Lecture Notes in Networks and Systems

Volume 530

AF870101

Series Editor

Janusz Kacprzyk, Systems Research Institute, Polish Academy of Sciences, Warsaw, Poland

Advisory Editors

Fernando Gomide, Department of Computer Engineering and Automation—DCA, School of Electrical and Computer Engineering—FEEC, University of Campinas—UNICAMP, São Paulo, Brazil

Okyay Kaynak, Department of Electrical and Electronic Engineering, Bogazici University, Istanbul, Turkey

Derong Liu, Department of Electrical and Computer Engineering, University of Illinois at Chicago, Chicago, USA
Institute of Automation, Chinese Academy of Sciences, Beijing, China

Witold Pedrycz, Department of Electrical and Computer Engineering, University of Alberta, Alberta, Canada
Systems Research Institute, Polish Academy of Sciences, Warsaw, Poland

Marios M. Polycarpou, Department of Electrical and Computer Engineering, KIOS Research Center for Intelligent Systems and Networks, University of Cyprus, Nicosia, Cyprus

Imre J. Rudas, Óbuda University, Budapest, Hungary

Jun Wang, Department of Computer Science, City University of Hong Kong, Kowloon, Hong Kong

The series "Lecture Notes in Networks and Systems" publishes the latest developments in Networks and Systems—quickly, informally and with high quality. Original research reported in proceedings and post-proceedings represents the core of LNNS.

Volumes published in LNNS embrace all aspects and subfields of, as well as new challenges in, Networks and Systems.

The series contains proceedings and edited volumes in systems and networks, spanning the areas of Cyber-Physical Systems, Autonomous Systems, Sensor Networks, Control Systems, Energy Systems, Automotive Systems, Biological Systems, Vehicular Networking and Connected Vehicles, Aerospace Systems, Automation, Manufacturing, Smart Grids, Nonlinear Systems, Power Systems, Robotics, Social Systems, Economic Systems and other. Of particular value to both the contributors and the readership are the short publication timeframe and the world-wide distribution and exposure which enable both a wide and rapid dissemination of research output.

The series covers the theory, applications, and perspectives on the state of the art and future developments relevant to systems and networks, decision making, control, complex processes and related areas, as embedded in the fields of interdisciplinary and applied sciences, engineering, computer science, physics, economics, social, and life sciences, as well as the paradigms and methodologies behind them.

Indexed by SCOPUS, INSPEC, WTI Frankfurt eG, zbMATH, SCImago.

All books published in the series are submitted for consideration in Web of Science.

For proposals from Asia please contact Aninda Bose (aninda.bose@springer.com).

More information about this series at https://link.springer.com/bookseries/15179

José M. Cascalho · Mohammad Osman Tokhi ·
Manuel F. Silva · Armando Mendes ·
Khaled Goher · Matthias Funk
Editors

Robotics in Natural Settings

CLAWAR 2022

Springer

Editors
José M. Cascalho
Faculty of Sciences and Technology
University of the Azores
Ponta Delgada, Portugal

Mohammad Osman Tokhi
School of Engineering
London South Bank University
London, UK

Manuel F. Silva
School of Engineering
Polytechnic Institute of Porto
Porto, Portugal

Armando Mendes
Faculty of Sciences and Technology
University of the Azores
Ponta Delgada, Portugal

Khaled Goher
Mechanical, Materials and Manufacturing
Engineering (M3), Faculty of Engineering
University of Nottingham
Nottingham, UK

Matthias Funk
Faculty of Sciences and Technology
University of the Azores
Ponta Delgada, Portugal

ISSN 2367-3370 ISSN 2367-3389 (electronic)
Lecture Notes in Networks and Systems
ISBN 978-3-031-15225-2 ISBN 978-3-031-15226-9 (eBook)
https://doi.org/10.1007/978-3-031-15226-9

© The Editor(s) (if applicable) and The Author(s), under exclusive license
to Springer Nature Switzerland AG 2023
This work is subject to copyright. All rights are solely and exclusively licensed by the Publisher, whether the whole or part of the material is concerned, specifically the rights of translation, reprinting, reuse of illustrations, recitation, broadcasting, reproduction on microfilms or in any other physical way, and transmission or information storage and retrieval, electronic adaptation, computer software, or by similar or dissimilar methodology now known or hereafter developed.
The use of general descriptive names, registered names, trademarks, service marks, etc. in this publication does not imply, even in the absence of a specific statement, that such names are exempt from the relevant protective laws and regulations and therefore free for general use.
The publisher, the authors, and the editors are safe to assume that the advice and information in this book are believed to be true and accurate at the date of publication. Neither the publisher nor the authors or the editors give a warranty, expressed or implied, with respect to the material contained herein or for any errors or omissions that may have been made. The publisher remains neutral with regard to jurisdictional claims in published maps and institutional affiliations.

This Springer imprint is published by the registered company Springer Nature Switzerland AG
The registered company address is: Gewerbestrasse 11, 6330 Cham, Switzerland

Foreword

Climbing and Walking Robots (CLAWAR) started with a six months exploratory phase in 1996 by four European organizations, namely University of Portsmouth, Royal Military Academy, FZI and RISO with the view to identify robotic stakeholders across Europe. The outcome was initiation of the CLAWAR thematic network of excellence supported by the European Commission over two phases, namely CLAWAR1 under the EC Brite Euram program during 1998–2002 and CLAWAR2 under the EC GROWTH program during 2002–2005.

CLAWAR Association was established by the end of 2005 to continue the activities of the CLAWAR Network globally, with the mission to advance robotics for the public benefit. The Association was registered in March 2006 with the Companies House in the UK as a non-profit making limited company by guarantee and in 2012 with the Charities Commission in the UK as a charitable organization.

The CLAWAR annual conference series is one of the main activities of CLAWAR Association. The first nine issues of the conference starting from 1998 were held in locations across Europe and further issues in various countries worldwide. The COVID-19 pandemic has had an impact on mode of participation in the conference, and while issues to 22 (2019) were held in physical participation mode, virtual participation mode has been exercised for issues 23 (2020, Russian Federation) and issue 24 (2021, Japan). The CLAWAR conference series has established itself as a popular and high profile platform for networking and dissemination of research and development findings in the area of mobile robotics and associated technologies.

Preface

CLAWAR 2022 is the twenty-fifth edition of International Conference series on Climbing and Walking Robots and the Support Technologies for Mobile Machines. The conference is organized by CLAWAR Association in collaboration with the University of the Azores in Ponta Delgada, Portugal, during September 12–14, 2022.

CLAWAR 2022 brings new developments and new research findings in robotics technologies within the framework of "robotics for natural settings." The topics covered include wearable assistive robotics from augmentation to full support for those with mobility disorders, innovative designs of components and full systems and application-specific robotic solutions.

The CLAWAR 2022 conference includes a total of 51 regular and special submission articles, from research institutions worldwide. This number has been arrived at through rigorous review of initial submissions, where each paper initially submitted has received at least three reviews. The conference further features four plenary presentations;

Developing biomedical devices
Cristina P Santos, University of Minho, Portugal

It's alive! From bioinspired to biohybrid robots
Victoria A. Webster-Wood, Carnegie Mellon University, USA

Heterogeneous teams of UGVs and UAVs for multipurpose operations
Pedro U. Lima, Universidade de Lisboa, Portugal

Key steps toward development of humanoid robots
Xie Ming, Nanyang Technological University, Singapore

It is believed that this book will serve as a source of inspiration and further innovation in research and development in the rapidly growing area of mobile service robotics.

Acknowledgements

The editors would like to thank members of the International Scientific Committee and National Organizing Committee of CLAWAR 2022 for their efforts in reviewing the submitted articles:

Ahmad, S., Malaysia
Alkan, B., UK
Almeshal, A., Kuwait
Amar, F. B., France
Armada, M., Spain
Banfield, I., Panama
Belter, D., Poland
Berns, K., Germany
Birla, R. P., India
Bonsignorio, F., Italy
Bridge, B., UK
Briskin, E., Russia
Burlacu, A., Romania
Chevallereau, C., France
Chrysostomou, D., Denmark
Chugo, D., Japan
Costa, M. T., Portugal
Darus, I. M., Malaysia
Dias, A., Portugal
Dillmann, R., Germany
Doroftei, I., Romania
El Youssef, E. S., Brazil
Ermolov, I., Russia
Faina, A., Denmark
Farias, P., Brazil
Fernandez, R., Spain
Ferreira, P., Portugal
Friebe, A., Sweden
Funk, M., Portugal
Gallegos Garrido, G., UK
Garcia, D., Panama
Goher, K. M., UK
Grand, C., France
Guedes, P., Portugal
Hassan, M. K., Malaysia
Hwang, K.-S., Taiwan
Ion, I., Romania
Jamali, A., Malaysia
Kaur, A. P., UK
Kottege, N., Australia
Lefeber, D., Belgium
Leon-Rodriguez, H., Colombia
Lera, F. R., Luxembourg
Li, G., UK
Manoonpong, P., Thailand
Marques, L., Portugal
Massoud, R., Syria
Mendes, A., Portugal
Mohamed, Z., Malaysia
Molfino, R., Italy
Mondal, S., UK
Monje, C. A., Spain

Montes, H., Panama
Moon, S., Korea
Muscato, G., Italy
Nakamura, T., Japan
Okui, M., Japan
Papadopoulos, E., Greece
Paraforos, D., Germany
Petry, M., Portugal
Qian, F., USA
Reina, G., Italy
Ribeiro, M., Portugal
Rincon, L. M., Brazil
Rocha, R. P., Portugal
Rodríguez, H., Panama
Sathyan, A., USA
Sattar, T. T., UK
Semini, C., Italy
Seong, H., Korea
Sequeira, J., Portugal
Silva, M., Portugal
Skrzypczynski, P., Poland
Tokhi, M. O., UK
Visser, A., The Netherlands
Wu, J., China
Xie, M., Singapore
Yigit, A., Kuwait
Yokota, S., Japan
Zaier, R., Oman
Zhao, Z., UK
Zielinska, T., Poland

Contents

List of Contributors

Al Bassit Lama Université Clermont Auvergne, INRAE, UR TSCF, Clermont-Ferrand, France

Artemyev K. S. Department of Theoretical Mechanics, Volgograd State Technical University, Volgograd, Russia

Asaoka Yuya Research Institute of Electrical Communication, Tohoku University, Sendai, Japan;
Department of Electrical Engineering, Tohoku University, Sendai, Japan

Asawalertsak Naris Vidyasirimedhi Institute of Science and Technology, Rayong, Thailand

Badri-Spröwitz Alexander Dynamic Locomotion Group, Max Planck Institute for Intelligent Systems, Stuttgart, Germany

Barrientos Antonio Centro de Automática y Robótica, Consejo Superior de Investigaciones Científicas, Universidad Politécnica de Madrid, Madrid, Spain

Behnke Sven Institute for Computer Science VI, University of Bonn, Bonn, Germany

Behr André Universidade dos Açores, FCT - UAc, Ponta Delgada, Portugal

Benist Bram TU Delft, Delft, Netherlands

Berns Karsten Department of Computer Science, Technische Universität Kaiserslautern, Kaiserslautern, Germany

Besselmann Marvin Grosse Department of Interactive Diagnosis and Service Systems (IDS), FZI Research Center for Information Technology, Karlsruhe, Germany

Biondo Alessia Dipartimento di Ingegneria Elettrica Elettronica e Informatica, University of Catania, Catania, Italy

Bjelonic Marko Robotic Systems Lab, ETH Zürich, Zürich, Switzerland

Borgese Andrea Dipartimento di Ingegneria Elettrica Elettronica e Informatica, University of Catania, Catania, Italy

Briskin E. S. Department of Theoretical Mechanics, Volgograd State Technical University, Volgograd, Russia;
Center for Technology Components of Robotics and Mechatronics, Innopolis University, Innopolis, Russia

Buchmann Alexandra Applied Mechanics, Department of Mechanical Engineering, TUM School of Engineering and Design, Technical University Munich, Garching, Germany

Buettner Timothee FZI Research Center for Information Technology, Karlsruhe, Germany

Carvalho Simão P. Center for MicroElectroMechanical Systems (CMEMS), University of Minho, Braga, Portugal;
LABBELS – Associate Laboratory, Braga/Guimarães, Portugal

Cascalho José Universidade dos Açores, FCT - UAc, Ponta Delgada, Portugal;
Grupo de Robótica e Inteligência Artificial da Universidade dos Açores GRIA-LIACC, Ponta Delgada, Portugal

Cepolina Elvezia Maria University of Genoa, Genoa, Italy

Cepolina Francesco University of Genoa, Genoa, Italy

Chakarov Dimitar Institute of Mechanics, Bulgarian Academy of Sciences, Sofia, Bulgaria

Chen Chen Tsinghua University, Beijing, China

Chowdhury Abhra Roy Indian Institute of Science, Bengaluru, India

Chugo Daisuke Kwansei Gakuin University, Sanda, Hyogo, Japan

Coba Gabriel R. Rivera Quito, Ecuador

Costa Paulo FEUP - Faculty of Engineering, University of Porto, Porto, Portugal;
INESC TEC - Institute for Systems and Computer Engineering, Technology and Science, Porto, Portugal

Costa Pedro INESC TEC - Instituto de Engenharia de Sistemas e Computadores, Tecnologia e Ciência, Porto, Portugal;
FEUP - Faculty of Engineering of University of Porto, Porto, Portugal

Danilov Vladimir Voltbro LCC, Moscow, Russia;
Institute of Control Problems of Russian Academy of Sciences, Moscow, Russia

Degani Amir Technion-Israel Institute of Technology, Haifa, Israel

del Cerro Jaime Centro de Automática y Robótica, Consejo Superior de Investigaciones Científicas, Universidad Politécnica de Madrid, Madrid, Spain

Diane Sekou Institute of Control Problems of Russian Academy of Sciences, Moscow, Russia

Dillmann Ruediger FZI Research Center for Information Technology, Karlsruhe, Germany

Dillmann Rüdiger Department of Interactive Diagnosis and Service Systems (IDS), FZI Research Center for Information Technology, Karlsruhe, Germany

Domínguez David Centro de Automática y Robótica, Consejo Superior de Investigaciones Científicas, Universidad Politécnica de Madrid, Madrid, Spain

Duan Fang School of Engineering, London South Bank University, London, UK

Eguchi Amy University of California San Diego, San Diego, USA

Faiña Andres Robotics, Evolution and Art Lab (REAL), IT University of Copenhagen, Copenhagen, Denmark

Fauroux Jean-Christophe Université Clermont Auvergne, Clermont Auvergne INP, CNRS, Institut Pascal, Clermont-Ferrand, France;
MecaBotiX, Toulouse, France

Ficht Grzegorz Institute for Computer Science VI, University of Bonn, Bonn, Germany

Figueiredo Joana Center for MicroElectroMechanical Systems (CMEMS), University of Minho, Braga, Portugal;
LABBELS – Associate Laboratory, Braga/Guimarães, Portugal

Fujikawa Chikage Chuo University, Tokyo, Japan

Fukuhara Akira Research Institute of Electrical Communication, Tohoku University, Sendai, Japan

Funk Matthias G. Universidade dos Açores, FCT - UAc, Ponta Delgada, Portugal;
Grupo de Robótica e Inteligência Artificial da Universidade dos Açores GRIA-LIACC, Ponta Delgada, Portugal

Gaathon Adar Technion-Israel Institute of Technology, Haifa, Israel

Gorb Stanislav N. Kiel University, Kiel, Germany

Guastella Dario C. Dipartimento di Ingegneria Elettrica Elettronica e Informatica, University of Catania, Catania, Italy

Hashimoto Hiroshi Advanced Institute of Industrial Technology, Shinagawa, Tokyo, Japan

Hermes Luca Machine Learning Group, Bielefeld University, Bielefeld, Germany

Huang Geng Tsinghua University, Beijing, China

Hutter Marco Robotic Systems Lab, ETH Zürich, Zürich, Switzerland

Ikeda Masahiro Graduate School of Information Science and Technology, The University of Tokyo, Bunkyo-ku, Tokyo, Japan;
School of Science and Technology, Meiji University, Tama-ku, Kanagawa, Japan

Inagaki Shinkichi Nanzan University, Nagoya, Japan

Ishiguro Akio Research Institute of Electrical Communication, Tohoku University, Sendai, Japan

Ishii Yusuke Precision Mechanics, Chuo University, Tokyo, Japan

Ishikawa Masato Department of Mechanical Engineering, Osaka University, Suita, Japan

Ito Fumio Precision Mechanics, Chuo University, Tokyo, Japan

Jatsun Sergey Southwest State University, Kursk, Russia

Jørgensen Jonas University of Southern Denmark, Odense M, Denmark

Kato Kanta Department of Mechanical Engineering, National Institute of Technolog, Kisarazu College, Kisarazu, Chiba, Japan

Kato Ryota SHINMEI Industry Co., Ltd., Toyota, Aichi, Japan

Katsarov Georgi Institute of Mechanics, Bulgarian Academy of Sciences, Sofia, Bulgaria

Kaur Aman School of Engineering, London South Bank University, London, UK

Kaushik B. Carnegie Mellon University, Pittsburgh, USA

Kim Mi-Sook Kangwon University-Industry Cooperation Foundation, Chuncheon, South Korea

Kiss Bernadett Dynamic Locomotion Group, Max Planck Institute for Intelligent Systems, Stuttgart, Germany

Kirchgeorg Steffen ETH Zürich, Zürich, Switzerland;
WSL, Birmensdorf, Switzerland

Klemm Victor Robotic Systems Lab, ETH Zürich, Zürich, Switzerland;
ASL, ETH Zürich, Zürich, Switzerland

Koutsoukis Konstantinos School of Mechanical Engineering, National Technical University of Athens, Athens, Greece

Kovalev Alexander Kiel University, Kiel, Germany

Kubota Takashi Japan Aerospace Exploration Agency (JAXA), Kanagawa, Japan

Kulich Miroslav Czech Institute of Informatics, Robotics, and Cybernetics, Czech Technical University in Prague, Prague, Czech Republic

Kuniyoshi Yasuo Graduate School of Information Science and Technology, The University of Tokyo, Bunkyo-ku, Tokyo, Japan

Kurumaya Shunichi Precision Mechanics, Chuo University, Tokyo, Japan

Lee Joonho Robotic Systems Lab, ETH Zürich, Zürich, Switzerland

Lenain Roland Université Clermont Auvergne, INRAE, UR TSCF, Clermont-Ferrand, France

Li Teng Tsinghua University, Beijing, China

Lima José Research Centre in Digitalization and Intelligent Robotics (CeDRI), Instituto Politécnico de Bragança, Bragança, Portugal;
INESC TEC - Institute for Systems and Computer Engineering, Technology and Science, Porto, Portugal;
Centre for Robotics in Industry and Intelligent Systems, INESC TEC, Porto, Portugal

Lima Pedro U. Instituto Superior Técnico (IST), Universidade de Lisboa, Lisbon, Portugal;
Intelligent Robots and Systems Group, Institute for Systems and Robotics, Lisbon, Portugal

Malchikov Andrei Southwest State University, Kursk, Russia

Mannhart Dominik ASL, ETH Zürich, Zürich, Switzerland

Manoonpong Poramate Vidyasirimedhi Institute of Science and Technology, Rayong, Thailand;
University of Southern Denmark, Odense M, Denmark

Maree Hugo European Space Agency, ESTEC, Noordwijk, The Netherlands

Marques Vitor London South Bank University, London, UK

Martinez Leon Andres Santiago Southwest State University, Kursk, Russia

Mastrogeorgiou Athanasios School of Mechanical Engineering, National Technical University of Athens, Athens, Greece

Matos Diogo INESC TEC-Instituto de Engenharia de Sistemas e Computadores, Tecnologia e Ciência, Porto, Portugal;
Centre for Robotics in Industry and Intelligent Systems, INESC TEC, Porto, Portugal

Mavridis Anestis European Space Agency, ESEC-Galaxia, Transinne, Belgium

Medeiros Paulo Universidade dos Açores, FCT - UAc, Ponta Delgada, Portugal;
Grupo de Robótica e Inteligência Artificial da Universidade dos Açores GRIA-LIACC, Ponta Delgada, Portugal

Mendes Armando B. Universidade dos Açores, FCT - UAc, Ponta Delgada, Portugal;
Grupo de Robótica e Inteligência Artificial da Universidade dos Açores GRIA-LIACC, Ponta Delgada, Portugal

Mendes João Research Centre in Digitalization and Intelligent Robotics CeDRI, Instituto Politécnico de Bragança, Bragança, Portugal

Mikula Jan Czech Institute of Informatics, Robotics, and Cybernetics, Czech Technical University in Prague, Prague, Czech Republic;
Faculty of Electrical Engineering, Czech Technical University in Prague, Prague, Czech Republic

Mintchev Stefano ETH Zürich, Zürich, Switzerland;
WSL, Birmensdorf, Switzerland

Miyazaki Yuya Kwansei Gakuin University, Sanda, Hyogo, Japan

Moreira João FEUP - Faculty of Engineering, University of Porto, Porto, Portugal

Moro Seiya Department of Mechanical Engineering, National Institute of Technolog, Kisarazu College, Kisarazu, Chiba, Japan

Muramatsu Satoshi Tokai University. Hiratsuka, Kanagawa, Japan

Murata Yuki SHINMEI Industry Co., Ltd., Toyota, Aichi, Japan

Muscato Giovanni Dipartimento di Ingegneria Elettrica Elettronica e Informatica, University of Catania, Catania, Italy;
Istituto di Calcolo e Reti ad Alte Prestazione, Consiglio Nazionale delle Ricerche, Palermo, Italy

Muslimov Tagir Ufa State Aviation Technical University, Ufa, Russian Federation

Nakamura Taro Precision Mechanics, Chuo University, Tokyo, Japan

Naruse Yuta Precision Mechanics, Chuo University, Tokyo, Japan

Niiyama Ryuma Graduate School of Information Science and Technology, The University of Tokyo, Bunkyo-ku, Tokyo, Japan;
School of Science and Technology, Meiji University, Tama-ku, Kanagawa, Japan

Nishihama Rie Research and Development Initiative, Chuo University, Tokyo, Japan

Norita Atsushi Research Institute of Electrical Communication, Tohoku University, Sendai, Japan;
Department of Electrical Engineering, Tohoku University, Sendai, Japan

N'chot Denis Université Clermont Auvergne, Clermont Auvergne INP, CNRS, Institut Pascal, Clermont-Ferrand, France

Oba Kiyotaka Chuo University, Tokyo, Japan

Okui Manabu Chuo University, Tokyo, Japan

Oliveira André Universidade Tecnológica Federal do Paraná (UTFPR), Curitiba, Paraná, Brazil

Or Keung Graduate School of Information Science and Technology, The University of Tokyo, Bunkyo-ku, Tokyo, Japan;
School of Science and Technology, Meiji University, Tama-ku, Kanagawa, Japan

Papadopoulos Evangelos School of Mechanical Engineering, National Technical University of Athens, Athens, Greece

Papatheodorou Aristotelis School of Mechanical Engineering, National Technical University of Athens, Athens, Greece

Park Hong Seong Kangwon National University, Chuncheon, South Korea

Paul Noah Technion-Israel Institute of Technology, Haifa, Israel

Pedro Francisco Escola Básica Integrada de Rabo de Peixe, Ribeira Grande, Portugal;
Grupo de Robótica e Inteligência Artificial da Universidade dos Açores GRIA-LIACC, Ponta Delgada, Portugal

Pereira Ana I. Research Centre in Digitalization and Intelligent Robotics CeDRI, Instituto Politécnico de Bragança, Bragança, Portugal

Pereira Miguel School of Engineering, Polytechnic Institute of Porto, Porto, Portugal

Pinheiro Cristiana Center for MicroElectroMechanical Systems (CMEMS), University of Minho, Braga, Portugal;
LABBELS – Associate Laboratory, Braga/Guimarães, Portugal

Pinto Vítor H. FEUP - Faculty of Engineering, University of Porto, Porto, Portugal;
SYSTEC (DIGI2)-Research Center for Systems and Technologies (Digital and Intelligent Industry Lab), Porto, Portugal

Plasberg Carsten Department of Interactive Diagnosis and Service Systems (IDS), FZI Research Center for Information Technology, Karlsruhe, Germany

Renjewski Daniel Applied Mechanics, Department of Mechanical Engineering, TUM School of Engineering and Design, Technical University Munich, Garching, Germany

Rito Diana Center for MicroElectroMechanical Systems (CMEMS), University of Minho, Braga, Portugal;
LABBELS – Associate Laboratory, Braga/Guimarães, Portugal

Roennau Arne Department of Interactive Diagnosis and Service Systems (IDS), FZI Research Center for Information Technology, Karlsruhe, Germany

Rohrich Ronnier Universidade Tecnológica Federal do Paraná (UTFPR), Curitiba, Paraná, Brazil

Rosendo Andre ShanghaiTech University, Shanghai, China

Rönnau Arne Department of Interactive Diagnosis and Service Systems (IDS), FZI Research Center for Information Technology, Karlsruhe, Germany

Sanchez Elsa Maria Alfonso European Space Agency, ESTEC, Noordwijk, The Netherlands

Santos Cristina P. Center for Micro Electro Mechanical Systems (CMEMS), University of Minho, Braga, Portugal;
LABBELS - Associate Laboratory, Braga/Guimarães, Portugal

Sathappan Nagu School of Engineering, London South Bank University, London, UK

Sattar Tariq Pervez London South Bank University, London, UK

Sauerbaum Sven Department of Interactive Diagnosis and Service Systems (IDS), FZI Research Center for Information Technology, Karlsruhe, Germany

Schilling Malte Machine Learning Group, Bielefeld University, Bielefeld, Germany

Schneider Axel Institute of System Dynamics and Mechatronics, Bielefeld University of Applied Sciences, Bielefeld, Germany

Schwertfeger Olivia KIT Karlsruhe Institute of Technology, Karlsruhe, Germany

Sharonov N. G. Department of Theoretical Mechanics, Volgograd State Technical University, Volgograd, Russia;
Center for Technology Components of Robotics and Mechatronics, Innopolis University, Innopolis, Russia

She Jin-Hua Tokyo University of Technology, Hachioji, Tokyo, Japan

Shirkoohi Gholamhossein School of Engineering, London South Bank University, London, UK

Siegwart Roland ASL, ETH Zürich, Zürich, Switzerland

Silva Manuel F. School of Engineering, Polytechnic Institute of Porto, Porto, Portugal;
INESC TEC, Porto, Portugal

Simmering Janneke Machine Learning Group, Bielefeld University, Bielefeld, Germany

Siqueira André School of Engineering, Polytechnic Institute of Porto, Porto, Portugal

Smirnaya L. D. Department of Theoretical Mechanics, Volgograd State Technical University, Volgograd, Russia;
Center for Technology Components of Robotics and Mechatronics, Innopolis University, Innopolis, Russia

Soares Salviano Universidade de Trás-os-Montes e Alto Douro, Vila Real, Portugal;
IEETA-Instituto de Engenharia Electrónica e Telemática de Aveiro, Aveiro, Portugal

Soares Inês N. FEUP - Faculty of Engineering, University of Porto, Porto, Portugal;
SYSTEC (DIGI2)-Research Center for Systems and Technologies (Digital and Intelligent Industry Lab), Porto, Portugal

Sutera Giuseppe Dipartimento di Ingegneria Elettrica Elettronica e Informatica, University of Catania, Catania, Italy

Suzuki Shura Department of Mechanical Engineering, Osaka University, Suita, Japan;
Japan Society for the Promotion of Science, Tokyo, Japan

Suzuki Tatsuya Nagoya University, Nagoya, Japan

Takaya Kazuki Precision Mechanics, Chuo University, Tokyo, Japan

Talevi Monica European Space Agency, ESTEC, Noordwijk, The Netherlands

Tanada Kosei Nagoya University, Nagoya, Japan

Tao Bo Tsinghua University, Beijing, China

Tokhi Mohammad Osman School of Engineering, London South Bank University, London, UK

Tokoi Ryosuke Chuo University, Tokyo, Japan

Tomioka Hiroki Graduate School of Information Science and Technology, The University of Tokyo, Bunkyo-ku, Tokyo, Japan

Tomoaki Tsuji Chuo University, Tokyo, Japan

Toyama Wataru Chuo University, Tokyo, Japan

Uchida Hiroaki Department of Mechanical Engineering, National Institute of Technolog, Kisarazu College, Kisarazu, Chiba, Japan

Ulloa Christyan Cruz Centro de Automática y Robótica, Consejo Superior de Investigaciones Científicas, Universidad Politécnica de Madrid, Madrid, Spain

Valente António INESC TEC-Instituto de Engenharia de Sistemas e Computadores, Tecnologia e Ciência, Porto, Portugal;
Universidade de Trás-os-Montes e Alto Douro, Vila Real, Portugal

Vaslin Philippe Université Clermont Auvergne, Clermont Auvergne INP, EMSE, CNRS, LIMOS, Clermont-Ferrand, France

Venev Pavel Institute of Mechanics, Bulgarian Academy of Sciences, Sofia, Bulgaria

Veneva Ivanka Institute of Mechanics, Bulgarian Academy of Sciences, Sofia, Bulgaria

Vidašič Jan Faculty of Electrical Engineering, Czech Technical University in Prague, Prague, Czech Republic

Vonwirth Patrick Department of Computer Science, Technische Universität Kaiserslautern, Kaiserslautern, Germany

Watanabe Jun'ichi Pigeon Home Products Corporation, Shizuoka, Japan

Webster-Wood Victoria A. Department of Mechanical Engineering, Carnegie Mellon University (CMU), Pittsburgh, PA, USA;
CMU Biohybrid and Organic Robotics Group, Pittsburgh, USA

Xie Ming Nanyang Technological University, Singapore, Singapore

Yang Dongchao Tsinghua University, Beijing, China

Yatsun Andrey Southwest State University, Kursk, Russia

Yokota Sho Toyo University, Kawagoe, Saitama, Japan

Yu Chen ShanghaiTech University, Shanghai, China

Zeng Zecui Tsinghua University, Beijing, China

Zhao Zhanfang School of Engineering, London South Bank University, London, UK

Zielinska, Teresa Faculty of Power and Aerospace Engineering, Warsaw University of Technology, Warsaw, Poland

List of Abbreviations

3R	Three Revolute
ABS	Acrylonitrile–Butadiene–Styrene
ACoM	Augmented CoM
ADLs	Activities of Daily Living
ADC	Analog-to-Digital Converter
ADD	Actuator Damping Distribution
ADL(s)	Activitie(s) of Daily Living
AEP	Anterior Extreme Position
AFD	Actuator Force Distribution
AGC	Automated Guided Cart
AGV	Autonomous Guided Vehicles
AI	Artificial Intelligence
ALNS	Adaptive Large Neighborhood Search
AMCL	Adaptive Monte Carlo Localization
AMR	Autonomous Mobile Robots
ANOVA	Analysis of Variance
AP	Antero-Posterior
API	Application Programming Interface
APO	Ankle Push-Off
ARTU-R	A1 Rescue Tasks Unitree Robot
BBS	Berg Balance Scale
BCE	Before the Common Era
BLE	Bluetooth Low Energy
BO	Bayesian Optimization
CAD	Computer-Aided Design
CAM	Centroidal Angular Momentum
CAN	Controller Area Network
CD	Centroidal Dynamic
CFRP	Carbon Fiber-Reinforced Plastic
CIM	Common Information Model

CMA	Covariance Matrix Adaptation
CMP	Centroidal Moment Pivot
Co	Comprehension of the user's commands
CoG (COG)	Center of Gravity
CoM (COM)	Center of Mass
COP	Center of Pressure
COT	Cost of Transport
CP	Capture Point
CPG(s)	Central Pattern Generator(s)
CPU	Central Processing Unit
CRB	Composite Rigid Body
CRES	Center for Renewable Energy Systems
CS	Computer Science
DAC	Digital-to-Analog Converter
DAG	Directed Acyclic Graph
DARPA	Defense Advanced Research Projects Agency
DC	Direct Current
DE	Differential Evolution
DF	Duty Factor
DIY	Do It Yourself
DoF (DOF)	Degree of Freedom
DRL	Deep Reinforcement Learning
DT	Digital Twin
EMC	Electro-Magnetic Compatibility
EMG	ElectroMyoGraphy
EMS	Electrical Muscle Stimulation
EOS	End-of-Step
ES	Electrical Stimulation/Effectiveness in Sorting
ESA	European Space Agency
EU	European Union
EVA	Ethylene-Vinyl Acetate
FB	FeedBack
FBG	Fiber Bragg Grating
FCP	Follow-the-Contact Point
FEA	Finite Element Analysis
FEM	Finite Element Method
FES	Functional Electrical Stimulation
FRP	Fiber-Reinforced Polymer
GAS	GAStrocnemius
GK	Gutin-Karapetyan (algorithm)
GLKH	Generalized Lin–Kernighan–Helsgaun (algorithm)
GLNS	Generalized Large Neighborhood Search
GLS	Generalized Least Squares
GNSS	Global Navigation Satellite System
GP	Gaussian Process

GPS	Global Positioning System
GPU	Ground Power Unit/Graphical Processing Unit
GP-UCB	GP Upper Confidence Bound
GSM	Global System Mobile
GTSP	Generalized Traveling Salesman Problem
HAT	Head Arm Trunk segment
HL	High Level
HRI	Human Robot Interaction
HRIM	Hardware Robot Information Model
HSV	Hue-Saturation-Value (image format)
ICS	Inertial Coordinate System
ID	IDentification
IFR	International Federation of Robotics
IK	Inverse Kinematics
IMI	Intrinsic Motivation Inventory
IMU	Inertial Measurement Unit
IO	Input-Output
IoT	Internet of Things
IPQ	Igroup Presence Questionnaire
IQ	Inspiration Quotient
IR	InfraRed
ISO	International Organization for Standardization
JAXA	Japan Aerospace Exploration Agency
KD	K Dimensional
KPI	Key Performance Indexes
LaMDA	Language Models for Dialogue Applications
LED	Light-Emitting Diode
LF	Left Front
LIPFM	Linear Inverted Pendulum Plus Flywheel Model
LIPM	Linear Inverted Pendulum Model
LL	Low Level
LO	Lift-Off
LP	Liquefied Petroleum
LQR	Linear-Quadratic Regulator
Max	Maximum
MCP	Meta Carpo Phalangeal
MDP	Markov Decision Process
MFL	Magnetic Flux Leakage
Min	Minimum
MINLP	Mixed-Integer Nonlinear Problem
ML	MedioLateral/Mid-Level
NASA	National Aeronautics and Space Administration
NDT	Non-Destructive Testing
NIST	National Institute of Standards and Technology
NLS	Nimble Limbs System

NMES	Neuromuscular Electrical Stimulation
NMS	Neuromuscular Simulation
NP	NeuroProsthesis/Nondeterministic Polynomial
NPL	Non-Priority Luggage
NSIRC	National Structural Integrity Research Center
OLS	Ordinary Least Squares
PAX	Passenger (embarqued at airport)
PC	Personal Computer
PCB(s)	Printed Circuit Board(s)
PCU	Production Control Units
PD	Proportional-Derivative
PEP	Posterior Extreme Position
PID	Proportional-Integral-Derivative
PL	Priority Luggage
PLA	PolyLactic Acid
PPO	Proximal Policy Optimization
ProBO	Progressive Bayesian Optimization
PV	Photo Voltaic
PWM	Pulse Width Modulation
RAM	Random-Access Memory
RBD	Rigid-Body Dynamics
RDF	Resource Description Framework
RFID	Radio-Frequency IDentification
RGB	Red Green Blue (image format)
RGB-D	Red Green Blue-Depth
RH	Right Hind
RL	Reinforcement Learning
RMS	Root Mean Square
RMSE	Root Mean Square Error
RNN	Recurrent Neural Network
ROI	Return on Investment
ROS	Robot Operating System
ROV	Remotely Operated Vehicle
RPM	Revolutions Per Minute
RVIZ	ROS VIsualiZation
SAR	Search And Rescue
SCI	Spinal Cord Injury
SEA(s)	Series Elastic Actuator(s)
SEE	Series Elastic Element
SI	Success in correct Identification of candy
SLAM	Simultaneous Localization and Mapping
SLIP	Spring-Loaded Inverted Pendulum
SLR	Swing Leg Retraction
SOL	Shinmei-Original-Locomotion/SOLeus
SPI	Serial Peripheral Interface

SPS	Standby Power Supply
SRC	Social Robotics Collaboration
SRS	Single-Robot Systems
SS	Success in picking up the correct candy and Segregating it
STD	Standard Deviation
STEAM	Science, Technology, Engineering, Art, and Mathematics
STEM	Science, Technology, Engineering, and Mathematics
STL	Standart Tessellation Language
SU	Shearography Unit
SUH	Shearography Holder Unit
TCP	Tool Center Point
TD	TouchDown
TES	Transcutaneous Electrical Stimulation
TO	Trajectory Optimization
TPP	Touring Polygons Problem
TRM	Tension Release Mechanism
TSP	Traveling Salesman Problem
TSPN	Traveling Salesman Problem with Neighborhoods
TUG	Timed Up and Go
TV	TeleVision
UAV(s)	Unmanned Aerial Vehicle(s)
UDP	User Datagram Protocol
UI	User Interface
UK	United Kingdom
UML	Unified Modeling Language
URDF	Unified Robot Description Format
USB	Universal Serial Bus
VDB	Virtual Data Base
VIP	Very Important Person
VMC	Virtual Model Control
VR	Virtual Reality
VRE	Virtual Reality Environments
W3C	World Wide Web Consortium
WBC	Whole-Body Control/Whole-Body Controller
WCPG	Workspace-Based Central Pattern Generator
Whegs	Wheel-legs
WLS	Weighted Least Squares
WT	Wait Time/Wind Turbine
WTB	Wind Turbine Blades
XML	Extensible Markup Language
ZMP	Zero-Moment Point

Plenary Presentations

Developing Biomedical Devices

Cristina P. Santos[1,2](✉)

[1] Center for Micro Electro Mechanical Systems (CMEMS), University of Minho, Braga, Portugal
cristina@dei.uminho.pt

[2] LABBELS - Associate Laboratory, Braga/Guimarães, Portugal

Abstract. More advanced and reliable robotic technologies for the healthcare industry are required since many industrialized countries are suffering from limited resources for healthcare, and an ever-increasing number of disabled and elderly population. Existing assistive technologies are unsatisfactory in meeting personal needs and fulfilling the required functions at a reasonable cost while being used in hospitals, schools, and homes in therapeutic programs that monitor, encourage, and assist their users. Any innovation to improve the capability of personalization or reduce the ratio of price and performance will promote the applications of robotic technologies in healthcare significantly, addressing more functionalities and targeted patients.

This presentation summarizes developments of assistive technologies for the disabled population to identify research targets that have guided the author research towards tangible, and measurable progress. Thus, it discusses the demands for assistive technologies for the disabled population, identifying the advantages and disadvantages of rehabilitation robots as assistive technologies and the issues involved in their development. Further, it selects a few critical challenge tasks in developing advanced rehabilitation robots to demonstrate progress. Further, presents some case studies of the author.

Keywords: Biomedical devices · Disabled population · Assistive technologies

© The Author(s), under exclusive license to Springer Nature Switzerland AG 2023
J. M. Cascalho et al. (Eds.): CLAWAR 2022, LNNS 530, p. 3, 2023.
https://doi.org/10.1007/978-3-031-15226-9_1

It's Alive! From Bioinspired to Biohybrid Robots

Victoria A. Webster-Wood[1,2(✉)]

[1] Department of Mechanical Engineering, Carnegie Mellon University (CMU), Pittsburgh, PA, USA
[2] CMU Biohybrid and Organic Robotics Group, Pittsburgh, USA
vwebster@andrew.cmu.edu

Abstract. Bioinspired robotics serves a dual role of advancing technology for real-world applications and creating tools for basic science research. However, just as there are gaps between simulations and reality, there are gaps between synthetic bioinspired robots and the animals they abstract or mimic. For example, the complex ability for abstract decision making and behavioral flexibility of biological brains, the adaptability and self-healing of living muscle, and the natural compliance of many biological systems, are lacking in many bioinspired robotics approaches. In this talk, I will explore the role of bioinspired robotics as a tool for basic science and address how biohybrid robotic approaches may help bridge the gap between bioinspiration and true biomimetics. I will also present recent work from our lab towards addressing sustainability in robotics and conclude with the ethical considerations and challenges facing biohybrid robotics as the field grows.

Keywords: Bioinspired robotics · Biohybrid robotics · Sustainability in robotics

© The Author(s), under exclusive license to Springer Nature Switzerland AG 2023
J. M. Cascalho et al. (Eds.): CLAWAR 2022, LNNS 530, p. 4, 2023.
https://doi.org/10.1007/978-3-031-15226-9_2

Heterogeneous Teams of UGVs and UAVs for Multipurpose Operations

Pedro U. Lima[1,2](✉)

[1] Instituto Superior Técnico (IST), Universidade de Lisboa, Lisbon, Portugal
[2] Intelligent Robots and Systems Group, Institute for Systems and Robotics, Lisbon, Portugal
pedro.lima@tecnico.ulisboa.pt

Abstract. The development of heterogeneous teams of unmanned ground vehicles (UGVs) and unmanned aerial vehicles (UAVs) for different mission types has deserved increasing attention of researchers in recent years, as the technology enables building more complex and resilient robot systems. In this talk I will describe some of the work done by my group on developing a multipurpose, autonomous, mobile manipulator for building outdoor structures and for firefighting, developed to participate in the Mohamed Bin Zayed International Robotics Challenge (MBZIRC) 2020, but also as a first step towards a team of UAVs and UGVs designed to inspect and regularly monitor solar farms.

I will present with some detail the hardware and software architectures of the developed mobile manipulator, and the developed methods for building outdoor structures consisting of heterogeneous brick patterns, and entering buildings to locate and extinguish fires. Solutions were successfully deployed in the near-realistic arenas of the MBZIRC 2020 competition and resulted in the first-place award for the firefighting scenario. Then I will provide some insight on ongoing work concerning the use of this UGV and an UAV fleet to search for solar panel defects in large-area solar farms, where autonomy is an important step for future operation.

Keywords: Unmanned ground vehicles (UGVs) · Unmanned aerial vehicles (UAVs) · Mobile manipulator

© The Author(s), under exclusive license to Springer Nature Switzerland AG 2023
J. M. Cascalho et al. (Eds.): CLAWAR 2022, LNNS 530, p. 5, 2023.
https://doi.org/10.1007/978-3-031-15226-9_3

Key Steps Toward Development of Humanoid Robots

Ming Xie(✉)

Nanyang Technological University, Singapore 639798, Singapore
mmxie@ntu.edu.sg

Abstract. This paper addresses the question of how to develop humanoid robots which could demonstrate human-like skills and human-like intelligence. Traditionally, the answer to such question includes the keywords such as design, analysis, sensing, perception, planning, control, actuation, etc. In this paper, the attempt is to make a comparison between human beings and humanoid robots. Particularly, the body-brain-mind model is adopted in order to better represent the skill-related and intelligence-related aspects of human beings. In this way, the key steps toward development of humanoid robots are explained according to body-brain-mind model, in which the material flow, the energy flow, the signal flow, the motion flow, and the knowledge flow could be easily understood. The extra benefit of such clearer explanation is to enable us to identify remaining challenges faced by the development of humanoid robots. Also, the well-designed invention of human-like humanoid robots will help us to discover the secrets behind the mind of human beings.

Keywords: Body · Brain · Mind · Action · Reaction · Computation · Memory · Cognition · Recognition · Material flow · Energy flow · Signal flow · Motion flow · Knowledge flow

1 Introduction

Our missions on Earth include the understanding of the world and the improvement of the world. The former requires us to discover the knowledge from the world, while the latter expects us to invent smarter machines, smarter products or smarter systems. For example, we may want to discover the physical principles behind the design of human beings' body, brain, and mind. Clearly, the invention of humanoid robots will help us to gain better understanding about the skill-related and intelligence-related aspects of human beings. Also, the deployment of humanoid robots as robot workers or robot drivers will certainly improve the eco-systems on Earth.

So far, the best humanoid robot prototype developed by universities is the HUBO series from KAIST. The best models of humanoid robots developed by research institutes are the HRP series from AIST in Japan and iCube from IIT in Italy. However, the best prototypes of humanoid robots developed by industries include ASIMO from Honda Co., Atlas from Boston Dynamics Co., QRIO from SONY Co., NAO from Soft Bank Co. Tesla's humanoid robot will join the list soon.

© The Author(s), under exclusive license to Springer Nature Switzerland AG 2023
J. M. Cascalho et al. (Eds.): CLAWAR 2022, LNNS 530, pp. 6–18, 2023.
https://doi.org/10.1007/978-3-031-15226-9_4

1.1 Body-Brain-Mind Model

Despite the advances in science and engineering, it is interesting to take note that we have not yet gain a full understanding of ourselves on Earth. Particularly, many people still believe that brain is mind while mind is what brain does. Such belief actually undermines the progress in cognitive science which mainly focuses on the study of mind. Interestingly, if we examine the discovery made in religions, minds could exist without brains and biological bodies. Therefore, it makes sense for us to advocate the body-brain-mind model as shown in Fig. 1.

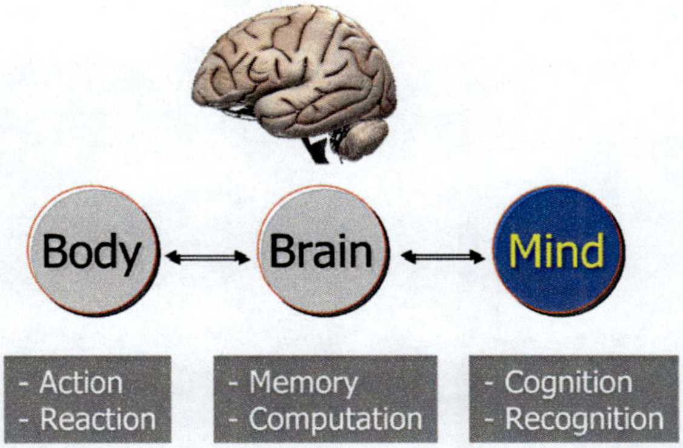

Fig. 1. Body-brain-mind model, in which: body is the provider of actions and reaction; brain is the provider of computation and memory; mind is the provider of cognition and recognition.

Refer to Fig. 1. The primary function of body is to provide actions and reactions. The primary function of brain is to provide computation and memory, while the primary function of mind is to provide cognition and recognition. Hence, these are three independent entities inside a human being, which could interface and work together.

1.2 Problem Statement

A humanoid robot is a machine which looks like a human being and could demonstrate human-like skills and human-like intelligence. Therefore, it is natural for us to ask this question of how to develop such human-like machines.

There are many ways to provide the answer to the above question. For example, people may say that the development of humanoid robots requires the applications of principles from design, analysis, sensing, perception, planning, control, actuation, etc. However, a better way to provide the answer is to adopt the body-brain-mind model. In this way, we could further raise these specific questions such as:

- What is the material flow inside a humanoid robot's body?
- What is the material flow inside a humanoid robot's brain?
- What is the energy flow inside a humanoid robot's body and brain?
- What is the signal flow inside a humanoid robot's body, brain and mind?
- What is the motion flow inside a humanoid robot's body?
- What is the knowledge flow inside a humanoid robot's mind?

We have developed a full-sized humanoid robot as shown in Fig. 2. According to our own research works on humanoid robots, the remaining part of this paper will provide partial answers to the above list of questions.

Fig. 2. A full-sized humanoid robot which was entirely developed by the author's teams at Nanyang Technological University between 2006–2012 with the grants from the Singapore government.

2 Key Steps Underlying Development of Humanoid Robot's Body

A humanoid robot could be considered as a network of rigid links which act and interact together for the purpose of providing motions intended by any task. Hence, the body of a humanoid robot contains material flow, energy flow and motion flow.

The material flow of a humanoid robot is manifested in the form of mechanism, structure and appearance. The design of mechanism, structure and appearance is a well understood topic in robotics.

In practice, two rigid links in a humanoid robot are connected by a link joint, which is coupled with a switch joint (i.e. clutch), a torque joint (i.e. a speed reducer) and a power joint (i.e. an actuator). This topic is also well understood in robotics. However, the conventional design is to do one-to-one coupling from a link joint to a power joint. But, this is not the case inside a human being's body because the failure of one muscle in a human being's body will not cause the entire body to fail. Therefore, the future challenge in the design of a humanoid robot's body is to achieve many-to-many coupling among link joints and power joints. In this way, the failure of one actuator (i.e. a power joint) will not cause a humanoid robot to malfunction.

Compared to industrial robots, the motion flow in humanoid robots is special. We know that in robotics, a local coordinate system is assigned to a link. Hence, when a humanoid robot has n links, the motion flow inside a humanoid robot includes all the kinematic chains which could be identified from the matrix shown in Fig. 3.

Particularly, there is a Jacobian matrix $J_{i \to j}$, which describes the motion kinematics $\dot{P}_{i \to j} = J_{i \to j}\dot{q}_{i \to j}$ of any possible kinematic chain from link i to link j, where $\dot{P}_{i \to j}$ is the linear/angular velocity vector of local coordinate system assigned to link j while $\dot{q}_{i \to j}$ is the velocity vector of the generalized joint coordinates from link i to link j.

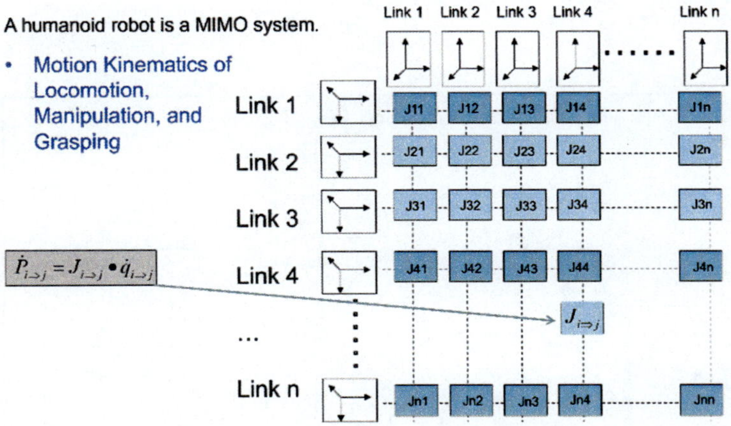

Fig. 3. A network of Jacobian matrices, which describe the motion kinematics of a humanoid robot with n rigid links. Here, notation Jij is equivalent to $J_{i \to j}$. MIMO stands for Multiple Inputs Multiple Outputs.

Another special consideration of a humanoid robot's motion flow arising from locomotion is the stability in sagittal plane (i.e. OZX plane) and coronal plane (i.e. OZY plane). In this case, dedicated controls must be put in place for achieving stable motion flow in both the sagittal plane and the coronal plane. Such dedicated controls may require the knowledge about the equations of dynamics of a humanoid robot's locomotion. So far, the popular models for deriving a humanoid robot's walking dynamics are:

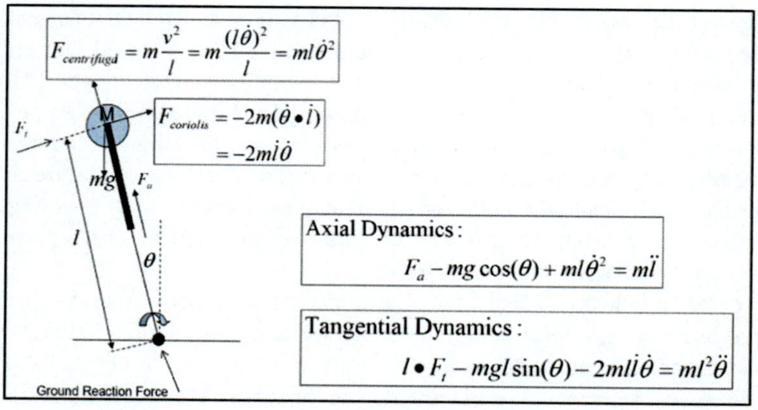

Fig. 4. Inverted pendulum model for deriving a humanoid robot's walking dynamics.

- Inverted Pendulum Model as Outlined in Fig. 4.
- Cart Table Model as Outlined in Fig. 5.
- Inverted Arm Model as Outlined in Fig. 6.

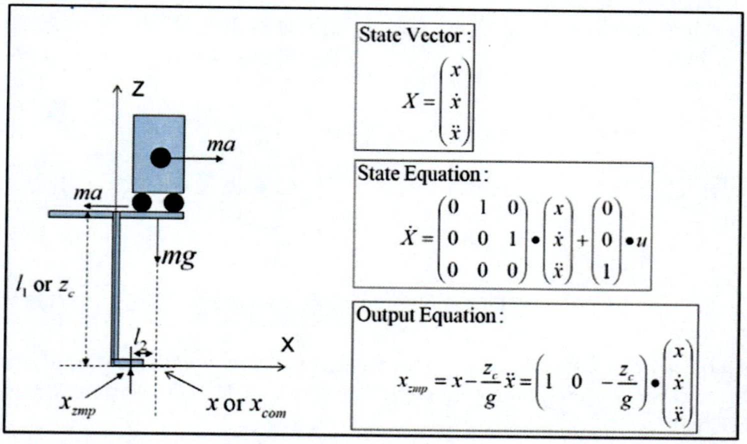

Fig. 5. Cart table model for deriving a humanoid robot's walking dynamics.

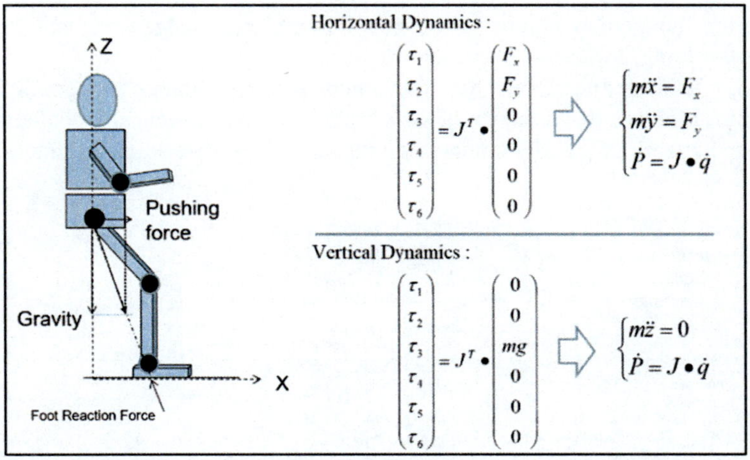

Fig. 6. Inverted arm model for deriving a humanoid robot's walking dynamics.

3 Key Steps Underlying Development of Humanoid Robot's Brain

One of the most important inventions made by human beings is microprocessor in which we have the full understanding about the following facts:

- A system of transistors allows us to implement logic gates such as AND gate, OR gate, XOR gate, etc.
- A system of logic gates allows us to implement register devices.
- A system of logic gates allows us to implement arithmetic devices.
- A system of register devices and arithmetic devices allows us to implement electronic brain such as the one shown in Fig. 7.

If we adopt today's microprocessor as the key component, we could propose the design solution for a humanoid robot's brain as shown in Fig. 8. In this way, we could say that a humanoid robot's brain is a network of transistors. Therefore, the material flow inside a humanoid robot's brain manifests in the form of a network of transistors, which undertakes computations with signal flow and knowledge flow.

Interestingly enough, the invention of microprocessor actually helps us to better understand our own brains. For example, in the literature, people have discovered that:

- A neural system could implement a logic AND gate.
- A neural system could implement a logic OR gate.
- A neural system could implement a logic XOR gate.

With these biologically implementable logic gates, our brains could actually be self-organized into complex neural systems for achieving the functionalities of doing computations (with signal flow and knowledge flow) as well as doing data storage.

Therefore, we could say that a human being's brain is a massive network of neurons which could be dynamically configured into two types of neural systems: the neural systems for doing computations (with signal flow and knowledge flow) and the neural systems for doing storages of learnt knowledge.

Also, it is interesting to take note that a human being's brain has over 100 billions of neurons. It means that a human being's brain could be dynamically configured into a massive network of parallel computing units as well as data storage units. This fact

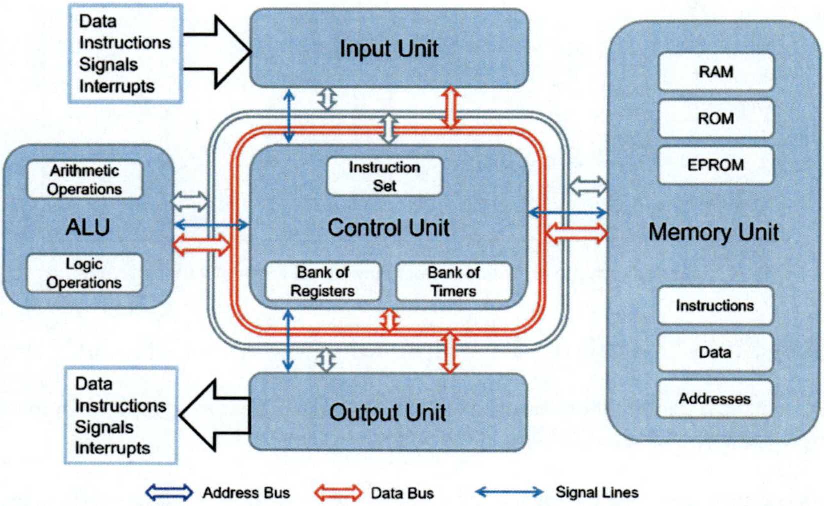

Fig. 7. Generic blueprint of microprocessor in which we could easily see not only the flow of energy, but also the flow of control signals as well as the flow of data signals.

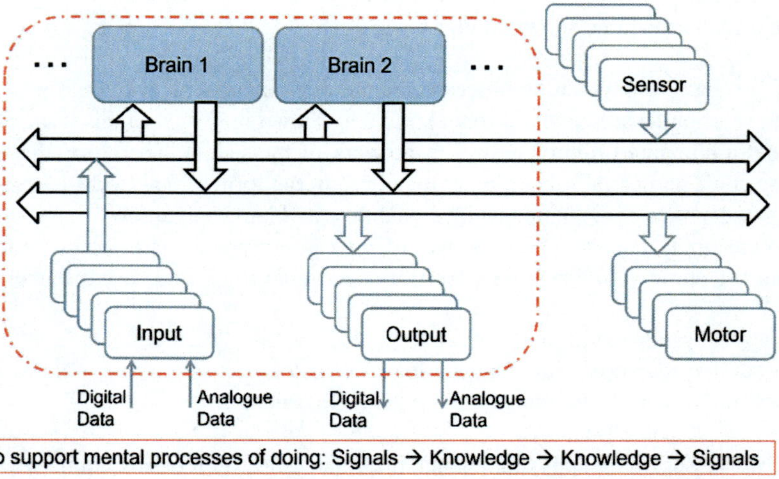

Fig. 8. Use of microprocessors to serve as brains inside a humanoid robot. In practice, more the number of brains is, better the performance will be.

explains why our brain is capable of processing massive amounts of signals coming from our vision system and other sensory systems.

This observation further leads us to suggest that today's microprocessor is not powerful enough in order to enable a humanoid robot's mind to achieve human-like intelligence. Hence, more research work needs to be done in this direction. This consists of another future challenge faced by the development of humanoid robots.

4 Key Steps Underlying Development of Humanoid Robot's Mind

Human beings are living inside the ocean of signals. The minds of human beings have the extraordinary capabilities of transforming signals captured by sensory systems into the cognitive states of knowing the meanings behind these signals. Similarly, humanoid robots are living inside the same ocean of signals. Hence, it is extremely important for us to develop artificial minds for humanoid robots so as to enable them to possess the mental capabilities which will be similar to the ones of human beings.

According to the ancient texts in Buddhism and repeatable spiritual experiments in today's religions, it is known to the public that minds are classified into three types which are coarse minds, subtle minds, and eternal minds. Most importantly, subtle minds and eternal minds belong to the category of immortal minds which are actually the so-called lives in layman terms (i.e. a very important answer to the big question of what life is). It also means that mind could exist without brain and biological body. This is an astonishing discovery made in religions. Therefore, the topics of mind should be considered as the most important study on Earth or even in the Universe.

Today, we do not know exactly what the blueprints (i.e. design drawings) underlying the mind systems of human beings are. Hopefully, we will know the answer after the journey of developing the mind systems for humanoid robots.

By default, we normally use the term Mind to refer to the coarse minds. Generally speaking, the two major functions of mind (i.e. coarse mind) are:

- The key steps toward achieving autonomous skills which are guided by sensing and perception as shown in Fig. 9, in which the outer loop of sensing, perception, planning, and control plays a vital role which transforms knowledge flow (i.e. task, action, motion) into flow of control signals. The inner loops are conventional feedback control systems which are well understood in control engineering. These feedback control systems could be position-based, velocity-based, force/torque based, or a combination of them (i.e. hybrid modes of control).
- The key steps toward achieving knowledge acquisition which transforms signals into the cognitive states of knowing the meanings behind these signals, as shown in Fig. 10. Here, we use visual signals as example to outline the key steps involved in the transformation. The reason of choosing visual signals is because if we rank the levels of importance among all the possible signals, visual signals will certainly occupy the first place of importance.

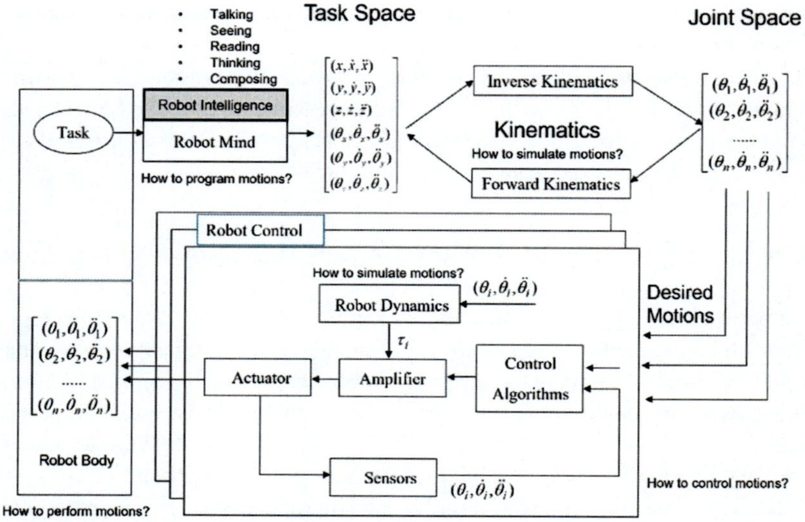

Fig. 9. Key steps toward achieving autonomous skills which are guided by sensing and perception. Particularly, the outer loop of sensing, perception, planning, and control plays a vital role.

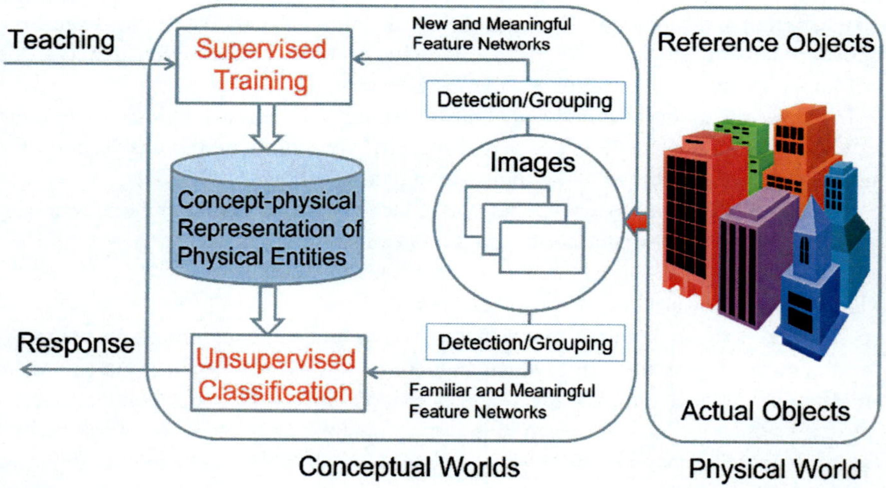

Fig. 10. Key steps toward transforming signals (i.e. signal flow) into the cognitive states of knowing the meanings (i.e. knowledge flow) behind these signals. Such mental capabilities are divided into two important categories which are cognition (i.e. supervised training) and recognition (i.e. unsupervised classification). Here, we use visual signals as example to outline these key steps involved in the transformation.

5 Highlights of Scientific Achievements

The full-sized version of a humanoid robot as shown in Fig. 2 was entirely developed by the author's teams at Nanyang Technological University with the grants from the Singapore government. Here, three significant scientific achievements are highlighted.

5.1 Achievement of Many-To-Many Coupling Scheme in Actuation

As mentioned earlier in the paper, the traditional way of implementing a robotic joint is to couple link joint, switch joint, torque joint, and power joint together by following the one-to-one coupling scheme. The drawback of such approach is the risk of total failure if one actuator (i.e. power joint) inside a humanoid robot's legs or arms fails.

However, if we examine the actuation of a human being's mechanism by the muscles, it is clear that the failure of one muscle will not cause the malfunction of a human being's body. This is because a joint inside a human being's body could be actuated by more than one muscle. Hence, the question will be whether it is possible to achieve many-to-many coupling scheme in the design of robotic joints inside a humanoid robot.

Actually, we have developed a workable solution which has been successfully demonstrated inside an industrial robotic arm, the upper body of a humanoid robot and a multiple-fingered robotic hand. The results are shown in Fig. 11.

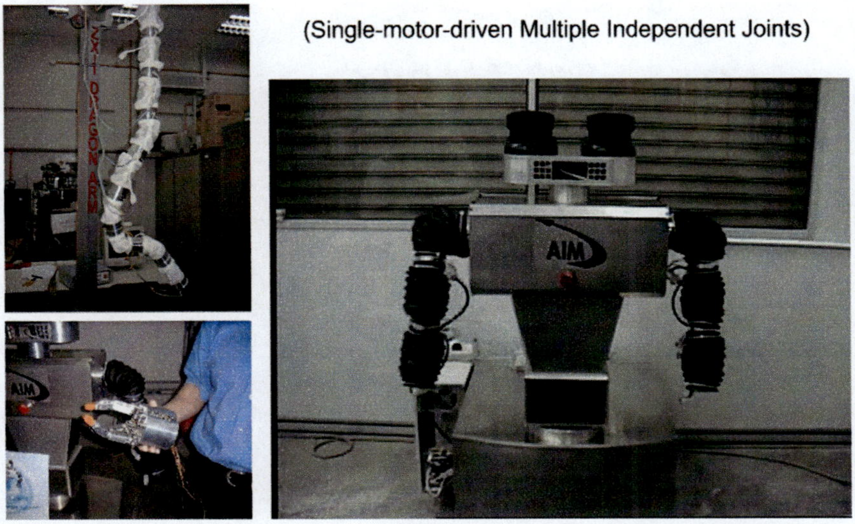

Fig. 11. Implementation of a novel many-to-many coupling scheme inside an industrial robotic arm, the upper body of a humanoid robot and a multiple-fingered robotic hand.

5.2 Achievement of Human-Like X-Eye Coordination

Human-like skills are mainly manifested in the form of X-Eye coordination where X may refer to hands, arms, and legs, etc. Most importantly, it is easy for us to observe that human-like skills heavily depend on the continuous running of the outer loop of sensing, perception, planning and control. Therefore, it is interesting for us to discover the working principle behind such X-Eye coordination which involves the use of binocular vision.

In the literature, there are many investigations under the topic of visual servoing. Most of them focuses on the use of monocular vision, which is not human-like vision system nor solution.

To the surprise of many researchers, with the use of human-like binocular vision, there exists a constant matrix which maps the disparity vectors observed from stereo images into the displacement vectors of hands, heads, or legs. The series of displacement vectors produced with the use of constant matrices allows the successful achievements of X-Eye coordination. We have validated this discovery with a robotic arm guided by a binocular vision as shown in Fig. 12. The demonstration has been broadcasted in the national TV channels in Singapore.

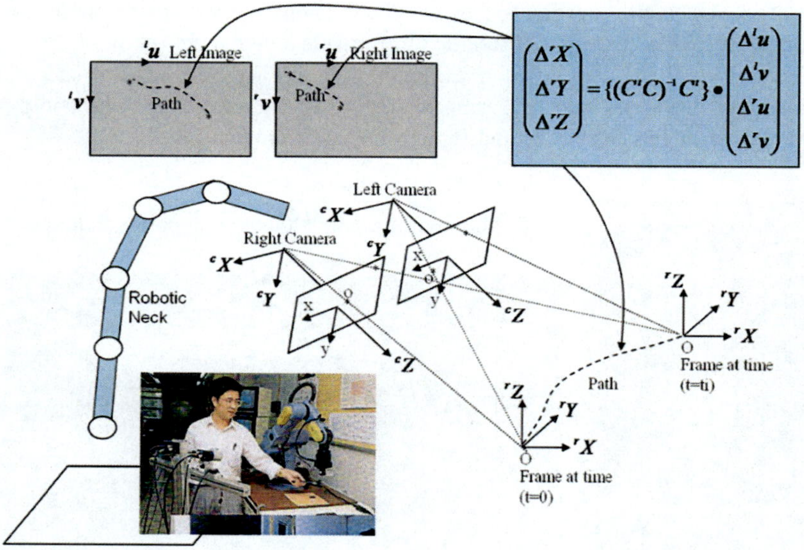

Fig. 12. Implementation of a novel X-Eye coordination solution which makes use of one or more constant matrices in order to map disparity vectors observed from the stereo images into displacement vectors of hands, heads or legs.

5.3 Achievement of Language-Based Conversational Dialogue

The flow of knowledge among the minds of human beings is transported by the texts and voices which are the outcomes due to the use of human languages. Hence, the mental capability of transforming audio signals into knowledge and vice versa is extremely important to human beings. Without such mental capability, human beings could not communicate at the cognitive level.

Similarly, if we want to use humanoid robots to serve as robot workers or robot drivers, it is necessary for future humanoid robots to master human languages. In this way, conversational dialogues with the use of human languages between human beings and humanoid robots will become feasible.

So far, great progresses have been in the areas of speech recognition (i.e. transforming audio signals into conceptually meaningful symbols) and speech synthesis (i.e. transforming meaningful symbols into audio signals). What is lacking is the transformation of conceptually meaningful symbols into physically meaningful symbols. For example, the sentence of "rabbit eats tiger" is conceptually meaningful, but not physically meaningful.

With a research grant from Singapore government, we have developed a meaning-centric framework in which a database of concept-physical knowledge has been developed. The use of such data allows us to filter out physically meaningless symbols (e.g. rabbit eats tiger). In this way, we have achieved the first version of a mind system which enables the conversational dialogues between human beings and humanoid robots as shown in Fig. 13.

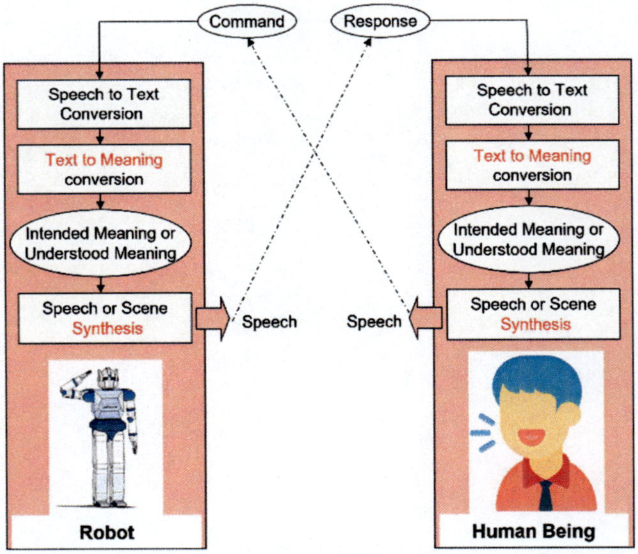

Fig. 13. A meaning-centric framework in which audio signals are transformed into meaningful symbols in any human language, and vice versa. In this way, conversational dialogue with the use of human languages becomes feasible between human beings and humanoid robots.

6 Conclusions

In this paper, I have shared our findings underlying the key steps toward the development of humanoid robots. Especially, I strongly advocate the adoption of body-brain-mind model in order to represent the skill-related and intelligence-related aspects of human beings. By following this model to develop humanoid robots, we see clearly that the interplay among the flow of materials, the flow of energy, the flow of signals, the flow of motions, and the flow of knowledge contributes nicely to the design of a humanoid robot's body, brain, and mind.

References

1. Xie, M., Chen, H., Hu, Z.C.: New Foundation of Artificial Intelligence. World Scientific Publishing Co., Singapore (2021)
2. Jayakumar, K.S., Xie, M.: Natural Language Understanding by Robots. LAP LAMBERT Academic Publishing Co., Republic of Moldova) (2010)
3. Xie, M.: Five Steps of Evolution from Non-life to Life-like Robots. Int. J. Humanoid Rob. **6**(2), 307–327 (2009)
4. Xie, M., Dubowsky, S., Fontaine, J.G., Tohki, O.M., Virk, G. (eds.): The Advances in Climbing and Walking Robots. World Scientific Publishing Co., Singapore (2007)
5. Xie, M.: Fundamentals of Robotics: Linking Perception to Action. World Scientific Publishing Co., Singapore (2003)

Biologically-Inspired Systems and Solutions

Power to the Springs: Passive Elements are Sufficient to Drive Push-Off in Human Walking

Alexandra Buchmann[1(✉)], Bernadett Kiss[2], Alexander Badri-Spröwitz[2], and Daniel Renjewski[1]

[1] Applied Mechanics, Department of Mechanical Engineering, TUM School of Engineering and Design, Technical University Munich, Boltzmannstr. 15, 85748 Garching, Germany
alexandra.buchmann@tum.de

[2] Dynamic Locomotion Group, Max Planck Institute for Intelligent Systems, Heisenbergstr. 3, 70569 Stuttgart, Germany
https://www.mec.ed.tum.de/am/

Abstract. For the impulsive ankle push-off (APO) observed in human walking two muscle-tendon-units (MTUs) spanning the ankle joint play an important role: Gastrocnemius (GAS) and Soleus (SOL). GAS and SOL load the Achilles tendon to store elastic energy during stance followed by a rapid energy release during APO. We use a neuromuscular simulation (NMS) and a bipedal robot to investigate the role of GAS and SOL on the APO. We optimize the simulation for a robust gait and then sequentially replace the MTUs of (1) GAS, (2) SOL and (3) GAS and SOL by linear springs. To validate the simulation, we implement NMS-3 on a bipedal robot. Simulation and robot walk steady for all trials showing an impulsive APO. Our results imply that the elastic MTU properties shape the impulsive APO. For prosthesis or robot design that is, no complex ankle actuation is needed to obtain an impulsive APO, if more mechanical intelligence is incorporated in the design.

Keywords: Ankle push-off · Gastrocnemius · Soleus · Energy storage · Catapult · Neuromuscular simulation · Bipedal robot

1 Introduction

Plantigrade bipedal walking, i.e. walking with heel and toes flat on the ground, distinguishes humans from most other animals. The complexity of human bipedal gait becomes visible when trying to technically replicate or restore natural human leg dynamics. Humanoid robots, lower limb prostheses or exoskeletons fall far short of human performance in terms of efficiency, versatility and robustness [1–3]. This can be partially attributed to technical limitations, but more importantly to a lack of understanding the underlying biomechanical and control principles of human walking.

© The Author(s), under exclusive license to Springer Nature Switzerland AG 2023
J. M. Cascalho et al. (Eds.): CLAWAR 2022, LNNS 530, pp. 21–32, 2023.
https://doi.org/10.1007/978-3-031-15226-9_5

A characteristic feature of human walking is the high power output at the ankle joint in late stance [4] which leads to an impulsive ankle push-off (APO) [5]. It has been shown that the peak power is substantially higher than maximum active power capacity of the plantar flexor muscle's Gastrocnemius (GAS) and Soleus (SOL) [6]. The power peak higher than the muscles' capacity indicates the use of elastic structures to store and rapidly release mechanical energy contributing to the energy efficiency of locomotion [7–9].

Hof *et al.* [10] were the first to label the impulsive APO in human walking a catapult. Catapult mechanisms are commonly used during locomotion in nature. Humans use a catapult during APO to propel their trailing leg into swing [6], as do galloping horses which use a catapult to achieve rapid limb protraction [11]. Locusts, frogs, fleas, and click beetles use a geometrical catapult for jumping [12–14].

Functionally, a catapult is characterized by slow storage of elastic energy followed by a rapid energy release with substantially higher power to accelerate a projectile. Three main components are needed to achieve this power amplification: an elastic element to store energy, a block or frame to take up forces arising during loading and a catch to hold the elastic energy and trigger its rapid release.

In all biological examples, elastic energy storage is facilitated by muscles-tendon-units (MTUs). During human APO, SOL and GAS MTUs provide up to 91% of the power from elastic energy [15]. During stance, GAS and SOL mostly work isometrically [16] while the Achilles tendon is stretched to store energy. The MTUs of GAS and SOL thus act like springs reducing the metabolic costs required for walking [15].

The observation that ankle kinematics and kinetics in human walking are mainly driven by the passive-elastic properties of ankle plantar flexor MTUs raises the following questions:

I Do humans need active ankle actuation for steady state walking with impulsive ankle push-off or would passive-elastic structures be sufficient?
II What are the individual contributions of the plantar flexor muscle-tendon-units, Soleus and Gastrocnemius, to the ankle push-off?

We investigate (I) and (II) using a 2D neuromuscular simulation (NMS) of human walking [17] and a bipedal robot for real world validation. First, we optimize the simulation for a robust, natural gait and then replace (1) GAS, (2) SOL and (3) GAS and SOL MTUs by passive linear springs without adapting neural control parameters. Finally, we implement NMS-3 on a bipedal robot with actuated knee and hip joints and passive elastic ankle joint, where springs connected to cables spanning the ankle joint represent GAS and SOL MTUs. NMS and robot were able to walk continuously in all trials showing an impulsive APO even in absence of active ankle actuation.

For prosthesis, exoskeleton, or bipedal robot design we state that more attention should be payed to an intelligent mechanical ankle design inspired by human morphology and biomechanical function. Reproducing natural leg dynamics with impulsive APO may be possible by utilizing passive-elastic elements for steady state bipedal walking instead of complex control or powerful actuator strategies.

2 Methods

2.1 Neuromuscular Simulation and Solver

We used a 2D neuromuscular simulation [17] with 9 degrees of freedom (trunk-to-world: 3, knee, ankle and hip 1 each) running on Matlab Simulink R2020b (The MathWorks Inc., Natick, Massachusetts, America). The NMS includes seven muscles groups per leg controlled by reflex feedback. Figure 1(a) shows the setup of the NMS-ref model and Fig. 1(b)–(d) NMS-1 to NMS-3 where we replaced GAS and/or SOL MTUs by springs.

The NMS is solved using a variable-order numerical solver *ode15s*[1] for stiff differential equations with a maximum step size of 0.1 s, a minimum step size of $1\,e^{-9}$ and relative and absolute tolerances of $1\,e^{-3}$ and $1\,e^{-4}$, respectively. The zero-crossing detection was set to be non-adaptive with a maximum number of 30 consecutive crossings before aborting the simulation.

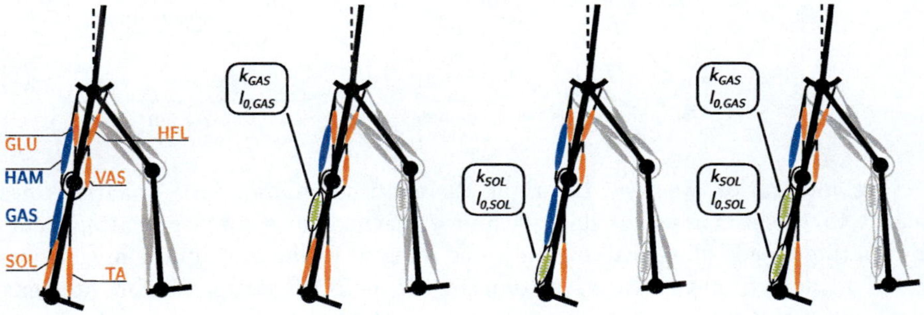

(a) NMS-ref. (b) NMS-1. (c) NMS-2. (d) NMS-3.

Fig. 1. Simulation experiments and model setup. (a) NMS-ref configuration with seven muscle groups per leg namely Gluteus *GLU*, Hamstrings *HAM*, Gastrocnemius *GAS*, Soleus *SOL*, Tibialis Anterior *TA*, Vastus *VAS* and Hip flexor muscles *HFL*. For details on the NMS see [17]. (b) to (d): three NMS trials where (1) GAS, (2) SOL and (3) GAS and SOL MTUs are replaced by linear springs with stiffness k and resting length l_0. Figure inspired by [18].

2.2 Optimization

To optimize the simulation for a natural, steady state walking we implemented a global optimization framework using a non-deterministic genetic algorithm *Matlab GA*[2] with a population size of 200 and a maximum of 50 stall generations as termination criterion. We ran the optimization on a 64-bit computer using a 12-core *Intel i9-10900K* CPU at 3.7 GHz with 64 GB RAM in parallel computing and a pre-compiled model in accelerator mode with fast restart.

1 https://www.mathworks.com/help/simulink/gui/solver.html.

2 https://www.mathworks.com/help/gads/ga.html.

We designed a custom cost function to optimize ten of the model's muscle reflex gains. The resulting gains are given in Table 2. Upper and lower bounds were selected according to boundary values for stable gait from [17, Table I, p. 269].

The cost function is inspired by Veerkamp *et al.* [19] and Falisse *et al.* [20]. We included the metabolic energy consumption, muscle fatigue expressed by muscle activation squared [21], ground reaction force jerks, the acceleration of the head arm trunk segment (HAT) and knee stop moments representing the passive structures in the knee, e.g. ligaments (Eq. (1)). We additionally included ankle stop moments to ensure the model stays within the physiologically range of motion and does not use ankle end stops. We calculated stop moments, muscle activations and ground reaction force jerks for both legs summed. The metabolic cost calculation follows [22].

$$J = \frac{1}{x_{end}} \int_{t_{step3}}^{t_{end}} \Big(\underbrace{w_1 \cdot E_{Met}}_{\substack{\text{metabolic}\\ \text{energy}}} + \underbrace{w_2 \cdot ACT^2}_{\substack{\text{muscle}\\ \text{fatigue}}} + \underbrace{w_3 \cdot GRF_{jerk}}_{\substack{\text{ground reaction}\\ \text{force jerks}}} + \underbrace{w_4 \cdot a_{HAT}}_{\substack{\text{HAT}\\ \text{acceleration}}} + \ldots$$

$$\ldots \underbrace{w_5 \cdot |T_K|}_{\substack{\text{stop moments}\\ \text{knee}}} + \underbrace{w_6 \cdot |T_A|}_{\substack{\text{stop moments}\\ \text{ankle}}} \Big) dt + \underbrace{w_7 \cdot (t_{max,sim} - t_{end})^2}_{\substack{\text{early fall}\\ \text{penalty}}} \quad (1)$$

For integration we used the right Riemann summation. All quantities are scaled to the maximum walking distance reached after 10 s simulation time. Evaluating steady state walking only, the integral in the cost function was evaluated starting from the third step onwards. If the NMS stopped before finishing three steps the entire simulation time was evaluated to have a smooth reward function also in absence of a good initial guess where the model falls before reaching steady state. If the NMS aborted before reaching 10 s a penalty was applied. The NMS stopped with an error, if the center of mass height fell below 1 m, the trunk's forward velocity decreased below zero or joint limits were exceeded reflecting the physiological range of motion.

To tune the weights of the cost function we ran the NMS once with the initial parameter set from Geyer and Herr [17] and calculated each contribution separately. We then tuned the weights based on the recommendations for relative weighting given in Veerkamp *et al.* [19, Table 3, p. 7]. For the ankle stop moment we selected a high weight compared to all other values to steer the optimization towards solutions avoiding ankle hard stops. Details see Table 1.

2.3 Muscle-Tendon-Unit Replacement

For NMS-1, 2, 3 we replaced the MTUs of GAS and/or SOL by linear mechanical springs as shown in Fig. 1*(b)–(d)*. To determine spring stiffness and resting length, we analyzed the force-length curves of both MTUs in NMS-ref (Fig. 2).

Table 1. Cost function weights. *Value* shows the final weighting, *Rel. w* the relative weights used for tuning from [19]. J_{Cont} and % are the absolute contributions after integration for one step and the contribution fractions to the reward function for NMS-ref.

Property	No.	Value	Rel. w	J_{Cont}	%
E_{CoT}	w_1	$5\,e^2$	5	239	69.9
ACT^2	w_2	10	0.1	2.9	0.9
GRF_{Jerk}	w_3	2	1	40	11.7
HAT_{acc}	w_4	$2\,e^4$	1.5	45	13.2
T_K	w_5	10	0.25	15	4.4
T_A	w_6	$1\,e^4$	n.a	0	–
Penalty	w_7	$1\,e^5$	n.a	0	–

Table 2. Muscle reflex gains for NMS-ref. Muscle gains of force feedback **(bold)** are normalized to F_{max} of the respective muscle.

Gain	NMS-ref
k_φ	4.25
G_{HAMHFL}	3
$\mathbf{G_{GAS}}$	1.10
G_{TA}	1.10
$\mathbf{G_{VAS}}$	1.15
$\mathbf{G_{SOLTA}}$	0.3
$\mathbf{G_{SOL}}$	1.2
G_{HFL}	0.35
$\mathbf{G_{HAM}}$	0.65
$\mathbf{G_{GLU}}$	0.4

We measured the MTUs length at minimum and maximum force in ascending force-length slope of stance phase and fitted these points coarse with a linear stiffness rounding to the next full ten N/mm resulting in $l_{0,GAS} = 0.440\,\text{m}$, $k_{GAS} = 60\,\text{N/mm}$, $l_{0,SOL} = 0.290\,\text{m}$, $k_{SOL} = 200\,\text{N/mm}$.

The joint torque resulting from F_s is modeled by $\tau_s = r_s(\varphi)F_S$ with $r_s = r_0 \cos\varphi - \varphi_{max}$, $r_0 = 50\,\text{mm}$ and $\varphi_{max} = 110°$ for GAS and SOL at the ankle and $r_0 = 50\,\text{mm}$ and $\varphi_{max} = 140°$ for GAS at the knee [17, Appendix III].

The spring replacements start loading when heel and ball are on the ground and are instantaneously switch off when the foot leaves the ground. We used the muscle gains given in Table 2 for all NMS trails, started the model in NMS-ref configuration with MTUs and turned on the springs after the third step when the model was already in a steady state.

2.4 Robotic Model

For validation we implemented NMS-3 on a child-size bipedal robot with an anthropomorphic ankle configuration [23]. SOL and GAS MTUs are represented by linear springs connected to cables acting on pulleys of fixed radii around the ankle- and knee joints. The SOL and GAS spring stiffness values are 4.5 N/mm and 1.4 N/mm, respectively with pulley-radii of 13 mm at the ankle for SOL and GAS springs, and 13 mm for the GAS spring at the knee.

The robot's hip and knee joints are actuated and controlled by an open-loop central pattern generator [24] with 1.0 Hz locomotion frequency which allows for a walking speed of 0.55 m/s on a treadmill. Control patterns roughly follow the NMS joint trajectories. The robot's ankle joints are not actively actuated, only SOL and GAS springs act on them. The designed anthropomorphic bipedal robot weights 2.22 kg, at a leg length of 0.35 m.

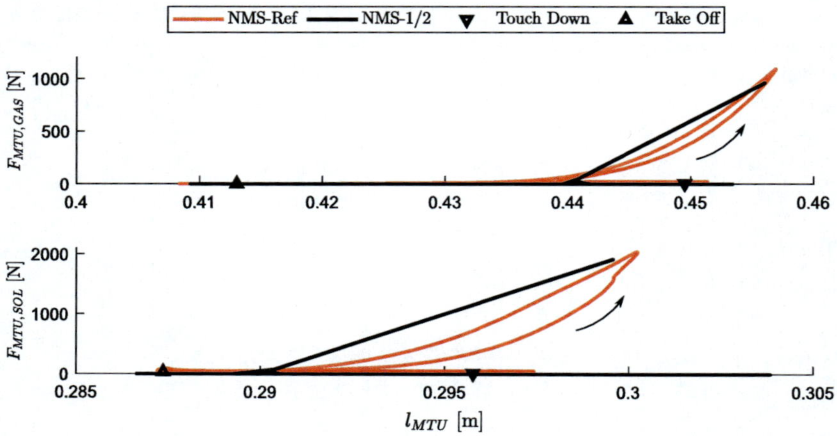

Fig. 2. Force length curves of GAS and SOL MTUs during one stride. *Note the different x- and y-axis for both plots.* The arrow indicates the direction of the orange work loops for the MTUs in NMS-ref for GAS at the top and SOL at the bottom. The black lines shows the force length curves when the MTU is replaced by a spring from NMS-1 in the top graph and NMS-2 for the bottom. The triangles denote take off and touch down.

The robot is equipped to measure joint angles and velocities. The ankle joint power is calculated from the stiffness values of the SOL and GAS springs when they are loaded according to:

$$P_A(\alpha, \dot{\varphi}) = \begin{cases} \dot{\varphi}_A \cdot (k_{SOL} \cdot r_{SOL}^2 \cdot \alpha_A + k_{GAS} \cdot r_{GAS}^2 \cdot (\alpha_A - \alpha_K)), & \text{if } \alpha_A \geq \alpha_K \\ \dot{\varphi}_A \cdot k_{SOL} \cdot r_{SOL}^2 \cdot \alpha_A, & \text{otherwise} \end{cases}$$

with P_A is the ankle power and $\dot{\varphi}_A$ the ankle angular velocity in rad/s. α_A and α_K are the ankle and knee angle in rad, k_{SOL} and k_{GAS} are the SOL and GAS spring stiffness values in N/m, and r_{SOL} and r_{GAS} are the pulley radii of the SOL and GAS spring-cables in m.

Both springs are loaded starting in plantarflexion at a fixed angle ($-22°$) as the ankle dorsiflexes. SOL's cable length only depends on the ankle angle while GAS's also depends on the knee angle. *For further details on design and experimental setup see* [23].

3 Results

3.1 Simulation

The NMS walks continuously for all experiments. Average forward velocity, stride time and duty factor show minor changes. Metabolic costs decrease from NMS-1 over NMS-2 to NMS-3 compared to NMS-ref by max. 20% for NMS-3 as fewer muscles consume energy when being replaced by a spring (Table 3).

Joint kinematics for all trials are shown in Fig. 3. The hip kinematics are not changing significantly. When replacing GAS in NMS-1 and NMS-3 only minor changes are visible in knee and ankle comparing NMS-1 to NMS-ref and NMS-3 to NMS-2. Replacing SOL shows impact on ankle and knee kinematics. NMS-2 and NMS-3 exhibit decreased knee flexion in stance and increased knee flexion in swing. The ankle plantar flexion is slightly decreased in stance. In swing the ankle dorsi-flexes rapidly and at 70% stride time hard stops (20°) engage. We assume the spring switch off at the end of stance described in Sect. 2.3 and thus the resulting immediate force drop to zero cause this artifact.

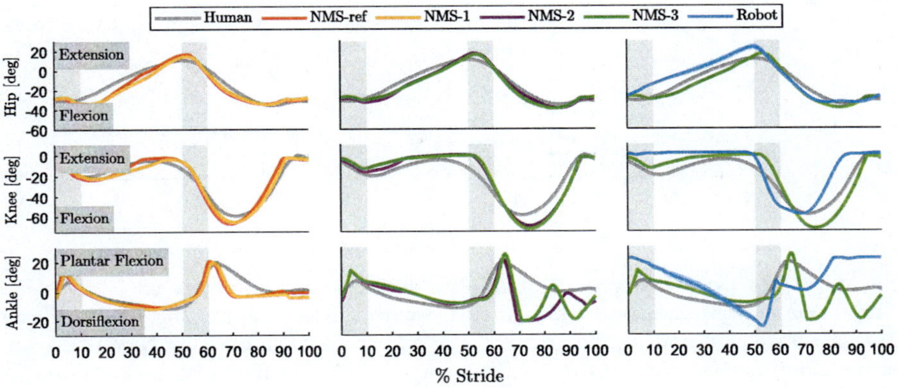

Fig. 3. Kinematics for hip, knee and ankle. Human data from [4] and sixth step for all NMS trials. Joint extension is always positive. Gray areas indicate human double support. *Left column:* NMS-ref and NMS-1 with very similar trajectories for all joints. *Middle column:* NMS-2 and NMS-3 with replacement of SOL MTU with deviations to human data at ankle and knee joint. *Right column:* NMS-3 and the robot. Knee and hip trajectories of the robot show similar shapes than NMS-3, the ankle shows clear deviations.

All trials show the characteristic APO (Fig. 4). In NMS-ref and NMS-1 the APO starts earlier than in humans (37% stride for NMS-ref, 40% for NMS-1 vs. 45% for human). In NMS-2 and NMS-3 the APO starts similar to human data and shows a double hump profile with interrupted unloading during the first half of the final double support phase. The total positive power output during APO with 15.9 J at 80 kg for NMS-ref is comparable to human data with 14.6 J at 1.3 m/s from [6] at 70.9 kg body mass. The power output increases for all experiments compared to NMS-ref by max. 16% in NMS-2. Negative power for loading and net energy at the ankle decrease from NMS-1 to NMS-3. We see a 1.8 times higher positive power peak when SOL's MTU is replaced by a spring compared to NMS-ref indicating that ankle peak power and thus power amplification (ratio of positive to negative peak power) are mainly determined by SOL. For more details see Table 3.

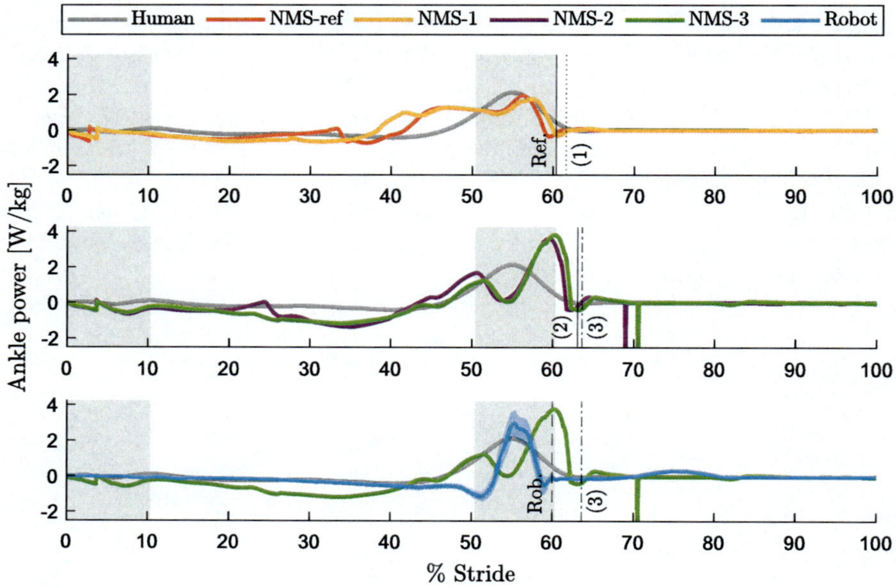

Fig. 4. Ankle power curves. Human data from [6], sixth step for all NMS trials. Gray areas indicate human double support, vertical lines take off. *Top:* NMS-ref and NMS-1. Ankle joint unloading starts earlier in NMS than in humans, for NMS-1 slightly earlier than for NMS-ref. *Middle:* NMS-2 and NMS-3. Loading with higher negative power compared to human data and faster unloading with higher positive power peak compared to NMS-1 and NMS-ref. Around 70% stride ankle hard stops engage. *Bottom:* NMS-3 and robotic data. The maximum peak for the robot is higher than for humans with similar peak timing but faster unloading. NMS-1, 2, 3 show a later take off than NMS-ref and humans.

3.2 Bipedal Robot

We successfully implemented NMS-3 on the bipedal robot. The robot walks continuously with a purely passive ankle joint at 0.55 m/s on a treadmill. With a leg length of 350 mm the robot walks at a *Froude number* of 0.09 compared to 0.16 for the NMS. Joint kinematics and ankle power curve of the robot, human, and NMS-3 are shown in Fig. 3 and 4. Hip and knee trajectories actively controlled by CPGs match NMS data reasonable well. Ankle kinematics however show clear differences. Especially dorsiflexion in swing and at the beginning of stance is missing. The ankle power curve shows the characteristic APO. With a positive peak power of $P_{max} = 3.0\,\text{W/kg}$ and a negative peak power of $P_{min} = -1.1\,\text{W/kg}$ the power amplification ratio $P_{PA} = 2.7$ is similar compared to NMS-ref, but smaller than in NMS-3 (Table 3). The peak timing at 55% stride is comparable to human data, but the unloading happens faster and at a higher peak power. *For detailed robotic results see* [23].

Table 3. Global gait measures and ankle power analysis. *GM:* average forward velocity of the upper body segment v_{HAT}, stride time t_s, duty factor DF and metabolic energy $\sum E_{metab}$ for the 6th step. *Ankle joint:* ankle power analysis for stance phase. ΔE^+ and ΔE^- denote the positive and negative power integral over one stride, ΔE_{tot} is their sum. P_{Max} and P_{Min} are the positive and negative peak values of the joint power. P_{PA} is the power amplification, i.e. the ratio of $|{}^{P_{Max}}/_{P_{Min}}|$.

		Unit	NMS-ref	MTU Replacement		
				NMS-1	NMS-2	NMS-3
GM	v_{HAT}	$[m/s]$	1.3	1.3	1.3	1.2
	t_s	[s]	1.2	1.2	1.2	1.1
	DF	[-]	0.6	0.6	0.6	0.6
	$\sum E_{metab}$	[W/kg]	5.9	5.5	5.3	4.7
Ankle joint	ΔE^+	[J]	15.9	18.6	18.8	17.8
	ΔE^-	[J]	−10.8	−13.1	−21.3	−23.1
	ΔE_{tot}	[J]	5.2	5.5	−2.5	−5.3
	P_{Max}	[W]	157.5	142.0	288.8	304.7
	P_{Min}	[W]	−57.3	−51.8	−109.4	−93.0
	P_{PA}	[-]	2.8	2.7	2.6	3.2

4 Discussion

Our initial research questions were **(I) Do humans need active ankle actuation for steady state walking** and **(II) What are the individual contributions of SOL's and GAS's MTUs to the APO in human walking**.

We used a 2D neuromuscular simulation and replaced (1) GAS, (2) SOL and (3) GAS and SOL MTUs by linear springs. We validated NMS-3 on a bipedal robot with active hips and knees and passive ankle joints.

All NMS trials and the robot walked stable showing an impulsive APO. With use of only elastic energy for ankle plantar flexion in NMS-3 and robot, continuous walking was possible supporting the hypothesis that humans mostly use the passive-elastic properties of GAS and SOL MTUs for walking [15,25].

In simulation we could reduce the metabolic costs by 20% from 5.9 W/kg in the NMS-ref to 4.7 W/kg in NMS-3. Comparable metabolic costs as in NMS-3 were observed for humans with 5–6 W/kg [26, Fig. 4(b)]. Ankle-powered gait is known to be very efficient [27] as efficiency of muscle work increases from proximal to distal [28]. Our results show, that therefore the APO and involved MTUs play a crucial role for the energy efficiency of human walking.

SOL is the stronger muscle with $F_{max,SOL} = 4000N$ compared to $F_{max,GAS} = 1500\,\text{N}$ [29], allowing for more elastic energy to be stored in the Achilles tendon. SOL moreover shows a larger influence on the reduction of metabolic energy than GAS (Table 3). SOL's strong influence on metabolic measures is in line with findings from Collins *et al.* [30] where an unpowered exoskeleton with a spring in parallel to the Achilles tendon reduced the metabolic energy consumption mainly by decreasing SOL's activity.

GAS shows less influence on the global joint kinematics when its MTU is replaced by a spring. Biarticular muscles have long tendons to support rapid recoil and the use of elastic energy [31] thus the MTUs behavior has a stronger elastic component.

Both muscles, GAS and SOL, have a passive role in human APO, isometrically holding the force when elastically loading the Achilles tendon [16]. Thus their MTUs can be replaced by fully passive springs as shown here. Further research is required to investigate the role of SOL and GAS in more detail, e.g. by eliminating GAS and/or SOL in the NMS, varying the stiffnesses for spring replacements in NMS-1, 2, 3 and investigating the power flow between knee and ankle.

NMS and robot only resemble sagittal plane motion and we used a coarse fitting for the spring replacement. Stance phase joint kinematics show good agreement with human data. Swing kinematics however do not match, especially for NMS-2 and NMS-3. The robot shows clear deviations from human data regarding ankle kinematics. Especially improvements on dorsiflexion capability are needed.

The well-known ankle power curve and the impulsive APO are preserved in both, robot and NMS. However, when SOL's MTU is replaced by a spring in NMS-2 and NMS-3, the APO comes with interrupted unloading; a case which has not been tested in the robotic model and for which no human data exists. We suspect the foot model has a yet unclear influence on this phenomenon. To investigate the foot model's influence in more detail we plan further studies with simulation and robotic hardware.

We conclude, that (I) steady state walking with impulsive APO does not require active ankle actuation. Elastic structures to store and release energy in a catapult-like fashion are sufficient in simulation and on a bipedal robot. For matching other gait phases or different gaits however, active muscles or more complex passive structures may be necessary.

(II) SOL is responsible for providing the necessary peak force and influences the APOs power amplification. GAS is responsible for coordination and energy transfer between knee and ankle using the elastic properties of its long tendons. Both muscles mainly hold the forces arising during elastic loading of their MTUs without active contraction.

Our research shows, that simulation in combination with a robotic model can help to improve the understanding of biomechanics and control principles in human walking. Simulation and robot facilitate decomposing complex structures and verifying findings under real-world conditions.

For designing prostheses, exoskeletons, or humanoid robots we conclude that no complex ankle actuation is needed to functionally reproduce the governing dynamic behavior of the ankle during stance. Incorporating more mechanical intelligence at the ankle could be a promising approach to restore and replicate human lower limb function without complex control or actuation to improve gait efficiency and reduced design complexity.

Acknowledgements. The authors thank Emre Cemal Gonen and An Mo for their help with the robot data. The project is funded by the Deutsche Forschungsgemeinschaft (DFG, German Research Foundation) - 449427815.

References

1. Sygulla, F., Rixen, D.: A force-control scheme for biped robots to walk over uneven terrain including partial footholds. Int. J. Adv. Rob. Syst. **17**, 172988141989747 (2020). https://doi.org/10.1177/1729881419897472
2. Lechler, K., et al.: Motorized biomechatronic upper and lower limb prostheses-clinically relevant outcomes. PM&R **10**, S207–S219 (2018). https://doi.org/10.1016/j.pmrj.2018.06.015
3. Cardona, M., Solanki, V.K., García Cena, C.E.: Exoskeleton Robots for Rehabilitation and Healthcare Devices. Springer, Singapore (2020). https://doi.org/10.1007/978-981-15-4732-4
4. Perry, J., Burnfield, J.M., Cabico, L.M.: Gait Analysis: Normal and Pathological Function. SLACK (2010). ISBN 9781556427664
5. Meinders, M., Gitter, A., Czerniecki, J.M.: The role of ankle plantar flexor muscle work during walking. Scand. J. Rehabil. Med. **30**, 39–46 (1998). https://doi.org/10.1080/003655098444309
6. Lipfert, S.W., Günther, M., Renjewski, D., Seyfarth, A.: Impulsive ankle push-off powers leg swing in human walking. J. Exp. Biol. **217**, 1218–1228 (2014). https://doi.org/10.1242/jeb.097345
7. McNeill, A.R., Bennet-Clark, H.C.: Storage of elastic strain energy in muscle and other tissues. Nature **265**, 114–117 (1977). https://doi.org/10.1038/265114a0
8. Alexander, R.M.: Energy-saving mechanisms in walking and running. J. Exp. Biol. **160**, 55–69 (1991). https://doi.org/10.1242/jeb.160.1.55
9. Alexander, M.R.: A model of bipedal locomotion on compliant legs (1992)
10. Hof, A.L., Geelen, B.A., van den Berg, J.: Calf muscle moment, work and efficiency in level walking; role of series elasticity. J. Biomech. **16**, 523–537 (1983). https://doi.org/10.1016/0021-9290(83)90067-2
11. Wilson, A.M., Watson, J.C., Lichtwark, G.A.: A catapult action for rapid limb protraction. Nature **421**, 35–36 (2003). https://doi.org/10.1038/421035a
12. Gronenberg, W.: Fast actions in small animals: springs and click mechanisms. J. Comparat. Physiol. A **178**, 727–734 (1996). https://doi.org/10.1007/bf00225821
13. Bennet-Clark, H.C.: The energetics of the jump of the locust schistocerca gregaria. J. Exp. Biol. **63**, 53–83 (1975). https://doi.org/10.1242/jeb.63.1.53
14. Nishikawa, K.C.: Neuromuscular control of prey capture in frogs. Philos. Trans. Roy. Soc. London Ser. B Biol. Sci. **354**, 941–954 (1999). https://doi.org/10.1098/rstb.1999.0445
15. Cronin, N.J., Prilutsky, B.I., Lichtwark, G.A., Maas, H.: Does ankle joint power reflect type of muscle action of soleus and gastrocnemius during walking in cats and humans? J. Biomech. **46**, 1383–1386 (2013). https://doi.org/10.1016/j.jbiomech.2013.02.023
16. Ishikawa, M., Komi, P.V., Grey, M.J., Lepola, V., Bruggemann, G.-P.: Muscle-tendon interaction and elastic energy usage in human walking. J. Appl. Physiol. (Bethesda Md.: 1985) **99**, 603–608 (2005). https://doi.org/10.1152/japplphysiol.00189.2005

17. Geyer, H., Herr, H.: A muscle-reflex model that encodes principles of legged mechanics produces human walking dynamics and muscle activities. IEEE Trans. Neural Syst. Rehabil. Eng.: Publ. IEEE Eng. Med. Biol. Soc. **18**, 263–273 (2010). https://doi.org/10.1109/TNSRE.2010.2047592
18. Geyer, H., Seyfarth, A.: Neuromuscular control models of human locomotion. In: Goswami, A., Vadakkepat, P. (eds.) Humanoid Robotics: A Reference, pp. 979–1007. Springer, Dordrecht (2019). https://doi.org/10.1007/978-94-007-6046-2_45. ISBN 978-94-007-6046-2
19. Veerkamp, K., et al.: Evaluating cost function criteria in predicting healthy gait. J. Biomech. **123**, 110530 (2021). https://doi.org/10.1016/j.jbiomech.2021.110530
20. Falisse, A., et al.: Rapid predictive simulations with complex musculoskeletal models suggest that diverse healthy and pathological human gaits can emerge from similar control strategies. J. R. Soc. Interface **16**, 20190402 (2019). https://doi.org/10.1098/rsif.2019.0402
21. Crowninshield, R.D., Brand, R.A.: A physiologically based criterion of muscle force prediction in locomotion. J. Biomech. **14**, 793–801 (1981). https://doi.org/10.1016/0021-9290(81)90035-X
22. Umberger, B.R., Gerritsen, K.G.M., Martin, P.E.: A model of human muscle energy expenditure. Comput. Methods Biomech. Biomed. Eng. **6**, 99–111 (2003). https://doi.org/10.1080/1025584031000091678
23. Kiss, B., et al.: Investigation on a bipedal robot: why do humans need both Soleus and Gastrocnemius muscles for ankle push-off during walking? 3 March 2022
24. Ijspeert, A.J.: Central pattern generators for locomotion control in animals and robots: a review. Neural Netw. **21**, 642–653 (2008). https://doi.org/10.1016/j.neunet.2008.03.014
25. Lichtwark, G.A., Bougoulias, K., Wilson, A.M.: Muscle fascicle and series elastic element length changes along the length of the human gastrocnemius during walking and running. J. Biomech. **40**, 157–164 (2007). https://doi.org/10.1016/j.jbiomech.2005.10.035
26. Umberger, B.R.: Stance and swing phase costs in human walking. J. R. Soc. Interface **7**, 1329–1340 (2010). https://doi.org/10.1098/rsif.2010.0084
27. Zelik, K.E., Huang, T.-W.P., Adamczyk, P.G., Kuo, A.D.: The role of series ankle elasticity in bipedal walking. J. Theor. Biol. **346**, 75–85 (2014). https://doi.org/10.1016/j.jtbi.2013.12.014
28. Sawicki, G.S., Lewis, C.L., Ferris, D.P.: It pays to have a spring in your step. Exerc. Sport Sci. Rev. **37**, 130–138 (2009). https://doi.org/10.1097/JES.0b013e31819c2df6
29. Yamaguchi, G.T., Sawa, A.G.U., Moran, D.W., Fessler, M.J., Winters, J.M.: A survey of human musculotendon actuator parameters (1990)
30. Collins, S.H., Wiggin, M.B., Sawicki, G.S.: Reducing the energy cost of human walking using an unpowered exoskeleton. Nature **522**, 212–215 (2015). https://doi.org/10.1038/nature14288
31. Schumacher, C., Sharba, M., Seyfarth, A., Rode, C.: Biarticular muscles in light of template models, experiments and robotics: a review. J. Roy. Soc. Interface **17**, 20180413 (2020). https://doi.org/10.1098/rsif.2018.0413

Grasping Characteristics of Flexible Propulsion Unit Using Braid Mechanism for Lunar Exploration Robot

Chikage Fujikawa[1](✉), Ryosuke Tokoi[1], Wataru Toyama[1], Manabu Okui[1], Tsuji Tomoaki[1], Taro Nakamura[1], and Takashi Kubota[2]

[1] Chuo University, Tokyo, Japan
{c_fujikawa,r_tokoi,w_toyama,m_okui}@bio.mech.chuo-u.ac.jp, {tsuji,nakamura}@mech.chuo-u.ac.jp

[2] Japan Aerospace Exploration Agency (JAXA), 3-1-1 Chuo-ku, Sagamihara-shi, Kanagawa 252-5210, Japan

Abstract. In this study, the authors model and conduct experiment on the braid propulsion unit of an excavation robot for lunar exploration. In existing studies, it is difficult for robots to create sufficiently big holes and achieve deep excavations. Therefore, in this study, we developed a lunar earthworm-type excavation robot "LEAVO," which attempts to excavate boreholes for environmental sensors and collect samples in particular layers. LEAVO has already succeeded in a vertical excavation of more than 938 mm and we have verified its usefulness in the first target mission. However, the existing propulsion unit of LEAVO cannot sufficiently bend and grip curved or uneven walls. In this research, we proposed a novel propulsion unit using braid to achieve curving excavation for the second target mission, modeled the gripping force of the proposed mechanism, and measured the gripping force of the prototype using the mechanism. The experiment demonstrated the validity of the modeling and the usefulness of the proposed mechanism.

Keywords: Braid · Peristaltic motion · Gripping force

1 Introduction

Various environmental surveys have been conducted on the Moon; however, several unknowns remain about the lunar sub-surface environment such as ground activity and subsurface constituents. Underground exploration is expected to help elucidate the origin of stars and discover new resources. Therefore, the Japan Aerospace Exploration Agency (JAXA) has set the installation of buried environmental survey sensors and collection of regolith samples as exploration goals [1]. Unmanned small excavation robots are required for these explorations owing to the transportation costs and the risk of ground collapses, as well as other hazards.

© The Author(s), under exclusive license to Springer Nature Switzerland AG 2023
J. M. Cascalho et al. (Eds.): CLAWAR 2022, LNNS 530, pp. 33–41, 2023.
https://doi.org/10.1007/978-3-031-15226-9_6

Existing excavation robots can be broadly classified into three types, according to their method of ground penetration: impact penetration-, screw drilling-, and biomimicry-type excavation robots. The Moon/Mars underground mole developed by Stoker et al. [2], an impact penetration-type robot, has a small body of diameter 40 mm and encounters difficulties in expanding boreholes. The robotic screw explorer developed by Nagaoka et al. [3] and mole-like robot developed by Yoshida et al. [4] are susceptible to soil pressure due to contact with the bodies and walls during excavation and are limited by their shallow depth of reach.

Therefore, in this study, we focused on the peristaltic motion of earthworms as a propulsion method that is less susceptible to soil pressure and easier to discharge soil; accordingly, we developed a lunar earthworm-type excavation robot (LEAVO) for lunar explorations [5]. LEAVO has successfully drilled a 938-mm vertical borehole and validated its usefulness with respect to borehole generation for buried sensors [6]. In addition, drilling tools and distributed driving systems (DDS) have been developed for bending and horizontal drilling, to collect samples at the same depth and handle underground obstacles [7, 8]. Threshold control using current sensors was introduced to the control method of the propulsion unit, which used to be the solely dependent on the rotation angle control of the motors for driving the propulsion unit. The threshold control enables autonomous drilling that responds to changes in borehole diameter. Furthermore, It was demonstrated that the grasping state can be measured by measuring the rotation angle and current of the motors for driving the propulsion unit [9]. However, in bending excavation, an existing propulsion unit is difficult to bend owing to its insufficient flexibility. In addition, the propulsion unit cannot adequately grasp a borehole that is prone to deformation.

Therefore, as an initial step, this study proposes a flexible propulsion unit using braid mechanism to address the problem posed by the reduced gripping force of the propulsion unit in bending excavation. Furthermore, we modeled the proposed unit and conducted a grasping force characteristic experiment using a prototype of the proposed unit to demonstrate the usefulness of the prototype and verify the validity of the modeling.

2 Lunar Earthworm-Type Excavation Robot "LEAVO"

2.1 Composition of the Robot

An overview of LEAVO comprising excavation and propulsion units is presented in Fig. 1. The excavation unit rotates an excavation tool called an auger to excavate the ground and transport the excavated soil backward. The propulsion unit comprises three propulsion subunits, and each subunit extends and expands in a peristaltic motion to grip the wall and move into the borehole. As illustrated in Fig. 2, LEAVO is propelled while excavating in cycle of (a)–(f).

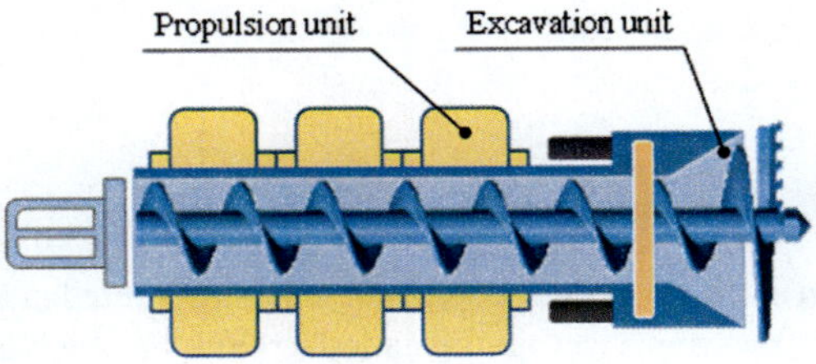

Fig. 1. Overview of LEAVO concept

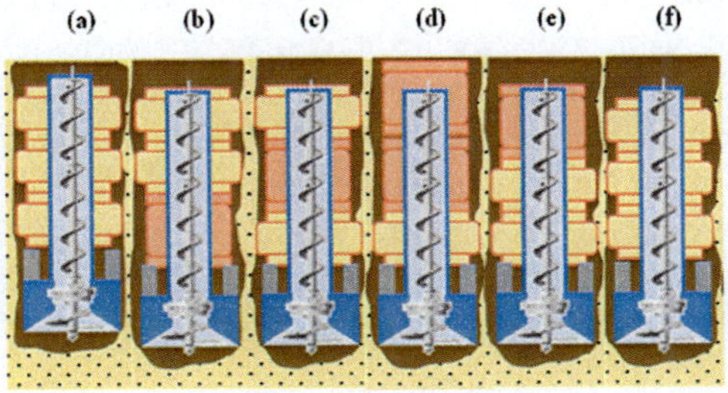

Fig. 2. Excavation process of LEAVO

2.2 Challenges for the Existing Robot in Curved Excavation

The conventional propulsion unit cannot bend owing to the insufficient flexibility of the gripping portion. In addition, as illustrated in Fig. 3, the gripping portion cannot adequately grasp curved or uneven wall surfaces because the rigid portion expands concentrically. Hence, to perform flexure drilling, these problems need to be addressed.

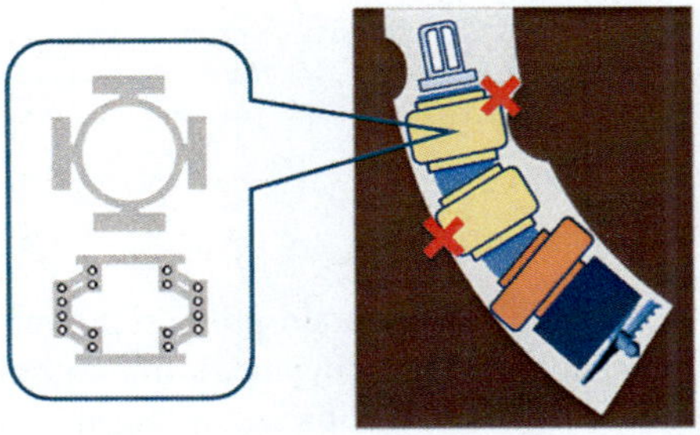

Fig. 3. Lack of gripping force of the existing propulsion unit

3 Braid Propulsion Unit

3.1 Requirements Specification

The following specifications are required for the propulsion unit to achieve flexural drilling:

- Ability to bend along the borehole
- Ability to grip curved or uneven wall surfaces

3.2 Braided-Inspired Propulsion Unit

A schematic diagram of the proposed propulsion unit is presented in Fig. 4. The grasping part of the proposed unit adopts a structure called a braid, in which the fibers are woven into a lattice pattern. Braiding has been utilized for the outer sleeves of McKibben artificial muscles [10]. The braid mechanism is also used in worm robots such as the Mesh-body Worm Robot [11]. Each fiber is helical and contracts axially as the diameter of the helix increases. Conversely, as the helix diameter decreases, it elongates in the axial direction. We believe that peristaltic motion can be reproduced by this thick short and thin long deformations, and that grasping is achieved in response to changes in the bore diameter. In addition, the braid propulsion unit can flexibly deform to adapt to the environment because each fiber is unglued. Therefore, the use of fibers with moderate flexibility and rigidity enables sufficient gripping of wall surfaces while accommodating bending and uneven wall surfaces.

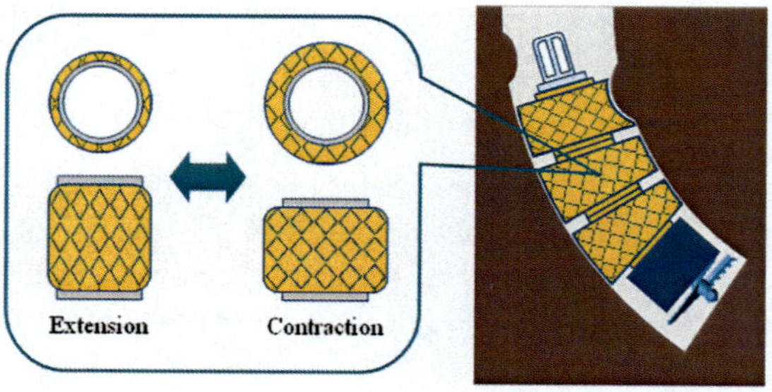

Fig. 4. Overview of flexible propulsion unit

3.3 Static Model of Gripping Force

Regarding the design and control of the braid propulsion unit, the proposed unit is modeled as a thin-walled cylinder. The gripping model of the proposed unit is illustrated in Fig. 5, and the parameters used in this section are presented in Table 1. The propulsion unit with initial height h and initial diameter d expands due to an applied compressive force F_{com} and receives external pressure p from the wall. Assuming that this causes the height to decrease by δ and the diameter to increase by Δd, Eq. (1) and (2) are obtained from the relationship between stress and strain.

$$-\frac{\delta}{h} = -\frac{F_{\mathrm{com}}}{\pi Edt} + \nu\frac{dp}{2tE} \tag{1}$$

$$\frac{\Delta d}{d} = -\frac{dp}{2tE} + \nu\frac{F_{\mathrm{com}}}{\pi Edt} \tag{2}$$

When $p = 0$, namely, no external pressure from the wall surface, Eq. (3) and (4) hold from Eq. (1) and (2).

$$\frac{F_{com}}{\pi dt} = E\frac{\delta}{h} \tag{3}$$

$$\frac{\Delta d}{d} = \nu\frac{\delta}{h} \tag{4}$$

Young's modulus and Poisson's ratio are required to derive the grasping force described below. Therefore, compressive forces are applied to the propulsion unit with $p = 0$, and the two parameters are experimentally obtained from the measured dimensional changes and Eq. (3) and (4). The external pressure p from the wall surface is obtained from Eq. (1) and (2) and can be expressed as:

$$p = \frac{2}{\pi d^2}(\nu F_{com} - \pi Et\Delta d) \tag{5}$$

The gripping force of the propulsion unit is illustrated in Fig. 6. If the vertical component of the force that the half-wall receives from the propulsion unit is defined as the gripping force, the grip force F_{grip} is given by Eq. (6).

$$F_{grip} = hd\int_0^{\pi} p \sin\theta d\theta = 2phd \tag{6}$$

Therefore, the gripping force can be derived from the compressive force and the change in the dimensions and diameter of the propulsion unit.

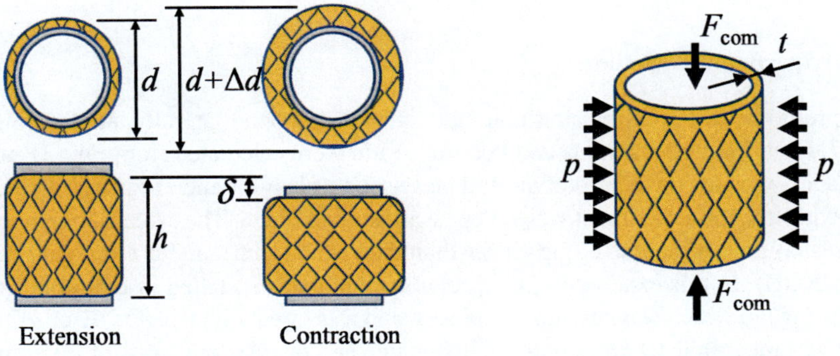

(a) Flexible propulsion unit (b) Thin-walled cylinder

Fig. 5. Static model of the flexible propulsion unit

4 Gripping Force Characteristic Experiment of Prototype

4.1 Objective of Experiment

In this chapter, the grasping force characteristic experiment of a prototype (initial diameter: 156.3 mm) made of Polypropylene bands (TRUSCO) is conducted to evaluate the modeling's validity and the usefulness of the proposed propulsion unit.

Table 1. Parameters of the model

h	Initial height	m	F_{com}	Compressive force	N
δ	Height displacement	m	E	Young's modulus	Pa
d	Initial diameter	m	ν	Poisson's ratio	
Δd	Diameter displacement	m	p	Pressure by soil	N/m^2
t	Thickness of gripping part	m	F_{grip}	Gripping force	N
π	Pi		θ	Angle of grip range	Rad

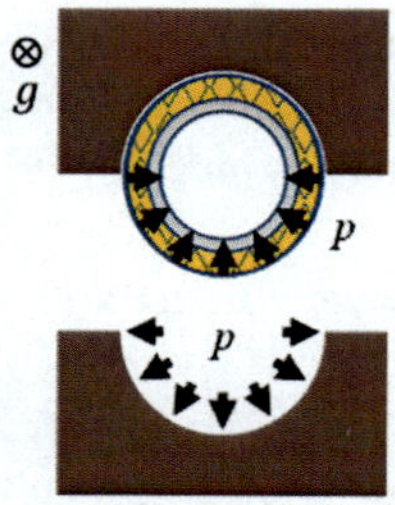

Fig. 6. Gripping force of the propulsion unit

4.2 Experimental Method

The prototype of the propulsion unit can contract and bend laterally, as illustrated in Fig. 7. First, Young's modulus and Poisson's ratio were calculated from Eq. (3) and (4) by placing weights on the prototype and measuring its heights and expansion diameters. Subsequently, the experiment was set up, as illustrated Fig. 8. The system comprises the propulsion unit, a measuring jig (inner diameter: 162.2 mm), and a force gauge (RZ-20, AIKOH). In this experiment, the propulsion unit was installed in the center of the measuring jig, which was cut into two pieces, and was fitted with weights. Each gripping force was measured 10 times with a force gauge and the average value of the gripping force at each compressive force was calculated.

4.3 Results and Discussion

The dimensions of the proposed propulsion unit at each compressive force are presented in Table 2. From presented the data, the Young's modulus and Poisson's ratio of the prototype were $E = 0.458$ MPa and $\nu = 0.936$, respectively. The theoretical and experimental values of gripping force at each compression force are presented in Fig. 9. Here, the experimental radial variation is $\Delta d = 0.006$ m. The experimental values of the gripping force increased as the compressive force increased. The obtained results indicate that the experimental values generally follow the same trend as the theoretical values.

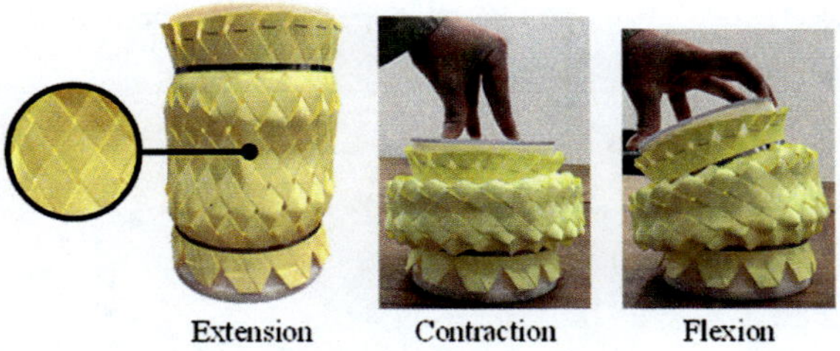

Fig. 7. Prototype of the propulsion unit

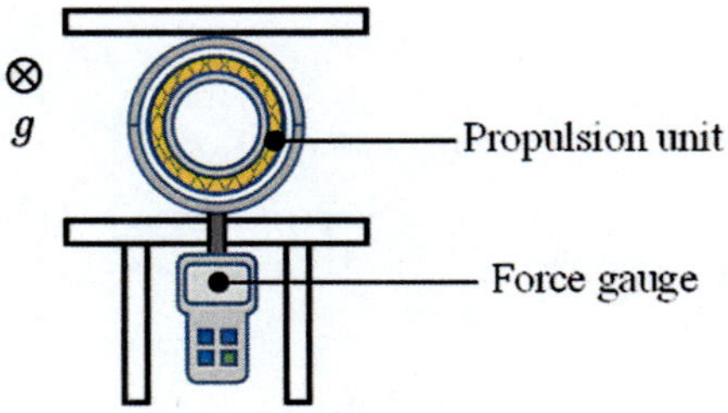

Fig. 8. Experimental setup of the gripping force properties

After the propulsion unit expanded to the inner diameter of the jig, the gripping force exhibited a corresponding decline in its magnitude. The uneven manual weaving of the fibers causes the body to expand while tilting, as illustrated in Fig. 10. We consider that the tilt of the body caused the contact with the wall to be non-uniform in the circumferential direction, thereby triggering a decrease in the gripping force. Therefore, we will reduce the inclination of the body by employing machine knitting and changing the fibers used.

Table 2. Values of compressive force and prototype dimensions

F_{com} [N]	δ [mm]	h [mm]	d [mm]	Δd [mm]	t [mm]
9.1	4.9			3.7	
13.0	8.2			7.5	
17.4	10.6	130.6	156.3	10.7	0.95
21.4	12.4			15.4	
25.8	16.3			19.3	

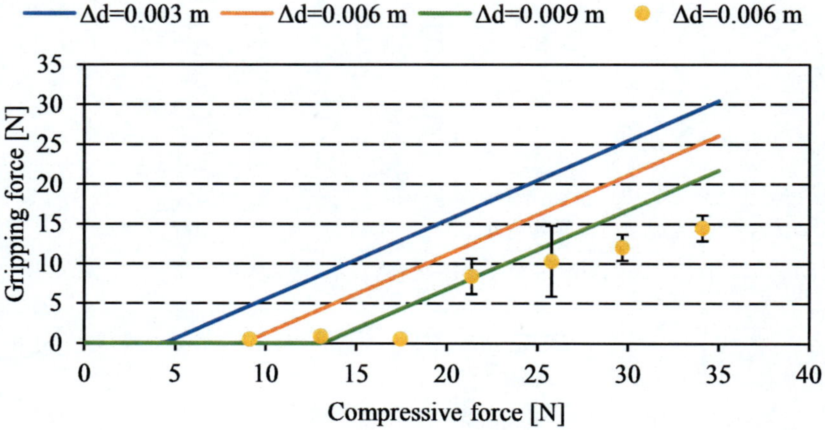

Fig. 9. Experimental results of gripping force properties (lines and markers indicate theoretical and experimental values, respectively)

Fig. 10. Tilt of the prototype

5 Conclusion

This study proposed a flexible propulsion unit with a braid mechanism to address the problem posed by the inadequate grip of the propulsion unit in bending excavation. Furthermore, the proposed mechanism was modeled and an experiment was conducted on gripping force characteristics using the prototype, to demonstrate the usefulness of the prototype and verify the validity of the modeling.

The values obtained from the experiments on the gripping force at each compressive force with a diameter change of $\Delta d = 0.006$ m generally exhibited the same tendency as the theoretical values. In addition, it was verified that the prototype is bendable.

In the future, we will conduct tests on gripping force characteristics for various hole diameters and examine the adopted fibers, fabrication method, and shape, to develop a braid propulsion unit that enables stable gripping in bending excavation.

References

1. Nagaoka, K., Kubota, T., Otsuki, M.: Experimental study on autonomous burrowing screw robot for subsurface exploration on the Moon. In: Proc. IEEE Int. Conf. on Intelligent Robots and Systems. Nice, France (2008)

2. Stoker, C.R., Gonzales, A., Zavaleta, J.R.: Moon/Mars Underground Mole. In: Proceedings of the NASA Science and Technology Conference (2007)
3. Nagaoka, K., Kubota, T., Otsuki, M., Tanaka, S.: Robotic screw explorer for lunar subsurface investigation: Dynamics modeling and experimental validation. In: Proc. IEEE Int. Conf. on robotics and Automation, pp. 1–6. Kobe, Japan (2009)
4. Yoshida, K., Kudo, T., Kawakatsu, Y., Yokoyama, T., Sonoyama M.: Development of a Mole-like Robot for lunar subsurface exploration. In: JSME Conference on Robotics and Mechatronics, 1P1–07. Tsukuba, Japan (2020)
5. Omori, H., Murakami, T., Nagai, H., Nakamura, T., Kubota, T.: Planetary subsurface explorer robot with propulsion units for peristaltic crawling. In: Proc. IEEE Int. Conf. Robotics and Automation. Shanghai, China (2011)
6. Nakatake, T., Konno, M., Mizushina, A., Yamada, Y., Nakamura, T., Kubota, T.: Soil circulating system for a lunar subsurface explorer robot using a peristaltic crawling mechanism. In: Proc. IEEE Int. Conf. on Advanced Intelligent Mechatronics, pp. 407–412. Banff, Canada (2016)
7. Fujiwara, A., et al.: Development of both-ends supported flexible auger for lunar earthworm-type excavation robot LEAVO. In: IEEE/ASME International Conference on Advanced Intelligent Mechatronics, pp. 924–929. Auckland, New Zealand (2018)
8. Watanabe, T., et al.: A development of front actuation-type excavating unit for lunar excavating exploration robot "LEAVO". In: JSME Conference on Robotics and Mechatronics, pp. 2P1–T06 (2019)
9. Tokoi, R., et al.: Propulsion unit control method for visualization of grasping state of earthworm-type lunar excavation robot. In: IEEE International Conference on Advanced Intelligent Mechatronics, p. 675681 (2021)
10. Chen, D.H., Ushijima, K.: Prediction of the mechanical performance of McKibben artificial muscle actuator. Int. J. Mech. Sci. **78**, 183–192 (2014)
11. Huang, Y., Kandhari, A., Chiel, H.J., Quinn, R.D., Daltorio, K.A.: Mathematical Modeling to Improve Control of Mesh Body for Peristaltic Locomotion. In: Mangan, M., Cutkosky, M., Mura, A., Verschure, P.F.M.J., Prescott, T., Lepora, N. (eds.) Living Machines 2017. LNCS (LNAI), vol. 10384, pp. 193–203. Springer, Cham (2017). https://doi.org/10.1007/978-3-319-63537-8_17

Bio-Inspired Imprecise Impedance Control of Muscle-Driven Robotic Limbs

Patrick Vonwirth(✉) and Karsten Berns

Department of Computer Science, Technische Universität Kaiserslautern, 67663 Kaiserslautern, Germany
{vonwirth,berns}@cs.uni-kl.de
https://agrosy.cs.uni-kl.de

Abstract. Breaking up with classical, precision-based control, the work at hand introduces a whole new bio-inspired control philosophy, the so-called *imprecise impedance control*. By combining technical and biological research observations, a very simple, but effective way of artificial, muscular control based on electrical series elastic actuators is developed. The presented control structure includes the whole drive train starting from abstract, digital control commands and ending with the electrical current that is given to the motors. Experimental evaluation demonstrates the proper functioning of the control as well as its stability and limitations.

Keywords: Bio-inspired · Impedance control · Imprecise control · Artificial muscles · Series elastic actuator · Humanoid robot

1 Introduction

Looking into the history of robotic locomotion, a relationship between the capabilities of actuation systems and robotic skills can be recognized. Today's industry robot arms are perfectly controlled by high precision motors with big gear ratios and different types of wheels have been invented for fast and robust mobility. Legged locomotion, however, only played a minor role in the past mobile robotic research. Mainly caused by the big problem of physical impact damage to stiff mechanical systems that, unfortunately, is inherent to legged locomotion. With the invention of soft actuation systems, e.g. pneumatic/hydraulic artificial muscles or series elastic actuators (SEAs), this area of research is now slowly coming back to interest. Recent progress on legged locomotion especially has been made in the context of quadruped robotic platforms, e.g. among others: *MIT Cheetah* [17], *HyQ* [16], *ANYmal* [4] or *Spot* [1]. However, comparing these against animals, it is obvious that there is still a lot to learn from nature.

A general proof-of-concept of bio-inspired bipedal locomotion has been given by Luksch [5] in 2010 with the development of a virtual bipedal robot control system, capable of walking over rough terrain. The first attempt to transfer

© The Author(s), under exclusive license to Springer Nature Switzerland AG 2023
J. M. Cascalho et al. (Eds.): CLAWAR 2022, LNNS 530, pp. 42–53, 2023.
https://doi.org/10.1007/978-3-031-15226-9_7

this pure virtual concept to a real machine then successfully has been demonstrated by Schütz in 2020 [13]. Following natural motorization principles, the planar, bipedal robot CARL (Fig. 4a) [15] has been built featuring both, mono- and biarticular muscular actuation using electrical SEAs [14]. As one aspect of impedance control, pure force control of muscular-driven robots has then first been analyzed in detail and successfully tested on CARL [10]. However, extensive experimental tries on controlling the other aspects of impedance (i.e. stiffness and damping) on CARL have failed. The growing complexity of too many nested and externally coupled control loops artificially lowered the stability limit of the overall system far below its physical capabilities.

Counteracting this, the paper at hand describes a newly developed, very simple, bio-inspired concept of *imprecise impedance control* based on the hardware of series elastic actuation systems. In contrast to other state-of-the-art approaches, the herein introduced concept does not require any dynamic model of the hardware. The innovative approach offers an especially simple, but sufficient strategy for compliant motion control of muscular-driven robotic limbs. Classical control goals like high precision are explicitly overruled by bio-inspired features and simplicity. Throughout this work, the whole actuation concept from single actuator control to abstract high-level limb control is sketched.

First, the design and control of a single actuation unit is sketched in Sect. 2. Section 3 then covers the high-level limb abstraction offering a simple work-space control API. Finally, the overall control approach is evaluated and concluded in the Sects. 4 and 5, respectively.

2 Bio-inspired Control of a Series Elastic Actuator

One of the biggest problems in legged locomotion is the necessity to compromise between controllability and impact tolerance. In 2015, Semini et al. hence proposed the principle of *stiff inside, soft outside* [16] as a recommended design guideline for such systems. They suggest building maximal stiff legs for high-precision control while environmental impacts are compensated by special, soft contact areas, e.g. rubber feet. This concept is quite common in legged robotics and has e.g. successfully been demonstrated in the design of the bipedal robot *DURUS* at the same time [12]. However, this approach limits the robot to mandatorily interact with nothing but the special contact areas. Furthermore, Semini et al. [16] propose to use pure active impedance control due to the better controllability compared to SEAs that suffer from their passive compliant elements inside. This statement is based on the assumption that SEAs are (classically) equipped with soft springs and controlled to act as an ideal, high-precision source of force.

Such a classical SEA control has been developed and successfully evaluated for the so-called RRLAB-SEA [14]. However, although successfully tested on a single actuator, the combined control of all five SEAs mounted on one leg of CARL uncovered big stability issues with unintended oscillations. Due to the pure amount of nested control loops and oscillation circuits at different frequencies, the overall system simply becomes too complex to be controllable the way it

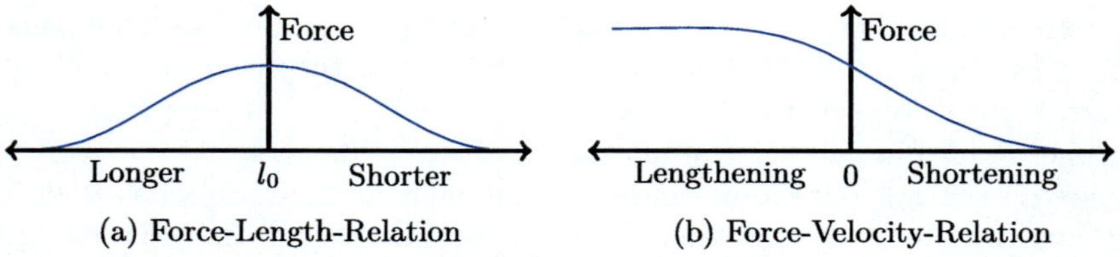

Fig. 1. Sketches of the principal impedance characteristics of natural muscles.

has been implemented originally. From low-level to high-level, mechanical and digital systems that are capable to oscillate are e.g.: The motor's mass against the physical spring; The by-passed PID spring-deflection controller; The overall actuator-impedance controller; The physical springs of several SEAs connected to one shared joint; The actuator force distribution strategy across all available actuators; The physical coupling of both legs; The high-level control strategy of the whole robot. Further information about these can be found in [8,13].

Looking into nature, a biological muscle can be described (strongly simplified) as a contracting motor part in series with a stretchable tendon. In between, the so-called *Golgi Tendon Organ* is located, sensing changes in the muscle-tendon tension. This organ is used to measure and control muscular tension, e.g. by triggering the commonly known muscle stretch reflex [6]. Following this natural construction, we propose to shift the principle of *stiff inside, soft outside* from limb-level to actuator-level. I.e., to use the mechanical structure of a SEA without controlling its elasticity. This way, the previously described stability issues with coupled SEAs are avoided at the cost of control precision. The elastic element, the *soft-outside*-part, hence reduces to only protect the completely stiff internal drive train from physical impacts. Due to the lack of feedback and control of the elastic element, external, long-term disturbances have to be handled by the actuator's transparency, i.e. its backdrivability, which becomes mandatory.

In 1985, Mussa-Ivaldi et al. [7] found out that the short-range end-point stiffness of a human arm has three important properties: First, measured stiffness matrices are nearly symmetric and can be assumed to be elliptical. Second, the general shape/orientation of the end-point stiffness ellipse is mainly influenced by the arm configuration, not by the muscles' activities. And third, muscular co-contraction strongly influences the overall size, i.e. the strength, of the stiffness ellipse. All these insights lead to the strong assumption that the stiffness is generated on a muscular level and not by some neuronal control loop. This hypothesis is further supported by recent studies about the effect of having natural (muscular) dynamics built-in inside actuators [3]. It has been found out that the pure existence of muscular output characteristics significantly reduces the amount of required central control commands. These properties build up a *low-level, zero-delay feedback system*, the so-called *Preflexes* [2].

It is well known that the force of an activated muscle is non-constant [2,20]. While the temporal force dependency can be ignored as an unnecessary limitation, force-length and force-velocity relationships form the build-in impedance

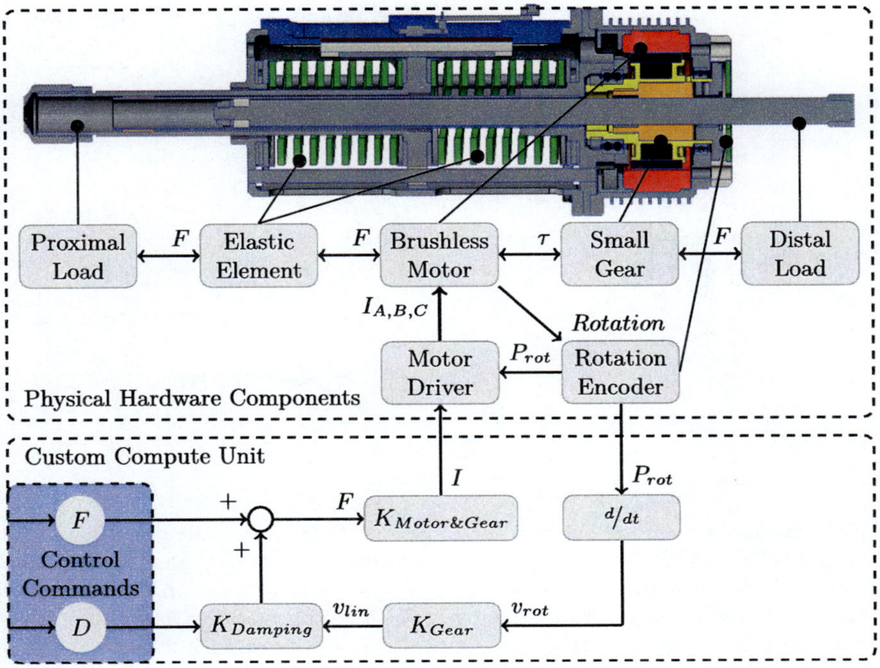

Fig. 2. Sketch of the bio-inspired control of one actuation unit. From top to bottom, the mechanical construction of the RRLAB-SEA with the respective physical drive train [14] and the novel, primitive, digital control.

of muscle. These two characteristics are sketched in principle within Fig. 1. The derivatives of both relationships can be interpreted as stiffness and damping respectively. While the stiffness contains positive and negative sections, the damping characteristic always stays non-negative. From the theoretical point of view, the muscular stiffness is hence partially unstable and therefore unsuitable for simple control. On the other hand, the muscular damping can offer local stability throughout all lengths and velocities.

Furthermore, damping is a non-directional property. While forces can be counteracted by antagonistic muscles, damping is always summed up. Hence, with the introduction of damping as the only secondary muscular control property next to force, an overall consistent concept of artificial muscular control is constructed. Intermuscular damping is hereby representing the co-contraction of a set of antagonistic muscles. Copying the natural separation of the contractile and the elastic element of a muscle, the overall required control logic of one actuator reduces to closely nothing. Figure 2 sketches the whole muscular control scheme from digital input commands to physical output at the example of the RRLAB-SEA hardware. It can be seen that the proposed approach is greatly simplified in comparison with traditional SEA control strategies, e.g. [11,14]. Muscular simulations by Brown and Loeb [2] that state a significant

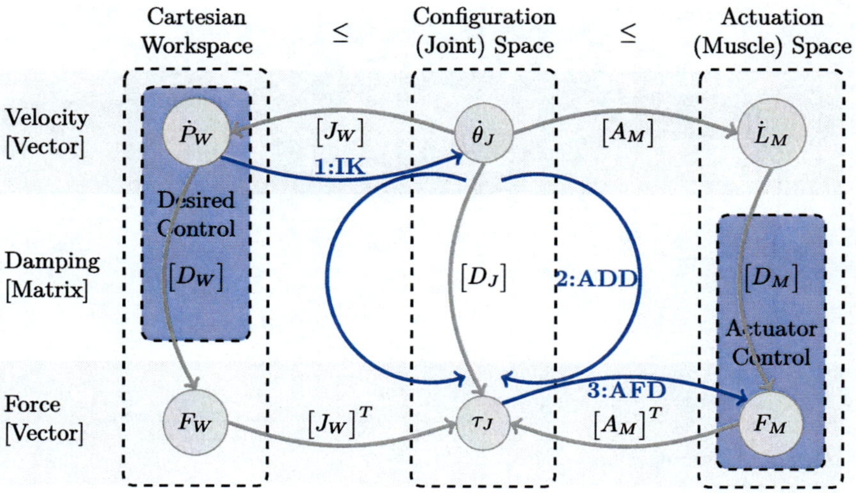

Fig. 3. Kinematic coupling between the Cartesian, the Configuration and the Actuation Space of a muscular-driven robotic limb. The two matrices $[J_W]$ and $[A_M]$ represent the *Position Jacobian* and the *Actuator Lever Arms* respectively [10]. All three matrices $[D_{W/J/M}]$ represent n-dimensional damping. The blue indicated control transformation is done in three steps from left to right: *1.IK* (Inverse Kinematics), *2.ADD* (Actuator Damping Distribution) and finally *3.AFD* (Actuator Force Distribution).

increasement of stability when introducing muscular damping while muscular stiffness has more or less no effect, further support the proposed muscular control concept.

3 Automatic Muscle Recruitment on Arbitrary Musculoskeletal Robotic Limbs

The kinematic coupling between the Cartesian workspace of a single limb, that is used as abstract control input and all actuator control values is sketched in Fig. 3. From left to right, the abstract Cartesian workspace, the configuration space and the actuation space are shown. The Cartesian workspace is the most simple abstraction offering an easy-to-use API for higher level limb control. In the middle, the configuration space represents the actual physical limb, defined by its joints. Finally, the so-called actuation space covers all actuators on the limb. Note, that the dimensionality of the three spaces might increase from left to right. The two matrices that describe the coupling between these three spaces, the *Jacobian* matrix $[J_W]$ and the *Lever Arm* matrix $[A_M]$ are hence in general not invertible. In general, the *Lever Arm* matrix $[A_M]$ contains the transmission ratios between the actuators and the joints of the robotic limb. Due to the background of using linear actuators, these transmission gears are classical lever arms. Further details about the kinematic coupling of actuators and the lever arms of CARL can be found in [10]. The CAD model of a single, motorized leg of CARL is given in Fig. 4b. The three damping matrices $[D_W]$, $[D_J]$ and $[D_M]$ all connect the velocity and force vectors inside their appropriate spaces. All three are symmetric and positive semidefinite in order to represent proper, non-negative, n-dimensional damping. The actuator damping $[D_M]$ is furthermore diagonal with the individual non-negative actuator damping values $\vec{d}_M$ on its diagonal: $[D_M] = diag(\vec{d}_M)$.

The flow of control data through the kinematic network is represented in Fig. 3 by the blue-colored elements. On the left side, the workspace control (velocity and damping) is entering the network. By three non-trivial, sequentially applied operations, these inputs are transformed into actuator controls (damping and force) on the right:

1. *IK:* Inverse Kinematics
2. *ADD:* Actuator Damping Distribution
3. *AFD:* Actuator Force Distribution

First, the abstract workspace velocity is translated into joint velocities by applying an inverse kinematic algorithm as e.g. presented in [19]. As a second step, all actuator damping controls are calculated using some ADD approach:

Since both, the Jacobian and the lever arm matrix are not invertible, it is not possible to simply calculate the actuator damping commands straight forward from the desired input. Furthermore, due to the structures of these two matrices, it is in general not even possible to reproduce the exact desired workspace damping using only the actuators' damping properties. Hence, a possible strategy to find suitable muscular damping values is to best fit *possible* joint space damping to the desired one.

$$\forall \dot{\theta}_J \in \{\dot{\theta}_J, \|\dot{\theta}_J\| = 1\} : \mathbf{A}_M^T \cdot \mathbf{D}_M \cdot \mathbf{A}_M \cdot \dot{\theta}_J \stackrel{!}{=} \mathbf{D}_J \cdot \dot{\theta}_J \tag{1}$$

This can be done by e.g. minimizing the integral of the squared torque difference over all possible unit joint velocities. Define:

$$\begin{aligned} &trace\,(\cdot) \text{ being the trace of matrix } [\cdot] \\ &row_i\,(\cdot) \text{ being the} i\text{-th row of matrix } [\cdot] \\ &row\,(\cdot) := \left(row_0\,(\cdot)\,,\ldots,row_i\,(\cdot)\right) \text{ i.e. being all rows concatenated} \end{aligned}$$

And derive the motor damping approximation formula:

$$\begin{aligned} &\min_{\mathbf{D}_M} \int_{\|\dot{\theta}_J\|=1} \left\| \left(\mathbf{A}_M^T \mathbf{D}_M \mathbf{A}_M - \mathbf{D}_J\right) \cdot \dot{\theta}_J \right\|^2 \qquad (2) \\ \Rightarrow \quad &\min_{\mathbf{D}_M} trace\left(\left(\mathbf{A}_M^T \mathbf{D}_M \mathbf{A}_M\right)^T \left(\mathbf{A}_M^T \mathbf{D}_M \mathbf{A}_M\right)\right) \\ &-2 \cdot trace\left(\left(\mathbf{A}_M^T \mathbf{D}_M \mathbf{A}_M\right)^T \mathbf{D}_J\right) \end{aligned}$$

$$\begin{aligned}
\Rightarrow \quad & \min_{\mathbf{D}_M} \frac{1}{2} row\left(\mathbf{A}_M^T \mathbf{D}_M \mathbf{A}_M\right) \cdot row\left(\mathbf{A}_M^T \mathbf{D}_M \mathbf{A}_M\right)^T \\
& \quad - row\left(\mathbf{A}_M^T \mathbf{D}_M \mathbf{A}_M\right) \cdot row\left(\mathbf{D}_J\right)^T \\
\Rightarrow \quad & \min_{\vec{d}_M} \frac{1}{2}\, \vec{d}_M^T \cdot \mathbf{L} \cdot \mathbf{L}^T \cdot \vec{d}_M - \vec{d}_M^T \cdot \mathbf{L} \cdot row\left(\mathbf{D}_J\right)^T
\end{aligned} \tag{3}$$

with

$$\mathbf{D}_J := \mathbf{J}_W^T \cdot \mathbf{D}_W \cdot \mathbf{J}_W \tag{4}$$

$$\mathbf{L} := \begin{bmatrix} row\left(row_0\left(\mathbf{A}_M\right)^T \cdot row_0\left(\mathbf{A}_M\right)\right) \\ \vdots \\ row\left(row_i\left(\mathbf{A}_M\right)^T \cdot row_i\left(\mathbf{A}_M\right)\right) \end{bmatrix} \tag{5}$$

Knowing the desired joint velocities and the actuator damping properties, the vector of all actuator forces $\vec{F}_M$ can be calculated.

$$\mathbf{A}_M^T \cdot \mathbf{D}_M \cdot \mathbf{A}_M \cdot \vec{\dot{\theta}}_J = \vec{\tau}_J \stackrel{!}{=} \mathbf{A}_M^T \cdot \vec{F}_M \tag{6}$$

In case of an overactuated limb as is the case on CARL, the lever arm matrix $\mathbf{A}_M$ is non-invertible and advanced *AFD* approaches have to be applied. Possible *AFD* algorithms, applied on CARL can be found in [9].

4 Imprecise Control Executed on the Planar Robotic Leg Carl

The design of the proposed muscular control is neither based on a concrete goal like achieving energy efficiency or precise force tracking, nor it is meant to be executed on a single SEA. As introduced above, the main goal of this work is to introduce and evaluate a bio-inspired, muscular control strategy for natural, imprecise movements. Against common habit, the work at hand hence does not cover an independent analysis of the performance of the proposed muscular control approach with only a single SEA being involved. Especially being designed simple and *imprecise*, standard rating methods cannot be applied to judge the herein presented studies. Instead, the introduced approach of *bio-inspired imprecise impedance control* is evaluated within the abstract Cartesian workspace on the planar, compliant robotic leg CARL [15] as pictured in Fig. 4.

Several compliant trajectories have been executed on the ankle of the right leg of CARL while keeping the desired foot orientation and impedance constant. A so-called *compliant trajectory* is hereby defined to be the equilibrium position and velocity of a virtual impedance element (spring/damper) that is virtually connected to the ankle of CARL. Figure 4c sketches the general principle of the compliant trajectory execution.

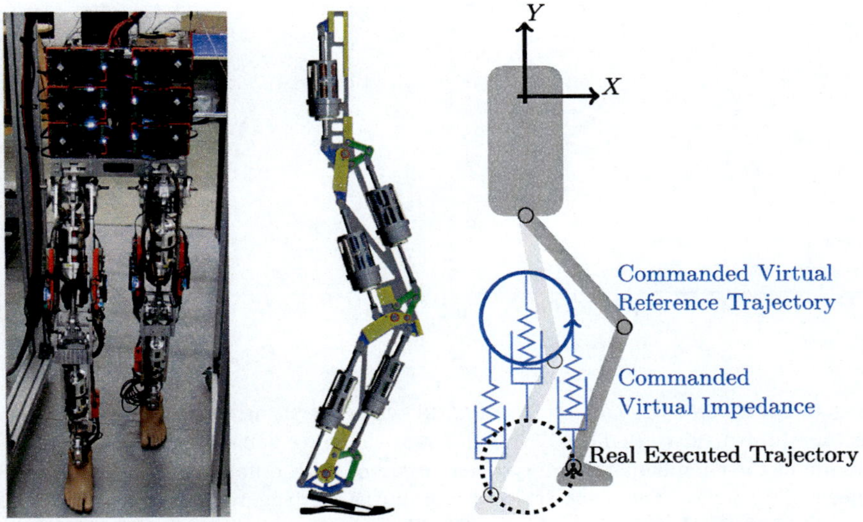

(a) Bipedal Robot CARL (b) CAD Model (c) Compliant Trajectory

Fig. 4. Bio-inspired Imprecise Impedance Control applied to the planar robotic leg CARL [15]. On the left, Figure (a) pictures the full bipedal version of CARL. Figure (b) shows the CAD model one leg of CARL, visualizing its physical joints and actuator mounting. On the right, the principle of a so-called *compliant trajectory* is sketched by Figure (c).

The virtual impedance element is composed of a spring and a damper in parallel. While the 2-dimensional damping value of the virtual damper can directly be used as the desired Cartesian damping control, the spring characteristic has to be mapped onto a desired Cartesian velocity (see Fig. 3). Since force has to be generated by a desired velocity with respect to a given damping, the virtual spring stiffness $\mathbf{K}$ has to be defined as follows

$$\mathbf{K} := \mathbf{D}_W \cdot \mathbf{K}_v \tag{7}$$

with $\mathbf{D}_W$ being the Cartesian Work Space Damping and $\mathbf{K}_v$ mapping position errors to desired velocities. In order to keep this virtual mass-spring-damper system stable, $\mathbf{K}_v$ is chosen to keep the system at least critically damped with respect to the given damping $\mathbf{D}_W$ and estimated mass M.

$$2 \cdot \sqrt{\mathbf{K} \cdot M} = \mathbf{D}_W \tag{8}$$

$$\sqrt{\mathbf{D}_W \cdot \mathbf{K}_v \cdot M} = \frac{1}{2}\mathbf{D}_W$$

$$\mathbf{K}_v = \frac{1}{4M}\mathbf{D}_W \tag{9}$$

As long as the estimated mass M is not underestimating the real mass, the virtual impedance will be over-, or at least critically damped. Due to the absence

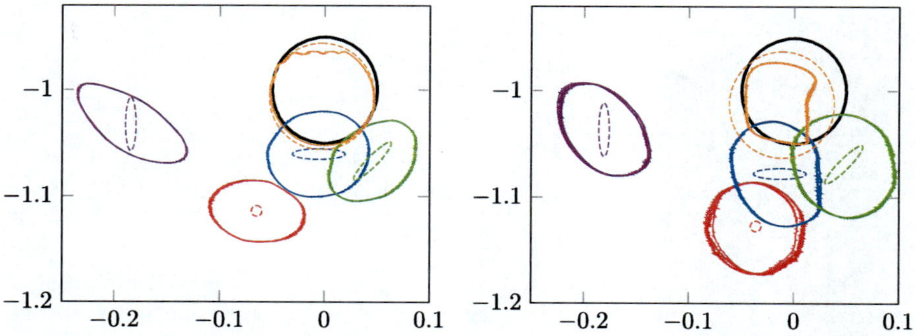

(a) At 30 RPM

(b) At 120 RPM

Fig. 5. Resulting ankle trajectories of CARL when executing a circular, compliant trajectory (drawn in thick black) with different damping values. Figure (a) shows the results of the compliant trajectory being executed with different damping values at a speed of 30 RPM. The same trajectory executions at identical damping values at a speed of 120 RPM is shown in Figure (b). The two-dimensional damping matrices are plotted as dashed ellipses of the same color as the appropriate ankle trajectories.

of any gravity (and inertia compensation) components, the real ankle trajectory is assumed to take place below the commanded one.

In order to verify and evaluate the functionality of the proposed imprecise control approach, two series of experiments have been performed. In all experiments, a virtual, circular compliant trajectory with a radius of 5 cm has been executed around the mid point $[0.0, -1.0]$ in the coordinate frame of the main body of CARL. The resulting real ankle trajectories at different desired workspace damping are shown for two different trajectory execution speeds of 30 RPM and 120 RPM in the Figs. 5a and 5b, respectively. The two-dimensional damping matrices that have been commanded to the system are plotted as dashed ellipses of the same color as the appropriate ankle trajectories. Their values are: *Red:* $\left[\begin{smallmatrix}50 & 0\\ 0 & 50\end{smallmatrix}\right]$, *Purple:* $\left[\begin{smallmatrix}50 & 0\\ 0 & 250\end{smallmatrix}\right]$, *Blue:* $\left[\begin{smallmatrix}250 & 0\\ 0 & 50\end{smallmatrix}\right]$, *Green:* $\left[\begin{smallmatrix}150 & 100\\ 100 & 150\end{smallmatrix}\right]$ and *Orange:* $\left[\begin{smallmatrix}500 & 0\\ 0 & 500\end{smallmatrix}\right]$ The trajectory plots are five consecutive compliant cycles that have been recorded at 1 kHz and are plotted raw, without any post-processing. I.e. for all 30 RPM experiments a total amount of 10000 positions per experiment have been recorded in each 10 s of execution. For all 120 RPM experiments, the execution duration and appropriately also the amount of recorded positions is quartered. All experiments have been executed with an estimated virtual mass of $M = 15.0\,\text{kg}$.

Both plots in Fig. 5 identically show the effect of changing the desired virtual impedance on the resulting real trajectory. As intended, increasing the virtual impedance increases the appropriate position accuracy.

The purple trajectory compared to the red one shows that the enlargement of the y-damping decreases the gravitational effect in this direction. However, since this leads to a stronger bended knee, the leg's Center of Mass is shifted front which, as can be seen, results in a stronger horizontal deflection as the leg rotates around the hip. In contrast, the blue trajectory represents increased horizontal damping which also leads to a better vertical alignment. This phenomenon can be explained by the physical structure of CARL, its motor placement, lever arm design and the ADD that distributes the workspace damping to the muscles. As analyzed by Nejadfard et al. in [10], the construction of CARL physically does not allow to generate horizontal-only damping. The ADD, introduced in Eq. (3), hence tries to minimize the horizontal damping error at the cost of over-damping the vertical component.

Comparing both evaluation plots with each other, it can be seen that the overall center positions of the executed trajectories are quite similar for both speeds. These displacements are mainly based on gravity and are hence, as expected, independent of the movement speed. However, the four times higher speed of the right execution plot results in much stronger dynamic, inertial effects. As a result, the trajectories in Fig. 5b occur at a much higher variety per execution cycle, as well as with stronger deformations of the resulting trajectory (especially compare the blue and orange ones). However, as can be seen on the red and green trajectories, the same dynamic effects can also help to compensate the mechanically given favor of the system to prefer horizontal movements.

On the top of the slow, orange trajectory in Fig. 5a, a repetitive hit of the physical boundary of the knee joint can be observed. Although the trajectory clearly visualizes the shank to unnecessarily bounce off the limit a little bit, the trajectory itself is both, deterministic and stable. Hence, the proposed algorithm for bio-inspired imprecise impedance control has been proven to be a valid and working control approach that offers a novel view on robot control by introducing desired imprecision.

5 Conclusion and Outlook

Previous works of Nejadfard et al. already offer great insights into the individual contribution of force-producing muscles on a robotic limb [9,10]. It has been shown that overactuating a limb with more motors than joints can be used to improve the energy efficiency of a whole limb [9]. While this observation is a nice feature from the energy point of view, it does not contribute to a robot's movement capability at all. Furthermore, it has been shown that the potential of saving energy by overactuation strongly depends on the actuators' mounting positions on the limb, the kinematic state of the limb, and the requested control. Energy efficiency hence has to be treated as probably not being the reason for the existence of overactuated limbs in nature.

By construction, every muscle has its own (Cartesian) direction of work that is purely dependent on the kinematic pose of the limb. Although only analyzed with respect to force by Nejadfard et al. [10], the contribution of muscular damping follows the exact same argumentation. Hence, the basic shape of possible workspace impedance, generated by muscular damping (and build-in stiffness) strongly depends on the kinematic pose of the limb. This phenomenon exactly reflects the biological findings of Mussa-Ivaldi et al. [7] as discussed in

Sect. 2. Obviously, in such a scenario, every single existing muscle adds flexibility and capability to the controlled limb to better approximate desired Workspace damping. Following this argumentation, to the best of our knowledge, the herein proposed muscular control strategy firstly offers a real beneficial reason for introducing overactuation on muscular-driven limbs.

The novel control structure that is introduced within the work at hand strongly reflects biological characteristics that have been observed in nature. By introducing the so-called *imprecise* control, the work at hand breaks with traditional robot control, establishing a new philosophy of non-perfect control. To the best of our knowledge, no other (legged) robot has been equipped and tested with such a type of muscular control system so far. The most similar control approach has been implemented on the so-called *BioBiped* which controls the mounting positions of joint-spanning springs [18].

With the provided experimental results of ten representative and comparable motion executions on the bipedal robot CARL, the novel *imprecise impedance control* strategy has been tested to be stable and functional. The approach has been evaluated especially without any dynamic or gravity compensation components to demonstrate its pure, unbiased functionality. Undesired perturbations by gravity and dynamic inertial forces are clearly shown. In future improvements, appropriate compensation techniques could be integrated into the proposed control structure. Overall, the proof-of-concept of the novel control philosophy introduced by the work at hand offers great confidence to be suitable and capable of further development towards complex motions like bipedal locomotion.

References

1. Bouman, A., et al.: Autonomous spot: long-range autonomous exploration of extreme environments with legged locomotion. In: 2020 IEEE/RSJ International Conference on Intelligent Robots and Systems (IROS), pp. 2518–2525. IEEE, Las Vegas (2020)
2. Brown, I.E., Loeb, G.E.: A reductionist approach to creating and using neuromusculoskeletal models. In: Winters, J.M., Crago, P.E. (eds.) Biomechanics and Neural Control of Posture and Movement, pp. 148–163. Springer, New York (2000). https://doi.org/10.1007/978-1-4612-2104-3_10
3. Haeufle, D.F.B., Wochner, I., Holzmüller, D., Driess, D., Günther, M., Schmitt, S.: Muscles reduce neuronal information load: quantification of control effort in biological vs. robotic pointing and walking. Front. Robot. AI **7**, 77 (2020)
4. Hutter, M., et al.: ANYmal - a highly mobile and dynamic quadrupedal robot. In: 2016 IEEE/RSJ International Conference on Intelligent Robots and Systems (IROS), pp. 38–44. IEEE, Daejeon (2016)
5. Luksch, T.: Human-like control of dynamically walking bipedal robots. RRLab Dissertations, Verlag Dr. Hut (2010)
6. Mihailoff, G., Haines, D.: Chapter 24 - Motor system i: peripheral sensory, brainstem, and spinal influence on anterior horn neurons. In: Fundamental Neuroscience for Basic and Clinical Applications, 5th edn., pp. 346–359.e1. Elsevier (2018)
7. Mussa-Ivaldi, F., Hogan, N., Bizzi, E.: Neural, mechanical, and geometric factors subserving arm posture in humans. J. Neurosci. **5**(10), 2732–2743 (1985)

8. Nejadfard, A., Schütz, S., Mianowski, K., Vonwirth, P., Berns, K.: Design of the musculoskeletal leg CARL based on the physiology of mono-articular and biarticular muscles in the human leg. Bioinspir. Biomimet. **14**(6), 066002 (2019)
9. Nejadfard, A., Schütz, S., Vonwirth, P., Mianowski, K., Berns, K.: Coordination of the biarticular actuators based on instant power in an explosive jump experiment. In: IEEE International Conference on Advanced Intelligent Mechatronics (AIM) (2018)
10. Nejadfard, A., Schütz, S., Mianowski, K., Vonwirth, P., Berns, K.: Moment arm analysis of the biarticular actuators in compliant robotic leg Carl. In: Vouloutsi, V., et al. (eds.) Living Machines 2018. LNCS (LNAI), vol. 10928, pp. 348–360. Springer, Cham (2018). https://doi.org/10.1007/978-3-319-95972-6_37
11. Paine, N., et al.: Actuator control for the NASA-JSC Valkyrie humanoid robot: a decoupled dynamics approach for torque control of series elastic robots. J. Field Robot. **32**(3), 378–396 (2015)
12. Reher, J., Cousineau, E.A., Hereid, A., Hubicki, C.M., Ames, A.D.: Realizing dynamic and efficient bipedal locomotion on the humanoid robot DURUS. In: 2016 IEEE International Conference on Robotics and Automation, pp. 1794–1801 (2016)
13. Schütz, S.: CARL - a compliant robotic leg designed for human-like bipedal locomotion. Doctoralthesis, University of Kaiserslautern, Kaiserslautern (2020)
14. Schütz, S., Mianowski, K., Kotting, C., Nejadfard, A., Reichardt, M., Berns, K.: RRLAB SEA – a highly integrated compliant actuator with minimised reflected inertia. In: 2016 IEEE International Conference on Advanced Intelligent Mechatronics (AIM), pp. 252–257. IEEE (2016)
15. Schütz, S., Nejadfard, A., Mianowski, K., Vonwirth, P., Berns, K.: CARL – a compliant robotic leg featuring mono- and biarticular actuation. In: IEEE-RAS International Conference on Humanoid Robots (2017)
16. Semini, C., et al.: Towards versatile legged robots through active impedance control. Int. J. Robot. Res. **34**(7), 1003–1020 (2015)
17. Seok, S., Wang, A., Chuah, M.Y., Otten, D., Lang, J., Kim, S.: Design principles for highly efficient quadrupeds and implementation on the MIT Cheetah robot. In: 2013 IEEE International Conference on Robotics and Automation, pp. 3307–3312. IEEE, Karlsruhe (2013)
18. Sharbafi, M.A., et al.: A new biarticular actuator design facilitates control of leg function in BioBiped3. Bioinspir. Biomimet. **11**(4), 046003 (2016)
19. Vonwirth, P., Berns, K.: Continuous inverse kinematics in singular position. In: Chugo, D., Tokhi, M.O., Silva, M.F., Nakamura, T., Goher, K. (eds.) CLAWAR 2021. LNNS, vol. 324, pp. 24–36. Springer, Cham (2022). https://doi.org/10.1007/978-3-030-86294-7_3
20. Winter, D.A.: Biomechanics and Motor Control of Human Movement, 4th edn. Wiley, Hoboken (2009)

Tegotae-Based Control for Non-animal-like Locomotion: A Case Study with Trident Snake

Shura Suzuki[1,2(✉)] and Masato Ishikawa[1]

[1] Department of Mechanical Engineering, Osaka University, Suita 565-0874, Japan
shura@eom.mech.eng.osaka-u.ac.jp

[2] Japan Society for the Promotion of Science, Tokyo 102-0083, Japan

Abstract. Tegotae-based control is a bio-inspired control scheme that can reproduce the behaviors of various animals with high resilience and robustnes. However, the robot application of Tegotae-based control has primarily focused on animal-like structures without considering applications with non-animal-like structures. In this study, we investigated the applicability of Tegotae-based control to a non-animal-like structure. This paper presents a case study of a trident snake with three actuated limbs radially, with each foot tip having a passive wheel. Robot experiments showed that rotational and translational movements emerged when using the Tegotae-based controller.

Keywords: Tegotae-based control · Bio-inspired robot · Non-bio-inspired robot · Trident snake

1 Introduction

The animal kingdom displays great diversity with various body structures, habitats, and locomotion modes [1]. However, regardless of the diversity, their motor control relies on distributed control systems in a similar manner [2]. Therefore, understanding their shared motor control principles helps to establish a robot controller with high applicability and effectiveness in the real world.

Central pattern generators (CPGs) and sensory feedback are the fundamental building blocks of the locomotor neural circuits in animals [2]. CPGs are responsible for generating basic locomotor patterns, which are modulated by sensory feedback. Previous studies have modeled these animal neural circuits and employed them to control robots [3]. However, these studies focused on understanding the musculoskeletal structure, motor control, and behaviors in a specific animal model. Therefore, shared motor control principles underlying animal locomotion have received less attention, and there is no systematic methodology for designing bio-inspired controllers.

A recently developed control scheme, Tegotae, is expected to provide resilient and robust locomotion skills [4].Tegotae is a Japanese concept that describes the

© The Author(s), under exclusive license to Springer Nature Switzerland AG 2023
J. M. Cascalho et al. (Eds.): CLAWAR 2022, LNNS 530, pp. 54–60, 2023.
https://doi.org/10.1007/978-3-031-15226-9_8

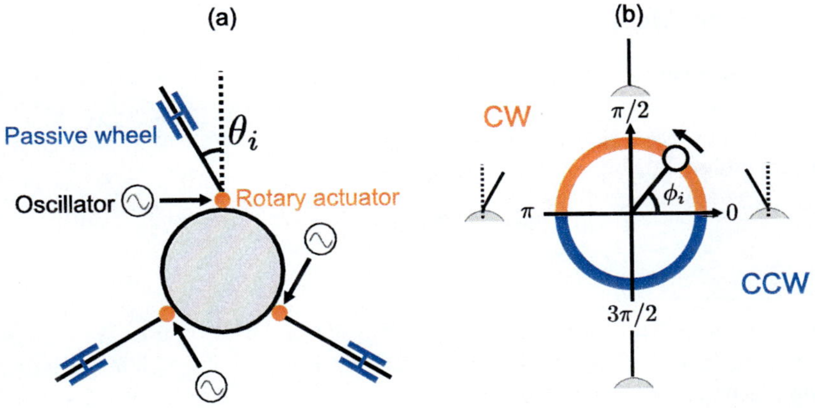

Fig. 1. Trident snake robot (a) model from the top view, (b) foot trajectory.

extent to which a perceived reaction matches one's expectations. Previous studies using Tegotae succeeded in reproducing various animal behaviors, such as the walking of insects [4], quadrupeds [5], humans [6], crawling of earthworms [7] and snakes [8], and swimming of eels [9], and all previous works have contributed to the understanding of animal motor controls. Here, we hypothesized that Tegotae-based control partially captures the common control principle in animal locomotion and helps to establish a design methodology for bio-inspired controllers. However, the robot application of Tegotae-based control has primarily focused on animal structures and not on non-animal-like structures; thus, the application scope of Tegotae remains unclear.

In this study, we investigated the applicability of Tegotae-based control to a non-animal-like structure. Specifically, we employed a trident snake robot [10] as an example of a nonanimal-like structure. The robot has three radially actuated limbs with each limb tip having a passive wheel, which was proposed from a purely engineering viewpoint considering the nonlinear control theory. We designed Tegotae-based control for a trident snake robot and performed tests using the robot. The results indicate that the proposed control algorithm generates rotational and translational movements. This suggests that Tegotae-based control is potentially applicable to non-animal-like structures as well as animal-like structures.

2 Model

2.1 Mechanical Structure

The trident snake consists of a central plate and three leg segments (Fig. 1). The central plate and each leg segment were connected via an actuated joint, and the leg tip had a nonslip passive wheel. Here, the leg angle θ_i is positive when the leg bends counterclockwise and the subscript i denotes the identifier of the leg segment.

2.2 Control Algorithm

The legs were controlled by proportional-derivative (PD) control, allowing the legs to approach their target positions. The target position was determined using a phase oscillator implemented in each leg (Fig. 1b). The time evolution of the oscillator phase can be described as follows:

$$\bar{\theta}_i = C_i^{amp} \cos\phi_i + C_i^{off}, \tag{1}$$

$$\dot{\phi}_i = \omega + f_i, \tag{2}$$

where $\bar{\theta}_i$ denotes the target angle of the ith leg and C_i^{amp} and C_i^{off} represent the amplitude and offset of the leg joint, respectively. The variable ϕ_i indicates the oscillator phase, ω represents the intrinsic angular velocity of the oscillator, and f_i symbolizes the feedback term.

Subsequently, we designed the feedback term f_i based on the Tegotae concept. Tegotae is the extent to which a perceived reaction matches the expectation (intention) of a controller [4]. For instance, when one is tired and wants to lean against a nearby wall, their intention is to lean against it. When the person leans against the wall, if the person receives a reaction force to support their body from the wall, then the person gets Tegotae. Note that Tegotae is not purely sensory feedback, and considers the match between reaction and intention. Tegotae-based control is a control scheme to enhance Tegotae, and it is realized by the Tegotae function quantifying the Tegotae. Generally, the Tegotae function is described in simple mathematical form as follows:

$$T_i(\phi_i, N) = C(\phi_i)S(N), \tag{3}$$

Function T_i denotes the Tegotae function. ϕ_i is a control variable (in this case, the oscillator phase) and N represents the obtained sensory information. The Tegotae function is expressed as the product of two functions $C(\phi)$ and $S(N)$, which describes the intention of the controller and the reaction from the environment, respectively. Here, we designed the feedback term f_i to increase the Tegotae function values. The oscillator model is simply expressed as the partial derivative of the Tegotae function with respect to the control variable ϕ_i as follows:

$$f_i = \frac{\partial T_i(\phi_i, N)}{\partial \phi_i}. \tag{4}$$

Here, we can systematically design decentralized controllers by designing only Tegotae functions. In this study, we define the Tegotae function as follows:

$$T_i(\phi_i, N) = \sigma(\sin\phi_i)(\tau_i), \tag{5}$$

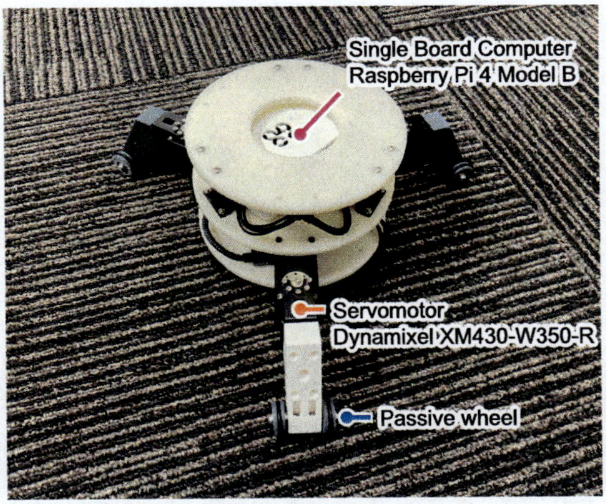

Fig. 2. Developed trident snake robot.

where the variable σ denotes the feedback gain and τ_i represents the reaction torque from the ground at the ith leg. A positive τ_i causes the joint to rotate counterclockwise. The Tegotae function has a positive value when the leg rotates clockwise ($\sin\phi_i > 0$) and obtains reaction torque from the ground ($\tau_i > 0$) or when the leg rotates counterclockwise ($\sin\phi_i < 0$) and obtains reaction torque from the ground ($\tau_i < 0$). In this study, we defined the match between the intention and the reaction as the Tegotae, and it modulates the oscillator rhythm that helps the leg obtain additional reaction torque from the ground.

Substituting Eqs. (3–5) into Eq. (2) yields the Tegotae controller as follows:

$$\dot{\phi}_i = \omega + \sigma\tau_i \cos\phi_i. \tag{6}$$

3 Results

3.1 Trident Snake Robot

Figure 2 presents the developed trident-snake robot. The main components were constructed using a 3D-printed ABS (acrylonitrile-butadiene-styrene) resin pieces. Rotary actuators were fabricated using servomotors (Dynamixel XM430-W350-R, ROBOTIS). The servomotor incorporates current and angle sensors. The current value was proportional to the generated torque; therefore, the current sensor was used as a torque sensor. A single-board computer (Raspberry Pi 4 Model B, 4 GB RAM, OKdo) was used as the controller. Furthermore, the robot was equipped with two lithium-ion batteries to power the motors and the on-board computer.

Table 1. Control parameter values

Common parameters	C_i^{amp} [rad]	$\{\pi/6, \pi/6, \pi/6\}$
	ω [rad/s]	1.5π
	σ [rad/(N·m·s)]	0.1
Rotational movement	C_i^{off} [rad]	$\{0.0, 0.0, 0.0\}$
Translational movement	C_i^{off} [rad]	$\{0.0, \pi/4, -\pi/4\}$

Fig. 3. Snapshots of rotational movement.

Fig. 4. Snapshots of translational movement.

3.2 Robot Experiment

We conducted robot experiments on flat terrain to verify the proposed control algorithm. First, we observed the resulting behavior of the trident snake robot using the Tegotae controller. We found that the robot generates two types of behaviors: rotation and translation. Furthermore, the resulting behavior changes depending on the leg offset angle C_i^{off}. Detailed control parameters are listed in Table 1. The initial oscillator phases were set to zero ($\phi_i = \{0.0, 0.0, 0.0\}$). Figure 3 illustrates snapshots during rotation. Evidently, the central plate hardly moved, but the three legs rotated counterclockwise around the central plate. Figure 4 shows snapshots of this translation. It is clear that the robot hardly rotates but moves downward, as shown in the figure. These results reveal that rotational and translational movements can emerge using Tegotae-based control.

Moreover, we examined the relationship between the oscillators during the resulting behavior. Figures 5 and 6 show the time evolutions of $\sin \phi_i$ during the

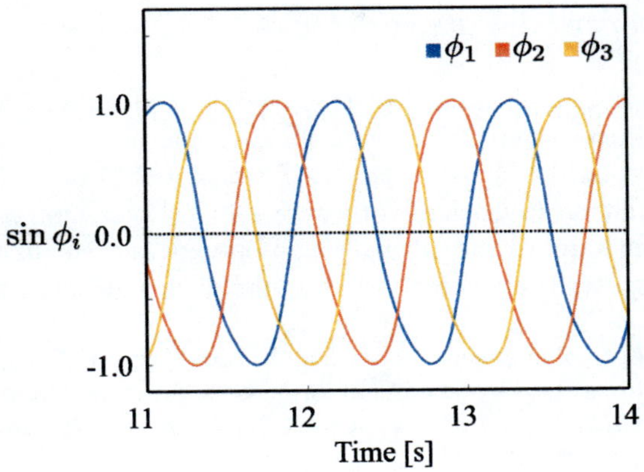

Fig. 5. Time evolution of the oscillator phases during rotational movement.

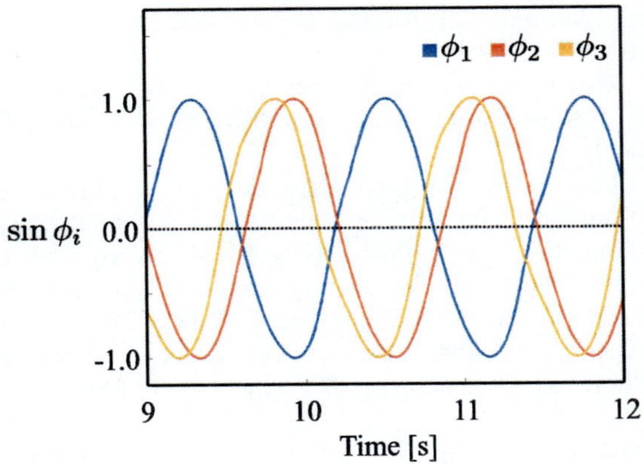

Fig. 6. Time evolution of the oscillator phase during translational movement.

rotation and translation, respectively. Note that the white leg corresponds to the first leg, and the leg number is counted clockwise. During rotation, the phase differences are approximately $2\pi/3$, and the coordinated patterns correspond to the rotation control based on the kinematic analysis in a previous study [10]. However, during translation, the first leg oscillator phase ϕ_1 is almost antiphase to the others, and ϕ_2 and ϕ_3 are almost in phase. These coordinate patterns were not predefined as inter-oscillator couplings and were self-organized by the Tegotae feedback term. Such self-organized behaviors depending on the control parameters are similar to Tegotae-based control using animal-like structures [4–8]. This suggests that the Tegotae-based controller can be applied to trident snake robots as well as animal-like robots.

4 Conclusion and Future Works

We investigated the applicability of the Tegotae control method to a trident snake robot to illustrate an example of a nonanimal-like structure. The experimental results revealed that rotational and translational behaviors emerged by Tegotae-based control. Remarkably, these two behaviors changed in response to the control parameters. In the future, we will investigate the underlying mechanisms of the bifurcation phenomenon using detailed mathematical analysis.

Acknowledgement. The authors thank Dr. Kotaro Yasui of Tohoku University for his helpful suggestions. This work was supported by a Grant-in-Aid for JSPS Research Fellows (No. 22J00076) and a JSPS Grant-in-Aid for Scientific Research (B) 21H01276.

References

1. McNeill, A.R.: Principles of Animal Locomotion. Princeton University Press, Princeton (2013)
2. Ijspeert, A.J.: Central pattern generators for locomotion control in animals and robots: a review. Neural Netw. **21**(4), 642–653 (2008). https://doi.org/10.1016/j.neunet.2008.03.014
3. Ijspeert, A.J.: Biorobotics: using robots to emulate and investigate agile locomotion. Science **346**(6206), 196–203 (2014). https://doi.org/10.1126/science.1254486
4. Owaki, D., Goda, M., Miyazawa, S., Ishiguro, A.: A minimal model describing hexapedal interlimb coordination: the tegotae-based approach. Front. Neurorobot. **11**, 29 (2017). https://doi.org/10.3389/fnbot.2017.00029
5. Owaki, D., Ishiguro, A.: A quadruped robot exhibiting spontaneous gait transitions from walking to trotting to galloping. Sci. Rep. **7**, 1–10 (2017). https://doi.org/10.1038/s41598-017-00348-9
6. Owaki, D., Horikiri, S., Nishii, J., Ishiguro, A.: Tegotae-based control produces adaptive inter-and intra-limb coordination in bipedal walking. Front. Neurorobot. **15**, 47 (2021). https://doi.org/10.3389/fnbot.2021.629595
7. Kano, T., Wakimoto, R., Sato, M., Shinohara, A., Ishiguro, A.: Designing higher Fourier harmonics of Tegotae function using genetic algorithm-a case study with an earthworm locomotion. Bioinspir. Biomim. **14**(5), 054001 (2019). https://doi.org/10.1088/1748-3190/ab2fab
8. Kano, T., Ishiguro, A.: Decoding decentralized control mechanism underlying adaptive and versatile locomotion of snakes. Integr. Comp. Biol. **60**(1), 232–247 (2020). https://doi.org/10.1093/icb/icaa014
9. Thandiackal, R., et al.: Emergence of robust self-organized undulatory swimming based on local hydrodynamic force sensing. Sci. Rob. **6**(57), eabf6354 (2021). https://doi.org/10.1126/scirobotics.abf6354
10. Ishikawa, M., Minami, Y., Sugie, T.: Development and control experiment of the trident snake robot. IEEE/ASME Trans. Mechatron. **15**(1), 9–16 (2010). https://doi.org/10.1109/TMECH.2008.2011985

Soft Gripper with Adjustable Microspines for Adhering to Tree Branches

Steffen Kirchgeorg[1,2](✉), Bram Benist[3], and Stefano Mintchev[1,2]

[1] ETH Zürich, 8092 Zürich, Switzerland
{skirchgeorg,smintchev}@ethz.ch
[2] WSL, 8903 Birmensdorf, Switzerland
[3] TU Delft, 2628 CD Delft, Netherlands

Abstract. Labour and resource intensive data collection methods drive the use of unmanned aerial vehicles in the field of environmental monitoring. UAV supported sensor deployment in forests can improve localized and continuous monitoring. To advance this field, we present a two fingered gripper with spines integrated on three phalanxes. The softness of the fingers combined with compliantly-supported microspines integrated into adjustable microspine-clusters allow the gripper to wrap and adhere to tree branches. With a differential drive actuation, microspine cluster adjustability as well as load-sharing between spine clusters is achieved. We characterize the bending behaviour of the soft fingers that adapt to curved and irregular objects. We show that the implementation of compliantly-supported and adjustable microspine-clusters increase holding force and that load-sharing between spine clusters is achieved with the differential drive actuation. The demonstration of UAV perching on a tree branch with the gripper shows that sensor deployment in these environments can be achieved.

Keywords: Mechanism design · Grasping · Gripping · Adhesion · Microspines

1 Introduction

Arboreal ecosystems gained fast growing scientific interest in the last decades as they are essential for global as well as local carbon, gas and water cycles [1]. The local, near-surface microclimates exhibited in these ecosystems depend on the global climate and vice-versa, resulting in an interdependence and complex relationship [2]. To date, monitoring is achieved with instantaneous manual sampling or deployment of sensors by either experienced climbers or on expensive fixed structures, such as canopy cranes [1]. With limited and difficult access, gaps remain in the localized and continuous sampling, but first robotic approaches for acquiring data have been proposed. Climbing robots that adhere and move on the tree structure have low power consumption [3], but handling complex manoeuvres, e.g. transitioning from the trunk to a branch, reliably remains an

© The Author(s), under exclusive license to Springer Nature Switzerland AG 2023
J. M. Cascalho et al. (Eds.): CLAWAR 2022, LNNS 530, pp. 61–74, 2023.
https://doi.org/10.1007/978-3-031-15226-9_9

Fig. 1. A UAV using the soft gripper with microspines for perching on a tree branch.

open research challenge [1]. Compared to climbing robots, aerial vehicles can easily reach forests and become of greater interest for continuous sampling and monitoring within canopies. UAVs have already demonstrated sensor deployment [4] or may serve as a sensor platform themselves by resting and perching on tree branches [5]. For either, the challenges lie in adhering with a sensor or the robot to the environment. Despite good performance on flat and artificial surfaces, wet- [6], dry- [7] and electro-adhesion [8] cannot yet handle the irregular bark of trees. Other approaches include using high-friction material [9] or grasping the whole object [10]. The most promising and employed approaches for adhering to tree barks are penetrating needles and microspines. Former can grasp and attach to branches in different orientations [11], but the large force required to penetrate the bark will push a gripper away from the surface unless it is kept in position by an external force during the attachment process. Microspines, on the other hand, have been demonstrated to adhere to small asperities on a variety of rough surfaces with low pressing or actuation force. [12,13] present rock grippers based on compliantly-supported microspines integrated into movable spine-clusters. This hierarchical design obtained by combining spine and cluster level adjustability allow the spines to travel along the surface and maximize the chance to find asperities. The grippers perform well on low curvature surfaces, but larger curvature obstacles pose a problem as the gripper bodies have limited [12] or no compliance [13] to adjust to curved objects. By taking advantage of a soft body, [14] developed a grapple that can conform to irregular overhanging surfaces, but does not include adjustability of the spines.

By taking inspiration from previous works, we combined a soft-robotics approach with compliantly-supported microspines integrated into microspine-clusters to develop a novel gripping strategy and a prototype gripper for curved rough surfaces as shown in Fig. 1. Our strategy exploits two soft fingers that adapt to curved and irregular objects, and two complementary approaches that

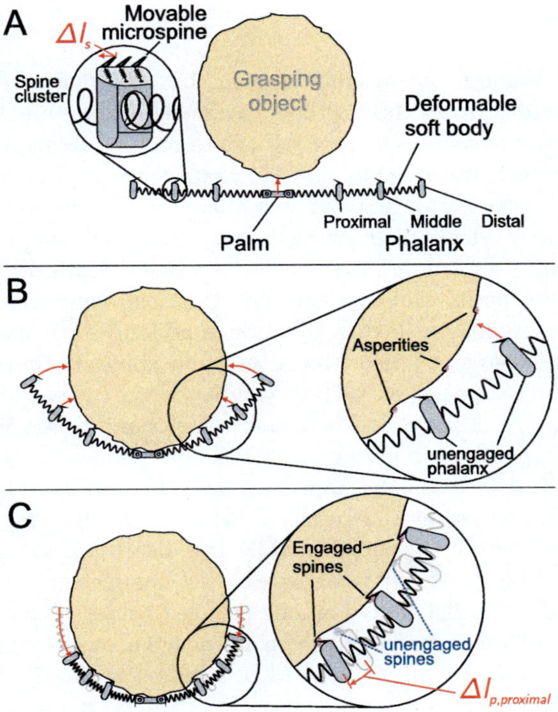

Fig. 2. Illustration of the gripping strategy. **A** The gripper uses two fingers with three phalanxes with multiple compliantly supported microspines (zoom-in) that are compliantly integrated (zoom-in), which enables them to move and adjust by Δl_s to find asperities. **B** Actuation of the fingers cause curling and brings phalanxes close to the grasping object. The spines start contacting the surface and look for asperities. **C** After full contact with the object, the fingers contract and the phalanxes travel along the surface, enabling the microspines to find and engage with asperities. Each finger can move independent of the other phalanxes within a small range by Δl_p.

increase spine engagement with the target surface by allowing the translation of both individual and clusters of spines. By granting each spine more freedom to move, the chance a spine engages is maximized. A differential drive actuation controls the closing of the fingers, enables spine cluster translation and load-sharing between clusters, such that the holding force, or payload, of the gripper is increased.

In the following section, we detail the developed gripping strategy and subsequently describe how it is translated into a mechanical design of a gripper in Sect. 3. We then characterize the gripper by analyzing its soft behaviour and the spine cluster and spine level adjustability as well as load-sharing in Sect. 4. Finally, we demonstrate the gripper on tree barks and integrat it on a UAV for perching on tree branches, simulating a possible sensor deployment scenario.

2 Gripping Strategy

Tree stems and branches are usually circular, but may have irregular features such as bumps and fissures due to their barks. Depending on the tree species, surface roughness can span from very smooth to rough. Microspines have shown strong carrying capacity and large surface versatility by holding on to small asperities in the surface [15]. However, they need to actively find suitable asperities, whose amount, depth and strength greatly depends on the material and surface roughness. Two failure mechanisms have been identified for microspines: slipping from an asperity and overload [16]. Overload causes damage either to the spine or the asperity, such that adhesion is no longer possible.

Our gripping strategy therefore aims to enable spines to find suitable asperities, thus avoiding slippage, as well as avoid overload by load-sharing between spines. This gripping strategy is illustrated in Fig. 2 and based on a palm-finger structure that exploits four principles: (i) A soft behaviour gives the fingers the ability to adjust to irregular shaped surfaces (Fig. 2B and C). Furthermore, microspines are employed in a hierarchical fashion with (ii) adjustability on the spine cluster-level, i.e. the proximal, middle and distal phalanxes and (iii) individual spine-level compliance. (iv) Load-sharing complements the strategy by allowing the load to be distributed among engaged spines. The two mechanisms for adjustability of the spines increase their engagement chances by giving the microspines two complementary degrees of freedom to find, adhere and continue adhering to asperities. By complementing them with a load-sharing ability (Sect. 3.2), the second failure mode, spine overload, can be reduced.

2.1 Spine Cluster Adjustability

On a spine cluster-level, each phalanx is connected through a soft connection that enables the fingers to not only curl around curved objects (Fig. 2B), but also to contract upon actuation (Fig. 2C) and thus allows the phalanxes and their spine clusters to travel along the surface of the object (Fig. 2C zoom-in). In this way, a phalanx with unengaged spines (Fig. 2C zoom-in) can translate by Δl_p and find suitable asperities on the target surface. The soft connection allows each phalanx to translate, within some bounds, regardless of the other phalanxes such that the engagement of spines on one phalanx does not hinder further translation and engagement of the other spine clusters.

2.2 Spine-level Adjustability

On a spine-level (Fig. 2A zoom-in), adjustability is achieved by compliantly integrating the microspines in the phalanxes. Each microspine within a spine cluster can therefore travel slightly by Δl_s with respect to the phalanx to adjust to minor changes in the surface. Therefore, multiple spines within a spine cluster can engage with different asperities.

3 System Description

In this section, we describe how the aforementioned gripping strategy with hierarchical compliance for both cluster and individual spine adjustability is translated into a mechanical design, and how system actuation and load-sharing is achieved.

3.1 Mechanical Design

The gripper consists of a palm and two soft fingers as illustrated in Fig. 3A. Three phalanxes with five microspines each are distributed along the finger length. A spring and a fiberglass reinforcement are used as a backbone for each finger. The spring allows the finger to bend in order to adjust to curved objects while the flat fiberglass reinforcement limits torsion within the finger. With this simple, yet effective design we achieve the soft behaviour, which forms the basis for our gripping strategy. The spring-based design was furthermore chosen to comply with the strategies' second principle: spine-cluster adjustability. By employing a spring, the phalanxes can travel along the spring's length through compression. The 3D printed phalanxes (Fig. 3B) are fixed in place on the spring by a blocking fiberglass sheet, but can slide on the fiberglass reinforcement, which therefore only limits torsion. Each phalanx represents a spine cluster made up of five microspines and by compression of the spring each cluster or phalanx translates (Fig. 3D). The phalanxes separate the spring sections such that each can compress independently. However, this requires a specific differential drive actuation with tendons, as described in detail in the next section. To ensure the principle of spine-level adjustability, the microspines are compliantly integrated in the phalanxes. Each spine is attached through an elastic material (3D printed Ninjaflex), which acts as a spring, to the phalanx. The motion of the spine is limited by a screw and a slot in the vertical direction (Fig. 3B and C). The five spines within a cluster can individually translate with respect to the phalanx and their neighbouring spines by a distance Δl_s. Once a spine engages with an asperity, it can translate with respect to the phalanx and remain attached to the found asperity while other spines continue translating with the phalanx. Two engaged, but vertically shifted spines in a single phalanx are shown in Fig. 3C.

Apart from the design considerations for the main working principles, slight adaptations were made to improve performance. A joint, much like the metacarpophalangeal (MCP) joint between the human finger and hand, is added between finger and palm to allow the finger to rotate and thus quickly adapt to the object before bending its finger. The fingers are also slightly rotated by φ to avoid colliding with each other when encircling smaller objects. Additionally, stabilizing tails are added to the palm to counteract a pitch-back moment that arises on inclined surfaces. As the center of mass (COM) is at a certain distance from the surface, a pitch-back moment occurs on inclined surfaces. Animals counteract it with tails [17], which is why two 3D printed tails are added whose shape and size may be adapted to the target application and surface.

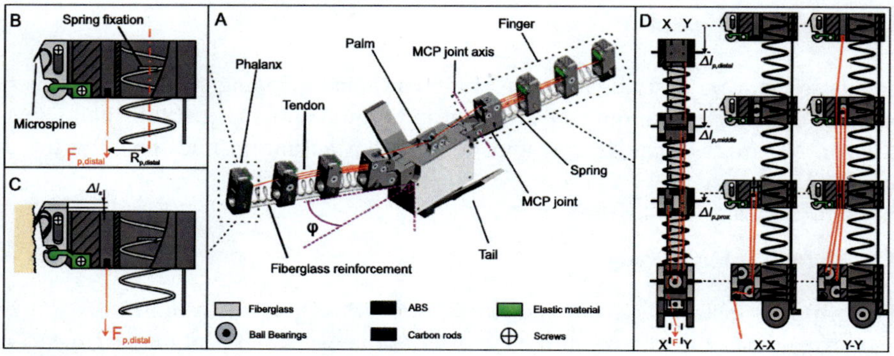

Fig. 3. Mechanical design and tendon routing: **A** Overall design with a palm and two fingers that can curl as well as rotate around the MCP joint axis. **B** Each phalanx has five microspines integrated with an elastic material such that **C** multiple spines can engage with a surface as each spine can translate by Δl_s. **D** Tendon routing allows for a differential drive actuation such that each phalanx can travel regardless of the other two by Δl_p.

3.2 Differential Drive Actuation and Load-Sharing

The fingers are actuated through tendons and a differential drive actuation, as shown in Fig. 3D is employed to ensure that phalanxes can travel independently. By passing the tendon from one phalanx back to the MCP joint before passing it to the next phalanx, the following forces F_p are applied:

$$\begin{aligned} F_{p,proximal} &= 2F, \quad R_{p,proximal} = 9\,\text{mm}, \\ F_{p,middle} &= 2F, \quad R_{p,middle} = 8\,\text{mm}, \\ F_{p,distal} &= F, \quad R_{p,distal} = 9\,\text{mm}, \end{aligned} \tag{1}$$

with F as the actuation force and R_p the moment arm between applied phalanx force F_p and the springs' central axis (Fig. 3B). The applied moments F_pR_p result in bending of the finger, with bending starting from the proximal phalanx and moving towards distal phalanx due to decreasing moments. This was chosen such that the gripper adjust to the surface's curvature from the proximal phalanx, as a bending initiated from the distal phalanx would result in a curvature smaller than that of the object. Once bending is restricted, an increase in actuation force compresses the spring and results in travel of the phalanxes. As each spring section before a phalanx is free to compress or even elongate, each phalanxes can travel independently even though their motion is constrained by the other phalanxes.

The tendon routing shown in Fig. 3D also enables load-sharing among spine clusters. An engaged spine restricts movement of the phalanx as it counteracts the applied force F_p on the phalanx. While the phalanx will remain in place, other phalanxes continue to translate until their spines engage to counteract F_p on their respective phalanx. Either the spines or the compression force of the spring will counteract F_p (applied by actuation force F according to eq. (1)) and once this is achieved for all phalanxes, the gripper stops bending and translating. Any increase in F, i.e. a payload, is transferred onto the phalanxes and then spines and thus shared between the engaged spines according to the previously presented distribution in Eq. (1).

The tendons are routed to a brushed DC motor in the palm via the MCP joint and when the tendons are actuated, the fingers will first rotate around the MCP joints as it's the path of least resistance. Once a phalanx gets in contact with an object, the motion is restricted and the finger starts bending. Lastly, the spring compresses until the spines are engaged and the actuation force is balanced out. In order to reduce friction forces arising from tendon redirection, the tendons are routed via carbon rods mounted in ball bearings. This eventually reduces the actuation force of the system.

For the control of the motor, an Arduino board (TinyDuino) and a battery are housed in the palm. The motor is connected through a worm-gear to the tendon spool, which automatically locks the position of the gripper when no power is supplied. The power of the motor can be adjusted such that it stalls when reaching a chosen activation force. The Arduino and thus the motor can be controlled remotely via Bluetooth Low Energy (BLE) to wind and unwind the tendons.

All in all, the mechanical design and differential drive actuation ensures that the amount of engaged microspines is maximized by allowing each spine to move independently, even when a spine within the same phalanx or within the same finger obtains a holding force by engaging with an asperity.

4 Results

The gripping strategy and derived mechanical design relies on fingers with a soft body and a maximization of spine engagement through hierarchical adjustability of spines and spine clusters. We first characterize the soft bending behaviour of the finger and compare it to model data in Sect. 4.1. We then prove that adjustability on the spine-level, Sect. 4.2, and cluster-level combined with load-sharing, Sect. 4.3, enable larger holding forces. Finally, we perform payload tests indoors for different sandpaper roughnesses in Sect. 4.4.

4.1 Characterization of Finger Curvature

The fingers need to adjust to the curvature of the gripping object. The force needed to bend the spring to achieve a certain curvature, i.e. a desired grasp diameter, was first modeled and subsequently experimentally verified. Here, we neglect the fiberglass reinforcement that prevents torsion due to its low bending stiffness. According to [18], the bending angle θ of a spring can be expressed as:

$$\theta = n\pi M R_s \left(\frac{1}{EI_a} + \frac{1}{GI_p} \right) \tag{2}$$

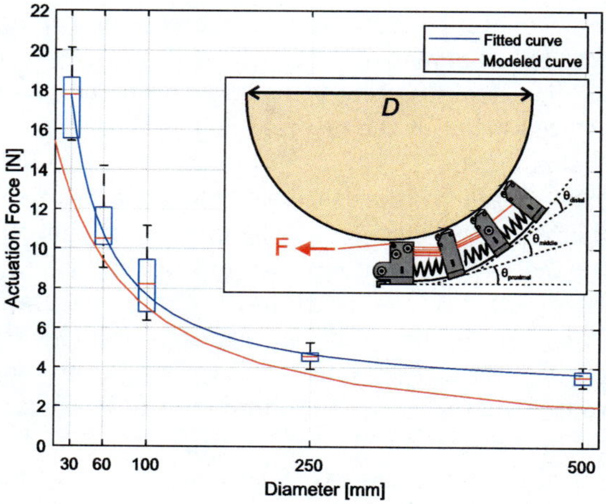

Fig. 4. Model and experimental data for the force required to bend a finger to achieve different diameters. Experimental data is based on ten measurements per diameter. Inset shows experimental setup and the angles θ required to calculate model data.

with n the amount of spirals in the spring, M the moment applied to the spring, R_s the radius of the spring, E the elastic modulus of the spring material, I_a the area moment of inertia, I_p torsional moment of inertia, and G the shear modulus. As the finger is segmented into three discrete parts, at which each end a phalanx is attached, we first compute the bending angles θ based on the moments M applied with the forces F_p and moment arms R_p according to Eq. 1. The bending angles for proximal, middle and distal phalanx are then converted to a diameter of curvature with:

$$D = \frac{C}{\pi}, \quad C = \frac{360°}{\theta_{proximal} + \theta_{middle} + \theta_{distal}}(L - \Delta H), \tag{3}$$

given L the spring length without any compression and ΔH the compression length. The diameter to which the spring adjusts to, given an actuation force F and a spring constant k, is then computed as:

$$D = \frac{360°}{\theta_{proximal} + \theta_{middle} + \theta_{distal}} \frac{(L - \frac{F}{K})}{\pi}. \tag{4}$$

Figure 4 shows the obtained model data as well as the experimental data acquired by applying a force F to a single finger and conforming to a diameter D with the setup shown in Fig. 4. While the experimental and model data share the same trend, a small difference in force is noticeable. Even though the mechanical design was optimized and ball bearings with carbon rods used to reduce friction, the tendon routing still resulted in noticeable friction that has to be overcome. The model does not account for friction, therefore predicting a smaller forces.

4.2 Spine-Level Adjustability

We integrated spines in each phalanx with a compliant 3D-printed material such that travel between adjacent spines is permitted. In order to validate that this enables higher holding forces, we compared the compliantly integrated spines to a rigid baseline. On a flat 80-grit sandpaper surface, we tested a finger with a single distal phalanx with five spines. The phalanx was engaged on the sandpaper using an actuation force of 10 N. Then, the sandpaper was pulled off in the opposite direction to simulate a payload. This force was measured and results for ten trials are shown in Fig. 5A. While the maximum holding force does not vary significantly, the compliantly integrated spines improve the mean holding force and reduce statistical variability compared to the baseline. In other words, the probability of all five spines finding an asperity is higher for the movable spines. This is also supported by the Wilcoxon Signed Rank Test with a value of $p \leq 0.05$, meaning that the compliant spines behave statistically significantly different from the rigid spines.

4.3 Spine Cluster Adjustability and Load Sharing

In order to achieve independent translation of each phalanx, a differential drive that distributes the actuation force to each phalanx according to Eq. 1. Thus, each phalanx can independently move, find asperities and halt translation to stay attached to the asperities. To ensure that the differential drive system achieves this behaviour, we performed experiments to compare the pulley-actuated system with two baselines, as shown in the insets in Fig. 5B. The first baseline is a finger with fixed phalanxes, where all phalanxes are constrained to move together and no independent movement is allowed. The second baseline is a distal phalanx actuated system, but the spring sections and thus the phalanxes can compress differently. In each experiment, the finger was engaged with a flat 80-grit sandpaper with an actuation force of 3.3 N for the fixed phalanx baseline and 10 N for the other experiments. The actuation force was reduced for the fixed phalanx experiment, as only the first spring element could compress. Subsequently, the sandpaper was pulled in the opposite direction to simulate a payload. The maximum force until failure for each system was recorded and evaluated for 10 measurements.

When comparing the fixed phalanx system to the final design, a twofold increase in holding force can be observed. As all phalanxes move together, less spines engage and even if multiple spines engage, the force does not seem to get distributed among them. The second baseline has a similar holding force as the first baseline. While all phalanxes can move, only the distal phalanx is actively actuated and therefore properly engages in the experiment. It becomes apparent that not only individual movement of the phalanxes must be allowed, but also each phalanx has to be actuated directly and the load preferably shared among engaged spines as mentioned in Sect. 3.2. Therefore, the differential drive actuated system engages more spines as well as shares the load between engaged

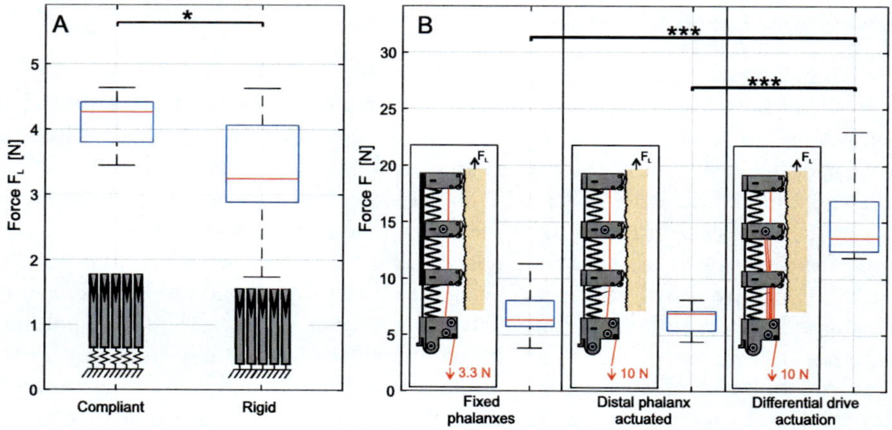

Fig. 5. Maximum holding force experiments on 80-grit sandpaper. **A** Force a phalanx with 5 compliant or 5 rigid spines can achieve with an activation force of 10 N before failure. **B** Holding force the fixed phalanxes, distal phalanx actuated and differential drive actuated system can achieve. Ten measurements were made each. Single and triple star indicate a probability of 5% and 0.1% of one the final systems performing worse than the baselines.

spines, allowing for a twofold increase over the baselines. The Wilcoxon Signed Rank Tests also suggests that the probability of the differential drive system performing worse than the other baselines is less than 0.1%.

4.4 Payload Tests

We have performed tests with a payload of 1 Kg for different surface roughnesses. The gripper was pressed vertically against cylindrical surfaces of different diameter as shown in the inset in Table 1. The tests were carried out with an activation force of 25 N split across two fingers. The successful total attempts are summarized in Table 1. K60–K120 describe the sandpaper rating, with K60 being the roughest and K120 the finest sandpaper tested. The gripper performs well for smaller diameters of 30 to 60 mm for all tested roughnesses. For a target with a 100 mm diameter, the gripper performs well for the roughest sandpaper but quickly fails for finer sandpaper. In this case, the gripper cannot exploit clamping as it spans less than half of the circumference of the obstacle. Nevertheless, with a rough sandpaper (K60) the gripper can still acquire normal forces from engaging with the asperities. For finer sandpapers, however, it becomes harder to engage and obtain holding forces from the small asperities. The microspines can easily slip from the smaller, shallower and rounded asperities in the fine sandpaper, causing the system to fail.

Table 1. Indoor tests with a payload of 1 Kg for different diameters and surface roughnesses. Low K-values refer to rough, while large K-values refer to smooth surfaces. Inset shows experpimental setup.

	Diameter [mm]		
Roughness	30	60	100
K60	5/5	5/5	4/5
K80	5/5	5/5	0/5
K100	5/5	5/5	0/5
K120	4/5	3/5	0/5

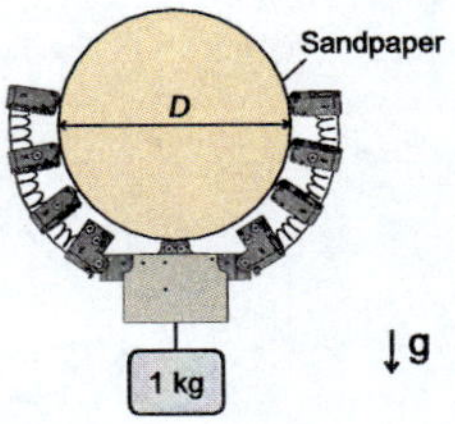

Fig. 6. Tests on different tree species with two successful and one failed engagement.

5 Demonstration

We have performed outdoor tests on several trees with a payload of 1 Kg as well as integrated the gripper on a UAV for perching. For the former, the gripper was pushed manually against the tree with its palm and the fingers were subsequently actuated. After reaching the maximum actuation force, the finger actuation stops and the manual support was released, such that the payload of 1 Kg was fully transferred onto the gripper. A selection of two successful and one failed example is shown in Fig. 6. For the successful attempts, Fig. 6A and B, the gripper properly encircled and adjusted to the tree branch curvature. All phalanxes are in contact with the tree branch, thus maximizing the amount of microspines that can engage with the surface. It also becomes apparent that the left tail counteracts the pitch-back moment in Fig. 6B, stabilizing the gripper. Figure 6C shows a more vertically oriented tree branch of the wild service tree, for which the gripping failed. Adhesion is more challenging due to the smoother surface and larger diameter compared to the himalayan spruce and the fir. Furthermore, the more vertical orientation of the branch may cause shear forces on the spines, which they do not handle well. Additionally, the pitch-back moment is large and the currently employed tail could not fully counteract it. A longer tail, or a three finger design would improve the gripping in this situation.

Fig. 7. Sequence of the UAV perching on a tree branch. A full clip is available: https://youtu.be/pTffRYXPBwQ

We integrated the gripper on the top of a UAV and demonstrate teleoperated perching of the system on a tree branch in Fig. 7. The UAV approaches the branch (Fig. 7A), aligns with it and makes contact with the palm (Fig. 7B) before actuating the gripper (Fig. 7C). The gripper closes and encircles the object, thus engaging the microspines for adhesion (Fig. 7E). The motors were fully turned off for maximum power saving. Afterwards, the gripper actuation was reversed and it disengaged from the branch, allowing the UAV to fly away as visible in Fig. 7F. The demonstration was achieved completely teleoperated without an additional control strategy. Once the gripper started closing around the branch, it stabilized the drone and eased teleoperation. However, it is necessary for the drone to maintain a stable hover for several seconds prior to gripper engagement. The demonstration is also comparable to sensor deployment as the gripper could simply be detached from the UAV after engagement.

6 Conclusion

A gripping strategy maximizing microspine engagement with adjustability on a spine and spine cluster level has been presented and translated into a mechanical design of a soft bodied gripper to grasp circular objects. By providing the microspines with two complementary degrees of adjustability, we improved spine engagement. We showed that the individual translation of spine clusters through a differential drive actuation coupled with the soft body increases the holding force of the mechanism two- to threefold. Furthermore, compliantly integrated spines within a spine cluster allow for independent micro adjustment of the microspines to asperities, further maximizing engagement and thus holding force.

The designed gripper successfully attached to a variety of controlled surface roughnesses and diameters as well as on tree barks outdoors. An integration of the gripper on a drone and perching on a branch showed that the use within a dynamically moving system is feasible. For larger diameters with low surface roughness, the addition of penetrating needles beside the microspines may improve adhesion of the system. Furthermore, a continuously soft finger from silicone could simplify the mechanical design and reduce actuation force, while also allowing the integration of microspines along the full body of the finger. A study on finger number and finger orientation for the desired application could also enhance capabilities.

With the proposed strategy and design, we hope to advance towards more versatile and reliable gripping solutions for robots in natural environments. Such solutions will improve and allow scaling of microclimate and ecosystem monitoring tasks in the future.

References

1. Cannon, C.H., Borchetta, C., Anderson, D.L., et al.: Extending our scientific reach in arboreal ecosystems for research and management. Front. For. Glob. Change **4**, 160 (2021)
2. Nakamura, A., Kitching, R.L., Cao, M., et al.: Forests and their canopies: achievements and horizons in canopy science. Trends Ecol. Evol. **32**(6), 438–451 (2017)
3. Tavakoli, M., Viegas, C.: Bio-inspired climbing robots. In: Biomimetic Technologies, pp. 301–320. Elsevier (2015)
4. Farinha, A., Zufferey, R., Zheng, P., et al.: Unmanned aerial sensor placement for cluttered environments. IEEE Robot. Autom. Lett. **5**(4), 6623–6630 (2020)
5. Hang, K., Lyu, X., Song, H., et al.: Perching and resting-a paradigm for UAV maneuvering with modularized landing gears. Sci. Robot. **4**(28), eaau6637 (2019)
6. Langowski, J.K., Dodou, D., van Assenbergh, P., van Leeuwen, J.L.: Design of tree-frog-inspired adhesives. Integr. Comp. Biol. **60**(4), 906–918 (2020)
7. Hawkes, E.W., Christensen, D.L., Cutkosky, M.R.: Vertical dry adhesive climbing with a 100× bodyweight payload. In: Proceedings - IEEE International Conference on Robotics and Automation, vol. 2015-June, no. June, pp. 3762–3769 (2015)
8. Prahlad, H., Pelrine, R., Stanford, S., et al.: Electroadhesive robots - wall climbing robots enabled by a novel, robust, and electrically controllable adhesion technology. In: Proceedings - IEEE International Conference on Robotics and Automation, pp. 3028–3033 (2008)
9. Kirchgeorg, S., Mintchev, S.: HEDGEHOG: drone perching on tree branches with high-friction origami spines. IEEE Robot. Autom. Lett. **7**(1), 602–609 (2021)
10. Zhang, H., Sun, J., Zhao, J.: Compliant bistable gripper for aerial perching and grasping. In: 2019 International Conference on Robotics and Automation (ICRA), pp. 1248–1253. IEEE (2019)
11. Xue, K., Qian, H., Sun, Z.: Design of an adaptive mini gripper for climbing robots. In: ICARM 2018 - 2018 3rd International Conference on Advanced Robotics and Mechatronics, pp. 618–623 (2019)
12. Parness, A., Frost, M., Thatte, N., et al.: Gravity-independent rock-climbing robot and a sample acquisition tool with microspine grippers. J. Field Robot. **30**(6), 897–915 (2013)
13. Wang, S., Jiang, H., Cutkosky, M.R.: A palm for a rock climbing robot based on dense arrays of micro-spines. In: IEEE International Conference on Intelligent Robots and Systems, vol. 2016-Novem, pp. 52–59 (2016)

14. Nguyen, H.N., Siddall, R., Stephens, B., et al.: A passively adaptive microspine grapple for robust, controllable perching. In: RoboSoft 2019 - 2019 IEEE International Conference on Soft Robotics, pp. 80–87 (2019)
15. Provancher, W.R., Clark, J.E., Geisler, B., Cutkosky, M.R.: Towards penetration-based clawed climbing. In: Armada, M.A., Santos, P.G. (eds.) Climbing and Walking Robots, pp. 961–970. Springer, Heidelberg (2005). https://doi.org/10.1007/3-540-29461-9_94
16. Jeffries, L., Lentink, D.: Design principles and function of mechanical fasteners in nature and technology. Appl. Mech. Rev. **72**(5) (2020)
17. Haynes, G.C., Khripiny, A., Lynch, G., et al.: Rapid pole climbing with a quadrupedal robot. In: 2009 Proceedings - IEEE International Conference on Robotics and Automation, pp. 2767–2772 (2009)
18. Yan, J., Shi, P., Xu, Z., Zhao, J.: Design and kinematics of cable-driven soft module coupled with spring. In: IEEE International Conference on Robotics and Biomimetics, ROBIO 2019, July 2021, pp. 2195–2200 (2019)

Check for updates

Exploiting Friction Anisotropy for Soft Robot Locomotion

Naris Asawalertsak[1], Stanislav N. Gorb[2], Alexander Kovalev[2], Jonas Jørgensen[3], and Poramate Manoonpong[1,3](✉)

[1] Vidyasirimedhi Institute of Science and Technology, Rayong, Thailand
{naris.a_s19,poramate.m}@vistec.ac.th
[2] Kiel University, Kiel, Germany
{sgorb,akovalev}@zoologie.uni-kiel.de
[3] University of Southern Denmark, Odense M, Denmark
jonj@mmmi.sdu.dk

Abstract. Animals can utilize the friction anisotropy of their skin to traverse various substrates efficiently. Inspired by this, we explore different bio-inspired sawtooth-like surface structures on a soft crawling robot. Our results show the robot can crawl in different directions by exploiting friction anisotropy on the substrates and using only a single source of pressure for its pneumatic actuation.

Keywords: Soft crawling robots · Bio-inspired robotics · Anisotropic friction

1 Introduction

Bio-inspired soft robotics research studies soft-bodied organisms with inherently adaptive behaviors in order to design robots. Several works have looked into worms with soft bodies and used them as inspiration for developing soft robots. To achieve worm-like crawling movement, multiple material and pneumatic actuators have often been used [1, 2]. In contrast to the conventional technique, we show a soft worm-like robot can crawl with only one type of material and one pneumatic actuator by exploiting friction anisotropy induced by an asymmetrical surface structure.

2 Materials and Methods

To fabricate the soft crawling robot for our study, we made three molds using a polylactic acid (PLA) filament with a 3D printer (Ultimaker S5 Pro, Ultimaker B.V., Singapore), and then used silicone rubber (Ecoflex 00–30, Smooth-On, Macungie, PA, USA) to mold the robot. Figure 1(a-c) shows the three molds that were used. We vacuumized the prepolymer silicone to remove air bubbles introduced by mechanical mixing using a vacuum chamber at 29 inches of mercury for 2–3 min, followed by pouring it into the molds to fabricate the body and base/pad of the robot. The silicone required five hours to cure at room temperature. Finally, we glued the body and pad by using uncured silicone. The robot body structure was designed based on the PneuNet actuator [3].

© The Author(s), under exclusive license to Springer Nature Switzerland AG 2023
J. M. Cascalho et al. (Eds.): CLAWAR 2022, LNNS 530, pp. 75–79, 2023.
https://doi.org/10.1007/978-3-031-15226-9_10

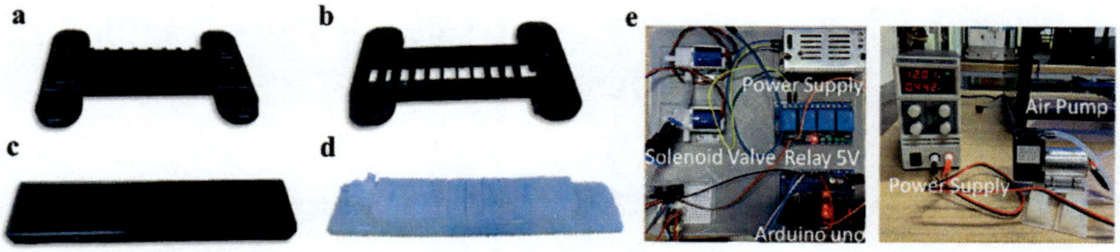

Fig. 1. The three molds used to fabricate the soft crawling robot. (a), (b) The molds for the body parts. (c) The mold for the base or pad. (d) Soft crawling robot after fabrication. (e) The electro-pneumatic actuation system used for the robot.

To investigate friction anisotropy for robot crawling, we created different pads with various sawtooth-like surface structures inspired by biological systems (e.g., shark skin, snakeskin, insect surfaces) [4]. For the preliminary study, three different pad surface structures were explored: Model 1: Oblique teeth to the right, Model 2: Oblique teeth to the left, and Model 3: Symmetrical teeth. The pad (Fig. 1(c)) was installed underneath the body of the robot. The robot was pneumatically actuated with one air tube connected to the body. The air distribution to the body was controlled by an *Arduino* board through a solenoid valve. The setup of the electronic system is shown in Fig. 1(e). The air tube was supported by a slide to so the robot could freely move in the x and y directions, as shown in Fig. 2(a).

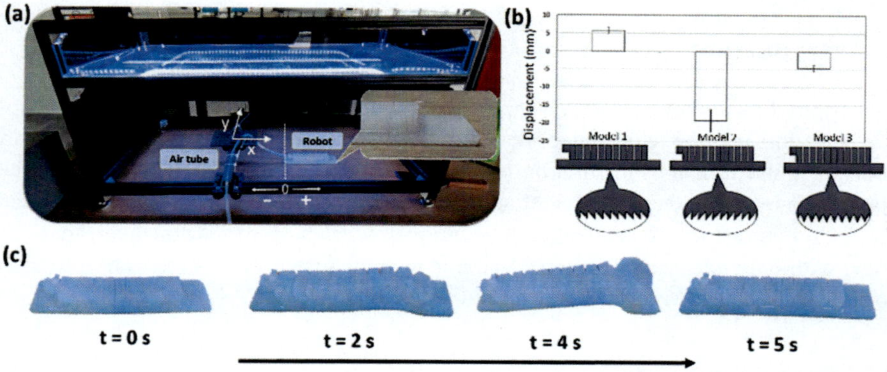

Fig. 2. (a) Experimental setup. (b) The distances traveled within 14 min. with respect to the pad models used. (c) One forward crawling step of the robot with the pad Model 1. A video of this experiment can be seen at www.manoonpong.com/CLAWAR22/Svideo.mp4.

3 Experiments and Results

In the experiments, the robot with different pads was placed on a slightly rough table surface (wood) and the air was distributed to drive the robot by turning the solenoid valve on and off with a frequency of 0.1 Hz [50% duty cycle] for 14 min. Figure 2(b) shows that the robot with the pad Model 1 could crawl forward (i.e., in the direction away from the air tube) with a speed of approximately 0.41 mm/min. In contrast, the robots with the pads Model 2 and Model 3 crawled backwards (i.e., in the direction of the air tube) with speeds of approximately 1.37 mm/min. and 0.32 mm/min., respectively. An example of the robot's crawling behavior is shown in Fig. 2(c).

To further analyze the three models with different scales/sizes, we performed a finite element analysis (FEA) simulation of a single tooth from each pad model with three sizes.

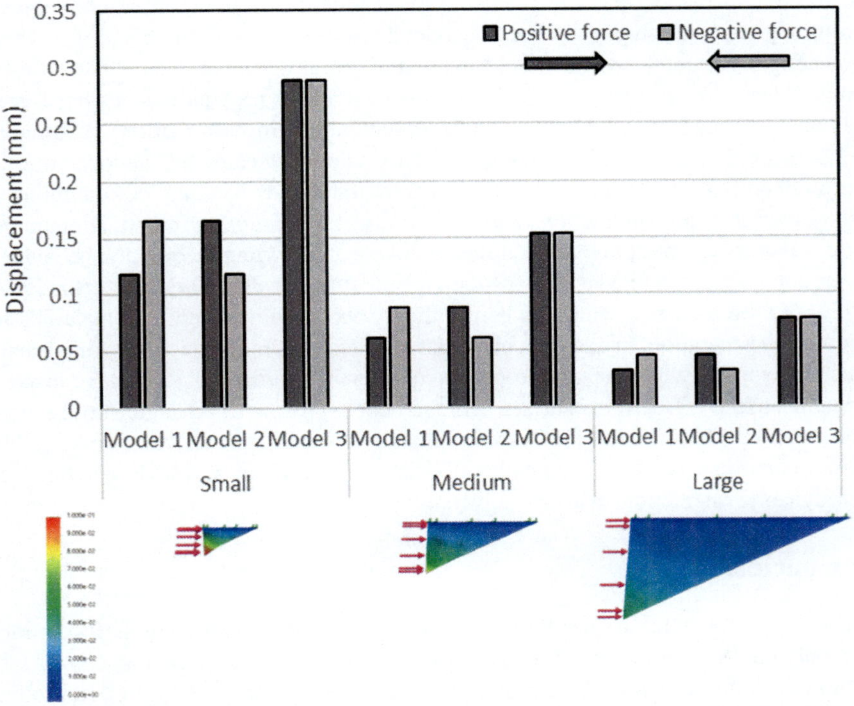

Fig. 3. The maximum displacement of a tooth of three pad models at different sizes when applying positive and negative forces (0.1 N). The FEA pictures show an example of applying a positive force, distributed to three sizes of a Model 2 tooth.

Medium size is the original one used in the locomotion experiments (Fig. 2(c)). Small size is scaled by 0.5 times from the medium size. Large size is scaled by 2 times from the medium size. Figure 3 shows the maximum deformation (displacement) of the tooth when force (0.1 N) is applied to both sides (positive and negative) of the tooth. It can be seen that under the same magnitude of force the three models show similar patterns of different magnitudes. For the positive force, Model 3 results in the highest deformation, followed by Model 2 and Model 1. For the negative force, Model 3 results in the highest deformation, followed by Model 1 and Model 2. While Model 3 has equal deformation for both positive and negative forces, Model 1 and Model 2 have deformations depending on direction of the force. This is because Model 3 has a symmetrical structure, whereas Model 1 and Model 2 have an asymmetrical structure. This shows that the design of the soft structure and the direction of the force can affect the deformation properties. Comparing all three sizes, the small size results in the highest deformation, followed by the medium and large sizes. This is because a larger size having a larger area can better absorb and resist force.

Comparing between the locomotion experiment and FEA results, the Model 1 robot crawled forward and had more tooth deformation with the negative force than with the positive force. The Model 2 robot crawled backward, with the negative force causing less tooth deformation than the positive force. While the tooth deformation of Model 1 with the negative force is similar to Model 2 with the positive force and vice versa, interestingly, Model 1 and Model 2 travelled different distances. The Model 3 robot crawled backward and had equal tooth deformation with the negative and positive forces. This analysis suggests that only tooth deformation (micro deformation) information cannot be used to predict robot locomotion. This is due to the fact that the locomotion is a result of both micro (tooth bending) and macro (robot body bending) deformations, as well as their dynamic interactions with the substrate. Thus, further investigation on this issue is required to be able to predict the complex soft crawling locomotion behavior.

In comparison to related state-of-the-art soft robots with a single actuator [5–7], our robot used a single material, whereas those robots required multiple materials for locomotion generation. In addition, our robot body exploits bio-inspired asymmetrically structured surface while the other robots take different approaches, e.g., snake-inspired kirigami surface [5], serrated surface (oriented hooks) [6], or tilted angle surface at the anterior and posterior parts [7]. Overall, our study provides an alternative approach for robot locomotion based on asymmetrically structured surface, soft body bending, and their dynamic interactions with the substrate.

4 Conclusion

Our study demonstrates that a soft robot can exploit friction anisotropy for its locomotion with only one type of material (i.e., Ecoflex 00–30) and one actuator (i.e., one air tube connected to the actuator). This can simplify the robot control. Instead of controlling multiple air tubes, we controlled only one tube. We found that the soft surface structure can affect locomotion efficiency. By having different surface structures, we can obtain different crawling directions on a slightly rough table surface. Moreover, the deformation of the soft structure depends on the direction of the force that acts on it.

Acknowledgements. This research was supported by the BrainBot project (I22POM-INT010) and the startup grant on bio-inspired Robotics of Vidyasirimedhi Institute of Science & Technology (VISTEC).

References

1. Wang, J., Min, J., Fei, Y., Pang, W.: Study on nonlinear crawling locomotion of modular differential drive soft robot. Nonlinear Dyn. **97**(2), 1107–1123 (2019). https://doi.org/10.1007/s11071-019-05035-0

2. Xie, R., et al.: A Pneumatic soft robot for locomoting like an inchworm. ICRA, pp. 3448–3453 (2018)
3. Soft Robotics Toolkit Homepage: https://softroboticstoolkit.com/book/pneunets-design. accessed 15 Nov 2019
4. Tramsen, H.T., et al.: Inversion of friction anisotropy in a bio-inspired asymmetrically structured surface. J. R. Soc. Interface c (2018)
5. Rafsanjani A., et al.: Kirigami skins make a simple soft actuator crawl. Sci. Robotics **3**(15) (2018)
6. Meder, F., Naselli, G.A., Sadeghi, A., Mazzolai, B.: Remotely light-powered soft fluidic actuators based on plasmonic-driven phase transitions in elastic constraint. Advanced Materials **31**(51) (2019)
7. Umedachi, T., Vikas, V., Trimmer, B.A.: Highly deformable 3-D printed soft robot generating inching and crawling locomotions with variable friction legs. In: IEEE/RSJ International Conference on Intelligent Robots and Systems, pp. 4590–4595 (2013)

Hybrid Legged-Wheeled Locomotion Robots

A Survey of Wheeled-Legged Robots

Marko Bjelonic(✉), Victor Klemm, Joonho Lee, and Marco Hutter

Robotic Systems Lab, ETH Zürich, 8092 Zürich, Switzerland
markob@ethz.ch

Abstract. The community in legged robotics focuses on bio-inspired robots, although there are some human inventions that nature could not recreate. One of the most significant examples is the wheel which has made our transportation system more efficient and faster. Inspired by this human-made evolution, we present a survey of wheeled-legged robots, allowing robotic systems to be efficient on flat as well as versatile on challenging terrain. This enhancement, however, comes at the cost of increased complexity due to additional degrees of freedom at the end-effector, which empowers motions along the rolling direction. The missing examples in nature make designing templates that capture the underlying locomotion principles cumbersome, making hybrid locomotion challenging. This paper reviews some of the novel locomotion frameworks overcoming these challenges.

Keywords: Wheeled-legged robots · Survey · Hybrid locomotion

1 Introduction

The fascination with legged machines has pushed the research community and private sector to aim for robotic creatures that can work next to us and improve our lives' quality. Since the first developments in 1967 and 1968 [1,2], research in legged robotics has been trying to outperform robots' natural counterparts. Today, in the year 2022, legged robots are finding their way into industries, primarily conducting inspection tasks [3–5] and exploration missions in underground environments [6,7]. We are becoming accustomed to these robots in media coverage, and due to the current pandemic, we even let robotic dogs enforce social distancing in a Singapore park [8]. It is only a matter of time before we see more of these robots in our daily lives. With this rapid advancement in the last decades, many companies worldwide are keen to include legged robotics solutions into their processes. We can see a temporal convergence of the robot's morphology towards four-legged creatures like *ANYmal* [9], *Spot* [10], *Aliengo* [11], quadrupeds from *DeepRobotics* [12], *VISION60 Q-UGV* [13], *MIT's Cheetah 3* [14] and *HyQReal* [15]. Bipedal robots, on the other hand, are still a research and development focus, but it is only a question of time until we see robots like *Atlas* [16], *HRP-5P* [17], *Digit* [18] and *Talos* [19] in industrial facilities around the world.

© The Author(s), under exclusive license to Springer Nature Switzerland AG 2023
J. M. Cascalho et al. (Eds.): CLAWAR 2022, LNNS 530, pp. 83–94, 2023.
https://doi.org/10.1007/978-3-031-15226-9_11

As shown in many impressive articles and videos, quadrupedal robots are getting more robust and skillful in their locomotion. This advancement enables them to explore any terrain like humans and animals, making them superior to their wheeled and tracked counterparts in challenging terrain. Compared to similar-sized flying machines, such as drones, legged robots can usually carry heavier payloads and have a longer battery life, enabling them to carry sensors and robotic arms. These advantages, however, come with a caveat: In terms of energy efficiency and locomotion speed, walking robots fall behind their wheeled counterparts when operating in flat environments.

Fig. 1. The *Swiss-Mile* robot [20]. This wheeled-legged robot reaches speeds of up to 6.2 m/s, overcomes obstacles [21,22], improves efficiency by 83% compared to legged systems [23], and transforms into a humanoid to press the button of the elevator [24].

As much as the research community in legged robotics likes to study bio-inspired robots, there are some human inventions that nature could not develop. It was sometime between 4000 and 3000 BCE when the human race invented the *wheel* [25], and since then, the wheel has continuously improved the way we live by making our means of transport more efficient in terms of travel time and energy consumption. The question of why animals do not have wheels is a fascinating research topic to answer. There might be a limitation to what kinds of biological structures can exist in nature, and thus it is simply not possible for evolution to develop wheels. Another theory claims that wheels are generally less effective than other modes of locomotion in terrestrial environments like forests, mountains, and other settings without flat terrains [26]. Because of the latter explanation, there is perhaps a little incentive for nature to invent wheels. Our urban environments, however, have changed drastically in the last centuries from terrestrial terrains to relatively flat and open terrain, making machines with wheels superior as a form of transportation. The key difference between a human on passive wheels, e.g., roller-skates, and a robot equipped with actuated wheels (see Fig. 1) is that the robot if appropriately engineered can exert control over its wheels with a nuance that humans can not match [27].

Adding wheels to the robot's legs allows us to be efficient on flat as well as agile on challenging terrain. With this combination, we overcome the trade-off between mobility and efficiency. While legged robots have already made their way into real-world applications, wheeled-legged robots like *Handle* [28], *Skaterbot* [29], *Ascento* [30], *RoboSimian* [31], *Roller-Walker* [32] and *Centauro* [33] remain only within research and development departments/labs. Their hybrid locomotion capabilities are less understood due to the additional complexity and the missing natural counterparts. The young research community addresses this by experimenting with novel methods that capture hybrid locomotion principles, i.e., simultaneous walking and driving. These additional barriers make the algorithm's design challenging, but if overcome, would allow robots to outperform nature. Our development over the past six years summarized in Fig. 1 shows the potential of these highly efficient, fast, and versatile machines. The robot's capability to stand up on two legs revolutionizes the way we categorize multi-limbed robots. Its humanoid-quadruped transformation makes the robot superior to traditional four-legged machines, allowing for a higher field of view, but also enabling it to perform practical skills such as calling an elevator, and in the future, grabbing packages and opening doors. A traditional quadruped without wheels struggles on two legs because of the limited degrees of freedom to balance.

In this work, we put wheeled-legged robots into context and give an outlook of where this relatively new domain is heading to. It also provides an overview of the leading research actors and presents some of the most notable robotic designs. We also discuss the underlying control strategies and the robots' applications. These materials can also be used as guidelines for future wheeled-legged robot designers and researchers.

2 Design and Control of Wheeled-Legged Robots

Locomotion has been thoroughly studied by a large community in legged robotics as well as computer graphics. In the last years, we have also seen a rise in novel hardware designs and control algorithms that contribute to wheeled-legged machines, which we discuss in the following sections. Furthermore, recent

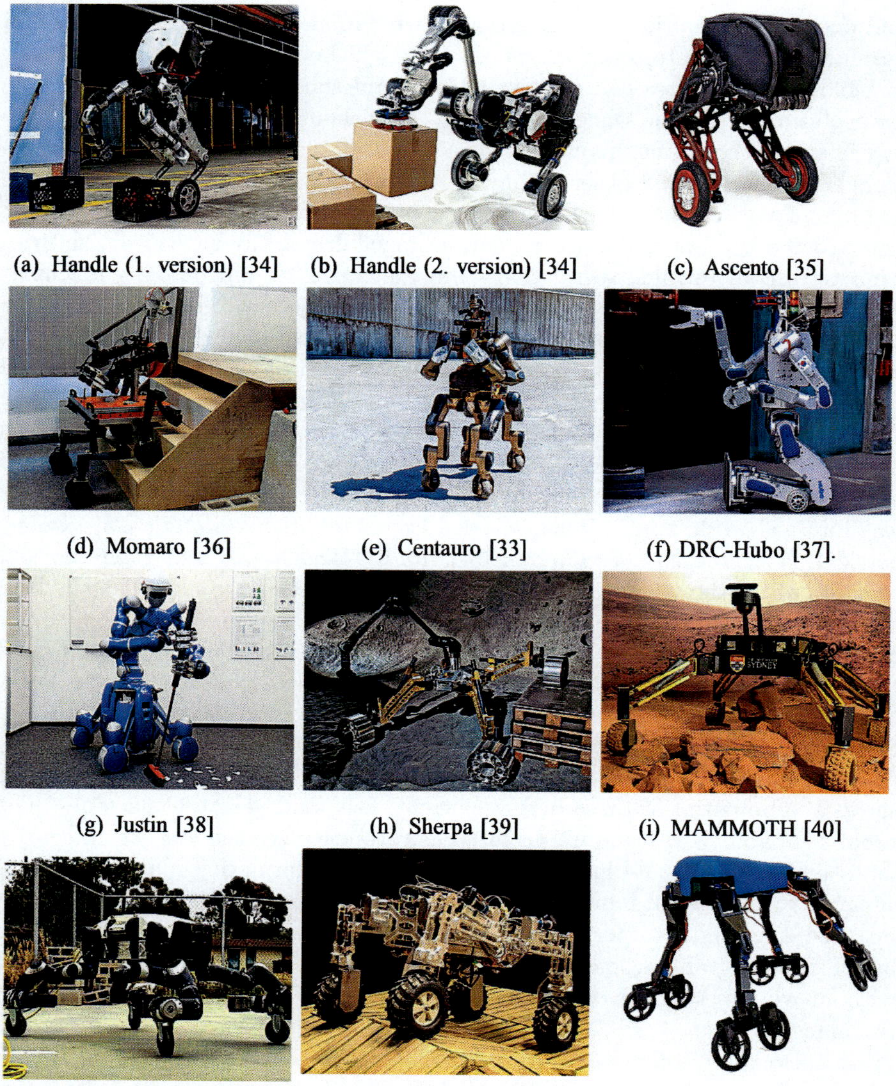

(a) Handle (1. version) [34] (b) Handle (2. version) [34] (c) Ascento [35]
(d) Momaro [36] (e) Centauro [33] (f) DRC-Hubo [37].
(g) Justin [38] (h) Sherpa [39] (i) MAMMOTH [40]
(j) RoboSimian [41] (k) Hylos [42] (l) Skaterbot [29]

Fig. 2. Research in the last years showed unique designs for wheeled-legged robots.

advancements in *Reinforcement Learning (RL)* had a significant impact on the space of locomotion, showcasing robust policies that can handle any terrain in the real world, as reviewed in the last section of this chapter.

2.1 Overview of Wheeled-Legged Robots

Combining walking and driving has been an active research topic in recent years. In particular, adding actuated wheels to walking robots increases the locomotion's efficiency while maintaining the rough terrain capabilities through the robot's legs. For example, *Boston Dynamics* shows impressive locomotion skills with their wheeled-legged platform *Handle* in Fig. 2a, jumping and driving at the same time over high obstacles [28]. Its newest version in Fig. 2b moves boxes in warehouses, and the company focuses less on the robot's walking abilities [34]. There are currently no published papers about Handle, and as such, no knowledge about Boston Dynamics' locomotion framework. Klemm et al. [30] introduce a smaller-sized version of Handle without arms called *Ascento*, as shown in Fig. 2c. *Momaro* in Fig. 2d, on the other hand, is a four-legged robot with wheels performing driving and stepping [36]. This wheeled quadrupedal robot uses a kinematic approach to drive and overcome obstacles like stairs and steps. Recently, the *Centauro* robot in Fig. 2e showed similar results over stepping stones and first attempts to overcome stairs while performing only slow static maneuvers [33,43–46]. *DRC-HUBO* in Fig. 2f is a wheeled humanoid robot that can switch between a walking and a driving configuration [37]. While driving, the robot is in a crouched position, and as such, the legs are not used for locomotion or balancing. With this novel idea, the transformer robot achieved first place at the Defense Advanced Research Projects Agency (DARPA) Robotics Challenge in 2015 due to its speed, adaptability, and reliability. Figure 2g shows the wheeled humanoid robot, *Justin* [38,47], a humanoid upper body attached to a reconfigurable base on wheels. To further reduce energy consumption, the authors of [48] introduce a leg design where the robot's gravitational force is completely compensated for in specific joint configurations.

Another class of wheeled-legged robots can be categorized as rovers [31,49, 50]. Most of these robots are not combining walking and rolling simultaneously, i.e., the legs typically act as a suspension system while driving. The *Sherpa* and *MAMMOTH* rover, shown in Fig. 2h and Fig. 2i, are used as case studies to investigate locomotion principles on extraterrestrial planets like Mars [39,40,51]. The version of *RoboSimian* with powered wheels performs impressive multimodal locomotion in first field trials on fractured, glacial ice [31]. It is a rover with 32 actively articulated degrees of freedom (DOF) designed to operate with articulated suspension whereby the normal load per wheel, body orientation, and available limb workspace are actively controlled (see Fig. 2j with passive wheels). A walking excavator showcases similar terrain adaptation results for autonomous excavation [52,53]. In Fig. 2k, the four-legged robot *Hylos* introduces a novel suspension system with 16 actuated DOF [42]. Each leg has two revolute parallel axes, and the end-effector is an active and steering wheel.

While most of the existing literature focuses on actuated wheels, locomotion with passive wheels poses a solution to decrease energy consumption when overcoming long distances over sufficiently even terrain with relatively low costs for hardware modifications. For example, the quadrupedal robot *RoboSimian* of NASA's Jet Propulsion Laboratory (JPL) in Fig. 2j uses six DOF to set each skate's pose [41] and there is no force feedback available during skating. The robot executes offline joint position trajectories tracked by a PID controller without considering unexpected changes in the contact states. By this means, the motion plan given by the joint trajectories is not updated during locomotion. Like RoboSimian, the *Roller-Walker* by Hirose and Endo et al. [32,54,55] performs skating motions without lifting its legs nor controlling the ground reaction force. These approaches do not explicitly deal with contact interactions and force control. Without force control, there is no guarantee that contact constraints are satisfied, resulting in less robust locomotion against unknown terrain irregularities. Humanoids on passive wheels are explored by some specialized research labs [56–58]. Most of the literature shows statically stable motions where both skates are always in contact, and the control approach is based on position-controlled joints. Iverach-Brereton et al. [59] show similar results for a small-sized humanoid robot that performs skating motions on ice. Takasugi et al. [60] generate skating motions with the human-sized humanoid robot *HRP-2* balancing on a skateboard. The robot uses a simplified approach for motion planning, where the robot controls the contact force on the ground to push with one leg and maintain balance. Impressive results are presented with *Skaterbots* in Fig. 21, which introduces a generic optimization to the planning of wheeled-legged robots' motions [29,61].

2.2 Optimization-Based Methods

Designing high-dimensional motions is elaborate, and, in most cases, it is not feasible to hand-craft due to the complexity of the physical constraints. One way to solve such a locomotion problem is through optimization-based methods. It appears that these methods can be separated from a practical point of view into *motion tracking* through inverse dynamics and optimal control technologies used for slow (<1 Hz) *offline optimization* (often referred to as trajectory optimization (TO)) and fast (10–100 Hz) *online optimization* (model predictive control (MPC)). Inverse dynamics approaches only look at one set point at a time, while TOs solve complex locomotion maneuvers over the task's time horizon. Approaches in the final category, MPC, contribute to the online generation of motions over a predefined look-ahead and explore task simplifications for computational reasons. It should be noted, however, that any attempt at categorization will result in an overlap.

Motion Control. While most wheeled-legged robots use a kinematic approach to generate velocity commands for the wheels, some promising research incorporates the robot's whole-body dynamics into motion control algorithms, computing torque commands for each of the joints, including the wheels. Sentis et al. [62] show a prioritized whole-body compliant control framework that generates motor torques for a humanoid robot's upper body attached to a wheeled base. Jeong et al. [63] incorporate the equations of motion, including the nonholonomic constraints, into the two-wheeled mobile robot's control framework.

Justin in Fig. 2g, a wheeled humanoid robot, creates torque commands for each of the wheels using an admittance-based velocity controller [38]. Each of these wheeled platforms, however, is unable to step due to the missing actuation on the legs, and as such, the robots are only performing wheeled locomotion. There is a research gap for wheeled-legged robots. Most of the robots using actuated wheels do not consider the whole-body's dynamic model, including the wheels. The lack of these model properties hinders these robots from performing dynamic walking and shaping the interaction with the environment. A wheeled-legged robot needs to carefully choose the reaction forces between its wheels and the terrain to control its motion. The switching of the legs' contact state, the additional constraint along the rolling direction of each wheel, and the reaction forces all need to be accounted for to reveal the full potential in speed and efficiency. So far, torque control for the wheels is only explored for some slowly moving wheeled mobile platforms. Without force or torque control, the friction constraints related to the no-slip condition cannot be fulfilled, and locomotion is not robust against unknown terrain irregularities. Building upon our proposed solution, Klemm et al. [35] introduce a hierarchical whole-body controller (WBC) leveraging the full rigid-body dynamics (RBD) of the wheeled bipedal robot Ascento in Fig. 2c.

Trajectory Optimization. Recently, research in hybrid locomotion showed a rise of new TO methods. Skaterbots in Fig. 2l show impressive results based on the centroidal dynamics (CD) model of robots with wheels and legs [29,61]. In contrast to the work of Medeiros et al. [64], the authors of Skaterbots include the leg's kinematics into the optimization problem, which becomes even more critical for wheeled-legged robots since the estimation of the rolling direction is required. Thus, Medeiros et al. [64] require a heuristic that approximates the rolling constraint. A whole-body, kinematic TO is introduced by Jelavic et al. [65] and showcases static motions for walking excavators in visualization only.

Model Predictive Control. The locomotion planning's high-dimensional problem can be decomposed into a separate torso and wheels planning [66]. Such a *decomposed task* approach is completed in two stages, where a heuristic is needed to connect the feet and torso planning stages. For example, a common method in legged locomotion designs foothold positions based on the heuristic described in [67] with a capture-point-based feedback term [68]. In contrast, a *single task approach* treats the continuous decision problem as a whole without breaking down the problem into several sub-tasks [21]. Here, the challenge is to solve the problem in a reasonable time so that online execution with feedback control becomes feasible. However, gait adaptation, i.e., switching between driving and walking, remains difficult to achieve in a holistic fashion.

2.3 Reinforcement Learning

The introduction of RL to traditional legged locomotion has enabled a massive increase in performance in terms of skills and robustness in a wide variety of challenging terrains [69, 70]. With this advancement, we believe that many of the underlying problems of hybrid locomotion can be solved [20, 24]. Switching between walking and driving requires time-consuming computational resources due to mixed-integer optimizations. Our preliminary results show that such a mechanism does not need heuristics when learning end-to-end.

3 Conclusions and Outlook

The combination of legs and wheels overcomes the trade-off between mobility and efficiency. By incorporating the best of both worlds, the robot can go anywhere efficiently and fast. Augmenting the robot's end-effector with wheels, however, comes with a drawback since an additional hardware component is added, and thus, the price of such a robotic creature increases. Tasks like large-scale infrastructure monitoring and emergency response require a robot to overcome long distances on one battery charge or reach navigation goals in a short time. Optimizing the morphology of a wheeled-legged robot is a fascinating avenue for future work that could, for example, enable higher payloads for logistics tasks. This paper reviews the locomotion problem of wheeled-legged robots on flat terrain and over challenging obstacles. Autonomous navigation in an urban environment, however, requires further development of perception and navigation tools. The most promising future directions are in navigation planning, incorporating a wheeled-legged robot model since the optimal navigation route of traditional legged robots or wheeled platforms will most likely not be the optimal solution for a wheeled-legged robot.

References

1. Frank, A.A.: Automatic control systems for legged locomotion machines. Technical report. University of Southern California Los Angeles Electronic Sciences Lab (1968)
2. McGhee, R.B.: Finite state control of quadruped locomotion. Simulation **9**(3), 135–140 (1967)
3. Hutter, M., et al.: Towards a generic solution for inspection of industrial sites. In: Hutter, M., Siegwart, R. (eds.) Field and Service Robotics. SPAR, vol. 5, pp. 575–589. Springer, Cham (2018). https://doi.org/10.1007/978-3-319-67361-5_37
4. Bellicoso, C.D., et al.: Advances in real-world applications for legged robots. J. Field Robot. **35**(8), 1311–1326 (2018)
5. Gehring, C., et al.: ANYmal in the field: solving industrial inspection of an offshore HVDC platform with a quadrupedal robot. In: Ishigami, G., Yoshida, K. (eds.) Field and Service Robotics. SPAR, vol. 16, pp. 247–260. Springer, Singapore (2021). https://doi.org/10.1007/978-981-15-9460-1_18
6. Tranzatto, M., et al.: Cerberus: autonomous legged and aerial robotic exploration in the tunnel and urban circuits of the DARPA subterranean challenge. Field Robot. (2021)
7. Bouman, A., et al.: Autonomous spot: long-range autonomous exploration of extreme environments with legged locomotion. In: IEEE/RSJ International Conference on Intelligent Robots and Systems (IROS), pp. 2518–2525 (2020)

8. BBC News: Coronavirus: robot dog enforces social distancing in Singapore park. https://www.bbc.com/news/av/technology-52619568
9. Hutter, M., et al.: ANYmal-toward legged robots for harsh environments. Adv. Robot. **31**(17), 918–931 (2017)
10. Boston Dynamics: Spot. https://www.bostondynamics.com/spot
11. Unitree Robotics: Aliengo. https://www.unitree.com/products/aliengo
12. DeepRobotics: DeepRobotics quadrupedal robots. http://www.deeprobotics.cn/
13. Ghost Robotics: Vision60 Q-UGV. https://www.ghostrobotics.io/
14. Bledt, G., Powell, M.J., Katz, B., Di Carlo, J., Wensing, P.M., Kim, S.: MIT Cheetah 3: design and control of a robust, dynamic quadruped robot. In: IEEE/RSJ International Conference on Intelligent Robots and Systems, pp. 2245–2252 (2018)
15. Semini, C., et al.: Brief introduction to the quadruped robot HyQReal. In: Italian Conference on Robotics and Intelligent Machines (I-RIM), pp. 1–2 (2019)
16. Boston Dynamics: More parkour atlas (2019). https://youtu.be/_sBBaNYex3E
17. Kaneko, K.: Humanoid robot HRP-5P: an electrically actuated humanoid robot with high-power and wide-range joints. IEEE Robot. Autom. Lett. **4**(2), 1431–1438 (2019)
18. Agility Robotics: Digit - advanced mobility for the human world. https://www.agilityrobotics.com/robots#digit
19. Stasse, O., et al.: TALOS: a new humanoid research platform targeted for industrial applications. In: IEEE-RAS International Conference on Humanoid Robotics (Humanoids), pp. 689–695 (2017)
20. Marko Bjelonic: Swiss-mile. https://www.swiss-mile.com/
21. Bjelonic, M., Grandia, R., Harley, O., Galliard, C., Zimmermann, S., Hutter, M.: Whole-body MPC and online gait sequence generation for wheeled-legged robots. In: IEEE/RSJ International Conference on Intelligent Robots and Systems (2021)
22. Bjelonic, M., et al.: Offline motion libraries and online MPC for advanced mobility skills. Int. J. Robot. Res. (2022, under review)
23. Bjelonic, M., et al.: Keep Rollin' - whole-body motion control and planning for wheeled quadrupedal robots. IEEE Robot. Autom. Lett. **4**(2), 2116–2123 (2019)
24. Vollenweider, E., Bjelonic, M., Klemm, V., Rudin, N., Lee, J., Hutter, M.: Advanced skills through multiple adversarial motion priors in reinforcement learning. In: IEEE/RSJ International Conference on Intelligent Robots and Systems (IROS) (2022, under review)
25. Bulliet, R.W.: The Wheel: Inventions and Reinventions. Columbia University Press, New York (2016)
26. LaBarbera, M.: Why the wheels won't go. Am. Nat. **121**(3), 395–408 (1983)
27. Ackerman, E.: Wheels are better than feet for legged robots: ANYmal demonstrates how hybrid mobility can benefit quadrupedal robots. https://spectrum.ieee.org/automaton/robotics/robotics-hardware/wheels-are-better-than-feet-for-legged-robots
28. Boston Dynamics: Introducing handle. https://youtu.be/-7xvqQeoA8c
29. Geilinger, M., Poranne, R., Desai, R., Thomaszewski, B., Coros, S.: Skaterbots: optimization-based design and motion synthesis for robotic creatures with legs and wheels. ACM Trans. Graph. (TOG) **37**(4), 160 (2018)
30. Klemm, V., et al.: Ascento: a two-wheeled jumping robot. In: ICRA International Conference on Robotics and Automation (ICRA), pp. 7515–7521 (2019)
31. Reid, W., Emanuel, B., Chamberlain-Simon, B., Karumanchi, S., Meirion-Griffith, G.: Mobility mode evaluation of a wheel-on-limb rover on glacial ice analogous to Europa terrain. In: IEEE Aerospace Conference, pp. 1–9 (2020)

32. Endo, G., Hirose, S.: Study on roller-walker-adaptation of characteristics of the propulsion by a leg trajectory. In: IEEE/RSJ International Conference on Intelligent Robots and Systems (2008)
33. Klamt, T., et al.: Remote mobile manipulation with the centauro robot: full-body telepresence and autonomous operator assistance. J. Field Robot. **37**(5), 889–919 (2020)
34. Boston Dynamics: Handle. https://www.bostondynamics.com/handle
35. Klemm, V., et al.: LQR-assisted whole-body control of a wheeled bipedal robot with kinematic loops. IEEE Robot. Autom. Lett. **5**(2), 3745–3752 (2020)
36. Klamt, T., Behnke, S.: Anytime hybrid driving-stepping locomotion planning. In: IEEE/RSJ International Conference on Intelligent Robots and Systems, pp. 4444–4451 (2017)
37. Lim, J., et al.: Robot system of DRC-HUBO+ and control strategy of team KAIST in DARPA robotics challenge finals. J. Field Robot. **34**(4), 802–829 (2017)
38. Dietrich, A., et al.: Whole-body impedance control of wheeled mobile manipulators. Auton. Robot. **40**(3), 505–517 (2015). https://doi.org/10.1007/s10514-015-9438-z
39. Reid, W., Pérez-Grau, F.J., Göktoğan, A.H., Sukkarieh, S.: Actively articulated suspension for a wheel-on-leg rover operating on a Martian analog surface. In: IEEE International Conference on Robotics and Automation (ICRA), pp. 5596–5602 (2016)
40. Cordes, F., et al.: An active suspension system for a planetary rover. In: Proceedings of the International Symposium on Artificial Intelligence, Robotics and Automation in Space (i-SAIRAS), pp. 17–19 (2014)
41. Bellegarda, G., van Teeffelen, K., Byl, K.: Design and evaluation of skating motions for a dexterous quadruped. In: IEEE International Conference on Robotics and Automation (ICRA), pp. 1703–1709 (2018)
42. Grand, C., Benamar, F., Plumet, F.: Motion kinematics analysis of wheeled-legged rover over 3D surface with posture adaptation. Mech. Mach. Theory **45**(3), 477–495 (2010)
43. Klamt, T., et al.: Supervised autonomous locomotion and manipulation for disaster response with a centaur-like robot. In: IEEE/RSJ International Conference on Intelligent Robots and Systems, pp. 1–8 (2018)
44. Laurenzi, A., Hoffman, E.M., Tsagarakis, N.G.: Quadrupedal walking motion and footstep placement through linear model predictive control. In: IEEE/RSJ International Conference on Intelligent Robots and Systems, pp. 2267–2273 (2018)
45. Kamedula, M., Kashiri, N., Tsagarakis, N.G.: On the kinematics of wheeled motion control of a hybrid wheeled-legged centauro robot. In: IEEE/RSJ International Conference on Intelligent Robots and Systems, pp. 2426–2433 (2018)
46. Klamt, T., et al.: Flexible disaster response of tomorrow: final presentation and evaluation of the CENTAURO system. IEEE Robot. Autom. Mag. **26**(4), 59–72 (2019)
47. Giordano, P.R., Fuchs, M., Albu-Schaffer, A., Hirzinger, G.: On the kinematic modeling and control of a mobile platform equipped with steering wheels and movable legs. In: IEEE International Conference on Robotics and Automation (ICRA), pp. 4080–4087 (2009)
48. Pinto, V.H., Soares, I.N., Rocha, M., Lima, J., Gonçalves, J., Costa, P.: Design, modeling, and control of an autonomous legged-wheeled hybrid robotic vehicle with non-rigid joints. Appl. Sci. **11**(13), 6116 (2021)
49. Squyres, S.W., et al.: Exploration of Victoria crater by the mars rover opportunity. Science **324**(5930), 1058–1061 (2009)

50. Cordes, F., Kirchner, F., Babu, A.: Design and field testing of a rover with an actively articulated suspension system in a mars analog terrain. J. Field Robot. **35**(7), 1149–1181 (2018)
51. Reid, W., Göktogan, A.H., Sukkarieh, S.: Moving mammoth: stable motion for a reconfigurable wheel-on-leg rover. In: Proceedings of Australasian Conference on Robotics and Automation, pp. 1–10 (2014)
52. Hutter, M., et al.: Towards optimal force distribution for walking excavators. In: IEEE International Conference on Advanced Robotics (ICAR), pp. 295–301 (2015)
53. Jud, D., Hottiger, G., Leemann, P., Hutter, M.: Planning and control for autonomous excavation. IEEE Robot. Autom. Lett. **2**(4), 2151–2158 (2017)
54. Hirose, S., Takeuchi, H.: Study on roller-walk (basic characteristics and its control). In: IEEE International Conference on Robotics and Automation, vol. 4, pp. 3265–3270 (1996)
55. Endo, G., Hirose, S.: Study on roller-walker-improvement of locomotive efficiency of quadruped robots by passive wheels. Adv. Robot. **26**, 969–988 (2012)
56. Xu, Z., Lü, T., Tian, H., Xu, Z., Song, L.: Dynamic analysis of the biped ice-skater robot of passive wheel type. J. Shanghai Jiaotong Univ. (Sci.) **3**, 122–128 (2008). https://doi.org/10.1007/s12204-008-0122-8
57. Ziv, N., Lee, Y., Ciaravella, G.: Inline skating motion generator with passive wheels for small size humanoid robots. In: IEEE/ASME International Conference on Advanced Intelligent Mechatronics (AIM) (2010)
58. Matsumoto, O., Kajita, S., Komoriya, K.: Flexible locomotion control of a self-contained biped leg-wheeled system. In: IEEE/RSJ International Conference on Intelligent Robots and Systems (2002)
59. Iverach-Brereton, C., Baltes, J., Anderson, J., Winton, A., Carrier, D.: Gait design for an ice skating humanoid robot. Robot. Auton. Syst. **62**(3), 306–318 (2014)
60. Takasugi, N., Kojima, K., Nozawa, S., Kakiuchi, Y., Okada, K., Inaba, M.: Real-time skating motion control of humanoid robots for acceleration and balancing. In: IEEE/RSJ International Conference on Intelligent Robots and Systems (2016)
61. Geilinger, M., Winberg, S., Coros, S.: A computational framework for designing skilled legged-wheeled robots. IEEE Robot. Autom. Lett. **5**(2), 3674–3681 (2020)
62. Sentis, L., Petersen, J., Philippsen, R.: Implementation and stability analysis of prioritized whole-body compliant controllers on a wheeled humanoid robot in uneven terrains. Auton. Robot. **35**(4), 301–319 (2013). https://doi.org/10.1007/s10514-013-9358-8
63. Jeong, S., Takahashi, T.: Wheeled inverted pendulum type assistant robot: design concept and mobile control. Intel. Serv. Robot. **1**(4), 313–320 (2008). https://doi.org/10.1007/s11370-008-0024-5
64. Medeiros, V.S., Jelavic, E., Bjelonic, M., Siegwart, R., Meggiolaro, M.A., Hutter, M.: Trajectory optimization for wheeled-legged quadrupedal robots driving in challenging terrain. IEEE Robot. Autom. Lett. **5**(3), 4172–4179 (2020)
65. Jelavic, E., Hutter, M.: Whole-body motion planning for walking excavators. In: IEEE/RSJ International Conference on Intelligent Robots and Systems, pp. 2292–2299 (2019)
66. Bjelonic, M., Sankar, P.K., Bellicoso, C.D., Vallery, H., Hutter, M.: Rolling in the deep - hybrid locomotion for wheeled-legged robots using online trajectory optimization. IEEE Robot. Autom. Lett. **5**(2), 3626–3633 (2020)
67. Raibert, M.H.: Legged Robots that Balance. MIT Press, Cambridge (1986)
68. Pratt, J., Carff, J., Drakunov, S., Goswami, A.: Capture point: a step toward humanoid push recovery. In: IEEE-RAS International Conference on Humanoid Robots (Humanoids), pp. 200–207 (2006)

69. Miki, T., Lee, J., Hwangbo, J., Wellhausen, L., Koltun, V., Hutter, M.: Learning robust perceptive locomotion for quadrupedal robots in the wild. Sci. Robot. **7**(62) (2022)
70. Lee, J., Hwangbo, J., Wellhausen, L., Koltun, V., Hutter, M.: Learning quadrupedal locomotion over challenging terrain. Sci. Robot. **5**(47), eabc5986 (2020)

Performance Evaluation of an Innovative Suspension System for Obstacle Crossing

Denis N'chot[1](✉), Jean-Christophe Fauroux[1,2], Lama Al Bassit[3], Roland Lenain[3], and Philippe Vaslin[4]

[1] Université Clermont Auvergne, Clermont Auvergne INP, CNRS, Institut Pascal, 63000 Clermont-Ferrand, France
denis.n-chot@sigma-clermont.fr

[2] MecaBotiX, At Home/La Cité, 55 Av. Louis Bréguet, 31400 Toulouse, France

[3] Université Clermont Auvergne, INRAE, UR TSCF, 63000 Clermont-Ferrand, France

[4] Université Clermont Auvergne, Clermont Auvergne INP, EMSE, CNRS, LIMOS, 63000 Clermont-Ferrand, France

Abstract. Unmanned ground vehicles designed for exploring unstructured unknown environments usually move at very low speed because they can encounter different types of obstacles. For some tasks, however, they need to move faster, which increases the risk of tipping over and taking off the ground. This paper compares a new suspension, build with both vertical and longitudinal dampers (Susp-4D), to a classical vertical-only damping suspension (Susp-3D) through geometric and kinematic models, and experimental tests. The rover equipped with Susp-4D endured a lower deceleration from the shock with the obstacle, and lost less speed and less kinetic energy. The vehicle with Susp-4D did not pitch down against the obstacle, but rather pitched up, which should allow it to cross any obstacle easier than with a classical suspension.

Keywords: Obstacle crossing · Suspension evaluation · Vehicle kinematics · Geometric modelling

1 Introduction

All-terrain vehicles are generally used for rolling in unstructured environment and can encounter different types of obstacles depending on the ground surface continuity (Fig. 1(a) [1]). Obstacles generally have a front surface of which the normal vector $\overrightarrow{N}$ may be more or less inclined. When a wheel hurts an obstacle, the orientation of $\overrightarrow{N}$ causes a reaction force that has a horizontal ($\overrightarrow{F}_H$) and a vertical ($\overrightarrow{F}_V$) component (Fig. 1(b)). The vehicle trajectory is greatly impacted by such reaction forces. Besides, a previous experimental work [2] showed that the longitudinal impact force had the same intensity as the vertical one and that there is a stability domain of impact speed and obstacle heights out of which the vehicle tips over (Fig. 1(c)). Indeed, this horizontal impact force can create

© The Author(s), under exclusive license to Springer Nature Switzerland AG 2023
J. M. Cascalho et al. (Eds.): CLAWAR 2022, LNNS 530, pp. 95–106, 2023.
https://doi.org/10.1007/978-3-031-15226-9_12

a high pitch rate and generate an important pitch momentum, which could lead to tipping over when, for a given obstacle height, the vehicle speed increases or when, for a given vehicle speed, the obstacle becomes higher. Many locomotion systems have been developed to improve obstacle-crossing capabilities and stability of rovers such as legged [3], wheel-legged [4], tracked [5] or tracked-legged systems [6]. But these systems are only convenient for slow motion and do not damp the horizontal component of the impact force. From experiment [2] arose a new concept of 2-DoF suspension [7] patented in [8] that allows damping motions along both X_V and Z_V directions in the vehicle reference frame (X_V, Y_V, Z_V) when crossing obstacles that have at least C^0 continuous surfaces (Fig. 1(a)).

This new concept is represented in 2D in Fig. 2. It is composed of two suspension limbs, of which neutral axes are coplanar and orthogonal when the vehicle is in a reference position. Each suspension limb $\{S_l\}$ (with $l = 1$ for horizontal and $l = 2$ for vertical direction) is made of two rigid solids: a $\{piston\}_l$ and a $\{cylinder\}_l$ connected by a cylindric joint in parallel with one deformable solid $\{spring\}_l$. The horizontal (respectiveley vertical) limb is connected to the chassis by a revolute joint in O_1 (respectively O_2) and to the hub carrier by a revolute joint in C that is the intersection point of both limbs. This suspension system called 'Susp-4D' [8], allows horizontal and vertical damping (along X_V and Z_V, respectively) that constitute the first 2-DoF, and allows two additional DoF: rotation of the wheel around the transversal axis Y_V, and steering rotation of the wheel around the vertical axis Z_V. This mechanism is expected to extend the stability limit while crossing obstacles due to the horizontal damping (Fig. 1(c): green dashed curve). This paper is focused on the analysis of obstacle-crossing capabilities of the Susp-4D compared to Susp-3D, which is Susp-4D without the longitudinal damper. As a first approach, Sect. 2 presents the 2D geometric model used for comparing both suspensions workspaces. Section 3 introduces the experimental environment and protocol. Section 4 presents experimental results and evaluation criteria, and the performances of both suspensions are discussed in Sect. 5.

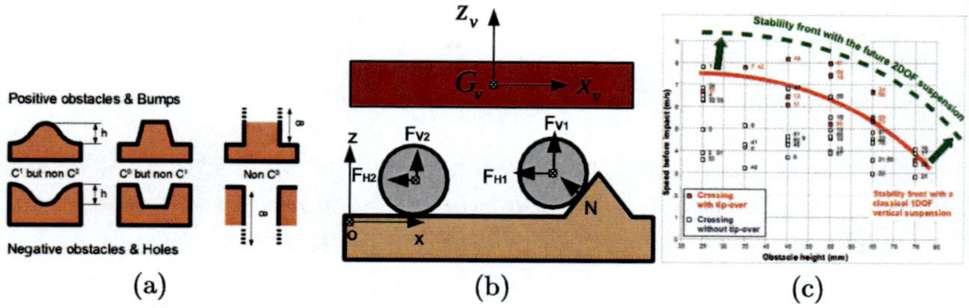

Fig. 1. (a) Obstacles classification. (b) Effort components generated while crossing an obstacle. Adapted from [7] (c) Experimental results of four-wheeled vehicles, equipped with 1-DoF suspension, when crossing obstacles of increasing height at various speeds – Red curve: actual stability boundary with 1-DoF suspension - Green dashed curve: expected stability boundary with 2-DoF suspensions [2].

2 Determination and Characterization of the Mechanism Workspace

For that purpose, it was assumed that the wheel is carried by a fixed platform representing the front left quarter of the vehicle. The wheel centre (F) is confounded with the intersection (C) of both limbs (Fig. 2) and can move from right to left with respect to the chassis [9]. The following reference frames have been thus defined: $R\,(O, \overrightarrow{x}, \overrightarrow{y}, \overrightarrow{z})$, the terrestrial Galilean reference frame; $R_G\,(G, \overrightarrow{x_G}, \overrightarrow{y_G}, \overrightarrow{z_G})$, a vehicle reference frame placed at the centre of gravity (G) of the front left quarter-vehicle; $R_1\,(O_1, \overrightarrow{x_1}, \overrightarrow{y_1}, \overrightarrow{z_1})$ and $R_2\,(O_2, \overrightarrow{x_2}, \overrightarrow{y_2}, \overrightarrow{z_2})$, the reference frames linked to $\{S_1\}$ and $\{S_2\}$, with $\overrightarrow{x_1}$ and $\overrightarrow{x_2}$ oriented from O_1 to C and O_2 to C, respectively; $R_C\,(C, \overrightarrow{x_C}, \overrightarrow{y_C}, \overrightarrow{z_C})$, the reference frame at the intersection of both limbs C with its unit vector $\overrightarrow{x_C}$ collinear to vector $\overrightarrow{O_1O_2}$. The angle $\phi = (\overrightarrow{x_G}, \overrightarrow{x_C})$ represents the orientation of R_C with respect to R_G and remains fixed when C moves. Both limbs work in compression, but not in traction. When the mechanism is at rest without load, both limbs are at their maximum lengths ($S_{1M} = S_{2M}$) and the wheel centre F is at its initial position C_0. The second defined position, C_p, is the pre-constrained position of F, confounded with C_0 on Fig. 2. This position is reached when the quarter-vehicle is in static equilibrium under its own weight. In this position, limbs lengths are pre-constrained values (S_{1p} and S_{2p}, respectively).

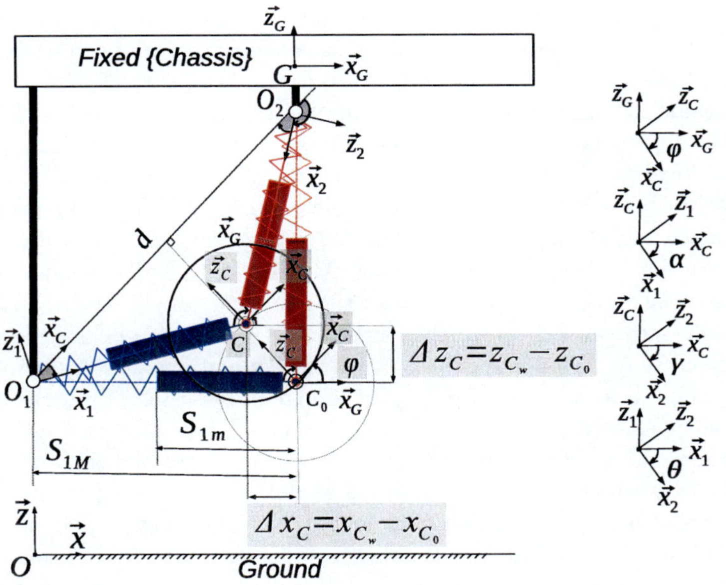

Fig. 2. Reference frames (R, R_G, R_1, R_2, R_C) and typical positions of front left wheel centre in the front left quarter-vehicle system. In this figure, C_0 and C_p are confounded.

2.1 2D Geometric Model of Susp-4D and Susp-3D Suspensions

The purpose of this model is (i) to define the input-output law of the suspension only based on its geometry and (ii) to determine its workspace (W). The Susp-4D suspension is assimilated to a plane parallel mechanism with two degrees of freedom [10] allowed by two linkages ($\{piston\}$ + $\{cylinder\}$) (Fig. 2). In Susp-3D, the longitudinal DoF is blocked. The mechanism inputs are the lengths O_lC (S_l) of the suspension limbs $\{S_l\}$ and the output is the position vector of C in ($\overrightarrow{x_G}$, $\overrightarrow{z_G}$) of R_G. The distance (d) between points O_1 and O_2, and the angle ϕ are fixed. The direct geometric modelling links the length vector $\overrightarrow{S}\begin{pmatrix} S_1 \\ S_2 \end{pmatrix}_{/\mathbf{R_G}}$ to the position vector $\overrightarrow{GC}\begin{pmatrix} x_C \\ z_C \end{pmatrix}_{/\mathbf{R_G}}$.

The resulting equations are circle equations of centres O_1 and O_2 and of radii S_1 and S_2, respectively. The workspace (W) is the trace [11] of point C when suspension lengths S_l vary between min (S_{lm}) and max (S_{lM}) values.

2.2 Global (W) and Useful (W_u) Workspaces of both Suspensions

These workspaces (Fig. 3) have been determined using the geometric modelling open source software *GeoGebra* [12]. The global workspace is an area [13] bounded by four circles: $\mathcal{C}_{(O_1;S_{1m})}$, circle of centre O_1 and radius S_{1m}; $\mathcal{C}_{(O_1;S_{1M})}$, circle of centre O_1 and radius S_{1M}; $\mathcal{C}_{(O_2;S_{2m})}$, circle of centre O_2 and radius S_{2m}; $\mathcal{C}_{(O_2;S_{2M})}$, circle of centre O_2 and radius S_{2M}. Taking into account the joints used for fastening the suspensions with the chassis and the wheel, overall limbs lengths are: $S_{1m} = S_{2m} = 9.25$ cm and $S_{1M} = S_{2M} = 13.5$ cm.

The global workspace is a double arrowhead surface that is symmetrical with respect to O_1O_2 axis (Fig. 3(a)). The intersection point C of both limbs has the theoretical possibility to reach the apex C' (Fig. 3(b)), but in this position the attachment point O_1 would touch the ground, which would prevent the suspensions to fulfil their function. To avoid this undesirable theoretical position and to prevent C from passing over O_1O_2 axis, limbs lengths must meet the following constraint: $S_{1m} + S_{2m} > d$. In the practical case of Traxxas vehicle E-Maxx 3903 (see Sect. 3.1), the displacement of C is limited because it is attached to the upper shaft of the hub carrier [7]. In the simulations, the minimum length of both limbs were set at: $S_{1m} = S_{2m} = 9.6$ cm. This value of minimum length sets the suspension ratio between full (S_M) and compressed (S_m) length at 1.4 in order to keep the same geometrical behaviour as in reality. (W) is then reduced to a useful workspace (W_{u4}), which is the area below O_1O_2 axis (Fig. 3(b): brown hatched area). (W_{u4}) has a rhombus shape of which longest diagonal C_0A is oriented at $\phi = -45°$ around Y_G-axis with respect to Z_G-axis. For Susp-4D, the useful workspace (W_{u4}) coincides with the whole W_{u4}. For Susp-3D, the useful workspace (W_{u3}) was obtained when $\{S_1\}$ was blocked at the maximum value S_{1M} and corresponds to the edge of W_{u4} between C_0 and D, which is an arc of circle $\mathcal{C}_{(O_1;S_{1M})}$. In the case of a collision between the wheel and an obstacle, the trajectory of C remained on this arc of circle.

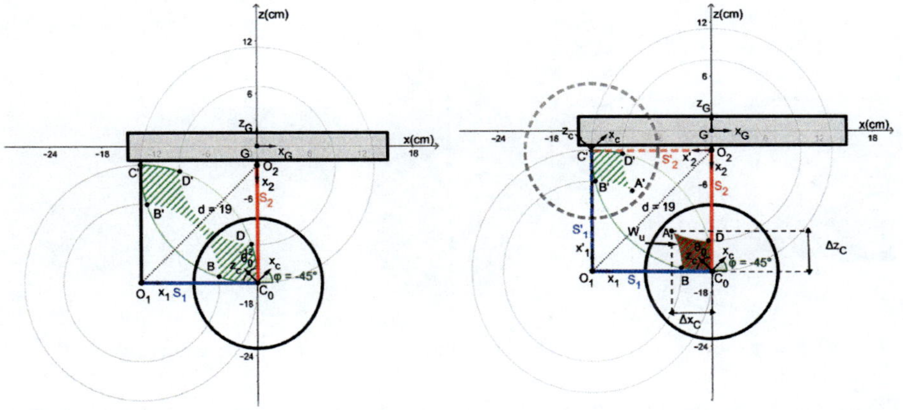

(a) (b)

Fig. 3. (a) Global workspace (W) (green hatched area) of Susp-4D suspension. (b) Useful workspace (W_u) (dark hatched area). Solid line: position of the wheel centre at apex C_0 of W_u; Dashed line: its theoretical position at point C'. Δx_C and Δz_C are the maximum possible strokes of C in the horizontal and vertical directions, respectively.

3 Materials and Methods

3.1 Experimental Materials

The performances of Susp-3D and Susp-4D have been compared by recording and processing videos of an obstacle crossing by the same experimental vehicle [14]. For practical reasons, we decided to use the small-scale all-terrain Traxxas E-Maxx 3903 vehicle (Fig. 4, Table 1) that can reach a maximum speed of 20 m/s, on which we mounted the suspensions (Fig. 4(c)). As both limbs had to be perpendicular to each other at C_p, the distance (d) between attachment points O_1 and O_2 was set to:

$$d = \|\overrightarrow{O_1O_2}\| = \sqrt{S_{1M}^2 + S_{2M}^2} = \sqrt{13.5^2 \times 2} \simeq 19.1\,\text{cm} \qquad (1)$$

Susp-4D parts have been manufactured in aluminum and two Susp-4D were mounted on the front axle (Fig. 4(c)). A plexiglass cover was also added to protect the rover in case it fell on the roof after ballistic phases. With all these elements, the total weight of the rover used in this experiment reached 7.2 kg.

The obstacle was a rectangular steel shaft fixed on the ground with two vertical rods (Fig. 5(b)). It was 60-mm high, which corresponded to 80 % of Traxxas wheel radius. A few meters before the obstacle, two parallel lines were materialized on the ground with adhesive tape, perpendicularly to the obstacle, in order to help the pilot to drive the rover along a linear path, so that both front wheels hurt the obstacle at the same time. When this condition was fulfilled, the rover performed a 2D crossing and all its movement remained in the sagittal plane, before and after the crossing (Fig. 5(b)). The rover speed was progressively

(a) Front view (b) Side view (c) front wheel equiped

Fig. 4. (a) Front and (b) Side views of the quarter of Traxxas E-Maxx 3903 vehicle (adapted from [15]). Legend: Chassis (1); Suspension limbs (2); Damper (3); Spring (4); Wheel (5) see dimensions in Table 1(c) Prototype with two Susp-4D suspensions on front wheels.

increased from each trial to the other until it tipped over. The rover was first tested with Susp-4D that was then changed to Susp-3D by replacing longitudinal limbs with aluminum rods.

Table 1. Dimensions and components of the Traxxas E-Maxx 3903 rover.

E-Maxx3903	Wheel	Spring	Damper
Mass: 5.16 kg	Rim diameter: 96.5 mm	Reference: 4957/4957R	Reference: 3766A
L × W × H	Hub size: 14 mm	Compressed: 30.48 mm	Retracted: 69.85 mm
518 × 419 × 242 mm	Tire diameter: 149 mm	Uncompressed: 82.55 mm	Extended: 102 mm

As the prototype was not yet equipped with inertial sensors, all trials have been recorded using an AOS S-Motion high speed camera [16] (1000 frames per second, 1280 × 512 pixels) placed near the obstacle, with its optical axis oriented perpendicularly to the sagittal plane of the vehicle.

For the purpose of the 2D kinematic analysis, two reference frames have been defined: $R\,(O, \overrightarrow{x_0}, \overrightarrow{y_0}, \overrightarrow{z_0})$, the terrestrial reference frame linked to the environment, and $R_v\,(O_v, \overrightarrow{x_v}, \overrightarrow{y_v}, \overrightarrow{z_v})$, the mobile reference frame linked to the vehicle. The system was composed of three solids: the chassis $\{S\}$, the rear $\{S_r\}$ and the front $\{S_f\}$ wheels. Thus, five targets were stuck on the vehicle: two on the centres of front (F) and rear (R) wheels, one on the chassis centre of gravity (G_{ch}), one on the origin (O_v) of R_v, and another on a fixed point (M) on the chassis (Fig. 5(a)).

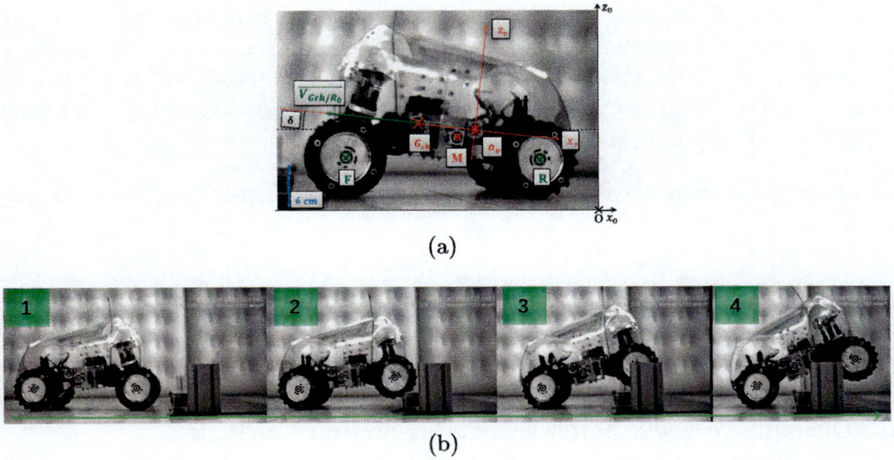

Fig. 5. (a) Reference frames and targets definition.(b) Obstacle crossing over.

3.2 2D Kinematic Analysis

Among the numerous trials performed in this experiment, only two video recordings of successful trials have been processed using the freeware motion tracking Kinovea [17]: one with Susp-3D and one with Susp-4D. The 2D coordinates of all targets in $R\,(O, \overrightarrow{x_0}, \overrightarrow{y_0}, \overrightarrow{z_0})$ have been digitized frame by frame on both trials (measurement uncertainty $= 0.1\,\text{cm}$). Positions coordinates, linear speed components and pitch rate have been filtered using a Savitzky-Golay filter (eleven elements, 3rd order polynomial) before computing the following parameters:

- The relative displacement of the wheel centre with respect to the chassis has been used for defining the collision phase, which began just before the front wheel hurt the obstacle (wheel centre at C_p position) and ended when the front wheel centre returned to C_p position during the airborne phase that followed the initial shock;
- Linear velocity of the chassis: this parameter has been computed all along the crossing over for each trial, and its values at the beginning and at the end of the collision phase have been used for computing the relative speed loss;
- Deceleration of the chassis: this parameter has only been computed during the collision phase of each trial;
- Kinetic energy of the chassis: this parameter has been computed all along the crossing over for each trial, and its values at the beginning and at the end of the collision phase have been used for computing the relative kinetic energy loss;
- Potential energy of the chassis: this parameter has been computed all along the crossing over for each trial, and its values at the beginning and at the end of the collision phase have been used for computing the slope of the potential energy during the collision phase.

- Pitch rate of the chassis: this parameter has been computed all along the crossing over using the coordinates of the targets fixed on the chassis. As a consequence of the chosen axes, when the vehicle pitched up, its pitch rate was positive, whereas a pitch down movement of the vehicle was expressed by a negative pitch rate.

All these parameters have been used for comparing Susp-4D vs. Susp-3D suspensions.

4 Experimental Results

In both trials selected for Susp-3D and Susp-4D, the rover movement remained in the sagittal plane all along the crossing over and could thus be analyzed in 2D. Quantitative results of this analysis are presented in Table 2.

The crossing over was divided into three phases: (i) before the collision between front wheels and the obstacle, the rover was rolling on a horizontal plane with all wheels on the ground; (ii) during the collision, the front wheel centre moved in the useful workspace W_u from C_p position and went back to C_p abscissa; (iii) after the collision, the front wheel centre fell down with respect to the chassis and reached C_0 position. At the end of the crossing over, in both trials rear wheels hurt the obstacle, which had some unintended consequence on the airborne phase, but that was out of the purpose of this study. In the trial with Susp-4D, the collision phase was 2.8 times longer (0.065 s) than with Susp-3D (0.023 s). In this latter trial, the relative movement of the wheel centre F with respect to the chassis during the collision went out of W_u (Fig. 6(a)) because in the prototype F is lower than the intersection C of both limbs (Fig. 4(c)).

Table 2. Kinematic and kinetic parameters of the vehicle with Susp-4D vs. Susp-3D suspensions during the collision phase. Means and standard deviations (SD) have been calculated during the collision duration with each suspension. Absolute and relative variations have been computed between initial and final collision time of each trial.

	Susp-3D	Susp-4D
Collision duration	0.023 s	0.065 s
Vehicle initial speed	3 m/s	3.4 m/s
Chassis deceleration (SD)	−6.1(4.3) g	−3.4(2.2) g
Relative speed loss	−26%	−10%
Relative kinetic energy loss	−43.9%	−14.0%
Potential energy variation	0.24 J	2.01 J
Pitch rate (SD)	+41.8(124.4)°/s	+245.1(100.7)°/s

This trajectory was not linear and showed a maximum difference of 3 mm (Fig. 6(a)), with respect to that of C, in the longitudinal direction, which let suppose a bending of the upper shaft supporting both limbs during the collision. In the trial with Susp-4D (Fig. 6(b)), the trajectory of F in W_u was confounded with that of C, and it was almost linear during the collision phase.

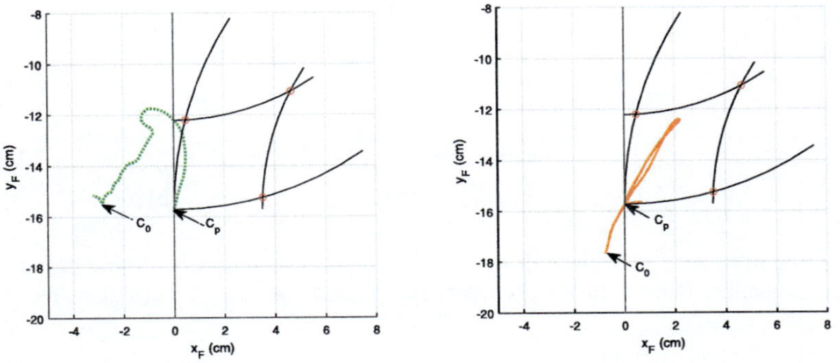

(a) (b)

Fig. 6. Trajectories of wheel centre with Susp-3D (a) and Susp-4D (b) suspensions in (W_u).

In both trials, the rover reached the obstacle within the same range of initial speeds (Susp-3D: 3 m/s; Susp-4D: 3.4 m/s), which were however rather low with regard to its maximum speed (20 m/s). The mean deceleration endured by the chassis during the collision phase was 44.3% less with Susp-4D (3.4 g) than with Susp-3D (6.1 g). The relative speed loss during the collision phase was more than twice lower with Susp-4D (−10%) than with Susp-3D (−26%). After the collision phase, the vehicle speed showed a sudden increase (+0.6 m/s) with Susp-3D (Fig. 7(a)) whereas it remained steady with Susp-4D, until it dropped again in both trials at the end of the crossing over, when the rear wheel hurt the obstacle.

The relative kinetic energy loss during the collision phase (Fig. 8(a)) was 3.1 times lower with Susp-4D (−14%) than with Susp-3D (−43.9%). After the collision, the kinetic energy increased ten times more with Susp-3D (7.3 J) than with Susp-4D (0.74 J), before it dropped again when the rear wheel hurt the obstacle. In both trials, the potential energy progressively increased from the beginning to the end of the crossing over (Fig. 8(b)). This increase was greater and more linear with Susp-4D (31 J/s) than with Susp-3D (10.8 J/s), which means that the vehicle centre of mass reached a higher altitude at the end of the trial with Susp-4D than with Susp-3D.

During the collision phase, the pitch rate ($\dot{\delta}$) was first negative (pitch down) and then positive (pitch up) with Susp-3D, whereas it was always positive and reached a maximum value of 360°/s with Susp-4D (Fig. 7(b)). On average, the pitch rate was 5.9 higher with Susp-4D (245.1°/s) than with Susp-3D (41.8°/s). After the collision, it decreased in both trials and converged to - 450°/s.

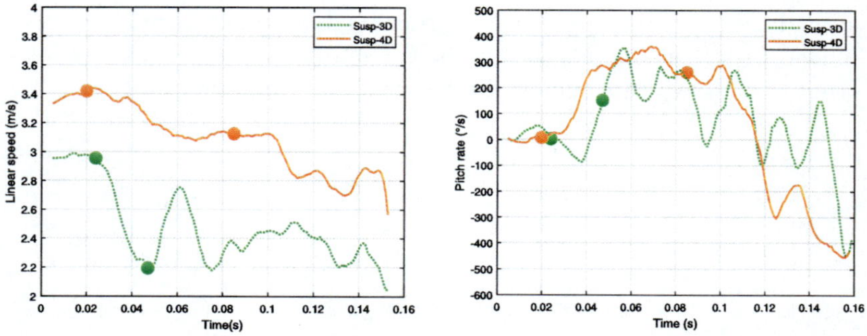

(a) (b)

Fig. 7. Linear speed (a) and pitch rate (b) of the chassis during obstacle crossing with both suspensions. Dots indicate the beginning and the end of the crossing over of each trial.

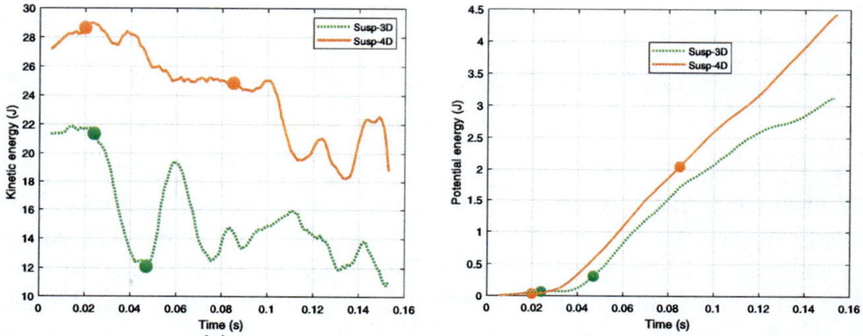

(a) (b)

Fig. 8. Kinetic (a) and potential (b) energy of the chassis during obstacle crossing with both suspensions. Dots indicate the beginning and the end of the crossing over of each trial.

5 Discussion

The results of the 2D kinematic analysis of these two trials showed differences in the vehicle behaviour depending on the type of suspension mounted on its front wheels. Susp-4D is a new kind of suspension allowing a larger longitudinal displacement of the wheel centre, and thus a greater workspace than Susp-3D - a suspension close to a classical one.

When it was equipped with Susp-3D, and although its initial speed was lower (3 m/s < 3.4 m/s), the rover was more decelerated and thus endured a greater bump during the crossing over than with Susp-4D. Moreover, this latter suspension allowed the vehicle to dissipate less kinetic energy than the classical one.

During the collision phase, the rover equipped with Susp-3D was first subject to a pitch down followed by a late pitch up movement, just before front wheels took off from the obstacle. Such a behaviour increases the risk of tipping over when speed increases: in this case, indeed, the suspension may not have enough time to raise the vehicle nose, which would thus take off and perform a partial or complete somersault, with the risk of falling on the roof.

On the opposite, the rover equipped with Susp-4D showed a positive pitch rate immediately after the contact with the obstacle and until it took off. At low speed, this behaviour should help the vehicle to cross any obstacle [2], and at high speed, it should provide it a positive angular momentum around the pitch axis. Moreover, the increase in potential energy of the rover equipped with Susp-4D after the crossing over suggests that the time of flight is increased. Both these latter results are interesting for the attitude control in ballistic phase [18].

6 Conclusion

This study compared two suspension systems, one with a vertical damper only (Susp-3D) and one with both a vertical and a longitudinal damper (Susp-4D), mounted on a small-scale all-terrain vehicle crossing over an obstacle of 60-mm height. This comparison was based on selected parameters computed from a 2D kinematic analysis of the rover during and after the collision with the obstacle: deceleration, relative speed loss, relative kinetic energy loss, pitch rate. When it was equipped with Susp-4D, the rover endured a lower deceleration from the shock with the obstacle, and lost less speed and less kinetic energy than when it was equipped with Susp-3D. During the collision phase, the vehicle with Susp-4D did not pitch down against the obstacle, which is a cause of tipping over and landing on the roof. On the opposite, the rover pitched up during and after the collision, which should allow it to cross any obstacle more easily than with a classical suspension. Furthermore, the rover potential energy increased linearly after the collision, which may increase its time of flight and provide an additional delay for an eventual attitude correction, in case of a ballistic phase. Overall, Susp-4D improved the global safety of the rover during and after crossing over an obstacle and it should be particularly convenient for an off-road vehicle rolling at high speeds on uneven grounds.

Acknowledgment. This work has been sponsored by the French government research program "Investissements d'Avenir" through the IDEX-ISITE initiative 16-IDEX-0001 (CAP 20-25) and the IMobS3 Laboratory of Excellence (ANR-10-LABX-16-01).

References

1. Fauroux, J.C., Dakhlallah, J., Bouzgarrou, B.C.: A new concept of fast mobile rover with improved stability on rough terrain. In: 8th International Advanced Robotics Programme (IARP), no. 26, pp. 1–16 (2010). http://jc.fauroux.free.fr/PUB/ARTICLES/2010_HUDEM_Fauroux_Dakhlallah_Bouzgarrou_DRAFT_A_New_Concept_of_FAST_Rover_with_Improved_Stability_on_Rough_Terrain.pdf

2. Fauroux, J.C., Bouzgarrou, B.C.: Dynamic obstacle-crossing of a wheeled rover with double-wishbone suspension. In: Proceedings of 14th International Confrence on Climbing and Walking Robots, Field Robotics, pp. 642–649 (2011). http://jc.fauroux.free.fr/PUB/ARTICLES/2011_CLAWAR_Fauroux_Bouzgarrou_DRAFT_Dynamic_obstacle-crossing_of_a_wheeled_rover_with_double_wishbone_suspension.pdf
3. Charlet, M., Gosselin, C.: Reorientation of free-falling legged robots. ASME Open J. Eng. **1**, 1–10 (2022)
4. Wang, Y., Yu, X., Chi, R., Tang, W.: Obstacle-climbing performance simulation analysis of wheel-legged rover. In: 2nd International Conference on Computer Engineering and Technology, vol. 1, pp. 74–79 (2010)
5. Cui, D., Gao, X., Guo, W.: Mechanism design and motion ability analysis for wheel/track mobile robot. Adv. Mech. Eng. **8**(11), 1–13 (2016)
6. Guarnieri, M., Debenest, R., Inoh, T., Fukushima, E., Hirose, S.: Development of Helios VII: an arm-equipped tracked vehicle for search and rescue operations. In: RSJ: International Conference on Intelligent Robots and Systems, vol. 1, pp. 39–45. IEEE (2004)
7. Fauroux, J.C.: Synthesis of spatial parallel mechanisms for a vertical and longitudinal all-terrain suspension. In: Joint XI International Conference on Mechanisms and Mechanical Transmissions (MTM) and International Conference on Robotics (Robotics 2012). Applied Mechanics and Materials, vol. 162, pp. 19–28 (2012). http://jc.fauroux.free.fr/PUB/ARTICLES/2012_MTM-ROBOTICS_Fauroux_DRAFT_Synthesis_of_spatial_parallel_mechanisms_for_a_vertical_and_longitudinal_all-terrain_suspension.pdf
8. Fauroux, J.C., Cousturier, R.: Wheel suspension device and vehicle fitted with at least one wheel equipped with such a suspension device. Brevet PCT WO2013/045634, 36 p. (2012). https://worldwide.espacenet.com/publicationDetails/biblio?DB=EPODOC&II=8&ND=3&adjacent=true&locale=frEP&FT=D&date=20130404&CC=WO&NR=2013045634A1&KC=A1#
9. Prairie, E.: Dynamic kinematic and compliance testing. MTS Systems Corporation, pp. 1–2 (2011). http://jc.fauroux.free.fr/PUB/2011_MTS_Dynamic_Kinematic_and_Compliance_Testing.pdf
10. Arsenault, M., Boudreau, R.: Synthesis of planar parallel mechanisms while considering workspace, dexterity, stiffness and singularity avoidance. J. Mech. Des. **128**, 69–77 (2006)
11. Gosselin, C.M., Jean, M.: Determination of the workspace of planar parallel manipulators with joint limits. Robot. Auton. Syst. **17**(3), 129–138 (1996)
12. GeoGebra: for teaching and learning math. https://www.geogebra.org/
13. Gosselin, C.M., Guillot, M.: The synthesis of manipulators with prescribed workspace. J. Mech. Des. **113**, 451–455 (1991)
14. Romain, J., Mathieu, P.: Fabrication et expérimentation d'une suspension innovante à deux degrés de liberté pour le franchissement dynamique d'obstacles. Master dissertation, IFMA, 44 p. (2013)
15. Traxxas: E-Maxx (3903) (2014). https://traxxas.com/products/models/othermodels/3903emaxx2011
16. AOS technologies: S-MOTION, 13 April 2022. https://www.aostechnologies.com/high-speed-cameras/self-contained/non-high-g-rated/s-motion
17. Kinovea 0.9.5: Free motion tracking. https://www.kinovea.org/
18. Vaslin, P., Petrila, S., Davis, M., Ju, L., Gouinaud, C., Fauroux, J.C.: Experimental evidence of pitch control of an all-terrain *AGV* during a ballistic phase. In: MCG: 5th International Conference on Machine Control & Guidance (2016). https://hal.uca.fr/hal-01380990

Autonomous Wheeled Locomotion on Irregular Terrain with Tactile Sensing

Hiroki Tomioka[1(✉)], Masahiro Ikeda[1,2], Keung Or[1,2], Ryuma Niiyama[1,2], and Yasuo Kuniyoshi[1]

[1] Graduate School of Information Science and Technology, The University of Tokyo, 7-3-1 Hongo, Bunkyo-ku, Tokyo, Japan
{tomioka,ikeda,or,niiyama,kuniyosh}@isi.imi.i.u-tokyo.ac.jp

[2] School of Science and Technology, Meiji University, 1-1-1 Higashimita, Tama-ku, Kanagawa, Japan
{ikeda_ma,or,niiyama}@meiji.ac.jp

Abstract. Adopting the concept of the tactile wheel, which considers the interaction between the wheel and the ground, this paper simulates reinforcement learning to show the usefulness of tactile sensing for autonomous wheeled robots on irregular terrain and to clarify the characteristics of the information to be acquired. A wheeled robot model with a wheel-on-leg structure is created and tested on two types of irregular terrain. The tactile information from each wheel is used as part of the reinforcement learning state. The average return and sample efficiency respectively increase by factors of 1.18 and 2.21 on uneven terrain. On fractal terrain, they increase by factors of 1.31 and 2.51 times, respectively, confirming the usefulness of tactile information. Tactile wheels using analog tactile information perform better in terms of adaptability to unknown terrain.

Keywords: Irregular terrain · Tactile wheel · Reinforcement learning

1 Introduction

Mobile robots have long been actively researched and are highly practical. The wheeled robot is the most common mobile robot and is widely used because of its high travel speed, simple structure and easy control. However, the wheeled robot faces problems in traveling over irregular terrain [1]. Crawler-type and leg-type hybrid robots have been developed to address the issue of traversability over irregular terrain [2]. However, the unlimited complexity of the structure and control of hybrids are undesirable in that the simplicity of wheeled robots is lost and downsizing is difficult. There is thus a need for technology that improves performance of mobile robots traveling over irregular terrain while maintaining a simple structure. Scenarios in which mobile robots are used include work in unknown environments that are inaccessible to humans. Under such scenarios, it is difficult for operators to accurately grasp the robot's surroundings from a

© The Author(s), under exclusive license to Springer Nature Switzerland AG 2023
J. M. Cascalho et al. (Eds.): CLAWAR 2022, LNNS 530, pp. 107–118, 2023.
https://doi.org/10.1007/978-3-031-15226-9_13

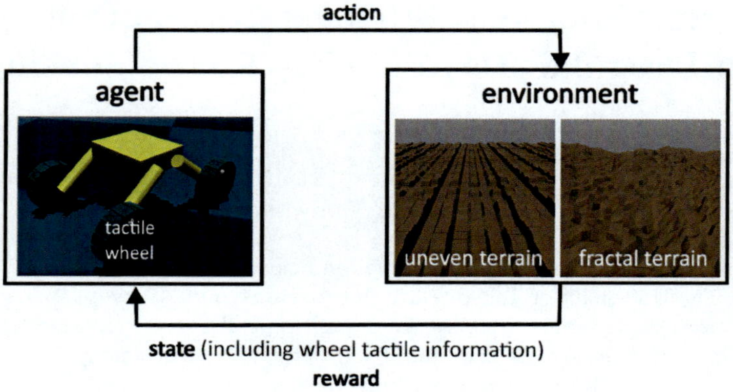

Fig. 1. Conceptual diagram

remote location, and remote control is often undesirable in terms of the operation delay and communication environment. In addition, predesigned control does not support adaptability to irregular terrain that may have various patterns. It is therefore important that the robot autonomously selects actions.

As an approach focusing on the interaction between the wheel and the environment, the use of tactile wheels has been studied for the direct acquisition of wheel tactile information [3–6] and is expected to provide information that cannot be obtained only visually. Despite the simplicity of the idea of giving the wheel a sense of touch, the approach is expected to provide richer information of the terrain and is therefore compatible with irregular terrain having a wide range of characteristics.

This paper presents the usefulness of tactile sensing for autonomous wheeled robots on irregular terrain and clarifies the characteristics of the information to be acquired. We create a wheel-on-leg robot model with tactile sensors attached to the wheels in a simulator. Reinforcement learning is adopted for autonomous behavior generation because diverse and adaptive behaviors are considered to be required for traversing irregular terrain. The wheeled robot with tactile wheels is used as an agent, two types of irregular terrain are considered for the environment, and the tactile information transmitted between the ground and wheels is taken as the state to learn the forward action (Fig. 1).

2 Related Works

The tactile wheel has been studied to take advantage of the specialized features of wheels, which are the only components of a wheeled robot that directly interact with the environment. There are diverse sensing methods. Lauria et al., for example, applied the concept of tactile wheels to the wheeled robot Octopus having eight wheels and 15 degrees of freedom (DOFs) using infrared sensors [3]. The tactile wheel developed by Nagatani et al. can directly measure the normal stress distribution using a pressure sensor array, and a study was conducted to estimate the traction force on soft soil by estimating terramechanical parameters [4]. Additionally, wheels that sense contact through resistance have been developed. A sensor shield that detects the exact contact point contributed to the estimation of robot stability [5], and a sensor system that detects multiple contact points successfully scanned the terrain without visual information [6]. The cited studies show the affinity between tactile wheels and irregular terrain.

The range of applications of tactile wheels is gradually expanding, and tactile wheels have been used in machine learning. Marchetti et al., for example, manually generated features from measurements of a two-dimensional pressure grid placed on a wheel and adopted stochastic gradient boosting to estimate the wheel slip and detect the presence of rocks [7]. Another study studied the performance of a similar wheel in terrain classification using graph convolutional networks [8]. However, these studies focused on measuring and classifying terrain by calculating the wheel slip and soil properties, and the scope of their application was narrow. Among these studies, there was no application of tactile information obtained from tactile wheels to autonomous motion control on irregular terrain through machine learning.

3 Experimental Setup

3.1 Model of the Wheeled Robot

As a model that has a simple structure and can generate various actions, we adopt a wheel-on-leg structure and create a robot model with four swing arms and four wheels as shown in Fig. 1. We use the physics engine Multi-Joint dynamics with Contact (MuJoCo) [9] for simulation. Each wheel is attached to the tip of the swing arm and each swing arm has one DOF. All components in the simulation are rigid and OpenAI Gym is used to create the model. The concept of a wheel-on-leg robot with four swing arms is applied to some robots [10,11]. The robot of simple structure can travel rapidly by means of its wheels and can overcome obstacles by swinging its arms.

One end of each of the four arms is connected to one side of the body by a rotary joint and is driven by a position servo that is controlled through an angle command. Each wheel is connected to the end of an arm via a rotary joint and is driven through torque control using an infinitely rotating motor. As shown in Fig. 2a, the angular range of motion of the arms is set to $7\pi/18$. When the front and rear arms are opened to the maximum extent, the front wheels, body, and rear wheels are aligned, and the robot's center of gravity is at its lowest, which is the most stable posture. Meanwhile, when the front and rear arms are closed to the maximum extent (i.e., the front and rear wheels are closest to each other), the wheels do not collide with each other.

3.2 Tactile Wheels

Each wheel of the robot has 16 tactile sensors mounted along its circumference, making for a total of 64 tactile sensors (Fig. 2b). Tactile sensors are equally

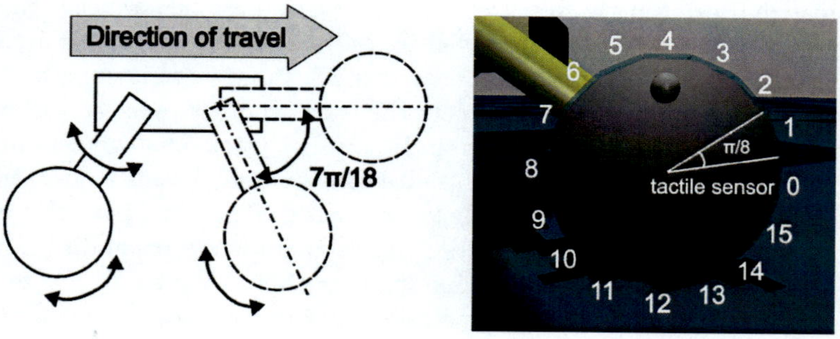

(a) (b)

Fig. 2. Wheeled robot model: **a** A side view of the model. The angular range of motion of the arms is $7\pi/18$. A gray arrow indicates the direction of travel and black arrows indicate the DOFs of rotation. **b** The placement of tactile sensors in MuJoCo. Sixteen tactile sensors are equally spaced at intervals of $\pi/8$.

spaced at intervals of $\pi/8$ and sense the wheel surface. The number of sensors mounted on the wheels of Octopus [3] is 16, and we define the number of sensors as that number from the viewpoint of feasibility. In the implementation within MuJoCo, if a contact point falls within the sensed area, the contact force is included in the sensor output. The output is a non-negative scalar computed by summing the normal forces from all included contacts.

We prepare two types of robot to investigate the learning contribution of wheel tactile information; one acquires analog tactile information (robot-SA) and the other acquires digital tactile information (robot-SD). These are collectively called robot-S. As analog values, the normal force obtained by contact is clipped to [0, 500]. As digital values, all values other than 0 are converted to 1. We conduct learning experiments on irregular terrain using a robot without tactile sensors (robot-N) in addition to robot-SA and robot-SD, and we compare the experimental results.

3.3 Irregular Terrain

Focusing on the geometric features of the terrain, we select two types of irregular terrain, uneven terrain (Fig. 3a) and fractal terrain (Fig. 3b), for the experiments. Uneven terrain is relatively artificial and represented as a line of randomly generated convex parts with a specific unevenness frequency and maximum height. The upper limit of the number of convex parts per fixed area is set by the unevenness frequency, and the number is randomly generated within the allowable range. The height is randomly determined up to the maximum height. Fractal terrain is a more natural type of terrain and randomly generated using the diamond-square algorithm with a specific roughness and maximum height. As the roughness value increases, the ground in a certain area is subdivided

Fig. 3. Two types of irregular terrain: **a** uneven terrain and **b** fractal terrain.

into smaller pieces and its height is determined. The maximum elevation of the terrain is determined by the value of maximum height.

The terrain in the simulation is newly generated for each episode of reinforcement learning. Because the terrain is generated using random values, such as in the uneven-terrain experiment, training is performed on terrain comprising different groups of convex parts in each episode.

3.4 Settings for Reinforcement Learning

Algorithm: Reinforcement learning is adopted in this experiment as the method of action generation, and we adopt the soft actor critic algorithm. All hyperparameters and training procedures are detailed in [12]. The training is performed using the PFRL [13], a reinforcement learning library.

State: The state of robot-N comprises 26 dimensions (i.e., the velocity of the body, angular velocity of the body, orientation of the body, angle of each arm joint, angular velocity of each arm joint, angle of each wheel and angular velocity of each wheel). The only difference between robot-N and robot-S is the length of the state vector that the agent obtains as input at each time step. In contrast with the state of robot-N, the state of robot-S also includes 64-dimensional wheel tactile information, and the state vector thus has 90 dimensions.

Action: The robot used in the simulation has eight DOFs of the four swing arms and four wheel rotations. Each DOF is driven independently, the actions learned in reinforcement learning are thus eight dimensional, and the learned policy outputs an action vector comprising eight floating point numbers for a given environmental state. Note that the joint angle for the posture shown in Fig. 1 is defined as zero, and rotation such that the wheels move backward in the direction relative to the body is considered positive. Of the arms drawn with the dotted lines in Fig. 2a, the one raised is at an angle of $-7\pi/36$ and the other is at an angle of $7\pi/36$. The torque applied to the wheels is expressed as a positive value for wheel rotation torque in the direction that the robot moves forward and a negative value for torque in the opposite direction.

Reward: For the reward function used in reinforcement learning, the forward velocity is selected as a measure of traversability. As the traversability improves, the forward velocity should increase proportionally. Another important aspect of a mobile robot is that the robot avoids flipping over while traveling. We therefore define a reward that penalizes flipping over as

$$r(t) = v_{forward}(t) - C_{flip-over}(t), \tag{1}$$

where $v_{forward}(t)$ is the forward velocity. The reward term $C_{flip-over}(t)$ represents the penalty for flipping over and is defined by

$$C_{flip-over}(t) = \begin{cases} 100 \text{ (if robot flips over)} \\ 0 \quad \text{(else)} . \end{cases} \tag{2}$$

Episode End Condition: The time from the start to the end of a task performed in an environment is referred to as an episode, and the agent robot maximizes the return for the entire episode. In this experiment, the upper limit for the number of steps in the learning episode is set to 1000, and the episode is terminated when 1000 steps are reached. One-thousand learning steps, which is the upper limit of the episode length, corresponds to 10 s. If the robot flips over, it cannot be said that it has succeeded in traversing, and no further progress can be expected. Therefore, in this situation, the episode is terminated compulsorily. In other words, if the robot flips over, the episode ends with a penalty value of 100 deducted from the return, and no further reward is earned. The falling of the robot is determined by the fact that the inner product of the z-axis of the world coordinate system and the z-axis of the local coordinate system of the robot body becomes negative.

Others: One-million steps of training are performed for each terrain. During the training, the performance is evaluated by performing 10 deterministic test rollouts for every 5000 steps and calculating the return value as the average for these rollouts. The experiments are conducted with eight different random seeds. The same action is applied for five environmental steps before returning control to the agent (where the agent action frequency is 100 Hz).

4 Experimental Results

We focus on the average return (i.e., accumulated forward speed minus fall penalties) and sample efficiency to verify that reinforcement learning with tactile information improves the traversability of the wheeled robot over irregular terrain. We consider that a smaller number of learning steps required to earn a certain return in the learning process corresponds to higher sample efficiency. In the experiment, we use nine settings for uneven terrain generated using three levels of the unevenness frequency and three levels of the maximum height. Additionally, three levels of the roughness and three levels of the maximum height, making for a total of nine experiments, are used for the fractal terrain.

Figure 4 presents two examples of the learning curves obtained in the experiments. We can see that the presence of tactile information brings high reward. Figure 5 presents the behaviors of the trained robot on uneven and fractal terrain. For both types of terrain, the trained robot moves forward with walking-like locomotion not only using its wheels but also rotating its arms. Such action is particularly notable when overcoming the convex parts of uneven terrain. Tables 1 and 2 show that tactile information improves agent performance based on the evaluation indicators used in [14]. To compare the performances of robot-S and robot-N, the average return over the last 50 epochs of the experiments with robot-S is divided by the average return of the experiments with robot-N (Table 1). In the experiments, we find that the return with tactile information is higher on average by a factor of 1.18 for uneven terrain and by a factor of 1.31 for fractal terrain. Mostly, there is no significant difference in terms of whether the tactile information is analog or digital for both uneven and fractal terrain. Nonetheless, the use of digital tactile information results in a particularly high average return for a specific uneven terrain (having a high unevenness frequency and mid-range maximum height) shown in Fig. 4a. Table 2 shows an increased ratio of sample efficiency when tactile information is used. To compare the sample efficiency with and without tactile information, we count the number of learning steps required to reach the average return of robot-N in a given environment. The increase in the ratio of sample efficiency is expressed as the number of required steps for robot-N divided by that for robot-S. As an example, for fractal terrain with low roughness and low maximum height parameters (Fig. 4b), robot-N requires 2.01 times the learning steps required by robot-S to obtain the average return of robot-N over the last 50 epochs, which is 642.3 (Table 2). When tactile information is available, the reaching the average return of robot-N is faster by a factor of 2.21 on average for uneven terrain and 2.51 on average for fractal terrain.

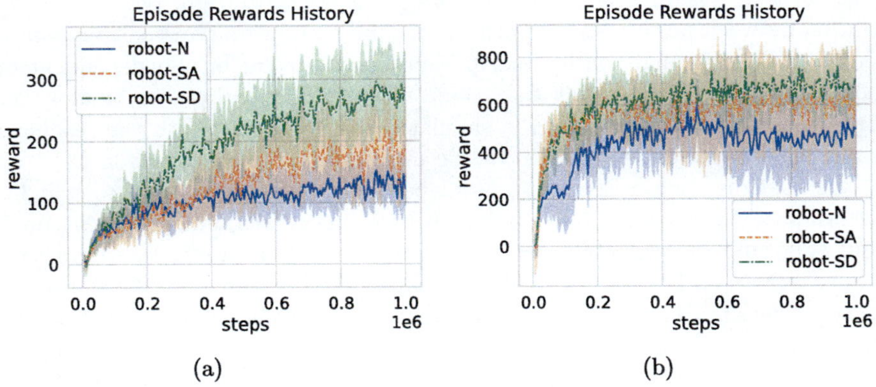

Fig. 4. An example of the learning curve for each terrain. The shaded region represents the standard deviation of the trials: **a** uneven terrain and **b** fractal terrain.

Fig. 5. Examples of traversing actions of the trained robot. The top figures show actions on uneven terrain. The bottom figures show actions on fractal terrain.

5 Discussion

The experimental results show a performance improvement on both relatively artificial uneven terrain and natural fractal terrain, indicating that tactile information from the wheels is useful for driving on irregular terrain. Table 1a shows that the average rate of increase in the average return increases as the unevenness frequency increases, revealing positive correlation between the unevenness frequency and the improvement in traversability due to tactile sensing. This implies that wheel tactile information is more useful when there are more contact points between the wheels and ground. Table 2a shows that the rate of increase in the sample efficiency tends to decrease as the unevenness frequency and maximum height increase. This infers that there is a tradeoff between the increasing ratio of the average return and sample efficiency. For fractal terrain, there is no correlation between performance improvement and terrain roughness or maximum height, indicating that tactile information is not a contribution dependent on these parameters. The result that there is no difference between analog and digital information in most of the simulations is consistent with the results of reinforcement learning studies on tactile manipulation of robot hands [14].

We next discuss the unique results where digital information gives a higher performance than analog information on uneven terrain under certain conditions (Fig. 4a). We evaluate the adaptability to unknown terrain by running the trained agents on terrain not used for training (i.e., on terrain unknown to the agent). There are 17 types of unknown terrain, including eight types of uneven terrain and nine types of fractal terrain. Table 3 shows the average return of robot-S divided by the average return of robot-N when the trained agent runs on each terrain 10 times. The numbers in parenthesis are standard deviations, and conditions for which a significant difference is confirmed using Steel's method with a significance level of 5% are shown in bold. There are significant differences for 13 of the 17 types of terrain for robot-SA and three types of terrain for

Table 1. Average return increase ratio of learning with tactile information relative to learning without tactile information. The ratio is calculated as the average return over the last 50 epochs of the experiments with robot-S divided by that of the experiments with robot-N: **a** uneven terrain and **b** fractal terrain.

(a)

		Unevenness frequency						
		Low		Medium		High		Average
		Analog	Digital	Analog	Digital	Analog	Digital	
Maximum height	Low	1.06	1.05	1.12	1.08	1.18	0.87	1.06
	Medium	1.08	1.09	1.14	1.13	1.40	2.09	1.32
	High	1.17	1.16	1.34	1.32	1.10	0.85	1.16
	Average	1.10		1.19		1.25		1.18

(b)

		Roughness						
		Low		Medium		High		Average
		Analog	Digital	Analog	Digital	Analog	Digital	
Maximum height	Low	1.20	1.36	1.22	1.16	1.24	1.26	1.24
	Medium	1.29	1.47	1.29	1.18	1.13	1.29	1.28
	High	1.37	1.56	1.32	1.27	1.36	1.53	1.40
	Average	1.38		1.24		1.30		1.31

Table 2. Sample efficiency of learning with tactile information relative to learning without tactile information. The indicator is calculated as the number of required steps to reach the average return of robot-N for robot-N divided by that for robot-S: **a** uneven terrain and **b** fractal terrain.

(a)

		Unevenness frequency						
		Low		Medium		High		Average
		Analog	Digital	Analog	Digital	Analog	Digital	
Maximum height	Low	3.18	3.05	3.04	2.59	2.22	2.24	2.72
	Medium	3.02	3.30	1.85	1.99	0.96	1.44	2.09
	High	2.75	2.95	1.35	1.22	1.26	1.31	1.81
	Average	3.04		2.00		1.57		2.21

(b)

		Roughness						
		Low		Medium		High		Average
		Analog	Digital	Analog	Digital	Analog	Digital	
Maximum height	Low	2.01	2.01	3.04	2.75	2.56	2.32	2.45
	Medium	3.19	2.46	2.71	2.86	2.57	2.41	2.70
	High	2.61	1.61	2.01	2.75	2.77	2.47	2.36
	Average	2.30		2.69		2.51		2.51

robot-SD. Regarding robot-SD, two of the three types of terrain for which there is a significant difference are uneven terrain with the same unevenness frequency as the terrain used for learning. This indicates that robot-SA is highly adaptive to unknown terrain. In addition, it can be interpreted that an agent will readily overlearn with respect to the unevenness frequency if it uses digital tactile information. It is therefore possible that the agents may acquire a higher return on terrain having the same unevenness frequency. This may be due to the analog

Table 3. Average return increase ratio for unknown terrain with tactile information relative to learning without tactile information. The condition that unevenness frequency is high and maximum height is medium is excluded because the terrain is used for learning: **a** uneven terrain and **b** fractal terrain.

(a)

		Unevenness frequency					
		Low		Medium		High	
		Analog	Digital	Analog	Digital	Analog	Digital
Maximum height	Low	**1.10** (14.52)	0.92 (18.91)	**1.18** (39.01)	0.99 (66.07)	**1.62** (164.25)	**1.53** (34.22)
	Medium	**1.17** (39.20)	0.99 (24.75)	1.12 (179.50)	1.44 (147.16)	–	–
	High	**1.23** (171.25)	1.02 (148.07)	1.44 (191.49)	1.53 (191.45)	1.28 (71.81)	**1.88** (104.23)

(b)

		Roughness					
		Low		Medium		High	
		Analog	Digital	Analog	Digital	Analog	Digital
Maximum height	Low	**1.17** (62.16)	0.98 (42.85)	**1.16** (99.45)	0.97 (47.14)	**1.34** (70.80)	1.10 (59.20)
	Medium	**1.59** (148.77)	1.41 (79.53)	1.05 (187.77)	0.99 (127.12)	**1.39** 253.11	1.23 (197.62)
	High	**2.36** (236.28)	**2.08** (301.61)	**1.80** (218.44)	1.27 (248.86)	**2.33** (301.61)	2.02 (181.33)

information being richer because analog information reflects the magnitude of force whereas digital information only involves contact detection.

6 Conclusions

This study aimed to improve the traversability of a mobile robot on irregular terrain through reinforcement learning using wheel tactile sensing. Simulation experiments were conducted to learn forward motion. The wheeled robot model used in the experiments was a wheel-on-leg structure with wheels at the ends of four arms and 16 tactile sensors along the circumference of each wheel. To investigate the learning contribution of wheel tactile information, both analog and digital tactile information was used.

We observed that on both uneven and fractal terrain, regardless of the unevenness frequency, terrain roughness, or maximum height, the average return and sample efficiency obtained in learning were higher. Furthermore, it was found that sensing using analog and digital values made no difference to the contribution to learning in most cases. When we examined a special case of uneven terrain, where particularly high performance was achieved for digital information, it was clear that the use of digital tactile information may be prone to a kind of overlearning with respect to the unevenness frequency. In addition, the results show that the agents' traversability increased on unknown terrain when analog tactile information was used.

We therefore conclude that incorporating tactile information from wheels into learning is beneficial to improving autonomous traversability on irregular terrain. In particular, when designing a robot using tactile wheels, it is considered desirable to use analog tactile information for highly adaptability to unknown terrain.

Future work will include real-world experiments to demonstrate the usefulness of tactile sensing, taking into account not only the geometric features of the terrain but also the soil characteristics. Moreover, it is worth investigating the impact of the changes in the structure of the robot and the number of tactile sensors.

Acknowledgements. This work was supported by JSPS KAKENHI Grant Number 18H05466.

References

1. Oliveira, J., Farçoni, L., Pinto, A., Lang, R., Silva, I., Romero, R.: A review on locomotion systems for RoboCup rescue league robots. In: Akiyama, H., Obst, O., Sammut, C., Tonidandel, F. (eds.) RoboCup 2017. LNCS (LNAI), vol. 11175, pp. 265–276. Springer, Cham (2018). https://doi.org/10.1007/978-3-030-00308-1_22
2. Bruzzone, L., Quaglia, G.: Review article: locomotion systems for ground mobile robots in unstructured environments. Mech. Sci. **3**(2), 49–62 (2012)
3. Lauria, M., Piguet, Y., Siegwart, R.: Octopus: an autonomous wheeled climbing robot. In: Proceedings of the Fifth International Conference on Climbing and Walking Robots (CLAWAR 2002), pp. 315–322 (2002)
4. Nagatani, K., Ikeda, A., Sato, K., Yoshida, K.: Accurate estimation of drawbar pull of wheeled mobile robots traversing sandy terrain using built-in force sensor array wheel. In: IEEE/RSJ International Conference on Intelligent Robots and Systems, pp. 2373–2378 (2009)
5. Mardani, A., Ebrahimi, S.: Simultaneous surface scanning and stability analysis of wheeled mobile robots using a new spatial sensitive shield sensor. Rob. Auton. Syst. **98**, 1–14 (2017)
6. Ebrahimi, S., Mardani, A.: A new resistive belt sensor for multipoint contact detection of robotic wheels. Iran. J. Sci. Technol. Trans. Mech. Eng. **43**(1), 399–414 (2018). https://doi.org/10.1007/s40997-018-0166-9
7. Marchetti, Y., Lightholder, J., Junkins, E., Cross, M., Mandrake, L., Fraeman, A.: Barefoot rover: a sensor-embedded rover wheel demonstrating in-situ engineering and science extractions using machine learning. In: IEEE International Conference on Robotics and Automation (ICRA), pp. 6000–6006 (2020)
8. Chen, Y., Marchetti, Y., Gel, Y.R.: Deepening the sense of touch in planetary exploration with geometric and topological deep learning. In: Proceedings of the AAAI Conference on Artificial Intelligence, vol. 35, no. 17, pp. 15,278–15,285 (2021)
9. Todorov, E., Erez, T., Tassa, Y.: MuJoCo: a physics engine for model-based control. In: IEEE/RSJ International Conference on Intelligent Robots and Systems, pp. 5026–5033 (2012)
10. Wilcox, B.H., Jones, R.M.: The MUSES-CN nanorover mission and related technology. In: IEEE Aerospace Conference Proceedings (Cat. No.00TH8484), vol. 7, pp. 287–295 (2000)

11. Zhao, J., Han, T., Wang, S., Liu, C., Fang, J., Liu, S.: Design and research of all-terrain wheel-legged robot. Sensors **21**(16), 5367 (2021)
12. Haarnoja, T., et al.: Soft actor-critic algorithms and applications (2018). arXiv preprint arXiv:1812.05905
13. Fujita, Y., Nagarajan, P., Kataoka, T., Ishikawa, T.: ChainerRL: a deep reinforcement learning library (2019). arXiv preprint arXiv:1912.03905
14. Melnik, A., Lach, L., Plappert, M., Korthals, T., Haschke, R., Ritter, H.: Tactile sensing and deep reinforcement learning for in-hand manipulation tasks. In: IROS Workshop on Autonomous Object Manipulation (2019)

Control of Wheeled-Legged Quadrupeds Using Deep Reinforcement Learning

Joonho Lee(✉), Marko Bjelonic, and Marco Hutter

Robotic Systems Lab, ETH Zürich, 8092 Zürich, Switzerland
joleee@ethz.ch

Abstract. The control problem of wheeled-legged locomotion is still an open problem in the robotics community. Each leg has multiple discrete control modes (rolling, point-foot mode, swing phase), which results in highly nonlinear system dynamics. Most existing works rely on model-based control approaches, and they reduce the complexity of the problem by introducing handcrafted contact sequences or simplified dynamics models. In this work, we attempt to develop a locomotion controller for a wheeled-legged robot using model-free Reinforcement Learning (RL). We train a control policy in simulation, where we simulate the full dynamics of the system and random external disturbances. We then deploy the trained policy on the real robot. Like recent state-of-the-arts in legged locomotion using RL, our preliminary results show that RL is a promising framework for wheeled-legged robots. The policy learns to dynamically switch between driving mode and walking mode in response to the user command and terrain.

Keywords: Wheeled-legged robots · Reinforcement learning · Hybrid locomotion

1 Introduction

Legged robots are promising mobility platforms for challenging environments that are inaccessible or too dangerous for humans, such as disaster areas and the surfaces of other planets. Unlike traditional wheeled or tracked robots, legged systems have the freedom to select support surfaces and can walk over unstructured terrain with steep slopes, steps, and gaps. Researchers have made notable progress in this field [1–5] and several commercial platforms are being deployed in the real world [6–9]. However, despite the versatility and superior traversability over rough terrains, legged systems are inherently limited in speed and endurance compared to wheeled systems because the walking motion requires the joint actuators to continuously work.

Wheeled-legged robots are alternative concepts that leverage actuated wheels to lower the cost of transport (COT) [10] while taking advantage of legged locomotion over challenging obstacles [11]. Additional actuation, however, results in increased complexity in determining when to step and drive, i.e., hybrid locomotion requires discrete decision making. Traditional control approaches struggle to find a solution and require handcrafted heuristics for online gait generation [10].

© The Author(s), under exclusive license to Springer Nature Switzerland AG 2023
J. M. Cascalho et al. (Eds.): CLAWAR 2022, LNNS 530, pp. 119–127, 2023.
https://doi.org/10.1007/978-3-031-15226-9_14

Fig. 1. Swiss-Mile robot traversing stairs.

1.1 Related Work

Most existing works in wheeled-legged locomotion controllers rely on model-based control methods. Our previous work [10] utilized the model predictive control (MPC) formulation of traditional quadrupedal robot [12]. To generate gait, the work uses the heuristic approach of [13] to adapt the MPC's gait timing a priori. However, such a handcrafted motion pattern does not scale to rough terrains and lacks robustness against unperceived obstacles. Trajectory optimization methods can be used to generate gait timings offline, and the MPC can track these offline trajectories on the real robot [11]. This approach, however, cannot adapt the gait plan in response to environmental changes and disturbances and computationally expensive. A sampling-based method [14] showcases a variety of static motions for wheeled-legged excavators and our wheeled-legged robot [11]. However, the optimization can not be computed in real-time on the robot and cannot handle dynamic case. The reduced models that model-based approaches rely on cannot accurately model wheel slippage or uncertainty in the observation, and the behavior is largely biased by handcrafted templates [13]. Specifically, they require human-engineered gait patterns [10,12] which do not change in react to the environment.

To tackle the aforementioned limitations of the conventional approaches, model-free RL has been introduced to this field. Lee et al. [4] proposed an RL formulation for legged robots to generate gaits and controlling motions at the same time. The work demonstrated that RL agents can discover walking behavior without predefined gait patterns and demonstrated remarkable robustness in practical setups where model-based controllers often become fragile.

1.2 Contribution

In this paper, we extend the RL-based perceptive locomotion controller by Miki et al. [15] with additional gait randomization, and present preliminary results on the wheeled-legged robot shown in Fig. 1.

The motion of our controller is shown in the supplementary video. We achieve higher robustness and maximum speed over uneven terrains compared to existing model-based controllers [10].

2 Method

We develop a command-conditioned locomotion controller using model-free RL framework. Command-conditioned locomotion is a key requirement for many applications of mobile robots such as navigation or exploration [16], and a command can be defined in various ways: target direction [4], target velocity [15], or goal position tracking (PointNav) [17]. In this work, we use target velocity as the command for our controller.

2.1 Architecture

The control architecture is shown in Fig. 2. The architecture is similar to the perceptive locomotion controller of Miki et al. [15]. The output of the RNN policy modulates Central Pattern Generators (CPGs) for the leg joints, corrects the joint position targets with residual actions, and commands wheel velocities. The 'joint position controller' is replaced by the actuator network [3] during the training in simulation to improve the accuracy of the simulator. The wheel velocity controller is implemented as a simple proportional controller, i.e., $\tau_t = D(\dot{q}_t^{desired} - \dot{q}_t)$. On the robot, the joint PD controllers run at 400 Hz.

Gait Generation. The CPGs are generating periodic foot trajectories defined by a 6 dimensional vector $\langle \phi_s^1, \phi_s^2, \phi_s^3, \phi_s^4, DF, \dot{\phi} \rangle$. Each leg's trajectory is generated by a periodic phase $\phi^{n \in \{0,1,2,3\}} \in [0, 1)$ that cycles at the frequency $\dot{\phi}$, i.e., $(\phi^n)_{t+dt} = wrap_{01}((\phi^n)_t + \dot{\phi} \cdot dt)$, where $wrap_{01}(x) := x - \lfloor x \rfloor$. The swing phase starts at ϕ_s and ends at $\phi_s + DF$, where DF represents the duty factor. For example, $\langle 0.0, 0.5, 0.5, 0.0, 0.5, 1.0 \rangle$ represents a trot gait. The last three entities are target horizontal velocity and yaw rate defined in body frame. During the swing phase, the target foot position is shifted by 0.2 m upward. The definitions of gaits are given in Table 1. Instead of directly sampling each ϕ_s^n, we first sample ϕ_s^0 from a uniform distribution $U(0.0, 1.0)$ and then sampled phase offset for each leg. We used the gaits from Belicoso et al. [18] and added random offsets.

Gait Randomzation. The gait parameter is randomized during the training to enable the policy to generalize to different gait patterns and to efficiently explore distinct behaviors. The gait parameter is then fixed during the deployment.

Table 1. Gait definitions

	$\phi_s^1 - \phi_s^0$	$\phi_s^2 - \phi_s^0$	$\phi_s^3 - \phi_s^0$	Duty factor	Frequency
Trot	$U(0.4, 0.6)$	$U(0.4, 0.6)$	$U(-0.1, 0.1)$	$U(0.3, 0.5)$	$U(0.8, 1.6)$
Walk	$U(0.2, 0.3)$	$U(0.45, 0.55)$	$U(0.7, 0.8)$	$U(0.2, 0.25)$	$U(0.6, 1.0)$
Driving	$U(0.0, 1.0)$	$U(0.0, 1.0)$	$U(0.0, 1.0)$	$U(0.0, 0.0)$	$U(0.0, 1.0)$

Similarly to [15] and [4], we fix the gait parameter to the "trot" gait during the test time. However, this does not limit the capability of the policy because the policy directly acts on the gait parameter (see Gait param. offsets in Fig. 2) and can switch to different gait. Through RL, the policy learns to switch between different gaits in response to the command velocity and the terrain.

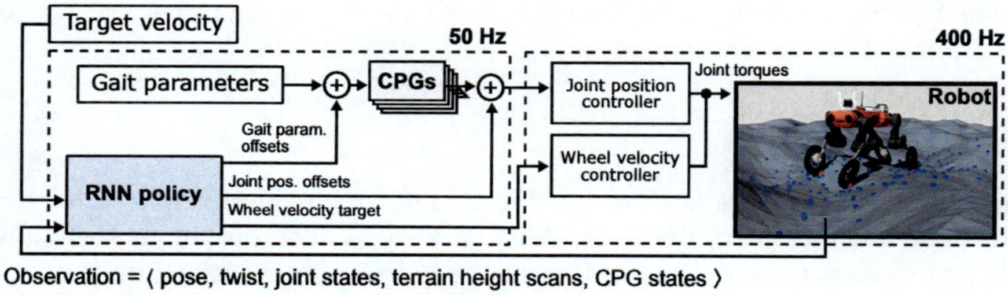

Fig. 2. Control architecture

2.2 Reinforcement Learning

In RL framework, a control problem is modeled as a Markov Decision Process (MDP). MDP is a mathematical framework for modeling a stochastic control process commonly used in RL. At each timestep, an agent observes a state $s_t \in \mathcal{S}$ from the environment and outputs an action $a_t \in \mathcal{A}$. The environment then evolves into s_{t+1} via the state transition function $p(s_{t+1}|s_t, a_t)$, gives a reward $r_t \in \mathcal{R} : \mathcal{A} \times \mathcal{S} \rightarrow \mathbb{R}$ and a new observation to the agent. The agent may act according to a stochastic policy $\pi_\theta(a_t|s_t)$ parametrized by θ. RL algorithms update θ to maximize the cumulative discounted rewards $\mathbb{E}[\sum_{t=k}^{\infty} \gamma^t r_t]$ through environmental interactions, where k is the current timestep and γ is the discount factor.

MDP Formulation. An MDP is defined by a tuple of state space ($\mathcal{S}$), action space ($\mathcal{A}$), $r_t(s_t, s_{t+1})$, and $p(s_{t+1}|s_t, a_t)$ Our state definition inherits from our previous work [15]. The policy observes ground height scans around the robot and proprioceptive measurements from IMU and joint encoders. As described above, the action is defined as a 21 dimensional vector that consists of gait

Table 2. Reward functions. (The velocities are defined in the base frame)

Reward terms	
Linear velocity	$1.5\exp(-3(v^{xy}-\hat{v}^{xy})^2)$
Angular velocity	$1.0\exp(-3(\omega^{z}-\hat{\omega}^{z})^2)$
Base motion penalty	$0.8\exp(-((v^{z})^2)+\exp(-(\omega^{x,y})^2)$
Torque penalty	$-10^{-6}\|\|\tau\|\|^2$
Joint speed penalty	$-0.05\sum_{i=0}^{12} max(\|\dot{\phi}_i\|-\dot{\phi}_{jslim},0)^2$
Action smoothness penalty	$-0.05\|a_{t-2}-2a_{t-1}+a_t\|$
Symbols	
$\dot{\phi}_{jslim}$	Maximum joint speed
τ	Vector of joint torques
fc	Foot contact state
$\hat{(\cdot)}$	Target quantity
$(\cdot)_g$	Sub-goal quantity
$\mathbb{1}_{(\text{condition})}(\cdot)$	Indicator function

parameter offsets (5), joint position offsets (12), and wheel velocity commands (4). The reward function is defined by the sum of the reward terms given in Table 2. The state transition follows the simulated rigid by dynamics, and each episode terminates when the main body of the robot touches the ground or when it reaches joint torque and speed limits.

Training. We follow the privileged training method by Lee et al. [4]. We first train a teacher policy with noiseless simulated states (privileged information), using an on-policy RL algorithm [19]. In other words, we append s_t with additional simulation states including ground friction coefficient, ground reaction forces, and the contact state of each link. The privileged information enables the teacher policy to learn terrain-adaptive behaviors. Thanks to the noise-free and accurate simulation states, the teacher policy converges fast to a highly performant policy. The teacher policy training requires about 8 h on a desktop PC using PPO algorithm [19].

Then we train a student policy for the real-world deployment. The student policy imitates the teacher without the privileged information and with observation noise. With imitation learning [20], it learns to construct an internal representation of the world from the sequence of the noisy real-world observations. A policy trained in this way has proven to be more adaptive and robust in real-world settings with high disturbances and noisy observations [4].

2.3 Implementation Details

Our training environment is implemented using Raisim [21] simulator. We used actuator network [3] to simulate the series elastic actuators at the joints, and the wheel actuators are simulated as ideal torque sources. The wheel is simulated as a rigid disk, and we did not introduce any tire models [22]. Instead, we randomized the wheel diameter and applied random damping during the training. The randomization parameters are observed by the teacher policy.

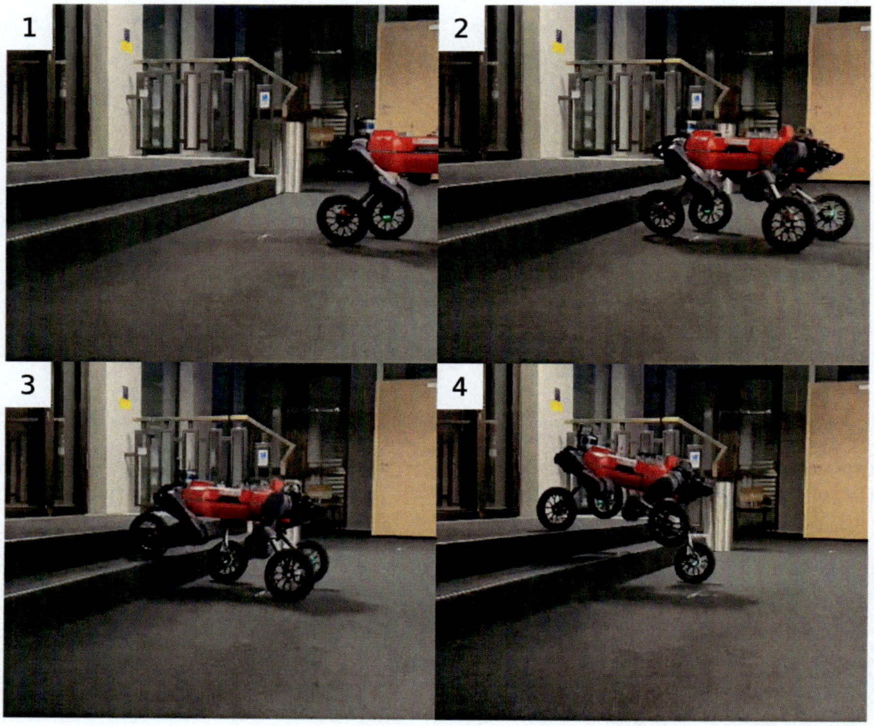

Fig. 3. Gait transition from Driving to Trot.

3 Qualitative Results

We tested our controller in various urban environments. The motion is shown in the supplementary video. Notably, the neural network policy learns to switch between Driving gait and Trot gait. When the command velocity is given in x-direction (longitudinal direction with respect to the main body) or commanded zero velocity, the policy does not step. It starts stepping when commanded in other directions or when it faces an obstacle as shown in Fig. 3.

Additionally, we tested the controller for the collision avoidance in simulation. As shown in Fig. 4, we command the robot into an obstacle. The robot steps to the side to avoid the collision. Our controller decides when and where to step based on the terrain height information around the robot (shown by blue dots in Fig. 4A). Figure 4B shows the trajectory of the robot's main body in the world frame, and Fig. 4C shows the foot contact sequence.

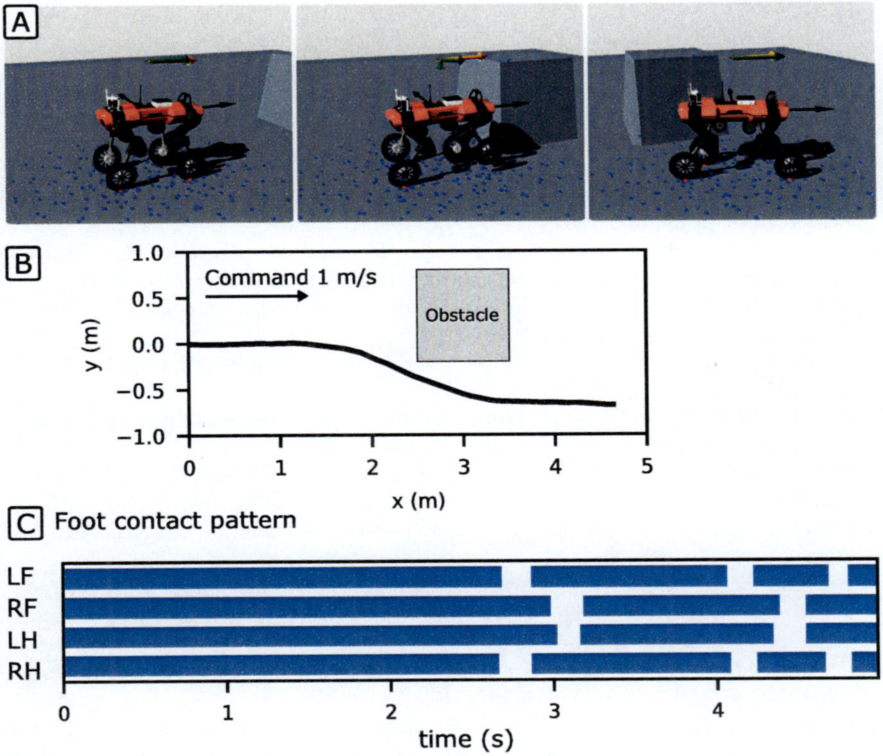

Fig. 4. Obstacle avoidance behavior. (A) Motion sequence of the robot during the obstacle avoidance. (B) Base trajectory. (C) Foot contact sequences during the experiment. LF means Left Front, RH means Right Hind leg.

4 Future Work

As shown by the qualitative results, existing RL-based locomotion strategy [4,15] for point-foot quadrupedal robots transfers well to the wheeled-legged systems. However, handcrafted gait switching and gait randomization is still required to encourage the policy to explore different gait patterns during the training. In the future work, we will investigate hierarchical RL approaches [23,24] to introduce a learned agent to propose gait parameters in response to environmental changes and reduce the handcrafting effort in the development.

References

1. Raibert, M., Blankespoor, K., Nelson, G., Playter, R.: BigDog, the rough-terrain quadruped robot. IFAC Proc. Vol. **41**(2), 10,822–10,825 (2008)
2. Katz, B., Di Carlo, J., Kim, S.: Mini Cheetah: a platform for pushing the limits of dynamic quadruped control. In: 2019 International Conference on Robotics and Automation (ICRA), pp. 6295–6301. IEEE (2019)
3. Hwangbo, J., et al.: Learning agile and dynamic motor skills for legged robots. Sci. Robot. **4**(26), eaau5872 (2019)
4. Lee, J., Hwangbo, J., Wellhausen, L., Koltun, V., Hutter, M.: Learning quadrupedal locomotion over challenging terrain. Sci. Robot. **5**(47) (2020)
5. Park, H.W., Wensing, P.M., Kim, S.: Jumping over obstacles with MIT Cheetah 2. Robot. Auton. Syst. **136**, 103,703 (2021)
6. Boston Dynamics: Spot (2021). https://www.bostondynamics.com/spot. Accessed Mar 2021
7. Agility Robotics: Robots (2021). https://www.agilityrobotics.com/robots. Accessed June 2021
8. Unitree Robotics: A1 (2021). https://www.unitree.com/products/a1/. Accessed Mar 2021
9. Ghost Robotics: Vision 60 (2021). https://www.ghostrobotics.io/. Accessed June 2021
10. Bjelonic, M., Grandia, R., Harley, O., Galliard, C., Zimmermann, S., Hutter, M.: Whole-body MPC and online gait sequence generation for wheeled-legged robots. arXiv preprint arXiv:2010.06322 (2020)
11. Bjelonic, M., et al.: Offline motion libraries and online MPC for advanced mobility skills. Int. J. Robot. Res. 02783649221102473 (2022)
12. Grandia, R., Farshidian, F., Ranftl, R., Hutter, M.: Feedback MPC for torque-controlled legged robots. In: 2019 IEEE/RSJ International Conference on Intelligent Robots and Systems (IROS), pp. 4730–4737. IEEE (2019)
13. Boussema, C., Powell, M.J., Bledt, G., Ijspeert, A.J., Wensing, P.M., Kim, S.: Online gait transitions and disturbance recovery for legged robots via the feasible impulse set. IEEE Robot. Autom. Lett. **4**(2), 1611–1618 (2019)
14. Jelavic, E., Farshidian, F., Hutter, M.: Combined sampling and optimization based planning for legged-wheeled robots, pp. 8366–8372 (2021)
15. Miki, T., Lee, J., Hwangbo, J., Wellhausen, L., Koltun, V., Hutter, M.: Learning robust perceptive locomotion for quadrupedal robots in the wild. Sci. Robot. **7**(62), eabk2822 (2022)
16. Tranzatto, M., et al.: Cerberus: autonomous legged and aerial robotic exploration in the tunnel and urban circuits of the DARPA subterranean challenge. J. Field Robot. (2021)
17. Anderson, P., et al.: On evaluation of embodied navigation agents. arXiv preprint arXiv:1807.06757 (2018)
18. Bellicoso, C.D., Jenelten, F., Gehring, C., Hutter, M.: Dynamic locomotion through online nonlinear motion optimization for quadrupedal robots. IEEE Robot. Autom. Lett. **3**(3), 2261–2268 (2018)
19. Schulman, J., Wolski, F., Dhariwal, P., Radford, A., Klimov, O.: Proximal policy optimization algorithms. arXiv preprint arXiv:1707.06347 (2017)
20. Ross, S., Gordon, G., Bagnell, D.: A reduction of imitation learning and structured prediction to no-regret online learning. In: Proceedings of the Fourteenth International Conference on Artificial Intelligence and Statistics, pp. 627–635. JMLR Workshop and Conference Proceedings (2011)

21. Hwangbo, J., Lee, J., Hutter, M.: Per-contact iteration method for solving contact dynamics. IEEE Robot. Autom. Lett. **3**(2), 895–902 (2018)
22. Pacejka, H.B., Bakker, E.: The magic formula tyre model. Veh. Syst. Dyn. **21**(S1), 1–18 (1992)
23. Nachum, O., Tang, H., Lu, X., Gu, S., Lee, H., Levine, S.: Why does hierarchy (sometimes) work so well in reinforcement learning? arXiv preprint arXiv:1909.10618 (2019)
24. Jain, D., Iscen, A., Caluwaerts, K.: Hierarchical reinforcement learning for quadruped locomotion. In: 2019 IEEE/RSJ International Conference on Intelligent Robots and Systems (IROS), pp. 7551–7557. IEEE (2019)

Design Optimization of a Four-Bar Leg Linkage for a Legged-Wheeled Balancing Robot

Victor Klemm[1,2](✉), Dominik Mannhart[1], and Roland Siegwart[1]

[1] ASL, ETH Zürich, 8092 Zürich, Switzerland
{vklemm,dominikm,rsiegwart}@ethz.ch
[2] RSL, ETH Zürich, 8092 Zürich, Switzerland

Abstract. Balancing legged-wheeled robots have gained popularity in recent years due to their locomotive efficiency while still being able to conquer rough terrain and obstacles. Furthermore, as this type of robot maintains ground contact with its wheels for most of the time, passive gravity compensation mechanisms can greatly minimize power consumption. Various designs with different leg configurations have emerged, whereby a 1-DOF mechanism per leg already showed sufficient compliance to adapt to most outdoor terrain. We propose a design optimization procedure for a 1-DOF four-bar leg linkage to ensure minimum pitch angle correction of the robot's base while varying the leg extension. Gravity compensation is further achieved through an optimized torsional spring. Finally, we evaluate the performance of the leg linkage and gravity compensation mechanism on real hardware.

Keywords: Design optimization · Gravity compensation · Balancing robots · Legged-wheeled robots

1 Introduction

The emergence of legged-wheeled robots in the last five years had a disruptive effect on the robotics scene [1]. With principles of legged robotics and new actuator types becoming readily available, suddenly, numerous systems showed astounding capabilities which rival the ones of purely legged systems while having a faster and more efficient way of locomotion [2]. Especially bipedal legged-wheeled systems gained popularity with their versatility [3], [4] and simplistic design [5]. While Boston Dynamics' Handle [6] shows impressive driving, jumping, and manipulation performance, their system has an extremely high degree of complexity.

Our previous work [7] shows that even a 1-DOF leg per wheel can allow for swift traversal of uneven terrain, as vertical compliance to the wheel is critical. A straightforward approach for achieving a basic level of vertical compliance is linear actuators. While they allow for a straight up and down motion, linear

© The Author(s), under exclusive license to Springer Nature Switzerland AG 2023
J. M. Cascalho et al. (Eds.): CLAWAR 2022, LNNS 530, pp. 128–139, 2023.
https://doi.org/10.1007/978-3-031-15226-9_15

Fig. 1. Different possible leg linkage designs are visualized for the *Ascento* robot by varying the link lengths and angles. The design optimizer's objective is to find a solution that allows keeping the robot's base as level as possible over the entire operating range of the leg linkage.

Fig. 2. A real-world *Ascento* prototype with the optimized linkage and spring is shown.

actuators are either known to be slow or require additional onboard compressors. The *Ascento* robot [8], on the other hand, uses rotative actuators with the combination of a four-bar linkage to achieve a similar motion of the wheel end-effectors. We thereby optimize the leg linkage to move the wheel up and down in a straight line. The aim behind this objective is to decouple the driving and leg dynamics, especially as the said robot is designed for jumping. This comes with the following drawback: When changing its base height, the center of mass of the robot shifts away from its equilibrium point, and the robot needs to adjust its pitch angle through a significant driving maneuver to regain balance, which is undesirable. This can also be detrimental to sensor performance due to rapid accelerations during the maneuver or the resulting static offset of their field of view. Unfortunately, the mentioned effects are amplified for the new *Ascento* concept, see Fig. 1. The new design has a significantly higher lateral shift of the total center of mass when changing height. This is due to the all-metal leg structure, which is heavier relative to the base, and the new placement of the motor, which is in the knee joint, for better leverage. In this work, we aim to solve these issues through design optimization of the linkage geometry, see Fig. 2. Design optimization has recently gained popularity, especially in legged robotics. In [9] a toolbox is presented, allowing to optimize leg designs for a specific objective, e.g. minimum effort walking. In this work, we present a custom procedure, as our objective differs from typical approaches due to the treated parallel kinematic four-bar linkage.

An additional distinctive feature of the *Ascento* robot, which was reconsidered in the re-design, is the gravity compensation mechanism. Known bipedal legged-wheeled systems use hydraulic [10], or electric actuation, which requires significant power consumption even in an idle state. Especially when standing with bent knees, which is necessary for compliance, the static torque of the knee motors can be immense. To alleviate this, *Ascento* uses a spring for gravity compensation: This technique is beneficial for legged-wheeled robots, which mostly have ground contact while driving. In contrast, for legged locomotion, this only works to a certain extent, as during a step, the spring force needs to be overcome. Typical mechanical gravity compensation mechanisms, such as in [11], [12], are either focused on manipulators or actuators. We further highlight [13], where the parallel kinematic leg structures of the quadruped are supported by linear parallel springs to store energy for jumping maneuvers and [14], where cams with springs are used to achieve a specific compliance behavior of the end-effector. In this work, we present an optimization procedure to find the spring parameters of a gravity compensating parallel torsional spring for the optimized four-bar linkage. The spring parameters of the previous *Ascento* prototype were found by curve fitting the spring torque to the residual motor torque in a least-squares fashion. In the following, we improve on that by introducing a worst-case optimization that we deem more suitable for our case and outline the spring optimization for the first time in detail.

The two design optimization procedures of the linkage geometry and the spring parameters form the contribution of this work. Similarly, the paper is structured into two parts; Sect. 2 describes the linkage design optimization, and Sect. 3 describes the gravity compensation mechanism. To our knowledge, this is the first time a structured approach to design optimization for parallel-kinematics mechanisms regarding their statics properties is shown.

2 Linkage Design Optimization

In the following, our goal is to determine the mentioned parameters through constrained optimization. We choose this approach because it allows us to define a complex objective while respecting several hard constraints necessary for a feasible hardware design. We keep the base fixed for the following optimization problem and formulate the problem out of the base's reference frame, which leads to an algebraically simpler formulation. We start by fully parametrizing the linkage by its geometric properties, see Fig. 3. This includes the link lengths a, b, c, d of the four-bar linkage, the length e of the shank where the wheel is attached, the angle β between the forward direction of the base and the head linkage link and the angle γ between the shank and the knee link. The linkage kinematics is now fully defined, whereby varying the hip angle α results in the extension and retraction of the linkage. Its upper and lower limits are denoted as α_{min} and α_{max}, respectively. Finally, we parametrize the x position of the center of mass of the base, as it can be varied easily by modifying the battery placement in the base's hardware design.

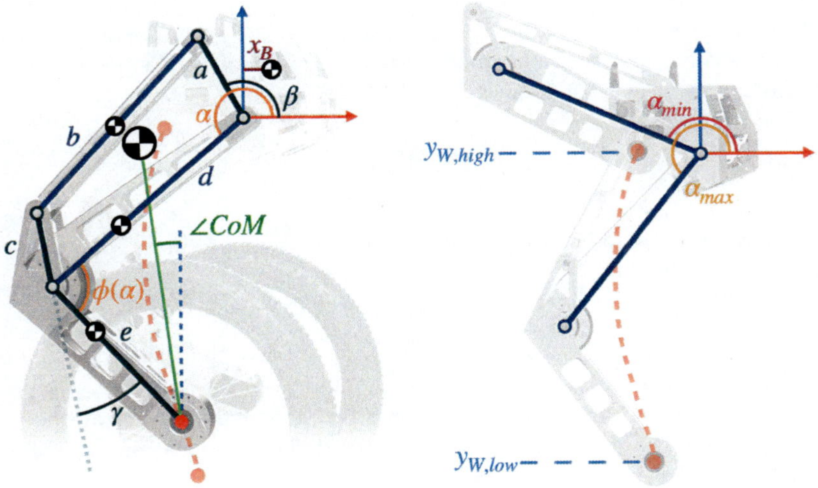

Fig. 3. *Left:* The parametrized leg linkage is shown. The leg can be extended and retracted by varying the hip angle α, whereby the wheel center follows the red dashed line. The current $\angle CoM$ angle between the base vertical and the total center of mass w.r.t. the wheel center is marked light green. *Right:* The limits of the hip angle α_{min}, α_{max} are shown, indicating the maximally retracted and extended positions of the linkage. Furthermore, the corresponding upper and lower wheel heights $y_{W,low}$, $y_{W,high}$ are marked with dashed light blue lines, indicating the range that the wheel's center must cover when moving the leg angle from one limit to another.

2.1 Problem Formulation

The objective of the optimization can be written as

$$\underset{a,b,c,d,e,\beta,\gamma,x_{base},\alpha_{min},\alpha_{max}}{\text{minimize}} \int_{\alpha_{min}}^{\alpha_{max}} f_{\angle CoM}\left(\alpha, a, b, c, d, e, \beta, \gamma, x_{base}\right) d\alpha, \tag{1}$$

where we integrate the angle between the vertical of the base and the total center of mass of the inverted pendulum structure[1] w.r.t. the wheel center. Minimizing this $\angle CoM$ angle over the entire operating range of the linkage will result in a linkage design which tries to keep the base as level as possible at the balance point for every linkage extension. The angle can be defined as

$$f_{\angle CoM} = \arctan\left(\frac{r_{CoM,x} - r_{W,x}}{r_{CoM,y} - r_{W,y}}\right) \tag{2}$$

where $\boldsymbol{r}_{CoM}$ denotes the position of the total center of mass and $\boldsymbol{r}_W$ the position of the wheel center. To determine these two quantities based on the linkage parameters, we need to find expressions for the positions of the joints and centers of mass for each link, respectively. This can be done by considering a kinematic solution for the parallel kinematics leading to a function for the knee angle $\phi(\alpha)$, which can be derived from [15] without much effort. Next, the desired quantities can be obtained by forward kinematics, see e.g. [16]. Finally, it's possible to find the total center of mass as follows:

$$\boldsymbol{r}_{CoM}(\alpha, a, b, c, d, e, \beta, \gamma, x_{base}) = \frac{\sum_{i=1}^{4} m_i\, \boldsymbol{r}_{i,CoM}\left(\alpha, a, b, c, d, e, \beta, \gamma, x_{base}\right)}{\sum_{i=1}^{4} m_i}. \tag{3}$$

The required masses of the links are calculated by using a linear law

$$m_i = m_{fixed,i} + l[i]\, m_{varying,i}, \quad \text{where} \quad l \in \{a, b, c, d, e\}, \tag{4}$$

to account for changing link mass when varying the length of the corresponding link. The law consists of a fixed part $m_{fixed,i}$ and a varying part $m_{varying,i}$ and is approximately valid over the allowed length range of the optimization.

Additionally, the lower and upper bounds of the optimization variables are enforced with

$$\begin{bmatrix} a_{lb} \\ b_{lb} \\ c_{lb} \\ d_{lb} \\ e_{lb} \\ \beta_{lb} \\ \gamma_{lb} \\ x_{B,lb} \\ \alpha_{min,lb} \\ \alpha_{max,lb} \end{bmatrix} \leq \begin{bmatrix} a \\ b \\ c \\ d \\ e \\ \beta \\ \gamma \\ x_B \\ \alpha_{min} \\ \alpha_{max} \end{bmatrix} \leq \begin{bmatrix} a_{ub} \\ b_{ub} \\ c_{ub} \\ d_{ub} \\ e_{ub} \\ \beta_{ub} \\ \gamma_{ub} \\ x_{B,ub} \\ \alpha_{min,ub} \\ \alpha_{max,ub} \end{bmatrix}, \tag{5}$$

[1] The inverted pendulum structure includes all the bodies except the wheels.

which allow to specify a range that is still feasible for the subsequent hardware design. A further linear constraint is added to specify the order of the hip angle limits:

$$\alpha_{min} \leq \alpha_{max}. \tag{6}$$

Finally, nonlinear constraints are required to enforce the desired height range $[y_{W,low}, y_{W,high}]$ of the linkage:

$$y_{W,high} \leq f_{W,y}\left(\alpha_{min}, a, b, c, d, e, \beta, \gamma\right) \tag{7}$$

$$f_{W,y}\left(\alpha_{max}, a, b, c, d, e, \beta, \gamma\right) \leq y_{W,low}. \tag{8}$$

2.2 Optimization Setup

We implement our approach in MATLAB and solve the resulting constrained optimization problem using the `fmincon` function. With `fmincon` we choose a gradient-based method, as we expect our problem to be smooth in the area of interest. We initialize the optimization with a scaled version of the linkage parameters of the previous prototype. The optimization takes an average time of around 12 seconds to converge to a local minimum with the default solver settings.[2]

Table 1. Optimized linkage parameters and the resulting link masses.

a	b	c	d	e	β	γ	x_B	α_{min}	α_{max}	m_a	m_b	$m_{c,e}$	m_d
0.158 m	0.408 m	0.126 m	0.437 m	0.322 m	2.03 rad	0.621 rad	−0.00981 m	2.52 rad	4.05 rad	4.01 kg	0.408 kg	1.85 kg	0.930 kg

2.3 Results

The resulting optimized linkage parameters satisfy the required constraints and achieve the desired objective by several orders of magnitude better than the initial guess. Despite mass uncertainties, qualitative results on real hardware show minimal tilt correction for different leg extensions. A total of 14 evenly spaced leg extensions are chosen, and the corresponding base pitch angle offsets are averaged over several seconds. A quantitative evaluation is given in Fig. 4. The resulting parameters are given in Table 1.

3 Gravity Compensation Mechanism

This section aims to optimize a spring-based gravity compensation mechanism for the previously optimized leg linkage. We limit the solution space to torsional springs, as their ease of implementation and compactness keep the mechanical complexity of the system low. Regarding the spring placement, the four joints

[2] On an Intel i7-8809G processor.

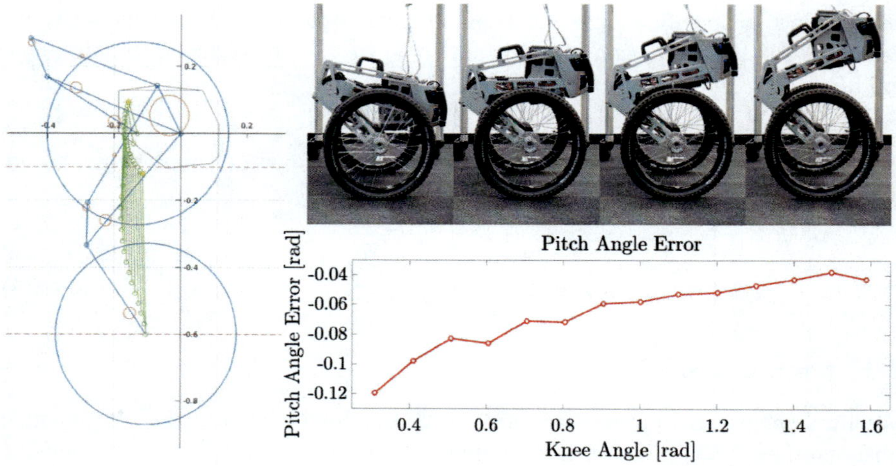

Fig. 4. *Left:* The optimal solution of the linkage parameters is highlighted. The green lines connecting the wheel center and the total center of mass are close to vertical over the entire operating range of the linkage while fulfilling the constraint on the demanded height range (dotted lines). The size of the orange circle drawn at the center of mass of each part indicates their relative masses. *Right:* The real-world *Ascento* prototype with the optimized linkage extends its legs in the top image sequence. Below, the base's corresponding measured pitch angle error is shown in steady-state. We note the similarity to the calculated behavior, i.e., the pitch angle needs less correction when the linkage is extended, and the maximal pitch angle deviation is well below 10°.

of the four-bar linkage are all viable solutions. As the knee joint covers the highest absolute range during the leg mechanism's full up and down motion, an additional torque at this joint exhibits the highest leverage benefit. We, therefore, choose the design shown in Fig. 5, allowing for the compact implementation of motor and spring at the same joint.

In the following, the spring parameters are to be optimized. We choose the typical parameters for torsional springs, which are stiffness k and preload angle ϕ_0, and assume a linear spring law for its torque

$$\tau_S(\alpha) = k\ (\phi(\alpha) - \phi_0)\,. \tag{9}$$

3.1 Problem Statement

To determine the optimal spring parameters, we must define a metric beforehand. The spring should bring the residual knee motor torque required to hold the robot in equilibrium to zero for every leg configuration for an ideal gravity compensation mechanism. As this is not possible due to the high nonlinearities of the linkage, we define a cost function. In contrast to the least-squares

Fig. 5. The final motor-spring module design is highlighted in the left knee of the robot. The torsional spring is wound around the motor and presses against the shank and the lower thigh to support the upper body's weight, reducing the required static torque of the motor substantially.

optimization, which was used for the previous *Ascento*, we present a worst-case optimization by using a bi-level optimization objective

$$\underset{k,\phi_0}{\text{minimize}} \left\| \begin{bmatrix} \min\limits_{\alpha_{min} \leq \alpha \leq \alpha_{max}} & f_{\tau,K}(\alpha, k, \phi_0) \\ \max\limits_{\alpha_{min} \leq \alpha \leq \alpha_{max}} & f_{\tau,K}(\alpha, k, \phi_0) \end{bmatrix} \right\|_2, \tag{10}$$

where $f_{\tau,K}$ denotes the residual knee torque to keep the system in equilibrium. We derive it using the principle of virtual work in combination with differential kinematics of the expressions from Sect. 2.

We motivate the choice of cost function as follows: A least-squares fit optimizes overall values equally, potentially leaving high costs at some places for the benefit of others. But in our case, it is desirable to keep the peak torque over the range of the leg linkage minimal, as the robot should be able to drive for extended amounts of time at any leg height. The chosen optimization objective has the interesting effect that the boundary points of the torque function, which is roughly parabolic in shape, come to lie on the same torque level. This behavior can be intuitively explained: The objective squeezes the torque function together until equality between the two boundary points is reached.

Similar to Sect. 2, the optimization variables are bounded due to design constraints:

$$\begin{bmatrix} k_{lb} \\ \phi_{0,lb} \end{bmatrix} \leq \begin{bmatrix} k \\ \phi_0 \end{bmatrix} \leq \begin{bmatrix} k_{ub} \\ \phi_{0,ub} \end{bmatrix}. \tag{11}$$

3.2 Optimization Setup

Similar to the optimization setup for the leg linkage optimization, we implement our approach in MATLAB and solve the resulting constrained optimization problem using the `fmincon` function.

As only two optimization variables are to be determined, we can plot the objective function as a 2D surface. Furthermore, the problem has no nonlinear constraints, unlike the leg linkage optimization. Only explicit bounds on the optimization variables are required, rendering the problem structure significantly simpler. We initialize the optimization with a scaled version of the spring parameters of the previous prototype. The optimization takes an average time of around 2 s to converge to a local minimum with the default solver settings.

3.3 Results

The resulting optimal spring parameters can be identified as the global minimum of the objective function, see Fig. 6. In our case, the respective values are −3.12 Nm/rad for the stiffness k and 5.80rad for the preload angle ϕ_0. With the optimized spring parameters, the robot's average theoretical holding torque is reduced from 17.8 Nm to 2.4 Nm, leading to substantial power savings and allowing to select motors with significantly lower nominal torques.

We conduct an experiment on the real *Ascento* robot prototype by measuring the residual steady-state motor torques while balancing on different leg heights. The torque measurements are inferred from the motor currents, whose relation has been carefully calibrated and is reasonably deterministic for the pseudo-direct drives in question. A total of twelve evenly spaced leg extensions are chosen, and the torques of the left and right knee motor are averaged over several seconds. We measure each leg extension twice, once by extending and once by retracting the leg mechanism. This ensures that direction-dependent effects, such as joint hysteresis can be identified. With a maximum[3] of just below 4 Nm the spring is significantly reducing the required power of the motor. We assign the differences in the calculated values to the mass uncertainties of the real-world prototype and the non-linearities and manufacturing tolerances (> 10%) of the torsional springs. Qualitatively, the springs render the robot's body practically weightless when manually moved. Due to stiction effects, the robot also remains passively on a constant height when released.

[3] Averaged over the up and down motion to remove hysteresis effects.

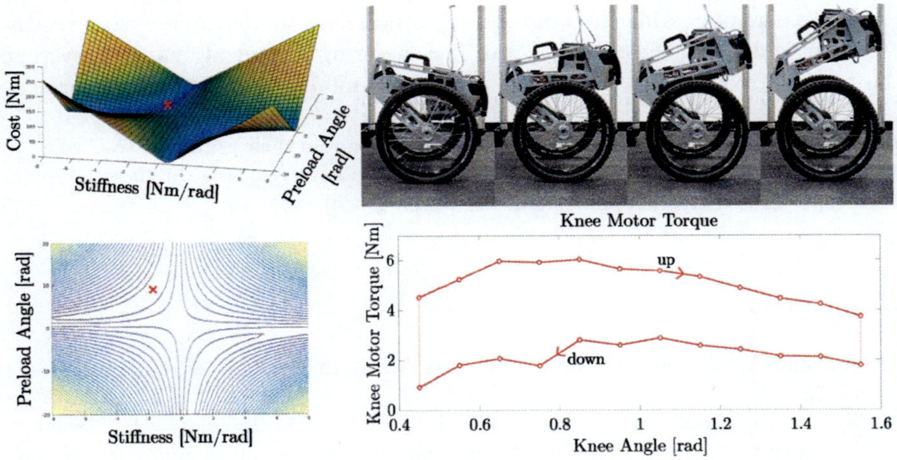

Fig. 6. *Left:* In the top graph the cost function of the spring parameter optimization is plotted over the feasible ranges of stiffness and preload angle values. The level sets of said function are further outlined in the graph below. *Right:* In the bottom graph the residual motor torques of the knee motors are plotted for a full extension and retraction cycle of the leg linkage. The above image sequence shows the corresponding extension, whereby each data point was measured in steady-state. The torque differences between up and down motions can be explained by joint hysteresis friction effects. We note that due to the spring, the knee motors are far from their nominal torque of 13 Nm, causing significant power savings and providing higher margins for maneuvers.

4 Conclusions and Future Work

In this work, we show an optimization procedure for designing a 1-DOF four-bar leg linkage with the application of a wheeled balancing robot in mind. The approach successfully produces leg linkage parameters that minimize correction of the robot's base pitch angle over the entire operating range of the leg linkage. In a second step, the spring parameters of an additional parallel spring are determined. The spring's functionality is to compensate for the proper weight of the robot, reducing the power consumption of the knee motor drastically. The optimized results for both pitch angle correction and gravity compensation are validated on real hardware.

We note that the leg linkage optimization is highly sensitive to the initial guess, as only local convergence can be ensured. To alleviate this, we propose to use global optimization methods, which we leave as future work. In that case, a re-implementation utilizing a framework that allows for faster function and derivative evaluation would be highly beneficial, e.g., CasADi [17], as the current approach is severely limited by the evaluation time of the objective function and constraints. However, local convergence was more than sufficient for the targeted accuracy.

A further interesting direction to explore is how the optimal spring relates to dynamic criteria, which might give rise to spring optimizations that leverage the natural frequency of the system during common locomotion maneuvers.

Acknowledgements. The authors would like to express their gratitude to Ciro Salzmann, Lionel Gulich, Alessandro Morra for their support in realizing the shown *Ascento* prototype, as well as all supporters of the *Ascento* project.

References

1. Bjelonic, M., Klemm, V., Lee, J., Hutter, M.: A survey of wheeled-legged robots (2022)
2. Bjelonic, M., Bellicoso, C.D., de Viragh, Y., Sako, D., Tresoldi, F.D., Jenelten, F., Hutter, M.: Keep rollin'-whole-body motion control and planning for wheeled quadrupedal robots. IEEE Robot. Autom. Lett. **4**(2), 2116–2123 (2019)
3. Wang, S., et al.: Balance control of a novel wheel-legged robot: design and experiments. In: 2021 IEEE International Conference on Robotics and Automation (ICRA), pp. 6782-6788. IEEE (2021)
4. Vollenweider, E., Bjelonic, M., Klemm, V., Rudin, N., Lee, J., Hutter, M.: Advanced skills through multiple adversarial motion priors in reinforcement learning. arXiv preprint. arXiv:2203.14912 (2022)
5. Zambella, G.: Dynamic whole-body control of unstable wheeled humanoid robots. IEEE Robot. Automa. Lett. **4**(4), 3489–3496 (2019)
6. Boston Dynamics: Introducing Handle. https://youtu.be/-7xvqQeoA8c, YouTube (2017)
7. Klemm, V., et al.: LQR-assisted whole-body control of a wheeled bipedal robot with kinematic loops. IEEE Robot. Autom. Lett. **5**(2), 3745–3752 (2020)
8. Klemm, V., et al.: A two-wheeled jumping robot. In: 2019 International Conference on Robotics and Automation (ICRA), pp. 7515-7521. IEEE (2019)
9. Chadwick, M., Kolvenbach, H., Dubois, F., Lau, H.F., Hutter, M.: Vitruvio: an open-source leg design optimization toolbox for walking robots. IEEE Robot. Autom. Lett. **5**(4), 6318–6325 (2020)
10. Boston Dynamics: Handle Robot Reimagined for Logistics. https://youtu.be/5iV_hB08Uns, YouTube (2019)
11. Ulrich, N., Kumar, V.: Passive mechanical gravity compensation for robot manipulators. In: 1991 IEEE International Conference on Robotics and Automation, pp. 1536-1541. IEEE Computer Society, Los Alamitos (1991)
12. Ham, R.V., Sugar, T.G., Vanderborght, B., Hollander, K.W., Lefeber, D.: Compliant actuator designs. IEEE Robot. Autom. Mag. **16**(3), 81–94 (2009)
13. Arm, P., et al.: SpaceBok: a dynamic legged robot for space exploration. In: 2019 International Conference on Robotics and Automation (ICRA), pp. 6288-6294 (2019)
14. Mannhart, D., Dubois, F., Bodie, K., Klemm, V., Morra, A., Hutter, M.: CAMI-analysis, design and realization of a force-compliant variable cam system. In: 2020 IEEE International Conference on Robotics and Automation (ICRA), pp. 850-856 (2020)

15. Hartenberg, R.S., Danavit, J.: Kinematic Synthesis of Linkages, pp. 295–297. McGraw-Hill Book Company, New York (1964)
16. Denavit, J., Hartenberg, R.S.: A kinematic notation for lower-pair mechanisms based on matrices. Trans. J. Appl. Mech. (23), 215-221 (1955)
17. Andersson, J.A.E., Gillis, J., Horn, G., Rawlings, J.B., Diehl, M.: CasADi: a software framework for nonlinear optimization and optimal control. Math. Prog. Comput. **11**(1), 1–36 (2018). https://doi.org/10.1007/s12532-018-0139-4

Hybrid Legged-Wheeled Robotic Platforms: Survey on Existing Solutions

João Moreira[1(✉)], Inês N. Soares[1,2], José Lima[3,4], Vítor H. Pinto[1,2], and Paulo Costa[1,4]

[1] FEUP - Faculty of Engineering, University of Porto, Porto, Portugal
{up201506628,insoares,vitorpinto,paco}@fe.up.pt
[2] SYSTEC (DIGI2)-Research Center for Systems and Technologies (Digital and Intelligent Industry Lab), Porto, Portugal
[3] Research Centre in Digitalization and Intelligent Robotics (CeDRI), Instituto Politécnico de Bragança, Bragança, Portugal
jllima@ipb.pt
[4] INESC TEC - Institute for Systems and Computer Engineering, Technology and Science, Porto, Portugal

Abstract. This survey analyses and compares ten different robots capable of hybrid locomotion in an attempt to elucidate the readers on several aspects of importance when designing and implementing a legged-wheeled vehicle. With this purpose in mind, the robots are compared based on their goals, kinematic configurations, joint specifications and overall performance. In this text, their variety and versatility is presented, justifying their use in real-world scenarios.

Keywords: Hybrid robotic locomotion · Legged-wheeled platforms · Survey · Kinematic configurations

1 Introduction

In the last couple of decades, there has been an increase in the development of hybrid mobility vehicles, *i.e.* those that move using more than one mechanism. This emergence is justified by the broad range of applications covered by these platforms. These include, but are not limited to, search and rescue missions [1–3], indoor inspection [1,4], surveillance [5] and planetary exploration [6,7].

This wide variety of use cases is related to these vehicles' capabilities. Automation has inserted itself in our way of living decades ago, both digitally and physically. However, the physically automated tasks that robots currently perform are mostly done in controlled and structured environments like factory floors and warehouses. With legged vehicles, the promise of operating in unstructured scenarios largely increases the utility of mobile robots.

On the other hand, wheeled vehicles still largely outperform legged ones in speed and efficiency when used in flat terrain. Therefore, instead of attempting

© The Author(s), under exclusive license to Springer Nature Switzerland AG 2023
J. M. Cascalho et al. (Eds.): CLAWAR 2022, LNNS 530, pp. 140–151, 2023.
https://doi.org/10.1007/978-3-031-15226-9_16

to push legged platforms beyond their limits, many researchers have been focusing on combining different types of locomotion into one vehicle in an attempt to create broad purpose robots capable of traversing a wide variety of terrain effectively.

Concurrently, a lot of effort has been put into the research of autonomous aerial vehicles. However, despite their ability to quickly traverse over any kind of terrain, their applications are limited by their maximum time of operation, as well as reduced capability to carry any payload. Given this discrepancy in the abilities of ground and aerial vehicles, research on both fields is to be expected in the upcoming years. Therefore, it is safe to assume that the future of automation in unstructured environments will, to a certain extent, depend on hybrid locomotion vehicles.

Considering the emergence and fast evolution of this field, the authors collected and filtered information on several projects that are considered of great relevance to the advances on the area of legged-wheeled locomotion. This information was then compiled in this document in an attempt to enlighten any readers who might be interested on developing or improving their own legged-wheeled robot, or even just becoming acquainted with the recent developments of the field. In that sense, the authors hope that this survey serves as a tool for quickly updating any researchers on the state of the art of this domain, and consequently improving the convenience and time of implementation of new solutions.

The remainder of this document is structured as follows. Section 2 motivates the choice of the discussed robotic platforms. Section 3 compares the different kinematic configurations of these vehicles while providing some insight into the advantages and disadvantages of each design. In Sect. 4 the main hardware attributes and performance metrics are described. Subsequently, the actuating mechanisms of the robots are reviewed. Lastly, some final remarks are included in Sect. 5.

2 Robotic Platforms

The physical aspects of mobile robots are essential aspects to be analysed, considering that they define the range of their abilities. Specifically, the different categories of hybrid mobile robots will be explored in this section. Subsequently, a set of robotic platforms will be chosen and compared based on their actuating components, kinematics and overall performance.

The main actuating mechanisms with which a vehicle moves on ground are wheels, legs and tracks. Any vehicle can be considered to have hybrid locomotion capabilities as long as it has more than one mechanism by which it moves in space. Therefore, the best known types of hybrid locomotion include legged-wheeled, wheeled-tracked, legged-tracked and legged-wheeled-tracked locomotion. From these categories, the legged-wheeled paradigm appears to be the most capable in terms of speed, energy efficiency and mobility in unstructured terrain, while only suffering from the ability to traverse soft terrains like sand [8]. Therefore, this revision will focus solely on legged-wheeled locomotion vehicles.

Within the legged-wheeled category, there is still a wide variety of configurations that robots can follow in an attempt to better explore their mobility capabilities. For instance, there are countless examples in which the wheels are attached to the ends of the legs, such as ANYmal [9], Momaro [2] and CENTAURO [3]. A number of different prototypes, like Hytro-I [10] and Wheeleg [11], separate the wheel and leg modules. The Quattroped [12] and Wheel Transformer [13] robots have reconfigurable limbs that can serve as either legs or wheels. Finally, Asguard [5] and RHex [14] are equipped with limbs that can either be considered non-circular wheels or legs with one degree of freedom (DoF). Despite this wide range of configurations, this text will only review platforms following the first paradigm, *i.e.* the one including robots with wheels at the end of their legs. This is due to both the gargantuan amount of existing legged-wheeled robots, which cannot be accurately and realistically represented within one survey, and the current trend in research and commercial use of robots of the selected configuration. It should also be noted that, unlike with the other paradigms, robots with wheels at the end of their legs are capable of not only driving and walking but also performing some kind of hybrid locomotion that makes use of legs and wheels at the same time.

Taking the previous aspects into consideration, a list of robots was chosen as a representative subset of the state of the art on hybrid locomotion vehicles. These are:

- ANYmal [1,9];
- Momaro [2];
- CENTAURO [3];
- The work of V. H. Pinto *et al.* [15], which for simplicity purposes will be referred to as University of Porto robot throughout this paper;
- Complios [16];
- Ascento [4];
- Cassino Hexapod III [17];
- SherpaTT [6];
- MAMMOTH [7];
- BIT-NAZA-II [18].

3 Kinematic Configurations

Despite the aforementioned design restrictions imposed on the selected list of robots, there is still a wide variety of geometric configurations available to choose from. These will largely define the way in which different robots attain different goals.

For example, the number of legs is an important factor when maintaining balance. Cassino Hexapod III [17] and BIT-NAZA-II [18] are equipped with six legs, which allow them to lift three legs at the same time during a walking motion while maintaining static balance using the other three legs. However, it is still possible for a robot to actively balance itself with less feet. For example, Ascento [4] only has two legs.

However, most platforms were designed to have four legs. These include ANYmal [1], Momaro [2], CENTAURO [3], Complios [16], the University of Porto robot [15], SherpaTT [6] and MAMMOTH [7]. This design choice is more stable than the humanoid configuration, and allows for static gaits while moving just one leg at a time. This safe mode of locomotion is used mostly to traverse difficult terrain. Despite being slow, this does not constitute a significant limitation since these robots were designed to spend most of their time driving or using hybrid locomotion methods. The sufficient utility of this number of legs, along with the good design practice of trying to maintain systems as simple as possible, has made four the most often used number of legs in legged-wheeled vehicles.

Besides the number of legs, the number of DoFs of each of them, as well as the orientations of their axes of rotation, influence the kind of locomotion that can be performed. For a quick review of these aspects, the reader can consult Table 1.

Table 1. Leg kinematics for the different robots

Robot	# DoFs per leg[a]	Rotation axes[b]
ANYmal	3	roll-pitch-pitch
Momaro	4	pitch-pitch-pitch-yaw
CENTAURO	5	yaw-pitch-pitch-pitch-yaw
Complios	1	pitch
University of Porto robot	3	yaw-pitch-pitch
Ascento	1	pitch
Cassino Hexapod III	2	pitch-pitch
SherpaTT	4	yaw-pitch-pitch-yaw
MAMMOTH	3	yaw-pitch-yaw
BIT-NAZA-II	6	not applicable

[a] without the wheels
[b] from body to feet

Ascento has only one actuator per leg that moves it along the pitch axis [4]. However, it has four joints linked in such a way that keep the center of the wheel roughly below the CoM for a wide range of motion of the actuator. This geometric characteristic was chosen to allow decoupling of the driving and jumping motions.

Complios is also limited to one pitch DoF per leg [16] that is also used to control mostly the wheel's height. Similarly to Ascento, Complios has more joints than actuators in each leg. These are connected by springs, which allow the wheels' horizontal positions to be controlled by their speed differential. This means that, unlike most robots mentioned here, foothold positioning is only controllable when the corresponding wheel is on the ground. Furthermore, the position of the body is not controllable for a given foothold pattern, since it is predetermined by the balance in forces applied to the springs.

A slightly more complex leg is that of the Cassino Hexapod III, which has two pitch DoFs [17], allowing for full control of the position of the wheel in the sagittal plane.

However, in order to place a wheel in any position in space within the workspace of a leg, at least three DoFs are necessary. Examples of robots with this ability are ANYmal and the University of Porto robot. Their main difference is in the way they allow the wheel to escape the sagittal plane. One of ANYmal's hip joints is oriented along the roll axis, providing the leg with a range of motion so high that two kinematic configurations are made available, the O-type and the X-type [1]. On the other hand, the University of Porto robot has a hip joint oriented along the yaw axis, which allows it to drive omnidirectionally and control the height and orientation of the body independently [19]. However, omnidirectional driving is only available as long as the active wheels are kept along the corresponding yaw axis of the actuator. An additional problem with the University of Porto robot configuration is the lack of compliance along the roll axis of its legs, which causes lateral impacts in the feet to apply a direct torque on the whole body of the robot.

The difference between these last two configurations is often referred to as mammal type (ANYmal) and spider type (University of Porto robot). The mammal-like configuration grants higher mobility capabilities, and exploits this advantage mostly when the legs are roughly straight. The spider-like configuration is better suited for robots with a larger support polygon and a lower CoM, providing it with the ability to perform high power tasks such as payload carrying. The reader is referred to [20] to find out more about the advantages of each configuration.

Note that most robots with more than two DoF mentioned here follow the spider-like configuration. This is a natural consequence of the design process of these vehicles. Since the use of wheels is preferable whenever it is possible to use them, given their superior speed and efficiency, many of these platforms use the legs only rarely to overcome difficult terrain, quickly returning to wheeled locomotion once that has been accomplished. In fact, SherpaTT, which follows a similar configuration, rarely refers to its legs as such, and often uses the term *active suspension system* [6], which emphasises their purpose of *aiding* in locomotion. This makes the ability to perform dynamic motions less important. In contrast to this design philosophy, ANYmal follows the mammal-like configuration probably since it started out as a purely legged robot [1].

Another three DoF configuration is that of MAMMOTH, which uses only two DoF to place each foot and uses the last one to steer the wheel. Its pitch DoF is controlled with a linear actuator in such a way that the plane of the wheel is always perpendicular to the plane of the body, which removes the necessity of an extra DoF to align its wheels with a flat ground.

Moving further along the complexity spectrum, Momaro has four DoFs per leg following a pitch-pitch-pitch-yaw configuration [2]. The third pitch actuator is used to maintain the plane of the wheel perpendicular to the driving surface, allowing omnidirectional driving regardless of the wheels' positions. However, considering that the actuator aligned with the yaw axis is the one closest to the wheel, it confines its position to the sagittal plane.

SherpaTT uses a configuration similar to that of MAMMOTH with an extra pitch DoF using a linear actuator [6]. This extra DoF gives it the ability to choose feet placement in three-dimensional space.

CENTAURO has five DoFs following a yaw-pitch-pitch-pitch-yaw configuration [3]. This means that each leg has full control of the position and orientation of each wheel. It should also be noted that the choice of the yaw-pitch-pitch axes for the first three joints was made to follow the spider-like configuration in order to reduce the required gravity compensation torques [20].

Finally, BIT-NAZA-II follows a completely different leg configuration. Instead of using serial actuators, it controls the position and orientation of each wheel using a Stewart platform [18]. The six DoF legs can position and orient the wheels in any way within the limits of the joint space.

4 Hardware

While the kinematics of a robot limit its theoretical capabilities, the hardware defines the range of its practical ones. In the case of mobile robots, the most desired traits are high speed, low power consumption, high maximum payload weight and low cost. However, aspects like size and weight also relate directly to a robot's functionality, and are usually defined according to their objective. Notably, not all of these metrics are available in the literature, and some of them had to be deduced from other available data. This is not surprising considering the different purposes with which these vehicles were built. Nevertheless, Table 2 displays a collection of the available data, leaving unavailable data marked by the letters NA, signifying "not available".

ANYmal's main design criteria include robustness, high mobility, fast locomotion and simple maintenance [1]. Initially, this robot had no wheels, but improvements were noted both in efficiency and mobility once these were added. With them, the speed record for this vehicle became of 4 m/s [21]. Other specifications were obtained from the ANYbotics website [22], namely size, its maximum payload capacity of 15 kg and a battery runtime of roughly 90 min.

Momaro was built in the context of the DARPA Robotics Challenge of 2015 [2], and therefore it was designed to complete the set of tasks defined by the competition. Some guiding principles used to improve its utility and robustness include a large support polygon, a lightweight design, and omnidirectional driving capabilities for easy adjustment of the workspace of the manipulators. This robot was equipped with a 22.2 V 16 Ah battery that is capable of powering the whole platform for roughly 1.5 to 2 h.

The creators of CENTAURO realised that a part of the need for legged-wheeled robots is related to heavy duty tasks, such as carrying payloads or handling heavy objects. With this in mind, CENTAURO was designed to have weight and dimensions similar to those of humans, as well as high strength and high power to be able to perform such heavy duty tasks [3]. More specifically, the robot is able to squat with 60 kg of weight attached to the pelvis or carry objects of up to 12 kg with the manipulators. Additionally, this vehicle was also built

Table 2. Main metrics for the selected robots

Robot	Length × Width × Height (m)	Weight (kg)	Power consumption (W)
ANYmal	$0.8 \times 0.6 \times 0.7$	30–40[c]	13.7–156[d]
Momaro	$0.8 \times 0.7 \times$ NA	58[a]	176–237[e]
CENTAURO	$0.61 \times 0.615 \times 1.706$[a b]	92[a]	576–864[e]
Complios	$0.655 \times 0.38 \times 0.34$	15	NA
University of Porto robot	$0.85 \times 0.465 \times 0.505$[b]	18	NA
Ascento	$0.45 \times 0.36 \times 0.66$[b]	10.4	37[e]
Cassino Hexapod III	$0.4 \times 0.3 \times 0.3$[b]	3.1	4.11–9.25[d]
SherpaTT	$2.6 \times 2.6 \times$ NA[b]	166	200–250[d]
MAMMOTH	NA	80	NA
BIT-NAZA-II	$1.8 \times 1.8 \times 1.4$	400	NA

[a] includes upper body with manipulators
[b] with legs fully stretched
[c] estimated using "cost of transport" [9]
[d] depending on locomotion mode
[e] estimated from battery parameters and operation time

with mobility in mind, and can achieve 1.6 m/s while driving. However, such capabilities come at a cost. This robot is powered by a 48 V battery and drains from 12 A of current when standing still to 18 A when using legged locomotion.

Unlike the previous examples, Complios was not designed to use legged or hybrid locomotion. Instead, it uses the actuators in its legs to adapt the height of the wheels to the geometry of the floor [16]. This way, it can keep the base balanced and control the weight distribution between the wheels. This is the vehicle's approach to dealing with unstructured terrain.

Of all the previous designs, none put a special concern in cost. This far too real restriction was one of the leading criteria in the development of the University of Porto robot [15]. The authors attempted to design a cost-effective, low power platform without giving up on the mobility advantages of hybrid locomotion vehicles. In the end, the whole robot's cost was estimated to lie between 1899€ and 2516€.

Unlike the initial prototypes mentioned, Ascento was built for operation in indoor environments [4]. Therefore, some of its design guidelines included compact size and high maneuverability. Its maximum speed of 2.20 m/s confirms this last ability. Its 11.1 V 5 Ah LiPo battery can power the wheels and joints for roughly 1.5 h, which serves as sufficient autonomy for many indoor inspection tasks.

The Cassino Hexapod III robot was also designed with inspection in mind. Despite its similarity in size with Ascento, its main purpose is more focused on outdoor inspection, specifically, operation in archeological sites [17]. This goal sets some requirements for the platform. Namely, it was intended to be able to carry up to 1 kg of surveying equipment, it was designed to be operatable by non-experts, and to operate wirelessly. Additionally, just like all the above robots, it is supposed to be able negotiate different obstacles. This small robot is capable of reaching a speed of 0.2 m/s while driving omnidirectionally [23].

On the larger platforms side, SherpaTT was designed to be a rover capable of traversing extraplanetary terrain [6]. This robot weighs 166 kg and can carry up to 80 kg. Its large size and weight makes it difficult and dangerous to move at high speeds. The hardware allows it to reach 0.7 m/s. However, the software limits it to 0.2 m/s. This massive system can be powered externally through a tether or using its two on-board 44.4 V 10 Ah batteries.

With a similar purpose, MAMMOTH weighs roughly half of what SherpaTT weighs [7]. This rover is powered by twelve 95 Wh Li-Ion cells.

On the very end of the weight spectrum, BIT-NAZA-II has a mass of 400 kg and a maximum payload capacity of 350 kg, which allows it to carry people. At such a massive size, this hexapod is capable of reaching 5.6 m/s using wheeled locomotion and 1.2 m/s with legged locomotion. Despite the weight and achievable speeds, the robot has an autonomy of 8 h. This machine is clearly in another category when compared to the previously mentioned robots in terms of capabilities and goals.

4.1 Actuators

An important part on the realization of these systems is their actuation mechanisms, since a robot's dynamic locomotion skills or payload carrying capabilities are strongly tied to its building blocks. For the joints, some important metrics are displayed in Table 3.

ANYmal's joints are made of DC motors with harmonic gears and a rotational spring at the output [24]. The use of an elastic element in the joint is a common setup for legged vehicles known as Series Elastic Actuation (SEA). Its main purposes are to increase mechanical robustness by reducing its reflected inertia, and improve energetic efficiency during locomotion. Once again, given that ANYmal's original version did not have wheels, dynamic locomotion was of high importance to travel fast and efficiently. This design guideline is visible in its maximum joint speed of 12 rad/s, which is the second fastest speed in Table 3, being only slower than CENTAURO's wheel steering joint. One last noteworthy aspect is the modularity of the joint, which is important for robot maintenance and large scale production. These joints were also used in Ascento.

Momaro has flexible links formed by carbon fiber springs, which act as a form of compliant mechanism. Regarding the joints, they are made up of DC motors with a dedicated microcontroller each [25–27]. This is the case for most robots presented here.

CENTAURO's heavy duty capabilities require high maximum torques, specially for the joints closest to the hip. For the joints closer to the wheels, torque was replaced by smaller sizes and weights, making the legs lighter and the wheel steering movements faster. The large number of different joints is a philosophy opposing to that of ANYmal's modularity choice, making this robot an example of the trade-offs between ease of use and mechanical strength.

Table 3. Main Metrics for the Robots' Joints

Joint's robot	Joint locations	Speed[a]	Continuous-peak torque/force[a]	Continuous-peak power (W)	Length × Width × Height (mm)	Weight (g)
ANYmal	Everywhere	12 rad/s	15 Nm - 40 Nm	180^b–480^b	93 × 80 × 80	900
Momaro	Hip and Knee	3 rad/s	44.7 Nm - NA	223^c - NA	126 × 54 × 54	855
	Ankle (pitch)	3 rad/s	25.3 Nm - NA	132^c - NA	108 × 54 × 54	732
	Ankle (yaw)	3 rad/s	5.1 Nm - NA	36^c - NA	84 × 42 × 42	340
CENTAURO	Hip (yaw)	8.8 rad/s	92 Nm - 268 Nm	778–2096	166 × 98 × 98	1730
	Hip (pitch) and Knee	6.6 rad/s	123 Nm - 314 Nm	778–2096	166 × 98 × 98	1730
	Ankle (pitch)	6.1 rad/s	46 Nm - 147 Nm	259–820	150 × 93 × 93	1280
	Ankle (yaw)	20.4 rad/s	9 Nm - 28 Nm	167–518	120 × 82 × 82	870
Complios	Knee	4.7 rad/s	NA - 8.4 Nm	NA - $39.5^{b\,c}$	NA	NA
University of Porto robot	Hip (yaw)	1.75 rad/s	NA - 19.2 Nm	36 - NA	124.5 × 84.5 × 185	400
	Hip (pitch) and Knee	2.7 rad/s	NA - 12.4 Nm	36 - NA	110.5 × 85 × 85	1000
Ascento	Knee	12 rad/s	15 Nm - 40 Nm	180^b–480^b	93 × 80 × 80	900
Cassino Hexapod III	Everywhere	7.5 rad/s	NA - 1.47 Nm	NA - $11^{b\,c}$	NA	60
SherpaTT	Hip (yaw)	0.12 rad/s	433 Nm - NA	$52^{b\,c}$ - NA	NA	NA
	Hip (pitch) and Knee	12 mm/s	3500 N N - NA	$42^{b\,c}$ - NA	NA	NA
	Ankle	3.1 rad/s	60 Nm - NA	$186^{b\,c}$ - NA	NA	NA
MAMMOTH	Hip (yaw) and Ankle	NA	24.8 Nm - NA	NA	NA	NA
	Hip (pitch)	NA	1500 N N - NA	NA	NA	NA
BIT-NAZA-II	Whole leg	NA	NA	NA	NA	NA

[a] units depend on whether the actuator is rotational or linear
[b] only mechanical useful power, not electrical consumed power
[c] estimated using speed and torque parameters

Another leg setup that relies on SEA is that of Complios. With springs in both of its links, this vehicle adapts to uncertain terrain not only using its joints, but also using the compliance. In the previous examples, the main goals of including a compliant element were safety and energy efficiency. For Complios, the springs, specially the one in the thigh, are responsible for setting the position of the wheels by actuating the wheels themselves. This way one DoF can be removed from the leg while keeping control of the wheel in the sagittal plane.

The University of Porto robot serves as yet another example of the use of compliance mechanisms in legged-wheeled vehicles [28]. Recognizing its advantages, this robot includes one non-rigid joint. The remaining joints are not compliant to reduce cost. In order to protect the yaw joint's motor axis from impacts, the motor was decoupled from the output using an extra external gear [15].

Given the small size and load requirements of Cassino Hexapod III, it is no surprise that its motors are the ones with the lowest torque capacity [29]. In turn, the small torque requirement means that the maximum speed is relatively high, making this six legged vehicle more capable of fast legged locomotion.

The joints of all these robots are rotational. However, there are advantages to using linear joints. SherpaTT and MAMMOTH use these. This way, as was mentioned above, the planes of the wheels remain perpendicular to the plane of the robot's body, reducing complexity in the kinematic design and control algorithm. However, this reduces drastically their range of motion, even though it is more than sufficient for their motion requirements. With a different configuration, BIT-NAZA-II also uses linear actuators in its Stewart platform legs. Even though there are not enough relevant specifications available for the actuators of BIT-NAZA-II, it is evident from these last three robots that linear actuators are preferred for large scale applications.

5 Final Remarks

In this survey, ten different robots with hybrid locomotion capabilities were analysed. Given they're recent popularity, legged-wheeled vehicles with the wheels equipped at the end of the legs were the only ones considered here.

From these vehicles, a large number of differences was found. On the kinematics side, the number of legs varies between two to six. Additionally, the great number of joint configurations shows the wide variety of platforms that can be built. Nevertheless, some of them can be categorized and compared. Namely, the spider and mammal types of configurations were visible in several robots and their advantages were discussed and related to the purpose of each system implementing them.

Many other aspects of wide variety include the hardware itself and some performance metrics. In this text, size, weight, power consumption, dynamic motion skills and payload capacity were compared.

Regarding the actuating mechanisms, the series elastic actuation paradigm is often included, and it proves to be a useful addition to many of these vehicles. Even for joints, some design choices are still not clearly better than others.

Modularity over requirement specific designs are still aspects to consider when developing a robot according to its purpose. Additionally, the use of linear joints over revolute ones is also an aspect to take into consideration. From the systems discussed here, linear actuation seems to be preferable in large scale solutions.

Acknowledgment. This work has been supported by FCT - Fundação para a Ciência e Tecnologia within the Project Scope: UIDB/05757/2020 and Project Warehouse of the Future (WoF), with reference POCI-01-0247-FEDER-072638, co-funded by FEDER, through COMPETE 2020.

References

1. Hutter, M., et al.: Anymal-toward legged robots for harsh environments. Adv. Robot. **31**(17), 918–931 (2017)
2. Schwarz, M., et al.: Nimbro rescue: solving disaster-response tasks with the mobile manipulation robot momaro. J. Field Robot. **34**(2), 400–425 (2017)
3. Kashiri, N., et al.: Centauro: a hybrid locomotion and high power resilient manipulation platform. IEEE Robot. Autom. Lett. **4**(2), 1595–1602 (2019)
4. Klemm, V., et al.: Ascento: A two-wheeled jumping robot. In: 2019 International Conference on Robotics and Automation (ICRA), pp. 7515–7521. IEEE (2019)
5. Eich, M., Grimminger, F., Kirchner, F.: A versatile stair-climbing robot for search and rescue applications. In: 2008 IEEE International Workshop on Safety, Security and Rescue Robotics, pp. 35–40. IEEE (2008)
6. Cordes, F., Kirchner, F., Babu, A.: Design and field testing of a rover with an actively articulated suspension system in a mars analog terrain. J. Field Robot. **35**(7), 1149–1181 (2018)
7. Reid, W., Pérez-Grau, F.J., Göktoğan, A.H., Sukkarieh, S.: Actively articulated suspension for a wheel-on-leg rover operating on a martian analog surface. In: 2016 IEEE International Conference on Robotics and Automation (ICRA), pp. 5596–5602. IEEE (2016)
8. Bruzzone, L., Quaglia, G.: Locomotion systems for ground mobile robots in unstructured environments. Mech. Sci. **3**(2), 49–62 (2012)
9. Bjelonic, M., Bellicoso, C.D., Tiryaki, M.E., Hutter, M.: Skating with a force controlled quadrupedal robot. In: 2018 IEEE/RSJ International Conference on Intelligent Robots and Systems (IROS), pp. 7555–7561. IEEE (2018)
10. Lu, D., Dong, E., Liu, C., Xu, M., Yang, J.: Design and development of a leg-wheel hybrid robot "HyTRO-I". In: 2013 IEEE/RSJ International Conference on Intelligent Robots and Systems, pp. 6031–6036. IEEE (2013)
11. Lacagnina, M., Muscato, G., Sinatra, R.: Kinematics, dynamics and control of a hybrid robot wheeleg. Robot. Auton. Syst. **45**(3–4), 161–180 (2003)
12. Chen, S.-C., Huang, K.-J., Chen, W.-H., Shen, S.-Y., Li, C.-H., Lin, P.-C.: Quattroped: a leg-wheel transformable robot. IEEE/ASME Trans. Mechatron. **19**(2), 730–742 (2013)
13. Kim, Y.-S., Jung, G.-P., Kim, H., Cho, K.-J., Chu, C.-N.: Wheel transformer: a wheel-leg hybrid robot with passive transformable wheels. IEEE Trans. Robot. **30**(6), 1487–1498 (2014)
14. Saranli, U., Buehler, M., Koditschek, D.E.: RHex: a simple and highly mobile hexapod robot. Int. J. Robot. Res. **20**(7), 616–631 (2001)

15. Pinto, V.H., Soares, I.N., Rocha, M., Lima, J., Gonçalves, J., Costa, P.: Design, modeling, and control of an autonomous legged-wheeled hybrid robotic vehicle with non-rigid joints. Appl. Sci. **11**(13), 6116 (2021)
16. Bouton, A., Grand, C., Benamar, F.: Design and control of a compliant wheel-on-leg rover which conforms to uneven terrain. IEEE/ASME Trans. Mechatron. **25**(5), 2354–2363 (2020)
17. Tedeschi, F., Carbone, G.: Design of a novel leg-wheel hexapod walking robot. Robotics **6**(4), 40 (2017)
18. Chen, Z., et al.: Control strategy of stable walking for a hexapod wheel-legged robot. ISA Trans. **108**, 367–380 (2021)
19. Pinto, V.H., Gonçalves, J., Costa, P.: Design, modeling, and control of a single leg for a legged-wheeled locomotion system with non-rigid joint. In: Actuators, vol. 10, p. 29. Multidisciplinary Digital Publishing Institute (2021)
20. Kashiri, N., Ajoudani, A., Caldwell, D.G., Tsagarakis, N.G.: Evaluation of hip kinematics influence on the performance of a quadrupedal robot leg. ICINCO **1**, 205–212 (2016)
21. Bjelonic, M., et al.: Keep rollin'-whole-body motion control and planning for wheeled quadrupedal robots. IEEE Robot. Autom. Lett. **4**(2), 2116–2123 (2019)
22. Anymal - autonomous legged robot - anybotics. https://www.anybotics.com/anymal-autonomous-legged-robot/#specs. Accessed 27 April 2022
23. Orozco-Magdaleno, E.C., Gómez-Bravo, F., Castillo-Castañeda, E., Carbone, G.: Evaluation of locomotion performances for a mecanum-wheeled hybrid hexapod robot. IEEE/ASME Trans. Mechatron. **26**(3), 1657–1667 (2020)
24. Anydrive. https://echord.eu/public/wp-content/uploads/2018/01/D3.3-Integrated-Electronics-MODUL.pdf. Accessed 28 April 2022
25. H54-200-S500-R. https://emanual.robotis.com/docs/en/dxl/pro/h54-200-s500-r/. Accessed 28 April 2022
26. H54-100-S500-R.https://emanual.robotis.com/docs/en/dxl/pro/h54-100-s500-r/. Accessed 28 April 2022
27. H42-20-S300-R. https://emanual.robotis.com/docs/en/dxl/pro/h42-20-s300-r/. Accessed 28 April 2022
28. Pinto, V.H., Gonçalves, J., Costa, P.: Towards a more robust non-rigid robotic joint. Appl. Syst. Innov. **3**(4), 45 (2020)
29. DS RDS3115MG. https://alexnld.com/product/ds-rds3115mg-15kg-large-torque-180-degree-biaxial-digital-servo-for-rc-robot/. Accessed 28 April 2022

Legged Locomotion

Bridging the Reality Gap via Progressive Bayesian Optimisation

Chen Yu(✉) and Andre Rosendo(✉)

ShanghaiTech University, Shanghai, China
{yuchen,arosendo}@shanghaitech.edu.cn

Abstract. The recent rapid development of deep-learning-based control strategies has made the reality gap a critical issue at the forefront of robotics, especially for legged robots. We propose a novel system identification framework, Progressive Bayesian Optimisation (ProBO), to bridge the reality gap by tuning simulation parameters. Since dynamic locomotion trajectories are usually harder to narrow the reality gap than their static counterpart, we train a Gaussian process model with the easier trajectory data set and make it a prior to start the learning process of a harder one. We implement ProBO on a quadruped robot to narrow the reality gaps of a set of bounding gaits at different speeds. Results show that our methods can outperform all other alternatives after training the initial gait.

Keywords: Reality gap · Sim2Real · Bayesian optimisation · Simulator

1 Introduction

Simulators have accelerated the impact of deep learning on Robotics by allowing faster and highly-scalable data collection. However, a critical issue of deep-learning-based control strategies is the transfer of controllers learned in simulation to reality. This is especially the case for legged robots as they are highly dynamic and hard to accurately model. For example, Reinforcement Learning (RL) has shown promising results for a range of control problems in legged robots, such as planning [1] and locomotion [2]. Deep RL policies are often trained in a simulated environment due to the training time, cost, and safety concerns: These RL-based methods often need tens of millions of data samples to learn a meaningful policy, which can take months of data collection on a real robot [3]. However, capabilities demonstrated by simulated agents can hardly be realised by their physical counterparts because of the inaccurate representation of the real environment in the simulator, i.e., the *reality gap* [4]. Traversing this reality gap without any intermediary adaptation is usually based on two methods: domain randomisation and system identification [5].

Domain randomisation exposes controllers to a wide range of noises in simulation to produce a robust final policy in the target real-world environment [6].

© The Author(s), under exclusive license to Springer Nature Switzerland AG 2023
J. M. Cascalho et al. (Eds.): CLAWAR 2022, LNNS 530, pp. 155–166, 2023.
https://doi.org/10.1007/978-3-031-15226-9_17

The randomised parameters vary from robotic vision inputs [7,8] (such as object textures, colour, shading, camera position, etc.) to physical dynamics settings [4] (such as masses, damping, friction, inertial moments of robot links, etc.). Tan et al. [9] randomly sample a set of physical parameters at the beginning of each episode for locomotion training of a quadruped robot Minitaur in the simulator. These parameters are carefully chosen to prevent overly conservative locomotion gaits. Hwangbo et al. [10] robustify the controller by training with 30 different ANYmal models with stochastic inertial properties (the centre of mass positions, the masses of links, and joint positions) of the links. Siekmann et al. [11] uses dynamics randomisation for the training of a human-scale bipedal robot, Cassie. They show that controllers trained without dynamics randomisation are unable to consistently transfer to hardware, while the same policy, when trained with dynamics randomisation, are able to consistently transfer to the real world.

System identification utilises real-world data to tune the simulation parameters to increase the fidelity of the simulator [12,13]. Chebotar et al. [14] adapt the distribution of simulation parameters using real-world roll-outs data interleaved with policy training. By closing the loop of optimising the simulator and control policy, they improve the policy transfer by matching the policy behaviours in simulation and the real world. Yu et al. [15] not only train a policy on the distribution of simulated parameters, but also train to predict the simulated parameters from the current states and actions, and use the prediction as an additional input to the policy. Ramos et al. [13] propose a method to compute the posterior distribution of simulation parameters with data of trajectories obtained with a physical robot.

Here we focus on system identification since this can benefit the transference of more simulated controllers from arbitrary training processes. We propose a novel system identification method, especially for legged robot locomotion, based on Bayesian Optimisation (BO). BO refers to a class of methods for global optimisation of a black-box function that can only be accessed through point evaluations, which uses observations to form the posterior distribution over the objective function [16]. BO has been used as a promising probabilistic framework for robot learning in the last decade [17,18].

In this work, we propose Progressive Bayesian Optimisation (ProBO), for the optimisation of simulation parameters towards the goal of achieving real-world performance. This framework is designed for legged robot locomotion that utilises the difference in dynamic levels of different locomotion modes. In our setting, we wish to narrow the reality gaps of different locomotion modes (i.e., different locomotion gaits). Since trajectories of more dynamic locomotion (i.e., running at a higher speed) is harder to narrow the reality gap and trajectory of more static locomotion (i.e., walking slowly) is easier, we train a Gaussian process (GP) model with the easier trajectory data set and make it a prior as the other harder trajectory training process.

In the experiment, we implement our methods and evaluate the performance on a quadruped robot Mini Cheetah [19]. We use the bounding gaits at different frequencies as a proxy of locomotion of different dynamic levels. We record the trajectories of a quadruped robot bounding at different frequencies in both the simulators and real-world (Fig. 1). Rather than using differential evolution (DE) [20] to tune the simulation parameters as in [21], we use BO to optimise these parameters. Of benefit from the backend GP of BO, we use the training results from the data of lower bounding frequency as the prior of the case of bounding at a higher frequency. We show that our method can outperform other alternatives in minimising the error between real-world trajectories and simulated ones.

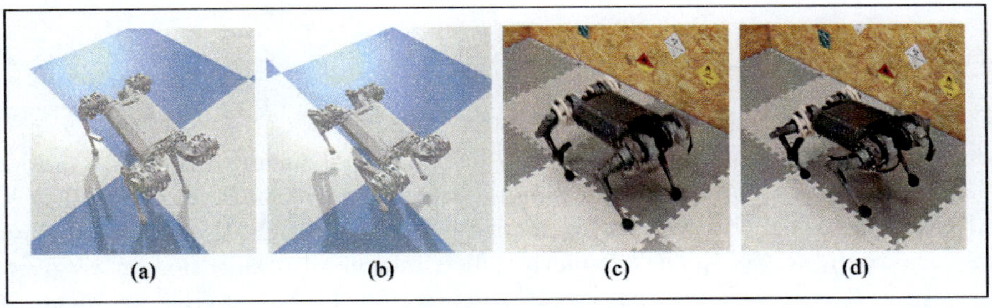

Fig. 1. Quadruped robots bounding in the simulation and real world. We show the snapshots of our robot bounding in the PyBullet simulator (a, b) and in the real world (c, d).

2 Progressive Bayesian Optimisation

The main goal of this work is to minimise the behaviour difference between the simulation and the real-world in arbitrary locomotion cases. In the following, we provide the preliminary mathematical tools and intuition over which our progressive training strategy is constructed.

2.1 Problem Formulation

We use system identification to bridge the reality gap between simulation and real world. We model this process as an optimisation problem:

$$x_\star = \arg\min_{x \in \mathcal{X}} f(x) \tag{1}$$

where the objective $f : \mathcal{X} \rightarrow \mathbb{R}$ is a black-box function that maps a vector of simulation parameters $x \in \mathcal{X}$ to the reality gap. The choose of the simulation parameters x is detailed in the experimentation Sect. 3.1.

We model the trajectories produced by a controller following the Markov Decision Process (MDP), where the agent at every time step is at state $s_t \in \mathcal{S}$ and takes actions $a_t \in A$. A trajectory of length T is then

$$\tau = \{s_0, a_0, s_1, a_1, \ldots, s_{T-2}, a_{T-2}, s_{T-1}\}.$$

The reality gap f is quantified as the state-wise Root Mean Square Error (RMSE) between the simulated trajectories τ_{sim} and real-world implementation τ_{real} following the same action sequence. Hence, the reality gap f is defined as:

$$f = \frac{1}{T}\sum_{t=0}^{T-1}\sqrt{\frac{1}{N}\sum_{s_t \in \mathcal{S}}(\tau_{\text{sim}} - \tau_{\text{real}})^2} \tag{2}$$

where N is the dimension of the state space $\mathcal{S}$. The detailed setup of $\mathcal{S}$ is descried in Sect. 3.2.

2.2 Gaussian Process

To minimise the objective reality gap, we model it as a GP, $f \sim \mathcal{GP}(0, k(x, \hat{x}))$, with a covariance function $k : \mathcal{X} \times \mathcal{X} \rightarrow \mathbb{R}$ and prior mean of zero. We model the observations $y(x) = f(x) + \epsilon$ using additive Gaussian noise $\epsilon \sim \mathcal{N}\left(\varepsilon; 0, \sigma^2\right)$. After $n \leq N - 1$ observations of reality gap is collected: $\mathcal{D}_n^f = \{\boldsymbol{x}_n, \boldsymbol{y}_n\} = \{x_1, \ldots, x_n, y_1, \ldots, y_n\}$, the predictive distribution at a location x is given by $p\left(f \mid \mathcal{D}_n^f, x\right) = \mathcal{N}\left(f(x); \mu\left(x \mid \mathcal{D}_n^f\right), \sigma^2\left(x \mid \mathcal{D}_n^f\right)\right)$. The predictive mean is given by $\mu\left(x \mid \mathcal{D}_n^f\right) = \boldsymbol{k}_n^{\top}(x)\left[K_n + \sigma^2 I\right]^{-1}\boldsymbol{y}_n$, where vector $\boldsymbol{k}_n(x)$ has entries $\left[\boldsymbol{k}_n(x)\right]_i = k\left(x_i, x\right)$, the Gram matrix K_n has entries $\left[K_n\right]_{i,j} = k\left(x_i, x_j\right)$, and the vector of observations y_n has entries $\left[\boldsymbol{y}_n\right]_i = y_i$. The predictive variance is given by $\sigma^2\left(x \mid \mathcal{D}_n^f\right) = k(x, x) - \boldsymbol{k}_n^{\top}(x)\left[K_n + \sigma^2 I\right]^{-1}\boldsymbol{k}_n(x)$.

2.3 Progressive Training

The intuition behind our progressive training scheme is that trajectories of a more dynamics locomotion (i.e., running at a higher speed) are harder to narrow the reality gap, while trajectories of more static locomotion (i.e., walking slowly) are easier. Assume that we want to narrow the reality gap in L different locomotion modes. We denote the corresponding reality gaps of these locomotion modes as $f_0, f_1, \ldots, f_{L-1}$. We sort these reality gap objectives f_i by the dynamicity of the corresponding locomotion. It means that the indexed $f_i(x)$ are sorted in ascending order by their value, given initial simulation parameters x.

For f_0, its corresponding GP $\mathcal{GP}_0$ is trained by trajectory dataset $\mathcal{D}_n^{f_0}$, with predictive mean $\mu\left(x \mid \mathcal{D}_n^{f_0}\right)$ and predictive variance $\sigma^2\left(x \mid \mathcal{D}_n^{f_0}\right)$. For f_i with $i > 0$, we use the GP for the lower dynamic locomotion as a prior, and hence the GP is a regression on the previous observations of other more static locomotion and current observations: GP $\mathcal{GP}_i$ is a regression of trajectory dataset $\mathcal{D}_n^{f_{i-1}}$, with predictive mean

$$\mu\left(x \mid \mathcal{D}_n^{f_i} \cup \mathcal{D}_{N-1}^{f_{i-1}} \cup \cdots \cup \mathcal{D}_{N-1}^{f_0}\right)$$

and predictive variance

$$\sigma^2\left(x \mid \mathcal{D}_n^{f_i} \cup \mathcal{D}_{N-1}^{f_{i-1}} \cup \cdots \cup \mathcal{D}_{N-1}^{f_0}\right).$$

In the context of BO, the candidate simulation parameters are suggested using an *acquisition function*, whose value can be high at some positions that the predictive mean is high, or at some positions that is unexplored where uncertainty is high. We use the acquisition function of the GP upper confidence bound (GP-UCB) algorithm [22] in our setting. For f_0, we have

$$x_t = \arg\min_{x \in \mathcal{X}} \mu\left(x \mid \mathcal{D}_n^{f_0}\right) + \beta\sigma\left(x \mid \mathcal{D}_n^{f_0}\right) \tag{3}$$

where β is a hyper-parameter for the tradeoff between exploration and exploitation. For f_i with $i > 0$, the candidate simulation parameters are suggested based on both the previous observations of other more static locomotion and observations of the current locomotion mode:

$$\begin{aligned} x_t = \arg\min_{x \in \mathcal{X}} & \mu\left(x \mid \mathcal{D}_n^{f_i} \cup \mathcal{D}_{N-1}^{f_{i-1}} \cup \cdots \cup \mathcal{D}_{N-1}^{f_0}\right) \\ & + \beta\sigma\left(x \mid \mathcal{D}_n^{f_i} \cup \mathcal{D}_{N-1}^{f_{i-1}} \cup \cdots \cup \mathcal{D}_{N-1}^{f_0}\right) \end{aligned} \tag{4}$$

Overall, we first sort the locomotion modes that we want to optimise the reality gap objectives from static to dynamic. For the first locomotion, we use the traditional BO method with a GP-UCB acquisition function to find the optimum simulation parameters. For other locomotion, each training observation is appended to the previous training dataset: In this way, the GPs of the simpler trajectories are served as a prior of more dynamics trajectories. The overall process of ProBO is shown in Algorithm 1.

3 Reality Gap of Legged Locomotion

We implement our ProBO framework on a quadruped robot Mini Cheetah [19] and its counterpart in the simulator PyBullet [23]. We use the bounding gaits at different speeds as a proxy of locomotion of different dynamic levels. We detail our experimental setting in the following.

3.1 Simulation Parameters

We choose the simulation parameters and their ranges carefully so that the search space can be narrowed utilising known system specifications (i.e., the maximum speed reported by the datasheet of each joint motor) but also cover the optimum setting. Mass is a commonly used simulation parameter in both system identification and domain randomisation [24]. The control step may also vary because of the non-real-time system. Some quantities related to the motor are also important since they may change over time. For example, the maximum velocity and torque of motors can vary due to wear and tear [25], whose theatrical values are 300 rpm (31.4 rad/s) and 18 Nm. We use three parameters: contact friction, contact stiffness, and contact damping to jointly model the complicated dynamics between the rubber robot feet and the soft foam floor mat. As in [21], we also consider joint damping, angular damping, and linear damping. The complete list of the simulation parameters and their ranges is summarised in Table 1.

Algorithm 1. Progressive Bayesian Optimisation (ProBO) for Simulator Tuning

Input: $N, \beta, \mathcal{D}_0^{f_0}, \mathcal{D}_0^{f_1}, \ldots, \mathcal{D}_0^{f_{L-1}}$,
Output: $x_0^\star, x_1^\star, \ldots, x_{L-1}^\star$

Sort locomotion gaits i by dynamic levels
for $n = 0, \ldots, N-1$ iterations **do**
  $x_n \leftarrow \arg\max_a \alpha(x|\beta, \mathcal{D}_n^{f_0})$ (Eq. (3))
  Compute the reality gap according to Eq. (2)
  $\mathcal{D}_n^{f_0} = \{\boldsymbol{x}_n, \boldsymbol{y}_n\}$
  Update $\mathcal{GP}_0(x_t|\mathcal{D}_n^{f_0})$.
end for
$x_0^\star \leftarrow \arg\min_{x\in\mathcal{X}} \mu(x|\mathcal{D}_{N-1}^{f_0})$
for $i = 1, \ldots, L-1$ locomotion **do**
  for $n = 0, \ldots, N-1$ iterations **do**
    $x_n \leftarrow \arg\max_a \alpha(x|\beta, \mathcal{D}_n^{f_i} \cup \mathcal{D}_{N-1}^{f_{i-1}} \cup \cdots \cup \mathcal{D}_{N-1}^{f_0})$ (Eq. (4))
    Compute the reality gap according to Eq. (2)
    $\mathcal{D}_n^{f_i} = \{\boldsymbol{x}_n, \boldsymbol{y}_n\}$
    Update $\mathcal{GP}_i(x_t|\mathcal{D}_n^{f_i} \cup \mathcal{D}_{N-1}^{f_{i-1}} \cup \cdots \cup \mathcal{D}_{N-1}^{f_0})$.
  end for
  $x_i^\star \leftarrow \arg\min_{x\in\mathcal{X}} \mu(x|\mathcal{D}_n^{f_i} \cup \mathcal{D}_{N-1}^{f_{i-1}} \cup \cdots \cup \mathcal{D}_{N-1}^{f_0})$
end for

3.2 State Space

The state space of the robot is designed similar to a classic simulated quadruped agent Ant in the RL literature [26], which covers the data from the IMU and the encoder of each joint motor. Specifically, it contains the x displacement, y displacement, z displacement (i.e., the height), x velocity, y velocity, z velocity, roll, pitch, yaw, roll velocity, pitch velocity, yaw velocity, and 12 joint positions. For the real robot, these data are read and recorded using LCM [27].

3.3 Bounding Gait

We use the bounding gaits at different speeds of our quadruped robot as a proxy of locomotion of different dynamic levels.

We design a set of position-control bounding gaits that only involves a low-level PD controller. These bounding gaits vary from different speeds, and hence different length of periods: 130, 140, 150, 160, 170, and 180 time steps. Since the control frequency of the robot is 500 Hz, the bounding frequencies for these gaits are 3.85 Hz, 3.57 Hz, 3.33 Hz, 3.12 Hz, 2.94 Hz, and 2.78 Hz.

For each speed, we record a trajectory that has a length T of 6000 time steps, which is averaged over 5 trials, as shown in Fig. 2.

Table 1. Physical simulation parameters for optimisation.

Parameter	Lower bound	Upper bound	Description
Torso mass	5 kg	10 kg	Mass of the robot body
Control step	1 ms	3 ms	Control time step
Maximum torque	5 Nm	20 Nm	Maximum torque of the motors
Maximum velocity	10 rad/s	30 rad/s	Maximum velocity of the motors
Contact friction	0	1.5	Friction factor between feet and the floor
Contact stiffness	1000	8000	Stiffness of the floor
Contact damping	0	2000	A property paired with the contact stiffness
Joint damping	0	0.5	Reduction of the joint velocity
Angular damping	0	0.5	Reduction of the angular velocity
Linear damping	0	0.5	Reduction of the linear velocity

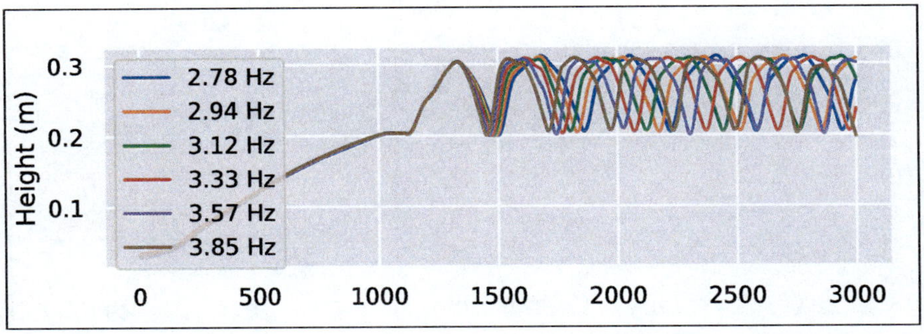

Fig. 2. Real-world recorded trajectory. We visualise the height dimension of the real-world trajectory data with different bounding speeds. Only half of each trajectory is shown.

4 Experimental Results

We compare ProBO with standard BO, Covariance matrix adaptation (CMA) evolution strategy [28], differential evolution (DE) [20], and random search (RS). Standard BO uses independent GPs for each locomotion gait. CMA is a derivative-free optimisation algorithm, where candidate solutions are sampled based on a distribution whose covariance matrix is updated based on the best-performing individuals and the "evolutionary path" that stores the update direc-

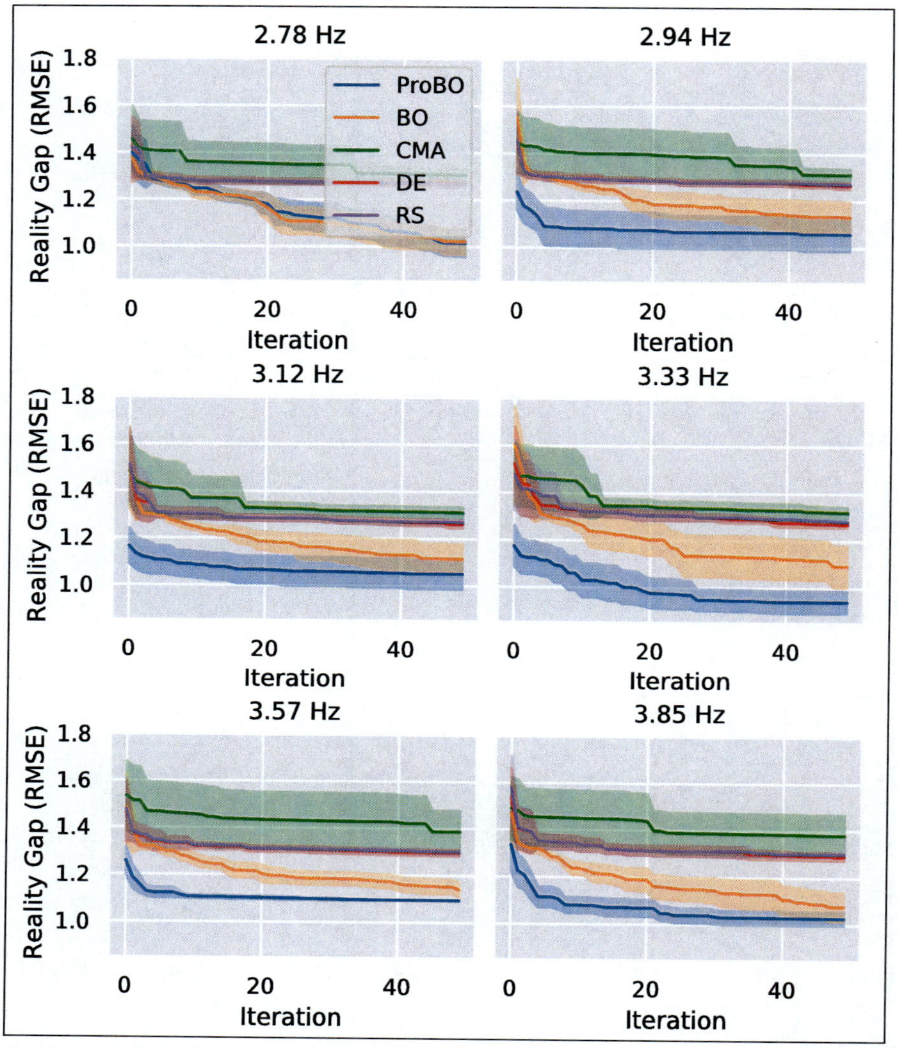

Fig. 3. Training curves of our Progressive Bayesian Optimisation (ProBO) framework and other alternatives. In our setting, we want to minimise the reality gaps in 6 different locomotion gaits: bounding at different frequencies, from slow to fast. Except for the initial gait of bounding at 2.78 Hz, ProBO makes the data of slower locomotion as the prior of the GP for optimising the simulation of more dynamic bounding. It consistently outperforms all other methods in these cases. All the results are averaged over 10 trials.

tion of generations. Previous works have shown its efficiency in control [29]. DE is another evolution strategy that improves candidates by introducing mutations into the population. It has been successfully used on simulator tuning in previous works [21,30]. We also adopt RS as a baseline where candidate solutions are sampled in every iteration independently of the prior iterations.

We show the training curves of our ProBO framework and other alternatives on various bounding gaits (different frequencies) in Fig. 3. We allow a training budget T of 50 for each gait and all the results are averaged over 10 trials.

Results show that ProBO consistently outperforms all other methods in all cases, except in the case of bounding at 2.78 Hz, where ProBO shows almost the same results as standard BO since both of them are using the same randomly initialised GP. ProBO can successfully extract the information from previous observations of other more static locomotion to provide a better initial guess and more efficient exploration. CMA shows the worst performance in all cases, which shows the inefficiency of CMA in simulator tunning. DE and RS show very similar performances in all cases, which are all better than CMA. Their results are converged to a local minimum.

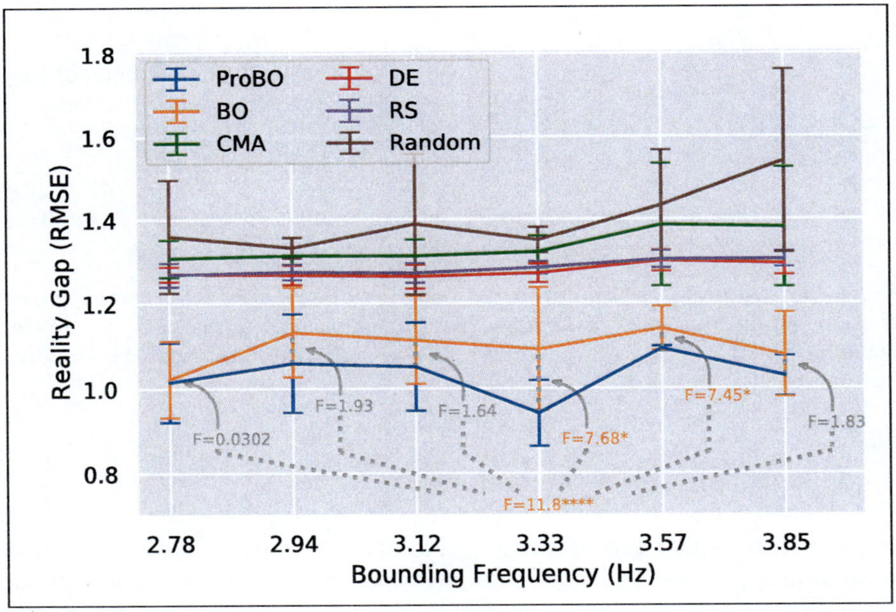

Fig. 4. The mean and standard deviation of training results from each method and the initial reality gap (Random). The initial reality gaps (the purple line) show that generally, the faster the bounding gait is, the wider the initial reality gap would be. However, ProBO can still statistically significantly outperform BO in bounding gaits of 3.33 Hz and 3.57 Hz, and in an overall performance comparison between the two (ANOVA: ****, $p < 0.001$, or *, $p < 0.05$).

We compare the training result from each method and the initial reality gap of each bounding gaits in Fig. 4. The initial reality gaps (the purple line) show that generally, the faster the bounding gait is, the wider the initial reality gap would be. After the first locomotion gait, ProBO can clearly outperform all other methods consistently. We further perform an Analysis of Variance (ANOVA) test on these training results between ProBO and BO. The F statistic is also shown in Fig. 4 where **** stands for $p < 0.001$ and * stands for $p < 0.05$. ProBO shows a statistically significant superiority over BO in bounding gaits of 3.33 Hz and 3.57 Hz, and in an overall performance comparison between the two.

We visualise an example simulated trajectory after training by ProBO compared with its initial data and real-world data in Fig. 5. The pitch value of the trajectories shows that the robot is flipped upside down while bounding with the initial simulation parameters. The error between the real and simulated trajectories is greatly narrowed after training.

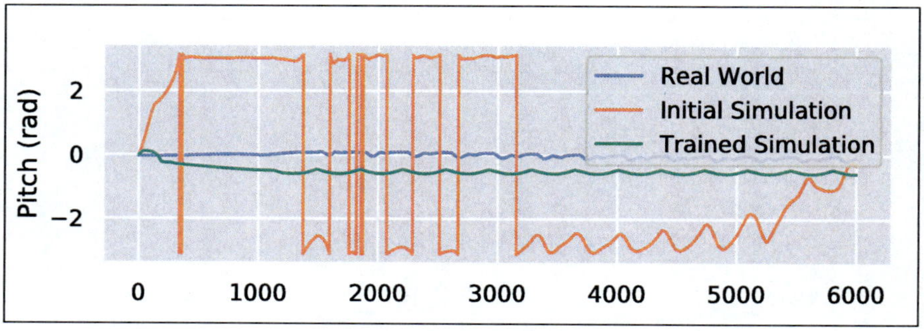

Fig. 5. An example of simulated trajectories before and after training and the corresponding real-world data. The pitch dimension of the state space of the bounding gait at 2.78 Hz is shown. The error between the real and simulated trajectories is narrowed after training.

5 Discussion and Conclusion

Narrowing the reality gaps of the legged locomotion is harder than other tasks such as manipulation [21] since the differences in physical parameters of the simulator can lead to a significant change of the final dynamic performance, as we can see from Fig. 5. Encouragingly, results shown in Fig. 3 suggest that our ProBO framework can successfully utilise the data of other more static locomotion data while training each locomotion gait, and hence constantly outperform other methods after the training of the initial locomotion gait.

The experimental results of ProBO show a promising application in designing deep-learning-based controllers for legged robots: Although legged robots are highly dynamic and hard to model accurately, our method can still progressively traverse the reality gap starting from some static locomotion gaits. We will apply ProBO on training more dynamic locomotion gaits of walking robots such as galloping or bipedal walking in our future work.

References

1. Schrittwieser, J., Hubert, T., Mandhane, A., Barekatain, M., Antonoglou, I., Silver, D.: Online and offline reinforcement learning by planning with a learned model. In: Advances in Neural Information Processing Systems, vol. 34 (2021)
2. Song, S., et al.: Deep reinforcement learning for modeling human locomotion control in neuromechanical simulation. J. Neuroeng. Rehabil. **18**(1), 1–17 (2021)
3. Ibarz, J., Tan, J., Finn, C., Kalakrishnan, M., Pastor, P., Levine, S.: How to train your robot with deep reinforcement learning: lessons we have learned. Int. J. Robot. Res. **40**(4–5), 698–721 (2021)
4. Peng, X.B., Andrychowicz, M., Zaremba, W., Abbeel, P.: Sim-to-real transfer of robotic control with dynamics randomization. In: 2018 IEEE international conference on robotics and automation (ICRA), pp. 3803–3810. IEEE (2018)
5. Collins, J., Brown, R., Leitner, J., Howard, D.: Follow the gradient: crossing the reality gap using differentiable physics (realitygrad). arXiv preprint arXiv:2109.04674 (2021)
6. Tobin, J., Fong, R., Ray, A., Schneider, J., Zaremba, W., Abbeel, P.: Domain randomization for transferring deep neural networks from simulation to the real world. In: 2017 IEEE/RSJ International Conference on Intelligent Robots and Systems (IROS), pp. 23–30. IEEE (2017)
7. James, S., Davison, A.J., Johns, E.: Transferring end-to-end visuomotor control from simulation to real world for a multi-stage task. In: Conference on Robot Learning, pp. 334–343. PMLR (2017)
8. Borrego, J., Figueiredo, R., Dehban, A., Moreno, P., Bernardino, A., Santos-Victor, J.: A generic visual perception domain randomisation framework for gazebo. In: 2018 IEEE International Conference on Autonomous Robot Systems and Competitions (ICARSC), pp. 237–242. IEEE (2018)
9. Tan, J., et al.: Sim-to-real: learning agile locomotion for quadruped robots. arXiv preprint arXiv:1804.10332 (2018)
10. Hwangbo, J., et al.: Learning agile and dynamic motor skills for legged robots. Sci. Robot. **4**(26), eaau5872 (2019)
11. Siekmann, J., et al.: Learning memory-based control for human-scale bipedal locomotion. In: M. Toussaint, A. Bicchi, T. Hermans (eds.) Robotics: Science and Systems XVI, Virtual Event/Corvalis, Oregon, USA, 12–16 July 2020 (2020). https://doi.org/10.15607/RSS.2020.XVI.031
12. Kolev, S., Todorov, E.: Physically consistent state estimation and system identification for contacts. In: 2015 IEEE-RAS 15th International Conference on Humanoid Robots (Humanoids), pp. 1036–1043. IEEE (2015)
13. Ramos, F., Possas, R., Fox, D.: BayesSim: adaptive domain randomization via probabilistic inference for robotics simulators. In: Proceedings of Robotics: Science and Systems. FreiburgimBreisgau, Germany (2019). https://doi.org/10.15607/RSS.2019.XV.029
14. Chebotar, Y., et al.: Closing the sim-to-real loop: adapting simulation randomization with real world experience. In: 2019 International Conference on Robotics and Automation (ICRA), pp. 8973–8979. IEEE (2019)
15. Yu, W., Tan, J., Liu, C.K., Turk, G.: Preparing for the unknown: learning a universal policy with online system identification. In: Amato, N.M., Srinivasa, S.S., Ayanian, N., Kuindersma, S. (eds.) Robotics: Science and Systems XIII, Massachusetts Institute of Technology, Cambridge, Massachusetts, USA, 12–16 July 2017 (2017). https://doi.org/10.15607/RSS.2017.XIII.048. http://www.roboticsproceedings.org/rss13/p48.html

16. Snoek, J., Larochelle, H., Adams, R.P.: Practical Bayesian optimization of machine learning algorithms. In: Advances in Neural Information Processing Systems, vol. 25 (2012)
17. Antonova, R., Rai, A., Li, T., Kragic, D.: Bayesian optimization in variational latent spaces with dynamic compression. In: Conference on Robot Learning, pp. 456–465. PMLR (2020)
18. Müller, S., von Rohr, A., Trimpe, S.: Local policy search with Bayesian optimization. In: Advances in Neural Information Processing Systems, vol. 34 (2021)
19. Katz, B., Di Carlo, J., Kim, S.: Mini cheetah: a platform for pushing the limits of dynamic quadruped control. In: 2019 International Conference on Robotics and Automation (ICRA), pp. 6295–6301. IEEE (2019)
20. Storn, R., Price, K.: Differential evolution-a simple and efficient heuristic for global optimization over continuous spaces. J. Global Optim. **11**(4), 341–359 (1997)
21. Collins, J., Brown, R., Leitner, J., Howard, D.: Traversing the reality gap via simulator tuning. arXiv preprint arXiv:2003.01369 (2020)
22. Srinivas, N., Krause, A., Kakade, S., Seeger, M.: Gaussian process optimization in the bandit setting: no regret and experimental design. In: Proceedings of the 27th International Conference on International Conference on Machine Learning, ICML 2010, pp. 1015–1022. Omnipress, Madison (2010)
23. Coumans, E., Bai, Y.: Pybullet, a python module for physics simulation for games, robotics and machine learning (2016)
24. Rajeswaran, A., Ghotra, S., Ravindran, B., Levine, S.: EPOpt: learning robust neural network policies using model ensembles. In: 5th International Conference on Learning Representations. OpenReview.net (2017)
25. Tan, J., et al.: Sim-to-real: learning agile locomotion for quadruped robots. In: Proceedings of Robotics: Science and Systems, Pittsburgh, Pennsylvania (2018). https://doi.org/10.15607/RSS.2018.XIV.010
26. Tassa, Y., Erez, T., Todorov, E.: Synthesis and stabilization of complex behaviors through online trajectory optimization. In: 2012 IEEE/RSJ International Conference on Intelligent Robots and Systems, pp. 4906–4913. IEEE (2012)
27. Huang, A.S., Olson, E., Moore, D.C.: LCM: lightweight communications and marshalling. In: 2010 IEEE/RSJ International Conference on Intelligent Robots and Systems, pp. 4057–4062. IEEE (2010)
28. Auger, A., Hansen, N.: Tutorial CMA-ES: evolution strategies and covariance matrix adaptation. In: Proceedings of the 14th Annual Conference Companion on Genetic and Evolutionary Computation, pp. 827–848 (2012)
29. Le Goff, L.K., et al.: Sample and time efficient policy learning with CMA-ES and Bayesian Optimisation. In: ALIFE 2020: The 2020 Conference on Artificial Life, pp. 432–440 (2020). https://doi.org/10.1162/isal_a_00299
30. Lim, V., et al.: Planar robot casting with real2sim2real self-supervised learning. In: 2022 International Conference on Robotics and Automation (ICRA). IEEE (2022)

A Review of Current Approaches to Configuration Detection in Modular Legged Robots

Timothee Buettner[1(✉)], Olivia Schwertfeger[2], Arne Roennau[1], and Ruediger Dillmann[1]

[1] FZI Research Center for Information Technology, Karlsruhe, Germany
{buettner,roennau,dillmann}@fzi.de
[2] KIT Karlsruhe Institute of Technology, Karlsruhe, Germany
oliv.schwert@gmx.de

Abstract. In this paper, we review monolithic and reconfigurable modular robotic systems, focusing on configuration detection and legged locomotion. The focus of the review lies on the realization of a robotic system consisting of autonomous robotic legs with integrated PC and power units and grants a specialized look at embodiment for such robots. We further dissect the challenges of disembodied modular legged systems and propose how they may be realized using configuration detection. This is necessary in case the configuration as well as the payload is unknown to the robot. Using the state of the art approaches, we propose a approach for the realization of a modular disembodied legged robot consisting of a list of methods focussing on individual issues in the realization of such autonomous legs. We close with an outlook on future challenges for such a system and what research fields will need to be exploited.

1 Introduction

The ability to walk is crucial for a robot to execute a given task in an environment designed for humans. The human environment includes stairs and other kinds of unevenness that prove difficult to surpass for other ways of locomotion, such as classic wheeled systems [1,2]. Implementing some form of walking pattern as well as self detection is particularly difficult in modular robots due to them requiring highly reliable inter-module communication, synchronization, and coordination. This paper reviews the state of the art of mobile, modular robots, on the one hand, on the other hand it proposes an approach for configuration detection in disembodied modular robots, specifically Nimble Limbs as introduced by Buettner et al. in [3,4]. The term depicts any kind of modular robot with individual, autonomous legs as their modules. A Nimble Limbs system (NLS) is comprised of an arbitrary number of independent, identical leg modules that can be attached to arbitrary objects, such as furniture, containers, or sensors, and enable these objects to walk [4]. Therefore, the most prominent features of a NLS are its reconfigurability as well as the autonomy of its legs.

© The Author(s), under exclusive license to Springer Nature Switzerland AG 2023
J. M. Cascalho et al. (Eds.): CLAWAR 2022, LNNS 530, pp. 167–180, 2023.
https://doi.org/10.1007/978-3-031-15226-9_18

Fig. 1. 3D rendered concept of a group of autonomous intelligent legs attached to a transport box. The concept and motivation of Nimble Limbs is highlighted in detail in *Nimble Limbs - Intelligent attachable legs to create walking robots from variously shaped objects* [4].

Nimble Limbs are motivated by the desire to turn any payload into a capable walking robot. The current paradigm in walking robotics consists of strapping whatever needs transportation to a monolithic robotic platform. Our approach, however tries to find a middle ground between monolithic and modular systems to simply attach the legs (in appropriate quantity) to whichever system one wishes to transport. The legs then synchronize themselves and automatically detect positioning, center-of-mass and leg number to generate a stable walking pattern. This concept generates a list of requirements to be fully realized. The requirements are discussed in detail in [4] and are as follows. **"Each leg module..."**

- **is identical in size, shape and equipment**
- **houses three actuators, optical encoders, an inertial measurement unit (IMU), a PCU, and a battery**
- **uses only Wifi and Bluetooth to communicate (no direct physical connection)**
- **can be fixed to the payload at the first shoulder joint**

An example of the legs carrying a simple crate can be seen in Fig. 1 . Since the modules of a NLS are only physically connected by their common payload and because we assume this payload to be unknown, the non-trivial question that arises is how a NLS will be able to walk stably without any information about its physical presence. We will address this question by discussing agnostic as well as proprioceptive ways of learning.

Agnostic learning is similar to the processes found in nature. Learning in nature is realized through imitation, experience, conditioning, and understanding [5]. Wanting to walk seems to stem from instinct and imitation, however, learning to walk requires practice which in turn builds experience. A study conducted by

Adolph et al. shows that it takes infants multiple attempts over a period of months to learn how to walk. Infants practiced walking in variable patterns and contexts at seemingly random times. This may seem ineffective for robotic learning, however, it has also been shown that practice in random environments leads to more generalization and adaptability in locomotion skills of robots [6].

Our main goal with proprioceptive learning, in contrast, is to provide the NLS with a sense of its physical presence, i.e. its configuration. This requires the leg modules to communicate in order to exchange information. Such information may include sensory feedback or geographic positioning.

We will see that proprioceptive learning has some major advantages over agnostic learning when applied to robotics: the ability to detect its configuration allows the NLS to choose a locomotion gait and calculate its stability since it provides the robot with knowledge of the payload and its mass, of its Center of Mass (CoM), its number of legs and their mounting positions [4]. In turn, configuration detection allows for great adaptability and generalization. However, navigation and control of a disembodied system, such as a NLS, also requires precise synchronization of its modules. We will address these challenges because overcoming them will allow the NLS to learn how to walk, thus providing it with the ability to execute a given task [7,8].

We will discuss how different monolithic as well as reconfigurable, modular robotic systems may work as an inspiration to realizing a NLS. We have selected the approaches presented in this paper by developing a system of classification shown in Table 1. As a reference, the NLS has also been added to Table 1.

This paper is structured as follows: in Sect. 2, we will review the state of the art of monolithic and reconfigurable, modular robots with a focus on embodiment, communication between modules, localization, and legged locomotion. We will look at agnostic as well as proprioceptive approaches to learning to walk. In Sect. 3, we will discuss how the approaches from Sect. 2 may be applied to a NLS and which challenges remain. This discussion will be followed by an outlook on subsequent steps towards readying a NLS to execute a given task. We will conclude with a reflection on the feasibility of the proposed approaches.

2 State of the Art

In this state of the art we aim at highlighting the individual issues of realizing intelligent robotic legs. We start off with a general taxonomy of the field to generate an overview of robotic types and their topology. This serves to later classify which reference systems are relevant for NLS.

Figure 2 provides an overview of the field of monolithic and reconfigurable, modular robots and how we classify a NLS in this context: Monolithic robots are designed as a single physical unit. They are often built with similar morphology as their organic counterparts. This sets them apart from a NLS. Monolithic robots are represented by a variety of walking robots, such as the humanoid robot Atlas [9], the quadruped Spot [13] or the six-legged LAURON V [10], as some examples. They are not directly related to the presented modular legs but

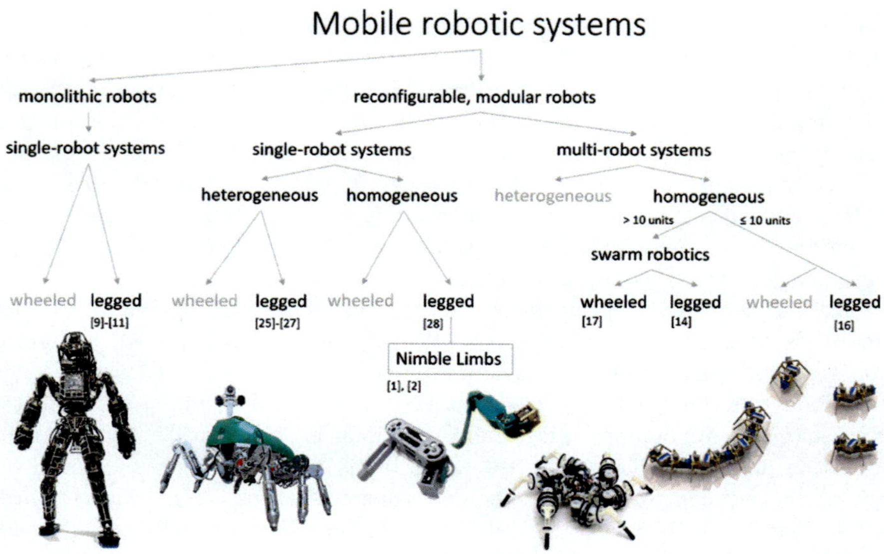

Fig. 2. Subject classification of Nimble Limbs among monolithic and reconfigurable, modular robots - gray: not reviewed in this paper. F.l.t.r.: Atlas [9], LAURON V [10], Nimble Limbs [3], Snapbot [11], Swarm robots [12]

provide various solutions in the fields of configuration detection and adaptive walking, as mentioned in later sections.

Since a NLS is comprised of leg modules with no designated main body, however, we will focus mainly on a comparison to reconfigurable, modular robots. For the purposes of this paper, we further split reconfigurable, modular robots into single-robot systems (SRS) and multi-robot systems (MRS) [7,8]. Both SRS and MRS are comprised of several modules which can either be identical (homogeneous) or not (heterogeneous) [7,8]. This division is sensible because there are two ways to look at a NLS: (1) we can think of each leg as a module. Hence, when multiple legs are attached to an object, the whole forms a unit, which is what we see in heterogeneous, reconfigurable SRS. (2) we can view each leg as an independent robot. Hence, multiple identical legs collaborate to transport an object, which is similar to task execution in homogeneous MRS.

Furthermore, configuration detection and walking require a unit that receives and processes data and makes decisions. Such a decision-maker has its organic counterpart in the brain. In monolithic and heterogeneous robots, a predefined control unit or module usually functions as such. In homogenous robots, however, any module can take control. This problem is often addressed through a concept called “leader election”, which is seen both in homogeneous SRS and MRS [12, 14,15], but not in swarms [16].

For data to be received in a modular system, communication between different modules is crucial. Since there is generally no physical contact between the legs of a NLS, the communication has to be wireless, which, as we will see, is rarely the case in the state of the art.

A variety of locomotion styles can be found in monolithic and modular robots, including legs [12,17] and wheels [18]. In this paper, we focus on legged locomotion in SRS and MRS because it is best suited for traversing through a human environment and because it can be better applied to a NLS.

In the following, the state of the art is categorized by the above-mentioned features. An overview of this categorization is depicted in Table 1.

2.1 Topology Detection in Monolithic Robots

Even though the internal model of a monolithic robot, such as masses and inertias, is known in most cases, topology detection[1] is necessary to operate in unknown and hazardous environments. This is because monolithic robots, like all robots in the physical world, are prone to suffer hardware damage: The detection of different failure modes is therefore crucial for autonomous missions and allows robots to finish their task or proceed to repair [19,20]. Damage detection can be compared to topology detection since the monolithic robot's original hardware has been altered. In order to enable a legged robot to detect different kinds of damages and recover from them autonomously, Bongard et al. simultaneously trained a simulated and a physical robot which update each other using a two-stage evolutionary algorithm [19,20]. The goal is to iteratively increase the accuracy of the simulation which requires the initial topology to be known. The authors show that this approach enables a monolithic legged robot to recover from different kinds of damages - including unanticipated damage. In its recovered state, the robot sometimes even exceeds the previous performance. Shortcomings of this approach are addressed and improved by Koos et al. [21]. These include long run times and the occurrence of a "reality gap". The latter describes a gap between the simulated robot's behavior and the real robot's behavior [21], often caused by simplifications or a lack of understanding of interdependences [22]. Koos et al. state that this gap is the main reason the algorithm by Bongard et al. sometimes predicted inaccurate self-models. Koos et al. improved damage recovery by implementing a T-Resilience algorithm which takes the reality gap into account as described in [23], and by making a compromise between speed and accuracy of the self-model [21].

2.2 Configuration Detection in Heterogeneous Robots

As stated earlier, a NLS can be seen as heterogeneous robotic system (1). Therefore we look into different approaches to configuration detection with regard to heterogeneous systems.

In [24], Bongard et al. introduce means to detect the configuration of a heterogeneous modular legged robot without prior information on its topology as

[1] In the case of monolithic robots, we will speak of topology detection rather than configuration detection, since they are not reconfigurable.

a whole. Only the topology of each module is assumed to be known. The main body has tilt sensors and is always the same, however, there are different types of leg and foot modules with included motors. To detect how the modules are connected to each other, a parallel hill climber is used. An evolutionary algorithm indirectly guesses the configuration using only the robot's tilt information. After multiple runs, the configuration the model has predicted is compared to the real configuration. This way the objective error is calculated.

Bongard et al. used the same robot to show an alternative approach to configuration detection [25]: first, the initial self-model is randomly generated by the algorithm. Then a random motor action is performed and recorded by a sensor. Through observing actuation patterns and resulting model disagreement, the configuration is iteratively detected.

Bongard et al. found out that the error is minimized by changing only a few parameters at once to learn about causal dependencies, and by optimizing a fitness function that rewards model disagreement [25] as well as reliability [24].

In [11], Kim et al. introduce Snapbot, a reconfigurable legged robot. Snapbot is comprised of a main body that a maximum of six legs can be attached to magnetically. The main body holds most of the computing power, including the locomotion algorithm. The possible configurations are known beforehand, yet their number (700) requires automatic detection. This is realized by scanning each servo's specific ID and thereby checking which of the predefines connection ports in the core body are occupied by which leg. Gim et al. continue this work with Snapbot V2 [26]. Snapbot V2 uses a external camera for configuration detection. A color detection algorithm distinguishes between orange- and blue-colored leg components in order to decide which legs are connected to which ports. This influences the choice of locomotion style which is predefined for each configuration. However, this requires the external camera to see all ports at all times, which cannot be guaranteed in a NLS. In our case, the premise of the NLS states that we neither want to use external systems, nor do we want to core body to contain any specific hardware (such as connection ports or magnets).

WalkingBot introduced by Wang et al. [27] is comprised of a main body and a set of building blocks, such as servos, sound and range sensors. Each servo motor has a designated ID. Configuration detection is realized through the exchange of predefined IDs, which is similar to [11]. The main body scans all IDs and sends them to an external PC via Bluetooth. This PC functions as a control panel for the robot to generate motion through pre-determined footfall patterns.

An approach that teaches a heterogeneous modular robot to walk without detecting its configuration first is the EMBER approach introduced by Jacob et al. It is based on modular reinforcement learning [28]. Under the assumption that locally similar actions induce locally similar states, Jacob et al. train a modular walking robot to walk in two steps: first, the legs are trained to walk individually. Second, the robot is assembled and now able to improve its walking abilities acquired in each leg before. Reinforcement learning which gives a reward for avoiding slip is applied to improve the robot's walking ability. It relies solely on the information that one leg can receive through its sensors and actors. This allows for a great ability to generalize and is particularly helpful because, during the learning phase of an individual leg, the specific kind of system the leg will be integrated into is unknown. Furthermore, Jacob et al. show that pre-training the legs in isolation increases the learning speed of the assembly.

Table 1. Examples from related work for legged robots.

Ref.	Hom/het/mon	Locomotion	MRS/SRS	Physical contact	Communication	Leader	Configuration detection
[3,4]	hom	legged	SRS	no	Bluetooth, WIFI	no	Geometric approximation
[19,21]	mon	legged	SRS	yes	n.a	no	minimize prediction error
[24,25]	het	legged	SRS	yes	n.a	no	minimize prediction error
[11,26]	het	legged	SRS	yes	n.a	no	ID exchange / camera
[27]	het	legged	SRS	yes	Bluetooth, IIC data bus	no	ID exchange
[28]	het	legged	SRS	yes	n.a	no	no
[14,15,29]	hom	legged	SRS	yes	IR	yes	ID exchange, roles
[17]	hom	legged	MRS	yes	CAN bus	no	ID exchange
[12]	hom	legged	MRS	both	LEDs, touch	yes	no

2.3 Configuration Detection in Homogeneous Robots

Since the legs of a NLS are identical, we review homogeneous modular robots with regard to configuration detection in the following.

Baca et al. present a modular homogeneous robot, assembled in a chain configuration, that can be classified as a SRS. It autonomously detects its configuration through wireless ID exchange with its neighbors [14,15]. Exchanging IDs to derive one's configuration is similar to the heterogeneous approaches in [11,27] (Sect. 2.2). In contrast, however, to find the correct configuration, Baca et al. have their algorithm assess if the modules within the configuration range belong to the configuration. Afterwards, a leader is randomly elected by all the modules within the same configuration, using their microprocessor's clock value. If the leader gets damaged or stops sending activity reports, a new leader will be elected the same way as its predecessor.

On top of leader election, Støy et al. introduce the idea of additionally distributing a role to each module through neighbor-to-neighbor communication via infrared (IR) transmitters [29]. Once each module has a role, Støy et al. address the challenge of synchronizing the legs of a walking robot. This is particularly relevant in a NLS due to its disembodiment. Støy et al. define the delay between leg pairs relative to the leader. The advantage of this approach is that each module keeps its sub-modules synchronized, allowing for changes in configuration, such as adding legs. If the leader fails, the modules can also replace it, thereby becoming the new leader of each of their sub-modules. Unfortunately, this approach is limited by the power source and motor power and grows in complexity the larger the system becomes [29].

All monolithic, heterogeneous and homogeneous approaches introduced so far describe SRS. Now we will look at homogeneous MRS. The advantage of MRS over SRS lies in their ability to perform tasks that are too difficult for a single robot (this especially applies to small robots transporting heavier payloads or traversing gaps or stairs [12,17]). They are also more robust to individual failure, as each robot can be easily replaced [30].

Ohira et al. developed a concept for a three-legged robot that is able to crawl towards and connect with others of its kind [17], thereby forming a MRS. Together, they can navigate through an unknown environment. Initially, each robot is supposed to communicate its orientations and environments to an external PC which will then generate a motion plan for each robot. Once a robot connects to another robot using its third leg, they exchange controller information and motor IDs via a CAN bus, which is similar to [14,15].

As stated earlier, a NLS can be viewed as the collaboration of multiple homogeneous, independent robots. This is similar to a sub-group of homogeneous MRS called "swarm robotics" (Fig. 2)[2]. Robotic swarms imitate swarms seen in nature, such as bees and ants [30], meaning they collaborate in large groups of homogeneous robots in a decentralized manner [31]. Each of these robots has simple hardware and can be easily replaced, hence homogeneous swarms are robust to individual failure [30]. Due to a swarm's size and its lack of a leader, self-organization, self-adaptiveness and the emergence of group intelligence are vital to its success in executing tasks [16]. This requires the robots to communicate with each other either directly or indirectly. Nurmaini et al. state that self-localization is integral to collision-free navigation in swarms [16]. Range-based localization meets this requirement and is simpler than mapping approaches [32] because they do not require laser scanners: the decrease in signal strength with an increase in distance and propagation can be used to estimate distances [33,34]. This is realized through deriving the distance between a sender and receiver from the attenuation of signal strength. Adding a fuzzy logic algorithm improves the accuracy of this approach, depending on the usage of WiFi or Bluetooth. Bluetooth triangulation yields an accuracy of approx. 1m, which can be highly improved (up to 0.15m) when adding ultrasound [34], however, this does not meet the current hardware limitations of a NLS since measures with ultrasonic are inhibited by the core body of the robot. Another approach called instantaneous localization, which uses a landmark as a reference [35], is not applicable because it requires a camera. In a NLS, however, visual contact cannot be guaranteed.

A different approach that does not use localization or even communication is that by Ozkan et al. [12]. They work with quadruped robots that collaborate either through a physical connection, similar to [17], or by positioning themselves around an object they want to transport. They can be categorized as homogeneous MRS which collaborate in swarms, however, this approach does not target configuration detection and is more similar to agnostic learning, like the EMBER approach by Jacob et al. [28]. Initially, a group of robots is sent out to execute a task. Each robot is equipped with a red LED. When one of them encounters an insurmountable obstacle, it flashes its LED to attract its group members, thereby becoming the leader. If necessary, the robots can con-

[2] Swarm robotics are also seen in heterogeneous MRS (Fig. 2). Because the leg modules are homogeneous, however, we do not include heterogeneous swarms in our review.

nect. A connection can be sensed using the touch sensors each robot is equipped with. The robots are merely attracted by light sources, they have no means of higher communication. Hence when transporting an object, they are only able to receive feedback through how much closer their last action brought them to the target light source. The problem with this is that an external light source relevant to the task is needed, which is especially problematic when using the robots outside.

3 Discussion

Monolithic, heterogeneous, and homogeneous mobile robotic systems were summarized in the previous section (Table 1). In this section, we will discuss if and how these approaches can be applied to a NLS. For that purpose, we rule out approaches that impose requirements for the robot's surroundings, such as external PCs as seen in [17,27] or artificial light sources [12]. Furthermore, hardcoding by an operator will be avoided, in favor of autonomy.

We first introduced damage detection in monolithic robots as a possible approach to realize configuration detection (Sect. 2.1). Even though a monolithic robot's initial topology is known, it may be altered and become unknown due to damages. The approach to topology detection by Bongard et al. [19] aims at detecting the altered topology in order to compensate for damages. This is done by comparing the physical robot's neuro-controller with that of a simulated robot and iteratively reducing the error using an evolutionary algorithm. An alternative to evolutionary algorithms, in this case, are policy gradient methods, where a controller's parameters are iteratively optimized using a reward function [21]. It is worth noting that simulations always involve the risk of a reality gap [21,22]. Specifically when compared to a NLS, it is questionable whether a leg's computing power would suffice to train both a simulated and a physical neuro-controller in parallel. This is a direct result of the main difference between monolithic robots and a NLS, which lies in a physical unit with a main body compared to a disembodied group of modules.

The disembodiment also marks the importance of communication between the leg modules of a NLS. We proposed to equip them with Wifi and Bluetooth [4] to tackle this problem. Unfortunately, the state of the art of modular robots is mostly represented by modules that are physically connected to each other, using connectors and data busses [11,17,26,27] or IR [14,15,29] to exchange data, such as motor IDs [11,14,17,27,29]. In a NLS, unique IDs can also be assigned to each leg and be exchanged between neighbors via Wifi or Bluetooth. We will discuss how the disembodiment can be dealt with later. Circling back to Bongard et al. and the neuro-controllers, rewarding the maximum distance traveled [19,20] would require the NLS to have at least basic locomotion skills before knowing its configuration.

This problem may be addressed using the EMBER approach by Jacob et al. [28] introduced in Sect. 2.2. Here each leg is trained to walk individually before the whole robot is assembled. This may be another way of realizing a

NLS, by simulating each leg to walk "blindly" beforehand and then detecting the NLS's configuration. The simulation can be the same for each leg, since the future assembly is unknown anyway. This way each leg would still have all the advantages of knowing its configuration (such as calculating its stability) while also being able to quickly improve its walking ability. However, this may not be applicable because of a lack of synchronization in the legs. The reality gap must also be taken into consideration [22].

Another problem arising from the approaches of Bongard et al. [19,20] and Koos et al. [21] is that they use the initial undamaged robot as a reference to what the nominal state is. This is not applicable to a NLS, since the initial configuration and state is unknown (such as the size and the position of the legs). However, Bongard et al. also developed two approaches for configuration detection in a legged heterogeneous modular robot. Here, the error between the estimated configuration and the true configuration is also successfully reduced [24,25]. In both approaches by Bongard et al. [24,25], a random initial configuration is selected as a reference. The error is found through an evolutionary algorithm that has access to the robot's tilt information [24]. Since the legs in a NLS are equipped with an IMU, this technique can be applied to them as well. In [25], the error is detected through observing actuation patterns and resulting model disagreement. Although the type of connection is unknown, the algorithm knows which and how many modules the robot is comprised of. A NLS is different because the number of legs is neither limited nor known and the payload is unknown as well. It might still be possible to use one of these approaches, though.

First, each leg would have to find its neighbors, either through direct or indirect communication [30]. In a NLS, direct communication via Wifi and Bluetooth can be used to find neighbors within each leg's proximity, by exchanging IDs, as stated above. Bluetooth needs less power than Wifi and thus should be used above the latter. It also makes sense for one leg to be elected as the leader and record all IDs, because these can be used for configuration detection later, as seen in [14,15]. The leader module can be elected randomly or based on computational power, should the legs have different hardware setups. Model disagreement, as introduced by Bongard et al. [24,25], can be used to detect the configuration for an initial estimation of the NLS. Unless the legs will be equipped with GPS in the future and to speed up the configuration detection, each leg's position can be estimated relative to its closest neighbors and to the leader. This can be realized through range-based localization using Bluetooth or WiFi. As explained in Sect. sec2c, interpreting received signal strength allows for a distance estimation between modules [34].

Once each leg's position has been estimated, roles as introduced by Støy et al. [29] can be distributed relative to the leader, allowing to define front and rear pairs of legs, depending on the configuration. This serves as a starting point for synchronizing leg movement as seen in [29] and is necessary to avoid delays between legs. This approach is robust to damages and changes in configuration, the latter being particularly relevant in a NLS. Walking can be realized through central pattern generators (CPGs) as seen in nature [28,36], reinforcement learning [28], and evolutionary algorithms [24,25].

Having analysed the complete pipeline from configuration detection through leader election to walking pattern generation, the following points serve as a recap of proposed steps to realize a nimble limbs system:

1. Learning to walk in isolation: Prior to assembly, each leg learns to walk individually in simulations by using reinforcement learning, as seen in the EMBER approach [28]. However, the reality gap must be taken into consideration to produce realistic and reliable results. Therefore the algorithm used to train each leg to walk should use, for example, a t-resilience approach to mitigate simulation bias [21,23], as explained in Sect. 2.1. As stated before, prior training of individual modules significantly improves the learning speed of the assembled robot. This step only has to be executed once and is independent from the way the leg modules will be assembled in the future. Hence it is well suited for a NLS with its restricted hardware.
2. Inferring the number of modules: As explained earlier, a leader simplifies decision-making and configuration detection. This is why all legs are automatically assigned a random ID when they are switched on and a leader is randomly selected. Due to the disembodiment, all legs send their ID information out using Bluetooth because it is more energy-efficient than WiFi. The leader collects all IDs to find out how many legs the system is comprised of, which is similar to [11,14,15,27,29], which are explained in Sect. 2.
3. Localizing the modules: Each leg localizes itself relative to its neighbors and the leader using range-based localization. This can be realized with the current hardware of a NLS, unlike mapping or vision-based approaches, as explained in Sect. 2.3. Range-based localization is realized through measuring the attenuation in received signal strength [33,34].
4. Minimizing the prediction error: Accuracy in 3) is not essential since the exact positioning can be derived from an IMU-based tilt routine similar to [24]. This is possible given the hardware assumptions stated in Sect. sec2c.
5. Synchronizing leg movement: Due to the disembodied structure of a NLS, special attention must be paid to possible delays between the legs before learning to walk. This is why the leg movement is synchronized. We use the approach by Støy et al. [29] because it is robust to configuration changes and damages, as discussed before. The approach is applied to a NLS through distributing roles to each module to find pairs of legs and to synchronize them with a delay relative to the leader elected in 2).
6. Learning to walk as a unit: Like in 1), reinforcement learning is now used to train the entire system to walk, building on each leg's prior training in simulation, which is similar to [28]. The learning of the entire NLS is done without a simulation, however, because it is faster, less expensive computationally, and because each leg is already able to walk by itself, which reduces the likelihood of damages.

4 Conclusion and Outlook

In this paper, we have presented an overview of the state of the art of configuration detection in monolithic and reconfigurable modular robots using leader election and communication. The review focuses on possible solutions to "nimble limbs systems": modular robots comprised of several autonomous legs and an unknown main body, as introduced by Buettner et al. [4]. For this purpose, we have presented several related approaches with a focus on embodiment, locomotion styles and communication. We looked at why configuration detection is useful, but also took alternatives into consideration. A discussion on how the state of the art differs from or may be applied to a NLS followed, concluding with a suggestion as to how configuration detection in a NLS may be realized.

The provided solution serves as a foundation for future research and development which will include an examination and proof of concept, taking computing power, robustness, computing time, and miscellaneous disruptive factors into account. Current research as well as future works are motivated towards a simulated proof-of-concept based on ROS [37] (and ROS2 [38]) to realize self-detection, stable walking and center of mass estimation for arbitrary leg numbers. Beyond the scope of simple walking and embodiment, future works for NLS will have to look into task execution, path planning and human-robot interaction. Challenges regarding the hardware may include designing shoulder mounts that can be safely and reliably attached to any surface and object.

References

1. Hurst, J.: Walk this way: to be useful around people, robots need to learn how to move like we do. IEEE Spectr. **56**(3), 30–51 (2019)
2. Siciliano, B., Khatib, O., Kröger, T.: Springer Handbook of Robotics, vol. 200. Springer, Heidelberg (2008). https://doi.org/10.1007/978-3-540-30301-5
3. Buettner, T., Wilke, D., Roennau, A., Heppner, G. Dillmann, R.: A scalable, modular leg design for multi-legged stair climbing robots. In: International Conference on Climbing and Walking Robots and the Support Technologies for Mobile Machines (CLAWAR) (2018)
4. Buettner, T., Heppner, G., Roennau, A. Dillmann, R.: Nimble limbs-intelligent attachable legs to create walking robots from variously shaped objects. In: 2019 IEEE/ASME International Conference on Advanced Intelligent Mechatronics (AIM), pp. 376–381. IEEE (2019)
5. Roennau, A.: Verhalten. In: T-INFO-101351 - Biologisch Motivierte Robotersysteme, p. 28 (2021)
6. Adolph, K.E., et al.: How do you learn to walk? thousands of steps and dozens of falls per day. Psychol. Sci. **23**(11), 1387–1394 (2012)
7. Liu, J., Zhang, X., Hao, G.: Survey on research and development of reconfigurable modular robots. Adv. Mech. Eng. **8**(8), 1687814016659597 (2016)
8. Ahmadzadeh, H., Masehian, E., Asadpour, M.: Modular robotic systems: characteristics and applications. J. Intell. Robot. Syst. **81**(3–4), 317–357 (2016)
9. Koolen, T., et al.: Design of a momentum-based control framework and application to the humanoid robot atlas. Int. J. Humanoid Robot. **13**(01), 1650007 (2016)

10. Rönnau, A., Heppner, G., Nowicki, M., Dillmann, R.: Lauron v: a versatile six-legged walking robot with advanced maneuverability. In: 2014 IEEE/ASME International Conference on Advanced Intelligent Mechatronics, pp. 82–87. IEEE (2014)
11. Kim, J., Alspach, A., Yamane, K.: Snapbot: a reconfigurable legged robot. In: 2017 IEEE/RSJ International Conference on Intelligent Robots and Systems (IROS), pp. 5861–5867. IEEE (2017)
12. Ozkan-Aydin, Y., Goldman, D.I.: Self-reconfigurable multilegged robot swarms collectively accomplish challenging terradynamic tasks. Sci. Robot. **6**(56), eabf1628 (2021)
13. Biswal, P., Mohanty, P.K.: Development of quadruped walking robots: a review. Ain Shams Eng. J. **12**(2), 2017–2031 (2021)
14. Baca, J., Woosley, B., Dasgupta, P., Nelson, C.A.: Configuration discovery of modular self-reconfigurable robots: real-time, distributed, ir+ xbee communication method. Robot. Auton. Syst. **91**, 284–298 (2017)
15. Baca, J., Woosley, B., Dasgupta, P., Dutta, A., Nelson, C.: Coordination of modular robots by means of topology discovery and leader election: improvement of the locomotion case. In: Chong, N.-Y., Cho, Y.-J. (eds.) Distributed Autonomous Robotic Systems. STAR, vol. 112, pp. 447–458. Springer, Tokyo (2016). https://doi.org/10.1007/978-4-431-55879-8_31
16. Nurmaini, S., Tutuko, B.: Intelligent robotics navigation system: problems, methods, and algorithm. Int. J. Electr. Comput. Eng. **7**(6), 3711 (2017). (2088-8708)
17. Ohira, M., Chatterjee, R., Kamegawa, T., Matsuno, F. Development of three-legged modular robots and demonstration of collaborative task execution. In: Proceedings 2007 IEEE International Conference on Robotics and Automation, pp. 3895–3900. IEEE (2007)
18. Johnson, P.J., Bay, J.S.: Distributed control of simulated autonomous mobile robot collectives in payload transportation. Auton. Robot. **2**(1), 43–63 (1995)
19. Bongard, J.C., Lipson, H.: Automated damage diagnosis and recovery for remote robotics. In: 2004 Proceedings of the IEEE International Conference on Robotics and Automation. ICRA 2004, vol. 4, pp. 3545–3550. IEEE (2004)
20. Bongard, J.C., Lipson, H.: Automated robot function recovery after unanticipated failure or environmental change using a minimum of hardware trials. In: Proceedings 2004 NASA/DoD Conference on Evolvable Hardware, pp. 169–176. IEEE (2004)
21. Koos, S., Cully, A., Mouret, J.-B.: Fast damage recovery in robotics with the t-resilience algorithm. Int. J. Robot. Res. **32**(14), 1700–1723 (2013)
22. Mouret, J.-B., Chatzilygeroudis, K.: 20 years of reality gap: a few thoughts about simulators in evolutionary robotics. In: Proceedings of the Genetic and Evolutionary Computation Conference Companion, GECCO 2017, New York, NY, USA, pp. 1121-1124. Association for Computing Machinery (2017)
23. Mouret, J.-B., Koos, S., Doncieux, S.: Crossing the reality gap: a short introduction to the transferability approach (2013)
24. Bongard, J.: Action-selection and crossover strategies for self-modeling machines. In: Proceedings of the 9th Annual Conference on Genetic and Evolutionary Computation, pp. 198–205 (2007)
25. Bongard, J., Zykov, V., Lipson, H.: Resilient machines through continuous self-modeling. Science **314**(5802), 1118–1121 (2006)
26. Gim, K.G., Kim, J.: Snapbot v2: a reconfigurable legged robot with a camera for self configuration recognition. In: 2020 IEEE/RSJ International Conference on Intelligent Robots and Systems (IROS), pp. 4026–4031. IEEE (2020)

27. Wang, M., Su, Y., Liu, H., Xu, Y.: Walkingbot: modular interactive legged robot with automated structure sensing and motion planning. In: 2020 29th IEEE International Conference on Robot and Human Interactive Communication (RO-MAN), pp. 307–312. IEEE (2020)
28. Jacob, D., Polani, D., Nehaniv, C.L.: Legs that can walk: embodiment-based modular reinforcement learning applied. In: 2005 International Symposium on Computational Intelligence in Robotics and Automation, pp. 365–372. IEEE (2005)
29. Støy, K., Shen, W.-M., Will, P.: How to make a self-reconfigurable robot run. Proc. First Int. Joint conf. Auton. Agents Multiagent Syst. part **2**, 813–820 (2002)
30. Tuci, E., Alkilabi, M.H., Akanyeti, O.: Cooperative object transport in multi-robot systems: a review of the state-of-the-art. Front. Robot. AI **5**, 59 (2018)
31. Jeong, D., Lee, K.: Distributed communication and localization algorithms for homogeneous robotic swarm. In: Chong, N.-Y., Cho, Y.-J. (eds.) Distributed Autonomous Robotic Systems. STAR, vol. 112, pp. 405–418. Springer, Tokyo (2016). https://doi.org/10.1007/978-4-431-55879-8_28
32. Clemens, J., Reineking, T., Kluth, T.: An evidential approach to slam, path planning, and active exploration. Int. J. Approximate Reasoning **73**, 1–26 (2016)
33. Alrajeh, N.A., Bashir, M., Shams, B.: Localization techniques in wireless sensor networks. Int. J. Distrib. Sens. Netw. **9**(6), 304628 (2013)
34. Liu, H., Darabi, H., Banerjee, P., Liu, J.: Survey of wireless indoor positioning techniques and systems. IEEE Trans. Syst. Man, and Cybernetics. Part C (Appl. Rev.) **37**(6), 1067–1080 (2007)
35. Jaulin, L.: Mobile Robotics, pp. 121–126 (2019)
36. Cruse, H.: What mechanisms coordinate leg movement in walking arthropods? Trends Neurosci. **13**(1), 15–21 (1990)
37. Quigley, M., et al.: Ros: an open-source robot operating system. In: Proceedings of the IEEE International Conference on Robotics and Automation (ICRA) Workshop on Open Source Robotics, Kobe, Japan (2009)
38. Macenski, S., Foote, T., Gerkey, B., Lalancette, C., Woodall, W.: Robot operating system 2: design, architecture, and uses in the wild. Sci. Robot. **7**(66), eabm6074 (2022)

Design and Mixed-Reality Teleoperation of a Quadruped-Manipulator Robot for SAR Tasks

Christyan Cruz Ulloa(✉), David Domínguez, Antonio Barrientos, and Jaime del Cerro

Centro de Automática y Robótica, Consejo Superior de Investigaciones Científicas, Universidad Politécnica de Madrid, 28006 Madrid, Spain
{christyan.cruz.ulloa,antonio.barrientos,j.cerro}@upm.es, d.dominguezc@alumnos.upm.es

Abstract. Technological development within the robotics field has made it possible to generate great advances, developing complex bio-inspired systems, such as robots with legs. The same ones have great applicability within search and rescue (SAR) tasks. This type of robot stands out for its extraordinary ability to move within unstructured environments, overcome obstacles and adapt to different terrains. However, quadruped robots have gained space in the SAR-Tasks field over the last few years. Most implementations are limited to collecting information using different sensors, such as cameras, lasers, or microphones. This article seeks to simulate, implement and carry out the teleoperation through Mixed-Reality (M-R) of a quadruped robot equipped with a manipulator with six degrees of freedom, for which the ARTU-R robot (A1 Rescue Tasks Unitree Robot) has been used. One of the main contributions of this work focuses on improving the efficiency in executing tasks of handling and transporting medical equipment in post-disaster situations, using Mixed Reality with the ARTU-R robot and a robotic manipulator. To develop this proof of concept, Matlab has been used as a computational tool for optimizing the workspace of the integrated robot. Simulations have been carried out on Gazebo in reconstructed post-disaster environments to validate the robot's functionality. At the same time, the effectiveness of the M-R system has been verified with field tests executing medical assistance tasks with the robot. The main results show a 21% increase in the efficiency of performing complex handling tasks using the proposed M-R system compared to conventional interfaces and the efficiency of using quadruped robots with manipulators for medical assistance tasks.

Keywords: Quadruped robot · Mixed reality teleoperation · ROS · Arm manipulator · Walking robots · Search and rescue · Gazebo

This work has received funding from the RoboCity2030-DIH-CM Madrid Robotics Digital Innovation Hub "Robótica aplicada a la mejora de la calidad de vida de los ciudadanos, fase IV"; S2018/NMT-4331), funded by "Programas de Actividades I+D en la Comunidad de Madrid" and cofunded by Structural Funds of the EU. and TASAR (Team of Advanced Search And Rescue Robots), funded by "Proyectos de I+D+i del Ministerio de Ciencia, Innovacion y Universidades".

© The Author(s), under exclusive license to Springer Nature Switzerland AG 2023
J. M. Cascalho et al. (Eds.): CLAWAR 2022, LNNS 530, pp. 181–194, 2023.
https://doi.org/10.1007/978-3-031-15226-9_19

1 Introduction

The study of quadrupeds robots focused on search and rescue (SAR) tasks has evolved to the multiple advantages provided over conventional mobile robots (robots with caterpillar tracks, wheels) [1]. Mainly their ability to move through unstructured terrain stands out, with debris [2,3].

The importance of quadruped robots as intervention platforms in post-disaster environments has been increasing in recent years, carrying out highly relevant developments in search and rescue. An example is the one developed by the authors, which focuses on detecting buried victims with thermal images using the Unitree A1 platform as a means of displacement in the environment [4].

In this preliminary study phase, the advantages of using a quadruped robot in this environment have been analyzed. However, it has been limited to data capture, as in most developments in this field [5–10]. Therefore, in this new phase of development, integrating a manipulator with six degrees of freedom has been proposed to carry out complementary tasks such as assistance to victims with medical equipment [11]. For this implementation, a workspace optimization phase has been previously carried out using Matlab and Gazebo simulations to verify the prototype's functionality.

The control of these systems is complex to conventional interfaces, such as screens with a mouse and keyboard, mainly due to the remote operator's lack of perception of the environment or the simultaneous handling of robot variables. Some developments have sought to address this problem by providing teleoperation solutions. However, they are quite rudimentary and invasive for the operator, like the developments carried out in [12,13].

The main contribution of this article focuses on the implementation of a Teleoperation method of this newly implemented platform through a Reality-Rixta (MR) system where the operator becomes aware of the environment through a virtual model projected through Hololens glasses, being able to carry out controlled movements in the workspace of the articulating robot, through a visual mark on the end of the robot.

The system has been shown to improve the efficiency in the execution of first aid equipment transport tasks due to the contribution of Mixed Reality since the operator can identify the type of object and perhaps its possible fragility, deformability, etc. and carry out the task with greater confidence.

The validation tests have been carried out in post-disaster scenarios reconstructed according to the specifications of the NIST (National Institute of Standards and Technology).

This paper is structured as follow. Section 2 describes the most relevant works in state-of-the-art; Sect. 3 shows materials and methods in detail. Section 4 containing the experiments and results. Finally, Sect. 5 presents the main findings.

2 Related Works

Figure 1 shows the articles increase, with topics related to Quadruped Robots for Search and Rescue; in the web of knowledge, it shows an mean increase of around 13% of articles per year until 2018. However, in the last three years, the number of articles published has increased significantly, up to an incremental rate of up to 100 compared to 30 in previous years. This indicates the importance of this type of robot within search and rescue. In [14] reference is made to the growth of these quadrupeds, while in studies such as [15], a follow-up and justification of the evolution of this type of mechanisms is made up to the present day.

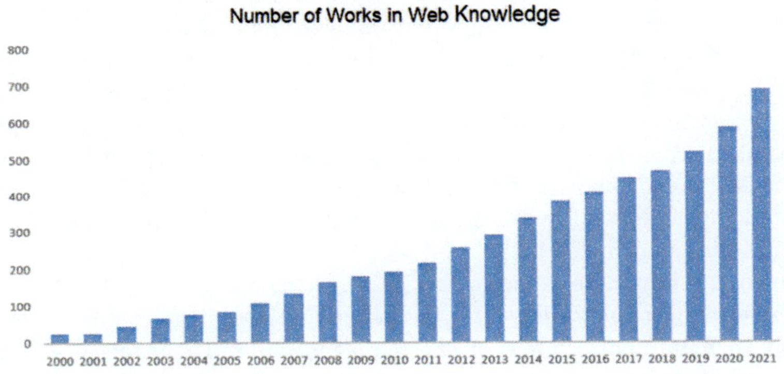

Fig. 1. Numbers of works in the web of knowledge including quadruped robots for search and rescue as topic.

Within the first phases of the study and analysis of quadrupedal robots, explorations and balance studies have been carried out [16], the behaviour of gait modes, trajectory planning, and commercial platforms such as ANYmal [17].

The variability of scenarios and complex tasks demand the remarkable capacity of robotic systems. It is convenient to operate with suitable actuators; depending on the field and the execution [18], there are differences between electrical, hydraulic, and pneumatic actuators. Regarding handling, in [19], it is about exposing the added difficulty that deformable objects have, a subject to study due to the uncertainty in SAR tasks and the variety of characteristics that can be found.

The teleoperation of the set is added as a complement to the control, also studied in this article, with references to previous works such as [20], where the problem of communication delays is also shown.

Implementations have been carried out with quadrupedal robots for search and rescue tasks focused on different objectives such as facilitating objects to the required place, picking up objects and taking them to a safe place or just obtaining data.

The environments that the robot faces can be very hostile. In addition to the topology of the terrain, they also affect its characteristics. Another studies works on an adaptive control to all types of surfaces, such as snow, water or leafy vegetation, making it difficult to perceive the scene [21].

Work times in the field can be very long, so good management and optimization of the energy consumed [22] is key to tackling SAR tasks. In case of having to cover much ground or collect very different types of data, it is interesting to study the use of fleets of robots. Studies have been carried out in terms of planning, assistance and communication with the human in these cases in which several quadrupeds are worked on simultaneously [23].

Among the most recent developments in this field are ALMA and RoLoMa, which use the ANYmal quadruped robot as a base, equipped with a 6 DoF manipulator ($Jaco^2$), on which different analyzes have been carried out, mainly dynamic and stability, as well as door opening tasks [24,25]. For all types of loads, it is necessary to have control over the influence of the forces generated by the arm when transporting them and also when carrying out the pick and place task, as studied in the article [26]. While other centaur-type robotic systems have focused on locomotion adaptation with heavy load [27]. For this type of task, implementations of hydraulic arms have been studied in conjunction with also hydraulic quadrupeds [28].

Due to the innumerable scenarios in which to carry out tasks, we have to minimize the uncertainty and errors that are made due to ignorance of the object we want to manipulate. Some of the main methods to control these complex systems use deep learning, as in the development carried out in [2], which uses the ALMA robot based on the ANYmal business model. Studies such as the one by P. Mitrano in the journal "ScienceRobotics" [29] deal with the problem of operating with deformable objects or with unexpected characteristics in order to know if the model implemented in the robot is valid.

There are many works on the control and modelling of manipulative quadrupeds, but the way of interacting and obtaining feedback is in a less developed phase. Remote control or touch control [30] are the most common. The main drawback in the development of teleoperation is the delays that may occur in the set of peripherals since they affect the model and it ceases to be robust. There is some study that integrates inertial sensors and virtual reality gloves [20], but it is the articles and field of study where, for the moment, less has been investigated.

3 Materials and Methods

3.1 Materials

The experimental development has been carried out in the facilities of the Automatic and Robotics Center of the UPM. The Robotics platform used for this development is a quadruped robot of the Unitree trademark, equipped with a sensory system and a robotic arm described in Table 1.

Table 1. Elements and components of the quadruped robot and M-R system.

Component	Amount	Description
Hololens (1st generation)	1	Mixed-reality glasses
Unitree A1	1	Quadruped robot
RP-Lidar	1	Lidar sensor
WX250S	1	6DoF robot arm
Nvidia Jetson Xavier-Nx	1	On-board embedded system
Real-sense	1	RGB-Deph sensor

3.2 Robot Design and Modeling

The modelling and design of the set have been developed in Matlab. The directing kinematic models have been obtained independently for the manipulator and one leg of the quadruped. In both cases, the Denavit-Hartenberg parameters are calculated; they are shown in the Tables 2 and 3 for the convenient reference system. The correct location of each leg of the model has been carried out using translations and rotations over Table 2.

The local reference systems of each leg and the robotic arm is located in the robot torso, as shown in Fig. 2(a). The entire assembly reference system has been placed in the centre of the A1 body, below the arm system.

Table 2. One leg Denavit-Hartenberg's parameters. 3DOF

Thetha (deg)	d (mm)	a (mm)	Alpha (deg)
q1 + 90	55	0	−90
q2	55	200	0
q3	0	215	0

Table 3. 6DOF manipulator's parameters

Thetha (rad)	d (mm)	a (mm)	Alpha (rad)
q1	104.43	0	pi/2
q2 + pi/2 − 0.2	0	254.95	0
q3 + 0.2	0	0	pi/2
q4	250	0	−pi/2
q5	0	0	pi/2
q6	131	0	0

Both the robotic arm and the height and position of the quadruped are predominant in determining the optimal workspace. This is created from iterating and saving the values (x, y, z) obtained from slight variations in the 6 degrees of freedom that the arm has.

Kinematic-ally there is access to a spherical but flattened shape, which covers from a radius of 650 mm to 740 mm. It is shown in red colour in Figs. 2(a)–(b). According to the manufacturer's recommendation regarding the optimal work zone, it is recommended to work at 70% of the maximum range; this is shown in the green zone.

There is a high risk of collision with the robot's head or rear. The different colours of the points in Fig. 2 were established to differentiate the manipulation zones that the robot can reach. Therefore, the green area is optimal for carrying out manipulation tasks. However, there are areas where it has not been considered for manipulation due to the manufacturer's recommendation or the generation of high torques. This is the case in the red zone.

Finally, and for stability reasons, it has been considered that the manipulator has more performance in the lateral areas than in the front and rear. An ellipsoid with its semi-major axis in the transverse direction of A1 is shown in blue.

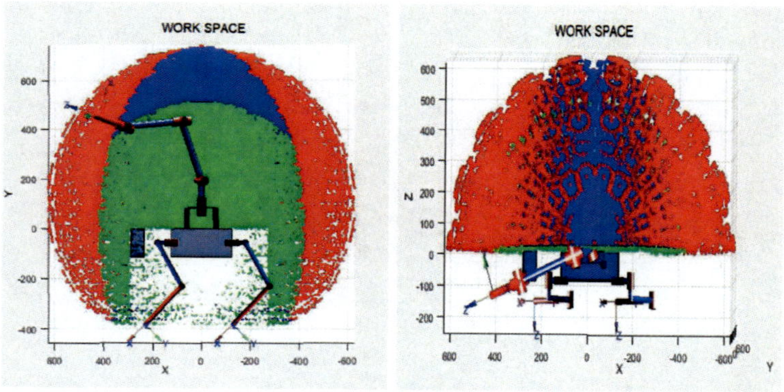

(a) Lateral Workspace view of the Model. (b) Upper Workspace view of the Model.

Fig. 2. Kinematic model and workspace of WX250S manipulator integrated to Unitree A1.

3.3 ROS-Gazebo Implementation and Simulation

Several test environments have been developed in ROS-Gazebo to test the movement of the entire kinematic assembly and the manipulator movement planning systems. Figures 3(a)–(c) show the simulated environments recreated based on post-disaster scenarios. These environments include small debris, unstructured soil, victims, and the quadruped robot model and the robotic arm integrated through an assembly carried out in *.xacro* files. Figures 3(b)–(d) correspond to the perception-planning systems of the respective Gazebo environments, where the robot model is shown in the current position (in black), the target position of the arm (in Orange).

(a) Simulation of the robotic system in Gazebo (indoors). (b) Planning of movements in MoveIt.

(c) Simulation of the robotic system in Gazebo (outdoors). (d) Planning of movements in MoveIt.

Fig. 3. Simulation of the quadruped robot together with the manipulator and Movement Planning (MoveIt) in different Gazebo environments for the execution of manipulation tests.

3.4 Mixed-Reality Teleoperation

Figure 4 shows the operating scheme of the teleoperation system. The operator uses Hololens glasses as a means of human-robot-environment interaction. In the glasses, both the physical environment and the hologram model of the robot are observed. It has a blue sphere at its end, which the user can manipulate through gestures to move the arm towards an area of interest. The joint positions of the simulated robot are sent to the controller of the real robot to execute the movement.

The movements of the fingers have been detected due to the functions of Holo-Toolkit, precisely the Manipulation Mode function. Which captures the union of the fingers showing the interaction with the virtual objects (end of the robot), then the inverse kinematics is calculated, and the positions are sent to the real robot via Rosbridge.

The mixed reality interface also has buttons that allow the gripper to open-close and execute movements to predefined positions (home and sleep) and a small screen where the image of the robot is transmitted outdoors. All these buttons can be manipulated by the joint gesture of the fingers (thumb-index) of the operator wearing the glasses.

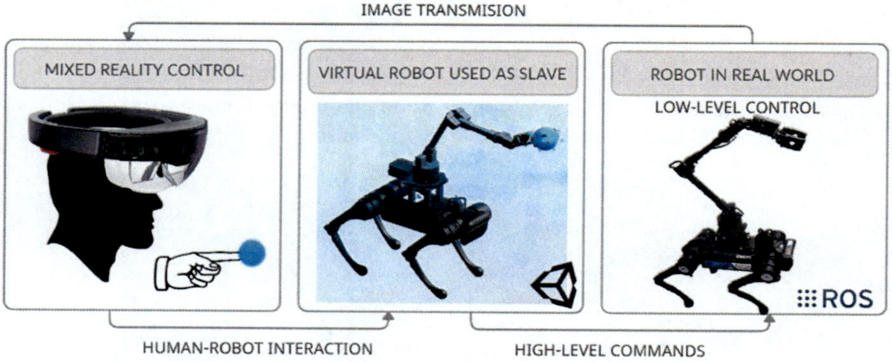

Fig. 4. Systematic scheme of teleoperation of the robot-manipulator through mixed reality.

The strategy contemplated for transporting first aid equipment using the implemented robot and the M-R system is based on an approach phase, where the robot with the arm in the collected position without load moves, minimizing its gravity centre towards point A. As long as it is not being manipulated, the ideal is to pick up the manipulator and work with a control model that dampens the oscillations generated in the quadruped body since it is the area where it joins the arm.

Once in the area, the user moves the arm remotely (at 5 m maximum distance) until taking the object with the gripper. The lateral posture is the most favourable to proceeding to load an object since the quadruped can adopt configurations with more stable joints to support the object's weight. This is due to the approach distances necessary to pick-place since they are longer longitudinally and generate more torque.

Once the object has been picked up, the arm acquires the transport position, reducing the gravity centre until the robot moves towards the victim's area (point B). Finally, there is teleoperated to deliver the equipment.

4 Experiments and Results

The development of tests to validate the functionality of the joint robotic system implemented (Quadruped Robot + Robotic Arm) and the functionality of the teleoperation system. The main tests to analyze the functionality of the robotic system have been based on manipulating objects with different weights, taking them from point A to point B.

The teleoperation system through mixed reality has been evaluated based on the versatility to control of the robotic system through teleoperation and has been contrasted with conventional interfaces. The performance tests can be seen in more detail in Appendix A.

4.1 Evaluation of the Mixed Reality System for Teleoperation

Tests have been developed with two types of interfaces. The first is the M-R Teleoperation system, and the second is a conventional interface (RVIZ); both have the same purpose, to perform movements on the manipulator robot of the robotic system, considering the workspace and the self-collision. Testers have used both interfaces to perform the same task, which is to move the end of the robot from point A to point B, marked with an X on the ground.

Figures 5(a)–(b)–(c) show the hand of the person taking the blue ball and performing the movements in space. These Figures show the superimposition of the robots' virtual model (grey colour) on the real robot (black colour); in such a way that as the blue ball moves, the virtual model adjusts the kinematic robots' positions to send them to the real robot, carrying out the teleoperation. In this first phase of experimentation, up to 200 ms in teleoperation has been delayed.

Figures 5(d)–(e) show the box diagrams of the data collected, where participants' experiences show that the implemented system generates a better experience for immersive situations by 15% and increases confidence when executing movements with the robotic arm by 23% compared to conventional interfaces.

4.2 Pick and Place Test Using M-R System

The next phase of experiments consisted of moving the robot's end from point A, taking a medical-type object and depositing it at point B, marked with an X on the ground. The transported objects are a first aid kit and a bottle of alcohol for healing.

Figure 6 [a–d] shows the different stages to transport a first aid kit, starting from $t = 0$ s. The first test object weights 100 g. The object has been reached and picked up in the first 10 s, while to reach the destination position and drop it, an additional 10 s has been required.

On the other hand, Fig. 6 [e–h] shows the steps to take from point A to point B, a bottle of alcohol weighing 150 g, unlike the previous test object, it is held in the first 7 s due to its cylindrical shape, but its transport to the destination point takes 12 s more because the operator requires greater precision in the movements to avoid the fall of the object with liquid that can damage the robot.

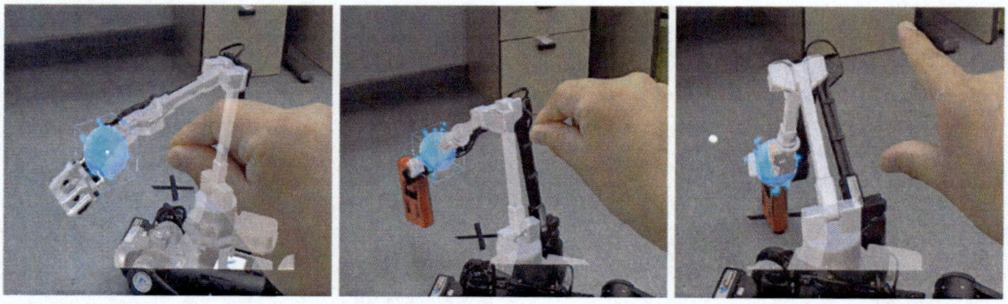

(a) Teleoperation in home position. (b) Teleoperation in 2nd Pose. (c) Teleoperation to final position.

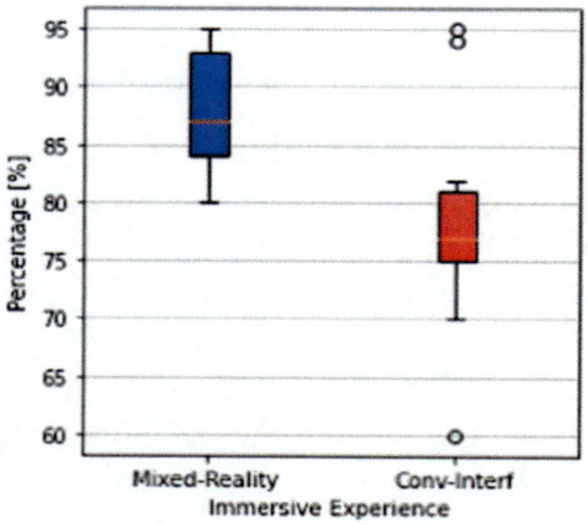

(d) Immersive experience analysis.

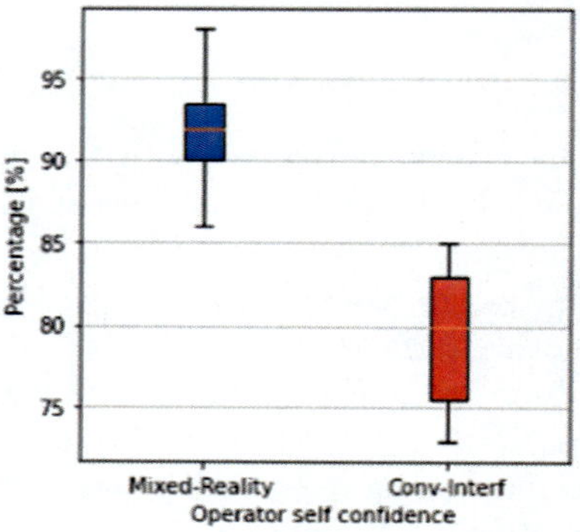

(e) Operator confidence analysis.

Fig. 5. Tests to evaluate the Mixed-Reality efficiency system, the operator holds the robots end (blue ball) and using his gestures, moves it towards the destination (X Mark).

Therefore, the greater the weight or, the more delicate it is, the operator will require more precise movements and the demand for time to carry out these tasks will increase proportionally to the weight.

4.3 Victim Assistance Test in Reconstructed Post-disaster Environments

The third phase of tests has been carried out in a test environment reconstructed according to NIST specifications, consisting of small debris and victims (mannequins and people). In this scenario, the task carries a first-aid kit from point A to point B (victim). Figures 7 [a–c] show the different stages of the trajectory executed by the robot. In the first instance, it approaches and takes the first aid kit, places the arm in the transport position and moves through the environment until reaching the victim. Finally, he extends his arm and drops the package near the victim. The entire process is teleoperated by the robot from the remote control and the arm from the M-R system.

(a) Home pose (b) Object capture (c) Object transport (d) Object deposit

(e) Approach Zone (f) Object capture (g) Object transport (h) Object deposit

Fig. 6. Tests of transportation medical assistance objects (First kit aid [a–d] and alcohol [e–h]), through M-R Teleoperation.

(a) Pose Robot tacking first kit aid. (b) Transporting to destination. (c) Kit delivery to the victim.

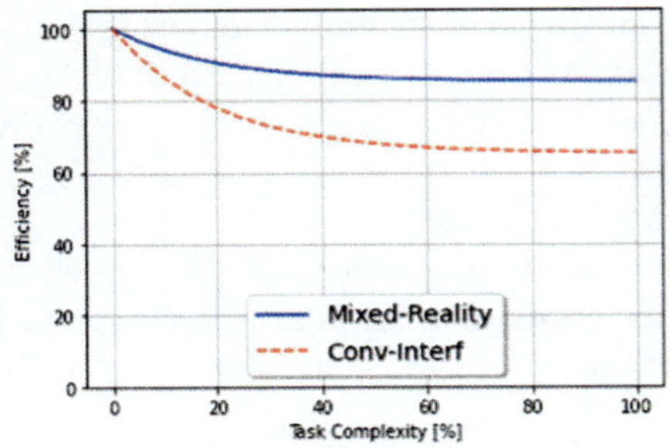

(d) Task complexly evaluation.

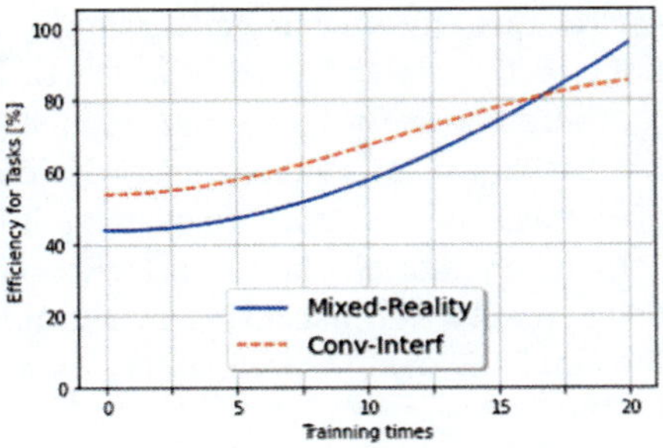

(e) Efficiency based on training.

Fig. 7. Teleoperation and metrics evaluation R-M teleoperation system for robot's assistance with medical equipment for a victim.

Based on the experiences collected by the operators, the following graphs have been established (Figs. 7(d)–(e)), which describe the efficiency of executing tasks based on the operators' complexity and previous experience.

Figure 7(d) shows the efficiency evolution based on the task complexity (defines if a task is easy to perform or not) for both cases, executions with Mixed Reality and conventional interface. Both show high efficiency for simple tasks like moving the arm from one point to another. However, for more complex tasks such as object manipulation, the mixed reality system is superior by 21%.

Figure 7(e) represents the evolution of the efficiency to execute tasks based on training with both types of interfaces. Initially, the conventional interfaces show to be better since they are intuitive. However, after a series of training sessions (20 tests), both interfaces are better. However, there is a turning point where mixed reality surpasses the conventional. This is due to several factors, one of which is the immersive experience generated by using these devices.

5 Conclusions

The implementation of the quadruped-manipulator joint kinematic modelling in Matlab has allowed calculating and optimizing the work area of the manipulator robot in such a way that an optimal zone has been established based on the kinematic characteristics and approach strategies of the robot for object manipulation.

The ROS simulation environment (Gazebo) has shown great potential and utility to recreate physical conditions and integrate the joint model designed. This has made it possible to validate the functionality before physical integration and the execution of movements without manipulators collision in the workspace, using the MoveIt packages.

Mixed-Reality interfaces as a means of teleoperation concerning conventional interfaces have shown potential advantages. Mainly in the increased confidence of operators for decision making and the significant improvement in efficiency to perform complex handling tasks with objects with variable geometry and weight.

Previous experience and training of operators with mixed reality greatly influence the efficiency with which tasks are executed. Initially, the interfaces used were better for inexperienced operators. Still, as a progressive training, the evolutionary results place the M-R interfaces as the most suitable solution to teleoperate complex robotic systems in search and rescue tasks.

In future works, the line is left open to evaluating other interfaces such as virtual reality and executing synchronized movements between the quadruped robot and the arm to perform the same tasks.

Appendix A

Video of Teleoperation tests through the Mixed-Reality Execution system for the implemented robot https://www.youtube.com/watch?v=_uN22XUTWjo.

References

1. Chai, H., et al.: A survey of the development of quadruped robots: joint configuration, dynamic locomotion control method and mobile manipulation approach. Biomim. Intell. Robot. **2**(1), 100029 (2022)
2. Ma, Y., Farshidian, F., Miki, T., Lee, J., Hutter, M.: Combining learning-based locomotion policy with model-based manipulation for legged mobile manipulators. IEEE Robot. Autom. Lett. **7**(2), 2377–2384 (2022)
3. Ubellacker, W., Csomay-Shanklin, N., Molnar, T.G., Ames, A.D.: Verifying safe transitions between dynamic motion primitives on legged robots. In: 2021 IEEE/RSJ International Conference on Intelligent Robots and Systems (IROS), pp. 8477–8484 (2021)
4. Ulloa, C.C., Sánchez, G.P., Barrientos, A., Del Cerro, J.: Autonomous thermal vision robotic system for victims recognition in search and rescue missions. Sensors **21**(21), 7346 (2021)
5. Cardona, G.A., Ramirez-Rugeles, J., Mojica-Nava, E., Calderon, J.M.: Visual victim detection and quadrotor-swarm coordination control in search and rescue environment. Int. J. Electr. Comput. Eng. (2088-8708) **11**(3), 2079 (2021)
6. Hoshino, W., Seo, J., Yamazaki, Y.: A study for detecting disaster victims using multi-copter drone with a thermographic camera and image object recognition by SSD. In: 2021 IEEE/ASME International Conference on Advanced Intelligent Mechatronics (AIM), pp. 162–167. IEEE (2021)
7. Subbayamma, B.V., Sushma, T.: Human detection robot in rescue operations. Int. Res. J. Innov. Eng. Technol. **5**(6), 122 (2021)
8. Tammana, A., Amogh, M.P., Gagan, B., Anuradha, M., Vanamala, H.R.: Thermal image processing and analysis for surveillance UAVs. In: Kaiser, M.S., Xie, J., Rathore, V.S. (eds.) Information and Communication Technology for Competitive Strategies (ICTCS 2020). LNNS, vol. 190, pp. 577–585. Springer, Singapore (2021). https://doi.org/10.1007/978-981-16-0882-7_50
9. Ilikci, B., Chen, L., Cho, H., Liu, Q.: Heat-map based emotion and face recognition from thermal images. In: 2019 Computing, Communications and IoT Applications (ComComAp), pp. 449–453. IEEE (2019)
10. Park, J., Chen, J., Cho, Y.K., Kang, D.Y., Son, B.J.: CNN-based person detection using infrared images for night-time intrusion warning systems. Sensors **20**(1), 34 (2019)
11. Ismail, R., Muthukumaraswamy, S.: Military reconnaissance and rescue robot with real-time object detection. In: Reddy, A.N.R., Marla, D., Favorskaya, M.N., Satapathy, S.C. (eds.) Intelligent Manufacturing and Energy Sustainability. SIST, vol. 213, pp. 637–648. Springer, Singapore (2021). https://doi.org/10.1007/978-981-33-4443-3_61
12. You, Y., et al.: State estimation for hybrid wheeled-legged robots performing mobile manipulation tasks. In: 2021 IEEE International Conference on Robotics and Automation (ICRA), pp. 3019–3025 (2021)
13. IIT: Robot teleoperative in collaboration with INAIL (2021)
14. Vyacheslav Lyashenko, M., Ahmad, A., Belova, N., Sotnik, S.: Modern walking robots: a brief overview. Int. J. Recent Technol. Appl. Sci. **3**(2), 32–39 (2021)
15. Chai, H., et al.: A survey of the development of quadruped robots: joint configuration, dynamic locomotion control method and mobile manipulation approach. Biomim. Intell. Robot. **2**(1), 100029 (2022)

16. Sun, W., Tian, X., Song, Y., Pang, B., Yuan, X., Qingyang, X.: Balance control of a quadruped robot based on foot fall adjustment. Appl. Sci. **12**(5), 2521 (2022)
17. Bellicoso, D., Jenelten, F., Gehring, C., Hutter, M.: Dynamic locomotion through online nonlinear motion optimization for quadrupedal robots. IEEE Robot. Autom. Lett. **3**, 2261–2268 (2018)
18. Hussain, K., Omar, Z., Wang, X., Adewale, O.O., Elnour, M.: Analysis and research of quadruped robot's actuators: a review. Int. J. Mech. Eng. Robot. Res. **10**(8), 436–442 (2021)
19. Zhu, J., et al.: Challenges and outlook in robotic manipulation of deformable objects. arXiv preprint arXiv:2105.01767 (2021)
20. Peers, C., Motawei, M., Richardson, R., Zhou, C.: Development of a teleoperative quadrupedal manipulator. In: UK-RAS21 Conference: "Robotics at Home" Proceedings, Leeds (2021)
21. Miki, T., Lee, J., Hwangbo, J., Wellhausen, L., Koltun, V., Hutter, M.: Learning robust perceptive locomotion for quadrupedal robots in the wild. Sci. Robot. **7**(62), eabk2822 (2022)
22. Liao, T., Ye, S., Chen, L., Sun, C., Zhang, A.: Energy efficient swing leg trajectory planning for quadruped robots walking on rough terrain. In: 2019 IEEE International Conference on Robotics and Biomimetics (ROBIO), pp. 2128–2133. IEEE (2019)
23. Kaufmann, M., et al.: Copilot mike: an autonomous assistant for multi-robot operations in cave exploration. In: 2021 IEEE Aerospace Conference (50100), pp. 1–9. IEEE (2021)
24. Ferrolho, H., Ivan, V., Merkt, W., Havoutis, I., Vijayakumar, S.: RoLoMa: robust loco-manipulation for quadruped robots with arms. arXiv preprint arXiv:2203.01446 (2022)
25. Dario Bellicoso, C., et al.: Alma - articulated locomotion and manipulation for a torque-controllable robot. In: 2019 International Conference on Robotics and Automation (ICRA), pp. 8477–8483 (2019)
26. Ewen, P., Sleiman, J.-P., Chen, Y., Lu, W.-C., Hutter, M., Vasudevan, R.: Generating continuous motion and force plans in real-time for legged mobile manipulation. In: 2021 IEEE International Conference on Robotics and Automation (ICRA), pp. 4933–4939. IEEE (2021)
27. Zhao, X., You, Y., Laurenzi, A., Kashiri, N., Tsagarakis, N.: Locomotion adaptation in heavy payload transportation tasks with the quadruped robot centauro. In: 2021 IEEE International Conference on Robotics and Automation (ICRA), pp. 5028–5034 (2021)
28. Semini, C., et al.: Design overview of the hydraulic quadruped robots. In: The Fourteenth Scandinavian International Conference on Fluid Power, pp. 20–22 (2015)
29. Mitrano, P., McConachie, D., Berenson, D.: Learning where to trust unreliable models in an unstructured world for deformable object manipulation. Sci. Robot. **6**(54), eabd8170 (2021)
30. Risiglione, M., Sleiman, J.-P., Minniti, M.V., Cizmeci, B., Dresscher, D., Hutter, M.: Passivity-based control for haptic teleoperation of a legged manipulator in presence of time-delays. In: 2021 IEEE/RSJ International Conference on Intelligent Robots and Systems (IROS), pp. 5276–5281. IEEE (2021)

On the Kinematic Characteristics of Walking Mechanisms of Mobile Robots with Walking Propulsion Devices

E. S. Briskin[1,2], N. G. Sharonov[1,2], L. D. Smirnaya[1,2(✉)], and K. S. Artemyev[1]

[1] Department of Theoretical Mechanics, Volgograd State Technical University, Lenin av., 28, Volgograd 400005, Russia
dtm@vstu.ru

[2] Center for Technology Components of Robotics and Mechatronics, Innopolis University, Universitetskaya str., 1, Innopolis 420500, Russia

Abstract. Various kinematic schemes of walking propulsion devices of mobile robots with different number of drives are analyzed. The errors in the implementation of the program movement of the propulsion device foot are estimated with the help transfer functions. With the use of transfer functions, matrices characterizing the specific mechanism of the mover are applied. The scalar indicator of the average accuracy of the realization of both position and speed of the foot in the program mode of movement is con-sidered. A specific "zoomorphic" propulsion device is considered and an estimate of its accuracy is given depending on the size of the links. The obtained dependences allow the formulation of the problem of optimization of geometric parameters of mechanisms.

Keyword: Mechanism. Propulsion device. Matrix of transfer functions. Scalar criterion

1 Introduction

Mobile robots with walking propulsion devices are known, which differ from each other both in the number of control drives and in their kinematic schemes.

Single-drive walking mechanisms, as a rule, are cyclic mechanisms, which are characterized by a well-defined trajectory of the supporting foot relative to the body of the walking robot. An example of such robots is the robot "Vosminog" [1] (Fig. 1) with a four-link walking mechanism equipped with a corrector gearbox.

The principle of operation of such a propulsion device and similar ones [2] is based on the control of the rotational movement of the driving crank. Therefore, despite the well-defined trajectory of the foot, it is possible to control the law of its movement of the trajectory by programmatically changing the angular velocity of the leading link - the crank. These methods can be provided both by using additional mechanisms included in the kinematic chain of the propulsion device drive [3], and by controlling the movement of the drive motor.

© The Author(s), under exclusive license to Springer Nature Switzerland AG 2023
J. M. Cascalho et al. (Eds.): CLAWAR 2022, LNNS 530, pp. 195–203, 2023.
https://doi.org/10.1007/978-3-031-15226-9_20

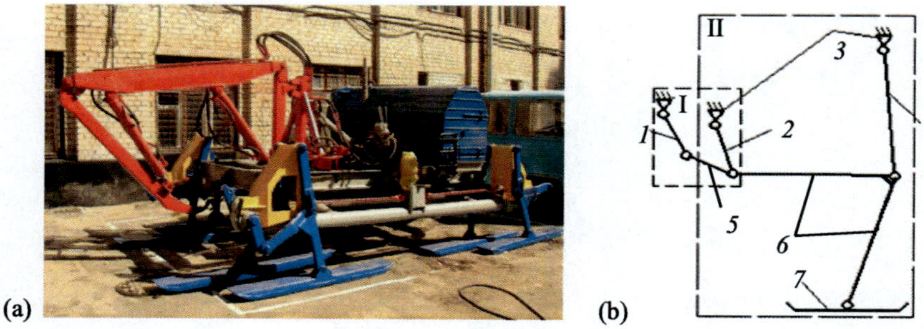

Fig. 1. Walking robot "Vosminog": (a) a prototype; (b) kinematic scheme of a four-link walking mechanism with a corrector gearbox in the form of a four-link mechanism: I - corrector gearbox, II - walking mechanism, *1* - driving crank of the gearbox, *2* - driving crank of the walking mechanism (driven crank of the corrector reducer), *3* - base, *4* - rocker, *5* - connecting rod of the corrector gearbox, *6* - crankshaft of the walking mechanism, *7* - foot

Another example of such propulsion devices are anchor-rope propulsion devices that provide propulsion for underwater pontoon robots with little positive buoyancy. The principle of operation of such "walking-like" propulsion devices is based on pulling the robot-pontoon to the anchors with their subsequent transfer to a new position [4]. However, two drive motors must be used for transfer.

Walking mechanisms are known, in which two drive motors are used in traction mode to carry out translational motion with rectilinear movement of the center of mass, and kinematic schemes correspond to zoomorphic (Fig. 2) [5] or orthogonal propulsion devices (Fig. 3) [6], which are distinguished by a series chain of drives.

Two-drive propulsion devices with a parallel chain of drives can also be used (Fig. 4). Features of such propulsion devices are simultaneous coordinated control of two drives.

Walking propulsion devices, driven by three drives to ensure the translational movement of the robot with the rectilinear movement of its center of mass [7] are also known. Among them, insectomorphic propulsion devices (Fig. 5) are most widely used, the supporting feet of which are controlled by three drives.

One of the most important characteristics of all propulsion devices, due to their kinematic scheme, is to ensure the program movement of the foot with a minimum positioning error and an error in the implementation of its program speed caused by an error in the drive control system.

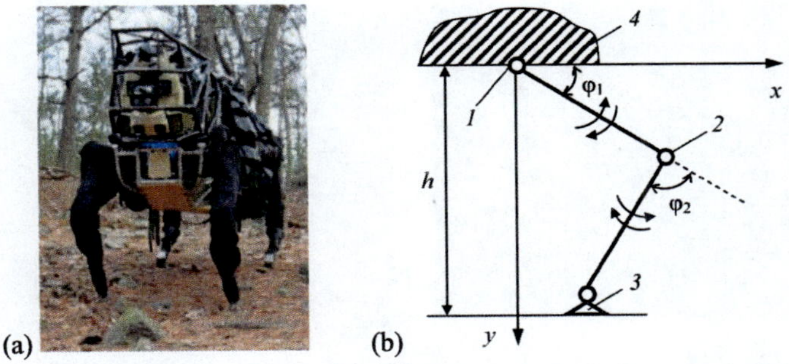

Fig. 2. Four-legged walking robot BigDog (USA, Boston Dynamics): (a) a prototype; (b) kinematic scheme of a zoomorphic propulsion device with a serial chain of drives for rotational movement: *1*, *2* - drives, *3* - foot, *4* - robot body.

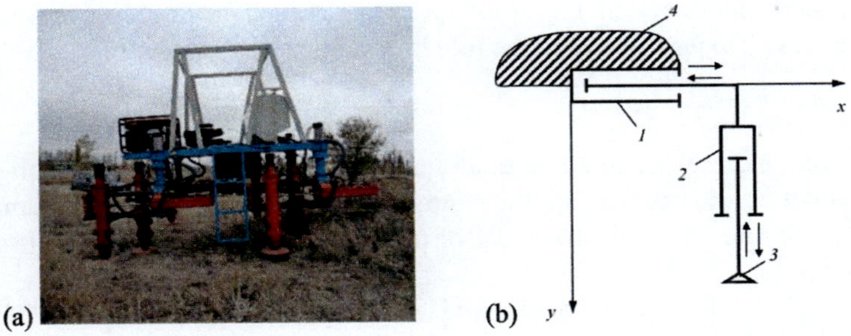

Fig. 3. Walking robot "Ortonog": (a) a prototype; (b) kinematic scheme of an orthogonal propulsion device with a serial chain of translational drives: *1*, *2* - drives, *3* - foot, *4* - robot body.

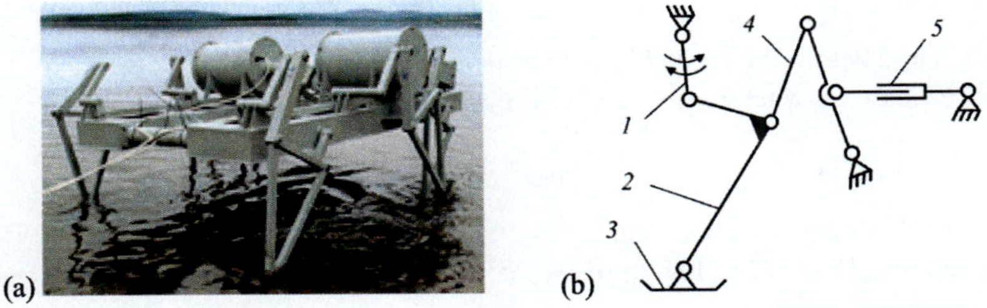

Fig. 4. Underwater walking robot "MAK-1": (a) a prototype; (b) kinematic scheme of the cyclic walking mechanism: *1* - driving crank, *2* - the curvilinear connecting rod of the walking mechanism, *3* - foot, *4* - rocker, *5* - drive to change the position of the rocker attachment

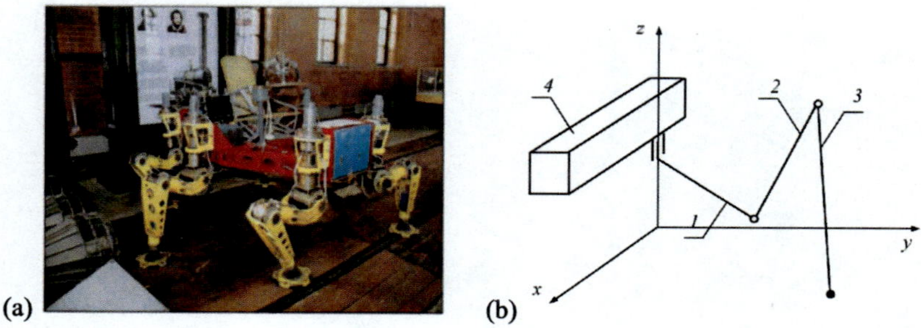

Fig. 5. NMSHA, IPM them. M.V. Keldysha: (a) a prototype; (b) kinematic diagram of the insectomorphic walking mechanism: *1, 2, 3* – links of the walking mechanism, *4* – robot body.

2 Formulation of the Problem

The propulsion control system must provide a programmed mode of foot movement. In the general case, the positioning of the foot is described by the functions

$$x^j = x^j(u^1, u^2 \ldots u^n), \quad j = 1, 2, 3; \quad n = 1 \ldots N, \tag{1}$$

where x^j are the Cartesian coordinates of the center of mass of the foot (its orientation in space is not considered), u^n are the control actions n of the drive, for example, the angle of rotation or linear displacement of the output element (crank or rod) n of the drive motor.

However, in addition to positioning, it is also required to provide the law of motion in time. Indeed, as is known [8], the state of a mechanical system is determined by the coordinates and velocities of its points. To do this, it is necessary to take into account the speed of the center of mass of the foot as a function of time.

$$\dot{x}^j = \frac{\partial x^j}{\partial u^n}\dot{u}^n. \tag{2}$$

In (2) and below, the repeating indices denote the summation over them.

For each drive, when implementing program control

$$u^k_{pr} = u^k_{pr}(t). \tag{3}$$

the error δu^k is taken into account.

Equations (1) and (2) allow us to estimate the error in the positioning δx^j of the foot and its program speed $\delta \dot{x}^j$

$$\delta x^j = \frac{\partial x^j}{\partial u^n}\delta u^n, \qquad \delta \dot{x}^j = \frac{\partial^2 x^j}{\partial u^n \partial u^k}\dot{u}^n \delta u^k + \frac{\partial x^j}{\partial u^n}\delta \dot{u}^n. \tag{4}$$

Thus, these errors that affect the stability of the program modes of motion of the walking propulsion device can be caused both by errors in the control system δu, $\delta \dot{u}$ and by the kinematic scheme of propulsion devices, which is described by the functions

$$\psi^j_n = \frac{\partial x^j}{\partial u^n}, \qquad \lambda^j_{nk} = \frac{\partial \psi^j_n}{\partial u^k} = \frac{\partial^2 x^j}{\partial u^n \partial u^k}. \tag{5}$$

Functions (5) for $j = n = k = 1$ are defined as transfer functions of the corresponding order and are introduced for cyclic mechanisms [9]. For walking propulsion devices controlled by several drives $N > 1$, (5) is a system of transfer functions that characterize the propulsion device mechanism.

Thus, the task is to assess the kinematic accuracy of the program foot movement of the walking mechanism based on transfer functions (5).

3 Solution Method

Functions (5) allow to estimate the integral errors in position I_{nk} and speed Φ_{nk} of the foot of the walking mechanism in the form of constant elements of square matrices with a dimension equal to the number of drives

$$\begin{cases} I_{nk} = \frac{1}{\tau}\int_0^{\tau} \chi_{nk} dt, \\ \Phi_{nk} = \frac{1}{\tau}\int_0^{\tau} \nu_{nk} dt, \end{cases} \quad (6)$$

where τ is the time of one cycle, χ_{nk}, ν_{nk} are local components of integral error elements.

$$\begin{cases} \chi_{nk} = \delta_{jl}\psi_n^j\psi_k^l, \\ \nu_{nk} = \delta_{jl}\{(\lambda_{nk}^j \dot{u}^r + \psi_n^j)(\lambda_{kq}^l \dot{u}^q + \psi_k^l), \end{cases} \quad r, q = 1, 2 \ldots N \quad (7)$$

δ_{jl} is the Kronecker function

$$\delta_{jl} = \begin{cases} 1 \text{ at } l = j \\ 0 \text{ at } l \neq j \end{cases} \quad (8)$$

A feature of the proposed estimates, in particular ν_{nk}, is the dependence on the speed of program control of drives $\dot{u}^r = \dot{u}^r(t)$.

The geometric and kinematic meanings of the introduced estimates I_{nk}, Φ_{nk} by analogy with the geometric space can be defined, respectively, as the integral components of the fundamental metric tensors [10] in the space of given controlled program movements of the drives $u^n(t)$ and program speeds $\dot{u}^n(t)$ or as the squared deviations in the position and speed of the foot, due to errors in the operation of n and k drives, depending on these kinematic schemes.

Evaluation of the kinematic schemes of propulsion devices according to these indicators is carried out by introducing a scalar function G, built on the basis of the estimates and errors in the program implementation of the movements of controlled drives δu^n, $\delta\dot{u}^n$.

Then

$$G = I_{nk}\delta u^n \delta u^k + \Phi_{nk}\delta \dot{u}^n \delta \dot{u}^k. \tag{9}$$

4 Kinematic Analysis of the Zoomorphic Propulsion Device During the Translational Movement of the Robot with Rectilinear Movement of Its Center of Mass at the Stage of Interaction of the Foot with the Supporting Surface

A zoomorphic propulsion device (Fig. 2) ($n = k = 1, 2$) with link lengths l_1 and l_2 and designed for operation in the "pushing" stable mode [11] is considered. Therefore, for the initial conditions, the equations take place

$$\begin{cases} x = x^1(0) = 0, \\ y = x^2(0) = h. \end{cases} \tag{10}$$

Then, considering the movement of the robot with a speed V_0 and that $\varphi_1 = u^1$ and $\varphi_2 = u^2$

$$\begin{cases} x^1 = l_1 \cos u^1 + l_2 \cos\left(u^1 + u^2\right) = V_0 t, \\ x^2 = l_1 \sin u^1 + l_2 \sin\left(u^1 + u^2\right) = h. \end{cases} \tag{11}$$

The control actions u^1 and u^2 in the time interval $0 < t < \tau$ are determined from (11)

$$\tau < \sqrt{\frac{(l_1 + l_2)^2 - h^2}{V_0^2}}, \tag{12}$$

where τ is the time spent by the foot in interaction with the supporting surface.

$$\begin{aligned} u^1 &= \arcsin\left(\frac{V_0^2 t^2 + h^2 + l_1^2 - l_2^2}{2l_1\sqrt{h^2 + V_0^2 t^2}}\right) - \arcsin\left(\frac{V_0 t}{\sqrt{h^2 + V_0^2 t^2}}\right), \\ u^2 &= \arcsin\left(\frac{V_0^2 t^2 + h^2 + l_2^2 - l_1^2}{2l_2\sqrt{h^2 + V_0^2 t^2}}\right) - \arcsin\frac{V_0^2 t^2 + h^2 + l_1^2 - l_2^2}{2l_1\sqrt{h^2 + V_0^2 t^2}}. \end{aligned} \tag{13}$$

The graph (Fig. 6) shows the corresponding characteristic dependencies.

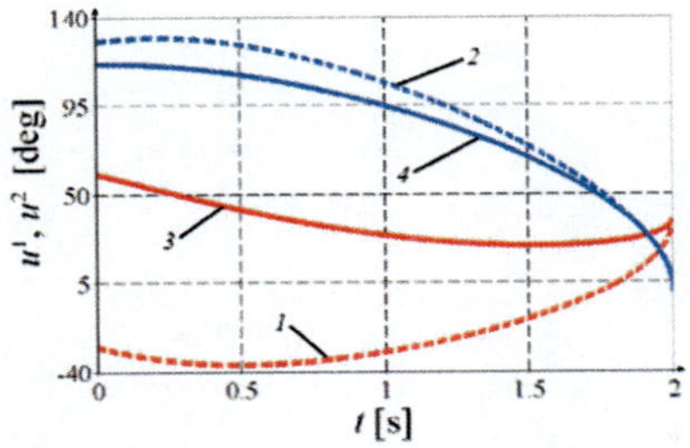

1 – u^1 at l_1=1/4 m, l_2=3/4 m;
2 – u^2 at l_1=1/4 m, l_2=3/4 m;
3 – u^1 at l_1=2/3 m, l_2=1/3 m;
4 – u^2 at l_1=2/3 m, l_2=1/3 m

Fig. 6. Dependences of program control actions u^1, u^2 on time at $h = 0.6$ m, $V_0 = 0.4$ m/s

The equations of motion (11) allow us to determine the functions (5)

$$\psi_1^1 = \partial x^1 / \partial u^1 = -l_1\left[\sin u^1 + sin\left(u^1 + u^2\right)\right];\quad \psi_2^1 = \partial x^1 / \partial u^2 = -l_1 sin\left(u^1 + u^2\right);$$
$$\psi_1^2 = \partial x^2 / \partial u^1 = l_2\left[\cos u^1 + l\cos\left(u^1 + u^2\right)\right];\quad \psi_2^2 = \partial x^2 / \partial u^2 = l_2\cos\left(u^1 + u^2\right);$$
$$\lambda_{11}^1 = \partial^2 x^1 / \partial u^1 \partial u^1 = -l_1\left[\cos u^1 + \cos\left(u^1 + u^2\right)\right];$$
$$\lambda_{12}^1 = \lambda_{21}^1 = \partial^2 x^1 / \partial u^1 \partial u^2 = -l_1\cos\left(u^1 + u^2\right);$$
$$\lambda_{22}^1 = \partial^2 x^1 / \partial u^2 \partial u^2 = -l_1\cos\left(u^1 + u^2\right);\quad \lambda_{11}^2 = -l_2\sin u^1 - l_2\sin\left(u^1 + u^2\right);$$
$$\lambda_{12}^2 = \lambda_{21}^2 = -l_2\cos\left(u^1 + u^2\right);\quad \lambda_{22}^2 = -l_2\sin\left(u^1 + u^2\right). \tag{14}$$

In turn, expressions (14) are the basis for calculating the elements of the matrices χ_{nk}, ν_{nk} (7), and integral estimates of the elements of these matrices I_{nk}, Φ_{nk} (6).

For instance,

$$\chi_{11} = l_1^2\left[\sin u^1 + \sin(u^1 + u^2)\right]^2 + l_2^2\left[\cos u^1 + \cos(u^1 + u^2)\right]^2,$$
$$\chi_{12} = \chi_{21} =$$
$$= l_1^2\left[\sin u^1 + \sin(u^1 + u^2)\right]\sin(u^1 + u^2) + l_2^2\left[\cos u^1 + \cos(u^1 + u^2)\right]\cos(u^1 + u^2),$$
$$\chi_{22} = l_1^2\sin^2(u^1 + u^2) + l_2^2\cos^2(u^1 + u^2). \tag{15}$$

The graphs (Fig. 7) show some dependences of the local components of the integral errors (15) for the considered mode of motion.

If we now ignore the speed $\delta\dot{u}^n$ errors, and assume the position errors are the same for all drives $\delta u^1 = \delta u^2$, then from (9) it follows

$$G = I_{nk}\delta u^n \delta u^k = (I_{11} + I_{12} + I_{21} + I_{22})\delta u^1 \delta u^2 \tag{16}$$

and up to a constant factor $\delta u^n \delta u^k$ in the first case, when $l_1 = 0.25$ m, $l_2 = 0.75$ m, $G_1 = 1.467$, and in the second when $l_1 = 0.666$ m, $l_2 = 0.333$ m, $G_2 = 1.895$. Therefore, according to the considered indicator, the first option is better than the second one.

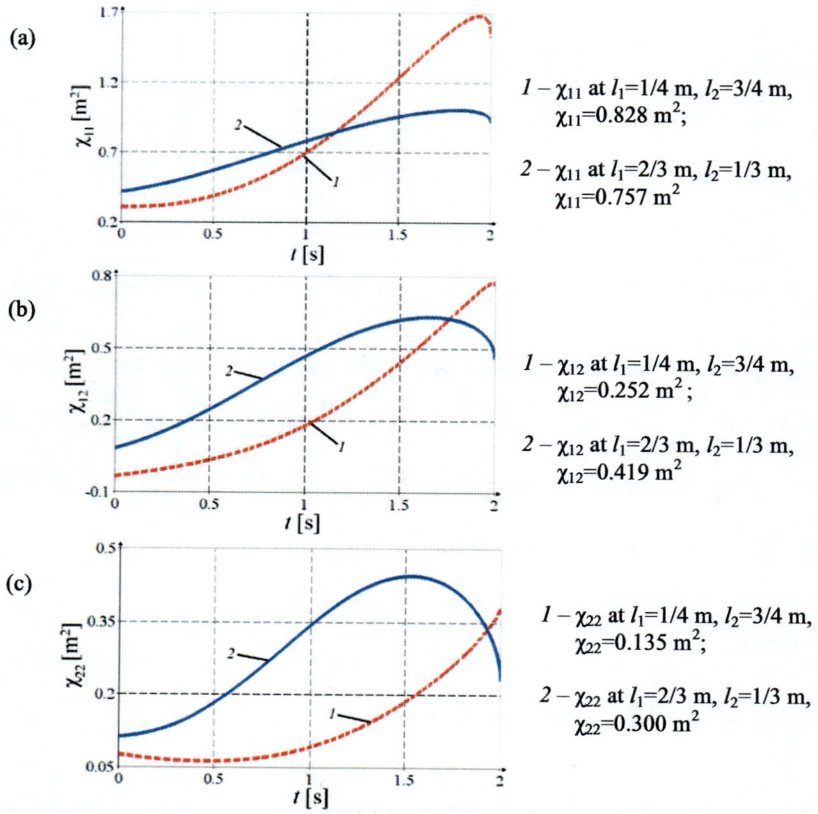

Fig. 7. Dependences of integral error components χ_{nk} on time: (a) $\chi_{11} = \chi_{11}(t)$; (b) $\chi_{12} = \chi_{12}(t)$; (c) $\chi_{22} = \chi_{22}(t)$.

However, the position changes if in some system of units $\delta u^1 = 3$, $\delta u^2 = 0.2$. Then in this system of units

$$G = 9I_{11} + 0.6I_{12} + 0.6I_{21} + 0.04I_{22} \tag{17}$$

Then in the first case $G_1 = 7.7598$ and in the second case $G_2 = 7.328$ and hence the second option is preferable.

5 Conclusion

The introduced scalar indicator of the accuracy of the kinematic scheme of the walking propulsion device mechanism makes it possible to compare mechanisms by this indicator. The introduced indicator is based on the transfer functions of the mechanism of the first and second order and their average values within the cycle. The geometric meaning of the scalar function G is the average square of the deviation of the foot in its actual movement from the program.

Acknowledgments. The study was supported by the Russian Science Foundation grant No. 22–21-00882.

References

1. Briskin, E.S., Chemyshev, V.V., Maloletov, A.V., Sherstobitov, S.V.: On Dynamics of Movement of Walking Machines with Gears on the Basis of Cycle Mechanisms. In: Morecki, A., Bianchi, G., Rzymkowski, C. (eds.) Romansy 13. ICMS, vol. 422, pp. 313–322. Springer, Vienna (2000). https://doi.org/10.1007/978-3-7091-2498-7_33
2. Bessonov, A.P., Umnov, N.V., Korenovsky, V.V., Silvestrov, E.E., Khoborkov, S.V.: Six Link Mechanisms for the Legs of Walking Machines. In: Morecki, A., Bianchi, G., Rzymkowski, C. (eds.) Romansy 13. ICMS, vol. 422, pp. 347–354. Springer, Vienna (2000). https://doi.org/10.1007/978-3-7091-2498-7_37
3. Briskin, E.S., Zhoga, V.V., Chernyshev, V.V., Maloletov, A.V.: Dynamics and motion control of walking machines with cyclic movers, p. 191. Mashinostroenie Publishers, Moscow (2009)
4. Briskin, E., Maloletov, A., Sharonov, N., Fomenko, S., Kalinin, Y., Leonard, A.: Development of Rotary Type Movers Discretely Interacting with Supporting Surface and Problems of Control Their Movement. In: Parenti-Castelli, V., Schiehlen, W. (eds.) ROMANSY 21 - Robot Design, Dynamics and Control. CICMS, vol. 569, pp. 351–359. Springer, Cham (2016). https://doi.org/10.1007/978-3-319-33714-2_39
5. Lucidarme, P., Delanoue, N., Mercier, F., Aoustin, Y., Chevallereau, C., Wenger, P.: Preliminary Survey of Backdrivable Linear Actuators for Humanoid Robots. In: Arakelian, V., Wenger, P. (eds.) ROMANSY 22 – Robot Design, Dynamics and Control. CICMS, vol. 584, pp. 304–313. Springer, Cham (2019). https://doi.org/10.1007/978-3-319-78963-7_39
6. Briskin, E., Kalinin, Y., Maloletov, A.: On Energetically Effective Modes of Walking Robots Movement. In: Arakelian, V., Wenger, P. (eds.) ROMANSY 22 – Robot Design, Dynamics and Control. CICMS, vol. 584, pp. 425–432. Springer, Cham (2019). https://doi.org/10.1007/978-3-319-78963-7_53
7. Russo, M., Ceccarelli, M., Cafolla, D., Matsuura, D., Takeda, Y.: An Experimental Characterization of a Parallel Mechanism for Robotic Legs. In: Arakelian, V., Wenger, P. (eds.) ROMANSY 22 – Robot Design, Dynamics and Control. CICMS, vol. 584, pp. 18–25. Springer, Cham (2019). https://doi.org/10.1007/978-3-319-78963-7_4
8. Landau, L.D., Lifshitz, E.M.: Theoretical physics. In: 10 volumes. Vol. I. Mechanics, p. 224. Fizmatlit, Moscow (2019)
9. Vulfson, I.I.: Dynamic calculations of cyclic mechanisms, p. 328. Mashinostroenie, Leningrad (1976)
10. McConnell, A.J.: Introduction to tensor analysis with applications to geometry, mechanics and physics, p. 411. Fizmatlit, Moscow (1963)
11. Briskin, E.S., Kalinin, Y.V., Artemyev, K.S.: On the stability of the plane movement of mobile robots with walking propulsion devices working in "Pulling" mode. Mekhatronika, Avtomatizatsiya, Upravlenie **22**(1), 28–34 (2021). (In Russ.) https://doi.org/10.17587/mau.22.28-34

Check for updates

Learning Energy-Efficient Trotting for Legged Robots

Athanasios Mastrogeorgiou(✉), Aristotelis Papatheodorou, Konstantinos Koutsoukis, and Evangelos Papadopoulos

School of Mechanical Engineering, National Technical University of Athens, Athens 15780, Greece
{amast,mc15407,kkoutsou,egpapado}@central.ntua.gr
https://csl-ep.mech.ntua.gr

Abstract. Quadrupedal locomotion skills are challenging to develop. In recent years, Deep Reinforcement Learning (DRL) promises to automate the development of locomotion controllers and map sensory observations to low-level actions. However, legged locomotion still is a challenging task for DRL algorithms, especially when energy efficiency is taken into consideration. In this paper, we propose a DRL scheme for efficient trotting applied on Laelaps II quadruped in MuJoCo. First, an accurate model of the robot is created by revealing the necessary parameters to be imported in the simulation, while special focus is given to the quadruped's drivetrain. Concerning, the reward function and the action space, we investigate the best way to integrate in the reward, the terms necessary to minimize the Cost of Transport (CoT) while maintaining a trotting locomotion pattern. Last, we present how our solution increased the energy efficiency for a simple task of trotting on level terrain similar to the treadmill-robot environment at the Control Systems Lab [1] of NTUA.

Keywords: Legged robots · Learning locomotion · Energy efficiency · Deep reinforcement learning

1 Introduction

Locomotion skills for quadruped robots require fast reactions, coordinated control of legs, precise manipulation of contact forces, and robust balance control. Setting up such controllers requires significant expertise and often tedious manual tuning. Data-driven methods, such as model-free DRL already have produced promising results showing that they can overcome the simplification of prior model-based approaches by learning effective controllers directly from experience [2–5]. According to [6], using a low-level toe trajectory planner, markedly reduces training time but requires an investigation on which gait parameters will be left for the agent to learn. Another aspect of DRL is the fact that learning based locomotion controllers usually focus on the target task to be

© The Author(s), under exclusive license to Springer Nature Switzerland AG 2023
J. M. Cascalho et al. (Eds.): CLAWAR 2022, LNNS 530, pp. 204–215, 2023.
https://doi.org/10.1007/978-3-031-15226-9_21

achieved without investigating the energy consumption during that task [7], or they just penalise high joint accelerations [3,8] without mechanical antagonism in mind [9]. Additionally, achieving a gait such as trotting, requires accurate orchestration of the robot legs, i.e., a task harder to train in joint space since it requires even more training data, while during training it may produce invalid configurations and thus, special care is needed to avoid this [8,10]. As a result, learning energy efficient locomotion skills for legged robots while maintaining a well orchestrated locomotion pattern has room for improvement especially when dynamic motions are involved. In this work, the trotting motion is investigated so that the quadruped robot Laelaps II [1] is able to move forward on level terrain, without diverging from its goal, while achieving reduced energy consumption.

1.1 Contribution and Overview

We propose learning efficient controllers directly in Cartesian space by selecting specific toe trajectory attributes while maintaining a trotting pattern as well as constant toe clearance from the ground. It is the first time that DRL with energy efficiency in mind is investigated for Laelaps II quadruped. As a result, it is shown that the Cost of Transport (CoT) is significantly reduced with the proposed reward function even when compared to similar approaches [3,8], since in this work actuator mechanical antagonism and electric loses are taken into account in the reward function.

The rest of the paper is organised as follows: Sect. 2 introduces Laelaps II quadruped and justifies all the choices towards creating an accurate model for training. Section 3 describes the developed training environment that produces the policies for energy efficient trotting. The final two sections present the results and discuss future research directions.

2 Laelaps II Quadruped

2.1 Robot Description

The Laelaps II quadruped, developed by the Legged Robots Team of the Control Systems Laboratory (CSL) [1] at the National Technical University of Athens (NTUA), is a 40 kg, state of the art quadruped that serves as a research platform for control, software development & DRL. Laelaps II legs consist of 3 main segments (femur, tibia, foot) and 3 revolute joints namely hip, knee and ankle joint. The hip and knee joints are actuated by actuation units mounted on the robot's body whereas the ankle joints are passive (a spring tendon connects foot and tibia), see Fig. 1a. Each actuation unit (see Fig. 1b) consists of the electric motor, the planetary gearbox, the encoder and the timing belt transmission (26/48 reduction ratio). For the hip, a Maxon EC45 250 W brushless motor is used, along with its GP52C planetary gearbox (with reduction ratio 8/343). The knees share the same family of gearboxes (with reduction ratio 12/637) with the hips, but they are actuated by Maxon RE50 200 W brushed motors. In the knees' case, a parallel mechanism (close kinematic chain consisting of the transmission arm, femur, tibia and rod) is responsible for the power transmission. A more detailed descriptions of the robot's architecture is presented in [11,12].

The objective of this work is to create an accurate model of the robot's dynamics, which will be used in DRL training, aiming to produce energy-efficient motions on flat terrain similar to the treadmill-robot setup at CSL [1]. Accurate modeling is a key requirement in extracting valid results and applying the trained policies to the real robot in the future. Towards this, the MuJoCo [13] rigid body dynamics simulator was chosen since it offers notable contact stability [14] and is able to deal with closed kinematic chains present in Laelaps II legs.

(a) (b)

Fig. 1. Laelaps II: (a) MuJoCo model, (b) Drivetrain.

2.2 Laelaps II MuJoCo Model

Our main aim during the modeling process was to create a MuJoCo model as realistic as possible. To this end, all the electrical and mechanical properties of the model (see Fig. 1a) correspond to the Laelaps II quadruped [9,11]. The properties were derived from the components' datasheets and/or were experimentally verified. Special mention should be made to the identification of the frictional parameters of each actuation unit, since they are directly connected to energy consumption.

In order to characterize the friction type we conducted various experiments with constant velocity. The sampling points were carefully chosen to be at the boundary of the breakaway torque as suggested in [15]. For a wide range of voltages, i.e. 0–48 V, applied at the motor's terminals, current consumption was measured using the high-precision FLUKE 289 True RMS multimeter[1]. Additionally, using a HEDL 5640 quadrature encoder[2] the joint's angular velocity was recorded. The current measurements were converted to torque at the actuation

[1] https://www.fluke.com/en/product/electrical-testing/digital-multimeters/fluke-289.

[2] https://docs.broadcom.com/doc/AV02-0993EN.

unit's output shaft using (1), with T and n being the actuation unit's output torque and reduction ratio respectively, K_T the motor's torque constant and i_m the applied current. The results of the described procedure are displayed in Fig. 2. Note that the sampling indicates that there were no significant deviations in the reverse direction. As a result, the same model is used for both directions of motion.

$$T = K_T i_m n \tag{1}$$

The linear model of Eq. (2) accounts for the static and linear viscous friction components, with τ_i and $\dot{q}_i$ representing the i_{th} sample of the frictional torque and angular velocity respectively and offers the right balance between accuracy and modeling complexity. The f_1 & f_2 are the unknown friction coefficients.

$$\tau_i = f_1\dot{q}_i + f_2 sign(\dot{q}_i) \tag{2}$$

This model is ideal for a model-based control design in terms of execution time, while at the same time it exhibits a good R^2 index. The regression problem is solved using linear least-squares, which produce the coefficients given in Figs. 2a and b for the hip's & knee's joints, respectively. Note that the torques induced by friction are relatively low, when compared to the maximum ones that the system can output at present, namely 70.21 Nm for the hips and 109.76 Nm for the knees. To avoid unrealistic motions, these limits are included in the developed MuJoCo model.

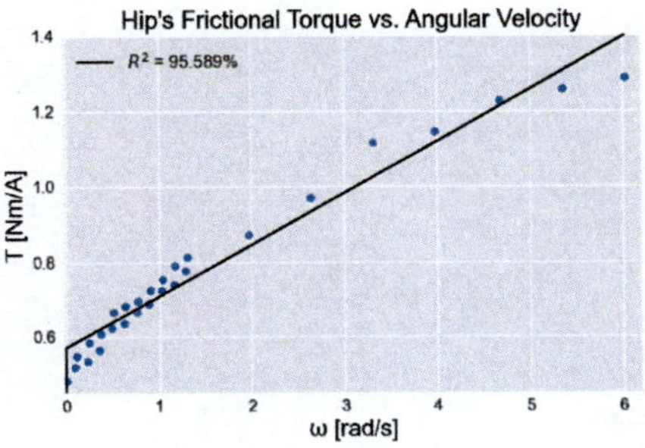

(a) $(f_1,\ f_2) = (0.138,\ 0.568)\ [Nm]$.

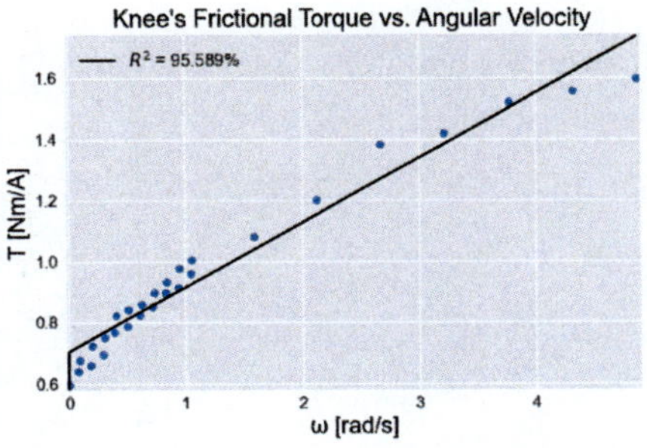

(b) $(f_1,\ f_2) = (0.211,\ 0.703)\ [Nm]$.

Fig. 2. (a) Hip experiment's friction model. (b) Knee experiment's friction model.

2.3 Laelaps II Planner

The planner introduced in [12] has been utilised here to allow the robot's model to move with constant stance velocity in MuJoCo, resulting in smoother gaits. This is a vital step towards the integration and testing of various motion modes, like trotting, originally investigated in [16]. Each leg follows a trajectory which includes separate formulations for swing and stance phases. The planner dictates the appropriate formula, by measuring the time progression of each step compared to the step's total period. By using the modulo operation, the whole process becomes independent of the step number. Moreover, a time phase shift is introduced, to enable different modes of motion (e.g. trotting, walking, etc.).

Let T_{step} be the total duration of a step, Δt_{phase} its phase shift, while T_{sw} and T_{st} the periods of the swing & stance trajectory phases, respectively. To make the trajectory invariant to the step's number, t_{traj} is introduced. Finally, the variable dir_{leg} is used to change the direction of leg's motion, with $dir_{leg} = 0$ for forward and $dir_{leg} = 1$ for backward motion. The resultant operational space trajectory $(x_{E,des},\ y_{E,des})$ is the half ellipse illustrated in Fig. 3, while its formulation is presented in Eqs. (3)–(5). Note that the ellipse's center is represented by $(x_0,\ y_0)$ while its horizontal and vertical radii with a and b, respectively. The operational-space trajectory is converted to a joint-space angle sequence that, in turn, is supplied to the low-level PV controllers of Laelaps II to produce the torques required by the desired motion.

$$t_{traj} = (t + \Delta_{phase}) \bmod (T_{step}), \quad with\ \ T_{step} = T_{sw} + T_{st} \tag{3}$$

Swing Phase

$$\begin{aligned} x_{E,des} &= x_0 + a\cos\left(\theta_{traj} + dir_{leg}\pi\right) \\ y_{E,des} &= y_0 + b\sin\theta_{traj}, \quad with\ \ \theta_{traj} = \frac{\pi}{2}\left(\cos\frac{\pi t_{traj}}{T_{sw}} + 1\right) \end{aligned} \tag{4}$$

Stance Phase

$$\begin{aligned} x_{E,des} &= x_0 + (1 - 2dir_{leg})\left(a - (t_{traj} - T_{sw})\,V_{st}\right), \quad V_{st} = \frac{2a}{T_{st}} \\ y_{E,des} &= y_0 \end{aligned} \tag{5}$$

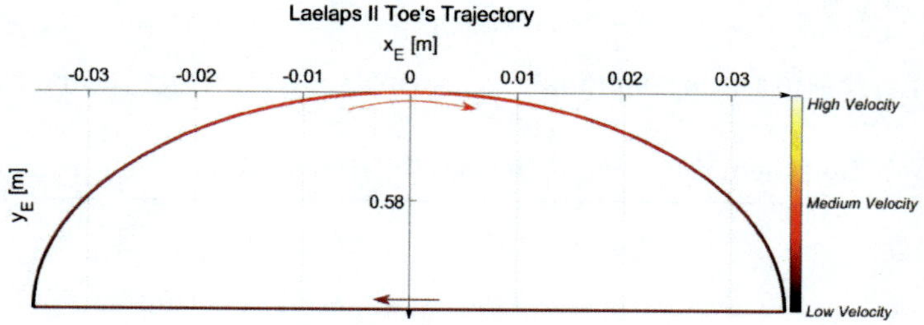

Fig. 3. Laelaps II toe's Cartesian path.

3 Training Environment

The fundamental constituents of every DRL training system is the task in context with an environment that the agent interacts. Here, the task is defined as a low-energy, low-frequency trotting in flat terrain and the environment consists of the MuJoCo model of Laelaps II traversing level terrain. The low-frequency trotting is promoted as the task, since experimental work has shown that it presents stability issues due to significant body-roll during motion; as a result a restrain mechanism was developed at CSL [1] (Laelaps II cannot perform hip abduction). For the environment, the OpenAI gym [17] together with Stable Baselines 3 [18] were utilised. According to [19], Soft Actor-Critic (SAC) [20] is a good match for training legged locomotion tasks and as a result it was employed in the current work. Concerning the hyperparameters, the buffer size was 10000, the learning rate was 0.0003 and the batch size was 256, while all other parameters were kept at the default values of [18].

3.1 Reward

To achieve efficient trotting, the energy consumption of the robot should be included in the reward function. The total actuation energy (E_{tot}) comes as the sum of the mechanical part of the actuation energy (E_{act}) and the electric losses (E_{el}) [9]. These quantities can be calculated by integrating the corresponding power formulas in (6). For the integration the Simpson 1/3 [21] was used for a given timestep (dt) and period ($\Delta t = t_2 - t_1$). Note that $\tau_{m,i}$, $\dot{q}_{m,i}$, $R_{m,i}$ and $K_{T,i}$ are the i_{th} motor's torque, angular velocity, windings' resistance and torque constant, respectively.

$$E_{act} = \int_{t_1}^{t_2} \sum_{i=1}^{8} |\tau_{m,i}\dot{q}_{m,i}| \, dt \ \ \& \ \ E_{el} = \int_{t_1}^{t_2} \sum_{i=1}^{8} \left[\left(\frac{\tau_{m,i}}{K_{T,i}} \right)^2 R_{m,i} \right] dt \tag{6}$$

The total energy is not a good metric for measuring the agent's performance and reward compared to the Cost of Transport ($CoT = E_{tot}/(mg\Delta x)$), which represents the energy required to move a robot of unit mass for one meter. In the reward function (7), a simplified version of CoT is used since the robot's mass (m) and the gravity's acceleration (g) are constants. Furthermore, the aforementioned reward accounts for the distance traversed from the beginning of the episode (Δx_{ep}), not only the one in the current step. Finally, the small constant ϵ is introduced to avoid division by $\sim$0 values, e.g.: when the robot stands still, which would destabilise the training process. A good value for the weight w_{en} has been found to be $3 \cdot 10^{-4}$.

$$rew_{en} = -w_{en} \frac{E_{tot}}{\Delta x_{ep} + \epsilon} \tag{7}$$

The second objective of the training task is to produce a policy that drives the quadruped forward without diverging from its goal. Laelaps II is an appropriate test-bed to evaluate the developed algorithm, since it cannot perform hip abduction which would help in stabilizing the robot in cases of increasing body roll/yaw angles. As a result, the terms (8) in the reward function are introduced to achieve the second objective. The weights w_x & w_y were set to 35 and 15, respectively, to promote forward motion. Finally, the total reward (rew_{tot}) which is calculated on every agent's timestep is presented in (9).

$$rew_x = w_x \left(|x_{now} - x_{previous}|\right) \;\; \& \;\; rew_y = -w_y \big| \, |y_{now}| - |y_{previous}| \, \big| \tag{8}$$

$$rew_{tot} = rew_x + rew_y + rew_{en} \tag{9}$$

3.2 Action Space

As showcased in Sect. 2.3, the quadruped's planners accept various parameters that form the desired trajectory of each toe. The high-level controller, in this case the DRL policy, is responsible for setting these parameters. However, to achieve a desired gait some of them need to remain unchanged. With that said, the vertical radius (b) remains constant during motion as it is related to the toe's clearance from the ground. For the current scenario its value is chosen to be 0.04 m, for all legs to avoid toe-collisions with the ground during their respective swing-phases, due to potential moderate body-roll/pitch induced by the whole-body motion. Moreover, the stance and swing phases are equal and constant, at 0.5 s each, to maintain a low-frequency trotting that the real robot can execute. Finally, the time-phase of either set of diagonal legs is the same and $T_{step}/2$ different from the other pair.

The rationale behind these choices is for the agent to learn parameters that effectively maximize the total reward of Eq. (9) and do not degrade the quality and synchronization of the desired gait. With this in mind, the agent selects the center (x_0, y_0) of the semi-ellipse and its horizontal radius (a). This center affects the body's pose and stability, while the horizontal radius is directly coupled to both the body's velocity and direction and thus to its kinetic energy. The resultant action space is given in Eq. (10), where L & R correspond to left and right legs while F & H to fore and hind ones, respectively.

$$\left[(x_0,\, y_0,\, a)_{RF},\, (x_0,\, y_0,\, a)_{RH},\, (x_0,\, y_0,\, a)_{LF},\, (x_0,\, y_0,\, a)_{LH}\right] \in \mathbb{R}^{12\times1}, \tag{10}$$

3.3 Observation Space

The observation space includes the standard outputs of the IMU [22] mounted on Laelaps II. Specifically, the linear (v_x, v_y, v_z) and angular ($\omega_x, \omega_y, \omega_z$) components of the body's velocity are included along with its roll-pitch-yaw angles ($\theta_x, \theta_y, \theta_z$). In accordance with [23] the observations time progress is also included in the observation space. In general to capture a motion, the sampling frequency is required to be at least two-times faster than the motion's frequency (see Nyquist-Shannon [24]). So, in the current work the observations are sampled ten-times faster than the fundamental frequency of the toe's motion (which is $f_{motion} = 2/T_{step} = 2\,\text{Hz}$), namely 20 Hz, to be on the safe-side and account for unexpected motion changes that may happen due to unmodeled dynamics. Then, the observation space is formed as in Eq. (11).

$$\left[(v_x, v_y, v_z)_{t-9\ldots now},\ (\omega_x, \omega_y, \omega_z)_{t-9\ldots now},\ (\theta_x, \theta_y, \theta_z)_{t-9\ldots now}\right] \in \mathbb{R}^{90\times 1} \tag{11}$$

3.4 DRL Control Architecture

The described architecture in Sect. 3 along with the created model presented in Sect. 2 is adumbrated in Fig. 4. The proposed architecture extends the standard OpenAI gym's framework by defining several callbacks and nested loops to execute multiple tasks in different frequencies. Specifically, the dynamics engine runs at 5 kHz to ensure convergence and avoid zero-crossing errors. At the same time, the planner along with the low-level PV leg-controllers run at 500 Hz. The agent chooses an action at a much slower rate, namely 2 Hz, i.e., on every toe's touchdown or liftoff, since each pair of diagonal legs have $T_{step}/2$ time-phase shift between them. Note that this is the ideal case, i.e., trotting motion. When the agent chooses significantly different attributes for each leg, the aforementioned assumption is invalid. However, this happens only when it corrects the robot's pose and direction, not during the steady-state forward motion. The framework of Fig. 4 is used for both training and testing.

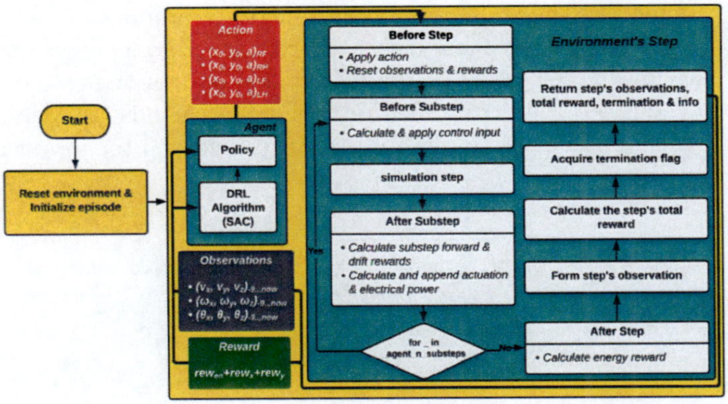

Fig. 4. Laelaps II DRL control architecture.

4 Results

The presentation of the results section focuses on three points: (a) the reduction of the *CoT* when using the energy reward term presented in Sect. 3.1, (b) the improvement in the quality of the produced motion with the proposed reward function, and (c) the fact that the produced footsteps tend to be within a specific area of the leg's workspace. The last conclusion coincides with work presented in [9] and opens new research directions, showing that the proposed DRL scheme is a way to validate analytic approaches on energy efficiency for the Laelaps II.

First, the energy levels are presented *without* and *with* the proposed energy term in the reward function in Figs. 5a and b, respectively. The lower energy consumed together with the longer distance covered (see Fig. 6a for the CoM's trajectory) in the second case results in a drop of ~37% in the CoT in favour of the developed reward function. Specifically, $CoT = 3.01$ in case of no energy term in the reward function compared to $CoT = 1.89$ when the energy term is introduced in the reward function).

The quality of motion is visualised in Fig. 6. Specifically, in Fig. 6a and b the CoM trajectory in the XY plane, i.e., top view in MuJoCo simulation environment, are presented for both simulation experiments. With the proposed reward function the robot moves almost in a straight line. In addition, the bounded yaw angle of the robot's body is shown in Fig. 6b. The yaw angle depicts the direction of the robot's body which using the proposed reward function is heading towards the goal with maximum yaw angle −0.05 rad.

Last, Figs. 7a and b illustrate the footsteps, i.e., ellipse centers and ellipse a radii that the DRL control scheme produces. The fact that the agent is trained to reduce energy consumption for the given task, results in smoother motion with bounded body angles as well footsteps that are within a limited area of the legs workspace. The latter coincides with analytical studies presented in [9]. Figure 8a summarises the training procedure and the produced CoT from each learned policy. Similar approaches have also tried to reduce energy consumption by penalising joint acceleration or the mechanical part of the actuation power during a gait, but not the drivetrain's total energy demands (6). After training and testing them on Laelaps II, our approach still achieved the lowest CoT, i.e., 1.89 (see Fig. 8b).

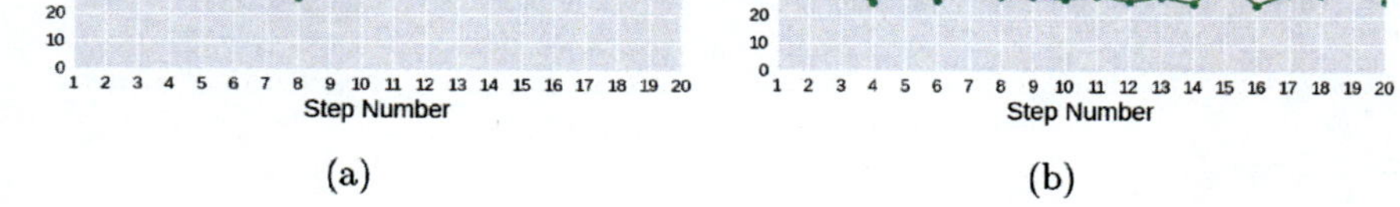

Fig. 5. Energy levels (a) *without*, (b) *with* the proposed reward function.

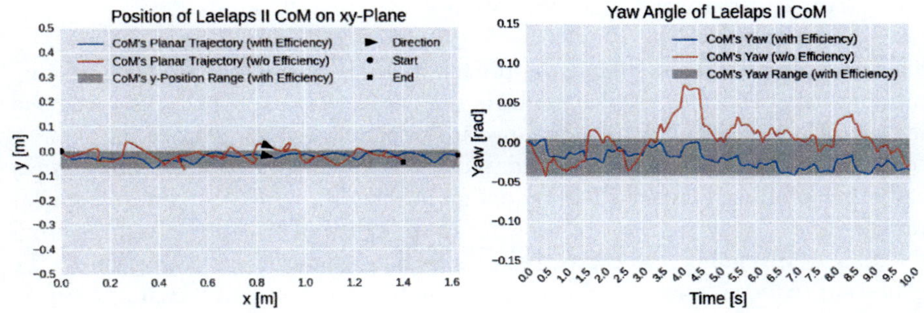

(a) (b)

Fig. 6. Motion quality. (a) CoM trajectories in top view. (b) Body yaw angle.

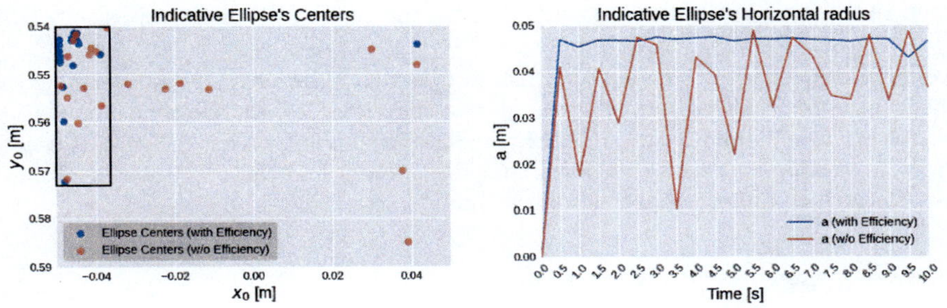

(a) (b)

Fig. 7. (a) The produced ellipse centers *with* the proposed reward function tend to be in specific area of the leg's workspace (black rectangle on the left). (b) The a radius also converges to a specific value.

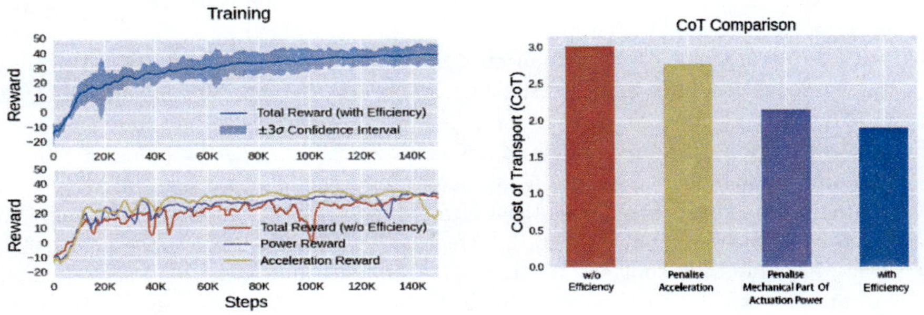

(a) (b)

Fig. 8. (a) All algorithms were trained until the mean reward converged so that the quadruped could successfully reach the goal on level terrain. (b) Using the proposed reward function the lowest CoT was achieved.

5 Conclusion and Future Work

This work described a systematic approach in producing energy efficient trotting motions using DRL. A detailed model of the Laelaps II quadruped was developed in MuJoCo and trained to move forward without diverging from its goal. A term for energy efficiency was introduced in the reward function resulting in (a) smoother produced motions, (b) reduction in the CoT by ~37%, and (c) smaller ranges in the body yaw angle, showing that the robot was not diverging from its goal. Additionally, the corrective actions needed during its transition to the goal were reduced. Last, the footsteps, i.e., ellipse centers and a radii, produced by the proposed DRL control scheme tended to be within a specific area of the legs workspace. Previous analytical studies reached similar conclusions [9].

Concerning future research directions, the gains of the low-level PV controllers can be also included in the DRL action space to investigate their effect on energy efficiency [11]. The training pipeline as well as the trained policies are open-source and available to download from the CSL's Bitbucket repository [25].

Acknowledgements. The authors wish to thank K. Machairas for co-developing the quadruped robot Laelaps II. This work was supported by the Hellenic Foundation for Research and Innovation (H.F.R.I.) under the "First Call for H.F.R.I. Research Projects to support Faculty members and Researchers and the procurement of high-cost research equipment grant" (Project Number: 2182).

References

1. https://nereus.mech.ntua.gr/legged/laelaps-2-2/. Accessed 15 Apr 2022
2. Peng, X.B., Coumans, E., Zhang, T., Lee, T.-W.E., Tan, J., Levine, S.: Learning agile robotic locomotion skills by imitating animals. Robot. Sci. Syst., July 2020
3. Haarnoja, T., Ha, S., Zhou, A., Tan, J., Tucker, G., Levine, S.: Learning to walk via deep reinforcement learning. Robot. Sci. Syst. (2019)
4. Xie, Z., Clary, P., Dao, J., Morais, P., Hurst, J., van de Panne, M.: Iterative reinforcement learning based design of dynamic locomotion skills for cassie. arXiv, preprint arXiv:1903.09537 (2019)
5. Lee, J., Hwangbo, J., Wellhausen, L., Koltun, V., Hutter, M.: Learning quadrupedal locomotion over challenging terrain. Sci. Robot. **5**(47), eabc5986 (2020)
6. Mastrogeorgiou, A., ElBahrawy, Y., Kecskeméthy, A., Papadopoulos, E.: Slope handling for quadruped robots using deep reinforcement learning and trajectory planning. In: Proceedings of the IEEE/RSJ International Conference on Intelligent Robots and Systems (IROS 2020), Las Vegas, NV, USA, 25–29 October 2020 (2020)
7. Iscen, A., Yu, G., Escontrela, A., Jain, D., Tan, J., Caluwaerts, K.: Learning agile locomotion skills with a mentor. In: 2021 International Conference on Robotics and Automation (ICRA) (2021)
8. Tan, J., et al.: Sim-to-real: learning agile locomotion for quadruped robots (2018). arXiv, doi: https://doi.org/10.48550/arxiv.1804.10332
9. Koutsoukis, K., Papadopoulos, E.: On the effect of robotic leg design on energy efficiency. In: Proceedings of the IEEE International Conference on Robotics and Automation (ICRA 2021), Xi'an, China, 30 May–5 June 2021 (2021)

10. Iscen, A., et al.: Policies modulating trajectory generators (2019). arXiv, doi: https://doi.org/10.48550/arxiv.1910.02812
11. Machairas, K., Papadopoulos, E.: An active compliance controller for quadruped trotting. In: 24th Mediterranean Conference on Control and Automation (MED 2016), Athens, Greece, 21–24 June 2016 (2016)
12. Papatheodorou, A.: Design & implementation of a real-time distributed EtherCAT-based motion control system for a multi-DoF quadruped robot. Diploma thesis, CSL, NTUA, Athens (2021). https://dspace.lib.ntua.gr/xmlui/handle/123456789/53738
13. Todorov, E., Erez, T., Tassa, Y.: MuJoCo: a physics engine for model-based control. In: IEEE/RSJ International Conference on Intelligent Robots and Systems, vol. 2012, pp. 5026–5033 (2012). https://doi.org/10.1109/IROS.2012.6386109
14. Collins, J., Chand, S., Vanderkop, A., Howard, D.: A review of physics simulators for robotic applications. IEEE Access **9**, 51416–51431 (2021). https://doi.org/10.1109/ACCESS.2021.3068769
15. Virgala, I., Kelemen, M.: Experimental friction identification of a DC motor. Int. J. Mech. Appl. **3**(1), 26–30 (2013)
16. Machairas, K., Papadopoulos, E.: An analytical study on trotting at constant velocity and height. In: IEEE/RSJ International Conference on Intelligent Robots and Systems (IROS), Madrid, Spain, October 2018, pp. 3279–3284 (2018)
17. Brockman, G., et al.: OpenAI Gym (2016)
18. Raffin, A., Hill, A., Gleave, A., Kanervisto, A., Ernestus, M., Dormann, N.: Stable-Baselines3: reliable reinforcement learning implementations. J. Mach. Learn. Res. **22**, 1–8 (2021). https://jmlr.org/papers/v22/20-1364.html
19. Mastrogeorgiou, A., ElBahrawy, Y., Machairas, K., Kecskeméthy, A., Papadopoulos, E.: Evaluating deep reinforcement learning algorithms for quadrupedal slope handling. In: 23rd International Conference on Climbing and Walking Robots and the Support Technologies for Mobile Machines (CLAWAR 2020), Moscow, Russian Federation, 24–26 August 2020 (2020)
20. Haarnoja, T., Zhou, A., Abbeel, P., Levine, S.: Soft actor-critic: off-policy maximum entropy deep reinforcement learning with a stochastic actor. In: International Conference on Machine Learning (2018)
21. Horwitz, A.: A version of Simpson's rule for multiple integrals. J. Comput. Appl. Math. **134**, 1–11 (2001)
22. https://www.xsens.com/products/mti-100-series
23. Gangapurwala, S., Geisert, M., Orsolino, R., Fallon, M., Havoutis, I.: RLOC: terrain-aware legged locomotion using reinforcement learning and optimal control (2020). arXiv, doi: https://doi.org/10.48550/arxiv.2012.03094
24. https://ocw.mit.edu/courses/2-161-signal-processing-continuous-and-discrete-fall-2008
25. https://bitbucket.org/csl_legged/laelaps_rl_mujoco/. Accessed 15 Apr 2022

Planning and Control

The Measure of Motion Similarity for Robotics Application

Teresa Zielinska[1(✉)] and Gabriel R. Rivera Coba[2]

[1] Faculty of Power and Aerospace Engineering, Warsaw University of Technology, Nowowiejska 24, 00-665 Warsaw, Poland
teresaz@meil.pw.edu.pl
[2] Quito, Ecuador
https://ztmir.meil.pw.edu.pl/web/eng/Pracownicy/prof.-Teresa-Zielinska

Abstract. The measure of motion similarity is defined. The measure is explained on the example of two different movement sequences, namely the step forward and backward and the reaction to the side push. The entered measure takes into account the displacements of body points. In this study, the relationship between the displacement of individual point masses of the body and the displacement of the resultant mass was investigated. It has been shown that the introduced measure of the similarity of movement allows to study the coordination of movements of human or humanoid robot body parts, which is useful for the synthesis of humanoid movement.

1 Introduction

There are many different methods of humanoid robot motion design and control. Some of them use models that represent in a simplified way the dynamics of the human body during locomotion [2]. Inverted pendulum models with a rigid rod or double inverted pendulum describe the processes of walking or stabilizing the posture, and pendulum models with a spring are used to model the running [4]. Unfortunately, typical models of inverted pendulums are of limited use, as the fixed length of the pendulum parts make it impossible to describe the posture with bent knees. With this in mind, in our previous work we proposed the concept of a variable double pendulum [18]. By testing various human movements, we have shown that such a model better reflects the dynamics of a human body than the typical inverted double pendulum. The inverted variable pendulum can be used to design humanoid postures. Generating humanoids motion applying simplified models, it is needed to know how to properly coordinate the movement of individual parts of the body. Neglecting this issue causes postural stabilization problems. Between others, the upper body swaying coordinated with the legs displacement, together with the swing of the upper limbs, support the dynamics of locomotion while maintaining the stability of the posture. Kinematic synergies are defined as coordinated joint movements that reduce the number of degrees of freedom [13]. Scientists often deal with kinematic synergies [8] and muscle synergies [10]. For example, the article [5] proposed a method of mimicking human

© The Author(s), under exclusive license to Springer Nature Switzerland AG 2023
J. M. Cascalho et al. (Eds.): CLAWAR 2022, LNNS 530, pp. 219–230, 2023.
https://doi.org/10.1007/978-3-031-15226-9_22

movement, a synergy of joint velocities (called first order synergy) was defined and using it, human like behavior was achieved when performing certain tasks. The authors of [1] emphasized that coordinated involvement of the ankle and hip joints is important for postural stabilization. However the motion synergies are investigated in a limited range. Such studies mainly concern the periodic movements, or the repetitive tasks and often are devoted to the diagnosing of illnesses [7]. In [8] the kinematic synergies taking into account the joints were studied. The focus was on the lower limbs. The Principal Component Analysis was used for this purpose. Relatively significant amount of works is devoted to the analysis of the synergies of the muscles work using EMG signals. The main purpose in this case is the development of exoskeletons [17] or the rehabilitation processes [12]. In this work the motion synergies aiming at non-periodical dynamical movements are studied. The overall purpose is to use this knowledge for humanoid robots motion synthesis. The motion similarity measure is proposed, and movement coordination of point masses of body parts is examined.

2 The Motion Similarity Measure

For creating the similarity measure the formula of Pearson's correlation coefficient was adapted:

$$r = \frac{\sum_{k=1}^{n} \left({}^{k}a - \bar{a}\right)\left({}^{k}b - \bar{b}\right)}{\sqrt{\sum_{k=1}^{n} \left({}^{k}a - \bar{a}\right)^{2} \sum_{k=1}^{n} \left({}^{k}b - \bar{b}\right)^{2}}} \tag{1}$$

where: r is the Pearson's correlation coefficient of the analyzed variables, ${}^{k}a$ is the k-th sample from the first data set, ${}^{k}b$ is the k-th sample from the second data set, and $\bar{a}$, $\bar{b}$ are the mean values respectively. The result r belongs to the range $\langle -1.00, +1.00 \rangle$, where:

- $+1$ means a perfect positive linear correlation between the variables,
- 0 means that there is no linear correlation between the variables,
- -1 means a perfect negative linear correlation between the variables.

The correlation coefficient provides the measure of linear dependence between two variables. It was used in our previous work [19]. Unfortunately it gives a single evaluation for a whole data set and only the linear dependency is tested. Defining the similarity measure for each time instant, the formula (1) was modified neglecting the sum in the numerator. The deviation for each measurement point was divided by the standard deviation component. Such formula made the similarity measure:

$$r_{ab}(i) = \frac{\left({}^{i}a - \bar{a}\right)\left({}^{i}b - \bar{b}\right)}{\sqrt{\sum_{k=1}^{n} \left({}^{k}a - \bar{a}\right)^{2} \sum_{k=1}^{n} \left({}^{k}b - \bar{b}\right)^{2}}} \tag{2}$$

where i denotes the i-th frame of the motion data record. Equation (2) enables the analysis of the similarities regardless of the amplitude differences and offsets in position (but not in a time). We used this formula producing the base curve created by the point mass of the whole human body (overall CoM) 'correlated' with itself and then we computed the similarity measure for the CoM and the point mass of the body segments. The x, y, z similarity measures were produced. The similarity measure for CoM with itself was used as the reference. In that case $a = b = x_{CoM}$, $a = b = y_{CoM}$ or $a = b = z_{CoM}$ respectively. If at some point the ${}^{i}a - \bar{a}$ deviation is significantly different than the deviation of overall CoM position ${}^{i}b_{CoM} - \bar{b}_{CoM}$ then the result $r_{ab_{CoM}}(i)$ is significantly different than $r_{b_{CoM}b_{CoM}}(i)$ – what is indicated as the difference in similarity measure. It means that the similarity measure curves have a similar shape when the motion of the considered quantities is similar.

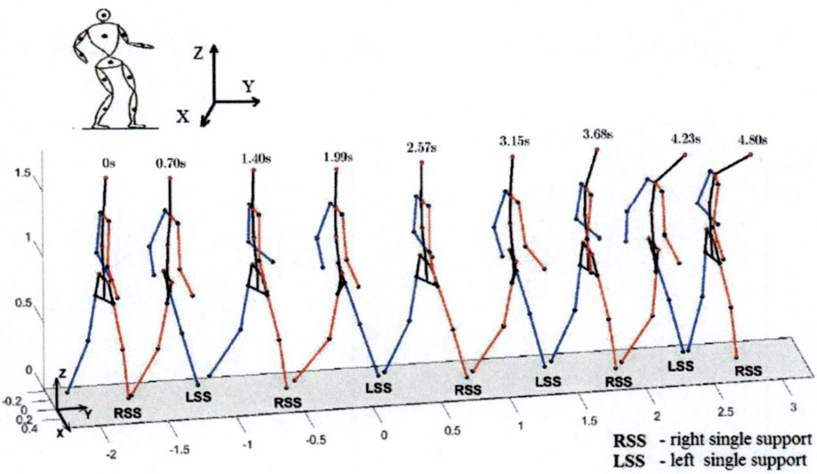

Fig. 1. Illustration of considered body segments and the stick diagram describing the walk.

3 The Research Method

The motion of a healthy, young female was recorded using the professional motion capture system (VICON). Several dynamic activities were studied. In this paper we describe two representative cases – the forward-backward stepping and – postural stability recovery after unexpected push from the side. Such selection enabled the comparison of slow dynamics (stepping) versus fast dynamics (push recovery).

The stick diagram of a human body consisting of 11 segments: – head & neck, – left upper arm, – right upper arm, – left forearm, – right forearm, – trunk, – pelvis, – left thigh, – right thigh, – left shank & foot, – right shank & foot was used for motion animation. Anthropometric data taken from [3,16] delivered the values and locations of a human body partial masses. In Fig. 1 the segmented model of a human body with point masses indicated by the dots, and the stick diagram reconstructed using the markers are illustrated. The markers position together with the added points used for reconstructing the pelvis and spine position are indicated. The motion animation in the 3D space using human motion data delivered the motion trajectories of the body point masses. Using these trajectories the similarity measure (2) of selected point masses was calculated. The measure obtained for overall CoM was drawn together with measures for the body points. The comparison of the curves enabled to conclude about the presence or lack of motion coordination (motion synergy).

4 Forward and Backward Stepping – Slow Dynamics

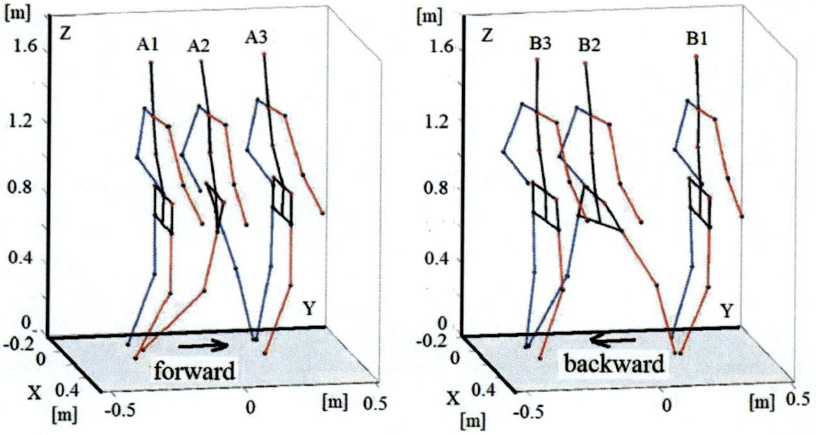

Fig. 2. Stepping. Visualization of events.

In a stepping forward and backward from a stand-still posture (double support) the following events were distinguished (Fig. 2):

- **A1** – 0.80 s; the single support of the right leg begins; the left leg starts its transfer to the front,
- **A2** – 1.75 s; the single support of the left leg begins; the right leg starts its transfer to the front,
- **A3** – 2.70 s; the double support begins,

- **B1** – 6.75 s; the single support of the right leg begins when stepping back; the left leg starts its transfer to the back,
- **B2** – 7.90 s; the single support of the left leg begins; the right leg starts its transfer to the back,
- **B3** – 8.70 s; the double support begins.

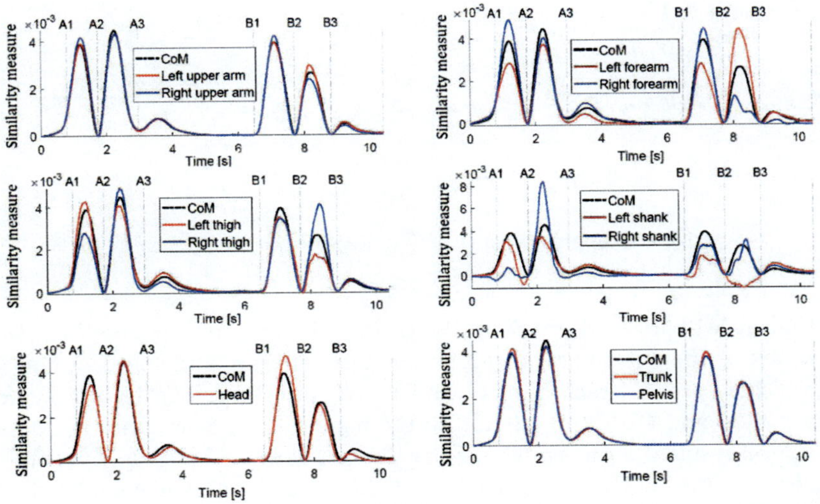

Fig. 3. Stepping. Trajectory of the similarity measure for the displacement along the X-axis.

Figure 3 presents obtained similarity measure for the motion along X axis. The black dashed line corresponds to the base curve created for the auto-similarity measure (the overall CoM with itself), while the other curves correspond to the similarity measure of the overall CoM with the other point masses. Comparing to Fig. 4 with the displacement trajectories of the body point masses, in Fig. 3 the differences and similarities in the displacements are much better seen. The best similarity to the reference trajectory (the overall CoM measure) is for the head, body, hips and shoulders. There is no similarity (i.e. lack of coordination of the character of the movement with reference trajectory $r_{x_{COM}x_{CoM}}$) for the right shank between events A1 and A2, and for the left shank between events B2 and B3. In the first case, it is the time interval in which the left leg is moving forward, and in the second case – when the right leg is moving backward. Remembering that in both of these cases, either the right leg or the left leg supports the body and the foot is stationary, while the rest of the body moves (with smallest displacement of the shank of the supporting leg), the graphs adequately reflect the real situation. A more detailed look at the graphs, and in particular on the trajectories obtained for the left and right lower leg and right arm, allows to notice lateral tilts. In the case of transferring the right leg forward (A2–A3)

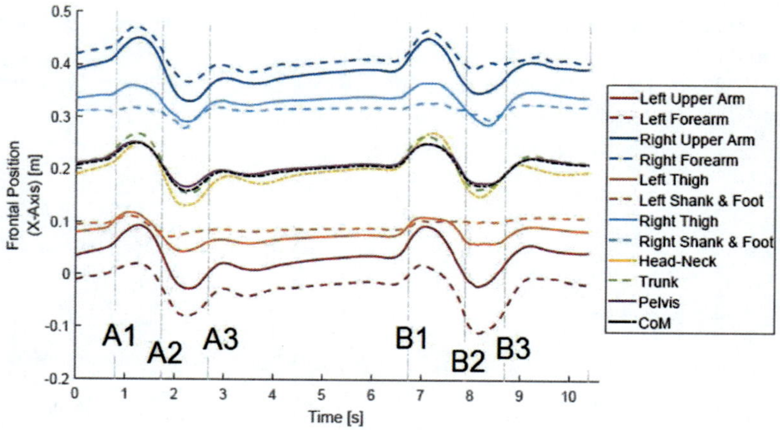

Fig. 4. Stepping. Displacement trajectories along X-axis

and the left leg backward (B1–B2), the trajectories of point masses of the thigh and shank of the left and right leg, respectively show a common coordination, however they are different from the reference trajectory. The lack of symmetry in A2–A3 and B1–B2 in relation to the regularities observed for the A1–A2 and B2–B3 results, in our opinion, from the fact that in these cases the full leg transfer (leading to a full step forward or backward) was not completed. The transferred leg was just placed next to the supporting leg, which resulted in that the lateral tilt of the body in these was smaller than that for a complete step.

When analyzing the similarity measures obtained for the displacements of point masses along the Y axis (Fig. 5), it can be concluded that the variability of all trajectories is similar, which means that all displacements are coordinated along the direction of movement. A slight shift and slightly smaller values towards the reference trajectory $r_{y_{COM}y_{CoM}}$ for the left shank in the A1–A2 result from that, that the left leg (leg just starting the forward transfer) and especially its foot and shank are behind the body. This situation is similar for the foot and shank of the right leg just before starting its backward transfer in B2–B3. In the A2–A3 interval, the transfer of the right leg is not so visible. We observe smaller values of the shank trajectory of transferred leg in relation to the reference trajectory in the B1–B2.

The greatest differences in the similarity measures for displacements along the Z axis (Fig. 6) concern both arms and legs, but in the case of upper limbs (arms and forearms) there is a similarity in the nature of the curves to the reference trajectory, which proves the coordination of displacements along the vertical axis. The different nature of the trajectories than $r_{z_{COM}z_{CoM}}$) of the left leg thigh and shank in the A1–A2 results from the fact that this leg is lifted, i.e. the nature of its movement is different than the movement of the rest of the body - it is especially visible for the shank. Lack of coordination with the reference trajectory for the left shank is also seen at the beginning of the A3

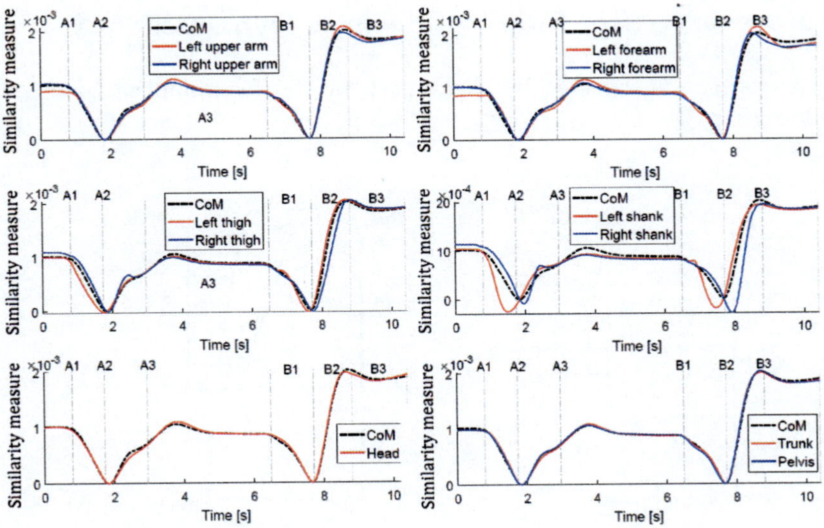

Fig. 5. Stepping. Similarity measure for the displacements along Y-axis.

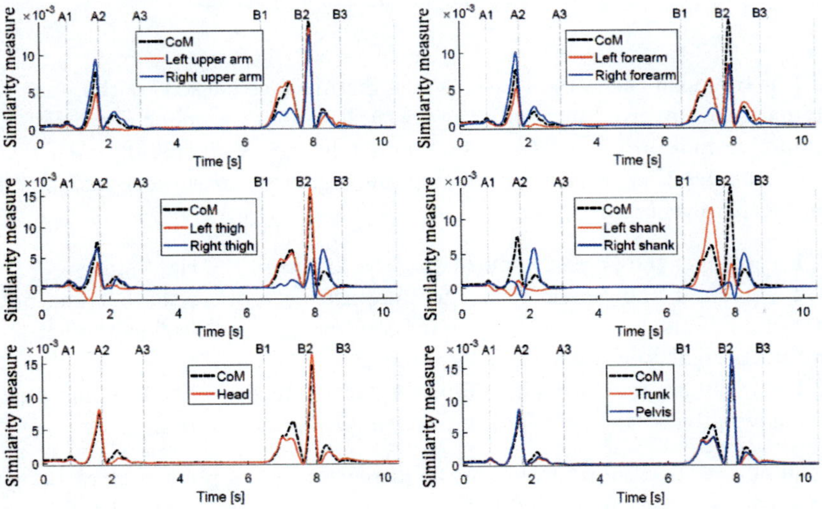

Fig. 6. Stepping. Similarity measure for the displacements along Z-axis.

interval, when the right leg starts the transfer, which may result from a slight flexion of the supporting leg (left leg). The trajectory of the similarity measure for the right shank has clearly greater values than the reference trajectory, in this interval.

5 Recovery After Side Push – Fast Dynamics

The person got unexpected but light push by the force applied to the right shoulder (Fig. 7).

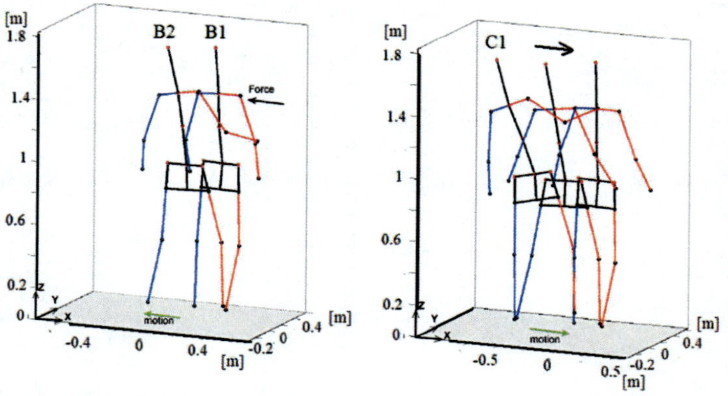

Fig. 7. Side push. Visualization of events.

After the push the person 'rejects' the disturbance caused by the push. The body parts are involved to regain the postural stability without any delay. First the reactive motion is performed by the left leg to avoid falling and next the person resumes the posture by moving the right leg to the original position. Main motion events are:

- **B1** – [6.25] s; the person is pushed to the left from the right side,
- **B2** – [6.70] s; the person reacts to the push by moving the left foot to the direction of the push in order to gain equilibrium, and reaches the left single support (right foot is above the ground),
- **C1** – [7.48] s; once in equilibrium, the person recovers the starting posture by moving back the right foot and placing it to the ground.

We can notice good coordination of the displacements of point masses along the X axis (push direction – lateral, Fig. 8). Only, after the push, the right forearms show less coordination of movement in relation to the reference trajectory $r_{xCOMxCoM}$. Small differences in the behavior of the forearms and shanks are also visible close to the moment C1, when the right leg is shifted towards the left leg to regain the postural equilibrium.

The measures of similarity for the displacements along the Y axis (Fig. 9) characterize well the process of equilibrium recovery. The trajectories of similarity measures for the upper body and pelvis are close to the reference trajectory $r_{yCOMyCoM}$, although here, too, some differences can be noticed for the trajectory of the pelvis just after the push.

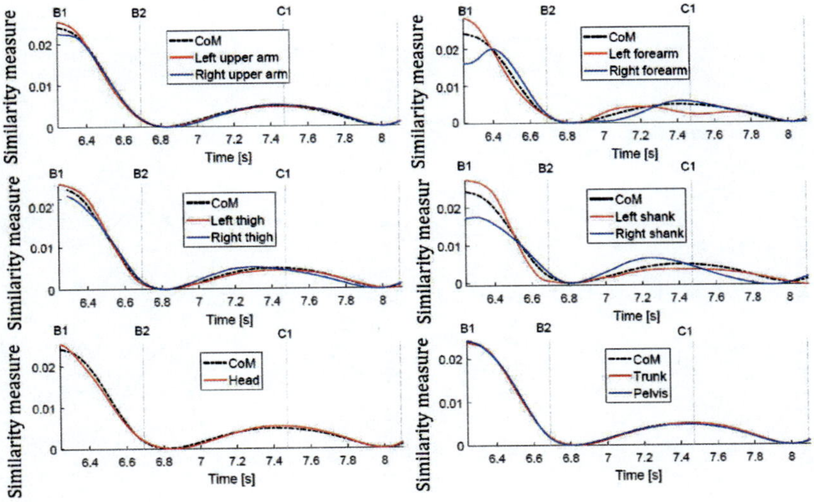

Fig. 8. Push recovery. Similarity measure for the displacement along X-axis.

Throughout the entire time interval, the similarity measure trajectories are mutually similar for the arm and forearm of the left limb and the arm and forearm of the right limb, respectively, but they are different for each limb and differ from the reference trajectory. So it indicates that each upper limb independently supports the process of regaining balance without changing its configuration in the XY plane. It can be also noticed that in the first phase of the movement, the left upper limb makes more voluminous movements (there are greater differences comparing to the reference trajectory), and in the second phase - it is a case for the right upper limb. The side shift of the left leg in the B1–B2 time interval is coordinated with the CoM movement (the trajectories of the thigh and shank of the left leg are close to the reference trajectory) and the placing the right leg next to the left restores the coordination with the reference trajectory (in the vicinity of the C1 point).

The similarity measures for the motion along the Z direction (Fig. 10) for the most extend are similar to the reference trajectory $r_{z_{COM} z_{CoM}}$. It is also seen a greater mutual similarity of the reference trajectories of the left forearm, head and left shank, which means greater coordination of vertical movements of these parts of the body. The purpose of introducing a measure of similarity is to use it to reconstruct the movement of human body parts using an inverted pendulum model with a variable structure. In our previous works [18,19], we introduced the concept of the variable double inverted pendulum, showing that such a model allows good representation of the motion dynamics in situations where the traditional inverted or double inverted pendulum models do not work (e.g. in postures with bent legs). In addition, the single inverted pendulum was used by us to design a gait of small humanoid that maintained an upright posture [15]. Currently, we are studying the transformation of human posture, the

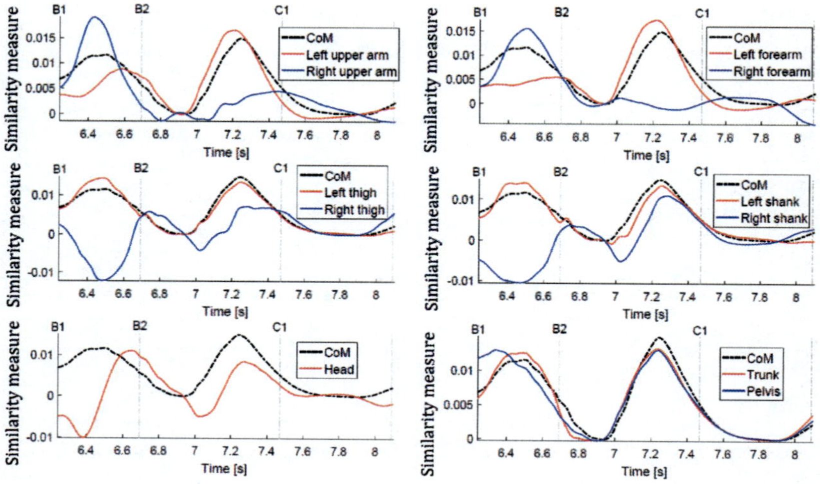

Fig. 9. Push recovery. Similarity measure for the displacement along Y-axis.

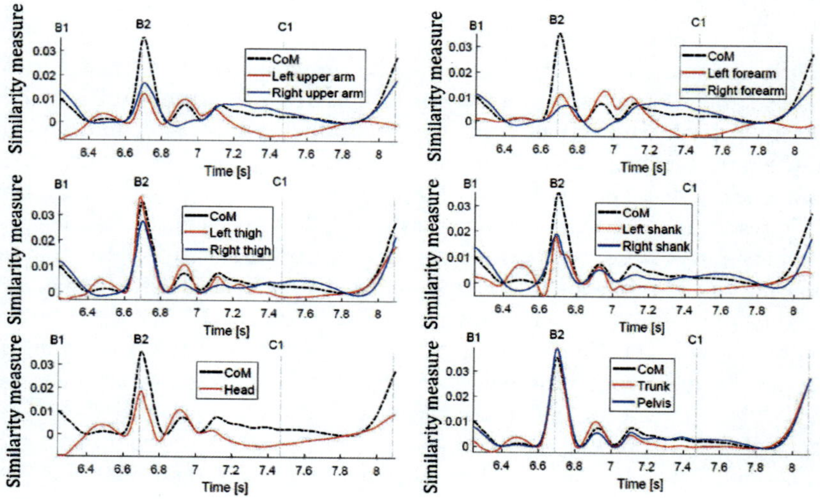

Fig. 10. Push recovery. Similarity measure for the displacement along Z-axis.

introduced measure of similarity makes it easier to identify the movement synergies of body point masses. This measure, unlike the Pearson coefficient, gives a score for each instant of time.

6 Conclusions and Discussion

The motion synergy depicts the relation between motions of the body segments. The similarity measure using normalized deviation provides information about motion similarities and differences in time domain. The similarity measure indicates well coordinated movements (stepping example) versus non-symmetrical (push reaction). As the measure in our case was expressed towards the CoM position deviation the result indicates if and when the motion deviation of the other point mass is bigger than the motion deviation of the CoM. This article describes two case studies (stepping and postural recovery after side push). The trajectory of the similarity measure of the body resultant mass point (CoM) was used as the reference trajectory. The next stage should be the extension of the analyzed examples and examination of the repeatability of the results. Bearing in mind the latter issue, it should be noted that the human motion was recorded in the hospital diagnostic laboratory. In addition to the VICON system, which is used to diagnose patients, and therefore must guarantee high accuracy, professional software was applied to eliminate reading errors. Thus, the data could be considered as sufficiently accurate. Nevertheless, the research concerned one person data. Therefore, in this article, we do not focus attention on the formulation of general conclusions regarding the coordination of body movements in specific movement situations, but rather analyze the information potentially carried by the introduced measure of similarity. When examining synergies, we were not interested in comparing the magnitude of displacements. The goal was to detect which parts of the body (their point masses) were moving in a similar way, especially in non-periodic events (such as postural balance recovery). The comparison of the trajectory of displacements and the trajectory of the introduced measure shows that this measure facilitates the identification of temporary synergies of displacements of individual parts of the body.

References

1. Alexandrov, A.V., Lippi, V., Mergner, T., Frolov, A.A., Hettich, G., Husek, D.: Human-inspired eigenmovement concept provides coupling-free sensorimotor control in humanoid robot. Front. Neurorobot. **11**, 1–13 (2017)
2. Liu, C., Ning, J., Chen, Q.: Dynamic walking control of humanoid robots combining linear inverted pendulum mode with parameter optimization. Int. J. Adv. Robot. Syst. **15**(1) (2018)
3. Contini, R.: Body segment parameters. Part II. Artif. Limbs Rev. Curr. Dev. **16**(1), 1–19 (1972)
4. Delattre, N., Lafortune, M., Moretto, P.: Dynamic similarity during human rRnning: about Froude and Strouhal dimensionless numbers. J. Biomech. **42**, 312–318 (2009). https://doi.org/10.1016/j.jbiomech.2008.11.010
5. García, N., Suárez, R., Rosell, J.: First-order synergies for motion planning of anthropomorphic dual-arm robots. IFAC-Papers On Line **5**(1), 2247–2254 (2017)
6. Gordon, C.C., Churchill, T., Clauser, C.E., Bradtmiller, B., McConville, J.T.: Anthropometric survey of US army personnel: methods and summary statistics. Technical report, Anthropology Research Project Inc Yellow Springs OH (1989)
7. Kogami, H., et al.: Analysis of muscle synergy and kinematics in sit-to-stand motion of hemiplegic patients in subacute period. Adv. Robot. **35**(13–14), 867–877 (2021)

8. Huang, B., Xiong, C., Chen, W., Liang, J., Sun, B.-Y., Gong, X.: Common kinematic synergies of various human locomotor behaviours. Royal Soc. Open Sci. (2021) http://doi.org/10.1098/rsos.210161
9. Liu, K., Xiong, C.-H., He, L., Chen, W.-B., Huang, X.-L.: Postural synergy based design of exoskeleton robot replicating human arm reaching movements. Robot. Auton. Syst. **99**, 84–96 (2018)
10. Liu, Y.X., Wang, R., Gutierrez-Farewik, E.M.: A muscle synergy-inspired method of detecting human movement intentions based on wearable sensor fusion. IEEE Trans. Neural Syst. Rehabil. Eng. **29**, 1089–1098 (2021). https://doi.org/10.1109/TNSRE.2021.3087135
11. McGrath, M., Howard, D., Baker, R.: The strengths and weaknesses of inverted pendulum models of human walking. Gait Posture **41**(2), 389–94 (2015). https://doi.org/10.1016/j.gaitpost.2014.10.023. Epub 2014 Oct 30. PMID: 25468688
12. Tang, S., Chen, L., Barsotti, M., Hu, L., Li, Y., et al.: Kinematic synergy of multi-DoF movement in uppar limb and its application for rehabilitation exoskeleton motion planning. Front. Neurorobot. **13**, article 99 (2019)
13. Aoi, S.: Gait generation and transition of a biped robot based on kinematic synergy in human locomotion. In: Habib, M.K. (ed.) Handbook of Research on Advancements in Robotics and Mechatronics. Kyoto University, Japan (2015)
14. Suzuki, Y., Nomura, T., Casadio, M., Morasso, P.: Intermittent control with ankle, hip, and mixed strategies during quiet standing: a theoretical proposal based on a double inverted pendulum model. J. Theor. Biol. **310**, 55–79 (2012)
15. Szumowski, M., Zurawska, M., Zielinska, T.: Preview control applied for humanoid motion generation. Arch. Control Sci. **29**(1), 111–132 (2019)
16. Winter, D.A.: Biomechanics and Motor Control of Human Movement. Wiley, Hoboken (2009)
17. Li, Z., Liu, H., Yin, Z., Chen, K.: Muscle synergy alternation of human during walking with lower limb exoskeleton. Front. Neuroscoence (2010). https://doi.org/10.3389/fnins.2018.01050
18. Zielinska, T., Rivera Coba, G.R., Ge, W.: Variable inverted pendulum applied to humanoid motion design. Robotica **39**(8), 1368–1389 (2021)
19. Magdalena, S.Z., Szumowski, M., Zielinska, T.: Reconfigurable double inverted pendulum applied to the modelling of humanoid robot motion. J. Autom. Mob. Robot. Intell. Syst. **11**(2), 12–20 (2017)

Design of a Minimalistic Torque Actuated Variable Rolling SLIP Leg for Robust Locomotion

Noah Paul(✉), Adar Gaathon, and Amir Degani

Technion-Israel Institute of Technology, Haifa, Israel
{noahpaul,adarg}@campus.technion.ac.il, adegani@technion.ac.il

Abstract. In this paper we investigate the use of an optimal control coupling applied to a continuously jumping robot utilizing a single spiral-shaped leg. Exploiting multiple controllers simultaneously can increase robustness over uneven terrain. However, this can be difficult to actualize. By designing a spiral-shaped leg, we can couple leg length and stiffness and optimize leg retraction rate and torque. Through this embodied design we can achieve a robust minimalistic running robot with a simple feedforward controller.

Keywords: Legged locomotion · Design optimization · Minimalistic control

1 Introduction

Dynamic legged locomotion outperforms wheeled and tracked locomotion over uneven and rough terrain [1]. However, controlling dynamic legged locomotion will often require complex closed loop control and fast accurate sensing to convey detailed information about the environment. Two popular minimalistic models used in modern dynamic legged locomotion are the Spring-Loaded Inverted Pendulum (SLIP) [2] and Swing Leg Retraction (SLR) [3] model. SLIP model reduces the leg to a point mass atop a springy leg and by controlling a set leg angle, achieves stable running over flat terrain in an energy conservative system with no perturbations. SLR is an extension of this method, which retracts the leg angle over time from a set angle at apex to add stability over varying terrain heights. This allows the velocity of the foot match the ground and promotes stability [3].

Applying other control methods can further increase stability [4–6] however, modifying multiple controllers is mechanically difficult to control and implement. Moreover, models such as SLIP and SLR function in energy conservative systems, losing the improvement in stability gained with energy dissipation [7].

© The Author(s), under exclusive license to Springer Nature Switzerland AG 2023
J. M. Cascalho et al. (Eds.): CLAWAR 2022, LNNS 530, pp. 231–241, 2023.
https://doi.org/10.1007/978-3-031-15226-9_23

In this paper we pursue a minimalistic approach involving a simplistic control scheme mixed with an "embodied" controller. Here an "embodied" controller, as depicted in Fig. 1. is a physical structure, optimized with coupled parameters, for a specific response. Leg length and stiffness are optimized and embodied in a spiral-shaped leg, where as the leg length increases the stiffness decrease. The torque is applied during stance and is controlled by the time of flight from apex. As a result, leg length, stiffness and torque are all controlled through a single actuator.

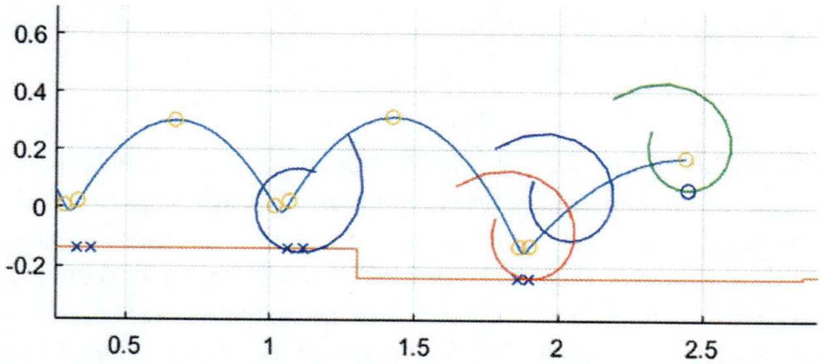

Fig. 1. An example of a torque actuated polynomial leg and its trajectory. The leg ends stance in one orientation, travels through flight phase and rotates further back during decent due to the step-down. The leg then travels through stance, applying torque, and jumps to the same y_{rel} at apex in reference to the terrain. Yellow circles represent state transitions (Touch Down, Stance, and Lift Off) and blue crosses represent foot TD and LO positions.

2 Related Work

Many existing models are based on the conservative planer SLIP model. Such models are often able to use little to no sensing and no closed-loop control. The SLR method is an open loop method that can increase robustness over moderate unknown terrain [3, 5, 8]. By turning the SLR parameters, the height above the last terrain level can be fixed as shown by Seyfarth et al. [8]. Furthermore, with proper SLR tuning the relative height (y_{rel}) above the terrain can be fixed in order to overcome unknown step-up and step-down perturbations as was shown by Shemer and Degani [9]. Previously, we showed these control parameters can be described by a second-degree polynomial [10].

Dynamically varying leg stiffness has also been shown to be advantageous. Changing the stiffness can control the natural frequency of the gait and increase stability [11]. Humans similarly adapt leg stiffness to compensate for terrain stiffness to maintain running mechanics [12]. Tunable stiffness models have been shown to reduce ground reaction forces [13], control forward velocity [14], and help prevent crashing at the end of stance [15]. Galloway et al. designed a tunable stiffness c-shaped leg that used an actuated sliding rigged spine for every leg to vary the stiffness of the leg while maintaining its shape [16].

Torque-actuated designs that consider rolling have also been explored in literature. Ankarali et al. proposed a hip torque actuated leg design with and inline spring and damper in the leg. They found the leg most stable with an energy regulating feedback law providing robust stability properties [17]. A Clocked-Torqued model was used on RHex in order to better capture the dynamics of animals exhibiting whole body motion [18]. Chia-Jui Hu et al. explored a rolling torqued SLIP model using a dampener added to the leg showing an increase in pitch disturbance rejection and running repeatability [19].

While c-shaped legs have been explored throughout literature, spiral-shaped legs are rarer, appearing in the works of Kovačić et al. and our previous work. There they are used as variable compliance foot on a 3-DOF leg. Kovačić et al.'s foot allowed the robot to compensate for surface compliance by modifying its foot position [20]. In this paper we aim to utilize the advantages of spiral legs found in our previous work [21]. There we leverage the coupled stiffness and leg length provided by the spiral design along with a minimalistic orientation-based controller to control for stability over terrain changes in an energy-conservative system. Here we aim to apply this minimalistic approach to non-conservative systems by considering dissipation as well as adding a hip torque for continuous jumping.

3 Models

3.1 Constant SLIP

The Spring-Loaded Inverted Pendulum (SLIP) model is one of the formative hybrid dynamical models for running used in robotics and the study of biology. The SLIP model imagines a point mass (m) atop a massless spring leg of length r_0 and stiffness k in an energy-conservative system. The leg may change its angle, relative to a reference position (straight down), by θ, its length can be compressed by r, and is bound in the sagittal plane such that no out of plane rotations may occur. Throughout the stride the leg passes through the repeating phases, a flight phase (composed of ascent and descent), touchdown (TD), stance phase, lift-off (LO), Fig. 2. During stance the foot is pinned and during flight the body follows ballistic dynamics. The stance phase equations are given by Eq. (1).

$$\begin{cases} m\ddot{r} = mr\dot{\theta}^2 + k(r_0 - r) - mgcos(\theta) \\ r\ddot{\theta} = -2\dot{r}\dot{\theta} + gsin(\theta) \end{cases} \tag{1}$$

Considering damping in the leg we can represent this as a non-conservative system and the stance equations become

$$\begin{cases} m\ddot{r} = mr\dot{\theta}^2 + k(r_0 - r) - mgcos(\theta) - mc_r\dot{r} \\ r\ddot{\theta} = -2\dot{r}\dot{\theta} + gsin(\theta) - rc_\theta\dot{\theta} \end{cases} \tag{2}$$

where c_θ and c_r are the damping constants.

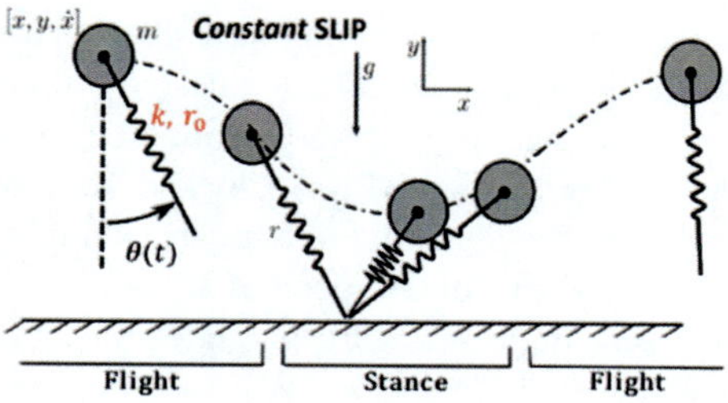

Fig. 2. The 2D SLIP model

3.2 Variable SLIP

Next, we couple the leg length and stiffness to the leg orientation to bind the parameters to our SLR controller, which we will explore below. Therefore, $r = r_0(\theta)$ and $k = k(\theta)$. We can then formulate the Lagrangian

$$\mathrm{L} = \mathrm{T} - \mathrm{V} = \frac{m\left(\dot{r}^2 + r^2\dot{\theta}^2\right)}{2} - \frac{k(\theta)}{2}(r_0(\theta) - r(t))^2 - mgr\cos(\theta) \tag{3}$$

Deriving this Lagrangian we find

$$\begin{cases} m\ddot{r} = mr\dot{\theta}^2 + k(\theta)\cdot(r_0(\theta) - r) - mg\cos(\theta) - mc_r\dot{r} \\ mr^2\ddot{\theta} = -2m\dot{\theta}r\dot{r} + mgr\sin(\theta) - \frac{p(\theta)(r_0(\theta)-r)^2}{2}\ldots \\ -k(\theta)q(\theta)(r_0(\theta) - r) - mr^2c_\theta\dot{\theta} \end{cases} \tag{4}$$

where, $p(\theta) = dk/d\theta$ and $q(\theta) = dr_0/d\theta$ and c_θ and c_r are the damping constants added to make the system nonconservative.

3.3 Variable Rolling SLIP

The Variable Rolling SLIP model is originally presented in Gaathon et al. [21]. It considers a leg shape defined by a polynomial as a physical way of binding leg orientation to leg length and stiffness. Due to the shape, the leg experiences rolling through the stance phase. Rolling is the continuous change in the ground contact point or Center of Pressure (COP) along the leg during stance denoted by θ. The rotation axis is at the hip and, similarly to most SLIP models, is the center of mass (COM). At each input angle α the leg the leg touches the ground at an angle θ that is dependent on the curvature of the leg. The angle of the COP relative to the input angle is denoted by β giving the relation $\alpha + \beta = \theta$, Fig. 3.

The Lagrangian can be constructed with state variables,

$$\mathrm{L} = \mathrm{T} - \mathrm{V} = \frac{\left(\dot{x}_{hip}^2 + \dot{y}_{hip}^2\right)m}{2} - \frac{k(\theta)}{2}(r_0(\theta) - r(t))^2 - mgr\cos(\theta) \tag{5}$$

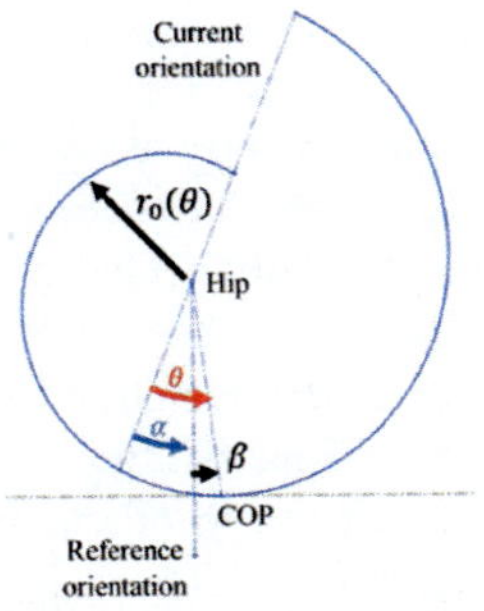

Fig. 3. An example of a polynomial leg shape and its related reference variables.

The derivation of the Lagrangian above and the equations of motion can be found in [21].

4 Methods

4.1 Control

In this paper we use a combination of flight phase control and stance phase control. During the long descent of the robot the leg angle is positioned according to the SLR parameters until TD when the stance phase controller takes over. During stance the robot rolls and the hip actuator adds a constant torque determined by the time from apex to touchdown, similarly to the SLR parameters controlling the leg angle in flight. This torque is used to recover any energy lost during stance. At the end of stance, the robot lifts-off, reaches apex and begins the process again. An example of this can be seen in Fig. 1.

During the descent phase the controller is based around the concepts explored in SLR control [3]. As the robot falls through descent, the leg is swept back at a rate from the time of apex. We control the input angle to the hip actuator through the polynomial $\alpha(t) = a_1t^2 + a_2t + a_3$. The stiffness and leg length are determined by a polynomial of the leg *angle*, the angle at which the leg touches the ground, determined by the COP. We assume then for the free leg length the form $r_0(\theta) = b_1\theta^2 + b_2\theta + b_3$ and stiffness $k(\theta) = c_1\theta^2 + c_2\theta + c_3$ where $-\pi \leq \theta \leq \pi$. The leg shape is in polar coordinates and able to describe a leg shape that varies between a spiral and an ellipse. The damping coefficients c_r and c_θ are also functions of θ and change with the leg angle. Therefor the flight phase SLR control is multi-purpose, controlling the leg, length, stiffness and damping all at the same time.

During stance we apply a constant torque. This constant torque is determined similarly to the input angle α and temporal. As the robot falls, the constant torque to be inserted during stance increases. This stance control takes the form of the polynomial $\tau(t) = d_1t^2 + d_2t + d_3$. This means for our controller there are twelve parameters that require optimization.

4.2 Single Jump Optimization

For the optimization we aim to find the optimal parameters for, SLR, free leg length, stiffness, and torque. To find the first three parameters the optimization launches trajectories from an apex while changing the initial velocity and terrain levels. The optimizer then finds three polynomial equations, for the temporal leg orientation, the leg stiffness and the free leg length that minimizes the deviation from a relative height (y_{rel}) over a *single* jump. This first optimization is done without damping and generates a leg shape that will be used to find the true damping coefficients in a physical model. We then further refine the SLR and find the optimal torque parameters through another round of optimization, keeping the leg shape, stiffness, and damping coefficients found from the first optimization. The second optimization is similar to the first in that the optimization launches trajectories from an apex for a *single* jump while changing the terrain levels, however, the initial velocity remains constant. The cost function attempts to minimize deviations from not only y_{rel} but also the deviation from the initial velocity at the next apex. We consider the forward velocity in addition to the deviation from y_{rel} since the goal of this optimization is to account for the energy lost from damping. This has the added benefit of after a single successful jump the robot appears to have returned to its initial conditions, as the robot has no sense of the upcoming terrain.

The first optimization process finds the optimal design parameter coefficients a_1- c_3, nine in total, minimizing deviation in apex height relative to the terrain, y_{rel}, over multiple terrain levels with differing initial forward velocity. We mathematically represent this as the double Euclidian norm of all deviations y_{rel} or

$$cost(y_{rel}) = \sqrt{\sum_{j=1}^{3}\left(\sqrt{\sum_{i=1}^{10}\left(y_{rel} - \left(y_{apex} - y_{ter(i)}\right)\right)^2}\right)^2} \tag{6}$$

where y_{ter} are the tested terrain deviations. The difference between y_{apex} and y_{ter} is the actual relative height above terrain level. We applied the constrained optimization, fmincon, in MATLAB® which uses the interior point algorithm. We applied the built-in MultiStart algorithm on each run with random initial conditions.

The second optimization process occurs after finding the true damping coefficients and making the system nonconservative. This optimization reevaluates the optimal design parameter coefficients a_1-a_3 and finds d_1-d_3, six parameters in total. The cost function is deviation from y_{rel} and forward velocity at apex or

$$\text{cost}\left(y_{rel}, \dot{x}_{inital}\right) = \left|y_{rel} - \left(y_{apex} - y_{ter(i)}\right)\right| + \left|\dot{x}_{inital} - \dot{x}_{apex}\right| \tag{7}$$

where y_{ter} are the tested terrain deviations and $\dot{x}_{apex}$ is the measured forward velocity at apex. We use a bounded search of the lowest cost function between feasible torque and angle values for each terrain level. We then plot the lowest cost torque and angle against the terrain deviation and use the polyfit function in MATLAB® to find the polynomial coefficients in terms of heights. Using basic trajectory equations, we can convert these polynomial coefficients to be in terms of time from apex giving us a SLR equation and torque varying in only in time.

4.3 Random Terrain Generation

To show the robustness of this powered curved leg design and single jump optimization process we chose to test the legs over multiple jumps on uneven terrain. In optimization the leg is made to account for terrain variations of no more than ± 5 cm. Therefore, the lower bound of terrain step height is five centimeters. The size of each step is randomly generated between one and three meters and the length of the terrain is fixed at 100 m. The max step height, or rise, is set by the user and step heights are randomly chosen within the bounds by MATLAB® with the default random number generator. The terrain also randomly steps up or down, allowing for multiple step-ups or step-downs, often a difficult task. The simulation tracks the center of mass and repots if a crash has occurred. A crash is considered as, the mass hitting the ground in stance or flight, mass hitting a wall in stance or flight, if the leg is fully compressed, and if the mass jumps backward. Only the point $r_0(\theta)$ on the leg is represented in the simulation and therefor there is the possibility a portion of the leg may collide with the wall while the mass does not and $r_0(\theta)$, and it will not be detected.

5 Results

We performed the optimization process to create three legs with three different relative hoping heights, $y_{rel} = 0.3$, 0.35, and 0.4 m. Each leg is optimized over a single jump using the single jump optimization process and is designed to jump from y_0 to y_{rel}. For simplicity we will call these, leg 1, leg 2 and leg 3, respectively.

We performed three simulated experiments. First, we simulated the leg over a single simulated jump to confirm that they have minimal deviation from their intended y_{rel}. Each leg hopped once, and we changed the terrain by 1cm between ± 5 cm for each jump for a total of 10 separate trials for each leg Fig. 4a. In Fig. 4b. we show a boxplot of the distribution of each leg, jumping from its intended y_0 to its intended y_{rel}. Leg 3, optimized for $y_0 = y_{rel} = 0.4$ had the lowest distribution with an interquartile range of 0.39939–0.39973.

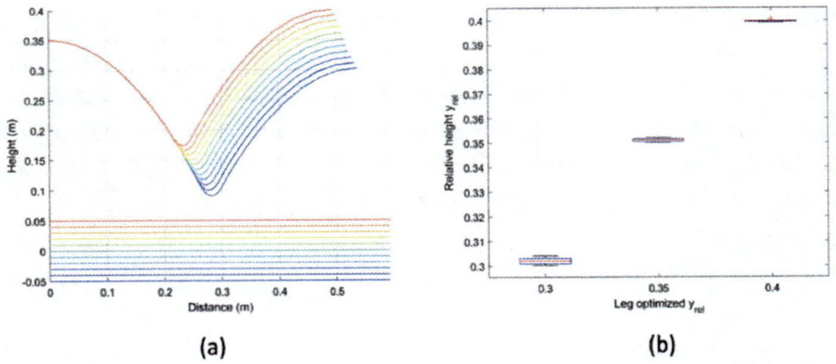

Fig. 4. Single hop tests and results. Plot a shows an example of the single jump test for leg 2. Each jump and their respective terrain deviations in the same color. Plot b shows boxplots of the deviation from the legs target y_{rel} in each test. Note no leg deviates from their y_{rel} by more than 0.5cm

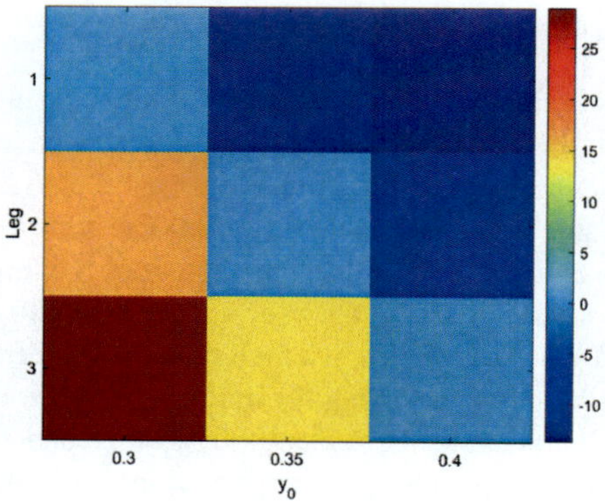

Fig. 5. Heatmap of percent deviation of leg from its staring y0

We then investigated if the optimized legs are truly optimized to jump to their intended relative height from their initial y_0, or if the initial y_0 influences the y_{rel}. We tested each leg over a single jump, $x_{terrain} = 0$, from the initial height of each other leg. By seeing which leg returns to the initial height we can confirm the leg is optimized for its relative height only, and is controlled by leg shape, SLR and torque, rather than start height y_0. In Fig. 5. we plot the matrix of percent deviation from y_0. Only the legs optimized to return to their initial height succeed in doing so, with a maximum percent deviation of 0.62%. Similarly, we see the legs that have been optimized for lower height undershoot y_0 and those optimized for higher height overshoot. In Fig. 5. positive numbers denote overshooting and negative numbers undershooting. The starting height is not the deciding factor of the jump height, rather the leg parameters control the height of the jump.

To explore the relationship between optimized y_{rel} and leg robustness we tested the leg on randomly generated rough terrain. We push the terrain roughness past the designed terrain variation of 0.05 m to test the robustness of each leg design. In our test terrains roughness is varied by $rise$ = [0.050.10.20.30.4] m. The legs are each tested on the same twenty randomly generated terrains at each roughness for a total of 50 tests on each leg. Each test the goal was to jump 50 jumps of the generated terrain. If successful, the leg scored a point. Figure 6. shows an example of a run, with each leg hopping on the same increasingly rough terrain. In Fig. 7. we see that leg 3, optimized for a y_{rel} = 0.4 m, succeeds more often than the other legs as roughness increases. The reason for the increased robustness of leg 3 may be attributed to its increase in jump height. This means the leg can jump over and ignore most potentially failure inducing step ups. However, if the starting energy for all three legs is the same, this would come at the cost of a slower hopping leg.

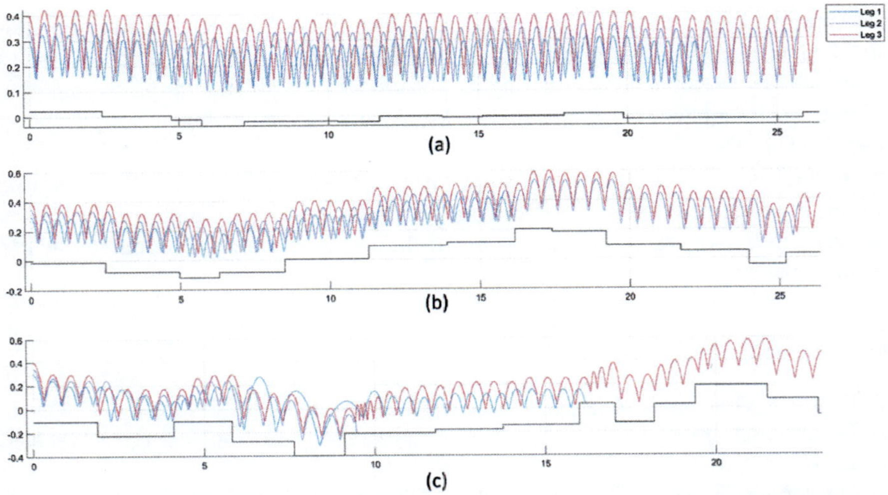

Fig. 6. An example of the legs hopping on simulated rough terrain. For visibility, the legs only jump a maximum of 30 times as opposed to in testing. In a) (rise = 0.05 m) all three legs successfully jump 30 times while in b) (rise = 0.2 m) only leg 1 and leg 2. In the roughest terrain, c), (rise = 0.4 m) only leg 3 completes the maximum jumps and begins showing instabilities like those seen right before 10 m.

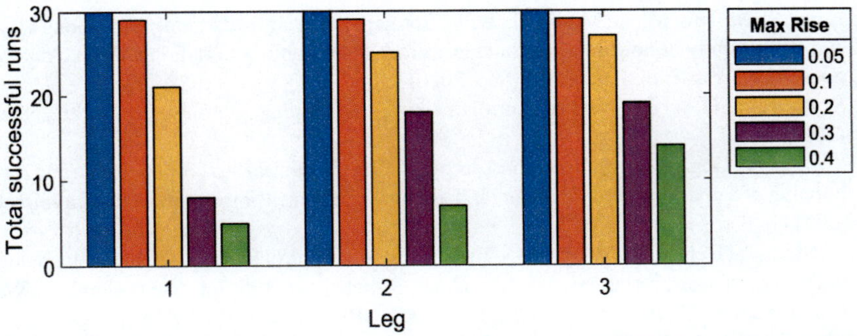

Fig. 7. Total successful runs over increasingly rough terrain. Bar graph shows as terrain level becomes more complex leg 3 outperforms against the other legs in part due to the higher jump height.

6 Conclusions and Future Work

In this work, we showed the integration of actuation into the variable rolling SLIP model and proved its functionality for multiple jumps over rough terrain in an energy non-conservative system. We showed the legs do not significantly deviate from their optimized y_{rel} heights and resistant to perturbation. We showed the previous single jump optimization method along with the addition of the new single jump optimization after damping can functionally translate to multi-hop scenarios with significant perturbations.

The optimization of these 12 leg parameters successfully allows for the use of a simplistic control scheme during jumping leveraging the effects of a spiral-shaped leg as a "embodied" controller. This coupling of stiffness, leg length and ultimately damping parameters in the physical structure of the leg along with the addition of torque allows robust and simple leg.

In the future will advance our simulation to better identify crashes the would be caused by portions of the leg that are not in contact with the ground during stance. Furthermore, we will transfer this simulation to the real world and reformat our real-world test bed to allow for torque control.

References

1. Raibert, M.H.: Legged robots that balance. MIT press (1986)
2. Blickhan, R.: The spring-mass model for running and hopping. J. Biomech. **22**(11), 1217–1227 (1989)
3. Seyfarth, A., Geyer, H., Herr, H.: Swing-leg retraction: A simple control model for stable running. J. Exp. Biol. **206**(15), 2547–2555 (2003)
4. Riese, S., Seyfarth, A.: Stance leg control: Variation of leg parameters supports stable hopping. Bioinspiration and Biomimetics **7**(1) (2012)
5. Karssen, J.G.D., Haberland, M., Wisse, M., Kim, S.: The optimal swing-leg retraction rate for running. In: Proc. - IEEE Int. Conf. Robot. Autom, pp. 4000–4006 (2011)
6. Jun, J.-Y., Clark, J.E.: Characterization of running with compliant curved legs. Bioinspir. Biomim. **10**(4), 46008 (2015). Jul.
7. Heim, S., Millard, M., Le Mouel, C., Badri-Spröwitz, A.: A little damping goes a long way: A simulation study of how damping influences task-level stability in running: A little damping goes a long way. Biol. Lett. **16**(9) (Sep. 2020)
8. Seyfarth, A., Geyer, H.: Natural control of spring-like running: Optimised selfstabilisation, 1 (2014)
9. Shemer, N., Degani, A.: A flight-phase terrain following control strategy for stable and robust hopping of a one-legged robot under large terrain variations. Bioinspiration and Biomimetics **12**(4) (2017)
10. Gaathon, A., Degani, A.: Adaptive swing leg retraction control for robust dynamic locomotion under large terrain variations. In: Adv. Coop. Robot. Proc. 19th Int. Conf. Climbing Walk. Robot. Support Technol. Mob. Mach. CLAWAR 2016, pp. 369–377 (2016)
11. Zhou, X., Bi, S.: A survey of bio-inspired compliant legged robot designs. Bioinspiration and Biomimetics **7**(4) (2012)
12. Ferris, D.P., Louie, M., Farley, C.T.: Running in the real world: Adjusting leg stiffness for different surfaces. Proc. R. Soc. B Biol. Sci. **265**(1400), 989–994 (1998)
13. Jun, J.Y., Clark, J.E.: Dynamic stability of variable stiffness running. In: 2009 IEEE International Conference on Robotics and Automation, pp. 1756–1761 (2009)
14. Ernst, M., Geyer, H., Blickhan, R.: spring-legged Locomotion on Uneven Ground: a Control Approach to Keep the Running Speed Constant. In: Mobile Robotics, pp. 639–644
15. Ernst, M., Geyer, H., Blickhan, R.: Extension and customization of self-stability control in compliant legged systems. Bioinspir. Biomim. **7**(4), 46002 (2012). Jul.
16. Galloway, K.C., Clark, J.E., Koditschek, D.E.: Design of a tunable stiffness composite leg for dynamic locomotion. In: Proc. ASME Des. Eng. Tech. Conf., vol. 7, no. Parts A and B, pp. 215–222 (2009)
17. Ankarali, M.M., Saranli, U.: Stride-to-stride energy regulation for robust self-stability of a torque-actuated dissipative spring-mass hopper. Chaos **20**(3) (2010)

18. Seipel, J., Holmes, P.: A simple model for clock-actuated legged locomotion. Regul. Chaotic Dyn. **12**(5), 502–520 (2007)
19. Hu, C.-J., Huang, C.-K., Lin, P.-C.: A torque-actuated dissipative spring loaded inverted pendulum model with rolling contact and its use as the template for design and dynamic behavior generation on a hexapod robot. IEEE International Conference on Robotics and Automation (ICRA) **2015**, 5177–5183 (2015)
20. Mutka, A., Kočo, E., Kovačić, Z.: Adaptive control of quadruped locomotion through variable compliance of revolute spiral feet. Int. J. Adv. Robot. Syst. **11**(10), 162 (2014)
21. Gaathon, A., Degani, A.: A Variable Rolling SLIP Model for a Conceptual Leg Shape to Increase Robustness of Uncertain Velocity on Unknown Terrain. RSS (2020)

Check for updates

Direct Centroidal Control for Balanced Humanoid Locomotion

Grzegorz Ficht(✉) and Sven Behnke

Institute for Computer Science VI, University of Bonn,
Endenicher Allee 19a, 53115 Bonn, Germany
{ficht,behnke}@ais.uni-bonn.de

Abstract. We present an integrated approach to locomotion and balancing of humanoid robots based on direct centroidal control. Our method uses a five-mass description of a humanoid. It generates whole-body motions from desired foot trajectories and centroidal parameters of the robot. A set of simplified models is used to formulate general and intuitive control laws, which are then applied in real-time for estimating and regulating the center of mass position and orientation of the multibody's principal axes of inertia. The combination of proposed algorithms produces a stretched-leg gait with naturally looking upper body motions. As only a 6-axis IMU and joint encoders are necessary for the implementation, the portability between robots is high. Our method has been experimentally verified using an igus® Humanoid Open Platform, demonstrating whole-body locomotion and push rejection capabilities.

Keywords: Humanoid robotics · Whole-body locomotion · Balancing

1 Introduction

Dynamic control of humanoid robots has been an active research area for several years. The high-dimensionality and underactuation of the system, coupled with environment unpredictability, signal noise, and non-deterministic actuation has led to the development of numerous approaches. When real-time control poses no issue, task-based optimal control methods are capable of producing motion plans exploiting the full dynamics with constraint awareness [1,19].

Supporting online application has been possible by utilising models capable of generalising a core part of the dynamics. For humanoid walking and balancing, a substantial line of work is based on tracking Center of Mass (CoM) and Zero Moment Point (ZMP) trajectories, generated with the Linear Inverted Pendulum Model (LIPM). Given desired ZMP locations, a CoM trajectory can be generated using preview control [16] and tracked in a state-space setting [18]. Computing optimal control gains—or even the complete trajectories over long horizons—can be done in a Linear-Quadratic Regulator (LQR) setting [28]. Another approach involves using Model Predictive Control (MPC) [8,31], also capable of handling large perturbations. Alternatively, some works build the control around the Capture Point (CP), reducing the system dynamics to first order [9,10].

© The Author(s), under exclusive license to Springer Nature Switzerland AG 2023
J. M. Cascalho et al. (Eds.): CLAWAR 2022, LNNS 530, pp. 242–255, 2023.
https://doi.org/10.1007/978-3-031-15226-9_24

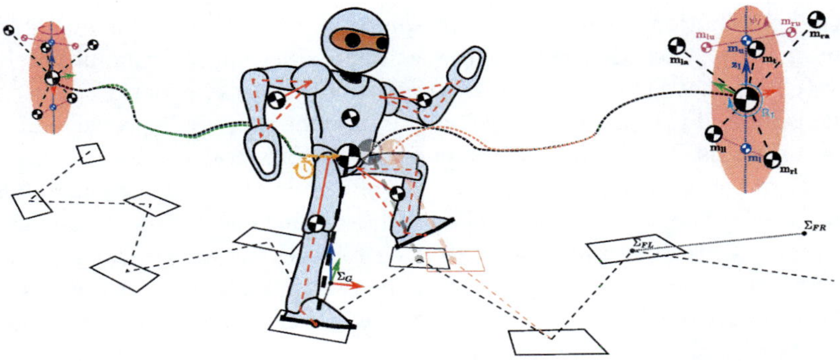

Fig. 1. Direct centroidal control, including CoM and CAM regulation with step size and timing adjustment. Our approach achieves walking with straight legs and angular momentum regulation using the tilt and yaw of the inertia principal axes. Five masses are combined into three dumbbels, whose relative movement translates between the full humanoid kinematics and its centroidal model.

The LIPM, while extremely useful, has two limitations which can hinder walking performance. The first one is constraining the CoM to move on a horizontal plane, which requires it's controllability in the vertical axis. Ensuring this usually results in the well-known, bent-knee walking to avoid kinematic singularities. Improving this issue has been the aim of multiple research works [13,15,20]. The second limitation stems from the assumption of zero angular momentum around the CoM. This is not an issue for generating the CoM motion, as humans are known to regulate the Centroidal Angular Momentum (CAM) close to zero when walking [26]. This ability however, needs to be included in the control and has led to the flywheel-extended LIPM [27]. The legs are then often assumed to be massless, while the torso acts as the flywheel [30]. This approximation is favored as so far there has been no clear and intuitive angular counterpart to the linear CoM motion. To the best of our knowledge, only angular excursions from whole body angular velocities so far have been used [26]. Simultaneous control of linear and angular momenta has been tackled also in more general approaches. Examples include the Resolved Momentum Control framework [17] and Whole-Body Control (WBC) for motion generation using centroidal dynamics [25], with improving online application being an active research topic [7,29].

A separate consideration to the control method is the capability of the hardware to execute these motion plans. On the hardware layer, this is often realized with backlash-free reducers and a high gain setting, to assure joint tracking performance. In lower-quality hardware, the limited joint torque and presence of compliance, backlash, and latency render many of these methods unapplicable. Also, ground contact and stability assessment is often assumed as given with force and/or torque sensors, which is not always the case [22]. In this regard, there exists a whole ZMP-less category of methods, which are designed around biologically-inspired Central Pattern Generators (CPG) [24]. As the trajectories are manually designed to produce a *self-stable*, open-loop gait, no precise modelling or foot sensors are required. Push rejection can be implemented on the joint level [3] or by modifying gait commands to alter the step parameters [23].

The contribution of this paper is an approach to whole-body humanoid robot locomotion by generating motion frames approximating the Composite Rigid Body (CRB) Inertia with five masses. To improve tracking performance, knee torque is limited through the inclusion of a straight-support-leg constraint. Reference CoM and ZMP values are combined with foot trajectories to determine reference orientations of the principle axes of inertia $\mathbf{R_I}$, which we introduce as an angular position equivalent to enhance the walking capabilities through CAM regulation. Control laws are then extended to track the centroidal parameters for both the linear and angular movement. Additionally, step size and timing feedback is developed using a closed-form prediction of the end-of-step state. As the approach requires only a 6-axis IMU and joint position feedback, it can be applied to a wide variety of robots.

2 Five-Mass Centroidal Description of a Humanoid

As shown in Fig. 1, our description of a humanoid robot is based on a non-uniform five-mass distribution model, where each limb and the torso are represented by their respective point mass. The torso mass is offset from the center of the pelvis, assumed to be the floating base of the robot Σ_B. Each limb mass is attached to the torso and parameterised through a triangle approximation, which ties in the kinematics and mass placement with a one-to-one mapping. The relative movement of these five masses shapes the system centroidal properties, e.g. it's CoM and system inertia $\mathbf{I_R}$, which we used to develop a whole-body motion generator in [12]. Our method can be thought of as a forward and inverse kinematics approach to whole-body humanoid control, where both the CoM and system inertia can be altered to achieve desired dynamic effects, e.g. concurrent linear and angular momentum regulation.

We compute the joint configuration $\mathbf{q}$ by providing CoM-relative foot frames Σ_{FL}, Σ_{FR}, constrained by set inertia principal moments $\mathbf{I_{PA}}$ and axes orientation $\mathbf{R_I}$ (Fig. 1, on the right). The upper body and lower body (leg) masses $\mathbf{m_u}$, $\mathbf{m_l}$ form a dumbbell, which tilts around the CoM. A virtual single leg evaluates the placement of the base frame from $\mathbf{m_l}$ with respect to $\mathbf{m_u}$ reachability. Having confirmed a viable dumbbell (inertia) tilt $\mathbf{z_I}$, the trunk is placed with its mass pointing towards $\mathbf{m_u}$ and aligned with the set inertia yaw angle ψ_I. This completes the trunk orientation $\mathbf{R_t}$ for Σ_B, which in conjunction with Σ_{FL} and Σ_{FR} specifies the leg configuration. The leg masses form a dumbbell, which rotates around $\mathbf{m_l}$ with a yaw angle ψ_l. A similar dumbbell is put at $\mathbf{m_u}$, yawing with ψ_u. Both of these dumbbells work to realise the set yaw rotation ψ_I. This top-down approach introduces more detail on each step of the pose generation scheme, which naturally resolves the upper body movement to satisfy the constraints driven by the feet. For a full description on the pose generation scheme, please refer to [12].

During the typical bent-knee walking, the majority of the weight is put on the supporting knee joint. Most tracking errors arise at this point due to the accumulation of non-negligible joint compliance, backlash, and limited torque.

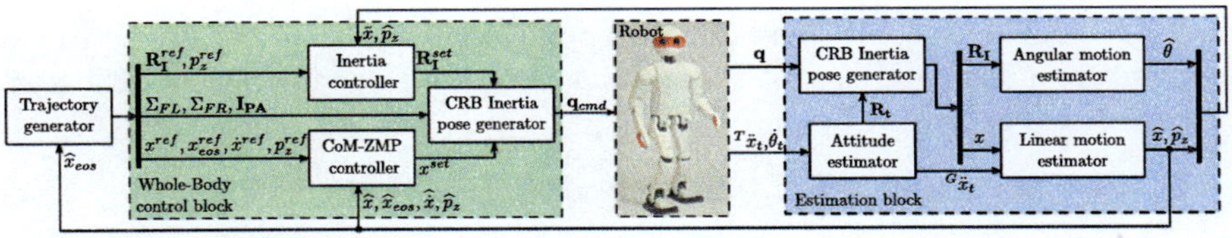

Fig. 2. Interaction between the framework modules.

We work around this issue by extending our pose generator [12] by including a straight-support-leg constraint. After computing the initial torso orientation and hip origins, we evaluate the distance to the supporting leg's ankle. If it is shorter than the sum of both leg links, the base frame position is shifted from the virtual leg origin by that difference. By doing so, the motion generator will resolve the CoM and inertia placement with a recomputed torso orientation $\mathbf{R_t'}$. The pose generation continues with Σ_B', which allows to maintain control over the CoM in the vertical axis by adjusting the upper body and arms.

3 Control Approach

Figure 2 gives an overview of our approach. For brevity, we limit the majority of the equation derivations to the planar case, three-dimensional coordinates are provided only when necessary.

3.1 Reference Trajectory Generation

Generating a frame of motion with the five-mass model from Sect. 2 requires specifying desired foot frames Σ_{FL}, Σ_{FR} relative to the CoM and desired system inertia $\mathbf{I_R}$ composed of the principal axes orientation $\mathbf{R_I}$ and moments $\mathbf{I_{PA}}$. In scenarios with plain single or double support balancing, the reference CoM and ZMP stay constant—unlike in walking—which requires specifying trajectories due to contact point exchange. In this regard, we follow the approach of Kajita [14] with the well-known planar Linear Inverted Pendulum Model (LIPM) dynamics and its analytical solution:

$$\begin{aligned} \ddot{x} &= \omega^2(x - p_z), \quad \omega = \sqrt{g/h} \\ x(t) &= x_0 \cosh(\omega t) + \dot{x}_0 \frac{1}{\omega} \sinh(\omega t), \\ \dot{x}(t) &= x_0 \omega \sinh(\omega t) + \dot{x}_0 \cosh(\omega t), \end{aligned} \tag{1}$$

where h is the constant CoM height, g is gravity, while x and p_z are the horizontal CoM and ZMP positions, respectively. Given a set of initial conditions $(x_0, \dot{x}_0)$, we can compute the state $(x, \dot{x})$ with respect to elapsed time t. Two independently generated planar solutions are combined into a three-dimensional ground-relative CoM coordinate $\mathbf{c}$. A given velocity v_g^{ref} can be achieved through multiple combinations of step size and applied gait frequency f_g. By setting a nominal frequency, we control the walking speed with stride length. The step progression is expressed by μ:

$$\mu \in [0,1), \qquad \mu[n+1] = \mu[n] + f_g \Delta t, \tag{2}$$

where Δt is the time step arising from the system control frequency. We let μ wrap around when crossing its border values, which triggers a sign flip of the supporting leg sign $\iota : \{\mathrm{L} = -1, \mathrm{R} = 1\}$. We compute the next initial step $\mathbf{s}_i$ coordinates x_s, y_s and heading ψ_s as follows:

$$\mathbf{s}_i \begin{cases} x_s & = v^{ref}_{g,x}/f_g \\ y_s & = \iota s_w + s_y = \begin{cases} 2v^{ref}_{g,y}/f_g & \text{if } \iota = \mathrm{sgn}(v^{ref}_{g,y}) \\ 0 & \text{otherwise} \end{cases}, \\ \psi_s & = v^{ref}_{g,\psi}/f_g \end{cases} \tag{3}$$

which is then refined into the applied footstep $\mathbf{s}$ by slight position adjustments to produce a single continuous trajectory according to [14]. When moving sideways, we first take a larger step, and then a trailing one with nominal step width s_w. Only three steps are necessary to be stored at any given time. The supporting foot stays at the current foothold $\mathbf{s}[k]$ while the swing leg travels from the previous footstep location $\mathbf{s}[k-1]$ to the next one $\mathbf{s}[k+1]$, reaching a step height of s_h with:

$$\Sigma_F \begin{cases} \mathbf{f} = \begin{cases} x_f = (1-\mu)\mathbf{s}^x[k-1] + \mu\mathbf{s}^x[k+1] \\ y_f = (1-\mu)\mathbf{s}^y[k-1] + \mu\mathbf{s}^y[k+1] \\ z_f = s_h \sin(\mu * \pi) \end{cases} \\ \psi_f = (1-\mu)\mathbf{s}^\psi[k-1] + \mu\mathbf{s}^\psi[k+1]. \end{cases} \tag{4}$$

For simplicity, the roll ϕ_f and pitch θ_f are set to zero to keep the feet parallel to the ground. When a foothold has been reached, the queue of steps with their coordinates are shifted, and a new step is generated. As the footholds are in reference to the ground, the final foot frame positions are offset by the set CoM position, to conform with the whole body pose generator. The final components required to define a whole body pose are the inertial properties $\mathbf{I_R}$. The definition for the inertia orientation $\mathbf{R}_\mathbf{I}^{ref}$ is obtained with Rodrigues' axis-angle formula from the inertia z-axis tilt $\mathbf{z}_\mathbf{I}^{ref}$ and the rotation angle around it ψ_I. A continuous *neutral* tilt trajectory of $\mathbf{z}_\mathbf{I}^{ref}$ should not result in generating CAM. This is achieved when the torque of a flywheel attached to a pendulum is zero, e.g. it does not rotate with respect to it. We therefore align $\mathbf{z}_\mathbf{I}^{ref}$ with the robot tilt defined as the vector originating at the feet of the robot, pointing towards the CoM and set ψ_I to align with the step progression μ:

$$\mathbf{z}_\mathbf{I}^{ref} = -(\mathbf{f_l} + \mathbf{f_r})/2, \quad \psi_I = (1-\mu)\mathbf{s}^\psi[k] + \mu\mathbf{s}^\psi[k+1]. \tag{5}$$

With this we keep CAM mostly independent from CoM and foot placement. Finally, nominal principal moments $\mathbf{I}_{\mathbf{PA}^{ref}}$ are kept constant and computed from an upright standing pose, with the pose generator adjusting their values to satisfy other constraints.

3.2 State Estimation

Estimating the robot's state is a necessity when it comes to precise control involving system dynamics. To maintain generality and a wide applicability, we require only joint encoders and a 6-axis IMU in the trunk to reconstruct the necessary state variables. The IMU is used to provide the torso (base) orientation $\mathbf{R_t}$ of Σ_B with an attitude estimator of choice [2,6,21]. In combination with the measured joint angles $\mathbf{q}$, we compute the foot frames Σ_{FL}, Σ_{FR} and the positions of the five masses. These are then used to reconstruct the system CoM and inertial properties $\mathbf{I_R}$, $\mathbf{I_{PA}}$. For synchronising the measured and control state we use a common reference *ground* frame Σ_G. This frame is laterally offset from the supporting foot by half of the step width from the last step taken (see Fig. 1). The supporting foot is chosen as the lower one of the two with a hysteresis applied to disable rapid switching of the frames, within which the robot is assumed to be in *semi double support.* By keeping track of every support exchange, we update the robot's odometry. To provide the robot with a notion of balance, a stability criterion is required. As we base our control around the LIPM concept, we use the ZMP which is a linear combination of the CoM position x and acceleration $\ddot{x}$ as seen in (1). We use a Kalman filter to estimate the complete state vector $\mathbf{x} = \begin{bmatrix} x \; \dot{x} \; \ddot{x} \end{bmatrix}$. Unlike work mentioned in [22], we do not explicitly use the LIPM state transition matrix in the filter design to not enforce it's dynamics in the estimation. Instead we use the linear mass model. Although the five-mass approximation does provide accurate x measurements, relying solely on them does not provide satisfactory acceleration estimates. ZMP estimates $\widehat{p}_z$ will be inoperative due to the non-negligible time delay before $\widehat{\ddot{x}}$ converges in response to actual changes in $\ddot{x}$. We supplement the measurement model $\mathbf{z}_k$ with the unrotated and unbiased for gravity g trunk acceleration ${}^G\ddot{x}_t$:

$$\begin{bmatrix} {}^G\ddot{x}_t \\ {}^G\ddot{y}_t \\ {}^G\ddot{z}_t \end{bmatrix} = \begin{bmatrix} {}^T\ddot{x}_t \\ {}^T\ddot{y}_t \\ {}^T\ddot{z}_t \end{bmatrix} \mathbf{R_t} - \begin{bmatrix} 0 \\ 0 \\ g \end{bmatrix}, \quad \mathbf{z}_k = \begin{bmatrix} 1 & 0 \\ 0 & 0 \\ 0 & 1 \end{bmatrix} \begin{bmatrix} x \\ {}^G\ddot{x}_t \end{bmatrix} + \mathbf{v}_k \tag{6}$$

Having sources stemming from two different state variable measurements allows for low-delay, robust estimation of $\mathbf{x}$. Even though ${}^G\ddot{x}_t$ only roughly coincides with $\ddot{x}_t$ only when CAM is zero, its fusion with CoM measurements provides a refined estimate $\widehat{\mathbf{x}}$ with fast $\ddot{x}$ convergence. The ZMP estimate $\widehat{p}_z$ is then simply:

$$\widehat{p}_z = \widehat{x} - \frac{\widehat{z}}{g}\,\widehat{\ddot{x}}. \tag{7}$$

A second Kalman filter is used for the angular motion, where angles θ of the inertia orientation $\mathbf{R_I}$ are provided to estimate $\dot{\theta}$ and $\ddot{\theta}$ of the multibody.

3.3 Center of Mass Controller

We start off with the control law proposed by Choi et al. [5], which realizes input-to-state stability by adjusting the velocity to steer the CoM x back to follow the nominal velocity $\dot{x}^{ref}$ and ZMP p_z^{ref} in the presence of errors:

$$\begin{aligned} \dot{x} = \dot{x}^{ref} + \dot{x}_{ctrl}, \qquad \dot{x}_{ctrl} = \dot{x}_{zmp} + \dot{x}_{com}, \\ \dot{x}_{zmp} = K_{zmp}(p_z^{ref} - p_z),\ \dot{x}_{com} = K_{com}(x^{ref} - x). \end{aligned} \tag{8}$$

We integrate its output to work with our position-based motion generator. We substitute the integral of $\dot{x}^{ref}$ with x^{ref}, while integrating $\dot{x}_{ctrl}$ with a *leaky* integrator to produce the sum x_i. The output position is then:

$$x^{set} = x^{ref} + x_i, \quad x_i[n] = \dot{x}_{ctrl}[n]\Delta t + (1-\alpha)x_i[n-1], \quad \alpha \in [0,1] \tag{9}$$

with α being an exponential decay factor. This approach already provides two benefits over a non-leaky integration approach. The first one being the ability to forget past errors, which propagate during walking from one footstep to another, while the second being a more aggresive tuning of the K_{zmp} and K_{com} gains. In a non-leaky integrator, using low gain values is necessary to prevent velocity build-up and overshooting, at the expense of insufficient error rejection capabilities and a long settling time. With the leaky integration however, the CoM velocity does not wind-up to levels which would first require the controller to slow it down.

This version of the controller is already capable of tracking CoM-ZMP trajectories with limited push rejection capabilities. However, the nature of perturbations during walking is erratic with concurrent impulse and constant forces acting on the robot, which disrupt the rhythm of the gait. To sustain a nominal gait cycle, we introduce two additional velocity terms $\dot{x}_{eos}$ and $\dot{x}_{vel}$ into $\dot{x}_{ctrl}$ from (8):

$$\begin{aligned} \dot{x}_{vel} = K_{vel}(\dot{x}^{ref} - \dot{x}), \\ \dot{x}_{eos} = K_{eos}(x_{eos}^{ref} - x_{eos}). \end{aligned} \tag{10}$$

The reasoning behind the $\dot{x}_{vel}$ component is to maintain the reference velocity across the whole step. In standing or walking-on-the-spot situations this term mostly contributes to decreasing the settling time of a transient response to a push. The second term $\dot{x}_{eos}$ uses the LIPM equations (1) with the remaining step time $t_r = (1-\mu)/f_g$ to compute reference and expected end-of-step (EOS) states. Given the currently estimated state and confining the ZMP to the support polygon, the computed $\dot{x}_{eos}$ steadily steers the CoM towards the nominal EOS position in preparation for the next support exchange, essentially providing the controller with predictive capabilities. While the CoM-ZMP regulator is responsible for instantaneous adjustments and keeping the ZMP within stability margins as per the definition in [5], the introduction of extra terms assures consistent long-term behaviour of the system. The final form of the extended control law, which steers the CoM position to stability is:

$$\dot{x}_{ctrl} = \dot{x}_{com} + \dot{x}_{zmp} + \dot{x}_{vel} + \dot{x}_{eos}. \tag{11}$$

3.4 Inertia Controller

The reference generator and position controller are designed around the Linear Inverted Pendulum (1) concept, with dynamics that make use of two assumptions: constant CoM height, and lack of Angular Momentum around the CoM.

In practice, this leads to making assumptions that a fixed torso height and orientation leads to zero Angular Momentum. Building control around this discards a set of dynamics, which—if properly utilised—can increase the stability region and performance of the system. There is no guarantee that the torso keeps still, especially with high joint compliance, backlash and external disturbances. Our inertia-based whole-body pose generation is designed to accomodate for this, along with foot swinging allowing for unified regulation of the centroidal system description.

The purpose of the inertia controller is to actively control the principle axes of inertia orientation $\mathbf{R_I}$ to maintain nominal CoM-ZMP trajectories. We do this by computing a complementary rotation $\mathbf{R_c}$ to the neutral one (5):

$$\mathbf{R_I}^{set} = \mathbf{R_c}\mathbf{R_I}^{ref}. \tag{12}$$

For this, we utilise the concept of *augmented CoM* (ACoM) $\tilde{x}$ introduced in [30]. The premise of the augmented CoM is to combine the linear and angular movement of the mass x and flywheel θ_f of a Linear Inverted Pendulum Plus Flywheel Model (LIPFM) within a single state variable, and control it with respect to an Augmented ZMP $\tilde{p}_z$, putting the LIPFM dynamics into LIPM form:

$$\tilde{x} = x + \frac{I\theta_f}{mh}, \quad \tilde{p}_z = p_z + \frac{I\theta_f}{mh}, \quad \ddot{\tilde{x}} = \frac{g}{h}(\tilde{x} - \tilde{p}_z), \tag{13}$$

where m and h are the system mass weight and height, respectively, I is the inertia moment, and θ_f the flywheel angle around the perpendicular axis. Our aim is to maintain zero angular velocities of the flywheel, by regulating its orientation to zero. To achieve this, we use a similar control law as in (8), and set the reference ACoM position to align with the output of the CoM-ZMP controller:

$$\begin{gathered} \dot{\tilde{x}}_{ctrl} = \dot{\tilde{x}}_{zmp} + \dot{\tilde{x}}_{com}, \\ \dot{\tilde{x}}_{zmp} = K_{zmp}^{aug}(\tilde{p}_z^{ref} - \tilde{p}_z), \quad \dot{\tilde{x}}_{com} = K_{com}^{aug}(\tilde{x}^{ref} - \tilde{x}). \end{gathered} \tag{14}$$

With the desired orientation kept at zero, the augmented and set state coincide ($\tilde{x}^{ref} \simeq x^{set}$), as well as the references for the regular and augmented ZMP ($\tilde{p}_z^{ref} \simeq p_z^{ref}$). Due to working in the space of the augmented CoM and its derivatives, the $\ddot{\theta}$ term of the Centroidal Moment Pivot (CMP) gets hidden within $\ddot{\tilde{x}}$. This benefits our control scheme by enabling regulation of the CMP towards the ZMP, without having to measure or estimate $\ddot{\theta}$ of our abstract flywheel. The control velocity $\dot{\tilde{x}}_{ctrl}$ is then integrated as in (9) into $\tilde{x}_i$, which holds the inertia rotation angles for $\mathbf{R_c}$:

$$\tilde{x}_i = \frac{I\theta_f}{mh} \Rightarrow \theta_f = \frac{\tilde{x}_i mh}{I}. \tag{15}$$

The rotation $\mathbf{R_c}$ can be expressed with any favorable representation. We use projected angles to avoid gimbal lock and maintain a set angle on two axes independently [11].

3.5 Stepping Controller

Conceptually, the LIPM assumes the dynamics can be split into two independent planar components. In reality, they are coupled by the cyclic nature of the gait. The lateral pendulum oscillations produced by instantaneous support foot exchanges define the rhythm of the gait. A wrongly timed foot exchange can quickly lead to balance loss. We introduce timing feedback, based on the nominal y_{eos}^{ref} and expected EOS y_{eos} lateral CoM error, by using the remaining step time with (1):

$$\tilde{f}_g = f_g + K_t(y_{eos}^{ref} - y_{eos}). \tag{16}$$

The timing adjusts the rate of step progression with $\tilde{f}_g$ in (2). Intuitively, if the CoM is on a trajectory to cross the lateral apex point, the progression will slow down and provide more time to both balance controllers to bring the CoM back. Similarly, overshooting towards the swing leg will speed-up the step exchange to maintain the nominal cycle.

The sagittal motion is less sensitive in the timing, as the velocity is set through the step size, and known frequency (3). However, its location needs to be set accordingly, given that with the current velocity it is possible to return to a nominal trajectory in a finite number of steps. The active next foothold $s[k+1]$ distance x is constantly recalculated (3) in response to the current step EOS CoM error, smoothly adapting the foot trajectory and walking speed:

$$x_{s[k+1]} = v_{g,x}^{ref}/f_g - K_s(x_{eos}^{ref} - x_{eos}). \tag{17}$$

4 Experimental Results

Verification of the proposed controller was performed on a 90 cm tall, igus® Humanoid Open Platform robot [4] possessing 20 position-controlled joints. The complete control pipeline was executed using the robot's on-board computer to produce the reference and stabilising motions. A single passthrough of the whole control loop with computing joint targets took on average 143 µs. As the control rate is limited to 100 Hz by the actuators, we are still left with more than 98% of idle time, to be spent on higher-layer tasks.

For the following experiments, we set the gait frequency f_g to 2.6 Hz which is quite high, but the underpowered ankles necessitate making smaller step sizes. The integration decay factor α of all controllers was set to 0.03, corresponding to a decay half time of roughly half the step time. The regular and augmented CoM-ZMP gains K_{zmp}, K_{com}, K_{zmp}^{aug}, K_{com}^{aug} were set according to [5], but with quite high gains due to the leaky integration. We have found that the long-term behaviour gains supplemented the controller best when set to $K_{vel} = 0.5$, $K_{eos} = 0.5K_{com}$, as the output would then equally mediate between the long and short-term control goals. Step timing and adjustment were set to $K_t = 10f_g$, $K_s = 1$, making a 10 cm EOS error in the lateral plane fully slow down, or double the gait progression. A similar error in the sagittal direction would result in adjusting the foot placement by 10 cm.

Table 1. Influence of the developed control techniques on walking quality. Control components are gradually added and tuned to achieve the best possible performance. The assessment is performed at highest achievable stable velocities with integral square errors of CoM, ZMP and velocity tracking, normalized by walking time and velocity.

Control mode	$v_{g,x}^{ref}\left[\frac{\mathrm{m}}{\mathrm{s}}\right]$	$v_{g,x}\left[\frac{\mathrm{m}}{\mathrm{s}}\right]$	e_c	e_z	e_v
Open-Loop	0.025	0.1304	5.3877	9.3537	76.734
Straight leg constraint	0.050	0.0970	2.3155	4.3287	18.473
Closed-Loop	0.005	0.0583	35.722	56.002	214.34
Leaky integration	0.125	0.2162	0.6266	1.2438	11.886
Extended control terms	0.350	0.4403	0.3921	1.0304	5.8843
Inertia orientation control	0.400	0.4916	0.2925	0.8572	3.9287
Step timing regulation	0.400	0.5166	0.4269	0.9225	4.8888
Step size adjustment	**0.485**	**0.5292**	**0.1895**	**0.7522**	**1.8463**
Without inertia control	0.350	0.3972	0.2676	0.7988	3.0245

To display the influence and justify the usage of each proposed component, we perform a series of tests where a highest achievable, stable and constant walking velocity is requested. The quality is assessed in each run by comparing the integral square errors of the relevant variable vector $\mathbf{v}$ normalized by a representatively sufficient walking time t_w and requested forward velocity $v_{g,x}^{ref}$:

$$e_* = \frac{1}{v_{g,x}^{ref} t_w} \int_0^{t_w} (\mathbf{v}^{ref} - \mathbf{v})^2 dt, \quad * : \begin{Bmatrix} c \,|\, \mathbf{v} = [x \; y] \\ z \,|\, \mathbf{v} = [p_{z_x} \; p_{z_y}] \\ v \,|\, \mathbf{v} = [\dot{x} \; \dot{y}] \end{Bmatrix}, \tag{18}$$

as higher velocities are much harder to maintain. The results of this assessment can be seen in Table 1. Due to the limited torque in the joints, coupled with backlash and compliance, the trajectories can barely be executed with frequent falls during open-loop walking. The straight-support-leg constraint reduced the necessary torque on the knees significantly to allow for better, more uniform operation. Closing the loop with a non-leaky integrator resulted in inoperability, where only a few steps caused a fall from self-excited instability. Introducing leaky integration provided a substantial gain in the control capabilities to the point where omnidirectional walking with varying velocities became possible. At moderate speeds, the walking velocity would fluctuate in a sinusoidal manner, eventually leading to a fall. The additional control terms solved this issue, increased the accuracy and more than doubled the maximum speed. Reaching the limits of CoM-ZMP control, regulating the inertia orientation provided another substantial increase in performance. The continuous adjustments to the torso and arm positioning generated sufficient angular momentum to complement and relieve the CoM-ZMP controller, increased the stability margin and allowed for higher walking speeds. Finally, step size and timing adjustments made the gait more accomodating against sudden slips and stumbles while also compensating

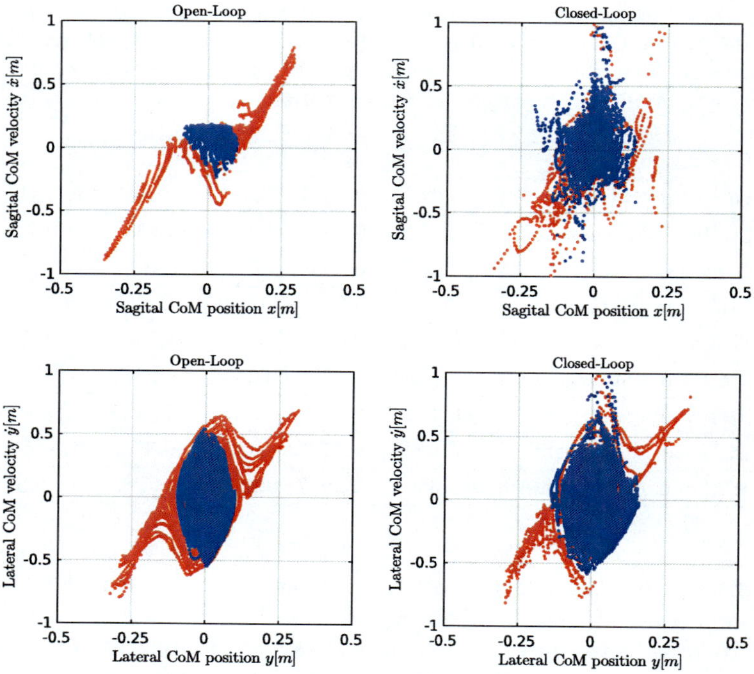

Fig. 3. Phase plots displaying the stability of open and closed-loop performance in the sagittal and lateral direction. Blue and red dots represent stable and unstable trajectory points respectively.

for visible velocity overshooting with a significant e_v reduction. Although often omitted in typical LIPM control scenarios, here we can clearly notice the influence that angular momentum has on walking performance. Turning off CAM regulation while keeping the step adaptation resulted in difficulties in maintaining higher velocities, where constant gait adjustment was necessary.

We assess the push recovery performance of the presented unified control scheme by aggregating several pushes into phase plots in Fig. 3, with examples of robot performance shown in Fig. 4. The Open-Loop system has a tendency to sag towards the front of the feet and innately dampened pushes of unsignificant strength in the sagittal direction. Closing the loop corrected the sagging and typically responded to a push with both leaning and stepping in it's direction.

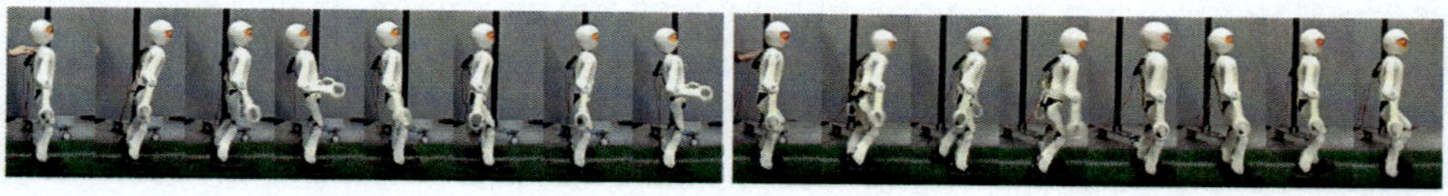

Fig. 4. A time evolution of the robot performing push recovery with the developed controllers. Left: using the controller while standing. Right: while walking.

Table 2. Number of withstood simulated pushes out of 20

Impulse [Ns]	1.2	1.5	1.8	2.0	2.2	2.4	2.6	2.8
Ours	20	20	20	18	17	15	13	10
Allgeuer [3]	20	20	17	16	14	9	7	4

Here, an undesired coupling was observed for pushes coming from the front, where rotating the trunk would limit the leg touchdown towards the back due to kinematic limits. Laterally, a push on the open-loop system lead to desynchronising the gait or tipping over the outer edge of the supporting foot. Enabling feedback smoothly adapted the gait progression to allow the CoM and inertia controllers to bring the state back onto the nominal trajectory with a combination of CoM shifting and corrective inertia roll angle. A substantial stability increase in both directions is clearly visible.

We have also evaluated the controller quantitatively in a Gazebo simulator. We compare our results to [3], which tackles balanced walking with a model-free, CPG-based gait. It is also the most recent work using the same robot and simulator for the evaluation. We replicate the benchmarking scenario of performing sets of 20 pushes from a random direction with increasing push strength. The result of this experiment is shown in Table 2, with a video available[1]. Our control over the centroidal state and step feedback produces clearly better push rejection capabilities than [3]. When measuring forward velocities, a maximum mean of 49.3 cm/s was achieved, which is also an improvement when compared to 45.7 cm/s from [3]. Not only is the robot able to walk faster, but is twice as likely to sustain a relatively strong push.

5 Conclusions

We presented a unified control approach to whole-body locomotion and balancing of humanoid robots and verified it experimentally. While state-of-the-art methods using simplified models are known to work, they require sufficient sensing, joint torque and tracking capabilities which lower quality hardware cannot guarantee. We have shown that such methods are also capable of operating on such hardware, given that supplementary measures are taken. Furthermore, we have increased the capability of an existing CoM-ZMP controller with additional feedback terms that operate on closed-form solutions. This provides the control with predictive capabilities and greatly increases performance in both gait speed and stability. Unlike in MPC, our scheme easily handles on-the-fly CoM height changes, as no preview matrices need to be computed. The Augmented CoM concept has been employed to control the orientation of the whole-body inertia principal axes, which proved to be a meaningful representation for centroidal angular momentum regulation. By employing a reduced five-mass description of a humanoid robot, we achieved a direct feedback control scheme on the centroidal

[1] https://youtu.be/MRQXE4ig0yA.

parameters of the system with a relatively small and intuitive set of gains. This is in contrast to current state-of-the-art optimisation-based approaches. By adopting minimal sensing and the centroidal system, our approach can be applied to a wide range of humanoids, even with existing hardware imperfections.

In future work, we would like to tackle issues that currently limit the hardware performance, such as performing the control on a latency-predicted state, anticipating kinematic limits, and modifying the swing foot trajectories to allow for more angular momentum compensation during a push.

References

1. Al Borno, M., De Lasa, M., Hertzmann, A.: Trajectory optimization for full-body movements with complex contacts. IEEE Trans. Visual Comput. Graph. **19**(8), 1405–1414 (2012)
2. Allgeuer, P., Behnke, S.: Robust sensor fusion for biped robot attitude estimation. In: Proceedings of 14th IEEE-RAS International Conference on Humanoid Robotics (Humanoids), Madrid, Spain (2014)
3. Allgeuer, P., Behnke, S.: Bipedal walking with corrective actions in the tilt phase space. In: 2018 IEEE-RAS 18th International Conference on Humanoid Robots (Humanoids), pp. 1–9 (2018)
4. Allgeuer, P., Farazi, H., Ficht, G., Schreiber, M., Behnke, S.: The igus humanoid open platform. KI-Künstliche Intelligenz (2016)
5. Choi, Y., Kim, D., Oh, Y., You, B.J.: Posture/walking control for humanoid robot based on kinematic resolution of com Jacobian with embedded motion. IEEE Trans. Robot. **23**(6), 1285–1293 (2007)
6. Cisneros, R., Benallegue, A., Chitour, Y., Morisawa, M., Kanehiro, F., et al.: Lyapunov-stable orientation estimator for humanoid robots. IEEE Robot. Autom. Lett. **5**(4), 6371–6378 (2020)
7. Dai, H., Valenzuela, A., Tedrake, R.: Whole-body motion planning with simple dynamics and full kinematics. In: 14th IEEE-RAS International Conference on Humanoid Robots (Humanoids) (2014)
8. Dimitrov, D., Sherikov, A., Wieber, P.B.: A sparse model predictive control formulation for walking motion generation. In: 2011 IEEE/RSJ International Conference on Intelligent Robots and Systems, pp. 2292–2299 (2011)
9. Englsberger, J., Ott, C., Albu-Schäffer, A.: Three-dimensional bipedal walking control based on divergent component of motion. IEEE Trans. Robot. **31**(2), 355–368 (2015)
10. Englsberger, J., Ott, C., Roa, M.A., Albu-Schäffer, A., Hirzinger, G.: Bipedal walking control based on capture point dynamics. In: 2011 IEEE/RSJ International Conference on Intelligent Robots and Systems, pp. 4420–4427 (2011)
11. Ficht, G., Behnke, S.: Online balanced motion generation for humanoid robots. In: IEEE-RAS 18th International Conference on Humanoid Robots (Humanoids) (2018)
12. Ficht, G., Behnke, S.: Fast whole-body motion control of humanoid robots with inertia constraints. In: 2020 IEEE International Conference on Robotics and Automation (ICRA), pp. 6597–6603 (2020)
13. Griffin, R.J., Wiedebach, G., Bertrand, S., Leonessa, A., Pratt, J.: Straight-leg walking through underconstrained whole-body control. In: 2018 IEEE International Conference on Robotics and Automation (ICRA), pp. 5747–5754 (2018)

14. Kajita, S.: Linear inverted pendulum-based gait. In: Humanoid Robotics: A Reference, pp. 905–922 (2019)
15. Kajita, S., et al.: Position-based lateral balance control for knee-stretched biped robot. In: 2019 IEEE-RAS 19th International Conference on Humanoid Robots (Humanoids), pp. 17–24 (2019)
16. Kajita, S., et al.: Biped walking pattern generation by using preview control of zero-moment point. In: IEEE International Conference on Robotics and Automation, ICRA 2003, vol. 2, pp. 1620–1626 (2003)
17. Kajita, S., et al.: Resolved momentum control: humanoid motion planning based on the linear and angular momentum. In: IEEE/RSJ International Conference on Intelligent Robots and Systems (IROS), vol. 2, pp. 1644–1650 (2003)
18. Kajita, S., et al.: Biped walking stabilization based on linear inverted pendulum tracking. In: 2010 IEEE/RSJ International Conference on Intelligent Robots and Systems, pp. 4489–4496 (2010)
19. Lengagne, S., Vaillant, J., Yoshida, E., Kheddar, A.: Generation of whole-body optimal dynamic multi-contact motions. Int. J. Robot. Res. **32**(9–10), 1104–1119 (2013)
20. Li, Z., Tsagarikis, N.G., Caldwell, D.G., Vanderborght, B.: Trajectory generation of straightened knee walking for humanoid robot iCub. In: 2010 11th International Conference on Control Automation Robotics and Vision, pp. 2355–2360 (2010)
21. Ludwig, S.A., Burnham, K.D.: Comparison of Euler estimate using extended Kalman filter, Madgwick and Mahony on quadcopter flight data. In: 2018 International Conference on Unmanned Aircraft Systems (ICUAS), pp. 1236–1241. IEEE (2018)
22. Masuya, K., Ayusawa, K.: A review of state estimation of humanoid robot targeting the center of mass, base kinematics, and external wrench. Adv. Robot. **34**(21–22), 1380–1389 (2020)
23. Missura, M., Bennewitz, M., Behnke, S.: Capture steps: robust walking for humanoid robots. Int. J. Humanoid Robot. **16**(06), 1950032 (2019)
24. Nielsen, J.B.: How we walk: central control of muscle activity during human walking. Neuroscientist **9**(3), 195–204 (2003)
25. Orin, D.E., Goswami, A., Lee, S.H.: Centroidal dynamics of a humanoid robot. Auton. Robot. **35**(2–3), 161–176 (2013)
26. Popovic, M., Hofmann, A., Herr, H.: Angular momentum regulation during human walking: biomechanics and control. In: Proceedings of the IEEE International Conference on Robotics and Automation, ICRA 2004, vol. 3, pp. 2405–2411 (2004)
27. Pratt, J., Carff, J., Drakunov, S., Goswami, A.: Capture point: a step toward humanoid push recovery. In: 2006 6th IEEE-RAS International Conference on Humanoid Robots, pp. 200–207 (2006)
28. Tedrake, R., Kuindersma, S., Deits, R., Miura, K.: A closed-form solution for real-time ZMP gait generation and feedback stabilization. In: 2015 IEEE-RAS 15th International Conference on Humanoid Robots (Humanoids), pp. 936–940 (2015)
29. Wensing, P.M., Orin, D.E.: Improved computation of the humanoid centroidal dynamics and application for whole-body control. Int. J. Humanoid Robot. **13**(01), 1550039 (2016)
30. Whitman, E.C., Stephens, B.J., Atkeson, C.G.: Torso rotation for push recovery using a simple change of variables. In: 2012 12th IEEE-RAS International Conference on Humanoid Robots (Humanoids 2012), pp. 50–56 (2012)
31. Wieber, P.B.: Trajectory free linear model predictive control for stable walking in the presence of strong perturbations. In: 2006 6th IEEE-RAS International Conference on Humanoid Robots, pp. 137–142 (2006)

Decentralized Control Mechanisms for Trunk, Head, and Tail–limb Coordination in Quadruped Running

Shura Suzuki[1,2](✉), Yuya Asaoka[3,4], Atsushi Norita[3,4], Akira Fukuhara[3], Masato Ishikawa[1], and Akio Ishiguro[3]

[1] Department of Mechanical Engineering, Osaka University, Suita 565-0874, Japan
shura@eom.mech.eng.osaka-u.ac.jp
[2] Japan Society for the Promotion of Science, Tokyo 102-0083, Japan
[3] Research Institute of Electrical Communication, Tohoku University, Sendai 980-8577, Japan
[4] Department of Electrical Engineering, Tohoku University, Sendai 980-8579, Japan

Abstract. Body–limb coordination, a well-coordinated behavior of the legs and other body parts such as the trunk, head, and tail, provides quadruped animals with notable locomotor skills. However, the control mechanisms responsible for body–limb coordination remain unclear because the animal behavior is highly dependent on body structure and varies with animal species. Here, we hypothesized the common control principle underlying trunk, head, and tail–limb coordination in quadruped running and investigated plausible control mechanisms through mathematical modeling and simulations. In this paper, we present a key control concept for body–limb coordination and simple two-dimensional simulation experiments. The simulation results revealed that the proposed mechanisms generate running behavior with trunk, head, and tail–limb coordination in response to body structure. The results suggest that the proposed mechanisms play an important role in body–limb coordination.

Keywords: Quadruped locomotion · Decentralized control · Body–limb coordination

1 Introduction

Quadruped animals exhibit high agility and adaptability to terrestrial environments. Such locomotion abilities are owing to the coordination between the legs and other body parts, such as the trunk, head, and tail, i.e., body–limb coordination. For instance, trunk flexion for the cheetah [1], head swing for the horse [2], and tail wagging for the lizard [3] improve the locomotion performance of these animals. These behaviors suggest that body–limb coordination plays an essential role in motor control in quadruped animals. Therefore, the decoding of

© The Author(s), under exclusive license to Springer Nature Switzerland AG 2023
J. M. Cascalho et al. (Eds.): CLAWAR 2022, LNNS 530, pp. 256–263, 2023.
https://doi.org/10.1007/978-3-031-15226-9_25

body–limb coordination mechanisms will help understand animal motor control and establish a design methodology for quadruped robots with high mobility.

The locomotion of quadruped animals is mainly controlled by a distributed neural network in the spinal cord called the central pattern generator (CPG), and sensory feedback in the peripheral nervous system [4]. However, the detailed mechanisms responsible for body–limb coordination remain elusive because motor behaviors result from intricate interactions between the CPG, sensory feedback, musculoskeletal structure, and environment. For this reason, a synthetic approach using simulations and robots is gaining importance as a technique for investigating complicated interactions and testing plausible control mechanisms [5]. Several studies focusing on body–limb coordination have discussed its usefulness and underlying control mechanisms [3,6–14]. However, these studies focused on specific animals, and the shared control principle for the body–limb coordination has received less attention.

In this study, we investigated the common control principle for body–limb coordination in quadruped running through mathematical modeling and simulations. We focused on the common role of trunk, head, and tail movements in quadruped running, and hypothesized that these movements increase the leg thrust. To verify this hypothesis, we designed a decentralized control algorithm and implemented it in a three-type of quadruped robot model with an actuated trunk, head, and tail. The simulation results revealed that each robot model could generate running behavior with self-organized body–limb coordination in response to the body structure. This suggests that the proposed mechanisms potentially play an important role in body–limb coordination.

2 Model

In this section, the key concept of the body–limb coordination mechanism is first presented. In quadruped running, the thrust force mainly originates from interactions between the legs and ground. Other body parts, such as the trunk, head, and tail, hardly experience thrust from the ground directly, with some exceptions, such as the kangaroo tail [15]. We hypothesized that these body parts would play a similar role in quadruped locomotion. Specifically, they help the legs to generate greater thrust forces. For example, when a thrust is generated in the hind legs, trunk extension contributes to increasing the thrust (Fig. 1a). Similarly, lowering the head and raising the tail also increases the leg thrust force (Figs. 1b and c). Based on this perspective, we designed a common control algorithm for trunk, head, and tail–limb coordination. The following section describes the control algorithm and quadruped robot models in detail.

2.1 Robot Structure

We introduce three quadruped robot models with an actuated trunk, head, and tail. These robot models employ a body to run in a sagittal plane on flat terrain, and the body dynamics are described as a mass-spring-damper system for simplification.

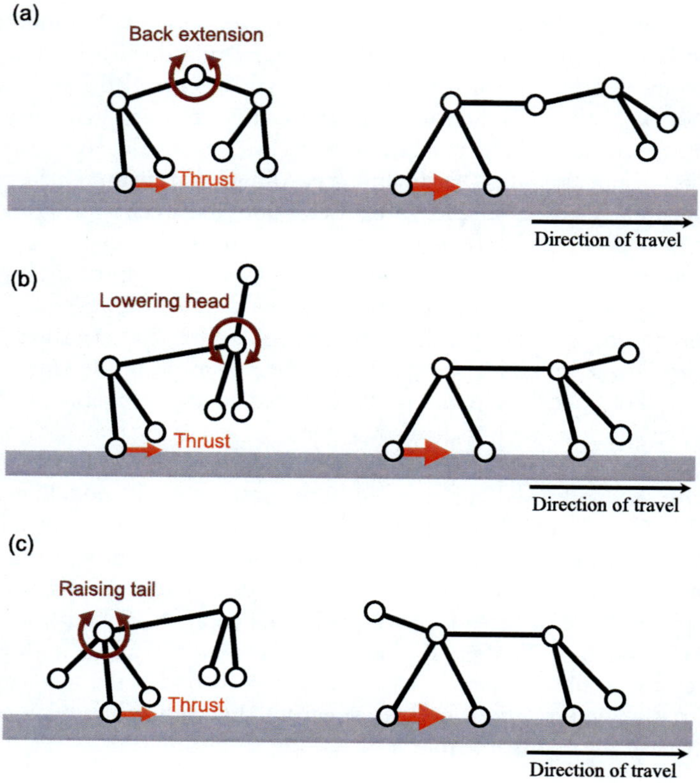

Fig. 1. Body–limb coordination patterns in quadruped running. These body movements increase the leg thrust in quadruped running. (a) Back extension lifts the forebody part up and the hind legs support a greater load. (b) Lowering the head leads to movement of the center of mass forward and provides the hind legs with more thrust. (c) Raising the tail makes the hind legs press against the ground, and a greater load is supported by the hind legs.

Figure 2a shows an overview of the quadruped robot model with an actuated trunk. The trunk has a single pitch joint (spine joint) and connects to the fore and hind legs with a single pitch joint (shoulder and hip joints). Each leg has two actuators, one for rotation and the other for linear motion. The point masses are located at the shoulder, spine, hip, fore, and hind feet. The linear springs between the shoulder, spine, and hip point masses are defined as stiff to ensure that the links remain rigid.

Figure 2b and c show an overview of the quadruped robot model with an actuated head and tail, respectively. The main components of these models are the same as those of the robot model with an actuated trunk, as shown in Fig. 2a. The robot model with an actuated head also comprises a head segment with a rotary actuator at the shoulder. The robot model with an actuated tail has a tail segment with a rotary actuator at the hip. The head and tail have point masses,

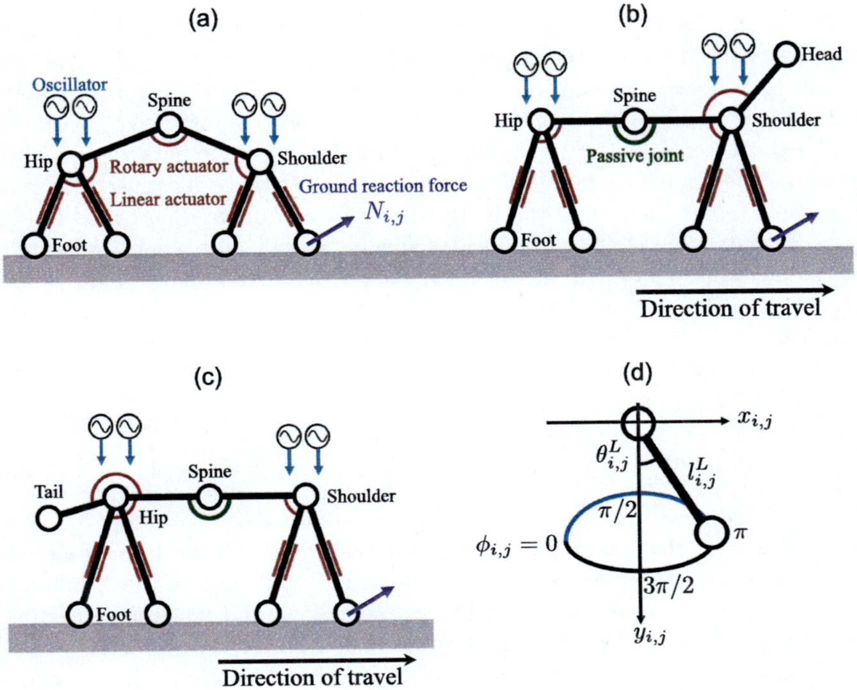

Fig. 2. Overview of the quadruped model (a) with an actuated trunk, (b) with an actuated head, (c) with an actuated tail, and (d) foot trajectory.

and the linear springs between the head and shoulder, and between the tail and hip are stiff to ensure that the links remain rigid. Furthermore, the spine joints of these two models are stiff to ensure that the trunk remains straight.

2.2 Control Algorithm

The legs are controlled by proportional-derivative (PD) control so that they approach their target positions. The target position is determined using a phase oscillator implemented in each leg (Fig. 2d). The time evolution of the oscillator phase can be described as follows:

$$\dot{\phi}_{i,j} = \omega - \sigma_{LL} N^V_{i,j} \cos\phi_{i,j} - \sigma_{BL} \dot{\psi}^B_i \sin\phi_{i,j} \tag{1}$$

where $\phi_{i,j}$ denotes the oscillator phase. The subscripts i and j are the indices of the fore and hind legs ($i = 1$: fore legs, $i = 2$: hind legs) and left and right legs ($j = 1$: left legs, $j = 2$: right legs), respectively. The variable ω represents the intrinsic angular velocity of the oscillators; σ_{LL} and σ_{BL} are the weights of the sensory feedback; $N^V_{i,j}$ represents the vertical component of the ground reaction force detected at the foot tip; and $\dot{\psi}^B_i$ is the angular velocity of the i side of the body trunk.

The second term on the right-hand side of Eq. (1) represents the local sensory feedback for the interlimb coordination proposed by Owaki et al. [16]. This feedback helps the loaded leg continue supporting the body. Specifically, it modulates the oscillator phase to $3\pi/2$ when the leg senses a normal force ($N_{i,j}^V > 0$).

The third term on the right-hand side of Eq. (1) represents the body-tilt feedback. The angular velocity of the pitch angle of the trunk reflects how the body joints (spine, neck, and tail joints) work. Thus, this feedback modifies the leg movements to exploit the body movements. Specifically, when the trunk is tilting downward ($\dot{\psi}_i^B < 0$), the leg oscillators are modulated toward π, and when the body is tilting upward ($\dot{\psi}_i^B > 0$), the leg oscillators are modulated toward 0.

The body (neck, spine, and tail) joints are controlled as follows:

$$\tau^X = -k_p(\theta^X - \theta_0^X) - k_d\dot{\theta}^X + \sigma_{LX}\sum_{i=1}^{2}\sum_{j=1}^{2}(-1)^{i+\epsilon^X}\max[0,\, N_{i,j}^H], \tag{2}$$

where variable X denotes the location of the body joint (N: neck, S: spine, and T: tail joint). The variables θ^X and θ_0^X denote the actual angle of the X joint and the offset angle of the X joint, and the variables σ_{LX} and ϵ^X are the feedback gains. Here, ϵ^X has a binary value ($\epsilon_S, \epsilon_T = 0$, $\epsilon_N = 1$). Variable $N_{i,j}^H$ is the horizontal component of the ground reaction force, and the thrust has a positive value. The first and second terms on the right-hand side of Eq. (2) originate from the passive spring and damper effects, respectively.

The third term on the right-hand side of Eq. (2) is a sensory feedback term. The feedback makes the body part bends to increase the leg thrust. For the spine joint ($\epsilon^X = 1$), when the thrust is generated in the hind legs, the spine joint bends to extend the trunk ($\tau^X > 0$). The generated torque lifts the forebody part, and consequently, the hind legs receive a greater ground reaction force from the ground (Fig. 1a). Similarly, in the case of the neck joint ($\epsilon^X = 0$), when the hind legs get thrust, the neck joint bends to lower the head ($\tau^X < 0$). The generated torque moves the center of mass forward and provides the hind legs with more thrust (Fig. 1b).

3 Simulation Result

We conducted simulation experiments and observed the gait patterns (Fig. 3) that show the time evolution of each body joint angle and the duration of the foot on the ground in each body structure.

In the footfall pattern for the model with an actuated trunk (Fig. 3a), the pair of contralateral feet is nearly synchronized, and there are two flight phases in the gait cycle. During the stance phase of the hind legs, the trunk extends ($\dot{\theta}^S > 0$), and vice versa. These features of the footfall and trunk–limb coordination pattern qualitatively correspond to those of a cheetah galloping [1].

In the case of a model with an actuated head (Fig. 3b), the footfall pattern is a four-beat in the order of H2, H1, F1, and F2 with a flight phase after the F2

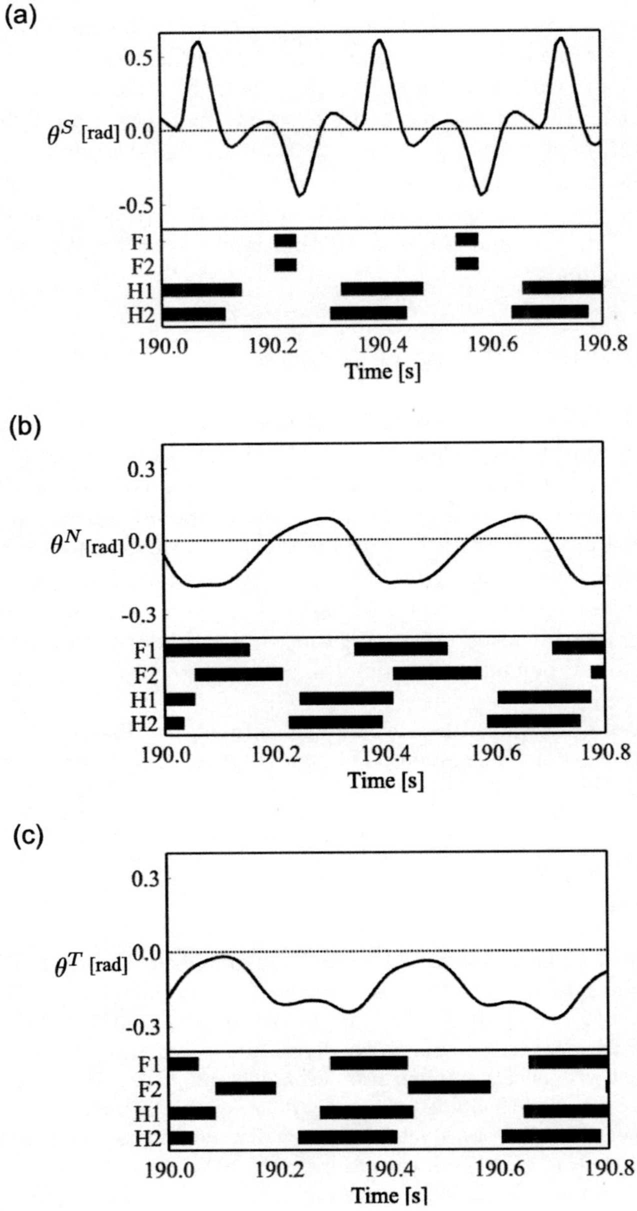

Fig. 3. The gait patterns of the robot model (a) with the actuated trunk, (b) with the actuated head, and (c) with the actuated tail. Time evolution of (top) each body joint angle and (bottom) gait diagrams. The colored region in the gait diagrams represents the stance phase of each leg ($N_{i,j} > 0$). F1 and F2 are the forelegs, and H1 and H2 are the hind legs, respectively.

stance phase. The neck joint lower the head ($\dot{\theta}^S < 0$) during the stance phase of the hind legs. These features of footfall and head–limb coordination patterns qualitatively correspond with those of a horse galloping [1].

For the model with an actuated tail (Fig. 3c), the footfall pattern is a three-beat in the order H2, H1 and F1, and F2, with a flight phase after the F2 stance phase. The tail joint raises the tail ($\dot{\theta}^T > 0$) during the stance phase of the hind legs. These features qualitatively correspond with those of a crocodile galloping [17]. Thus, the proposed control mechanism generated three types of body–limb coordination patterns, and the resulting behaviors were qualitatively similar to those of real animals.

4 Conclusion and Future Works

We investigated the common control principle underlying trunk, head, and tail–limb coordination in quadruped running. We focused on that trunk, head, and tail movements play similar roles, and designed a shared control algorithm for them. The simulation results revealed that the proposed control algorithm succeeded in generating these three types of body–limb coordination patterns. In the future, we will measure locomotion speed and cost efficiency, and investigate the extent to which body–limb coordination improves locomotion performance. Furthermore, we will also investigate the similarities and differences between simulated and real behaviors of animals.

Acknowledgement. The authors would like to thank Dr. Takeshi Kano of Tohoku University for his helpful suggestions. This work was supported by a Grant-in-Aid for JSPS Research Fellows (No. 22J00076) and Innovation Inspired by Nature" Research Support Program, SEKISUI CHEMICAL CO. LTD.

References

1. Bertram, J.E.A., Gutmann, A.: Motions of the running horse and cheetah revisited: fundamental mechanics of the transverse and rotary gallop. J. R. Soc. Interface **6**(35), 549–559 (2009). https://doi.org/10.1098/rsif.2008.0328
2. Loscher, D.M., Meyer, F., Kracht, K., Nyakatura, J.A.: Timing of head movement is consistent with energy minimization in walking ungulates. Proc. R. Soc. B: Biol. Sci. **283**(1843), 20161908 (2016). https://doi.org/10.1098/rspb.2016.1908
3. Libby, T., et al.: Tail-assisted pitch control in lizards, robots and dinosaurs. Nature **481**(7380), 181–184 (2012). https://doi.org/10.1038/nature10710
4. Ryczko, D., Simon, A., Ijspeert, A.: Walking with salamanders: from molecules to biorobotics. Trends Neurosci. **43**(11), 916–930 (2020). https://doi.org/10.1016/j.tins.2020.08.006
5. Ijspeert, A.J.: Biorobotics: using robots to emulate and investigate agile locomotion. Science **346**(6206), 196–203 (2014). https://doi.org/10.1126/science.1254486
6. Ijspeert, A.J., Crespi, A., Ryczko, D., Cabelguen, J.-M.: From swimming to walking with a salamander robot derven by a spinal cord model. Science **315**(5817), 1416–1420 (2007). https://doi.org/10.1126/science.1138353

7. Suzuki, S., Kano, T., Ijspeert, A.J., Ishiguro, A.: Spontaneous gait transitions of sprawling quadruped locomotion by sensory-driven body-limb coordination mechanisms. Front. Neurorobtics **15**, 645731 (2021). https://doi.org/10.3389/fnbot.2021.645731
8. Suzuki, S., Kano, T., Ijspeert, A.J., Ishiguro, A.: Sprawling quadruped robot driven by decentralized control with cross-coupled sensory feedback between legs and trunk. Front. Neurorobtics **14**, 607455 (2021). https://doi.org/10.3389/fnbot.2020.607455
9. Khoramshahi, M., Bidgoly, H.J., Shaflee, S., Asaei, A., Ijspeert, A.J., Ahmadabadi, M.N.: Piecewise linear spine for speed-energy efficiency trade-off in quadruped robots. Rob. Auton. Syst. **61**(12), 1350–1359 (2013). https://doi.org/10.1016/j.robot.2013.08.001
10. Zhao, Q., Nakajima, K., Sumioka, H., Yu, X., Pfeifer, R.: Embodiment enables the spinal engine in quadruped robot locomotion. In: 2012 IEEE/RSJ International Conference on Intelligent Robots and Systems, pp. 2449–2456. IEEE, Vilamoura (2012). https://doi.org/10.1109/IROS.2012.6386048
11. Fukuhara, A., Koizumi, Y., Baba, T., Suzuki, S., Kano, T., Ishiguro, A.: Simple decentralized control mechanism that enables limb adjustment for adaptive quadruped running. Proc. R. Soc. B Biol. Sci. **288**(1962), 20211622 (2021). https://doi.org/10.1098/rspb.2021.1622
12. Fukuhara, A., Suzuki, S., Kano, T., Ishiguro A.: Decentralized coordination mechanism between neck and limbs for efficient quadruped walking. In: 2019 IEEE/RSJ International Conference on Intelligent Robots and Systems, pp. 1715–1720. IEEE, Macau (2019). https://doi.org/10.1109/IROS40897.2019.8967619
13. Suzuki, S., Norita, A., Fukuhara, A., Kano, T., and Ishiguro, A.: Decentralized control mechanisms underlying neck–limb coordination in horse walking and trotting. In: The 9.5th International Symposium on Adaptive Motion of Animals and Machines, pp. 24–25. Adaptive Motion of Animals and Machines Organizing Committee, Ottawa (2021). https://doi.org/10.18910/84865
14. Heim, S.W., Ajallooeian, M., Eckert, P., Vespignani, M., Ijspeert, A.J.: On designing an active tail for legged robots: simplifying control via decoupling of control objectives. Ind. Rob. **43**(3), 338–346 (2016). https://doi.org/10.1108/IR-10-2015-0190
15. O'Connor, S.M., Dawson, T.J., Kram, R., Donelan, J.M.: The kangaroo's tail propels and powers pentapedal locomotion. Biol. Lett. **10**(7), 20140381 (2014). https://doi.org/10.1098/rsbl.2014.0381
16. Owaki, D.: Ishiguro: a quadruped robot exhibiting spontaneous gait transitions from walking to trotting to galloping. Sci. Rep. **7**(1), 1–10 (2017). https://doi.org/10.1038/s41598-017-00348-9
17. Renous, S., Gasc, J.-P., Bels, V.L., Wicker, R.: Asymmetrical gaits of juvenile Crocodylus johnstoni, galloping Australian crocodiles. J. Zool. **256**(3), 311–325 (2002). https://doi.org/10.1017/S0952836902000353

Adaptation of a Decentralized Controller to Curve Walking in a Hexapod Robot

Janneke Simmering[1], Luca Hermes[1], Axel Schneider[2], and Malte Schilling[1](✉)

[1] Machine Learning Group, Bielefeld University, Bielefeld, Germany
{jsimmering,lhermes,mschilli}@techfak.uni-bielefeld.de
[2] Institute of System Dynamics and Mechatronics, Bielefeld University of Applied Sciences, Bielefeld, Germany

Abstract. Curve walking is an important skill for multi-legged robot locomotion as it increases a robots' maneuverability. We investigate whether a decentralized system can explain even complex walking behaviors like curve walking. Based on an analysis of the curve walking capabilities of the decentralized control architecture Walknet, we propose a couple of simple but effective modifications with a main focus on the coordination between leg controllers: controlling the step length, shifting the AEP (transition point from swing to stance), and decreasing the step length for the legs on the inside of the curve. In simulation, the modified architecture demonstrated a significant improvement in the stability of the curve walking performance for tight curves and even allows a smooth transition to extremely tight curves and turning on the spot. Furthermore, the system is tested on a real robot and showed good qualitative results and robust curve walking behavior.

Keywords: Robot locomotion · Decentralized control · Biologically-inspired

1 Introduction

Mobile robots are a major research area in the field of robotics, as they are able to move around and solve tasks in the real world. Legged robots are of special interest, as they are able to potentially climb stairs or curbs, as well as deal with unpredictable environments in a highly adaptable way [1]. While walking straight ahead is a fundamental and well-studied behavior for animals as well as walking robots, only a few studies are dedicated explicitly to negotiating curves in legged robotics which is undoubtedly important for moving around.

Control of legged locomotion is usually divided into two general control problems: First, the movement of every single leg has to be controlled and, secondly, the movements of the different legs have to be coordinated to work together effectively. Curve walking affects both of these problems. On the one hand, the movements of the single legs relative to the body change in order to affect the rotation of the body. On the other hand, the coordination of the legs has to change, especially for tight curves during which the legs otherwise may collide.

© The Author(s), under exclusive license to Springer Nature Switzerland AG 2023
J. M. Cascalho et al. (Eds.): CLAWAR 2022, LNNS 530, pp. 264–275, 2023.
https://doi.org/10.1007/978-3-031-15226-9_26

The main focus of this article deals with the second, the coordination, problem (for the first problem, see control of stance movement in [20]).

The current state of the art in locomotion of multi-legged robot control is dominated by different approaches. Most can be sorted into one of three major categories. There are approaches based on central pattern generators (CPGs), learning approaches, and sensor-based approaches [6,7,10]. CPGs are systems of oscillators or in biological systems neural networks which produce rhythmic output without requiring patterned input [7]. Learning approaches optimize a function that takes in any sensory input and outputs a change in pose. The architecture discussed in this paper is called Walknet and it belongs to the third category of sensor-based approaches [14]. These have better explainability compared to learning-based approaches and are usually very adaptable to changing environments. Walknet is a decentralized control approach that is driven by proprioceptive sensory inputs, i.e. the selection of actions on the leg level depends on the current state of the body. It can deal with a wide range of environments and situations and reproduce many behaviors of the stick insect by which it is inspired [3,6,14]. The decentralized architecture simplifies the problem of legged locomotion control as modules of the system only deal with parts of the problem mostly based on local information. Walking behavior and gaits emerge out of the interaction of different controllers and the environment which proved to be quite adaptive. Such an approach should be well positioned to deal with curve walking because, in biological systems (e.g. stick insects), quite irregular gait patterns can be observed when negotiating curves which has been a challenge for approaches that employ fixed gait patterns [5,8]. This also motivated the extension of sersor-based approaches to control architectures [12].

The Walknet architecture was designed mostly based on the results from straight walking stick insects and usually does not allow the negotiation of tight curves. Previous Walknet versions are limited to a minimal curve diameter of about two body lengths [18], but stick insects can walk much smaller circles. This article examines whether a relatively simple decentralized system like the Walknet architecture can explain even complex walking behaviors like curve walking. We propose four simple modifications that aim at improving the curve walking capabilities while maintaining the simplicity of the original Walknet architecture and its straight walking performance. The resulting generalized solution to robotic legged locomotion contributes to building a complete system needed for legged mobile robots to accomplish robust and adaptable locomotion behavior able to deal with uncertain environments.

2 Methods

We evaluate the proposed modifications on a hexapod robot. The implementation is based on ROS and was tested in the Gazebo Simulation before being transferred onto real robotic hardware. In simulation a close approximation of the six legged PhantomX mark III was used while for the hardware simulation, the newer PhantomX robot in version mark IV—called WidowX—has been used. Both robots have six legs connected to a single body segment. Each leg consists of three revolute joints resulting in 18° of freedom (DoF). This section will briefly describe the baseline version of Walknet, before presenting the modifications for improved leg coordination during curve walking as the main contribution of this article.

2.1 Walknet Control Architecture

Walknet is a decentralized network architecture to control walking in hexapod robots with 18° of freedom (DoF), three for each leg, roughly modeling the leg anatomy of stick insects [14]. Multiple versions of Walknet exist and for our experiments we derived the implementation from the version by Schilling et al. which was implemented on the robot HECTOR [13,14,19]. That Walknet architecture is realized as a simple neural network to reproduce the walking behavior of stick insects [4]. The network parameters have been tuned manually.

In Walknet the movement of a single leg happens in two phases, the stance and the swing phase (see Fig. 1a). In the stance phase, the leg has contact to the ground and is moved backwards parallel to the robot's body. This pushes the robot forwards. The phase switches from stance to swing when the posterior extreme position (PEP) is reached. In the swing phase, the leg is lifted off the ground and moves along an arc towards the front. The phase switches back into stance phase when the anterior extreme position (AEP) is reached and the leg comes into contact with the ground. The exact trajectories of the motors of a leg for the stance phase are calculated by an internal explicit body model (for details refer to [20]). Broadly, the body model is realized as a system of linear equations and converts a pull vector at the front of the body into stance trajectories for the legs. Thus, the walking direction and speed can be controlled by changing this pull vector. The trajectories for the swing phase follow a simple Bézier curve that computes an arc from the PEP to the AEP.

Walknet implements six controllers, one for each leg. The controllers modulate the leg movement based on coordination rules depending on the state of neighboring legs (Fig. 1b). The coordination rules are based on studies of stick insects which found six coordination rules which act between neighboring legs and produce the stable walking behavior [2,14]. However, the first three are sufficient for reproducing stable gaits [4,9]. These are therefore implemented by changing the relative location of the PEP. In addition to the coordination rules, the legs are connected by the loop through the world, meaning that the controllers of the legs are influenced by one another through sensory feedback. In this decentralized structure, the walking behavior emerges in the form of a continuum of stable gaits, so-called free gaits, through this coordination of the legs on at least these two levels [14].

2.2 Adaptations of Control Architecture

The original Walknet is able to produce stable behavior when walking slight curves. This is detailed in the results for the baselines in Sec. 3.1. This section introduces our four proposed modifications to improve curve walking.

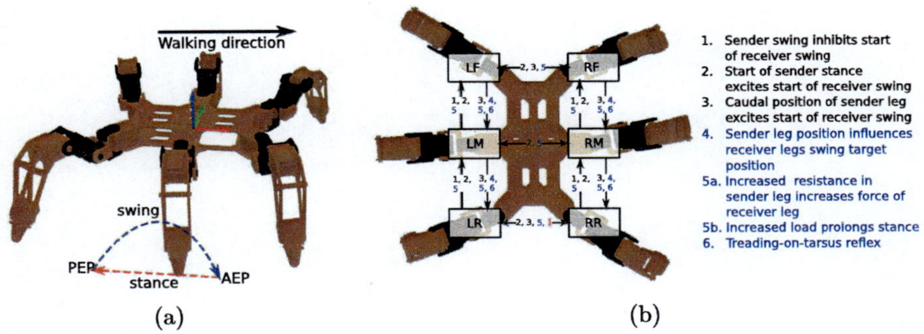

Fig. 1. (a) Visualization of swing (dashed blue arrow) and stance (dashed red arrow) with the corresponding walking direction (back arrow). (b) Leg controller and coupling via coordination rules (adapted from [6]). L and R–left, right side; F, M, and R–front, middle, and rear leg. Blue marked rules are not implemented.

Modification 1: Controlling the Step Length. The Walknet architecture defines the PEP as a scalar coordinate along the body axis [14]. It can be understood as a plane perpendicular to the body for each leg that is moved back and forth by the coordination rules and initiates the switch from stance to swing once the leg moves behind this plane. This is problematic in curve walking, when the trajectories of the legs no longer move parallel to the body but at an angle that increases with the tightness of the curve walked. As a result, some legs might never reach their PEP and therefore would never switch from stance to swing. To deal with this problem the length of the stance step independent of direction is used to determine the PEP instead of using the x-coordinate. If the step length increases beyond the PEP threshold, the leg switches from stance to swing phase. The step length threshold is influenced by the coordination rules, which change the length of the stance step instead of the position threshold. The PEP can now be considered as a sphere around the AEP whose radius is increased and decreased by the coordination rules (see Fig. 4).

Modification 2: Shifting the Anterior Extreme Position (AEP). As a second problem, legs on the inside of a curve are moved increasingly close to the body in curve walking. Studies of stick insects found, that the PEP and AEP are shifted in insects during curve walking. While the original Walknet architecture reproduces the tilted stance trajectories and therefore the shift of the PEP fairly well [4,9], the AEP is not automatically shifted. Shifting the AEP intuitively reduces the number of times the inner legs move too close to the body. It could also potentially improve the turning of the model, as the legs—if they are positioned farther away from the body—can contribute better to the rotation. As visualized in Fig. 2 we shift the AEP, so that following stance trajectories should intersect the mid point of a stance trajectory parallel to the body (dotted black line).

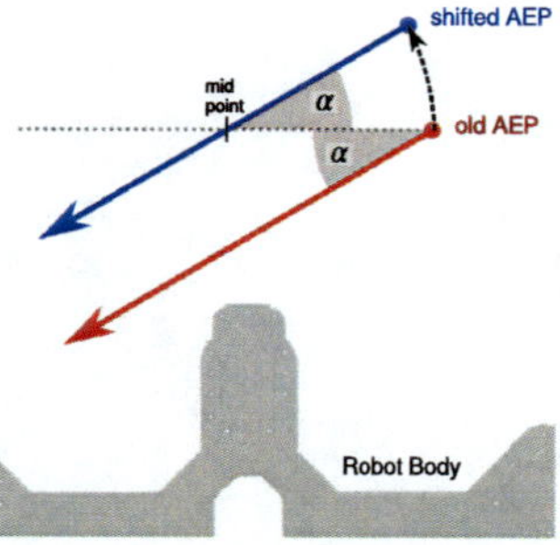

Fig. 2. AEP shift. Red arrow: Stance trajectory before AEP shift. Blue arrow: Stance trajectory after AEP shift. Dotted black line: Stance trajectory in straight walking.

Modification 3: Decreasing the Step Length of the Inner Legs. During curve walking, the legs on the inner side of the curve move at a reduced velocity compared to the legs on the outside of the curve. Walknet versions without a body model use this to achieve curve walking [4,9]. However, a decreased velocity of the legs on the inside of the curves means that these legs take longer to cover the same ground as the outside legs, potentially disrupting the overall coordination. Therefore, reducing the step length for these legs may improve leg coordination. A reduction of the step length of the inner legs can be found in the behavior of stick insects [8]. Reducing the step length of the inner legs could also reduce how often the legs come too close to the body. As the default step length is shortened, the modulation of the coordination rules had to be adapted accordingly.

Modification 4: Extending the Body Model. The stance movement of the legs is controlled through an internal body model [14,20]. Originally the turning of the robot is controlled by a single pull vector at the front of the body segment in the body model [20]. The rest of the body follows this pull based on the passive motion paradigm. As a modification, the body model is extended with an additional pull vector at the back of the body segment to help the robot turn which is similar to an approach proposed in [11].

3 Results: Simulation of Curve Walking

3.1 Baseline

We record the performance of the original Walknet architecture adapted to the PhantomX robot as a baseline. The system behaves as reported in realted work. Hence, walking straight shows a continuum of stable gaits that emerge depending on the target velocity. At low velocities a wave-gait emerges that transitions into a tetrapod-gait for velocities between 0.01 m/s and 0.02 m/s. For velocities greater than 0.02 m/s the tetrapod-gait fluently transitions into a tripod-gait where groups of three legs swing together. This velocity dependent continuum of gaits are the free gaits that also emerge in other Walknet implementations and reproduce the stick insect behavior.

3.2 Evaluation Methods

First, we analyze the emerging gaits for the baseline and our modified Walknet version with different velocities $v \in \{0.01, 0.02, .., 0.06\}$ m/s and different walking directions, rotating the pull vector by $\beta \in \{0.0, 0.2, .., 1.2\}$ radians. To intensify the evaluation scenario for our modified Walknet, we additionally perform evaluations for $\beta \in \{1.4, \frac{\pi}{2}\}$ radians. For each of the resulting conditions, we run 5 simulations.

To report quantitative results, we measure the static stability of the system. A pose of the robot is considered stable if the center of mass is lying inside the convex polygon spanned by feet that touch the ground. We report the stability of a trajectory as the percentage of stable poses.

We also use four different techniques to generate qualitative results. We report footfall patterns to examine the leg coordination. We report the stance trajectories for all six legs in a global frame of reference. This visualizes the work-space of the legs and indicates the direction of the whole body. We also report the trajectories of the whole robot to visualize the consistency of the curves. Lastly, we visualize the change in body height over time. Intuitively, a stable walking behavior should keep the body at a consistent height.

Curve Walking Static Stability. Table 1 shows the static stability for the baseline Walknet (Base) and Walknet with our modifications (Ours). These results indicate that the stability of the baseline is inversely proportional to both speed and angle and drops greatly for higher speeds and walking direction. In the simulations we observed that for tight curves, legs are often forced towards unreachable postures which contributes to the drops in performance. The trajectories of our modified Walknet have a stability above 99% in most cases. The stability drops below 99% only in four of the recorded 54 trajectories. The straight walking performance of the baseline Walknet is maintained.

Leg Coordination. Figure 3 shows a footfall pattern (a) and the work-space plots (b) and (c) of the baseline Walknet. It can be observed that the footfall patterns of the left legs are very irregular compared to the right legs. Inner legs stepped with a decreased frequency compared to the outer legs and with a decreasing frequency from rear to front. Furthermore, the stance trajectories in Fig. 3(b) and (c) of all the legs are oriented into the same direction. This leads to a more crab-like walking behavior, i.e. sideways walking, as we observed in our experiments.

The legs on the inside of the curve, mainly the front and middle leg, moved towards the body (Fig. 3b). This trend forces the legs into singular positions which resulted in unreachable postures and led to erratic behavior (Fig. 3c, for

Table 1. Static stability results of the original (Base) and modified (Ours) Walknet architecture. We report the worst run of each of the five trials. The (*) marks outlier trials for which the other runs were significantly ($\geq 5\%$) more stable than the reported run. Bold numbers denote a better result compared to the respective other model.

Speed [m/s]	0.01		0.02		0.03		0.04		0.05		0.06	
Model	Base	Ours	Base	Ours	Base	Ours	Base	Ours	Base	Ours	Base	Ours
0.00 rad	**100.0**	99.95	99.42	**99.90**	**100.0**	**100.0**	99.98	**100.0**	99.78	**100.0**	99.70	**99.92**
0.20 rad	**99.92**	99.82	**99.73**	99.10	**100.0**	**100.0**	**100.0**	99.98	**99.88**	**99.88**	98.27	**99.72**
0.40 rad	**99.88**	99.80	**99.67**	99.42	98.80	**100.0**	99.65	**100.0**	68.02*	**100.0**	83.85	**98.85**
0.60 rad	**100.0**	99.98	86.55*	**99.45**	97.22	**100.0**	97.27	**100.0**	76.24	**99.85**	51.79	**99.67**
0.80 rad	85.80*	**100.0**	80.94*	**100.0**	62.69	**99.92**	93.38	**99.85**	91.02	**99.53**	84.54	**99.92**
1.00 rad	98.65	**99.72**	95.68	**99.57**	87.57	**99.98**	73.99	**97.80**	74.37	**99.08**	75.34	**96.88**
1.20 rad	87.85	**100.0**	78.38	**99.73**	80.15	**99.58**	78.95	**100.0**	81.74	**100.0**	42.76	**100.0**
1.40 rad	–	99.10	–	100.0	–	100.0	–	99.30	–	99.18	–	99.90
1.57 rad	–	100.0	–	99.95	–	99.72	–	99.57	–	99.90	–	98.57

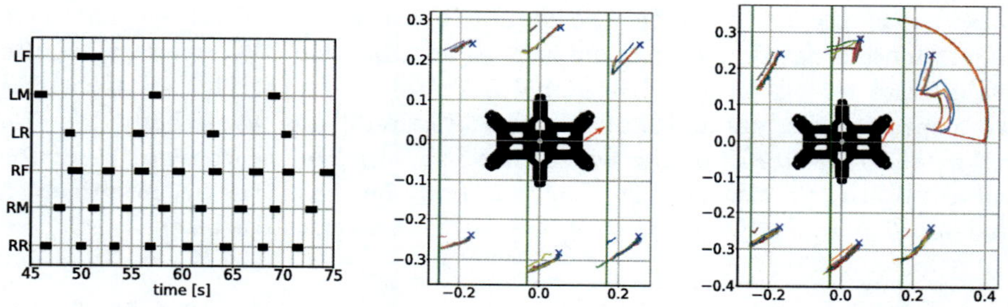

(a) 0.02 m/s 1.0 rad (b) 0.02 m/s 0.6 rad. (c) 0.02 m/s 1.0 rad.

Fig. 3. Results from the baseline Walknet control architecture: (a) Foot fall pattern of the emerging gait for tight curves. (b) and (c) workspace plots showing the development of the stance trajectories during curve walking. The green line represents the default PEP for the legs and the blue crosses the AEPs.

detailed results see [21]). For slight curves to the left, the emerging foot fall patterns were very similar to that of straight walking. Changes in the foot fall pattern of the right legs were less severe, but they also became more irregular for increasing velocities.

Figure 4(a)-(c) shows the footfall pattern of our modified Walknet. It can be observed that the gait patterns are comparably regular even for tight curves. The most obvious change over the tight curves (up to a walking direction of 1.2 rad, $\approx 69°$) is that the step frequency of the front and middle inner legs are reduced compared to that of other legs. For extremely tight curves, the frequencies of the left front and middle leg assimilated to that of the other legs, again, and the emerging step patterns appeared regular. In our experiments, we also observed that for straight walking and slight curves the emerging gaits were comparable to those of the initial implementation.

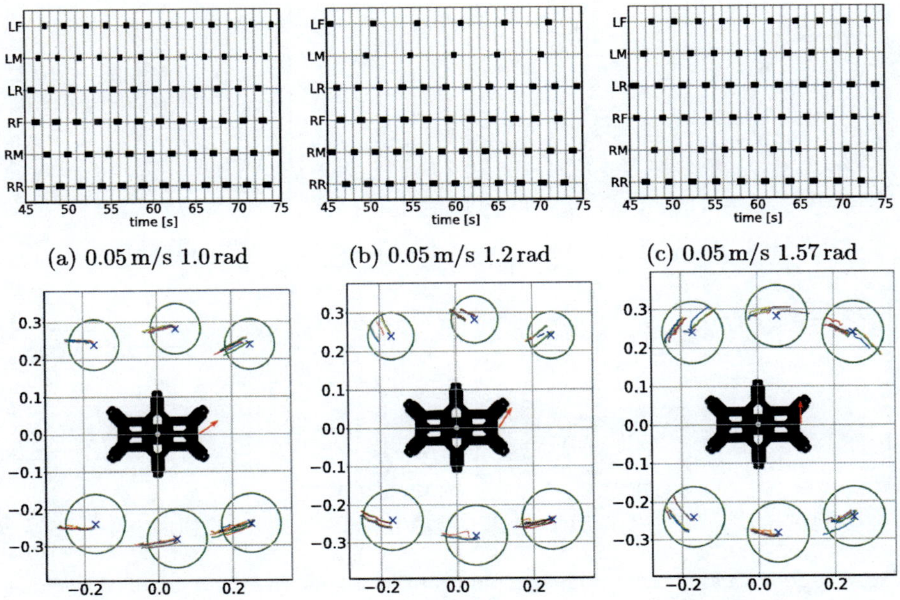

(a) 0.05 m/s 1.0 rad (b) 0.05 m/s 1.2 rad (c) 0.05 m/s 1.57 rad

(d) 0.03 m/s 0.6 rad (e) 0.03 m/s 1.0 rad (f) 0.03 m/s 1.57 rad

Fig. 4. Results from the modified Walknet architecture over the spectrum of walked curves: (a-c) Representative foot fall patterns for the development of the emerging gaits. (d-f) Workspace plots showing the development of the stance trajectories. The green circles represent the possible default PEPs for the legs and the blue crosses the AEPs. The default step length for the inner legs is reduced due to the third modification.

Figure 4(d)-(f) shows the work-space plots of our modified Walknet. The stance trajectories for slight curves are less tilted compared to those of the initial implementation. When comparing these plots to the results of the baseline above, the most apparent difference is that the stance trajectories of the six legs are not oriented into the same direction. Instead, the front legs are pulling into the curve and hind legs pushing into the opposite direction. This contributes to a turning behavior instead of a sideways walking. For extremely tight curves, stance trajectories lied in a circle around the robot with the left legs moving with reversed movement direction from a posterior AEP towards an anterior PEP (Fig. 4f).

Curve Walking Paths. Figure 5 shows walking trajectories as seen from above (first three columns) and plots of the body height (last column). The top and bottom rows show the results from the baseline and our modified Walknet, respectively. For all runs the target velocity was set to 0.06 m/s. The baseline shows stable behavior for slight curves (0.4 rad), but the trajectories become erratic for higher angles. This is also observable from the height plot in Fig. 5d.

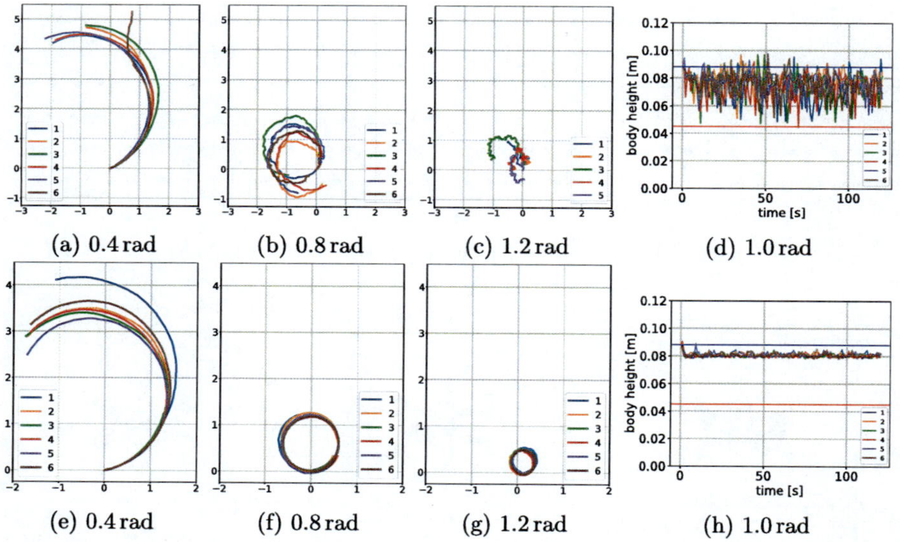

Fig. 5. Robot walking path and height plots for the original (a-d) and modified Walknet architecture (e-h) over five to six runs for each curvature. The target velocity is 0.06 m/s. For full results see [21]. In (d) and (h), the blue line marks the initial and overall desired body height, the red line marks half the initial body height.

After introduction of the modifications, the curve walking path plots show a clear improvement compared to the initial implementation. All trajectories show a smooth curvature. The height plot in Fig. 5h reflects this smoothness by showing little variance. It is also noticable that the circle in (a) is larger that the circle in (e), indicating that the translation of the pull vector into the final trajectory changes.

4 Discussion

Limitations of the Baseline Walknet. The results provide a unified picture of the curve walking capabilities of the baseline Walknet implementation. The system performs well for low to medium velocities and slight curves with walking directions up to 0.6 rad ($\approx 34°$). For tight curves with a higher walking direction angle, the performance rapidly decreases. Our results show that the leg coordination does not generalize to curve walking. In fact, the observed erratic behavior and decrease of statically stable trajectories can lead to hardware damage when executed on a robot.

Effect of Curve-Walking Modifications. The curve-walking modifications are meant to work in combination. Individually, none of them led to an overall satisfactory curve walking performance (results for the effect of the individual modifications are detailed in [21]). Adding the pull vector at the back improves the stability of the system. Controlling the step length allows the system to walk with any desired walking direction up to $\frac{\pi}{2} rad$ (90°) without the legs colliding. However, the circles for walking directions higher than 1.2 rad (≈ 69°) did not become much smaller. Shifting the AEP in combination with the controlled step length allows the system to actually walk circles matching the walking direction as shifting the AEP decreases the radius of the walked curves. Finally, decreasing the step length of the inside legs aims at increasing the differences between the legs for tight curves as is found in stick insects.

The final results were significantly improved compared to the original Walknet and compared to results for the individual modifications. The emerging gaits became regular and even for tight and extremely tight curves the workspace plots do not show any signs of erratic behavior as the legs were avoiding singularities and the joints were acting in a comfortable working range. Only 4 out of 54 trials were not considered statically stable. But even in these runs, the legs never moved too close to the body to damage any hardware. When using all modifications in combination the curvature increases for the extremely tight curves and for the walking direction of $\frac{\pi}{2} rad$ (90°) the robot basically turns on the spot. All these improvements were reached while maintaining the straight walking performance.

5 Application on the PhantomX Robot

To evaluate whether the system performs comparable to the simulation results on real robotic hardware, it was tested on the PhantomX mark IV robot (WidowX version). To prevent the robot from falling, a global check prevents four legs from being lifted at the same time. Secondly, swing starts are delayed if the current leg configuration without that leg would be unstable. Overall, in the recorded runs the stability enforcement worked as intended, only one trial showed static instability for a very brief moment.

To show the qualitative behavior of the WidowX robot, videos for different trials were recorded[1]. The videos show three different velocities that would produce different gaits in straight walking (wave-gait 0.01 m/s, tetrapod-gait 0.02 m/s, tripod-gait 0.05 m/s) each for five walking directions equally distributed between 0 rad and $\frac{\pi}{2} rad$ (90°).

The qualitative results for the WidowX robot were overall good and fairly comparable to the results in simulation. The high stability of the robot during walking suggests that the stability enforcement may not be necessary. The system produced the intended curves and these remained of stable curvature during a run and between runs. However, the WidowX moves slightly faster than the simulated PhantomX.

[1] Videos can be found in the publicly accessible git: https://gitlab.ub.uni-bielefeld.de/jsimmering/walknet-curve-walking-supporting-videos.git.

6 Conclusion

In this article, we introduced simple modifications to the decentralized Walknet architecture that significantly improve the stability of the system in curve walking, in particular for tight curves or when turning on the spot. The extended system has been tested in detail and evaluated in simulation. It has also been successfully applied on the hexapod robot WidowX. This demonstrates that such a decentralized system can explain even complex walking behaviors like curve walking which has shown to be difficult for many common control architectures.

Overall, improved stability for tight curves and the ability to walk extremely tight curves was achieved by four modifications. First, the stance trajectory generation is adapted as the rear of the body contributes to the rotation as well. It moves against the turning direction, thereby pushing the front of the body in turning direction. This adaptation also leads to less tilted stance trajectories which keeps the legs from moving too close to the body, therefore avoiding joint singularities. Secondly, the actual length of a step length is used to controlled switching from stance to swing. This enables the robot to walk extremely tight curves where the legs on the inside of the body move with reversed motion direction which overall stabilizes the system by improving the leg coordination for tight curves. The third modification shifts the anterior extreme position in the direction of the curve. This increases the curvature of the walked path, leading to smaller circles. Finally, the step lengths of the legs on the inside of the curve are decreased to increase the difference between the inner and outer legs. The effect of this modification is less obvious when tested in isolation but increased the stability of the system when combined with the other modifications.

As future work, we intend to examine the effect of perturbations, uneven terrain, and differing start positions on the curve walking performance. Therefore, the trajectory generation of the swing movement should be replaced with a more biologically inspired swing-net and the ground contact prediction should be exchanged with an actual ground contact detection. As the proposed system allows to freely control walking direction and speed, it could be utilized in navigation tasks. We want to connect it to our learning-based decentralized approach [15,17] and introduce a hierarchical structure for dealing with navigation [16].

Acknowledgement. This research was supported by the research training group "Dataninja" (Trustworthy AI for Seamless Problem Solving: Next Generation Intelligence Joins Robust Data Analysis) funded by the German federal state of North Rhine-Westphalia.

References

1. Buschmann, T., Ewald, A., von Twickel, A., Büschges, A.: Controlling legs for locomotion–insights from robotics and neurobiology. Bioinspiration Biomimetics **10**(4), 041001 (2015). https://doi.org/10.1088/1748-3190/10/4/041001
2. Cruse, H.: What mechanisms coordinate leg movement in walking arthropods? Trends Neurosci. **13**(1), 15–21 (1990)

3. Cruse, H., et al.: Walking: a complex behavior controlled by simple networks. Adapt. Behav. **3**(4), 385–418 (1995). https://doi.org/10.1177/105971239500300403
4. Cruse, H., Kindermann, T., Schumm, M., Dean, J., Schmitz, J.: Walknet—a biologically inspired network to control six-legged walking. Neural Netw. **11**(7), 1435–1447 (1998). https://doi.org/10.1016/S0893-6080(98)00067-7
5. Dürr, V., Ebeling, W.: The behavioural transition from straight to curve walking: Kinetics of leg movement parameters and the initiation of turning. J. Exp. Biol. **208**(12), 2237–2252 (2005). https://doi.org/10.1242/jeb.01637
6. Dürr, V., Schmitz, J., Cruse, H.: Behaviour-based modelling of hexapod locomotion: linking biology and technical application. Arthropod Struct. Dev. **33**(3), 237–250 (2004). https://doi.org/10.1016/j.asd.2004.05.004
7. Ijspeert, A.J.: Central pattern generators for locomotion control in animals and robots: a review. Neural Netw. **21**(4), 642–653 (2008)
8. Jander, J.P.: Untersuchungen zum Mechanismus und zur zentralnervösen Steuerung des Kurvenlaufs bei Stabheuschrecken (Carausius morosus). Ph.D. thesis, Universität Köln (1982)
9. Kindermann, T.: Behavior and adaptability of a six-legged walking system with highly distributed control. Adapt. Behav. **9**(1), 16–41 (2001)
10. Neftci, E.O., Averbeck, B.B.: Reinforcement learning in artificial and biological systems. Nature Mach. Intell. **1**(3), 133–143 (2019)
11. Paskarbeit, J.: Consider the robot-abstraction of bioinspired leg coordination and its application to a hexapod robot under consideration of technical constraints. Ph.D. thesis, Universität Bielefeld (2017)
12. Rosano, H., Webb, B.: A dynamic model of thoracic differentiation for the control of turning in the stick insect. Biol. Cybern. **97**(3), 229–246 (2007). https://doi.org/10.1007/s00422-007-0170-4
13. Schilling, M., Cruse, H.: ReaCog, a minimal cognitive controller based on recruitment of reactive systems. Front. Neurorobotics **11**, 3 (2017)
14. Schilling, M., Hoinville, T., Schmitz, J., Cruse, H.: Walknet, a bio-inspired controller for hexapod walking. Biol. Cybern. **107**(4), 397–419 (2013)
15. Schilling, M., Konen, K., Ohl, F.W., Korthals, T.: Decentralized deep reinforcement learning for a distributed and adaptive locomotion controller of a hexapod robot. In: 2020 IEEE/RSJ International Conference on Intelligent Robots and Systems (IROS), pp. 5335–5342. IEEE
16. Schilling, M., Melnik, A.: An approach to hierarchical deep reinforcement learning for a decentralized walking control architecture. In: Samsonovich, A.V. (ed.) BICA 2018. AISC, vol. 848, pp. 272–282. Springer, Cham (2019). https://doi.org/10.1007/978-3-319-99316-4_36
17. Schilling, M., Melnik, A., Ohl, F.W., Ritter, H.J., Hammer, B.: Decentralized control and local information for robust and adaptive decentralized deep reinforcement learning. Neural Netw. **144**, 699–725 (2021)
18. Schilling, M., et al.: A hexapod walker using a heterarchical architecture for action selection. Front. Comput. Neurosci. **7**, 126 (2013)
19. Schilling, M., Paskarbeit, J., Ritter, H., Schneider, A., Cruse, H.: From adaptive locomotion to predictive action selection-Cognitive control for a six-legged walker. IEEE Trans. Rob. **38**(2), 666–682 (2022)
20. Schilling, M., Paskarbeit, J., Schmitz, J., Schneider, A., Cruse, H.: Grounding an internal body model of a hexapod walker - control of curve walking in a biological inspired robot. In: Proceeding of International Conference on Intelligence Robots and Systems (2012)
21. Simmering, J.: Adapting the decentralized walknet architecture for improved curve walking in a six-legged phantomx. Master's thesis, Bielefeld University (2021)

CPG-Based Gait Generator for a Quadruped Robot with Sidewalk and Turning Operations

Vladimir Danilov[1,2](✉) and Sekou Diane[2]

[1] Voltbro LCC, Moscow, Russia
vldanilov90@gmail.com
[2] Institute of Control Problems of Russian Academy of Sciences, Moscow, Russia

Abstract. This article describes the quadruped robot gait generator algorithm based on the central pattern generator (CPG). The proposed architecture uses CPG as a phase signal generator for each leg, and a mapping function that builds the desired trajectory for the robot feet. The algorithm is able to change the gait type smoothly in real time, and also do the same with movement direction, frequency, height and length of the stride. In order to test the performance of the algorithm, experiments were carried out both in the simulation and on the real robot. The results show the efficiency of the algorithm and its ease of implementation.

Keywords: Quadruped robot · Central Pattern Generator (CPG) · Gait generator

1 Introduction

Walking robots are increasingly being used in various fields of human activity, as they have a greater cross-country ability, unlike wheeled vehicles. Over the past forty years there has been tremendous progress in this area. But the complexity of algorithms for controlling stable locomotion on different types of surfaces is still a serious problem.

To implement a stable static gait, the method of maintaining the center of mass point inside the reference polygon is used [1]. An extended version of this algorithm is the searching zero moment point (ZMP) [2, 3]. Its essence is to keep ZMP of the robot in the center of the reference polygon. This algorithm is well suited for six-legged robots [4] or slow four-legged ones [5]. This is due to the fact that the reference polygon is not presented during trot or gallop.

Another algorithm is CPG (Central Pattern Generator) inspired by research of mammalian locomotion. Neuroscientists have found that for the formation of rhythmic movements in the process of purposeful locomotion of animals, central pattern generators located in the spinal cord are used [6, 7]. A large number of implementations of this approach for mobile robots have been produced [8–11]. When applied to walking robots, usually each CPG neuron generates the desired joint angle [12–14]. The disadvantage of this method is that, in reality, joint trajectories are much more complicated than generated by CPG, and tedious manual adjustment of the parameters of each neuron is required for stable movement.

© The Author(s), under exclusive license to Springer Nature Switzerland AG 2023
J. M. Cascalho et al. (Eds.): CLAWAR 2022, LNNS 530, pp. 276–288, 2023.
https://doi.org/10.1007/978-3-031-15226-9_27

There is a large group of control algorithms based on dynamics of a robot mechanics: SLIP (Spring Loaded Inverted Pendulum), VMC (Virtual Model Control), MPC (Model Predictive Control). The SLIP method has been first proposed in [16] and it is based on the "virtual leg" concept [15]. Many works have been done [17–20], including Boston Dynamics' BigDog [19]. VMC has been successfully applied to HyQ [21] and StarlETH robots [22]. MPC is used in series of robots as Cheetah [23–27], ANYmal [28] and HyQReal [29]. As it can be seen, these methods are widely utilized for solving locomotion control problem. However, they require significant prior knowledge of the target robotic platform and the task [30], a very accurate dynamic robot model that is often difficult to obtain [31], and the need to manually adjust the parameters of the algorithms [32]. Moreover, separate controllers for each movement mode are used, which have a different architecture, and they are designed and configured separately [35].

Recently, model-free reinforcement learning (RL) algorithms have been widely advanced that can avoid the disadvantages described above. Although these algorithms have a large number of parameters, nevertheless, their configuration can be automated. However, there are two serious problems: a long training time and the difficulty of transition from simulation to hardware called the reality gap.

Solving the first problem, one can use parallel training of several thousand agents, which reduces the training time to few minutes [36]. Predefined gait generators are also used, and the goal of the agent is to optimize their trajectories [37]. The problem of the first implementation is that the robot could not change its movement speed, and in the second case, the robot moved only along the lateral plane and could not do gait transition online.

The solution of the second problem is combining accurate mathematical model of servo drives, the randomization of mechanical parameters and proprioceptive sensor data during training, as well as the imitation of external force disturbances acting on the robot [35, 38]. This algorithm provided stable and dynamic locomotion on such robots as Minitaur [30, 31, 33, 34, 37, 38], Unitree A1 [32] and ANYmal [35, 36].

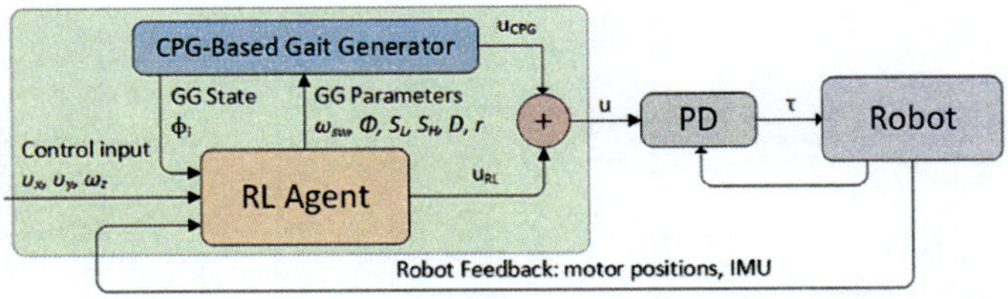

Fig. 1. Proposed robot locomotion control algorithm architecture

Based on the information above, the most interesting and promising approach is RL. The proposed scheme of the algorithm is shown in Fig. 1. It is based on the architecture obtained in [37] but it uses more efficient CPG as a gait generator. The CPG-based gait generator allows smoothly change the gait pattern, stride frequency, length and height, as well as additionally performs sidewalk and turning online. The goal of the trained agent is to set the input parameters for the CPG block and optimize its output trajectories of robot feet. GG State, Control Input and Robot Feedback signals form the agent observation vector. Owing to a predefined gait generator, the training time of the agent should be greatly reduced, and its CPG implementation with sidewalk and turning operations will also allow the agent to perform these actions.

In this paper, we focus on the CPG-based Gait Generator block. It must have the following properties:

- The network of CPG neurons must provide periodic clock signals for movement of a robot and smooth gait transition by adjusting only one parameter. The frequency of the signals should also vary smoothly based on a single parameter;
- Foot trajectories are calculated based on the rhythmic signals obtained by the CPG network;
- Trajectories must take into account both changes in stride height and stride length, as well as changes in direction of movement.

The algorithm will be tested in simulation and hardware quadruped robot model.

2 Gait Generator

The formation of the foot trajectories during locomotion is usually carried out using generators based on inverse kinematics [39]. The disadvantage of this method is that it does not allow both to do gait transition and to change stride frequency smoothly. The most attractive way is to use CPG, which is capable to generate trajectories of the foot positions [40, 41]. These publications describe development of such generators, but only to move forward. In this paper, we create a generator capable of form locomotion trajectories in all directions.

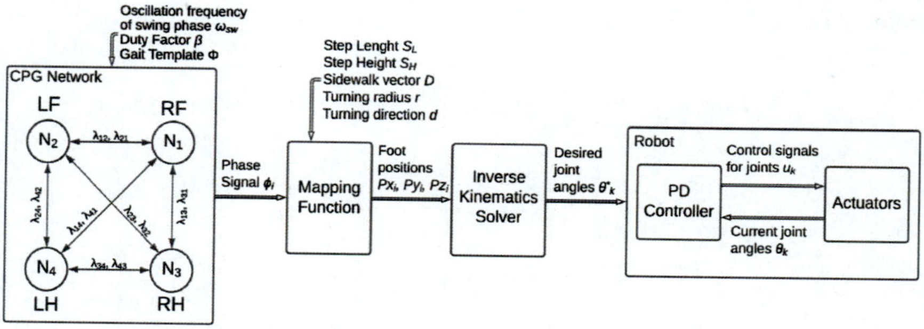

Fig. 2. Structure of the gait generator

2.1 Design Architecture

Figure 2 shows the general block diagram of the gait generator. The CPG Network block consists of four oscillators, each of which is responsible for formation of a phase trajectory of each leg. Stride frequency ω_{sw} and the vector Φ describing the gait type are taken as input. The output contains four phase signals ϕ_i, which are fed to the Mapping Function. The task of this block is to form stride cycles based on ϕ_i. The input parameters are: S_L - stride length, S_H - stride height, D – linear direction, r – turning radius and d - for turning direction (-1 - clockwise, 0 - straight, 1 - counterclockwise). The output parameters are desired foot position vectors P_{xi}, P_{yi}, P_{zi}. The Inverse Kinematics Solver block solves the inverse kinematics problem and, based on the desired positions, gives the desired joint angles, which are fed to the low-level part of the robot that controls the servos. This block is implemented by widely used matrix method [43–45].

2.2 CPG Network

CPG network consists of four symmetrically connected neurons, each of them is an oscillator that modulates the phase signal for a particular leg. There are several types of oscillators [42]. We choose the Hopf oscillator because it is the least computationally demanding, moreover it is able to provide fast switching between different gaits. The oscillator equations look like this:

$$\dot{x}_i = \alpha\left(\mu - r_i^2\right)x_i - \omega_i y_i + \eta\sum_{j=1}^{4}\left(x_j\cos\lambda_{ji} - y_j\sin\lambda_{ji}\right) \tag{1}$$

$$\dot{y}_i = \alpha\left(\mu - r_i^2\right)y_i + \omega_i x_i + \eta\sum_{j=1}^{4}\left(x_j\sin\lambda_{ji} - y_j\cos\lambda_{ji}\right) \tag{2}$$

$$r_i = \sqrt{x_i^2 + y_i^2} \tag{3}$$

$$\omega_i = \left(\frac{\beta}{(e^{-ay_i} + 1)} + \frac{1}{e^{ay_i} + 1}\right)\omega_{sw} \tag{4}$$

$$\phi_i = atan2(y_i, x_i), \tag{5}$$

where α – convergence rate, $\sqrt{\mu}$ – amplitude, i – oscillator number, ω_i – stride frequency, β – duty factor, ω_{sw} – swing phase frequency, a – alternation rate coefficient between swing and stance phases, $\lambda_{ji} = \varphi_i - \varphi_j$ – phase difference between legs i and j, $\varphi_i \in [0; 2\pi]$, x_i, y_i – internal oscillator state, η – coupling coefficient, ϕ_i – current phase of the oscillator, that is considered as a synchronization signal for the formation of the foot trajectories, $\phi_i \in [-\pi; 0)$– stance phase, $\phi_i \in [0; \pi]$– swing phase.

The phase differences φ_i for four different gaits are shown in Table 1. By setting these parameters for λ_{ji}, one can obtain different locomotion rhythms. The output values of the oscillators ϕ_i for gait patterns Φ from Table 1 are shown in Fig. 3. On the above graphs, one can observe a slight curvature of the lines. This is due to the value of β: the larger it is, the smaller the slope ϕ_i in the stance phase and the greater in the swing phase. By changing the parameters Φ, β and ω_{sw}, it is possible to perform a smooth change of ϕ_i. An example of a transition from walk to a more dynamic trot gait is shown in Fig. 4. Note, that gait transition occurs a little more than just one second.

Table 1. Phase relationship for different gait patterns.

Gait	RF φ_1	LF φ_2	RH φ_3	LH φ_4
Walk	π	0	0.5 π	1.5 π
Trot	π	0	0	π
Pace	π	0	π	0
Gallop	0	0	π	π

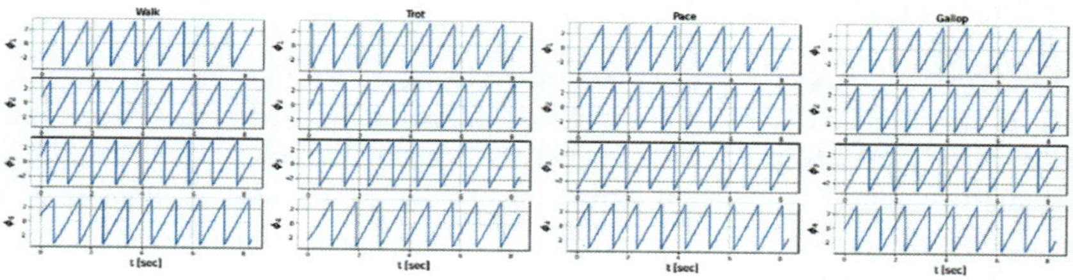

Fig. 3. CPG output curves ϕ_i for different gait patterns

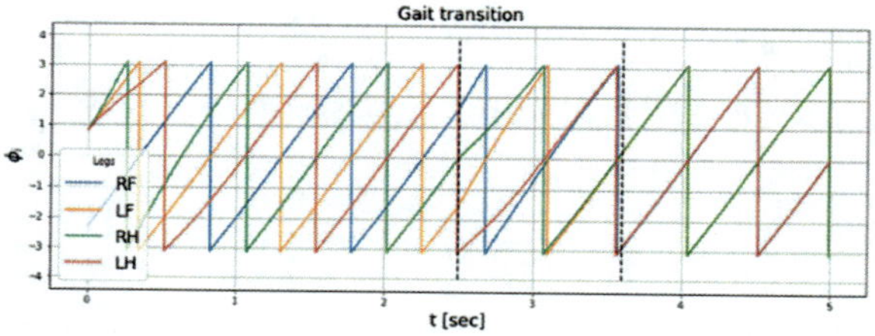

Fig. 4. Walk to trot transition. Dashed lines show the beginning and the end of the transition

2.3 Mapping Function

The task of the Mapping Function block is to form the desired trajectory of the robot feet based on the clock signal φ_i. Also, this block must take into account the desired stride height and stride length, and at the same time carry out turning and sidewalk. The algorithms used for the mapping function [40, 41] assume only forward movement. To implement other types of movement, the algorithm was refined; its block diagram is shown in Fig. 5. At the input, it takes the values of the clock signal ϕ_i, the stride length S_L, the stride height S_H, the two-dimensional velocity vector in the XY plane $D = \left[\upsilon_x \upsilon_y\right]^T$ and the turning radius r. Next, the desired foot positions are calculated. After that, the lateral deflection is sequentially calculated by the values sidewalk vector D, turning radius r and turning direction d. At the very end, the obtained trajectory is translated into the coordinate system of the robot foot relative to the center of mass of the body and the desired foot position P_x, P_y, P_z is calculated.

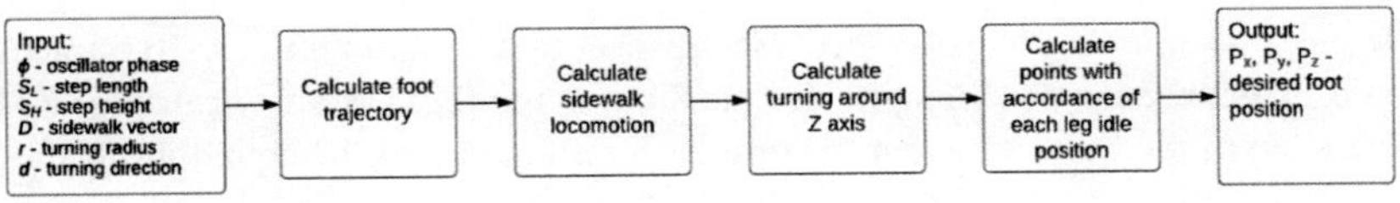

Fig. 5. Block diagram of Mapping Function algorithm

The basic calculation of the foot trajectory is carried out using the following equations:

If $\phi_i \in [-\pi; 0)$:

$$P_{x0} = -\left(\frac{1}{\pi}S_L\phi + \frac{S_L}{2}\right) \tag{6}$$

$$P_{z0} = 0 \tag{7}$$

$$P_{y0} = 0 \tag{8}$$

If $\phi_i \in [0; \pi]$:

$$P_{x0} = -\frac{1}{2\pi}S_L\sin(4\phi) + \frac{1}{\pi}S_L\phi + \frac{S_L}{2} \tag{9}$$

$$P_{z0} = \begin{cases} -\frac{1}{2\pi}S_H\sin(4\phi) + \frac{2}{\pi}S_H\phi, \phi \in \left[0; \frac{\pi}{2}\right] \\ \frac{1}{2\pi}S_H\sin(4\phi) - \frac{2}{\pi}S_H\phi + 2S_H, \phi \in \left[\frac{\pi}{2}; \pi\right] \end{cases} \tag{10}$$

$$P_{y0} = 0, \tag{11}$$

where P_{x0}, P_{y0}, P_{z0} – foot position, S_H, S_L – stride height and stride length respectively.

Sidewalk direction is calculated by determining the angle ψ (Fig. 6):

$$\psi = atan2(D_y, D_x) \tag{12}$$

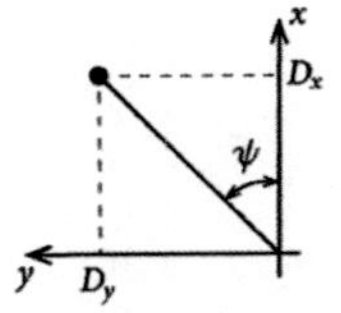

Fig. 6. Robot movement direction

And then the points of the generated trajectory are multiplied by the rotation matrix R_z:

$$P_1 = R_z(\psi)\cdot P_0 = \begin{bmatrix} \cos\psi & -\sin\psi & 0 \\ \sin\psi & \cos\psi & 0 \\ 0 & 0 & 1 \end{bmatrix}\begin{bmatrix} P_{x0} \\ P_{y0} \\ P_{z0} \end{bmatrix} = \begin{bmatrix} P_{x0}\cos\psi - P_{y0}\sin\psi \\ P_{x0}\sin\psi + P_{y0}\cos\psi \\ P_{z0} \end{bmatrix} \tag{13}$$

Turning of the robot is performed with respect to a certain radius r. To implement the turn, it is necessary to imagine the position of the legs, as shown in Fig. 7. The modification of the trajectories for the turn is carried out using the equations [46]:

$$\gamma = d * a\tan 2(h, 2r + d * b),\ \xi = d * a\tan 2(h, 2r - d * b) \tag{14}$$

$$P_2 = \begin{bmatrix} \cos\sigma_{\gamma\xi} & -\sin\sigma_{\gamma\xi} & 0 \\ \sin\sigma_{\gamma\xi} & \cos\sigma_{\gamma\xi} & 0 \\ 0 & 0 & 1 \end{bmatrix} \begin{bmatrix} P_{x1} \\ P_{y1} \\ P_{z1} \end{bmatrix}, \tag{15}$$

where d – turning direction, $\sigma_{\gamma\xi}$ - γ or ξ angles. $\sigma_{\gamma\xi} = \gamma$, if the leg is right, otherwise $\sigma_{\gamma\xi} = \xi$.

The final operation is shifting each vector according to the base position of each foot:

$$P = P_2 + P_{idle}, \tag{16}$$

where P_{idle} – the vector of the idle position of the foot relative to the center of mass of the robot body.

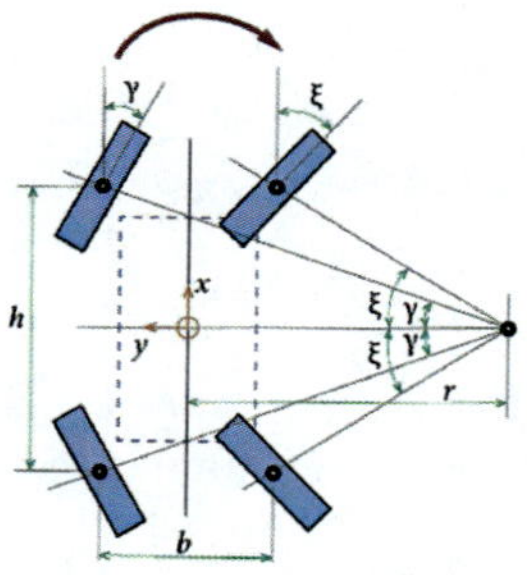

Fig. 7. Turning implementation

3 Results

This section presents the results of experiments on robot locomotion using the control algorithm described above.

3.1 Simulation Experiments

PyBullet is used as a simulation environment [46]. The robot model is described using URDF. Figure 8 shows a visual image of the robot in the PyBullet environment.

In all tests, the robot moves on a flat surface. The model processing frequency is 300 Hz. The parameters of the algorithm are presented in Table 2. The values of a, μ, α

Table 2. Parameters used in experiments for the CPG model and mapping function.

Parameter	Walk, trot, pace	Gallop
ω_{sw}	4π	6π
B	2	2
A	100	100
μ	1	1
a	100	100
S_H	0.06	0.08
S_L	0.06	0.08

did not lead to any major changes in the operation of the algorithm and were set as in [40].

The first experiment was the movement along the longitudinal plane of the robot, using all the gaits from Table 1. The simulation was performing for 5 s. Each gait showed the possibility of locomotion. Trot turned out to be the fastest gait, and in 5 s the robot covered ≈1.95 m. The slowest gait was walk. Using it, the robot covered only 1 m in the same time. Trot, pace and walk have minor deviations along the transverse Y axis, while the deviation of gallop was 0.2 m. Also, given the increased frequency of movement and stride length, the gallop was almost the slowest and least stable. Setting values, as in other gait types, the robot refused to make a steady movement. All this confirms the need to stabilize robot locomotion by the method presented in Chapter 1.

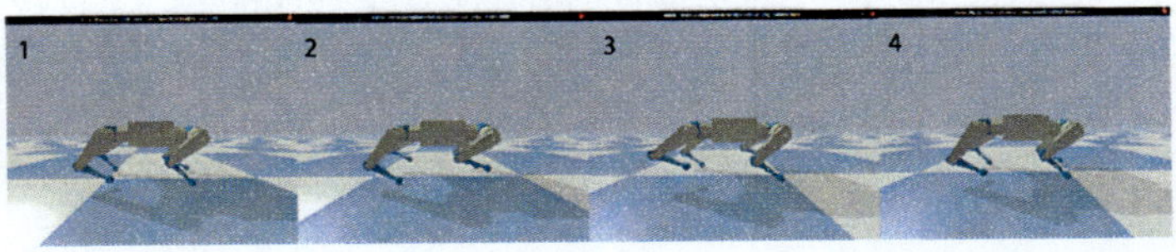

Fig. 8. Robot visualization in the PyBullet simulation environment while executing trotting gait

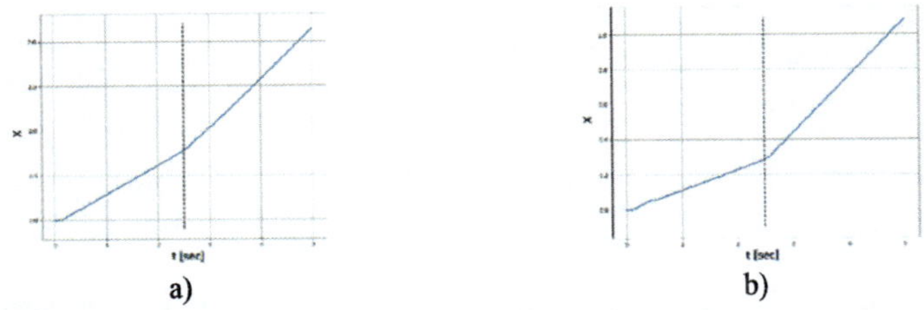

Fig. 9. Dependences of the CoG X position of the robot base on time with an increase in (a) stride frequency, (b) stride length

The second experiment is the effect of changing stride frequency and stride length. The robot walked using the trot gait, each simulation episode lasted 5 s. At the time t = 2.5 s, the following was performed: a) increase in step frequency from 4π to 6π with a step length of 0.03 m and b) increase in step length from 0.03 m to 0.08 m with a frequency of 4π. As can be seen in Fig. 9, the algorithm successfully changes these parameters, and the robot increases locomotion speed.

The third experiment is the transition between gaits during locomotion. Similar to previous experiments, the simulation lasted five seconds. At the beginning of the experiment, the robot moved with a walk gait, and at the time t = 2.5 s, the gait changed to a trot. As can be seen on the plot in Fig. 10, the robot successfully performed a gait transition in approximately 1 s, which can be seen from the change in the nature of the oscillations along the vertical Z axis.

The fourth experiment is testing robot turning capability. The simulation lasted 15 s, during which the robot made four turns. The experiment ended successfully, showing capability of implementing this operation. The results are presented in Fig. 11.

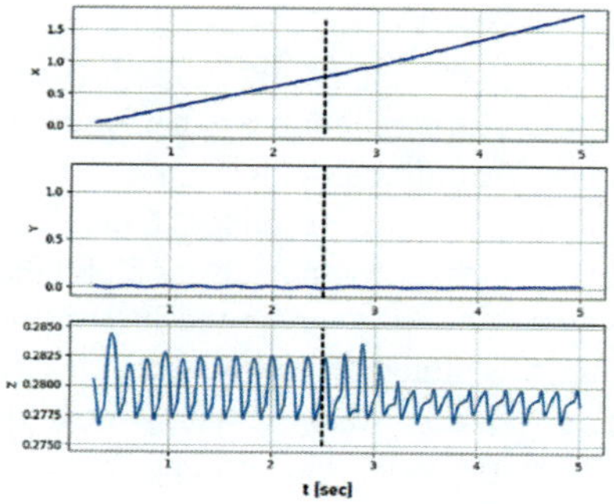

Fig. 10. Gait transition between walk and trot

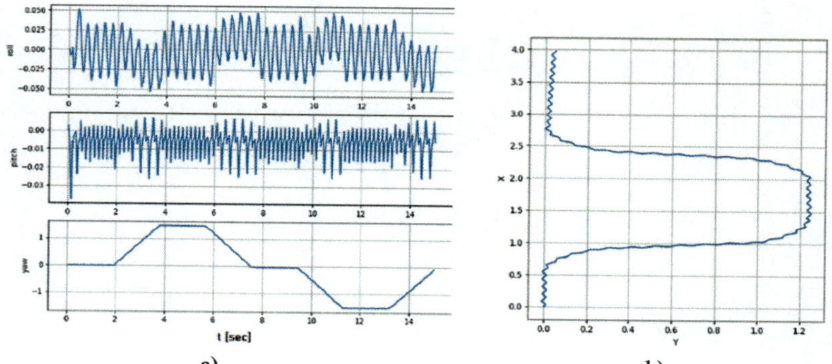

Fig. 11. Implementation of changing the direction of the robot. (a) CoG orientation angles, (b) CoG movement in the horizontal plane.

3.2 Hardware Experiments

While testing the performance of the algorithm on the hardware platform, the initial data from Table 2 was used. The robot moved on a flat surface.

As a result of the experiments, it was possible to move with all the proposed gait types with the ability to switch between them on the go and change other parameters (frequency, length and height of the step), even with the absence of feedback to stabilize the movement. The robot also coped with the task of changing the locomotion direction, executing both sidewalk and turning. Photographs of the robot movement during the experiments are shown in Figs. 12 and 13. Also you can watch the experiments in the supplementary video1[1].

Fig. 12. Trot gait experiments

Fig. 13. Changing moving direction experiments

4 Conclusions

In this paper, the Gait Generator algorithm is featured that combine CPG and mapping function with sidewalk and turning operations. All the gaits such as trot, pace, walk and gallop were executed both in simulation and real world. Moreover, this technique is able to change gait type online, locomotion direction, stride frequency, stride height and stride length. The results show the efficiency and ease of this approach.

Since the robot has to manage with rough terrain and stabilize its body under external disturbances, an additional work with feedback control is needed. Proposed approach is feet trajectory optimization with RL approach which is the aim of the feature work.

References

1. Lee, T.T., Liao, C.M., Chen, T.K.: On the stability properties of hexapod tripod gait. IEEE J. Robotics Autom. **4**, 427–434 (1988)

[1] https://youtu.be/bFAk3QXRYn4.

2. Vukobratovic, M., Borovac, B.: Zero-moment point - thirty-five years of its life. Int. J. Humanoid Rob. **01**(01), 157–173 (2004)
3. Li, J., Wang, J., Yang, S.X., Zhou, K., Tang, H.: Gait planning and stability control of a quadruped robot. Comput. Intell. Neurosci. **2016**, 9853070 (2016)
4. Xin, G., Deng, H., Zhong, G., Wang, H.: Gait and trajectory rolling planning for hexapod robot in complex environment. In: Mechanism and Machine Science, pp. 23–33. Springer Singapore, Singapore (2017)
5. Kalakrishnan, M., Buchli, J., Pastor, P., Mistry, M., Schaal, S.: Fast, robust quadruped locomotion over challenging terrain. In: 2010 IEEE International Conference on Robotics and Automation, pp. 2665–2670 (2010)
6. Delcomyn, F.: Neural basis of rhythmic behavior in animals. Science **210**(4469), 492–498 (1980)
7. Grillner, S., Zangger, P.: On the central generation of locomotion in the low spinal cat. Exp. Brain Res. **34**(2), 241–261 (1979)
8. Fukuoka, Y., Kimura, H., Cohen, A.H.: Adaptive dynamic walking of a quadruped robot on irregular terrain based on biological concepts. The Int. J. Roboti. Res. **22**(3–4), 187–202 (2003)
9. Grillner, S., Wallen, P., Brodin, L., Lansner, A.: Neuronal network generating locomotor behavior in lamprey: Circuitry, transmitters, membrane properties, and simulation. Annu. Rev. Neurosci. **14**(1), 169–199 (1991)
10. Ijspeert, A.J., Crespi, A., Ryczko, D., Cabelguen, J.M.: From swimming to walking with a salamander robot driven by a spinal cord model. Science **315**(5817), 1416–1420 (2007)
11. Conradt, J., Varshavskaya, P.: Distributed central pattern generator control for a serpentine robot. In: ICANN 2003 (2003)
12. Liu, C., Chen, Y., Zhang, J., Chen, Q.: CPG driven locomotion control of quadruped robot. In: Proceedings of the 2009 IEEE International Conference on Systems, Man and Cybernetics, SMC'09, p. 2368–2373. IEEE Press (2009)
13. Righetti, L., Ijspeert, A.J.: Pattern generators with sensory feedback for the control of quadruped locomotion. In: 2008 IEEE International Conference on Robotics and Automation, pp. 819–824 (2008)
14. Santos, C.P., Matos, V.: Gait transition and modulation in a quadruped robot: A brainstem-like modulation approach. Robot. Auton. Syst. **59**(9), 620–634 (2011)
15. Sutherland, I.E., Ullner, M.: Footprints in the asphalt. The Int. J. Roboti. Res. **3**, 29–36 (1984)
16. Cavagna, G.A., Heglund, N.C., Taylor, C.R.: Walking, running and galloping: mechanical similarities between different animals (1976)
17. Raibert, M.H., Chepponis, M., Brown, H.B.: Running on four legs as though they were one. IEEE J. Robotics Autom. **2**, 70–82 (1986)
18. Poulakakis, I., Papadopoulos, E., Buehler, M.: On the stability of the passive dynamics of quadrupedal running with a bounding gait. The Int. J. Roboti. Res. **25**, 669–687 (2006)
19. Raibert, M., Blankespoor, K., Nelson, G., Playter, R.: Bigdog, the rough-terrain quadruped robot. In: 17th IFAC World Congress **41**(2), 10822–10825 (2008)
20. De, A., Koditschek, D.E.: Vertical hopper compositions for preflexive and feedback-stabilized quadrupedal bounding, pacing, pronking, and trotting. The Int. J. Roboti. Res. **37**(7), 743–778 (2018)
21. Havoutis Ioannis, S.C.C.D.: Virtual model control for quadrupedal trunk stabilization. Dynamic Walking (2013)
22. Hutter, M., Gehring, C., Hopflinger, M.A., Blosch, M., Siegwart, R.: Toward combining speed, efficiency, versatility, and robustness in an autonomous quadruped. IEEE Trans. Rob. **30**(6), 1427–1440 (2014)

23. Hyun, D.J., Seok, S., Lee, J., Kim, S.: High speed trot-running: Implementation of a hierarchical controller using proprioceptive impedance control on the MIT Cheetah. The Int. J. Roboti. Res. **33**(11), 1417–1445 (2014)
24. Park, H.W., Wensing, P.M., Kim, S.: High-speed bounding with the MIT Cheetah 2: Control design and experiments. The Int. J. Roboti. Res. **36**(2), 167–192 (2017)
25. Bledt, G., Powell, M.J., Katz, B., Di Carlo, J., Wensing, P.M., Kim, S.: MIT Cheetah 3: design and control of a robust, dynamic quadruped robot. In: 2018 IEEE/RSJ International Conference on Intelligent Robots and Systems (IROS), pp. 2245–2252 (2018)
26. Di Carlo, J., Wensing, P.M., Katz, B., Bledt, G., Kim, S.: Dynamic locomotion in the MIT Cheetah 3 through convex model-predictive control. In: 2018 IEEE/RSJ International Conference on Intelligent Robots and Systems (IROS), pp. 1–9 (2018)
27. Katz, B., Carlo, J.D., Kim, S.: Mini Cheetah: a platform for pushing the limits of dynamic quadruped control. In: 2019 International Conference on Robotics and Automation (ICRA), pp. 6295–6301 (2019)
28. Sleiman, J.P., Farshidian, F., Minniti, M.V., Hutter, M.: A unified mpc framework for whole-body dynamic locomotion and manipulation. IEEE Roboti. Auto. Lett. **6**(3), 4688–4695 (2021)
29. Villarreal, O., Barasuol, V., Wensing, P.M., Caldwell, D.G., Semini, C.: Mpc-based controller with terrain insight for dynamic legged locomotion. In: 2020 IEEE International Conference on Robotics and Automation (ICRA), pp. 2436–2442 (2020)
30. Ha, S., Xu, P., Tan, Z., Levine, S., Tan, J.: Learning to walk in the real world with minimal human effort (2020)
31. Haarnoja, T., Ha, S., Zhou, A., Tan, J., Tucker, G., Levine, S.: Learning to walk via deep reinforcement learning (2018)
32. Kumar, A., Fu, Z., Pathak, D., Malik, J.: RMA: Rapid motor adaptation for legged robots (2021)
33. Yang, Y., Caluwaerts, K., Iscen, A., Zhang, T., Tan, J., Sindhwani, V.: Data efficient reinforcement learning for legged robots (2019)
34. Haarnoja, T., et al.: Soft actor-critic algorithms and applications (2018)
35. Hwangbo, J., et al.: Learning agile and dynamic motor skills for legged robots. Science Robotics **4**(26) (2019)
36. Rudin, N., Hoeller, D., Reist, P., Hutter, M.: Learning to walk in minutes using massively parallel deep reinforcement learning (2021)
37. Iscen, A., et al.: Policies modulating trajectory generators (2019)
38. Tan, J., et al.: Sim-to-real: Learning agile locomotion for quadruped robots (2018)
39. Lee, J.H., Park, J.H.: Turning control for quadruped robots in trotting on irregular terrain. In: Proceedings of the 18th International Conference on Circuits Advances in Robotics, Mechatronics and Circuits, pp. 303–308 (2014)
40. Wang, M., Tang, Z., Chen, B., Zhang, J.: Locomotion control for quadruped robot based on central pattern generators. In: 2016 35th Chinese Control Conference (CCC), pp. 6335–6339 (2016)
41. Liu, C., Chen, Q., Wang, D.: CPG-inspired workspace trajectory generation and adaptive locomotion control for quadruped robots. IEEE Transactions on Systems, Man, and Cybernetics, Part B (Cybernetics) **41**(3), 867–880 (2011)
42. Yu, J., Tan, M., Chen, J., Zhang, J.: A survey on CPG-inspired control models and system implementation. IEEE Trans. Neural Netw. Learn. Sys. **25**(3), 441–456 (2014). https://doi.org/10.1109/TNNLS.2013.2280596
43. Craig, J.J.: Introduction to robotics: mechanics and control. Pearson Educacion (2005)
44. Danilov, V., Diane, S., Gonchareko, V., Artamonov, A.: Algorithms for intelligent control of multi-link walking robots with self-learning capabilities. In: 2020 22th International Conference on Digital Signal Processing and its Applications (DSPA), pp. 1–5 (2020)

45. Danilov, V., Goncharenko, V.: Development and implementation of a six-legged walking robot prototype. In: 2020 13th International Conference "Management of large-scale system development" (MLSD), pp. 1–5 (2020)
46. Coumans, E., Bai, Y.: Pybullet, a python module for physics simulation for games, robotics and machine learning, http://pybullet.org (2016–2021)

Semi-autonomous Walking Control of a Hexapod Robot Based on Contact Point Planning and Follow-the-Contact-Point Gait Control

Kosei Tanada[1(✉)], Shinkichi Inagaki[2], Yuki Murata[3], Ryota Kato[3], and Tatsuya Suzuki[1]

[1] Nagoya University, Furo-cho, Chikusa-ku, Nagoya 464-8603, Japan
tanada.kosei.w3@s.mail.nagoya-u.ac.jp, t_suzuki@nuem.nagoya-u.ac.jp

[2] Nanzan University, Yamasato-cho 18, Syowa-ku, Nagoya 466-8673, Japan
inag@nanzan-u.ac.jp

[3] SHINMEI Industry Co., Ltd., Chokoji 8-20, Toyota, Aichi 471-0879, Japan
{murata y,kator}@shinmei.co.jp

Abstract. This paper proposes semi-autonomous walking control for a hexapod robot controlled by Follow-the-Contact Point (FCP) gait control. In the FCP gait control, each leg contacts the same contact point with which the front legs are in contact, so the motion planning of all legs is focused on the contact point planning of the foremost legs. The robot perceives the ground surface at each step with depth sensors and creates a grid map from the resulting point cloud. The operator inputs the robot's posture and height from the controller to a PC. The PC plans several steps for the foremost legs, and the robot moves the first of the planned steps. Contact point planning is based on a modified A* algorithm. Experiments in which the robot walked on uneven terrains and S-curve confirm the effectiveness of the proposed method.

Keywords: Hexapod robot · Semi-autonomous walking control · FCP gait control · Contact point planning

1 Introduction

Multi-legged robots are expected to play an active role in disaster sites where it is difficult for wheel robots to move. In unknown complex environments, multi-legged robots need to perceive the terrain and plan their movement online. In addition, considering the continuity of walking, several steps should be planned at once. However, due to the large degrees of freedom of multiple legs, real-time motion planning is a challenging task.

There are several approaches for real-time motion planning of multi-legged robots. The first is planning the movement of the body and the legs simultaneously. Hauser et al. [1] proposed a planning method based on a probabilistic

© The Author(s), under exclusive license to Springer Nature Switzerland AG 2023
J. M. Cascalho et al. (Eds.): CLAWAR 2022, LNNS 530, pp. 289–300, 2023.
https://doi.org/10.1007/978-3-031-15226-9_28

road map. The second is planning the movement of the body and the legs separately [2–4]. The trajectory of the body is roughly planned first, and then the proper movement of all the legs is planned, for quadruped robots [4] and hexapod robots [2,3]. The third is planning the movement of only the legs. Vernaza et al. [5] proposed planning a set of contact points of a quadruped robot based on static stability.

The above three approaches have computational difficulty because planning for all the legs is needed. Kishi and Inagaki [6] developed an algorithm to plan the contact points of only the foremost legs of a centipede-type multi-legged robot. In this algorithm, a series of contact points toward the desired walking direction is planned based on A* algorithm. The other legs are controlled to follow the fore legs based on Follow-the-Contact-Point (FCP) gait control [7,8]. In the FCP gait control, each leg except the two foremost legs is controlled to contact the same contact point on which its fore leg is contacting. As a result, planning the movement of all the legs is concentrated on planning the contact points of the foremost legs.

In this paper, an operating system of a hexapod robot based on the FCP gait control [7,8] and the contact point planning [6] is proposed. The operator inputs the desired body posture and height by a control pad, then the robot plans a suitable contact point in real-time. The developed control system allows the operator to determine the direction of travel and posture. Furthermore, the contact point planning and FCP compensate for the operator's difficulty in determining the contact points and foot movement. This reduces the operator's burden when walking on uneven terrain and enables safe walking. As a result, the operator's burden is reduced while the robot can walk over irregular terrains. The system is implemented on a hexapod robot and verified via experiments.

The rest of this paper is organized as follows: Sect. 2 introduces a hexapod mobile robot used in this paper and its system architecture. A user interface, perceiving a terrain surface and creating a grid map, and the FCP gait control, are also explained. In Sect. 3, contact point planning based on modified A* algorithm is proposed. In Sect. 4, experimental results are shown where the robot walk on uneven terrains and S-curve of even terrain, and the conclusion is given in Sect. 5.

2 Hexapod Robot and System Architecture

2.1 Hexapod Robot SOL and System Flow

In this research, a hexapod robot, Shinmei-Original-Locomotion (SOL), developed by SHINMEI Industry Co., Ltd. is used (Fig. 1). Each leg has three links and three servo motors. Three depth sensors are installed on the body front. Figure 2 shows the control system in this research. Three depth sensors are embedded in the robot, and images captured by the depth sensors are sent to a PC. The operator sees the images displayed on the PC screen and inputs body posture and height from a controller. The PC then plans a series of contact points and sends the first contact point in the plan and the body posture and

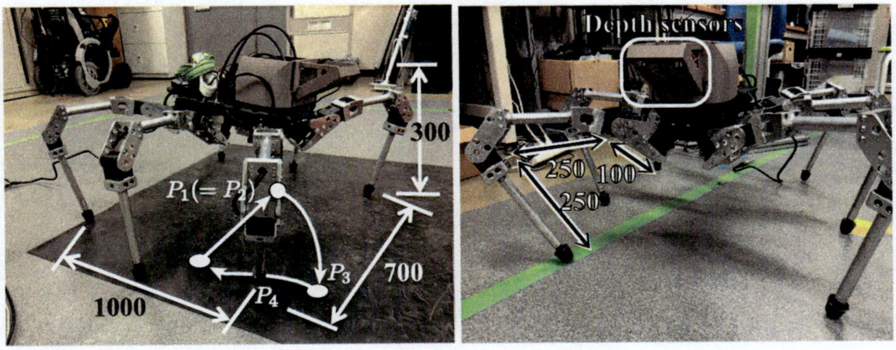

Fig. 1. Hexapod robot SOL

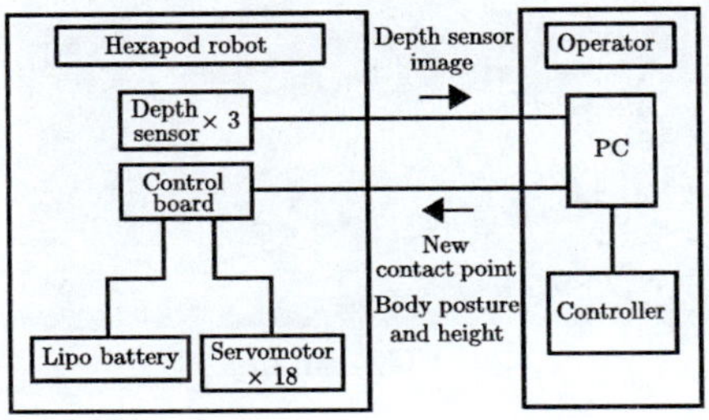

Fig. 2. System structure

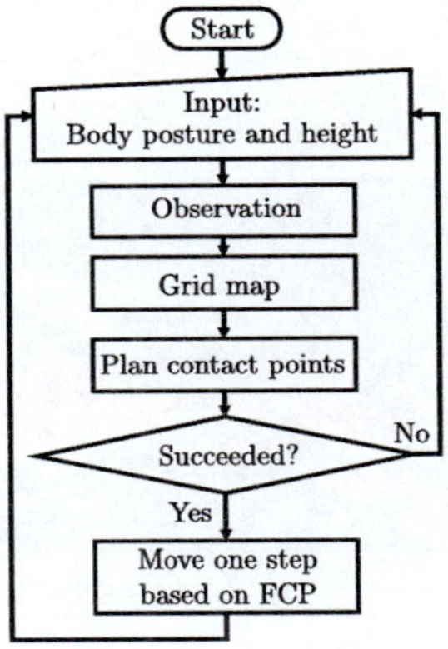

Fig. 3. System flow

the body height to the control board in the robot. Finally, the control board calculates the leg movements and controls the servo motors.

The flow of the semi-autonomous-walk control method in this research is shown in Fig. 3. First, the operator inputs the body posture and the body height by a controller while watching a display (Sect. 2.2). After that, the robot observes the front ground by the depth sensors, and a grid map is generated (Sect. 2.3). The robot plans a series of contact points ahead of the foremost legs based on an algorithm introduced in Sect. 3. The robot walks one step of the planned contact points based on the FCP gait control (Sect. 2.4).

2.2 Body Posture and Height Operation Using a Controller

The operator inputs the posture and the height of the robot body by a controller shown in Fig. 4. Once the body posture and height are input, the commands are sent to the robot, which immediately changes the posture and height of the body to reflect the input. Then, when the decision button is pressed, the robot

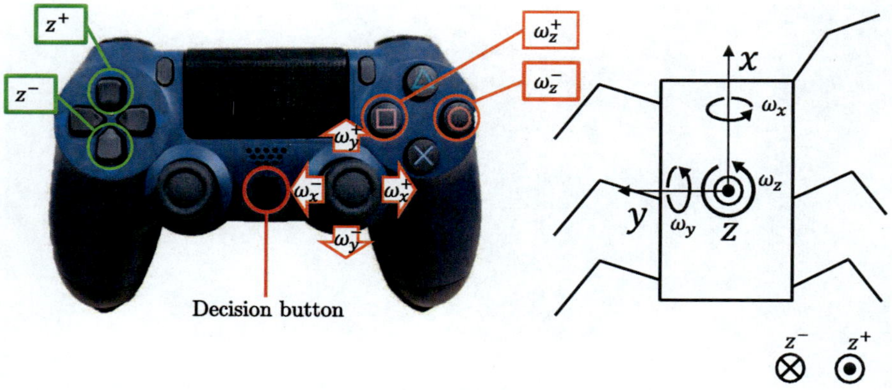

Fig. 4. Operation of body posture and height by a controller

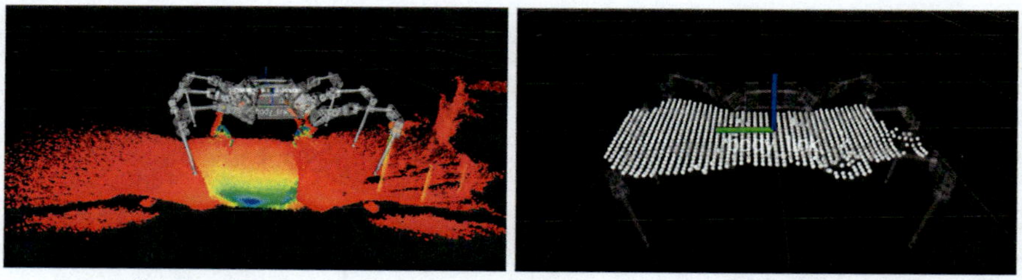

(a) Raw Point Cloud (b) Grid map

Fig. 5. Terrain surface observation by the depth sensors.

observes the terrain surface using its onboard depth sensors and begins planning the contact points. If contact point planning fails, the operator can change the robot's posture or height to get a better view and walk the robot again. In this paper, the height is set to 0.3 [m], and both the roll and the pitch angles are 0 [rad].

2.3 Observing Terrain Surface and Creating Grid Map

As shown in Fig. 1, three depth sensors are installed in front of the robot. These sensors observe the forward terrain surface using the infrared sensor and output it as point clouds (Fig. 5(a)). The image of the raw point cloud is shown in Fig. 5(a). It is difficult to deal with the raw point clouds directly because of the high computational burden. Therefore, a grid map is created by KD-Tree point search [9].

The grid is a lattice on xy plane in Σ_B, and the size of a cell is $c \times c$. In this paper, c is set at 0.03[m] as the same size as the robot leg tip. A grid map

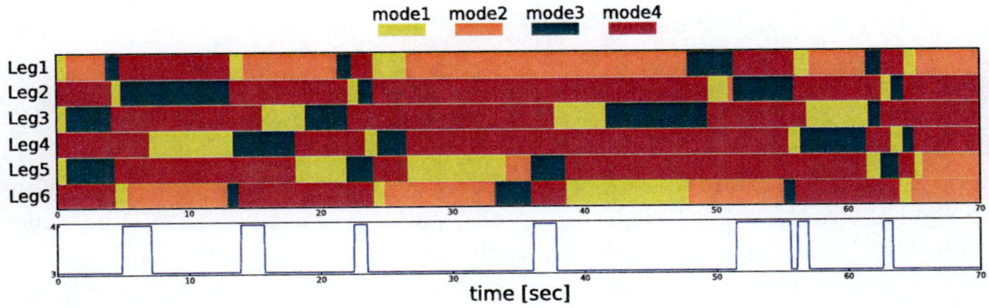

Fig. 6. Gait diagram.

is expressed as $M = \{(m, n)\}_{m \in [1,50], n \in [1,100]}$. Height information of a cell is calculated as follows:

$$h(m,n) = \begin{cases} \frac{1}{\#A(m,n)} \sum_{l \in A(m,n)} z_l & \text{if } \#A(m,n) \geq N \\ null & \text{otherwise} \end{cases}, \tag{1}$$

where $A(m, n)$ is a set of points whose xy coordinate is inside of a circle whose radius and center are r and the center of the cell (m, n) respectively, and $\#A(m, n)$ is the number of points in $A(m, n)$. The z coordinate value in Σ_B of point l is represented as z_l. At least N points are needed to give the height information of a cell. In this paper, N is set to 24 and r is 0.03 [m]. If the number of points belonging to a cell is less than N, the cell has null information and is not selected as a contact point when planning. The frame rates of the depth sensors and the grid map creation are 30 and 5 [fps], respectively.

2.4 FCP Gait Control

The Follow-the-Contact-Point (FCP) gait control is an event-driven control in which each leg has four control modes. In this section, the FCP gait control is explained briefly. See [7,8] for the details.

The modes have different target points of the leg tip P_j, ($j \in \{1, 2, 3, 4\}$). The movement of a leg tip are shown in Fig. 1, left, and FCP gait diagram is shown in Fig. 6. The leg tip moves from the ground to a midair point P_1 in Mode 1. When the leg tip reaches P_1, the mode changes to Mode 2. For the legs except the foremost ones, the leg tip keeps the position at $P_2(= P_1)$ until the leg tip of the fore leg enters its reachable area. In Mode 3, the leg tip moves to point P_3 where the fore leg is contacting the ground. For the foremost legs, the mode changes from 2 to 3 when the coordinate of a new contact point is given by the PC, and the leg tip moves to the new contact point P_3 in Mode 3. In Mode 4, the leg contacts the ground and propels the body together with other legs in Mode 4. The mode changes from 4 to 1 when conditions about the rear leg and the supporting polygon are satisfied. A gait is achieved by repeating this process in each leg.

3 Planning Contact Points

3.1 Searching Area for Next Contact Point

In the contact point planning method, each time the robot takes one step, it plans several steps ahead and adopts only the first step as a new contact point. This method allows the robot to achieve a robust gait against unknown environments since it will find a new location after each step.

A contact point is expressed as $C_i \in M$, $i \in \mathbb{Z}$. The foremost two contact points on which legs are contacting are represented as C_{-1} and C_0 as shown in Fig. 7(a). The contact points which are going to be planned are represented as C_1, C_2, ... in the contacting order. Hereafter, as shown in Fig. 7(b), when planning the contact points C_i $(i = 2, 3, \cdots)$, the direction of the body is assumed to be the same as at the start of planning. In addition, the body positions B_i $(i = 2, 3, \cdots)$ are assumed to be the same as the centroid of the triangle which is made from C_{i-1}, C_{i-2}, and C_{i-3}. Note that B_1 is equal to the origin of Σ_B.

When the robot whose body is located at B_i plans the contact point C_i, a cell (m, n) to be the candidate of C_i satisfies the following conditions:

1. The foremost legs can reach the cell (m, n).
2. When a contact point for the foremost right/left leg is planned, the cell (m, n) is on the right/left side of the body.
3. A triangle composed of (m, n), C_{i-1}, and C_{i-2}, has its centroid inside it with margin δ from the edges (Fig. 7(a)).
4. All the eight neighbor cells of (m, n) possess height information.
5. The cell (m, n) is not too high or too low compared to the surrounding cells:

$$h(m,n) - h(m+k, n+l) < c \max\{|k|, |l|\},\ \forall k \in [-2, 2],\ \forall l \in [-3, 2], \quad (2)$$
$$h(m,n) - h(m+k, n+l) > -c \max\{|k|, |l|\},\ \forall k \in [-2, 2],\ \forall l \in [-3, 5]. \quad (3)$$

3.2 Evaluation Function

To evaluate a contact point, an evaluation function of cell $c \in M$ is defined as follows:

$$f(c) = h(c) + g(c). \quad (4)$$

In this function, $h(c)$ is a heuristic function defined as follows:

$$h(c) = \omega_h \, |p(c) - p_h| \,. \quad (5)$$

where ω_h is a weight parameter. In this paper, $\omega_h = 2.0$. The heading point p_h is set ahead of the current body position as shown in Fig. 7(a). The term $g(c)$ is

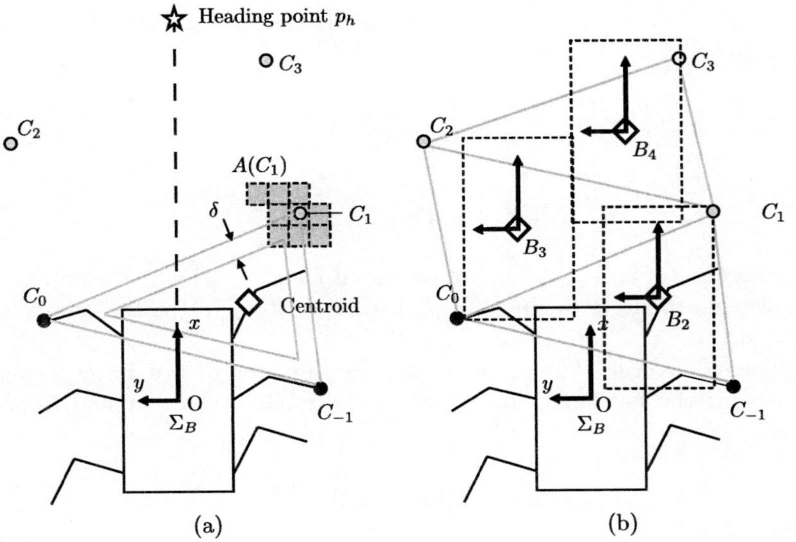

Fig. 7. Sequences of contact points (a); body position B_i when searching the next contact point C_i $(i = 1, 2, 3)$ (b)

the sum of the evaluation functions of all the steps from the contact point C_0 to cell c, so the following equations are satisfied:

$$g(c) = g\left(P(c)\right) + r(P(c), c),\ \ g\left(C_0\right) = 0 \tag{6}$$

where $P(c) \in M$ is the parent (previous contact point) of cell c, and $r(c, c')$ is a cost function in which the robot steps from c to c'.

The cost function $r(c, c')$ is composed of four cost functions: (i) landscape costs L_ϕ and $L_\sigma : M \to \mathbb{R}$, (ii) transition cost $T : M \times M \to \mathbb{R}$, and (iii) instability cost $I : M \times M \to \mathbb{R}$. Then, a cost function to evaluate the transition from cell $c \in M$ to $c' \in M$ is calculated as follows:

$$r(c, c') = w_\phi L_\phi(c) + w_\sigma L_\sigma(c) + w_T T(c, c') + w_I I(c, c') \tag{7}$$

where $w_\phi,\ w_\sigma, w_T, w_I \in \mathbb{R}$ are weight coefficients. In this paper, the weights are determined empirically as $(w_\phi, w_\sigma, w_T, w_I) = (1.0, 1.0, 5.0, 10.0)$. Each function is explained as follows:

(i) Landscape costs, $L_\phi(c)$ and $L_\sigma(c)$, evaluate the easiness for a leg to contact the cell. First, the least-squares plane is derived from $N \times N$ cells around cell c, $\mathcal{N}(c)$, in Σ_B:

$$z = \alpha(c)x + \beta(c)y + \gamma(c). \tag{8}$$

In this paper, we set $N = 5$. Then, the landscape functions are calculated as follows:

$$L_\phi(c) = \cos^{-1}\left(\frac{1}{\alpha^2(c) + \beta^2(c) + 1}\right), \tag{9}$$

$$L_\sigma(c) = \frac{1}{N}\sum_{c' \in \mathcal{N}(c)} |z(c') - (\alpha(c)x(c') + \beta(c)y(c') + \gamma(c))|, \tag{10}$$

where $L_\phi(c)$ is the gradient of terrain at c and $L_\sigma(c)$ is the variance of the terrain around c. In Eq. (10), $p(c) = (x(c), y(c), z(c))$ is the coordinate of cell c in Σ_B.

(ii) Transition cost, $T(c, c')$, evaluates the amount of the body movement toward the heading point p_h when the robot walks one step from c to c':

$$T(c, c') = (p(c) - p(c')) \cdot \frac{(p_h - p(c))}{|p_h - p(c)|}. \tag{11}$$

The target position is set 3 m straight ahead of the body in this paper.

(iii) The instability cost, $I(c, c')$, evaluates the narrowness between c and c':

$$I(c, c') = (x(c) - x(c')) - |y(c) - y(c')|. \tag{12}$$

Decreasing this cost function, the two cells c and c' become apart in y-axis direction of Σ_B, and c' is located in far front of c in the forward direction. is set so that the smaller the support polygon, the more unstable it is. Essentially, the instability cost should be calculated based on the support area created by the ground points. However, each contact point candidate c' has no information about the contact points other than the parent contact point c. Therefore, the alternative evaluation (12) is used.

3.3 Algorithm of Contact Point Planning

The algorithm to derive S steps from C_1 to C_S is shown in Algorithm 1. Each cell c has an evaluation function $f : M \to \mathbb{R}$, which means the cost when the robot starts walking from the initial step C_0 and finally reaches cell c. This algorithm is based on A* algorithm [10] with some modifications. In the original A* algorithm, once a cell is evaluated, the cell is never evaluated again. On the other hand, Algorithm 1 can re-evaluate the evaluation function even if the cell has been already evaluated. This re-evaluation allows the algorithm to search multiple branches of contact point sequences (underlined part of Lines 25 to 27).

In Line 6, the OPEN list and the cost function are initialized, where ":=" is a substitution operator. The OPEN list is the set of cells which have been evaluated but not selected as a candidate of contact point. When a cell is selected as a candidate of contact point from the OPEN list, it is moved into the CLOSE list.

Algorithm 1. Planning S contact points

1: Cost function of cell c: $f(c) = h(c) + g(c)$
2: Parent of cell c: $P(c)$ where $P(C_0) = C_{-1}$
3: Generation (# of steps) from C_0 to cell c: $G(c)$ where $G(C_0) = 0$
4: Searching area for the next contact point c' of cell c: $A(c')$
5:
6: Add C_0 to the OPEN list and $g(C_0) := 0$
7: **if** the OPEN list is vacant **then**
8:     Planning failed
9: **end if**
10: Find cell c that minimizes $f(c)$ from the OPEN list and move it to the CLOSE list
11: **if** $G(c) = S$ **then**
12:     Planning succeeded; $C_S := c$
13:     **for** $i = S, S-1, \cdots, 2$ **do**
14:         $C_{i-1} := p(C_i)$
15:     **end for**
16: **end if**
17: **for** $c' \in A(c')$ **do**
18:     $f'(c') := h(c') + g(c) + r(c, c')$
19:     **if** c' is neither included in the OPEN list nor the CLOSE list **then**
20:         Add c' to the OPEN list
21:         $f(c') := f'(c')$; $P(c') := c$
22:     **end if**
23:     **if** $f'(c') < f(c')$ **then**
24:         $f(c') := f'(c')$; $P(c') := c$
25:         **if** c' is in the CLOSE list **then**
26:             Move c' into the OPEN list
27:         **end if**
28:     **end if**
29: **end for**
30: Go to line 7

In Lines 7 to 9, the algorithm determines if the planning procedure failed. In Line 10, a candidate for the contact point which has a minimum value of cost function is selected from the OPEN list. When the candidate is S-th generation, that is S-th step, the planning procedure is determined as succeeded, and a sequence of contact points is generated by back-tracking the parents of the contact points.

The procedure in Lines 17 to 29 is conducted for each cell c' in the next contact point searching area of cell c. First of all, a prospective cost function $f'(c')$ is calculated in Line 18. If cell c' is neither included in the OPEN list nor the CLOSE list, it is added to the OPEN list and its cost function and parent cell are recorded (Lines 19 to 22). Otherwise, cell c' is in either the OPEN or CLOSE list. Then, if the value of prospective cost function $f'(c')$ is smaller than that of the cost function $f(c')$ which had been recorded in some previous procedure, $f(c')$ is updated to $f'(c')$ (Lines 23 to 28). Additionally, if cell c' is in the CLOSE list, it is moved back to the OPEN list (Lines 25 to 27).

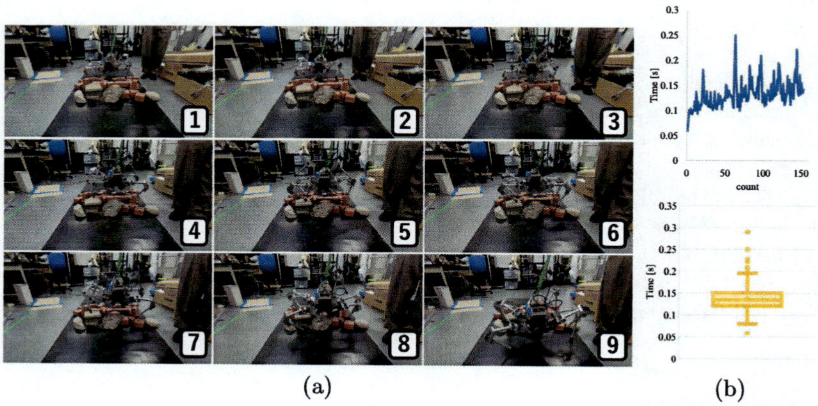

Fig. 8. Walking experiment on rough terrain: (a) Movement of the robot, (b) Computation time of the contact point planning

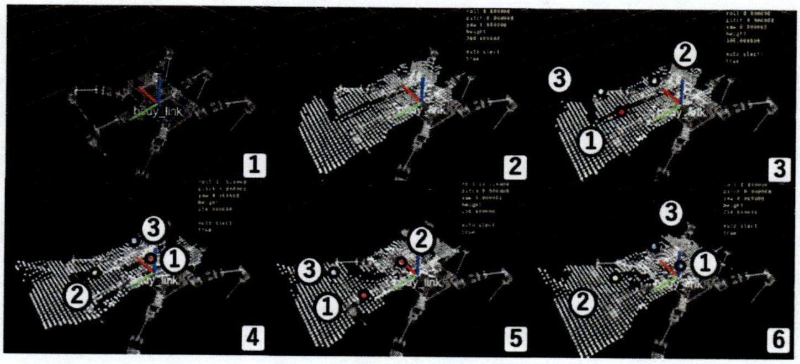

Fig. 9. Planned contact points when walking rough terrain: ①, ②, and ③, show the step number of planned contact points.

4 Experimental Results

Figure 8(a) shows scenes of an experiment where the robot walked on rough terrain with planning contact points. The numbers in the lower right corner of the images indicate the transition order. In this experiment, $S = 3$. Bricks and stones of up to 0.23 [m] height were placed as obstacles. The robot selected to step on the flat surface of the obstacles and climbed over them. The computation time to plan every three steps was approximately 0.14 [s].

Figure 9 shows the planned contact points when the robot walking on rough terrain in Fig. 8. The robot walks toward the upper left direction in each image. The numbers in the lower right corner of the images indicate transition order. The white points are the grid map. The number in a circle is the step number of the planned contact points.

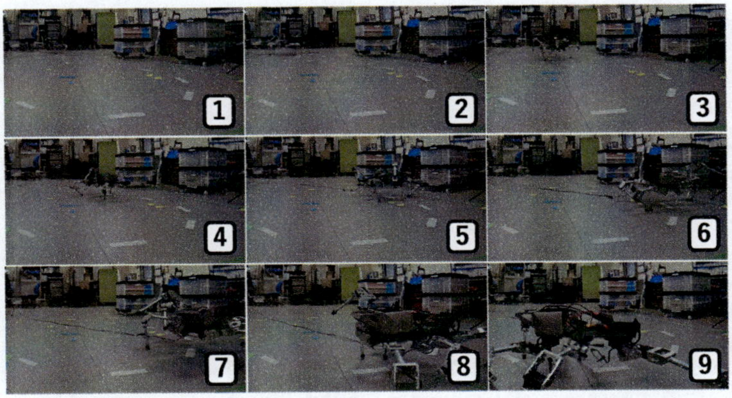

Fig. 10. Walking S-curve

Figure 10 shows scenes in which the robot walked along an S-curve. The S-curve was drawn on the room floor with tape. The operator controlled the posture of the robot so that it walked on the line while watching the display. The walking speed was approximately 0.3 [m/s], and the turning radius of the S-curve ranged from 0.8 [m] to 0.9 [m].

5 Conclusion

In this paper, the semi-autonomous walking control system for a hexapod robot using the FCP gait control is developed by implementing the contact point planning algorithm for the foremost legs. As the result, real-time contact point planning up to three steps ahead is achieved, and stable walking on both even and uneven terrains is achieved.

Acknowledgment. This research is subsidized by Grant-in-Aid for Scientific Research (19H02108).

References

1. Hauser, K., Bretl, T., Latombe, JC., Wilcox, B.: Motion planning for a six-legged lunar robot. In: Akella, S., Amato, N.M., Huang, W.H., Mishra, B. (eds) Algorithmic Foundation of Robotics VII. Springer Tracts in Advanced Robotics, vol. 47. Springer, Berlin, Heidelberg. https://doi.org/10.1007/978-3-540-68405-3_19
2. Eldershaw, C., Yim, M.: Motion planning of legged vehicles in an unstructured environment. In: Proceedings 2001 ICRA. IEEE International Conference on Robotics and Automation vol. 4, pp. 3383–3389 (2001). https://doi.org/10.1109/ROBOT.2001.933140
3. Belter, D., Skrzypczynski, P.: Integrated motion planning for a hexapod robot walking on rough terrain. IFAC Proc. Volumes **44**(1), 6918–6923 (2011)

4. Mastalli, C., Havoutis, I., Winkler, A.W., Caldwell, D.G., Semini, C.: On-line and on-board planning and perception for quadrupedal locomotion. In: 2015 IEEE International Conference on Technologies for Practical Robot Applications (TePRA), pp. 1–7 (2015). https://doi.org/10.1109/TePRA.2015.7219685
5. Vernaza, P., Likhachev, M., Bhattacharya, S., Chitta, S., Kushleyev, A., Lee, D.D.: Search-based planning for a legged robot over rough terrain. In: 2009 IEEE International Conference on Robotics and Automation, pp. 2380–2387 (2009). https://doi.org/10.1109/ROBOT.2009.5152769
6. Kishi, S., Inagaki, S.: Graph-search based footstep planning for multi-legged robots on irregular terrain by using depth-sensor. In: Mobile Service Robotics, pp. 417–424 (2014)
7. Inagaki, S., Niwa, T., Suzuki, T.: Follow-the-contact-point gait control of centipede-like multi-legged robot to navigate and walk on uneven terrain. In: IEEE/RSJ International Conference on Intelligent Robotics and Systems, pp. 5341–5346 (2010)
8. Murata, Y., Inagaki, S., Suzuki, T.: Development of an adaptive hexapod robot based on Follow-the-contact-point gait control and timekeeper control. In: 2019 IEEE/RSJ International Conference on Intelligent Robots and Systems, pp. 3321–3327 (2019)
9. Rusu, R.B., Cousins, S.: 3d is here: point cloud library (pcl). In: IEEE International Conference on Robotics and Automation, ICRA 2011, pp. 1–4 (2011). https://doi.org/10.1109/ICRA.2011.5980567
10. Hart, P.E., Nilsson, N.J., Raphael, B.: A formal basis for the heuristic determination of minimum cost paths. In: IEEE Transactions on Systems Science and Cybernetics, vol. 4, no. 2, pp. 100–107 (1968)

Omnidirectional Motion Control Method of Quadruped Robot Based on 3D-CPG Oscillator Group

Bo Tao, Dongchao Yang(✉), Geng Huang, Zecui Zeng, Chen Chen, and Teng Li

Tsinghua University, Beijing, China
{taob19,g-huang20}@mails.tsinghua.edu.cn, {ydc, liteng2020}@mail.tsinghua.edu.cn

Abstract. This paper focuses on the position control method of the quadruped robot, aims at the dynamic gait of the quadruped robot for Omnidirectional stable movement, combines the central mode generator to study the control method of the quadruped robot, and extends the CPG oscillator to 3 dimensions using the excellent dynamic characteristics and adaptive ability of CPG oscillator, the rhythmic foot movement trajectory for the trot gait is designed, and a universal quadruped robot is verified through dynamic simulation Motion trajectory planning methods.

Keywords: Online trajectory planning · Central pattern generator · Matlab simulation

1 Introduction

With the development of intelligent manufacturing industry, the robot presents more and more pivotal in industry, robots instead of humans to complete a large number of repetitive labor, can greatly improve production efficiency. And for dangerous work scenarios such as mine surveying, nuclear leakage accident handling, earthquake search and rescue, etc. The deployment of foot-type mobile robots with high kinematic performance and environmental adaptability will become a major trend in the future.

In the category of foot robots, quadruped robots have attracted great attention from research institutions around the world because they can take into account the characteristics of motion stability and high dynamic motion characteristics when dealing with most complex scenarios. In 2005, Boston Dynamics developed the Big Dog, a quadruped robot with a hydraulic drive solution [1] The Cheetah series [2] bionic quadruped robots released later by MIT can achieve fast dynamic gait movements, with greater flexibility on rough roads than the Big Dog based on quasi-static gaits. In addition, the research and development of quadruped robots in Europe [3], Japan [4] and other regions has also made remarkable progress. But so far, most of the research on quadruped robots is still in the laboratory stage, and the control scheme that has been productized worldwide tends to be single, that is, the body force control scheme based on model prediction control [5], which simplifies the dynamic equation of the whole machine, without considering the

© The Author(s), under exclusive license to Springer Nature Switzerland AG 2023
J. M. Cascalho et al. (Eds.): CLAWAR 2022, LNNS 530, pp. 301–312, 2023.
https://doi.org/10.1007/978-3-031-15226-9_29

impact of the movement of the swinging leg on the body. In order to reduce the amount of operation on the main control board to ensure sufficient force control frequency, the scheme is very sensitive to the inertia of the legs, and it is difficult to apply to the general structure of the quadruped robot.

The central mode generator [6] is a complex neural network composed of a series of neural oscillators and reflex circuits integrated together, which is an important part of the animal motion control system, capable of generating and controlling rhythmic movements. Barasuol et al. innovatively modified the extreme ring shape of cpg and directly corresponded it to the trajectory of the robot's foot, enabling simple and efficient gait programming, which was named WCPG [7] However, it does not fully take into account the change in the support phase time and the swing phase time generated after the introduction of the interception depth, because the support phase time of a certain leg has changed, which will reduce the coordination of the overall quadruped gait. And the oscillator can only be applied to the adaptive generation of the foot trajectory on the two-dimensional plane, which cannot meet the needs of the robot for steering movement. In view of these two problems, this paper introduces the foot contact judgment and the motion state machine to correct the angular frequency of the oscillator ω to ensure that the support phase time is consistent with the swing phase time, and designs the steering control strategy, introduces the lateral displacement of the foot y, and realizes the expansion of the oscillator from the two-dimensional plane to the three-dimensional space.

This study was conducted on a quadruped robot prototype platform shown in Fig. 1.

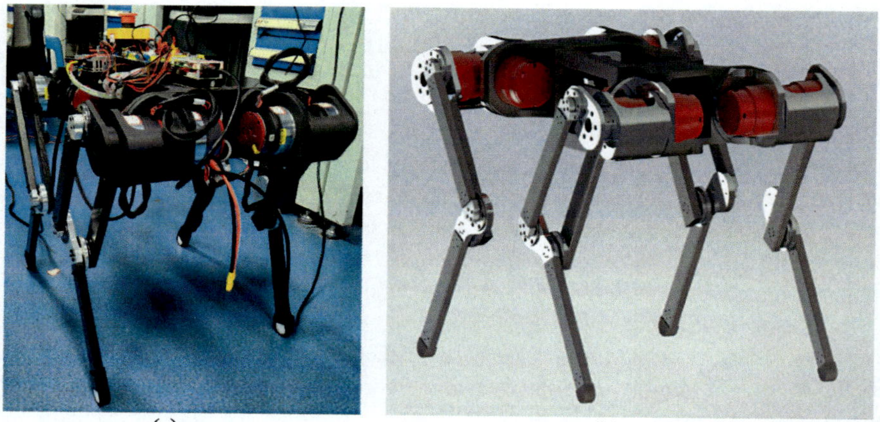

(a) (b)

Fig. 1. Quadruped robot prototype structure: solid prototype diagram (a), virtual prototype rendering diagram (b).

2 Control System Algorithm Framework

The quadruped robot motion control framework described in this article consists of three parts: a trajectory generator (CPG oscillator network), a motion state machine, and position control. Figure 2 shows the functions of each module of the robot's single leg and the data transfer between the modules.

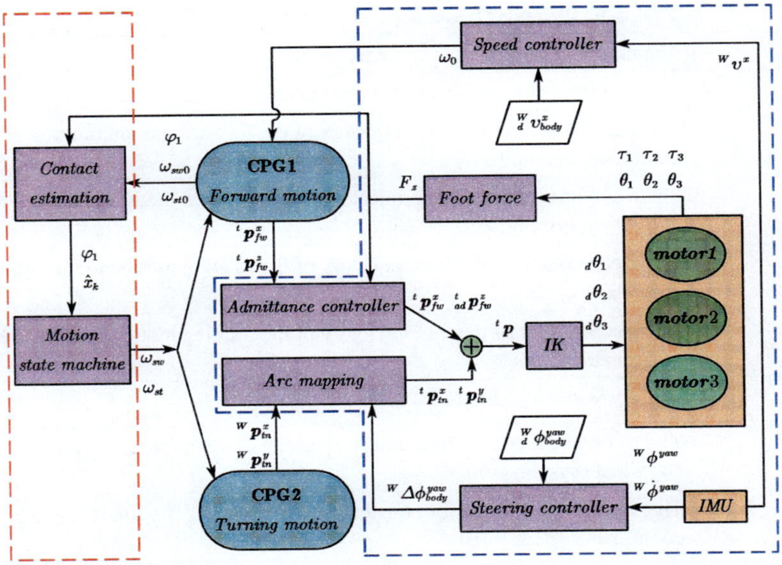

Fig. 2. One-leg control system algorithm framework

The blue rounded corner box represents an oscillator group with one leg, and the four oscillator groups are coupled to form the foot trajectory generation network of the whole machine; the orange part represents the robot's execution unit and inertial measurement element; and the purple rectangular box represents it Controls the specific nodes in the framework, and each purple box corresponds to a separate thread in the host computer program. The meaning of the data symbols in the Fig. 2 are shown in Table 1.

Table 1. Definition of each parameter.

Symbol	Meaning
$\tau_1\ \tau_2\ \tau_3$	The actual torque of the three motors (measured by current)
$\theta_1\ \theta_2\ \theta_3$	The actual angle of the 3 motors (measured by the encoder)
$_d\theta_1\ _d\theta_2\ _d\theta_3$	The desired joint angle solved by inverse kinematics
φ_1	The current phase angle of the main oscillator
ω_{sw0}	The initial angular frequency of the swing phase, when the foot falls on the flat ground
ω_{st0}	The initial angular frequency of the support phase when the foot falls on flat ground

(*continued*)

Table 1. (*continued*)

Symbol	Meaning
ω_0	The original angular frequency of the main oscillator
ω_{sw}	The modified swing phase angle frequency
ω_{st}	The corrected support phase angle frequency
${}^{t}\boldsymbol{p}_{fw}^{x}$	The x component of the foot trajectory of the forward motion in the coordinate system $\{traj\}$
${}^{t}\boldsymbol{p}_{fw}^{z}$	The z component of the foot trajectory of the forward motion in the coordinate system $\{traj\}$
${}^{W}\boldsymbol{p}_{tn}^{x}$	The x component of the foot trajectory of the rotating motion in the world coordinate system
${}^{W}\boldsymbol{p}_{tn}^{y}$	The y component of the foot trajectory of the rotating motion in the world coordinate system
${}^{t}_{ad}\boldsymbol{p}_{fw}^{z}$	${}^{t}\boldsymbol{p}_{fw}^{z}$ corrected by admittance controller
${}^{t}\boldsymbol{p}_{tn}^{x}$	The x component of the foot trajectory of the rotating motion in the coordinate system $\{traj\}$
${}^{t}\boldsymbol{p}_{tn}^{y}$	The y component of the foot trajectory of the rotating motion in the coordinate system $\{traj\}$
${}^{t}\boldsymbol{p}$	The final planned 3d trajectory in the foot coordinate system $\{traj\}$
$W\Delta\phi_{body}^{yaw}$	The circular angle corresponding to the foot track of single gait cycle in the world coordinate system
$W\phi^{yaw}$	Body yaw angle measured by IMU
$W\dot{\phi}^{yaw}$	Body yaw angular velocity measured by IMU
${}^{W}_{d}\phi_{body}^{yaw}$	Desired body yaw angle in the world coordinate system
${}^{F}v^{x}$	The body's forward velocity measured by IMU in the floating base coordinate system $\{F\}$
${}^{F}_{d}v_{body}^{x}$	The desired speed of the body in a floating base coordinate system

The foot contact estimation node indirectly solves the contact force of the robot foot by obtaining the joint torque in real time and calculating the Jacobian matrix of the leg, and fuses the contact force with the reference phase of CPG1 to achieve high accuracy contact judgment, and realizes the state transfer of the motion state machine according to the result of the contact judgment and the current phase of the leg; the admittive controller is paired by designing a virtual spring damping system between the foot and the hip joint The output of CPG1 is corrected in real time to achieve good foot force suppleness effect, and the arc mapping module combines the output of the steering controller and the output of CPG2 to plan the foot trajectory that drives the rotation movement of the robot. The adder superimposes the corrected outputs of the two oscillators to form a $\{traj\}$ three-dimensional foot trajectory of the robot's foot-end spatial coordinate system. The IK module is responsible for solving the joint angle and solving the position control input of the motor from the geometric relationship of the leg structure. The leg model and geometric parameters of the prototype are shown in Fig. 3.

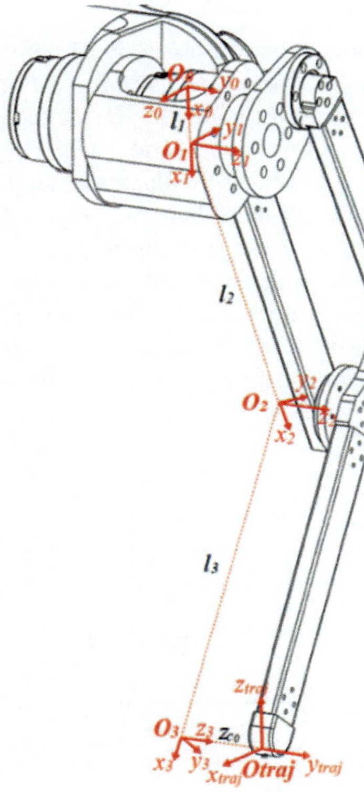

(a)

Single-leg DH parameter list

Joint i	l_i	α_i	d_i	θ_i
1	0	$-\frac{\pi}{2}$	0	q_1
2	250	0	0	$q_2-\frac{\pi}{6}$
3	250	0	0	$-q_3+\frac{\pi}{3}$

Dimension parameter list

Total mass	20kg
Body width W	444mm
Body length L	518mm
Height of CoM	493mm

(b)

Fig. 3. Quadruped robot prototype single-leg model: leg joint coordinate system (a), leg DH parameters and prototype size parameter list (b).

3 Control Framework Based on 3D-CPG Oscillator

The 3D-CPG oscillator that makes up the single-leg foot trajectory consists of a linear combination of two two-dimensional CPG oscillators, which are referred to as the master oscillator CPG1 and the slave oscillator CPG2, where CPG1 is The phase plane of the oscillator coincides with the XOZ plane of the foot coordinate system $\{traj\}$, the limit ring is semi-elliptical, and the CPG1 driver body realizes forward motion; the phase plane of the CPG2 oscillator coincides with the XOY plane of the world coordinate system $\{W\}$, and the limit ring is circular. The drive body achieves rotational motion.

The underlying oscillation unit of CPG1 is mathematically expressed as:

$$\begin{cases} \dot{x} = \alpha\left(1 - \frac{(x-x_0)^2}{L^2} - \frac{(z-z_0)^2}{H^2}\right)(x - x_0) + \frac{L}{H}\omega(z - z_0) \\ \dot{z} = \beta\left(1 - \frac{(x-x_0)^2}{L^2} - \frac{(z-z_0)^2}{H^2}\right)(z - z_0) - \frac{H}{L}\omega(x - x_0) \end{cases} \tag{3.1}$$

In Eq. (3.1), the origin coordinates of the oscillator in the two-dimensional phase plane, the differential equation system of (x_0, z_0) the CPG1 oscillator is composed of linear and nonlinear parts, and the solution of the linear part is an elliptic equation constituting a limit ring; the nonlinear part is equivalent to two first-order oscillation systems coupled to each other, and when a disturbance occurs, the oscillator can adaptively gradually converge from any position in the phase plane to the limit ring. Figure 4 is simulated, showing the transition from the origin of the phase plane to steady-state oscillation. α and β are the convergence coefficients of the oscillator in the x and Z directions respectively. The larger the convergence coefficients are, the faster the transition to steady state will be. ω is the frequency of the circle, and the greater the ω is the faster the point on the limit cycle moves. L and H are the major and minor axes of the ellipse respectively.

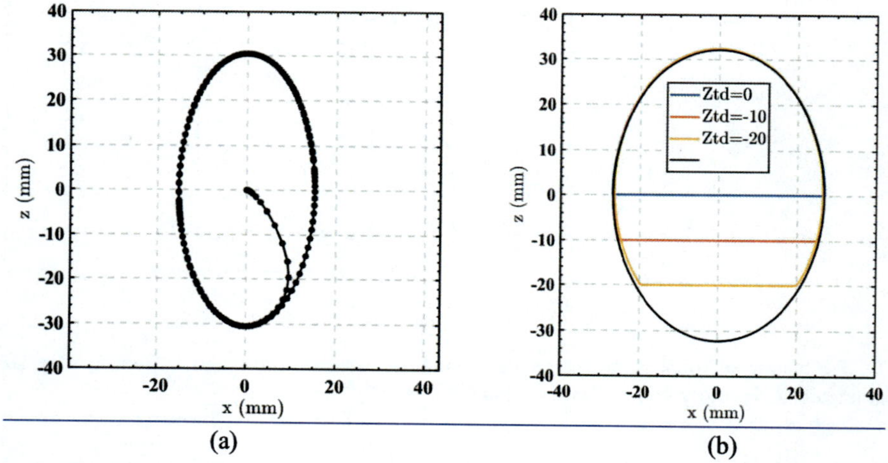

Fig. 4. The CPG1 oscillator properties: CPG1 converges from the origin to the limit cycle (a), Ztd Take the foot trajectory interception effect corresponding to different sizes

In order to make the limit ring shape of the CPG1 oscillator suitable for the online foot trajectory planning of the quadruped robot, this paper uses a smooth [7] to truncate the elliptic limit ring, the mathematical model of which is expressed as:

$$\begin{cases} \dot{z}_F = \left(\dot{z} + K_f(z - z_F)\right)\sigma_f(z) \\ \sigma_f(z) = \frac{1}{e^{-b_f\left(z - z_{kj}\right)} + 1} \end{cases} \tag{3.2}$$

In Formula (3.2), z_{td} is the interception depth, σ_f is the sigmoid function related to the interception depth, when $z > z_{td}$, the corresponding foot is in the swing phase, and the oscillator keeps the original output. When $z \leq z_{td}$, the corresponding foot enters the support phase, and z_f remains at the height of z_{td}, thus realizing the truncation of the limit cycle. K_f, b_f are parameters for adjusting the tracking characteristics of the output signal. However, when $z_{td} \neq 0$, if the angular frequency ω of the oscillator remains unchanged, the phase difference φ_{td} will lead to gait disorder, as shown in Fig. 5. Therefore, the angular frequency of the oscillator need to be segmapped, and the swing phase frequency ω_{sw} and support phase frequency ω_{st} of the leg in the next gait cycle are updated according to the current phase at the time of foot contact.

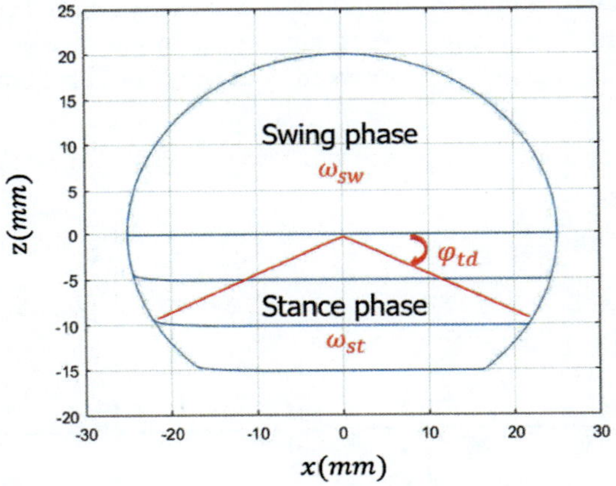

Fig. 5. Gait disturbance due to the introduction of the interception depth

Taking the diagonal gait studied in this paper as an example, the duty cycle coefficient $m = 0.5$, in order to ensure that the swing phase time is consistent with the support phase time, is updated as follows:

$$\begin{cases} (1-m)\frac{\pi-2\varphi_{td}}{\omega_{st}} = m\frac{\pi+2\varphi_{td}}{\omega_{sw}} \\ \frac{2\pi}{\omega_0} = \frac{\pi-2\varphi_{td}}{\omega_{st}} + \frac{\pi+2\varphi_{td}}{\omega_{sw}} \end{cases} \tag{3.3}$$

Solved by the system of equations of Eq. (3.3):

$$\begin{cases} \omega_{st} = \frac{\pi-2\varphi_{td}}{2\pi}\frac{\omega_0}{m} \\ \omega_{sw} = \frac{\pi+2\varphi_{td}}{2\pi}\frac{\omega_0}{1-m} \end{cases} \tag{3.4}$$

The corrected oscillator angle frequency is expressed as a function of the foot height z, and the frequency switching at commutation is achieved by the sigmoid function:

$$\begin{cases} \omega(z) = \omega_{st}\sigma_{dz1}(z) + \omega_{sw}\sigma_{dz2}(z) \\ \sigma_{dz1}(z) = \frac{1}{e^{-b_d(z-z_{dd})}+1} \\ \sigma_{dz2}(z) = \frac{1}{e^{b_d(z-z_{dd})}+1} \end{cases} \tag{3.5}$$

For the second problem mentioned in the introduction, the elliptical oscillator model is only a binary oscillation system with only x (forward direction) and z (leg lift direction) two outputs, in order to adapt the foot trajectory to the full range of motion scenarios. In this paper, a two-oscillator linear combination method is designed to extend the oscillator trajectory into three-dimensional space.

Robot body motion can be decoupled into two components: 1. Moving in the x-axis direction of the body coordinate system; 2. Rotating around the z-axis of a floating base coordinate system. The former drives the robot forward and backward, and the latter drives the robot to produce a deflection of the direction of motion. As shown in Fig. 6(a), there is a minimum external circle in the rectangle connected by the four hip vertexes on the robot body, taking the robot's right front leg as an example, to drive the body to rotate counterclockwise around the floating base Z axis in a gait cycle ${}^{W}\phi_{body}^{yaw}$ angle, you need to make the foot move along the circumference of the smallest external circle.In the world coordinate system, it's represented as moving Δx in the direction of x, Δy in the direction of y.

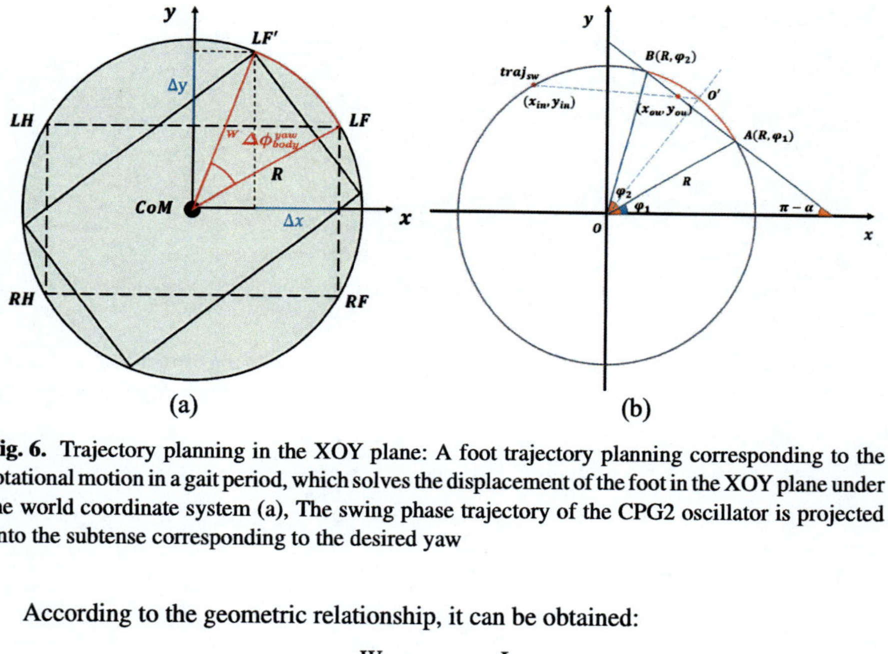

Fig. 6. Trajectory planning in the XOY plane: A foot trajectory planning corresponding to the rotational motion in a gait period, which solves the displacement of the foot in the XOY plane under the world coordinate system (a), The swing phase trajectory of the CPG2 oscillator is projected onto the subtense corresponding to the desired yaw

According to the geometric relationship, it can be obtained:

$$\begin{cases} \Delta x_{(RF,LH)} = \dfrac{W}{2}\sin \Delta\phi - \dfrac{L}{2}(1 - \cos \Delta\phi) \\ \Delta y_{(RF,LH)} = \dfrac{L}{2}\sin \Delta\phi + \dfrac{W}{2}(1 - \cos \Delta\phi) \\ \Delta x_{(LF,RH)} = -\dfrac{W}{2}\sin \Delta\phi - \dfrac{L}{2}(1 - \cos \Delta\phi) \\ \Delta y_{(LF,RH)} = \dfrac{L}{2}\sin \Delta\phi - \dfrac{W}{2}(1 - \cos \Delta\phi) \end{cases} \tag{3.6}$$

In this paper, it is stipulated that the position of the starting time of the robot foot coincides with the vertical projection point of the hip joint in the XOY plane of the floating base coordinate system, so under the premise that the foot is not slipped in the support phase stage, a foot curve and the WCPG in the XOZ plane are planned in the XOY plane The semi-elliptical trajectories are superimposed to achieve the overall rotation of the robot, which controls the yaw angle of the body, as shown in Fig. 6(a). The specific method is to use a PD controller to track the desired body yaw angle for the data feedback of the body IMU ${}^{W}_{d}\phi^{yaw}_{body}$:

$$ {}^{W}\dot{\phi}^{\mathrm{yaw}} = k\left({}^{W}_{d}\phi^{yaw}_{body} - {}^{W}\phi^{\mathrm{yaw}}\right) \tag{3.7} $$

The smooth transition of the trajectory curve is achieved by coupling a HOPF circular oscillator in the world coordinate system XOY plane, and the angular frequency of the HOPF oscillator is adjusted to continuously control the angular velocity of the body rotation, the system of differential equations for HOPF oscillators is as follows:

$$ \begin{cases} \dot{x} = \alpha\left(R^2 - r^2\right)x - \omega y \\ \dot{y} = \alpha\left(R^2 - r^2\right)y + \omega x \\ r^2 = x^2 + y^2 \end{cases} \tag{3.8} $$

In formula (3.8), R is the distance from the end point to the center of mass of the body at the starting time under the XOY plane of the floating base coordinate system, so the x and y controlling the rotation are coupled under the floating base coordinate system, and the angular frequency of the WCPG oscillator and the HOPF oscillator is adjusted at the same time. To ensure that the support phase time synchronization of the two oscillators.

The arc mapping module takes the desired yaw angle of the body within a single sampling period as the input, so as to truncate the inferior arc of the subtense corresponding to ${}^{W}_{d}\phi^{yaw}_{body}$ as the supporting phase trajectory of the foot in the XOY plane of the floating base coordinate system, and maps the optimal arc of the subtense to the subtense to obtain the swing phase trajectory of the rotation gait. The mapping method is shown in Fig. 6(b). OO' is the central perpendicular of subtense AB. According to the spherical mapping theorem, the line between O' point and any point on the optimal arc has an intersection with the subtense, and the trajectory formed by the intersection is the projection of the optimal arc on the subtense (Fig. 7).

4 Simulation Testing

In this paper, a virtual prototype model of a quadruped robot is established in the V-REP simulation software, and a virtual prototype control system is built in Simulink, and the accuracy of contact detection and the performance of the composite three-dimensional oscillator are tested through joint simulation. The control bandwidth in the simulation environment is 200 Hz, and during each sampling cycle, the Simulink model collects the body state and joint state, and the expected joint angles are solved in real time and sent to the virtual prototype in V-REP. This section tests the in-situ and forward steering of

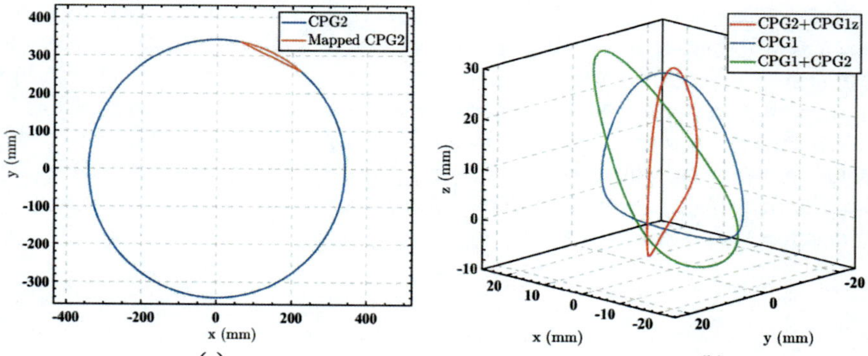

(a) (b)

Fig. 7. Matlab 30° simulation results: In the case of the expected amount of yaw in the body during a gait period, the CPG2 oscillator arc mapping simulation (a), C A three-dimensional foot trace curve (b) obtained by a linear combination of PG1 and CPG2

the robot in the simulation environment, and the test results are shown in Fig. 8. As can be seen from the figure, the steering controller with the 3D foot trajectory can achieve more accurate control of the yaw angle of the body, and the roll and pitch angle of the body are less variable.

The same CPG network parameters are used in the simulation environment and the physical environment, and the values of each parameter are shown in Table 2.

Table 2. CPG network parameters

Parameter	Value range
ω_0	4π
H	30
L	15
α	90
β	90
b_f	0.2
b_d	200
K_f	10

Next, the role of CPG1 oscillator's reference output in foot contact state estimation is shown. As shown in Fig. 9, the foot contact estimation module integrates CPG1 oscillator's reference phase information and foot force information, and realizes the maximum posterior estimation of foot contact state through Kalman filter algorithm [8]. On the basis of high sensitivity contact judgment, precise circular frequency switching of CPG oscillator group is realized.

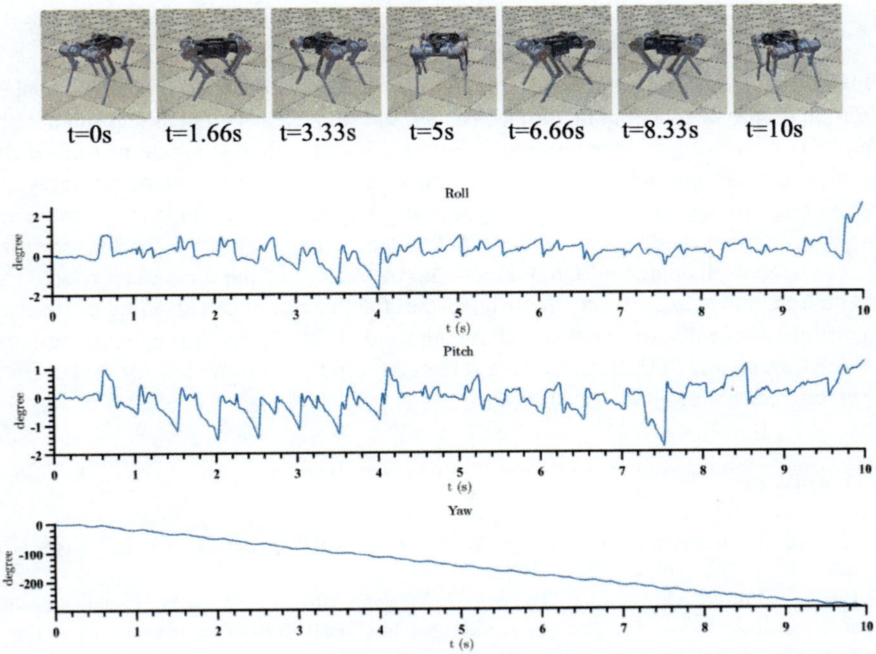

Fig. 8. In-situ steering simulation of the body Euler angle change curve

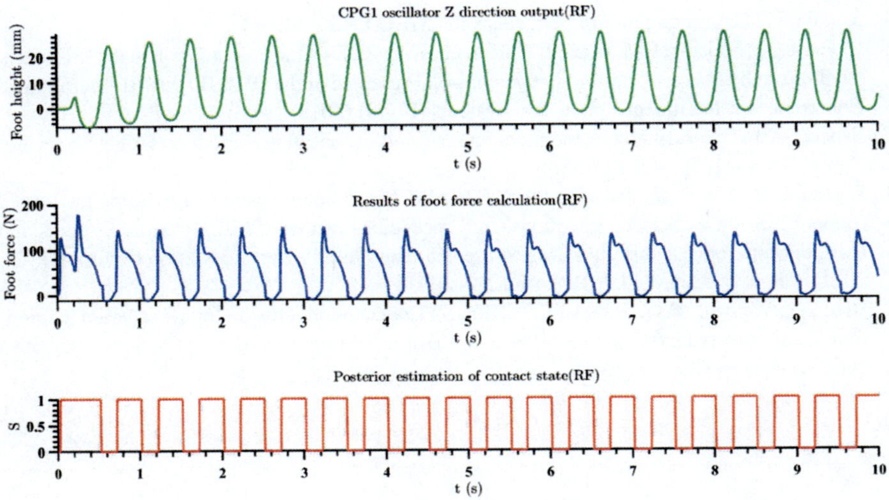

Fig. 9. High accuracy estimation of foot contact

5 Conclusion

This study focuses on the online gait planning problem of quadruped robots, establishes a virtual prototype model of quadruped robots, and studies the gait planning and control problems of quadruped robots through high-realism dynamic simulation. And in the quadruped robot physical prototype to build an attitude control system and integrate relevant control algorithms, the prototype can walk stably in the indoor environment. Drawing on bionic control theory, the 3D-CPG oscillator group is constructed to realize the online gait planning and Omnidirectional movement of the quadruped robot. The research content of this project is innovative, the effectiveness of gait planning and control algorithms has been fully verified in simulation and physical prototype experiments, and the relevant results of the project have reference value and promotion for the research of motion control of quadruped robots.

References

1. Raibert, M., Blankespoor, K., Nelson, G., Playter, R.: BigDog, the rough-terrain quadruped robot. IFAC Proc. Vol. **41**, 10822–10825 (2008)
2. Bledt, G., Powell, M.J., Katz, B., Di Carlo, J., Wensing, P.M., Kim, S.: MIT cheetah 3: design and control of a robust, dynamic quadruped robot. In: 2018 IEEE/RSJ International Conference on Intelligent Robots and Systems (IROS), pp. 2245–2252 (2018). (in English)
3. Hutter, M., et al.: ANYmal - a highly mobile and dynamic quadrupedal robot. In: 2016 IEEE/RSJ International Conference on Intelligent Robots and Systems (IROS), pp. 38–44 (2016)
4. Kemilä, K.: Generating motion sequences for AIBO ERS-7 (2006)
5. Di Carlo, J., Wensing, P.M., Katz, B., Bledt, G., Kim, S.: Dynamic locomotion in the MIT cheetah 3 through convex model-predictive control. Presented at the 2018 IEEE/RSJ International Conference on Intelligent Robots and Systems (IROS) (2018)
6. Ijspeert, A.J.: Central pattern generators for locomotion control in animals and robots: a review. Neural Netw. **21**(4), 642–53 (2008)
7. Barasuol, V., De Negri, V.J., De Pieri, E.R.: WCPG: a central pattern generator for legged robots based on workspace intentions. In: Proceedings of the ASME Dynamic Systems and Control Conference and Bath/ASME Symposium on Fluid Power and Motion Control (DSCC 2011), vol. 2, pp. 111–114 (2012). (in English)
8. Bledt, G., Wensing, P.M., Ingersoll, S., Kim, S.: Contact model fusion for event-based locomotion in unstructured terrains. Presented at the 2018 IEEE International Conference on Robotics and Automation (ICRA) (2018)

Evaluation of the Work Performance of a Paddy Field Weeding Robot Using Disturbance Observer

Seiya Moro, Hiroaki Uchida(✉), and Kanta Kato

Department of Mechanical Engineering, National Institute of Technolog, Kisarazu College, 2-11-1 Kiyomidai-higashi, Kisarazu 292-0041, Chiba, Japan
{sme21a14,sme22a08}@inc.kisarazu.ac.jp,
uchida@maple.m.kisarazu.ac.jp

Abstract. In this study, we first examine control methods to suppress the effects of disturbance on the work efficiency of a weeding robot with an automatic operation function, because disturbances caused by wind disturbance, rice growth, and uneven water depth in the environment where the paddy field weeding robot works affect the work efficiency of the robot. Next, a method for evaluating work performance is examined with the aim of constructing a work support system to support work in actual weeding operations. Specifically, we design a disturbance observer to suppress disturbance using a mathematical model of a paddy field weeding robot, and confirm its effectiveness through simulation. Next, a method to evaluate the work performance will be discussed. The work performance evaluation is performed by introducing an evaluation function.

Keywords: Paddy field weeding robot · Automatic driving system · Path planning · Disturbance observer

1 Introduction

In recent years, there has been a growing demand for herbicide-free paddy rice cultivation due to increasing consumer interest in food security and reduced burden on the environment. The authors have previously studied weeding methods that do not use herbicides, such as a controller-operated weeding robot that agitates the bottom of the paddy field to float weed seeds and muddy the water to prevent the growth of weeds [1]. The weeding robot is equipped with a propeller for propulsion and a propeller for steering, and operates two drive motors. We studied the automatic operation of this weeding robot by using a GNSS sensor to follow the set work path [2]. One of the problems of this system is that the disturbance caused by wind disturbance, rice growth, and unevenness of water depth adversely affects the performance of the system, such as deviation from the work path and changes in the work velocity. Therefore, it is necessary to construct an automatic driving system that takes into account the effects of disturbances.

Disturbance observer methods are known as disturbance suppression methods, and are diligently pursued in the fields of electric motor control and robot control. Since the

© The Author(s), under exclusive license to Springer Nature Switzerland AG 2023
J. M. Cascalho et al. (Eds.): CLAWAR 2022, LNNS 530, pp. 313–324, 2023.
https://doi.org/10.1007/978-3-031-15226-9_30

disturbance observer estimates the disturbance as part of the system state, it is possible to use the estimated disturbance to control the system and there are examples of introduction to human-assistive robots [3, 4]. Compared with large tractors, weeding robots are lightweight and are strongly affected by wind disturbance. And, in the case of weeding in paddy fields, weeding is expected to be affected by the terrain of the paddy field and by the wind during the work.

Furthermore, at present, methods for evaluating the work performance of agricultural robots have not been sufficiently established. Evaluation criteria for work performance differ depending on the user, weather conditions, and other factors. Currently, work plans and tasks are often carried out based on the experience of the operator, and it is desirable to construct a support system that assists the operator.

In this study, we first design a disturbance observer to suppress the effect of wind disturbance on work performance, with the aim of constructing a work support system for a paddy field weeding robot with an automatic operation function. Next, the effectiveness of the disturbance observer is verified by simulation using a mathematical model of a paddy field weeding robot. Finally, a method for evaluating the work performance of the weeding robot is studied using an evaluation function. Since wind disturbance acts from all directions in actual weeding work in paddy fields, the effect of wind disturbance on the work performance is verified using the proposed evaluation function.

2 Weeding Robot

2.1 Robot Appearance and Weeding Function

Fig. 1. Appearance of weeding robot

Table 1 shows the specifications of weeding robot, and Fig. 1 shows the appearance of weeding robot. Considering weeding work in paddy field, the body of the robot is used a body board. The material of the body is EVA (Ethylene-Vinyl Acetate). The reverse side of the body is made of plastic and flat. 12 plastic chains, in which one chain is a length of 700 mm and be fixed in a bar width of 900 mm, are attached to the rear of the body board. Since it is made of plastic, a metal hook is attached to the chain as a weight to prevent it from floating on water. The principle of weeding is that by stirring the paddy field with a chain, the water becomes muddy and the sunlight does not hit the weeds, and by floating the seeds of the weeds from the bottom of the paddy field, weeds are prevented from rooting.

2.2 Mathematical Model of a Weeding Robot

The mathematical mofdel of the weeding robot shown in Fig. 1 is represented in Fig. 2. The position of the robot's center of gravity is used as the origin of the robot coordinate

Table 1. Specifications of weeding robot

Weight	8.0 kg
Maximum height	450 mm
Maximum width	595 mm
Maximum vertical width	1200 mm

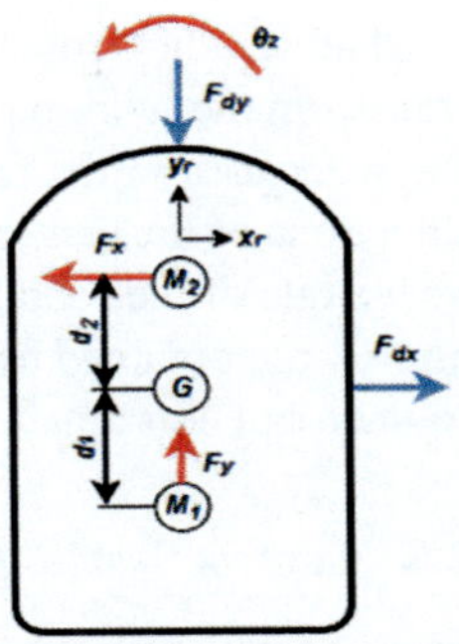

Fig. 2. Mathematical model of a paddy field robot

system, the longitudinal direction is y_r, the direction perpendicular to that axis is x_r, the rotational angle of the robot in the world coordinate system is θ_z. And M be the mass of the robot and I_z the moment of inertia around the center of gravity. M_1 and M_2 be the motors for propulsion and steering, respectively, and let F_y and F_x be the propulsive and steering forces generated by each. The direction of action of F_x is taken so that the rotational motion by the steering part of the robot is positive. The weeding robot for paddy fields is subjected to water, rice, weeds, and wind resistance during its work in the paddy field, so the disturbances are F_{dx}, F_{dy}, and τ_d in the x_r, y_r, and θ_z directions, respectively. The equations of motion for the robot are (1-a)–(1-c) below.

$$\begin{cases} M\dddot{x}_r + C_x\dot{x}_r = -F_x + F_{dx} & (1-a) \\ M\dddot{y}_r + C_y\dot{y}_r = F_y - F_{dy} & (1-b) \\ I_z\ddot{\theta}_z + C_z\dot{\theta}_z = F_x \times d_2 - \tau_d & (1-c) \end{cases}$$

In Eq. (1), letting the state variable vector $x = [x_r,\ y_r,\ \theta_z,\ \dot{x}_r,\ \dot{y}_r,\ \dot{\theta}_z]^T$, the state and output equations are as follows. Here the input vector is $u = [u_{steering}(= F_x), u_{propulsive}(= F_y)]^T$.

$$\begin{cases} \dot{x} = Ax + Bu + d & (2-a) \\ y = Cx & (2-b) \end{cases}$$

2.3 Work Path Planning

The work path is a series of target points at intervals of 500 mm. When the weeding robot judges that the target point has been passed, it updates the target point to the next value. When the target trajectories are set as $x_{ref}(i)$ and $y_{ref}(i)$, the robot angular velocity and speed in the direction of motion at the target are given by the following Eqs. (3).

$$\varphi_t = \tan^{-1}\frac{x_{ref}}{y_{ref}} \quad (3-a)$$

$$\omega_{ref} = \dot{\varphi} \quad (3-b)$$

$$v_{ref} = \sqrt{(\dot{x}_{ref})^2 + (\dot{y}_{ref})^2} \quad (3-c)$$

The weeding robot works in world coordinate space. Therefore, it is necessary to transform the work path to the robot coordinate system. Figure 3 shows the error with the work path in the robot coordinate system. e_{xr} and e_{yr} are the errors in the x and y directions of the center of gravity in the robot coordinate system, respectively. θ_z is the yaw angle of the robot's direction of travel relative to the world coordinate system. e_{xr} and e_{yr} are obtained by transforming the error obtained in the world coordinate system to the robot coordinate system using the rotation matrix θ_z.

2.4 Control System Design Considering Disturbance

From Sect. 2.3 work path planning, the control inputs become the inputs expressed by the following Eq. (4) [5].

$$\begin{cases} u_{steering} = \omega_{ref} + v_{ref}(K_x e_{xr} + K_\theta \sin e_\theta) & (4-\mathrm{a}) \\ u_{propulsive} = v_{ref} \cos e_\theta + K_y e_y + K_v v_{error} & (4-\mathrm{b}) \end{cases}$$

Here, ω_{ref} and v_{rer} are the robot's target angular velocity and speed in the direction of travel based on the work path planning, K_x, K_y, and K_θ are proportional gains, v_{error} is the velocity error, K_v is the velocity gain, e_θ is the difference between the robot's target posture angle and the angle error, between the current angle and the target point. The control input shown in Eq. (4) is based on the kinematic model of the weeding robot and the position deviation, and can't provide sufficient path-following performance in an environment where disturbances are acting. Therefore, we design a disturbance compensator considering a disturbance observer [6]. An overview of the disturbance compensator is shown in Fig. 4. The disturbance observer is a method that estimates the disturbance as well as the state x in Eq. (3). The estimated disturbance $\hat{D}(s)$ is multiplied by $1/k_n$ and added to the input $U(s)$ to generate enough input $U(s)$ to cancel the disturbance. This suppresses the disturbance $D(s)$.

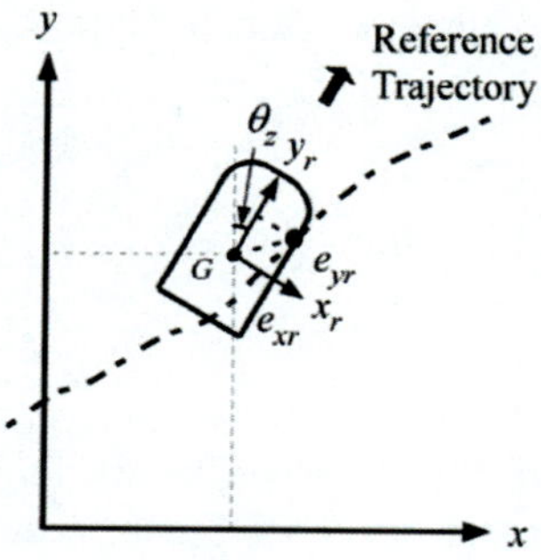

Fig. 3. Robot coordinate system

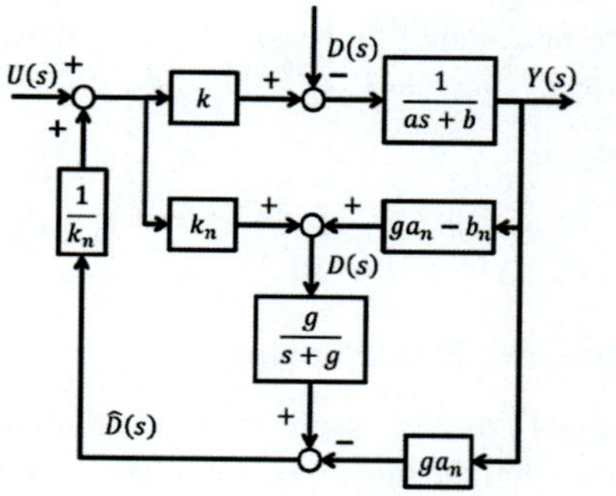

Fig. 4. Block diagram of disturbance observer

The performance of disturbance estimation is also determined by the pole g of $D(s)$. The control input considering the disturbance compensator is shown in Eq. (5). oU_1 and oU_2 are estimated disturbances, respectively.

$$\begin{cases} u_{steering} = \omega_{ref} + v_{ref}(K_x e_{xr} + K_\theta \sin e_\theta) + oU_1 & (5-\mathrm{a}) \\ u_{propulsive} = v_{ref} \cos e_\theta + K_y e_y + K_v v_{error} + oU_2 & (5-\mathrm{b}) \end{cases}$$

3 Examination of Automatic Driving Experimental Results by Feedback (FB) Controller

This section explains the results of experiments conducted using FB control method. These experiments were conducted in a paddy field. Figure 5 shows the target path using experiments in the paddy field. Figure 6 shows the target path and the work path of the weeding robot. The experiment was conducted under the condition that the wind was blowing at 4 m/s in the direction of the blue arrow in Fig. 5. From Fig. 6, it can be seen that the work path follows the target path, although it deviates from the path in some cases. Note that there is a section at point A that deviates from the target trajectory, but since the trajectory is tracked to the target trajectory afterwards, we do not consider this to be a problem for the work. Here, Fig. 7 shows the target path of the experiment conducted on a different day and the work path of the weeding robot. The experiment was conducted under the condition that the wind was blowing at 7 m/s in the direction of the blue arrow in Fig. 5. Figure 7 shows that the weeding robot is following the target path, but it is oscillating to the left or right during the run due to the wind effect. In addition, the experiment was terminated during the course of the experiment because the path deviated significantly from the target path due to the wind. Comparing Figs. 6 and 7, it can be seen that the wind increases, which in turn affects the automatic driving. Next, we consider the working time. In the case of Fig. 6, it took 654.7 s, and in the case of Fig. 7, it took 947.7 s. This indicates that the working time is also affected by the wind effect. Therefore, since weeding works by automatic operation for the weeding robot in actual paddy fields are affected by wind disturbance, it is considered necessary to construct a control system that suppresses the effects of wind disturbance. In this study, a disturbance observer is used as a method to suppress the effects of wind disturbance.

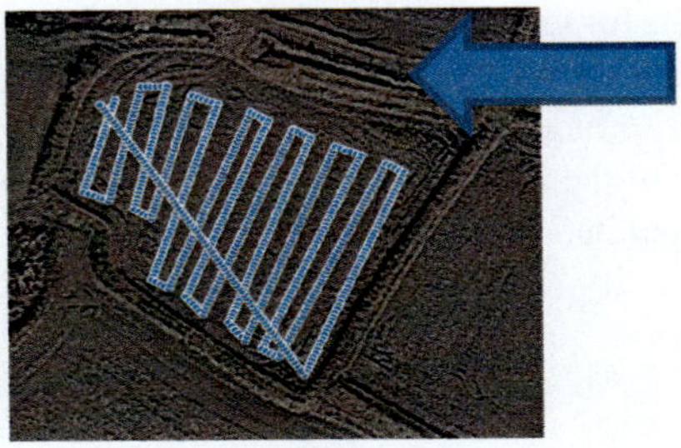

Fig. 5. Target trajectory

4 Simulations

Simulations were performed using numerical analysis software based on Sects. 2.2–2.4. In this simulation, the wind disturbance, which affects the velocity by about 0.1 m/s, is given as an external disturbance. The direction of the wind disturbance is assumed to be

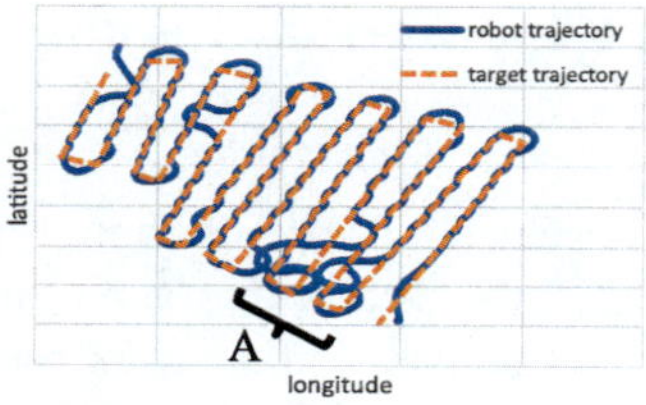

Fig. 6. Work results with 4 m/s wind

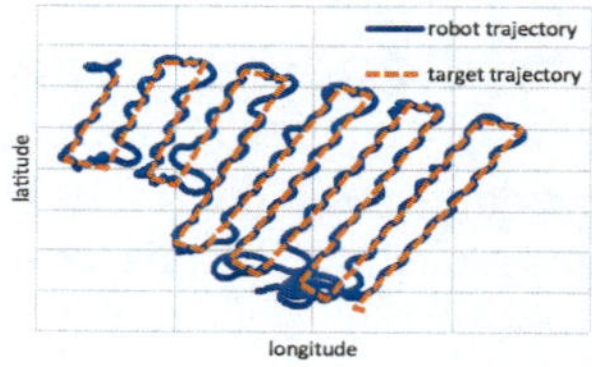

Fig. 7. Work results with 7 m/s wind

y-direction, where it is a tailwind when the robot moves in the positive direction of y-axis and a headwind when the robot moves in the negative direction of y-axis. We did three pattern simulations as ① the case with no disturbance, ② the case with a disturbance and no disturbance observer, and ③ the case with a disturbance and a disturbance observer. In the simulations, each parameter is set as $K_x = 0.5$, $K_y = 1$, $K_\theta = 20$, $K_v = 100$, $g_1 = 0.5$, $g_2 = 0.0025$. The simulation time is set to 450 s. The target path is a 20-m forward movement, a 2-m lateral movement, and five cycles of 20-m forward movement, given that the width of the weeding robot is about 1 m. The assumed direction of motion of the robot is assumed to be $-\pi/2$[rad] at 2 m lateral movement and $-\pi/2$[rad] at 20 m forward movement when the robot turns leftward from the target path, and $\pi/2$[rad] at 2 m lateral movement and $\pi/2$[rad] at 20 m forward movement when the robot turns rightward. The target moving speeds of the robot are set at 0.5 m/s for forward movement and 0.4 m/s for lateral movement, as requested by farmers.

4.1 Results of Simulation in Case ①

Figure 8 shows the results for the case ①. Figure 8(a)–(c) show the time response of the work path, posture angle of the robot, and the robot velocity, respectively. In the Figures, the solid blue and dotted orange lines indicate the robot's response and target values, respectively. The results of the robot's response to the target value confirm the validity of the control method used in Sect. 2.4.

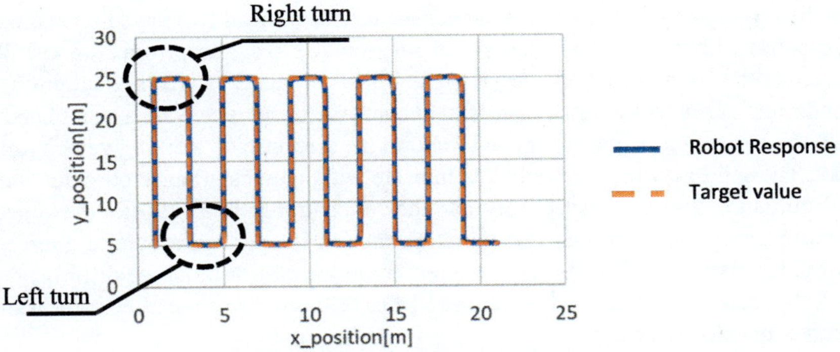

(a) Time response of working path of the robot

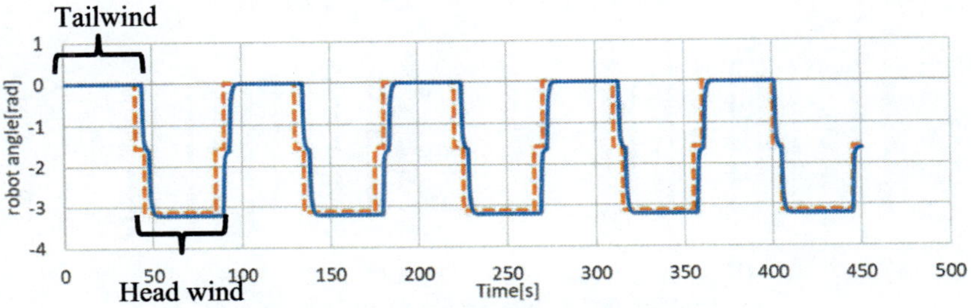

(b) Time response of posture angle of the robot

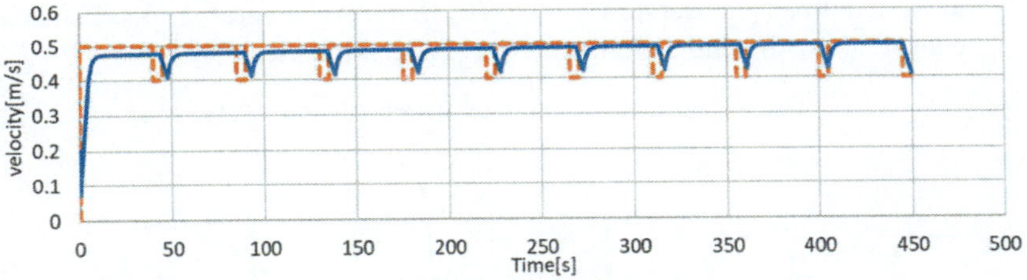

(c) Time response of working velocity of the robot

Fig. 8. Simulation results in case

4.2 Comparison with and Without Disturbance Observer in Case ②, ③

From Sect. 4.1, we can confirm that the simulation we are using works properly. Therefore, a comparison is made when a disturbance observer is introduced in condition with wind disturbance. Figure 9 shows the work path results. Figure 9 (a) shows the case without a disturbance observer and Fig. 9 (b) shows the case with a disturbance observer. From these figures, it can be confirmed that the work path with observer follows the target path in time. It is also confirmed that the work path with a disturbance observer reaches the target path in 13.6 s faster than the work path without the disturbance observer.

Figure 10 shows the response of posture angle of the robot. The solid gray line and blue line indicate without a disturbance observer and with a disturbance observer, respectively, and the dotted orange line indicates the target value. From the figure, we can see that the response without a disturbance observer is significantly delayed compared to the response with a disturbance observer from around 270 s. Also, from around 180 s, both results show errors from the target angle. This is because the errors occur during headwinds, and during tailwinds, the errors that could not be corrected are accumulated, although the correction is made. Figure 11 shows the response of working velocity of the robot. The solid gray line and blue line indicate without a disturbance observer and with a disturbance observer, respectively, and the dotted orange line indicates the target value. From the figure, it is seen that the work motion is operated around the target speed of 0.5 m/s with the disturbance observer, but in the case without the disturbance observer, the response is delayed from around 180 s. This confirms the effectiveness of the observer in velocity control.

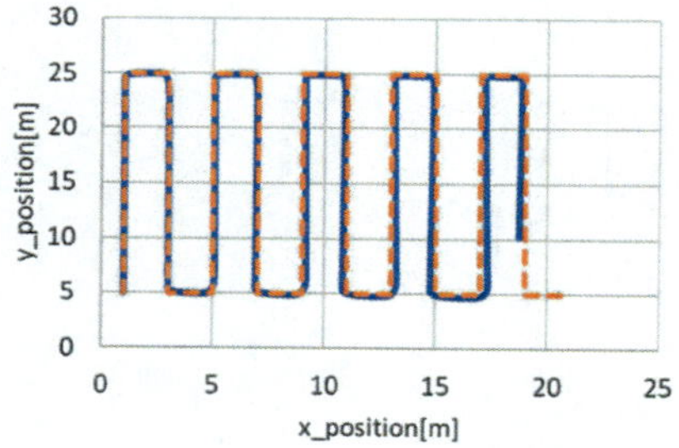

(a) No disturbance observer is considered with wind disturbance

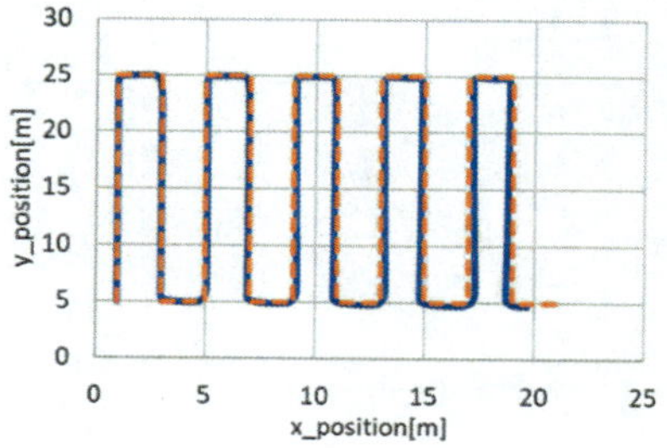

(b) Disturbance observer is considered with wind disturbance

Fig. 9. Working trajectory of the robot in case ②, ③

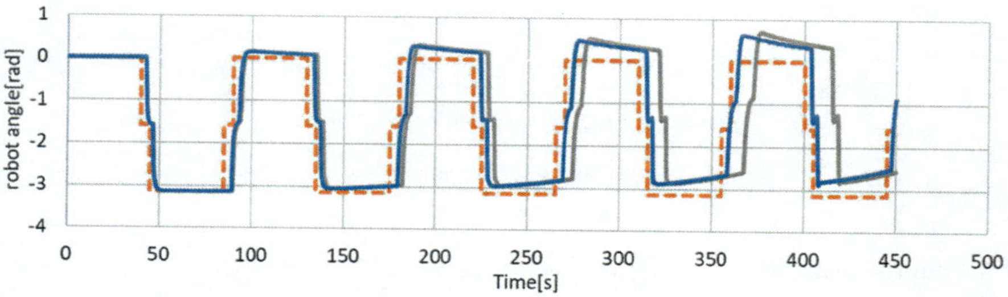

Fig. 10. Time response of posture angle of the robot in case ②, ③

Table 2 shows the total control inputs for the propulsion and steering motors. From the table, it can be confirmed that the amount of control input becomes large when a disturbance observer is introduced.

4.3 Evaluation Method for Paddy Field Weeding Robot

In this section, we discuss how to evaluate the work performance of the weeding robot. Since disturbances are always present in the working environment of the weeding robot,

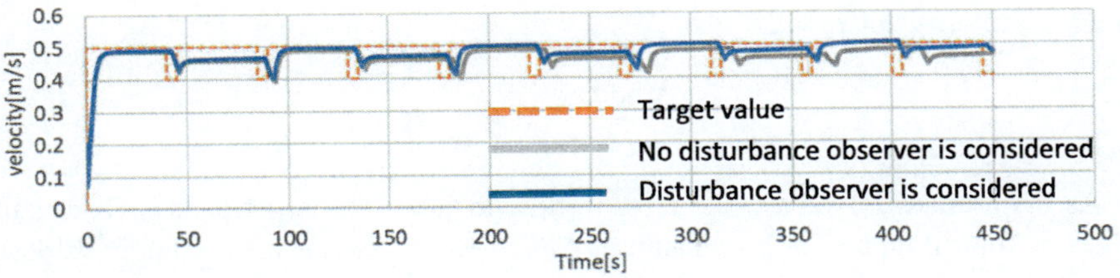

Fig. 11. Time response of working velocity of the robot in case ②, ③

Table 2. Total value of control inputs in case ②, ③

	No disturbance observer is considered	Disturbance observer is considered
Propulsive input [N]	33708.21	34663.11
Steering input [N]	1694.68	1843.10

it is assumed that a disturbance observer is used in the weeding work. First, five terms are set to evaluate the disturbance observer: (1) The error in time to reach the target path, (2) Sum values of control inputs for propulsion motors, (3) Sum values of target velocity error, (4) Sum values of target path errors, and (5) Sum values of target motion angle errors. The selection of the five terms is based on the consideration of battery energy efficiency and work efficiency during automated driving. The errors in these five terms are discussed in comparison with the case ② without a disturbance observer. Table 3 shows the difference of each term with and without disturbance observer. This difference indicates the difference between the value without the disturbance observer and the value with the disturbance observer. If the difference is positive, the error is small, and if it is negative, the error is large. Especially, the target path arrival time in Term (1) is set to 0 in the case with disturbance observer, while it shows + 13.6 s to reach the target path in the case without disturbance observer. Table 3 shows that four terms of five terms are positive. From these values, it confirms the superiority for the use of a disturbance observer.

Next, we consider the evaluation of the weeding performance by actual users. The evaluation of the five items is not necessarily equal, but depends on the user's request. Therefore, it is necessary to set the evaluation function that takes the five terms into account and make a judgment [7]. Here, the difference values of the five items are normalized (Min-Max) in order to set the evaluation function. For this reason, we simulate and analyze the case where the condition of the direction of the wind disturbance is changed. The wind disturbance directions set to 0°, 30°, 45°, 60°, 120°, 135°, 150°, and 180° as shown in Fig. 12. For each result, the difference from the case without the disturbance observer is calculated in the same way as in Table 3. The calculated differences are normalized from Eq. (6). Equation (6) shows $\dot{x}_i$ as the normalized value, x_i as the difference value, and $Max(x)$ and $Min(x)$ indicate the maximum and minimum values for each term, respectively.

$$\dot{x}_i = \frac{x_i - Min(x)}{Max(x) - Min(x)}(i = 1, 2, 3, 4, 5) \tag{6}$$

As the normalized values, the error in time to reach the target path is *Te(x)*, sum values of control inputs for propulsion motors is *Me(x)*, Sum values of target velocity error is *Ve(x)*, sum values of target path errors is *Pe(x)*, and Sum value of target motion direction errors is *De(x)*, respectively. Table 4 shows the results normalized by Eq. (6).

Table 3. Difference in values without and with disturbance observer

		No disturbance observer is considered	Disturbance observer is considered	Difference in value (x_i)
(1)	The error in time to reach the target path[s]	13.6	0	+13.6
(2)	Sum value of control inputs for propulsion motors [N]	33708.2	34663.1	−954.9
(3)	Sum value of target velocity error [m/s]	351.2	257.8	+93.4
(4)	Sum value of target path errors [m]	36650.5	15301.4	+21349.1
(5)	Sum value of target motion direction errors [red]	6226.6	2672.1	+2672.1

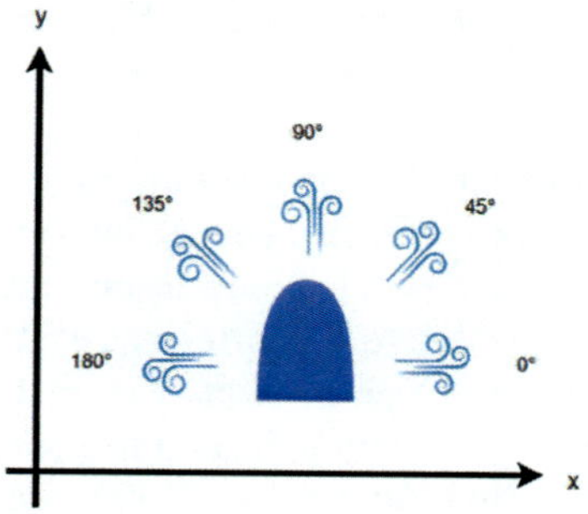

Fig. 12. The directions of the wind disturbance

Table 4. Normalized output due to direction of disturbance

	Direction of disturbance x	0°	30°	45°	60°	90°
(1)	The error in time to reach the target path [*Te(x)*]	1.00	0.90	0.91	0.91	0.84
(2)	Sum value of control inputs for propulsion motors [*Me(x)*]	0.00	0.20	0.46	0.69	0.95
(3)	Sum value of target velocity error [*Ve(x)*]	0.21	0.00	0.28	0.67	1.00
(4)	Sum value of target path errors [*Pe(x)*]	0.95	0.99	0.99	1.00	0.89
(5)	Sum value of target motion direction errors [*De(x)*]	0.24	0.21	0.31	0.51	0.93
	Direction of disturbance x	120°	135°	150°	180°	
(1)		0.67	0.54	0.39	0.00	
(2)		1.00	0.96	0.91	0.85	
(3)		0.89	0.88	0.87	0.87	
(4)		0.64	0.48	0.31	0.00	
(5)		1.00	0.85	0.59	0.00	

The range of values is from 0.00 to 1.00 and the closer the value is to 1.00, the smaller the error is. We define a evaluation function E(x) of the work performance for the weeding robot as Eq. (7).

$$E(x) = Te(x) * \alpha + Me(x) * \beta + Ve(x) * \gamma + Pe(x) * \delta + De(x) * \varepsilon \tag{7}$$

In Eq. (7), α, β, γ, δ, and ε are coefficients which determine the weight of evaluation terms, respectively. Here, the evaluation function $E(x)$ is used to examine the optimal target path depending on the direction of the wind disturbance.

5 Conclusion

We aim to construct a work support system for a paddy field weeding robot with automatic operation function, first, a disturbance observer was designed to reduce the effect of wind disturbance on the work performance. Simulation results using a mathematical model of the weeding robot confirmed that the disturbance observer functions effectively against wind disturbance and has control performance that reduces the effect of the disturbance. Furthermore, an evaluation method for the work performance of a paddy field weeding robot was studied, and an evaluation function was set up to take into account (1) arrival time to the target path, (2) control input of the propulsion motor, (3) work speed, (4) tracking error, and (5) angular error of the robot.

In the future, we plan to confirm the effectiveness of the disturbance observer and the work efficiency against the direction of wind disturbance in actual paddy field work by the weeding robot.

References

1. Uchida, H., Funaki, T.: Development of a remote control type weeding machine with stirring chains for a paddy field. In: 22nd International Conference on Climbing and Walking Robots and the Support Technologies for Mobile Machines, CLAWAR 2019 (2019)
2. Uchida, H., Moro, S., Nomura, K., Sekine, S.: Development of automatic operation system considering steering performance for a paddy field weeding robot. In: 24nd International Conference on Climbing and Walking Robots and the Support Technologies for Mobile Machines, CLAWAR 2021 (2021)
3. Chen, W.-H.: Disturbance observer based control for nonlinear systems. IEEE/ASME Trans. Mechatron. **9**(4), 706–710 (2004)
4. Mohammadi, A., Tavakoli, M., Marquez, H.J.: Disturbance observer based control of nonlinear haptic teleoperation systems. IET Control Theory Appl. **5**(18), 2063–2074 (2011)
5. Fierro, R., Lewis, F.L.: Control of a nonholonomic mobile robot: backstepping kinematics into dynamics. In: Proc. 34th IEEE Conf. Decision Contr., vol. 4, pp. 3805–3810 (1995)
6. Kouhei, O.: Special issue: advanced motion control. Robust motion control by disturbance observer. J. Robot. Soc. Japan **11**(4), 486–493 (1993)
7. Postmes, J., Botteghi, N., Sirmacek, B., Kamilaris, A.: A system for efficient path planning and target assignment for robotic swarms in agriculture. In: IoT'21, 8–12 Nov 2021

Application of Genetic Algorithm for Vector Field Guidance Optimization in a UAV Collective Circumnavigation Scenario

Tagir Muslimov(✉)

Ufa State Aviation Technical University, K. Marx Str., 12, 450008 Ufa, Russian Federation
tagir.muslimov@gmail.com

Abstract. Optimizing the trajectories of autonomous unmanned aerial vehicles (UAVs) in a decentralized cooperative tracking of a ground target is not a trivial undertaking. In this case, the UAV formation is a complex interconnected nonlinear system. This paper investigates a genetic algorithm for optimizing the trajectories of UAVs engaged in cooperative target tracking by means of vector field guidance, thus performing collective circumnavigation. Computational modeling shows that the genetic algorithm can effectively address trajectory optimization. Post-optimization reduction in the fitness function value is noted. Another finding is that it is necessary, when tuning the UAV heading controllers, to minimize not only the error of distance to the circular path around the target but also the relative inter-UAV distance error.

Keywords: Standoff target tracking · UAV formation · Target encirclement · UAV consensus · Flying robot

1 Introduction

Vector field guidance is commonly used on trajectory control of unmanned aerial vehicles (UAV). For instance, paper [1] describes using this method to enable UAVs to track a target. Of great interest is to develop approaches to UAV formation control, whether for centralized [2] or decentralized [3–5] formations. Vector field guidance has been successfully used to enable a consensus-based UAV formation to track a target, see [6]. As was shown there, the method complicates the tuning of formation controllers and adjusting their coefficients appropriately. This is due to both the interconnectedness of a decentralized formation system and the nonlinear dynamics of the UAVs themselves. The fact becomes even more obvious when using full 12-state, 6-degree-of-freedom models. Paper [6] describes using fuzzy model reference adaptive control (MRAC) to fine-tune the controllers. However, it does not question the optimality of transient trajectories that a UAV formation exhibits when making the preset arrangement.

Evolutionary algorithms are common in robot control research [7, 8] including trajectory control of autonomous UAVs [9, 10], task distribution [11, 12], and movement planning [13, 14]. In particular, genetic algorithms are quite common in UAV formation

© The Author(s), under exclusive license to Springer Nature Switzerland AG 2023
J. M. Cascalho et al. (Eds.): CLAWAR 2022, LNNS 530, pp. 325–333, 2023.
https://doi.org/10.1007/978-3-031-15226-9_31

control research [15, 16]. The goal hereof was to test a genetic algorithm to optimize UAV trajectories when controlling a vector field-guided formation. Paper [17], which investigates tuning the control system of a leader-follower UAV formation, is the example of most similar research, topic-wise. Closely related to our paper is [18], in which a genetic algorithm was used to tune the guidance gains in the UAV formation performing target observation. However, the number of publications on optimizing the trajectories for a consensus-based, vector field-guided UAV formation, is fairly limited. This is the subject matter of this paper.

2 Preliminary Remarks and System Models

This paper investigates a decentralized UAV formation, similarly [6]. Such formation is consensus-based, and its interaction can be described by an appropriate graph hereinafter referred to as the interaction graph. Key assumptions made for the models used are the same as in [6].

High-level guidance models are used herein for optimization by means of a genetic algorithm. Two factors make the use of such models plausible. First, they adequately describe autopilot-stabilized UAV flight dynamics; secondly, they are not computationally intensive, unlike full models. The simulation section uses such model from [19]:

$$\begin{aligned}
&\dot{p}_n = v_a \cos\psi \cos\gamma_a,\ \dot{p}_e = v_a \sin\psi \cos\gamma_a \\
&\dot{h} = v_a \sin\gamma_a,\ \dot{\psi} = \frac{g}{v_a}\tan\phi \\
&\dot{\gamma} = \frac{g}{v_a}(n_c \cos\phi - \cos\gamma) \\
&\dot{v}_a = b_{v_a}\left(v_a^c - v_a\right) \\
&\dot{\phi} = b_\phi\left(\phi^c - \phi\right),\ \dot{n} = b_n(n_c - n)
\end{aligned} \tag{1}$$

where p_n is the UAV coordinate on the northward axis of the inertial coordinate system (ICS); p_e is the UAV coordinate on the eastward axis of the ICS; v_a is the UAV airspeed; ψ is the heading angle; γ_a is the air-mass-referenced flight-path angle; h is the altitude;ϕ is the current roll angle; ϕ^c is the roll angle command; γ is the flight-path angle, whereby $\gamma_a = \gamma$ holds in the absence of wind; b_{v_a} and b_ϕ are positive constants that depend on the control loop implementation in the autopilot; n is the current load factor; n_c is the load-factor command.

3 Using a Genetic Algorithm for UAV Trajectory Optimization in Target Tracking

3.1 Interaction Model and UAV Control Laws

This paper uses the same control action vector calculation method and the same consensus-based interaction model as described in [6]. Let us further describe such consensus-based interaction; the description can also be found in [6].

Let $\mathbf{e}_\theta \in \mathbb{R}^{N\times 1}$ be the control vector, where $\mathbb{R}^{N\times 1}$ is a space of $N \times 1$-dimensional matrices with components from $\mathbb{R}$. To find this vector, use some elements of the vector of all possible relative phase shift angle errors $\overline{\mathbf{e}}_\theta = \left(\hat{e}_{i,j}\right) \in \mathbb{R}^{N(N-1)\times 1}$, where $\hat{e}_{i,j}$ is the value of error for the directly interacting ith and jth agents. The choice is dictated by the interaction architecture; in this research, the control action vector is set as such for open-chain interaction in the same manner as described in [6]:

$$\mathbf{e}_\theta = \begin{bmatrix} e_1 \\ \vdots \\ e_k \\ \vdots \\ e_N \end{bmatrix} = \begin{bmatrix} \hat{e}_{12} \\ \vdots \\ -\hat{e}_{k-1,k} + \hat{e}_{k,k+1} \\ \vdots \\ -\hat{e}_{N-1,N} \end{bmatrix} = \hat{\mathbf{M}}_\theta \breve{\mathbf{e}}_\theta + \mathbf{D}, \tag{2}$$

where $\mathbf{D} = -\mathbf{M}_\theta \mathbf{H}_\theta^{-1}\left(\left(\mathbf{P}_\theta^d\right)^{\mathrm{T}}, \hat{P}_\theta\right)^{\mathrm{T}}$ is a system control vector in the space of relative distances (a $(N-1)$-dimensional space generated by the interaction graph incidence matrix columns), $\mathbf{H}_\theta$ is a matrix that specifies the agents for agent-to-agent distance measurements, defined as follows:

$$\mathbf{H}_\theta = \begin{bmatrix} \mathbf{q}_1 \\ \mathbf{q}_2 \\ \vdots \\ \mathbf{q}_N \end{bmatrix}, \quad \mathbf{q}_i = \begin{bmatrix} \vdots \\ 1 \\ \vdots \\ -1 \\ \vdots \end{bmatrix}^{\mathrm{T}}, \quad i<N, \ \mathbf{q}_N = \begin{bmatrix} 1 \\ 1 \\ \vdots \\ 1 \end{bmatrix}^{\mathrm{T}},$$

whereby $\mathbf{H}_\theta \in \mathbb{R}^{N\times N}$, $\mathbf{q}_i \in \mathbb{R}^{1\times N}$, positions of "1" and "–1" in $\mathbf{q}_i$ are determined according to the structure of the interaction graph;

$\mathbf{P}_\theta^d \in \mathbb{R}^{(N-1)\times 1}$ is the vector of desired inter-UAV phase shift angles, $\hat{P}_\theta = \sum_{k=1}^{N} \varphi_k$ is the total of the current UAV phase angles in an inertial coordinate system;

$\breve{\mathbf{e}}_\theta = \left(\breve{e}_{i,i+1}\right)_{i=\overline{1,N-1}} \in \mathbb{R}^{(N-1)\times 1}$ is the vector of current phase shift angles for directly co-engaged agents, calculated by triple scalar product, e.g. when the final movement is directed clockwise:

if $\mathbf{n}\cdot(\mathbf{d}_i \times \mathbf{d}_{i+1}) \geq 0$, then $\breve{e}_{i,i+1} = \beta = \arccos\frac{(\mathbf{d}_i,\, \mathbf{d}_{i+1})}{\|\mathbf{d}_i\|\|\mathbf{d}_{i+1}\|}$, and $\breve{e}_{i,i+1} = 2\pi - \beta$ in other cases, where $\mathbf{d}_k$, $k \in \mathcal{N}$ is the vector of aircraft-to-moving-target distance at a given time, $\mathcal{N}$ is the set of all UAV agents in the formation, and $\mathbf{n} = (0,\ 0,\ 1)^{\mathrm{T}}$;

$\mathbf{M}_\theta \in \mathbb{R}^{N\times N}$ is an interaction matrix that in case of decentralized neighbor-neighbor interaction like herein is as follows:

$$\mathbf{M}_\theta = \begin{bmatrix} -1 & 1 & 0 & \cdots & 0 \\ 1 & -2 & \ddots & \ddots & \vdots \\ 0 & \ddots & \ddots & 1 & 0 \\ \vdots & \ddots & 1 & -2 & 1 \\ 0 & \cdots & 0 & 1 & -1 \end{bmatrix};$$

$\hat{\mathbf{M}}_\theta \in \mathbb{R}^{N\times(N-1)}$ is a matrix derived from the matrix $\mathbf{M}_\theta \mathbf{H}_\theta^{-1}$ by removing the Nth column.

The task delegated to UAVs in consensus-based interaction is named differently in different publications: target encirclement, standoff tracking, or collective circumnavigation. In essence, UAVs are tasked to maintain a preset distance to a moving or stationary target whilst flying around it. At the same time, they should also keep the preset distance to each other. The problem can be solved by vector field guidance. In this case, the following control laws $\mathbf{v}^c$ (cited from [6, 20]) will be used to set UAV speeds in order to perform cooperative target tracking:

$$\mathbf{v}^c = \begin{bmatrix} v_1 \; v_2 \dots v_N \end{bmatrix}^{\mathrm{T}} = v\mathbf{1}_N + \mathbf{L}, \tag{3}$$

where $\mathbf{1}_N = \begin{bmatrix} 1\,1 \dots 1 \end{bmatrix}^{\mathrm{T}} \in \mathbb{R}^{N\times 1}$ and the vector $\mathbf{L} = \left(v_f (2/\pi) \arctan\{k_\theta e_i + k_{\dot\theta} \dot e_i\}\right)_{i=\overline{1,N}} \in \mathbb{R}^{N\times 1}$ is found given (2), k_θ is the positive tuning coefficient, $k_{\dot\theta}$ is the positive tuning coefficient for the derivative signal, v_f is the maximum norm of the additional velocity vector that is to be adjusted for the constraints of the real UAV dynamics, v is the ultimate linear cruise speed of the UAVs provided that the target is stationary.

The following heading angle control laws $\boldsymbol{\chi}^c$ from [6] were used herein with minor changes:

$$\boldsymbol{\chi}^c = \left(\varphi_i + \lambda\left[\frac{\pi}{2} + \arctan\left(k_o^i(d_i - \rho) + k_{\dot o}^i \dot d_i\right)\right]\right)_{i=\overline{1,N}} \in \mathbb{R}^{N\times 1}, \tag{4}$$

where d_i is the ith UAV-to-target distance, $\dot d_i$ is the corresponding derivative signal, k_o^i is the tuning coefficient for the distance-to-circular-path signal for the ith UAV, $k_{\dot o}^i$ is the tuning coefficient for the distance-to-circular-path derivative signal for the ith UAV, ρ is the radius of the circular path that the UAV follows whilst encircling the target, φ_i is the phase angle of circumvolution around the target for the ith UAV.

3.2 Genetic Algorithm Application

Apparently, optimal values may vary depending on the initial position. Trajectory optimization can be practical if UAVs reach their designated points at some distance off the target and then engage in the tracking maneuver from those points.

For this research, we used `optimtool` from MATLAB2015b as well as the built-in `Genetic Algorithm Solver`. The following fitness function $F_{fitness}$ was used for minimization:

$$F_{fitness} = \int_0^{t_n} t \sum_{i=1,\ldots,4} (|d_i - \rho| + |e_i|)\, dt, \tag{5}$$

where t_n is the genetic algorithm runtime, the remaining parameters are defined in Eqs. (3) and (4).

Since this was a 4-UAV formation, we used 8 tuning variables, which are coefficients for the control law (4):

$$\mathbf{K} = [k_o^1\ k_{\dot{o}}^1\ \ldots\ k_o^4\ k_{\dot{o}}^4] \in \mathbb{R}^{1\times 8}. \tag{6}$$

The following constraints were imposed on this vector **K**, written as inequalities: $[0.1\ 0.2\ \ldots\ 0.1\ 0.2] \leq \mathbf{K} \leq [10\ 5\ \ldots\ 10\ 5]$.

The constraints were necessary to keep the entire UAV formation system stable. Upper and lower bounds can be selected by trial and error when running full nonlinear UAV models.

We chose the following `Initial Population`:

$$\mathbf{K}_{initial} = [1\ 2\ \ldots\ 1\ 2] \in \mathbb{R}^{1\times 8}. \tag{7}$$

We also chose:

- `Tournament` for the `Selection Function`
- `ConstraintDependent` for the `Mutation Function`
- `Arithmetic` for the `Crossover Function`
- `200` for the `Population Size`
- `Double vector` for the `Population type`
- `0.05*PopulationSize` for the `Elite count`
- `0.8` for the `Crossover fraction`

For `Stopping Criteria`, we only used 50 `Generations`, as we noted that using more would not significantly affect the quality of transient trajectories.

4 Numerical Simulation Results

4.1 Simulation Parameters

In the simulation, we used a formation of 4 UAVs that were tasked with tracking a ground target jointly. The high-level model (1) was used for computational modeling. Control laws (3) and (4) as well as genetic algorithm module parameters from Sect. 3.2 were applied. The control laws were enabled from the beginning of the simulation. Table 1 shows the simulation parameters. Thus, the parameters of the presented table fully specify the characteristics of the simulation.

Table 1. Simulation parameters

Parameter	Symbol	Values
Initial UAV heading vector, [rad]	$\boldsymbol{\chi}(0)$	$\begin{bmatrix} 0\,0\,0\,0 \end{bmatrix}^{\mathrm{T}}$
Initial UAV speed vector, [m/s]	$\mathbf{v}(0)$	$\begin{bmatrix} 13\ 13\ 13\ 13 \end{bmatrix}^{\mathrm{T}}$
Circular path radius, [m]	ρ	200
Initial target coordinates, [m]	$\mathbf{c} = \begin{bmatrix} c_e(0)\ c_n(0)\ h(0) \end{bmatrix}^{\mathrm{T}}$	$\begin{bmatrix} 500\ 500\ 100 \end{bmatrix}^{\mathrm{T}}$
Ultimate cruise speed of the formation in case of a stationary target [m/s]	v	13
Target speed, [m/s]	v_{target}	2
Target course angle, [rad]	χ_{target}	$\pi/4$
Vector of initial UAV coordinates in the ICS, [m]	$\mathbf{p}_1(0)$, $\mathbf{p}_2(0)$, $\mathbf{p}_3(0)$, $\mathbf{p}_4(0)$	$\begin{bmatrix} 190\ 920\ -100 \end{bmatrix}^{\mathbf{T}}$, $\begin{bmatrix} 255\ 250\ -100 \end{bmatrix}^{\mathbf{T}}$, $\begin{bmatrix} 720\ 145\ -100 \end{bmatrix}^{\mathbf{T}}$, $\begin{bmatrix} 920\ 770\ -100 \end{bmatrix}^{\mathbf{T}}$
Desired inter-UAV phase shift angles, $[^\circ]$	$\mathbf{P}_\theta^d$	$\begin{bmatrix} 270\ 260\ 290 \end{bmatrix}^{\mathrm{T}}$
Initial tuning coefficients	$k_\theta, k_{\dot\theta}, k_o^i, k_{\dot o}^i$	35, 15, 1, 2

4.2 Simulation Results and Discussion

Figure 1 shows the results of running the genetic algorithm. In this figure: top left shows how the value of the fitness function changed as the number of generations increased; top right shows the current value of the individual corresponding to the best fitness function value; bottom left shows the average distance between individuals for each generation; bottom right shows the genealogy of individuals (red lines show children of mutation, blue lines show children of crossover, black lines show elite individuals).

The following values were obtained for the elements of the vector **K** (6):

$$\mathbf{K}_{opt} = \begin{bmatrix} 0.6431\ 3.6207\ 0.5431\ 4.7572\ 1.6379\ 2.3851\ 1.8171\ 2.4721 \end{bmatrix}. \quad (8)$$

Figure 2 shows transient trajectories of the UAV formation: a) when using coefficients from Eq. (7), b) when using coefficients from Eq. (8). Fitness function $F_{fitness}(90)$ equaled 1.83e+05 before and approximately 1.5e+05 after optimization.

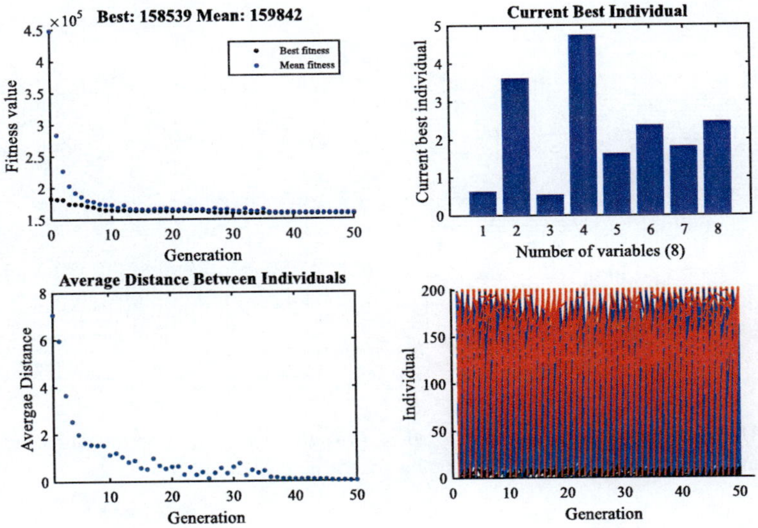

Fig. 1. Genetic algorithm results

A finding was made regarding the choice of a fitness function. Reason being, the coefficients selected for the control law (4) significantly affect the quality of convergence to the preset relative distances. Meanwhile, the coefficients were not tuned for the control law that governed the process of reaching the preset relative distances. Thus, the initially chosen fitness function as $\int_0^{t_n} t|d_i - \rho|\,dt$ can return poor results for this formation unless relative distance error is accounted for. Therefore, even for tuning the coefficients for the heading angle control law, one needs to bear in mind the decentralized, complexly interconnected formation structure. As a result, fitness function of choice must be (5) with adjustment for relative distance error.

Another simulation finding is that optimal coefficients in (8) have different values. Although for conventional trial-and-error tuning, coefficients for control laws were picked to have different values. This is quite expected, as they should depend on the relative positions and orientations of UAVs towards each other and the target as of engaging in collective circumnavigation. Notably, if trajectories need to be optimized in real time while the initial UAV positions are not known, online optimization by means of a genetic algorithm might only need to be done for a few time steps ahead.

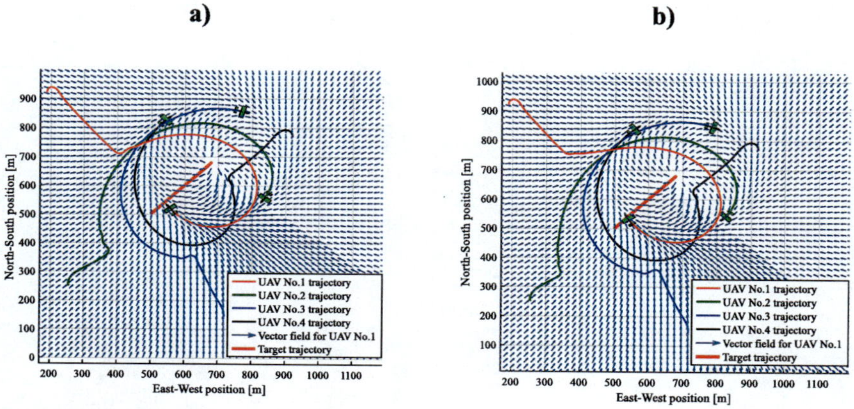

Fig. 2. UAV trajectories at the time $t = 90$ s. a) Trajectories before optimization by genetic algorithm; b) trajectories after optimization by genetic algorithm.

5 Conclusions

This paper shows that a genetic algorithm can be used to optimize the transient trajectories of a vector field-guided UAV formation. This is a decentralized consensus-based formation engaged in cooperative target tracking also known as collective circumnavigation. Simulations found the fitness function value to decrease. Other findings concern the choice of a fitness function that is adjusted for the interconnections in the UAV formation. Further research might seek to implement real-time trajectory optimization for initial UAV positions that are not known in advance.

Acknowledgements. This work was supported by the Ministry of Science and Higher Education of the Russian Federation (Agreement No. 075-15-2021-1016).

References

1. Harinarayana, T., Hota, S., Kushwaha, R.: Vector field guidance for standoff target tracking. Proc. Inst. Mech. Eng. Part G J. Aerosp. Eng. (2022). https://doi.org/10.1177/09544100211072320
2. Vishnevsky, V., Larionov, A., Zvyagin, M., Dinh, T.D., Ovchinnikov, A., Kirichek, R.: Software development for controlling a group of UAVs. ACM Int. Conf. Proc. Ser. (2020). https://doi.org/10.1145/3440749.3442614
3. Izhboldina, V., Lebedev, I., Shabanova, A.: Approach to UAV swarm control and collision-free reconfiguration. In: Ronzhin, A., Shishlakov, V. (eds.) Proceedings of 15th International Conference on Electromechanics and Robotics "Zavalishin's Readings." SIST, vol. 187, pp. 81–92. Springer, Singapore (2021). https://doi.org/10.1007/978-981-15-5580-0_6
4. Bassolillo, S.R., Blasi, L., D'Amato, E., Mattei, M., Notaro, I.: Decentralized triangular guidance algorithms for formations of UAVs. Drones 2022 **6**, 7 (2021). https://doi.org/10.3390/DRONES6010007

5. Ollervides-Vazquez, E.J., Rojo-Rodriguez, E.G., Garcia-Salazar, O., Amezquita-Brooks, L., Castillo, P., Santibañez, V.: A sectorial fuzzy consensus algorithm for the formation flight of multiple quadrotor unmanned aerial vehicles. Int. J. Micro Air Veh. **12**, 175682932097357 (2020). https://doi.org/10.1177/1756829320973579
6. Muslimov, T.Z., Munasypov, R.A.: Multi-UAV cooperative target tracking via consensus-based guidance vector fields and fuzzy MRAC. Aircr. Eng. Aerosp. Technol. **93**, 1204–1212 (2021). https://doi.org/10.1108/AEAT-02-2021-0058
7. Diveev, A., Shmalko, E.: Hybrid evolutionary algorithm for synthesized optimal control problem for group of interacting robots. In: 2019 6th Int. Conf. Control. Decis. Inf. Technol. CoDIT 2019, pp. 876–881 (2019). https://doi.org/10.1109/CODIT.2019.8820344
8. Konstantinov, S., Diveev, A.: Evolutionary algorithms for optimal control problem of mobile robots group interaction. In: Olenev, N.N., Evtushenko, Y.G., Jaćimović, M., Khachay, M., Malkova, V. (eds.) OPTIMA 2021. CCIS, vol. 1514, pp. 123–136. Springer, Cham (2021). https://doi.org/10.1007/978-3-030-92711-0_9
9. Wilburn, B.K., Perhinschi, M.G., Wilburn, J.N.: A modified genetic algorithm for UAV trajectory tracking control laws optimization. Int. J. Intell. Unmanned Syst. **2**, 58–90 (2014). https://doi.org/10.1108/IJIUS-03-2014-0002
10. Choutri, K., Lagha, M., Dala, L.: Multi-layered optimal navigation system for quadrotor UAV. Aircr. Eng. Aerosp. Technol. **92**, 145–155 (2019). https://doi.org/10.1108/AEAT-12-2018-0313
11. Darrah, M., et al.: A flexible genetic algorithm system for multi-UAV surveillance: algorithm and flight testing. Unmanned Syst. **3**, 49–62 (2015). https://doi.org/10.1142/S2301385015500041
12. Wu, X., Yin, Y., Xu, L., Wu, X., Meng, F., Zhen, R.: Multi-UAV task allocation based on improved genetic algorithm. IEEE Access. **9**, 100369–100379 (2021). https://doi.org/10.1109/ACCESS.2021.3097094
13. Gutierrez-Martinez, M.A., Rojo-Rodriguez, E.G., Cabriales-Ramirez, L.E., Reyes-Osorio, L.A., Castillo, P., Garcia-Salazar, O.: Collision-free path planning based on a genetic algorithm for quadrotor UAVs. In: 2020 Int. Conf. Unmanned Aircr. Syst. ICUAS 2020, pp. 948–957 (2020). https://doi.org/10.1109/ICUAS48674.2020.9213956
14. Pehlivanoglu, Y.V., Pehlivanoglu, P.: An enhanced genetic algorithm for path planning of autonomous UAV in target coverage problems. Appl. Soft Comput. **112**, 107796 (2021). https://doi.org/10.1016/j.asoc.2021.107796
15. Najm, A.A., Ibraheem, I.K., Azar, A.T., Humaidi, A.J.: Genetic Optimization-Based Consensus Control of Multi-Agent 6-DoF UAV System. Sensors 2020, vol. 20, pp. 3576 (2020). https://doi.org/10.3390/S20123576
16. Mondal, S., Tsourdos, A.: Autonomous addition of agents to an existing group using genetic algorithm. Sensors **20**, 6953 (2020). https://doi.org/10.3390/S20236953
17. Bożko, A., Ambroziak, L., Pawluszewicz, E.: Genetic algorithm for parameters tuning of two stage switching controller for UAV autonomous formation flight. In: Szewczyk, R., Zieliński, C., Kaliczyńska, M. (eds.) AUTOMATION 2021. AISC, vol. 1390, pp. 154–165. Springer, Cham (2021). https://doi.org/10.1007/978-3-030-74893-7_16
18. Kim, M.: Error dynamics-based guidance law for target observation using multiple UAVs with phase angle constraints via evolutionary algorithms. J. Control Autom. Electr. Syst. **32**(6), 1510–1520 (2021). https://doi.org/10.1007/s40313-021-00790-1
19. Beard, R.W., McLain, T.W.: Small Unmanned Aircraft: Theory and Practice. Princeton University Press (2012)
20. Muslimov, T.Z., Munasypov, R.A.: Coordinated UAV standoff tracking of moving target based on Lyapunov vector fields. In: 2020 International Conference Nonlinearity, Information and Robotics (NIR), pp. 1–5, IEEE (2020). https://doi.org/10.1109/NIR50484.2020.9290189

On the Travelling Salesman Problem with Neighborhoods in a Polygonal World

Miroslav Kulich[1(✉)], Jan Vidašič[2], and Jan Mikula[1,2]

[1] Czech Institute of Informatics, Robotics, and Cybernetics, Czech Technical University in Prague, Prague, Czech Republic
{miroslav.kulich,jan.mikula}@cvut.cz

[2] Faculty of Electrical Engineering, Czech Technical University in Prague, Prague, Czech Republic
http://imr.ciirc.cvut.cz

Abstract. The Travelling Salesman Problem with Neighborhoods (TSPN), as an extension of the broadly studied Travelling Salesman Problem, has many practical applications in robotics. Although several approaches to the problem have been introduced, they assume an environment without obstacles. In this paper, we introduce a method for the TSPN with polygonal neighborhoods taking polygonal obstacles into account. The method splits the problem into two subproblems: the Generalized Travelling Salesman Problem and the Touring Polygons Problem, which are solved sequentially. While the mature metaheuristic called Generalize Large Neighborhoods Search is employed to solve the first subproblem, the second one is solved by modifying the rubber-band algorithm. The experimental results show that the proposed approach outperforms a state-of-the-art algorithm modified for the environment with obstacles by 10–20% in most cases.

Keywords: Travelling Salesman Problem with Neighborhoods · Touring Polygons Problem · Combinatorial optimization · Path planning

1 Introduction

Assume a salesman that wants to meet several customers. Each customer specifies a compact set in a plane, his neighborhood, where he is willing to meet. These sets can be of arbitrary shape, for example, a disk or a polygon, and they are also allowed to intersect. The salesman aims to find a tour of the shortest length that intersects the neighborhoods of all the customers and returns to the initial point of departure.

This problem was first formulated as the Travelling Salesman Problem with Neighborhoods (TSPN) by Arkin and Hassin [4]. The TSPN is NP-hard as it extends the Travelling Salesman Problem (TSP), and it has recently received increasing attention in robotics and logistics. Example scenarios include sequencing of effective movements for robotic welding or cutting [1,2], vision-based inspection [7], or data collection in distributed wireless sensor networks [17].

© The Author(s), under exclusive license to Springer Nature Switzerland AG 2023
J. M. Cascalho et al. (Eds.): CLAWAR 2022, LNNS 530, pp. 334–345, 2023.
https://doi.org/10.1007/978-3-031-15226-9_32

Arkin and Hassin [4] studied the TSPN for various neighborhood types, including parallel unit segments, translates of a polygonal region, and circles. They derived lower bounds on the optimal tour lengths and presented simple heuristic procedures for tour construction. Gentilini et al. [7] formulated the problem as a non-convex Mixed-Integer Nonlinear Program (MINLP). They proposed a fast heuristics employing a Branch-and-Bound algorithm for convex MINLP to generate an initial good-quality solution. Furthermore, they devised an approach based on the global MINLP solver Couenne solving the instances with several neighborhoods up to 15 to optimality.

Several researchers take advantage of the availability of advanced solvers for similar optimization problems and use them for the TSPN. Mennel [10] suggests representing each neighborhood by a single point and finding the shortest tour visiting these points by a TSP solver. Optimization of the resulting tour with a fixed order of neighborhoods is consequently done by formulating it as the Touring Polygons Problem (TPP) [6]. TPP is an NP-hard problem of finding a point in each neighborhood such that the tour visiting all such points in the defined order is the shortest possible. Another approach splitting TSPN into subproblems was introduced in [1], where the insertion heuristic was applied to TSP and the rubber-band algorithm to TPP. The key distinction from the former approaches is that the two problems were solved simultaneously. Finally, the combination of 3-opt and the rubber-band algorithm was presented in [2].

The paper's contribution is twofold. First, we propose a decoupled approach for the TSPN consisting of three steps. Individual neighborhoods are sampled first, so each neighborhood is represented by a set of sample points. A sequence in which the neighborhoods are visited together with initial visiting points of individual neighborhoods is found as a solution to the Generalized Travelling Salesman Problem (GTSP) [16]. This sequence is subsequently refined by solving the TPP. Second, we propose a modification of the TPP, which takes polygonal obstacles in the environment into account and thus allows solving the TSPN in a polygonal world.

The rest of the paper is organized as follows. The problem is defined in Sect. 2, the proposed approach is introduced in Sect. 3, and the experimental results are presented in Sect. 4. Finally, Sect. 5 is dedicated to concluding remarks and future directions.

2 Problem Definition

Assume a set of disjoint polygons $O = \{O_i\}_{i=1}^{m}$ in plane representing obstacles and a set of possibly intersecting polygons $P = \{P_i\}_{i=1}^{n}$ representing neighborhoods. A neighborhood can touch an arbitrary obstacle (i.e., they have a common point(s) or line(s)) but do not intersect (do not share any area). The goal is to find a minimal-length cyclic tour $T = p_1, p_2, \dots p_{n+1}$ such that:

- $p_1 = p_{n+1}$
- For every neighborhood P_k, there exists a point $p_l \in T$ that lies in the neighborhood: $p_l \in P_k$.

- T is collision-free, i.e. it does not intersect with any obstacle.

The goal is thus to find the order in which the polygonal neighborhoods are visited together with points in these neighborhoods, which guarantee to visit the neighborhoods in an optimal way.

3 Approach

The proposed approach works in three phases, as illustrated in Fig. 1. The first phase starts by creating a triangular mesh of each polygon and pre-computing all-to-all distances between the center points of the triangles (Fig. 1a). The GLNS metaheuristic is employed on these points in the second phase (Fig. 1b) The solution that GLNS returns is a sequence of individual points, each representing a polygon of the instance, but this solution is generally far from the optimum for TSPN. Though, it gives a good approximation of the optimal order of the polygons in the tour. The third phase subsequently uses this order and employs the modified TPP algorithm, giving the output points and the order of polygons from the first phase as the input. The TPP algorithm's output forms the final solution, as shown in Fig. 1c. A triangular mesh is generated by the Delaunay refinement algorithm [15]. The rest of the algorithm is detailed in the following paragraphs.

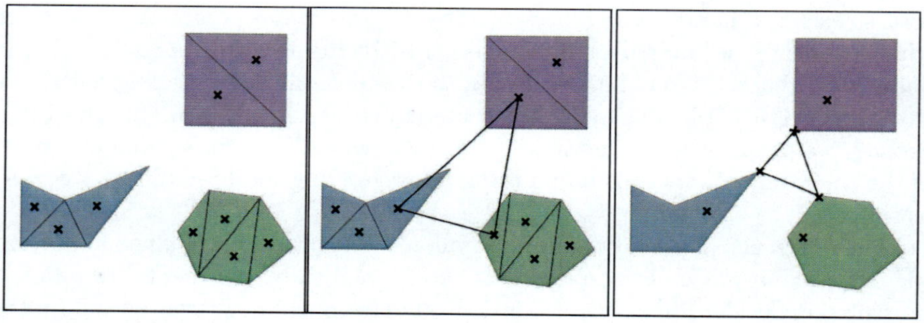

(a) (b) (c)

Fig. 1. General steps of GLNS-TPP. (a) The triangular mesh is created from the input polygons, the points in their centers are computed, and the distances between them are determined. (b) GLNS finds the approximate shortest GTSP tour among the sets (polygons) and their points. (c) TPP takes the tour found by GLNS and tries to optimize its points.

3.1 Generalized Large Neighborhood Search

The Generalized Large Neighborhood Search (GLNS) is a solver for the exactly one-in-a-set Generalized Travelling Salesman Problem (GTSP) [16]. It belongs to the state-of-the-art heuristics for solving the GTSP together with the GLKH [9], and GK [8] solvers. Experimental results show that it is competitive with the best-known algorithms solving GTSP, often finding better solutions given the same amount of time.

The general framework under which the GLNS operates is called Adaptive Large Neighborhood Search (ALNS) [12,13]. The main idea of the ALNS and, therefore, the GLNS is simple. An initial solution is found first by, e.g., a greedy approach, which is iteratively destroyed and repaired. It is accepted if a better solution than the current best one is found. The search is finished when a terminating condition is met. ALNS uses two sets of heuristics for removals and insertions of the solution. Each heuristic holds a weight representing how successful the heuristic is in improving the solution. This weight is updated as the algorithm runs; thus, the word adaptive in the ALNS. The more successful heuristics are selected more often than the less successful ones. GLNS also regularly optimizes the solution by local search.

The following section gives a simplified description of the original GLNS algorithm. More information can be found in [16].

Algorithm. The GLNS algorithm, shown in pseudocode in Algorithm 1 works in iterations. Each iteration i starts by constructing a random initial tour T (line 2) and making it a baseline best tour $T_{best,i}$ for this iteration (line 3). After that, the following cycle is repeated (line 4) until the stopping criterion is met. A single removal heuristic R and insertion heuristic I are selected first, according to their current weights (line 5). Next, a number N_r is uniformly randomly selected from between $1, \ldots, N_{max}$ (where N_{max} is a parameter of the algorithm), marking a number of vertices that are going to be removed (line 6). After creating a copy T_{new} of the current tour (line 7), we remove N_r vertices from T_{new} using the selected removal heuristic R (line 8). Then, using the selected insertion heuristic I, we insert a vertex into T_{new} for each removed vertex so that the tour T_{new} revisits all sets (line 9). The tour T_{new} is then locally optimized using ReOpt and MoveOpt heuristics. These two heuristics try to re-optimize the order of the sets and the vertex in each set (line 10). If the length of the tour T_{new} is smaller than the current best tour of the iteration $T_{best,i}$, then T_{new} becomes $T_{best,i}$ (line 11). The tour T_{new} is accepted or declined as a new tour T for this iteration based on the standard simulated annealing criterion (line 14). The stopping criteria of each iteration have two phases - initial descent and several warm restarts (line 18). The first phase, the initial descent, terminates after a fixed amount of non-improving iterations. Each warm restart starts with the best solution found in the second phase but a lower simulated annealing temperature. Each warm restart ends after several non-improving iterations.

Insertion Heuristics. The four insertion heuristics are used in GLNS – nearest, farthest, random and cheapest insertion. Each of these heuristics takes a GTSP instance (G, P_V) and a partial tour $T = (V_T, E_T)$ of graph $G = (V, E, w)$ as an input. $P_V = \{V_1, \ldots, V_m\}$ is a partition of V. The partial tour T is a cycle in G

Algorithm 1: GLNS(G,P_v)

Data: A GTSP instance (G, P_v)
Result: A GTSP tour on G

for $i = 1$ *to num_trials* **do**
  $T \leftarrow initial\_tour(G, P_v)$;
  $T_{best,i} \leftarrow T$;
  repeat
    Select a removal heuristic R and insertion heuristic I using the selection weights;
    Select the number of vertices to remove, N_r, uniformly randomly from $1, \ldots, N_{max}$;
    Create a copy of T called T_{new};
    Remove N_r vertices from T_{new} using R;
    For each of the N_r sets not visited by T_{new}, insert a vertex into T_{new} using I;
    Locally re-optimize T_{new};
    if $w(T_{new}) < w(T_{best,i})$ **then**
      $T_{best,i} \leftarrow T_{new}$;
    end
    if $accept(T_{new}, T)$ **then**
      $T \leftarrow T_{new}$;
      Record improvement made by R and I;
    end
  until *stop criterion is met*;
  Update selection weights based on improvements of each heuristic over trial;
end

such that each set in the partition P_V is visited at most once. The sets that are visited by the partial tour T are denoted $P_T \subseteq P_V$. The insertion heuristic then takes this partial tour T, inserts a vertex from an unvisited set into it (according to the heuristic-specific rules), and returns a tour that visits one more set than the input tour.

The heuristics use the set-vertex distance for their decision making about which set to insert (V_i in $P_V \backslash P_T$). The set-vertex distance is computed for each set V_i, $i \in \{1, \ldots, m\}$ and vertex $u \in V \backslash V_i$ at the beginning of the algorithm and is defined as:

$$dist(V_i, u) = min_{v \in V_i}\{min\{w(v, u), w(u, v)\}\}$$

Each heuristic has a specific way it selects the set to visit according to this distance. `Nearest Insertion` selects the set V_{opt}, which contains a vertex v at a minimum distance to any vertex on the partial tour T.

$$V_{opt} = \underset{V_i \in P_V \backslash P_T}{\text{argmin}} \; \min_{v \in V_T} dist(V_i, v)$$

`Farthest Insertion` selects the set V_i, that contains a vertex v which is at a maximum distance to any vertex on the partial tour T.

$$V_{opt} = \operatorname*{argmax}_{V_i \in P_V \setminus P_T} \min_{v \in V_T} dist(V_i, v)$$

`Random Insertion` selects the set V_i uniformly randomly from $P_V \setminus P_T$.

`Cheapest Insertion` selects the set V_i that contains a vertex v that minimizes the insertion cost.

$$V_{opt} = \operatorname*{argmin}_{V_i \in P_V \setminus P_T} \min_{(x,y) \in E_T, v \in V_i} \{w(x,v) + w(v,y) - w(x,y)\}$$

Removal Heuristics. GLNS uses three different removal heuristics – worst, distance, and segment removal. Each of these heuristics removes N_r vertices from a tour $T = (V_T, E_T)$ using a heuristic-specific strategy.

The goal of the `Worst Removal` heuristic is to remove the vertex that results in the most significant reduction in the tour length. Given a partial tour T we remove the vertex v_j that maximizes the removal cost

$$r_j = w(v_{j-1}, v_j) + w(v_j, v_{j+1}) - w(v_{j-1}, v_{j+1}).$$

`Distance Removal` aims to remove vertices that are close to each other. Given a complete tour T a random vertex is removed from T and added to a set $V_{removed}$. Then an iteration follows, where a seed vertex v_{seed} is uniformly randomly selected from $V_{removed}$ and for each $v_j \in V_T$ the removal cost r_j is computed as

$$r_j = min\{w(v_{seed}, v_j), w(v_j, v_{seed})\}.$$

The vertex with minimal r_j is removed. This process is repeated until N_r number of vertices is removed.

`Segment Removal` removes a continuous segment of the size N_r from the tour T. Given a complete tour T with vertices $V_T = \{v_1, \ldots, v_m\}$, a vertex v_j is uniformly randomly selected and then the vertices $v_j, v_{j+1}, \ldots, v_{j+N_r-1}$ are removed from the tour T.

Local Optimizations. Before evaluating the acceptance condition in an iteration of GLNS, the new tour is locally optimized. Two techniques are used for this optimization and they are called `ReOpt` and `MoveOpt`.

The goal of `ReOpt` is to optimize the choice of a vertex in each set, keeping the set order fixed. This goal is achieved by creating a directed acyclic graph (DAG) containing all vertices of the GTSP graph. The shortest path between each vertex of the first set of the tour and its copy in DAG is then determined, and the minimal shortest path becomes the optimized tour.

`MoveOpt` is a particular case of GLNS where only one set is removed and reinserted with minimum insertion cost. Given a complete tour $T = (V_T, E_T)$, a random vertex v_j is selected and removed from the tour. A vertex from the same set V_j with a lowest insertion cost $w(x, u) + w(u, y) - w(x, y)$ for all $u \in V_j$ and all $(x, y) \in E_T$ is reinserted afterwards. MoveOpt is repeated a specified number of times.

3.2 Touring Polygons Problem

The solution route found by the GLSN is refined by solving the Touring Polygons Problem (TPP). The TPP aims to find point p_i in each polygon P_i such that a closed collision-free route τ visiting all such points in a defined order is the shortest possible. We employ a rubber-band algorithm introduced by Pan et al. [11] for pairwise disjoint polygons in a plane. The algorithm takes a route found by the GLNS and a sequence of polygons in the order they are visited in the route and tries to shorten it sequentially. In each iteration, the algorithm goes over all triplets of consecutive points $\langle p_{i-1}, p_i, p_{i+j}\rangle$ and substitutes the middle point $p_i \in P_i$ by the one which minimizes the sum of distances to p_{i-1} and p_{i+1}, i.e.

$$p_i^{new} = \operatorname*{argmin}_{p \in P_i}(dist(p_{i-1}, p) + dist(p, p_{i+1})). \quad (1)$$

The process stops when the change of routes' lengths between two iterations is lower than a specified threshold.

To determine p_i^{new} in Eq. 1, we go over all edges of polygon P_i and compute a point minimizing the sum of the two distances to the edge. p_i^{new} is then the point with the minimal sum:

$$p_i^{new} = \operatorname*{argmin}_{p \in e}(dist(p_{i-1}, p) + dist(p, p_{i+1})) \quad \forall \text{edges } e \text{ of } P_i.$$

Computing p_i^{new} minimizing the sum of distances for a given edge can be realized numerically by Newton's method [14] or analytically using the law of reflection.

3.3 Adaptation to a Polygonal World

Several modifications of the described approach are needed to reflect the presence of obstacles in the environment. The modification of the first phase is straightforward. Instead of computing the Euclidean distance between all pairs of center points of the triangles, the visibility graph of the points in the environment with obstacles $\mathcal{W}$ is generated, and the shortest path distances are computed in this graph by the Floyd-Warshall algorithm [5]. Given the all-pairs distances reflecting obstacles, the GLNS algorithm can stay intact. The situation for the TPP is more complex. When looking for the optimal point $p_i \in P_i$ in regards to the points p_{i-1} and p_{i+1} in Eq. 1, it is necessary to take obstacles into consideration. The polygon P_i might be partially or entirely hidden from the view of the points p_{i-1} and p_{i+1}. The idea is to find the vicarious points p_{i-1}^V and p_{i+1}^V on the obstacle(s) for both p_{i-1} and p_{i+1} and then calculating the optimal point p in regards to p_{i-1}^V and p_{i+1}^V instead. The pseudocode of the modified algorithm is shown in Algorithm 2, while Fig. 2 illustrates its idea. The algorithm iterates

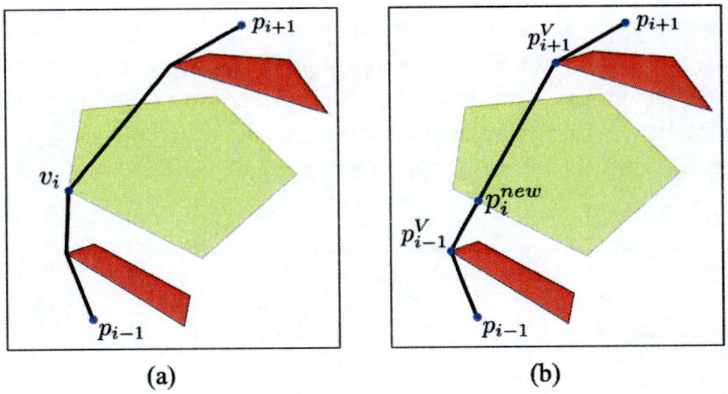

Fig. 2. The process of finding the optimal point in a polygonal world. (a) The shortest polyline connecting a vertex v_i and the points p_{i-1} and p_{i+1} is found. The grey polygons are the obstacles. (b) The vicarious points p_{i-1}^V and p_{i+1}^V are used to find the optimal point p_i^{new}. Obstacles are red, neighborhoods are yellow.

over all vertices $v \in P_i$ and computes the shortest polylines L^{-1} and L^{+1} in $\mathcal{W}$ from v to p_{i-1} and from v to p_{i+1} respectively. The vertex with a minimal sum of the lengths of L^{-1} and L^{+1} is selected, and the points directly before v on the polylines are chosen as the vicarious points. p_i^{new} is then determined using Eq. 1 with the vicarious points instead of p_{i-1} and p_{i+1}.

It should be noted that the presented approach is just suboptimal, as shown in Fig. 3. Moreover, another bottleneck is that the approach finds the optimal point on all edges of the input polygon, which leads to calling the time-consuming calculation of the shortest path many times. We thus work with the assumption that the vertex $v^* \in P_i$ minimizing length(L^{-1}) + length(L^{+1}) in Algorithm 2 is

Algorithm 2: Optimal point in a polygon through obstacles

```
Data: Point p_{i-1}, Point p_{i+1}, Polygon P_i
Result: Point p_i^new
minLength = infinite;
for vertex v ∈ P_i do
    polyline L^{-1} = shortest polyline from p_{i-1} to v ;
    polyline L^{+1} = shortest polyline from p_{i+1} to v ;
    currentLength = length(L^{-1}) + length(L^{+1});
    if currentLength < minLength then
        p_{i-1}^V = last point before v on the polyline L^{-1};
        p_{i+1}^V = last point before v on the polyline L^{+1};
        minLength = currentLength;
    end
end
p_i^new = argmin_{p∈P_i}(dist(p_{i-1}^V, p) + dist(p, p_{i+1}^V)) ;
Return p_i^new;
```

most likely around the area of the optimal point p_i^{new} that is determined at line 12. Instead of iterating over all edges of P_i, we try only the edge preceding and succeeding v^*. Of course, this leads to other inaccuracies in the optimal point determination. However, experimental results show that this suboptimality does not influence the quality of the whole TSPN solver much.

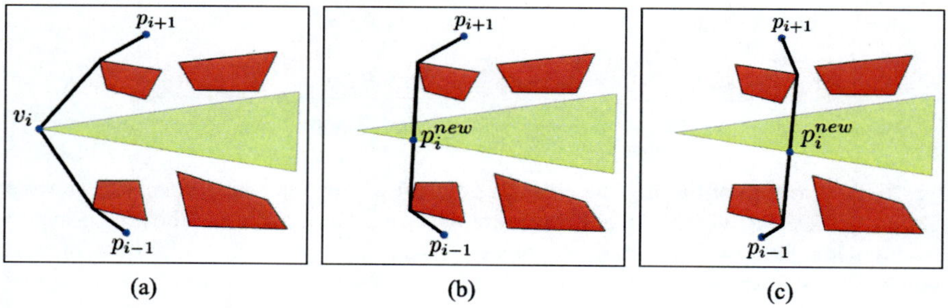

Fig. 3. Example of a solution that is not optimal found by Algorithm 2. (a) The shortest polyline connecting a vertex v_i and the points p_{i-1} and p_{i+1} is found. (b) Then the vicarious points p_{i-1}^V and p_{i+1}^V are used to find the point p_i^{new}. This solution is not optimal. (c) The correct optimal solution.

4 Experiments

The performance of the proposed approach in an environment with polygonal obstacles was statistically evaluated and compared with the state-of-the-art TSP-based approach, which solves the TSP on the centroids of the neighborhoods and refines the found route solving the TPP the same way as in the proposed approach. The TSP solver used is the Chained Lin-Kernighan heuristic from the Concorde package [3].

As there is no dataset for TSPN in a polygonal world, two sets of instances were created[1]. The first set A consists of six instances, with the same polygonal map of 59 obstacles and the varying number n of neighborhoods, $n \in \{12, 24, 33, 59, 99, 224\}$. The neighborhoods were generated to almost cover the whole map, so they have varying numbers of vertices and various shapes.

The second set B is based on the same map with 59 obstacles. We generated 150 random points in the map from which we computed visibility regions and cut them by regular decagons centered in these points to simulate a limited visibility range. 150 polygons of a similar size with around ten vertices were generated. We gradually removed polygons from this set of polygons to get instances with $n = \{10, 20, 30, 40, 50, 60, 70, 80, 90, 100, 110, 130, 150\}$ neighborhoods. Examples of the instances with found solutions are shown in Fig. 4.

[1] The generated instances can be found at http://imr.ciirc.cvut.cz/Datasets/TSPN.

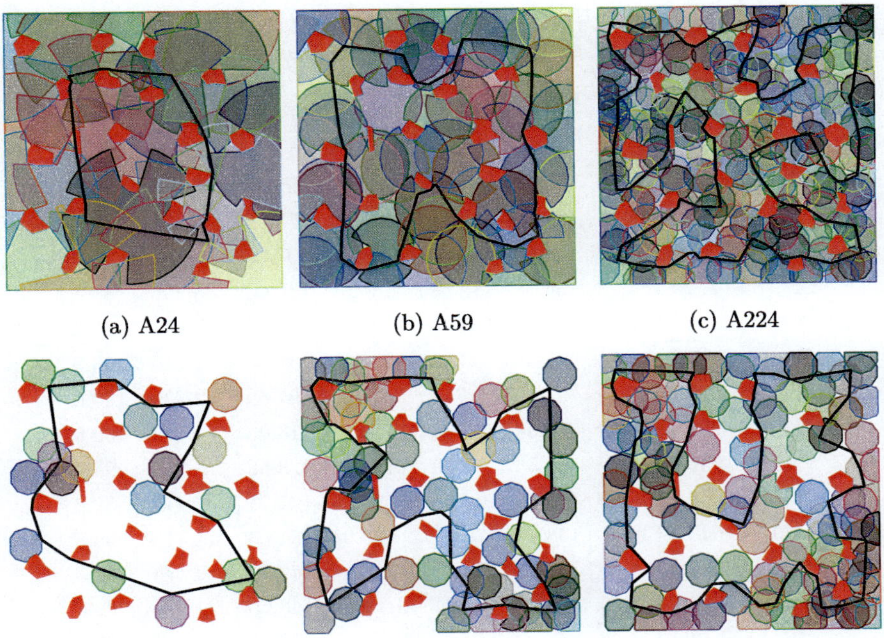

(a) A24 (b) A59 (c) A224

(d) B20 (e) B70 (f) B110

Fig. 4. Examples of instances and their solutions found by the proposed approach.

All experiments were performed within the same computational environment: a workstation with the Intel®Core^{TM} i7-6770 CPU at 3.4GHz running Ubuntu Linux with the kernel 5.4.0. The algorithms have been implemented in C++ and compiled with "-O2" flag. Fifty runs of both approaches were performed for each instance to provide statistically significant results.

The obtained results are summarized in Table 1, where *name* stands for the problem name, *size* is the number of generated triangles, and thus vertices in the GLNS/TSP, and *min* stands for the cost of the best solution found by the algorithm. Similarly, *mean* is the average cost over all 50 runs for the given problem, and *time* is the average time needed to find a solution. *Ratio* stands for the ratio of the minimal cost of the proposed and TSP-based approach.

We can state that the proposed algorithm finds solutions of better quality than the TSP-based approach for almost all instances. The only exception is the small *B*-instances where both algorithms found the same or very similar solutions. In the majority of instances, on the other hand, solutions found by the proposed approach are consistently better by 10–20%.

The main drawback of the GLNS-based approach is its computational time, which is up to four times higher than the TSP-based one. This is caused almost equally by (1) computation of all-pairs distances in the first phase, where the number of vertices is much higher for GTSP than for TSP, and (2) GLNS itself, which is slower than the TSP solver.

Table 1. Comparison of the proposed algorithm with the TSP-based approach.

Problem		TSP			proposed			Ratio
Name	Size	Min	Mean	Time [s]	Min	Mean	Time [s]	[%]
A12	988	26.46	26.46	7.359	11.33	11.37	16.569	42.8
A24	826	68.50	68.50	24.032	64.03	64.57	22.977	93.5
A33	857	87.60	87.60	22.931	74.72	75.02	25.527	85.3
A59	1204	133.16	133.16	22.286	105.58	107.26	43.145	79.3
A99	1472	154.73	154.73	28.758	136.67	139.03	83.295	88.3
A224	2096	227.03	227.03	48.817	194.77	200.24	387.651	85.8
B10	81	81.36	81.36	0.752	81.36	81.36	1.219	100.0
B20	177	91.69	91.69	2.775	91.69	91.69	2.830	100.0
B30	265	115.68	115.68	7.584	116.36	116.45	4.032	100.6
B40	359	125.36	125.36	3.970	123.46	123.56	6.753	98.5
B50	456	129.99	129.99	7.778	127.63	127.71	9.474	98.2
B60	543	138.47	138.47	7.617	130.96	131.03	14.513	94.6
B70	633	155.96	155.96	9.511	142.66	142.83	19.174	91.5
B80	724	171.06	171.06	10.291	147.66	148.69	24.727	86.3
B90	814	173.42	173.42	14.575	152.78	156.58	35.954	88.1
B100	901	185.58	185.58	17.596	154.66	155.75	44.165	83.3
B110	1000	190.52	190.52	17.330	163.85	166.82	55.679	86.0
B130	1203	207.67	207.68	20.288	177.68	182.01	82.415	85.6
B150	1403	206.66	206.66	29.073	182.80	186.46	118.986	88.5

5 Conclusion

Although the addressed Travelling Salesman Problem with Neighborhoods looks purely theoretically, it received the attention of many roboticists due to its applicability in robotic tasks. We studied the problem variant with polygonal obstacles present in the working environment, further increasing its application potential. We introduced a novel approach to the TSPN combining the solution of two subproblems: the GTSP and the TPP.

As the TPP is widely used in many other approaches incorporated not only in the TSPN, its proposed modification for a polygonal world is not limited to the TSPN only. However, it can also be used as a refinement procedure for similar problems, e.g., Watchman Route Problem or Close-Enough Travelling Salesman Problem. We will thus investigate possibilities of applying the presented principles to solve these problems. As mentioned, the weakest point of the method is its relatively high computational complexity. In future work, we thus want to focus on decreasing it by using advanced methods in the preprocessing phase and by optimizing GLNS.

Acknowledgement. This work was supported by the European Regional Development Fund under the project Robotics for Industry 4.0 (reg. no. CZ.02.1.01/0.0/0.0/15 003/0000470) and by the Grant Agency of the Czech Technical University in Prague, grant no. SGS21/185/OHK3/3T/37.

References

1. Alatartsev, S., Augustine, M., Ortmeier, F.: Constricting insertion heuristic for traveling salesman problem with neighborhoods. In: Proceedings of the 23rd International Conference on Automated Planning and Scheduling (ICAPS-2013). AAAI (2013)
2. Alatartsev, S., Mersheeva, V., Augustine, M., Ortmeier, F.: On optimizing a sequence of robotic tasks. In: Proceedings of the International Conference on Intelligent Robots and Systems (IROS). IEEE (2013)
3. Applegate, D., Bixby, R., Chvátal, V., Cook, W.: Concorde TSP Solver (2010)
4. Arkin, E.M., Hassin, R.: Approximation algorithms for the geometric covering salesman problem. Discrete Appl. Math. **55**(3), 197–218 (1994)
5. Cormen, T.H., Leiserson, C.E., Rivest, R.L., Stein, C.: Introduction to Algorithms, 3rd edn. The MIT Press (2009)
6. Dror, M., Efrat, A., Lubiw, A., Mitchell, J.S.B.: Touring a sequence of polygons. In: Proceedings of the Thirty-Fifth Annual ACM Symposium on Theory of Computing, STOC 2003, pp. 473-482. Association for Computing Machinery, New York, NY, USA (2003)
7. Gentilini, I., Margot, F., Shimada, K.: The travelling salesman problem with neighbourhoods: MINLP solution. Optim. Methods Softw. **28**(2), 364–378 (2013)
8. Gutin, G., Karapetyan, D.: A memetic algorithm for the generalized traveling salesman problem. Nat. Comput. **9**(1), 47–60 (2010). https://doi.org/10.1007/s11047-009-9111-6
9. Helsgaun, K.: Solving the equality generalized traveling salesman problem using the Lin-Kernighan-Helsgaun algorithm. Math. Program. Comput. **7**(3), 269–287 (2015). https://doi.org/10.1007/s12532-015-0080-8
10. Mennell, W.: Heuristics for solving three routing problems: close-enough traveling salesman problem, close-enough vehicle routing problem, sequence-dependent team orienteering problem. Ph.D. thesis, University of Maryland, USA (2009)
11. Pan, X., Li, F., Klette, R.: Approximate shortest path algorithms for sequences of pairwise disjoint simple polygons, pp. 175–178 (2010)
12. Pisinger, D., Ropke, S.: A general heuristic for vehicle routing problems. Comput. Oper. Res. **34**(8), 2403–2435 (2007)
13. Ropke, S., Pisinger, D.: An adaptive large neighborhood search heuristic for the pickup and delivery problem with time windows. Transp. Sci. **40**(4), 455–472 (2006)
14. Ryaben'kii, V.S., Tsynkov, S.V.: A theoretical introduction to numerical analysis. Chapman and Hall/CRC (2006)
15. Shewchuk, J.R.: Delaunay refinement algorithms for triangular mesh generation. Comput. Geom. **22**(1–3), 21–74 (2002)
16. Smith, S.L., Imeson, F.: Glns: an effective large neighborhood search heuristic for the generalized traveling salesman problem. Comput. Oper. Res. **87**, 1–19 (2017)
17. Yuan, B., Zhang, T.: Towards solving TSPN with arbitrary neighborhoods: a hybrid solution. In: Wagner, M., Li, X., Hendtlass, T. (eds.) ACALCI 2017. LNCS (LNAI), vol. 10142, pp. 204–215. Springer, Cham (2017). https://doi.org/10.1007/978-3-319-51691-2_18

Robotics and AI for Automated Infrastructure Inspection and Intervention

Confirmation of Variable Diameter TRM Operation

Yuta Naruse[1(✉)], Kazuki Takaya[1], Fumio Ito[1], Jun'ichi Watanabe[2], and Taro Nakamura[1]

[1] Precision Mechanics, Chuo University, Kasuga 1-13-27, Tokyo 112-8551, Japan
y_naruse@bio.mech.chuo-u.ac.jp

[2] Pigeon Home Products Corporation, Nakazato 2608-31, Shizuoka 417-0826, Japan

Abstract. This paper presents a variable-diameter TRM (Tension Release Mechanism) that can reduce the traction load for inspecting long-distance pipes. In this study, long, narrow, and complex pipes are inspected. The main objective of this study is to demonstrate the possibility of deflecting the traction wiring of a robot using a variable-diameter TRM and the possibility of increasing the inspection distance. Relevant studies include methods of attaching a braided sleeve or capsule to pulled wires and connecting multiple robots. However, these methods present some disadvantages, such as insufficient reduction in the traction load, TRM gets stuck in the piping, and insufficient space in the pipe. In this study, a prototype variable-diameter TRM is developed, its operation in a pipe is confirmed, and the push-in force is measured. It is shown that the variable-diameter TRM propagates in the piping and generates a pushing force of 40 N. Therefore, it is demonstrated that the traction load can be reduced by pushing.

Keywords: Pneumatic actuators · In-pipe inspection robot · Traction load

1 Introduction

In factories, complex narrow pipes are used for the long-distance transport of fluids to conserve space. Corrosion in these pipes can occur over long periods of time, resulting in the deterioration of fluid quality. Hence, periodic inspection of pipeline is required [1].

Push-in endoscopes [2] are used for pipe inspection. However, friction in the piping renders it difficult to inspect the entire pipeline using the device. Therefore, self-propelled endoscopy has been proposed as an alternative inspection method to push-in endoscopy.

In this paper, long-distance, narrow, and complex pipes in factories were investigated. In self-propelled endoscopy, a self-propelled robot must tow the wiring for pipe inspection and robot operation. Therefore, the robot must exert a significant amount of force during long-distance pipe inspections. In previous studies, wave-propagating robots [3, 4], such as inchworms and earthworms operated by fluid pressure, have been developed. Wave-propagating robots are considered suitable for propagating through complex tubes because of their flexible structures [5]. However, even in wave-propagating robots, the

© The Author(s), under exclusive license to Springer Nature Switzerland AG 2023
J. M. Cascalho et al. (Eds.): CLAWAR 2022, LNNS 530, pp. 349–357, 2023.
https://doi.org/10.1007/978-3-031-15226-9_33

structure becomes larger and more difficult to construct in narrow pipes when the traction force is increased. Therefore, to date, these robots are not able to inspect long-distance, narrow, and complex conduits. Hence, the traction load exerting on the robot must be reduced. The frictional force exerting on the wiring can be broadly classified into two factors:

(A) Friction between the inner wall of a straight pipe and the wiring.

(B) Friction generated by wiring against the inner wall of a curved pipe.

When inspecting long distances, (A) is the constant frictional force exerting on the pipeline length. However, for (B), the tension exerting on the wiring increases with the inspection distance, and the frictional force increases accordingly. Consequently, the force generated by (B) becomes dominant during the inspection of long-distance complex pipes and must be reduced.

Studies pertaining to traction load reduction include the installation of braided sleeves [6], capsules [7], and multiple robots [8]. However, braided sleeve and capsule attachment methods are limited in TRMs of traction load reduction because they do not directly reduce the wiring tension, which is the dominant source of frictional force in the inspection of long-distance complex pipes. In addition, a simple experiment conducted in the laboratory to investigate the capsule attachment method shows that the capsule cannot be extracted easily from the pipe because it is jammed when passing through a curved pipe. By contrast, the method of attaching multiple robots directly reduces wiring tension, which is a source of frictional force in long-distance complex pipes. It is considered an effective method for reducing the traction force required for robot propagation. However, when multiple robots are connected with the numerous wires required to operate them, the configuration of the robots becomes complex. Therefore, a system that can reduce the tension exerting on the wires while minimizing the number of wires required must be developed.

Therefore, the authors focused on TRMs (Tension Release Mechanism) [1], as shown in Fig. 1, which are used to clean pipelines. Generally, TRMs are cylindrical and have diameters larger than the inner diameter of the target pipeline. When fluid pressure is applied to the TRM from behind, it passes through the pipeline. Therefore, we assumed that attaching the TRM to the trailing wiring at the rear end of the robot and adjusting it would deflect the wiring and reduce the tension in the wiring even via a simple method. However, existing TRMs have a constant outer diameter, and the pipes in the subject pipeline have a constant inner diameter. This allows the TRM placed inside the pipe to be in constant contact with the inner wall of the pipe, thus sealing the pipeline. Furthermore, it allows the TRM to propagate through the conduit under fluid pressure. However, because the TRM is in constant contact with the inner wall of the pipe, the TRM cannot be removed easily from the pipe because of contact friction with the inner wall when the TRM is removed from the pipe, such as after inspection. In addition, even if the TRM can only be actuated by the robot after it is placed in the pipe, the constant contact of the TRM with the inner wall of the pipe results in a traction load. To prevent this phenomenon, a pushing pressure that requires a certain amount of air must be applied constantly. In our opinion, these issues can be resolved using a TRM with a variable outer diameter.

Therefore, a TRM with a variable outer diameter is proposed herein. In fact, the TRM expands when air pressure is applied and contracts when the air pressure is reduced. The variable outer diameter seals the inside of the tube during expansion, allowing the TRM to push in the wiring. However, during contraction, the TRM is not in contact with the inner wall, which implies no friction; this allows the TRM and robot to be extracted from the pipe. In this paper, characteristic tests of the force generated by the prototype variable-diameter TRM were conducted. The findings are expected to increase the inspection distance of in-pipe inspection robots. Hence, air pressure was applied to the developed TRM in advance. However, the TRM did not move. Therefore, the frictional force between the artificial muscle and inner wall of the canal absorbed the pushing force of the air pressure. When a low-friction material was attached around the high-friction artificial muscle in advance, it was confirmed that the developed variable outer-diameter TRM moved. Therefore, a comparison of the frictional forces before and after the installation of low-friction materials is presented herein. The next two chapters describe the prototype variable-diameter TRM. Chapter 3 describes the characteristic tests performed on the variable-diameter TRM, and Chapter 4 presents the conclusions.

The contributions of this study are as follows:

- A TRM with a variable outer diameter is proposed.
- It is demonstrated that a variable-diameter TRM can be used to seal the inside of pipeline.
- The variable-diameter TRM seals the inside of pipeline and demonstrates the possibility of deflecting the traction wiring of the robot by applying fluid pressure.

2 Variable Diameter TERM

This section presents the required specifications, principle of traction load generation, image of pushing by the TRM, and structure of the developed variable-diameter TRM.

2.1 Principle of Traction Load Generation

The principle of traction load generation and a model of the traction load tension will be explained. Friction on the contact surface with the pipe causes tension exerting on the wiring towed by the robot. In addition, as shown in Fig. 2, contact friction occurs in curved pipes as the wiring passes through the pipeline, which affects the traction load. Tension in a curved pipe can be expressed as shown in (1) [9]. The variables used in this study are listed in Table 1.

$$T_2 = T_1 + \mu N \tag{1}$$

T_1 and N increase with the penetration distance and number of curved pipes. Therefore, T_2 increases during long-distance complex pipe inspections, and when T_2 exceeds the traction force of the robot, it stops propagating. Therefore, a method to reduce the tension exerting on the traction wiring must be established for long-distance inspection by robots.

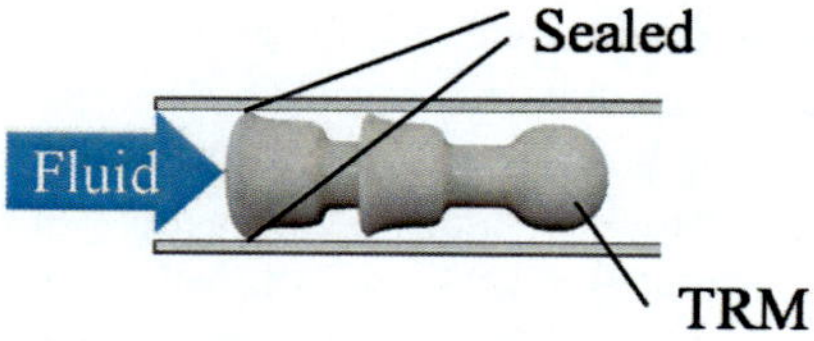

Fig. 1. TeRM advanced by fluid pressure.

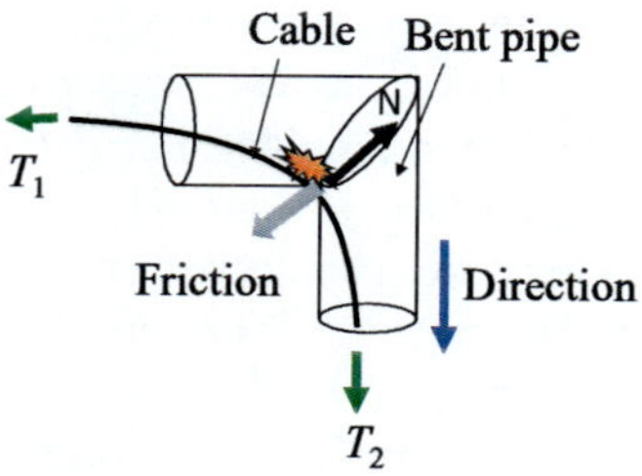

Fig. 2. Friction force model in elbow pipe.

Table 1. Symbols and nomenclature of variable-diameter TERM model.

T_1	Tension before passing through curved pipe	N
T_2	Tension after passing through curved pipe	N
M	Coefficient of friction	N/N
N	Pressing force against pipe inner wall	N
F_a	Compressed air force	N

2.2 Requirements

This section presents the requirements for the development of a TRM. The pipeline listed in Table 2 is selected for inspection because it is widely used in Japan. The pipeline to be inspected is narrow, which renders it difficult to use large robots. In addition, the proposed TRM expands and contracts with air pressure. Therefore, in this study, the robot used, which is an earthworm robot that can propagate inside a narrow pipe and adapts to the developed TRM, was operated pneumatically.

2.3 Concept

This chapter describes a traction load reduction method using TRMs. In this case, it is assumed that a TRM is attached to a wave propagation type mobile robot that consists of a pneumatic actuator. The robot is assumed to be driven by itself, and the robot is not driven when the TRM is used to push the robot. Figure 3 shows the operation pattern of the TRM.

Table 2. Specifications of pipeline inspected.

Inside diameter	35.5 mm
Material	Stainless steel
Properties length	20 m
Bend shape	90-degree bend
Number of bends	8
Wall shape	Circle
Wiring to tow	Thickness 15 mm (Five Φ4 nylon tubes and endoscope wiring covered with a braided sleeve) Length 20 m

① When the tension exerting on the wiring is greater than the traction force exerted by the robot, a variable-diameter TRM is attached to the wiring near the pipeline entrance after the robot stops propagating.

② The variable-diameter TRM is expanded, wiring is secured, and the inside of the pipe is sealed.

③ Air pressure is applied from the pipeline inlet, the variable-diameter TRM is pushed in, and the wiring is deflected.

④ When the wiring deflects, the tension exerting on the wiring is reduced, and the robot can proceed.

⑤ The variable-diameter TRM is reduced in size to secure the wiring and unseal the inside of the pipe.

⑥ The variable-diameter TRM is pulled back by the attached wire.

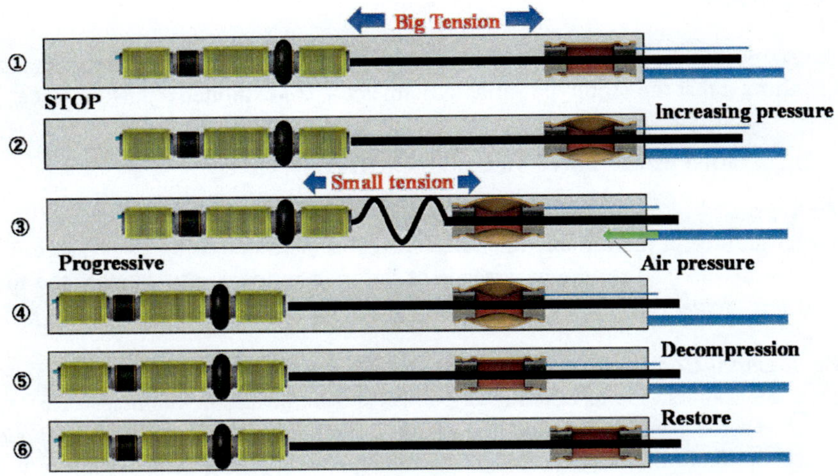

Fig. 3. Push-in pattern with variable diameter TRM.

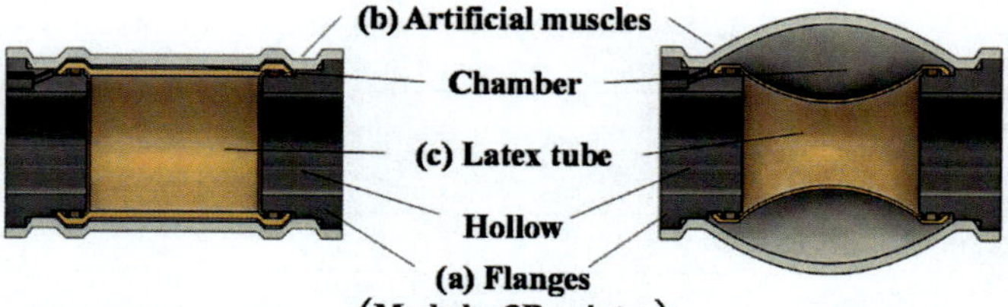

Fig. 4. Internal structure of TRM with variable outer diameter.

By repeating steps ① to ⑥, the inspection distance is expected to increase with the addition of only one actuator for long-distance pipes that were impossible to inspect previously. When wiring is pushed in by hand, the action is limited to a small distance near the pipe entrance, but with this method, wiring can be pushed in deeper into the pipe.

2.4 Structure of Variable-Diameter TRM

The internal structure of the developed variable-diameter TRM is illustrated in Fig. 4. The proposed variable-diameter TRM comprises two (a) flanges, (b) artificial muscles, and (c) latex tubes. The two (a) flanges are connected by (b) an artificial muscle and (c) a latex tube, and a chamber space exists between them. Pressurizing the chamber space via the (a) flange causes both (b) artificial muscle expansion and (c) latex tube expansion. Consequently, (b) the artificial muscle seals the inside of the pipe and (c) the latex tube supports the traction wiring in place. The TRM is pushed in by applying indentation pressure that is air pressure from the pipeline inlet.

3 Experiment

To realize push-in using the variable-diameter TRM, we measured the force generated and confirmed that the prototype variable-diameter TRM exhibited a push-in force.

3.1 Pneumatic Push-In Force Measurement Test

Test Purpose
The push-in force of the variable-diameter TRM was measured using air pressure. This was performed to confirm the occurrence of the push-in force, and whether the force exerted was greater than the frictional force against the inner wall of the pipe.

Environmental Conditions
The test environment is illustrated in Fig. 5. As shown, a variable-diameter TRM with an attached force gauge is installed in a stainless-steel pipe, and air pressure is applied. The variable-diameter TRM is pushed into the pipe by applying air pressure from the pipe inlet. A naphtholone sheet is attached to the variable-diameter TRM as a low-friction material. Measurements were performed thrice, and the mean values and standard deviations are shown in the graphs.

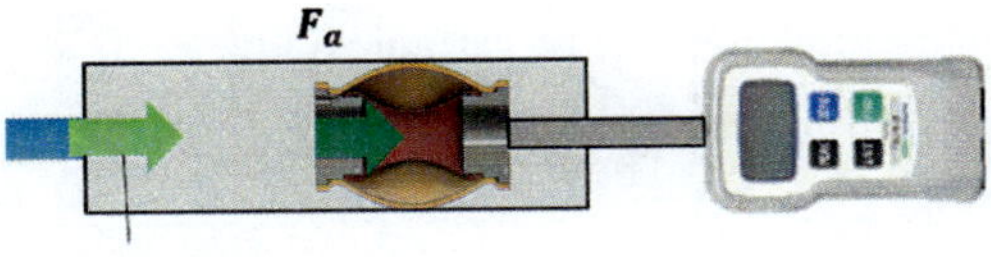

Fig. 5. Push-in force test method for TRM with variable outer diameter.

Test Result

The measurement results are shown in Fig. 6. A linear increase in the indentation pressure is observed at low pressures. However, this trend is not observed when the indentation pressure is increased. A maximum push-in force of 95.5 N is observed, but as the push-in pressure increased, a difference from the theoretical value is indicated. The maximum error generated is 61%, whereas the maximum standard deviation is 1.9.

Consideration

As the indentation pressure increases, the measured value deviates from the theoretical value, and no linear increase is observed because of the pressure loss caused by the air pressure arising from indentation leakage from the pipe. This occurred because the push-in pressure exceeded the TRM internal pressure, thereby causing the artificial muscle to be pushed and a gap to be created. Therefore, the internal pressure of the TRM was set to a low value. Meanwhile, the indentation pressure was set to a high value to achieve a large indentation force. However, the results indicate that the TRM should be operated with the internal pressure greater than the push-in pressure to prevent the TRM from denting. Consequently, the internal pressure of the TRM increases and it operates at a greater pressure than the push-in pressure. The standard deviation of 1.9 at the maximum value of this measurement is 2.0% smaller than the maximum value of 95.5 N. Therefore, the experimental data can be considered as highly reliable.

3.2 Summary

Based on the data presented in Sect. 3.1, $F_a = 40[N]$ when the indentation pressure is 50 kPa. However, by reducing the difference between the indentation pressure and TRM internal pressure, the loss of indentation pressure can be reduced, and a higher indentation force can be obtained. Therefore, we plan to implement the variable-outer-diameter TRM investigated in this study in future experiments after reducing the difference between the push-in pressure and internal pressure of the TRM.

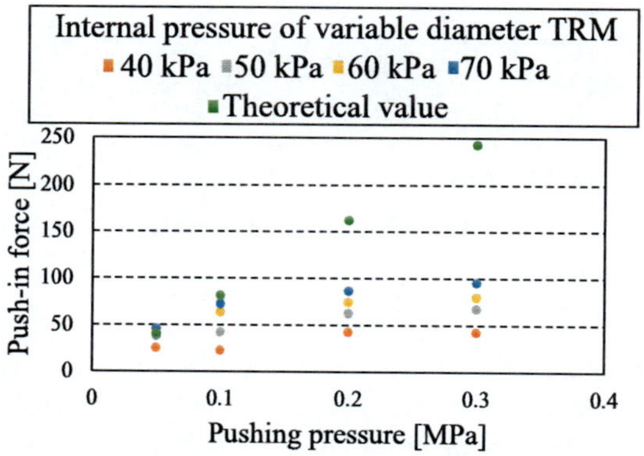

Fig. 6. Result of push-in force measurement.

4 Conclusion

In this study, a prototype variable-diameter TRM was developed for an in-pipe inspection robot to improve its mobility performance. In addition, a test was conducted to measure the force generated by the variable-diameter TRM. The results confirmed that a pushing force of 40 N was generated by the air pressure.

Future work will include measuring the frictional force between the variable-diameter TRM and the inner wall of the pipe and the force of the latex tube gripping the traction wiring. The variable- diameter TRM is then attached to the traction wiring of the worm-shaped robot to test the push-in assistance.

References

1. Mishra, D., Agrawal, K.K., Abbas, A., Srivastava, R., Yadav, R.S.: PIG [pipe inspection gauge]: an artificial dustman for cross country pipelines. Procedia Computer Science **152**, 333–340 (2019)
2. Yoda, S., Wakibe, Y., Imagawa, Y.: Inspection methods of piping. Bull Soc Sea Water Sci **68**, 57–62 (2014). (in Japanese)
3. Lim, J., Park, H., Moon, S., Kim, B.: Pneumatic robot based on inchworm motion for small diameter pipe inspection. In: Proceedings of the 2007 IEEE International Conference on Robotics and Biomimetics. Sanya, China, December 15–18 (2007)
4. Mano, Y., Ishikawa, R., Yamada, Y., Nakamura, T.: Development of contraction force control system of peristaltic crawling robot for sewer pipe inspection. In: Proceedings of the 2018 IEEE/ASME International Conference on Advanced Intelligent Mechatronics (AIM). Auckland, New Zealand, July 9–12 (2018)
5. Takayama, T., et al.: A twisted bundled tube locomotive device proposed for in-pipe mobile robot. IEEE/ASME Trans Mechatron **20**(6), 2915–2923 (2015)
6. Debenest, P., Guarnieri, M., Hirose, S.: PipeTron series - robots for pipe inspection. In: Proceedings of the 2014 3rd International Conference on Applied Robotics for the Power Industry, pp. 1–6 (2014). https://doi.org/10.1109/CARPI.2014.7030052

7. Arai, M., Hirose, S.: Improved reel mechanism to wind a cable uniformly in spherical trailer. In: 2007 IEEE International Workshop on Safety, Security and Rescue Robotics, pp. 1–6 (2007). https://doi.org/10.1109/SSRR.2007.4381257
8. Miyasaka, K., Kawano, G., Tsukagoshi, H.: Longmover: flexible tube in-pipe inspection robot for long distance and complex piping. In: 2018 IEEE/ASME International Conference on Advanced Intelligent Mechatronics (AIM), pp. 1075–1080 (2018). https://doi.org/10.1109/AIM.2018.8452690

Application of Radio Frequency Identification Underwater with Magnetic Flux Leakage Testing Technique

Nagu Sathappan(✉), Mohammad Osman Tokhi, Aman Kaur, Gholamhossein Shirkoohi, Zhanfang Zhao, and Fang Duan

School of Engineering, London South Bank University, London, UK
sathappn@lsbu.ac.uk

Abstract. Non-destructive testing (NDT) of engineering components and structures represents one of the strategic objectives of many industries in automation. It enables increased accuracy, precision and speed of inspection while reducing production time and associated labour costs in contrast to manual inspection. The use of robots can provide additional flexibility and autonomy to automated NDT. Mobile robot navigation has stood as an open and challenging problem over the last few decades. Despite the significant advances in this field, researchers have yet to reach a comfortable level of satisfaction. Radio-frequency identification (RFID) technology is now widely used in many commercial and industrial sectors, from object tracking to personal identification. Few studies have investigated the possibility of using RFID systems for underwater monitoring operations in fluvial environments. This paper presents the development of a device to collect and store data from a magneto resistance sensor for underwater corrosion monitoring using RFID. The findings show that RFID systems can be used to store data at near ranges and could be an option for underwater NDT applications. This paper also provides an investigation into RFID, transponders, and reader classification, as well as current underwater applications and their benefits in NDT.

Keywords: Radio-frequency identification · Non-destructive testing · Mobile robotics

1 Introduction

The use of wireless sensor networks in the underwater domain has a lot of potential for monitoring the health of rivers and oceans [1]. Corrosion monitoring in pipes is difficult and expensive in underwater environments; divers are limited in the hours and depths at which they can work, and they need a boat on the surface, which can be expensive to operate subject to weather conditions. A remotely operated vehicle (ROV) can carry a multitude of inspection tools and log a large amount of data gathered from the on-board sensors [2].

Compared to other communication techniques, RFID is omnidirectional, and this is its major advantage. Since electromagnetic fields are greatly affected by water, particularly saline water in an underwater environment, RFID, as a radio-frequency technology,

© The Author(s), under exclusive license to Springer Nature Switzerland AG 2023
J. M. Cascalho et al. (Eds.): CLAWAR 2022, LNNS 530, pp. 358–366, 2023.
https://doi.org/10.1007/978-3-031-15226-9_34

has few drawbacks, among them is the limited range of data transmission. RFID readers can read several tags simultaneously, store more data on the tag and it can be manipulated [3, 4]. The cost of implementing RFID technology is also decreasing, making commercial implementation with a higher Return on Investment (ROI) more feasible. Due to its ease of use, versatility, and low cost, recent initiatives in the field of mobile robot navigation have seen an increase in interest in the new RFID technology as a possible alternative strategy. Khubitz et al. [15] have presented an artificial landmark navigation system based on RFID tags. In recent years, significant research has been conducted on mobile robotics that incorporate several sensors and landmarks as navigation media in the environment.

2 RFID Technology

An active RFID system works on two main frequencies, namely 433 MHz and 2.5 GHz. Semi active tags are a combination of the characteristics of active and passive tags [5]. When the RFID tag is read within the reader's range, the energy from the radio-frequency (RF) waves is received by the tag's internal antenna as described in Fig. 1. Backscatter is a phenomenon that occurs in passive RFID tags in which a change in the RF wave is detected by the reader (via the antenna), which then interprets the data [5]. The coiled antenna inside a passive RFID tag generates a magnetic field as it encounters RF waves from the reader [6]. The RFID standards operate on specific frequency bands (e.g. UHF 860–915 MHz or HF 13.56 MHz) [6, 9]. In robotic applications, the robot is outfitted with an RFID reader that allows it to connect with the tags and estimate its location. RFID systems are primarily used for robot localization rather than navigation in mobile robot systems [5].

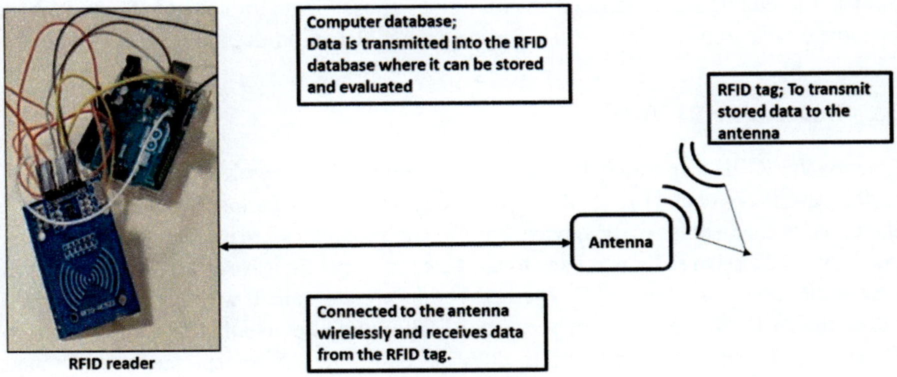

Fig. 1. RFID system

3 Underwater Communication Using RFID

Investigations have shown that the ability to transmit radio signals underwater is primarily determined by two factors: water conductivity and radio wave frequency. When RFID-based sensing is tested with a metal in a marine setting, the reading distance between the tag and reader is significantly reduced [10, 11]. Furthermore, the sensing region of RFID-based sensing is determined by the size of the tag's coil.

3.1 Underwater Radio Waves

In general, a tag can be placed as close to the reader's antenna range as practical. There are various variations in the RFID approach because it may be used in both salt and fresh water. Starting with seawater, the calculations suggest that only low frequency RFID may be employed for systems requiring a long reading distance (over 50 cm). The attenuation for a frequency of 13.56 MHz and an electrical conductivity of 4 S/m is given as

$$\Delta_{13.56\mathrm{MHz}} = \frac{1}{\sqrt{\pi f \mu \sigma}} = \frac{1}{\sqrt{\pi \times 13.56 \times 10^{6} \times 4\pi \times 10^{-7} \times 4}} = 0.0683\,\mathrm{m} \quad (1)$$

where Δ denotes attenuation, f denotes frequency in Mega Hertz, μ represents relative permeability, and σ denotes electrical conductivity in S per metre. The value in Eq. (1) is just below a low frequency device's maximum reading range, which is usually less than 1m. After determining the water conductivity value, the penetration depth and attenuation values can be estimated. Low-frequency RFID could therefore be used to distinguish submerged objects. While an efficient long-range RFID system might be realized at lower conductivity values, as the water conductivity increases, the penetration depth lowers in values [4], making this solution difficult, if not impossible, to implement. These sensors become nearly impossible to utilise for long-range detection as the frequencies grow.

3.2 Underwater RFID

Underwater RFID technology has been used to track pebbles, with displacement playing a vital part. Parks et al. [13] employed three different types of transponders to determine the range of the pebbles in the device. Passive transponders were implanted into pebbles and a core 125 kHz reader was used to emit electromagnetic waves to the transponder to determine the reading range [8]. The best results were obtained with a 125 kHz system rather than a 13.56 MHz system because the latter was attenuated. The range reduction for monopole antennas can be up to 80%, necessitating the transponder's attachment to the reader antenna. The reduction is lesser (about 30%) for low-frequency devices operating in the 125–134 kHz band, and a 50 cm reading range can still be reached using extended range antennas [3].

Pais and Symonds [4] conducted experiments using a typical USB HF reader to assess the reduction in read range when both devices (the reader and a transponder) are submerged.

4 Sensor Data Storage Using RFID Underwater

The data-on-tag concept is utilized in manufacturing for production control and maintenance documentation, as well as in healthcare for tracking patients and medical equipment [4]. Additional information for clothes has been stored in the tag in one case [4, 12]. RFID transponders have a large data storage capacity [4, 6]. A method to gather and store the data sent by the magneto resistance sensor is required. The data from underwater corrosion monitoring must be retrieved later, and a backup is required. Magnetic flux leakage (MFL) coil sensors are employed to monitor corrosion in pipes under normal environmental conditions. The (ferrite core) MFL coil sensor has been proven to be highly efficient for extreme temperature settings up to 200 °C [13]. RFID system provides data accuracy and involves shorter processes. The data sent by the magneto resistance sensor must be stored to detect corrosion. The RFID system ensures data integrity while also reducing the time it takes to complete the task. The tag functions like a USB where the data gets stored. The amount of information to be stored inside the tag depends on the type of tag to be used, and the storage capacity of each RFID tag varies.

4.1 Design of the System

Two different types of underwater environment, namely freshwater and saltwater, were considered in the study. The freshwater case involved placing the RFID tag (sealed) inside a small tank filled to a certain level with water and the RFID reader outside, as described in Fig. 2. Transponder waterproofing is important for many applications, and many devices have been developed that provide a high level of protection against water contact. While wristbands have been customized to be worn in the bath, plastic tags are waterproof by nature [7, 14]. However, none of these devices are intended to be read underwater and are only designed to survive water immersion. Moreover, to the authors' knowledge, no reader has ever been planned for use underwater. The specifications of the equipment used are as follows: RFID Reader, Tag – 28340, operating frequency = 125–134 kHz, maximum supply voltage 5.5 V. The RFID tag can be mounted anywhere within the tank that is within the reader's range. Table 1 shows the specifications of the components used for testing at room temperature and saltwater environment.

Table 1. Specification of the components

Component	Operating frequency	(app) Range
RFID Tag-28340	125–134 kHz	2–8 mm
RFID tag mifare1 S50	13.56 MHz	8–10 mm
RFID reader/writer-MFRC 522	13.56 MHz	

4.2 The Readable Transponder

With a frequency of 125 kHz, the RFID reader used in this experiment reads passive RFID transponder tags. With a maximum supply voltage of 5 V and a data rate of 2400 bits per second, the RFID reader is a powerful device. Using the experimental setup, the effective reading distance between the tag and the corresponding reader is measured directly. At a distance of 2 mm between RFID tag and reader antenna, signals (ID number) were obtained, with a maximum distance of 8 mm measured. Following initialization, the reader sends an electro-magnetic (radio-frequency) signal to the tag to energise it every 500 ms for 30 ms during a reading time, obtaining the tag's ID number. The scanned RFID tag determines the path length, demonstrating that RFID tags positioned in any location can be detected within the RFID reader's range as in Fig. 2.

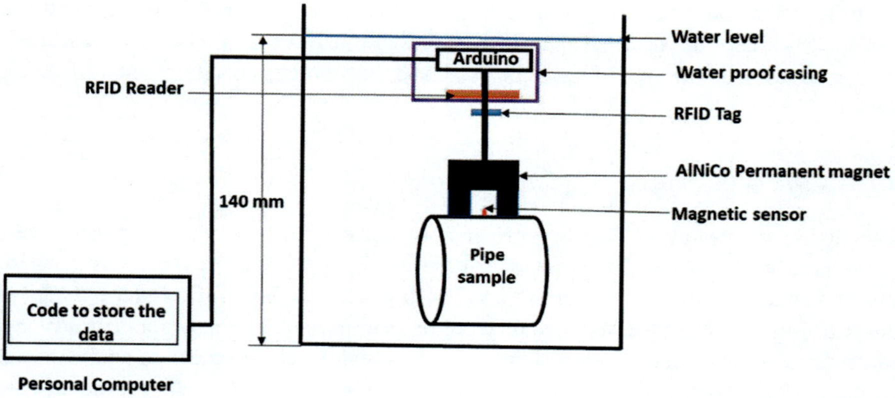

Fig. 2. Experimental setup with the RFID tag placed outside the casing

Measurement ranges were taken for various tags in order to define the maximum range obtained and tests were carried out in three environments as in Table 2.

Table 2. Reading ranges of different low frequency transponders at real conditions (All dimensions are in cm)

Tag-type	Frequency	Air	Fresh water	Salt water
PPS + Epoxy	125 kHz	10 cm	5.0 cm	3.0 cm
ABS	125 kHz	5.5 cm	5.3 cm	5.1 cm
Nylon	125 kHz	4.6 cm	4.2 cm	4.1 cm
PVC disc	125kHz	4.9 cm	3.6 cm	3.3 cm
Plastic	13.56 MHz	1.2 cm	5.0 cm	4.0 cm

4.3 The Rewritable Transponder

The experimental set-up was composed of the magnetic circuit immersed in the tank filled with water along with the RFID system, which comprises Arduino microcontroller connected to the reader along with the RFID tag placed inside a waterproof casing. The Arduino microcontroller was connected to the laptop through a serial port connector. Thus, the sensor data goes via the Arduino microcontroller to a rewritable RFID tag, and the reader reads the data and sends it to the laptop as described in Figs. 2 and 3. Arduino Uno Microcontroller has the output pin of the magneto resistance sensor connected to the A0 pin in the microcontroller through which the output is obtained. The sensor output was recorded along with range, and the result is shown in Fig. 4. The RFID tag was in close contact with the casing, and with the RFID tag submerged in water, the casing acts as a buffer between the reader and the tag, resulting in signal variation. As noted, with the sensor placed within the magnetic circuit, the sensor output varied over distance. The sensor resistance causes a drop in voltage, and due to the power utilization in the long run, the voltage should decrease since the RFID tag is placed inside the casing and not in direct contact with water, the voltage is increasing as in this case.

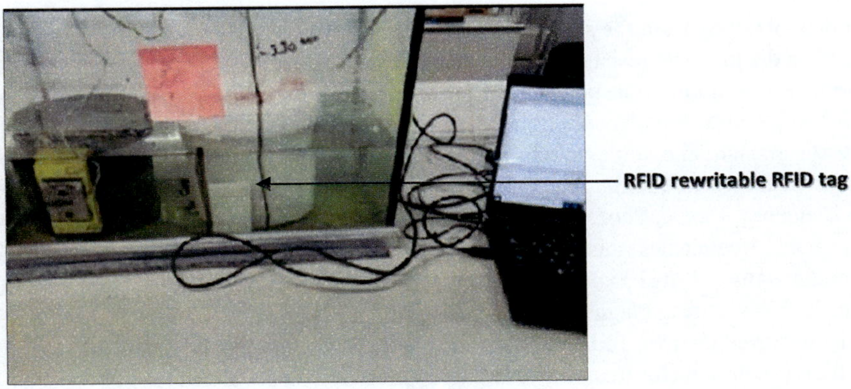

Fig. 3. RFID tag placed outside the casing

As noted, with the sensor placed within the magnetic circuit, the sensor output varied over distance.

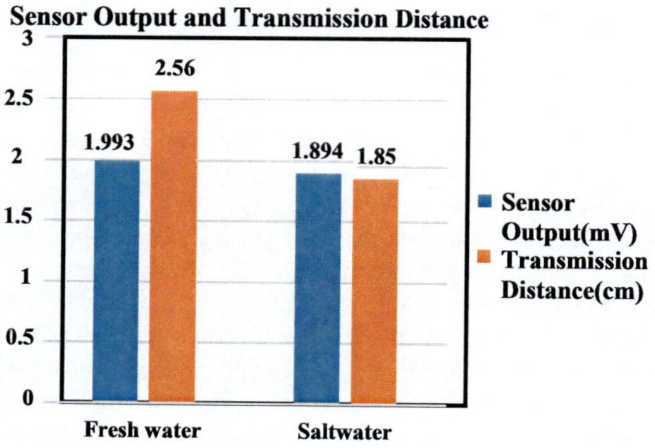

Fig. 4. Distance vs sensor output

5 Discussion

The data obtained from the sensor reveals that the developed RFID system can collect data from the magneto resistance sensor. Data were obtained using the developed RFID system with a unique transponder and reader. Regardless of the frequency range, the reader and the transponder should be placed in close contact in order to obtain the data from the sensor. The values obtained through the above tests refer to a short range communication while long range communication requires even lower frequencies. For low frequency, the depth of penetration is few metres. Due to limited bandwidth with lower range frequencies, this communication channel can be used to transmit data and not audio signals. Data has been obtained for a pipe sample thickness of 2 mm with the magnetic circuit using magneto resistance sensor and stored in the RFID system which can be retrieved later for further analysis. At higher frequencies, the chances of setting up efficient solutions for fresh water remain high. Based on the above results, especially for short-range applications, RFID becomes practically usable for fresh and salt water. However, the ability to use lower frequency bands has led to the advancement of certain RFID applications for purposes of both marine and fresh water environments.

6 Conclusion

The development of RFID sensor measurement has been presented and evaluated for underwater data collection. The experimental investigations show satisfactory results. The findings show that only at low frequencies can solutions for long reading ranges be created, both for salt and fresh water. The system has been developed for data storage at a range less than 5 mm with a single RFID tag whereas testing can be carried out with multiple tags as well. Future research involves creating RFID model to be used in simulations built from real- world data sampled from a real RFID system [17]. Multiple RFID systems can be used for collection of data at various ranges within the reader, and this will be a subject of future research. Furthermore, in mobile robotics, future work will consider several RFID tags mounted at unknown locations in 3-D space.

Acknowledgements. This publication was made possible by the sponsorship and support of Lloyd's Register Foundation. Lloyd's Register Foundation helps to protect life and property by supporting engineering-related education, public engagement and the application of research. The work was enabled through, and undertaken at, the National Structural Integrity Research Centre (NSIRC), a postgraduate engineering facility for industry-led research into structural integrity established and managed by TWI through a network of both national and international Universities.

References

1. Vasilescu, I., Kotay, D.K., Rus, L.D., Dunbabin, M., Corke, I.P.: Data collection, storage, and retrieval with an underwater sensor network. In: Proceedings of the 3rd International Conference on Embedded Networked Sensor Systems, pp. 154–165 (2005). https://doi.org/10.1145/1098918.1098936
2. Ho, M., El-Borgi, S., Patil, D., Song, G.: Inspection and monitoring systems subsea pipelines: a review paper. Sage J. **19**(2), 606–645 (2020). https://doi.org/10.1177/1475921719837718
3. Benelli, G., Pozebon, A.: RFID Under Water: Technical Issues and Applications (2013). https://doi.org/10.5772/53934
4. Pais, S., Symonds, J.A.: Data storage on a RFID tag for a distributed system. Int. J. UbiComp **2**(2) (2011). https://doi.org/10.5121/iju.2011.2203
5. https://www.atlasrfidstore.com/rfid-insider/active-rfid-vs-passive-rfid. Accessed 06 May 2020
6. Henrici, D., Kabzeva, A., Fleuren, T., Müller, P.: Data storage in RFID systems. In: Turcu, C. (ed.) Radio Frequency Identification Fundamentals and Applications Bringing Research to Practice (2010). ISBN: 978-953-7619-73-2. https://doi.org/10.5772/8002
7. Chichi, P.D., Kaur, P.: RFID technology and its applications: a review. Int. J. Electron. Commun. Instrum. Eng. Res. Dev. (IJECIERD) **2**(3), 109–120 (2012). ISSN 2249-684X
8. Ali, A.O., Aiyeoribe, M.A., Lucky, O.O.: The radio frequency identification principles and technical impacts in industrial sectors. AR J. Eng. Technol. **1**(1) (2020). ISSN Print: 2708-5155|ISSN Online: 2708-5163
9. Terman, E.F.: Radio Engineers Handbook, pp. 785–786. McGraw Hill, New York (1943)
10. Benelli, G., Pozzebon, A., Raguseo, G.: An RFID based system for the underwater tracking of pebbles on artificial coarse beaches. Sens. Technol. Appl. (2009). https://doi.org/10.1109/SENSORCOMM.52
11. Benelli, G., Bertoni, D., Sarti, G.: An analysis on the use of LF RFID for the tracking of different typologies of pebbles on beaches (2011). https://doi.org/10.1109/RFID-TA.2011.6068673
12. Melski, A., Thoroe, L., Caus, T., Schumann, M.: Beyond EPC – insights from multiple RFID case studies on the storage of additional data on tag. In: International Conference on wireless Algorithms, Systems and Applications, pp. 281–286 (2007)
13. Sathappan, N., et al.: Development of an MFL coil sensor for testing pipes in extreme temperature conditions. Sensors **21**(9) (2021). https://doi.org/10.3390/s21093033
14. Haas, M.L., Miller, J.R.: Transforming heterogeneous data with database middleware: beyond integration. In: Bulletin of the IEEE Computer Society Technical Committee on Data engineering, pp. 1–6 (1997)

15. Khubitz, O., Matthias, R.D., Berger, O., Perlick, M.: Application of radio frequency identification devices to support navigation of autonomous mobile robots. In: Proceedings of the IEEE Vehicular Technology Conference, pp. 126–130 (1997)
16. Gharpure, C., Kulyukin, V.: A robotic shopping assistant for the blind: a pilot study in a supermarket. Computer Science Department, Utah State University Logan, UT, Technical report. USU-CSATL-1-01-06 (2006)
17. Gueaieb, W., Miah, M.S.: An intelligent mobile robot navigation technique using RFID technology. IEEE Trans. Instrum. Meas. **57**(9), 1908–1917 (2008)

Robotic Deployment of Stabilized Shearography Unit for Wind Turbine Blade Inspection

Vitor Marques and Tariq Pervez Sattar(✉)

London South Bank University, London SE1 0AA, UK
sattartp@lsbu.ac.uk

Abstract. A wind turbine blade inspection system called Winspector has been developed by a European consortium to automate the in-situ non-destructive testing of wind blades. A robot platform is winched up a wind turbine tower to reach a blade locked into a 90° pitch angle. The blade region to be inspected is reached by a combination of an extension ladder and a 5-axis robot arm to place an end-effector onto the blade surface. The end-effector is critical to the success of the shearography technique used to detect subsurface defects when the blade is experiencing inevitable out-of-plane, in-plane, and root-to-tip vibrations. The paper describes the development of the end-effector which carries the shearography unit and ensures (with passive compliance and design based on analysis of aerodynamic blade shapes) that the unit remains at a constant distance from blade surface in the presence of blade vibrations. The system has been tested with three separate field trials in a wind farm near Athens in Greece. Sub-surface defect detection with shearography has been successfully demonstrated.

Keywords: Robotic in-situ inspection of wind blades · Measurement of wind blade vibrations · Shearography NDT of wind blades

1 Introduction

1.1 The Need to Inspect Wind Turbine Blades

Annual wind power installations in Europe increased from 3.2 GW in 2000, to 11.8 GW in 2014. At the end of 2014 the total installed capacity in the EU reached 128.8 GW [1] and 370 GW globally [2].

There are currently more than 700,000 wind turbine blades in operation, with approximately 243,600 wind turbine blades within the EU [3]. Given that an average of 3,800 blade failures occur annually, attributed to poor maintenance [4], costing between £70,000 and £700,000 each, preventative inspection every 3–4 months and maintenance every 6 months is necessary.

1726 accidents have been reported from 1970 to May 2015 with fatalities accounting for 45% of the total number [5]. There is a clear market need to develop automated inspection techniques that can meet the requirements of inspecting wind turbine installations so that structural failures and catastrophic consequences can be avoided.

© The Author(s), under exclusive license to Springer Nature Switzerland AG 2023
J. M. Cascalho et al. (Eds.): CLAWAR 2022, LNNS 530, pp. 367–379, 2023.
https://doi.org/10.1007/978-3-031-15226-9_35

1.2 Automation of the Non-destructive Testing (NDT) of Wind Turbine Blades

A solution to this problem is to utilize specially designed robot platforms that can climb WT towers, reach the blade and test wind turbine blades on site with non-contact non-destructive testing (NDT) methods. Systems to automate the NDT of wind turbine blades are in early stages of development with RadBlad [6] using robots to climb up a WT tower to reach a stationary wind turbine blade and use x-ray radiography to scan the blades. Another development [7] uses a climbing robot with permanent magnet adhesion to carry a long arm that can reach wind turbine blades. Other approaches [8–11] place a climbing robot carrying a shearography system directly on to a wind turbine blade using a crane.

2 The Winspector System

The Winspector project was funded by the European H2020 Fast Track to Innovation program [12] with WRS Marine in the Netherlands leading the project and Gamesa (Spain) and Iberdrola (Spain) providing wind turbine blade data and access to testing facilities. Winspector consists of a wheeled robotic platform developed by project partner Innora (Greece), see Fig. 1. The wheeled platform adheres to the wind turbine tower with two magnet systems at the top and bottom of the platform. A ladder arm is rotated to reach outwards towards a wind turbine blade. A five-axis robot arm (two prismatic joints, three rotational joints) mounted at the end of the ladder deploys the Digital Shearography Unit. The Shearography Unit is contained within an end-effector frame, termed the Shearography Unit Holder. The shearography unit was developed by project partner TWI (UK) while the Shearography Unit Holder was developed by London South Bank University (UK). The robot platform carries a winching system that uses two wire ropes from the nacelle to obtain up/down vertical motion of the robot.

The following sections describe the development of some aspects of the Winspector system with particular focus on overcoming problems caused by blade vibrations/oscillations due to wind effects and blade construction materials.

2.1 Development of the Winspector Robot

The prototype was developed for the middle range of the Gamesa family of wind turbines, G9x series. The design was optimized to be able to develop three sizes of this robot, targeting smaller or bigger wind turbines with respect to the G9x, with a tower height ranging from 50 m to 150 m.

A combination of climbing platform, ladder, robot arm and end-effector deploys the shearography equipment at a desired position along the blade chord while maintaining a steady distance of the shearography unit from the blade (despite blade movement due to wind gusts, etc.) and always at a normal orientation to the blade surface, see Fig. 2. The Winspector robot is designed to position the shearography system on both sides of a wind turbine blade while the wind turbine blade is stopped at a pitch angle of 90°. The end-effector, comprising of the Shearography Unit Holder and laser/optical unit is attached to the wind turbine blade surface during inspection.

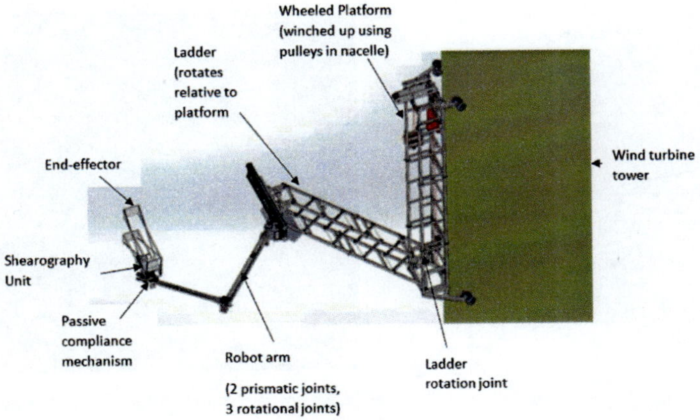

Fig. 1. Winspector robot

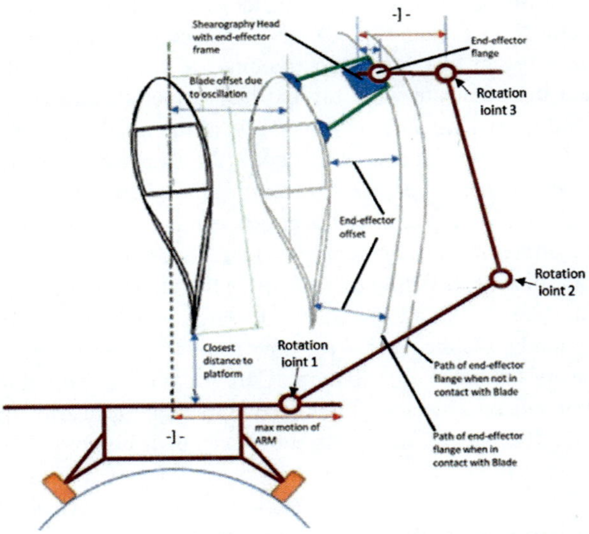

Fig. 2. Winspector schematic, platform, robot arm, end-effector with blade stopped at 90° pitch

This paper focusses on design of the end-effector and compliance mechanisms which are critical to enabling the shearography to detect blade flaws when the blade is vibrating in three dimensions.

2.2 The Winspector Shearography System

On-site shearography, a high-resolution inspection technique that can identify sub-surface defects, offers early detection of areas of risk to prevent costly blade failure incidents. A high-quality inspection carried out in-situ by examining up to 100 m^2/h,

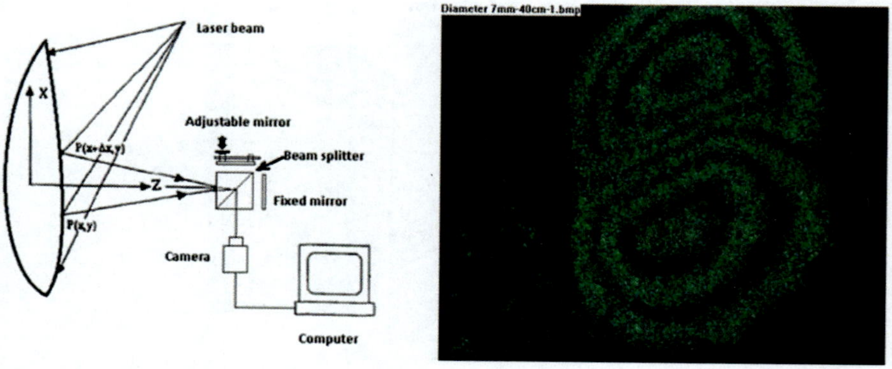

Fig. 3. Shearography system with a typical bull's eye fringe pattern indicating a subsurface defect

will reduce downtime by 25% and eliminate the need of other thorough techniques which require the dismantling of the blades and transfer to workshops.

The Winspector NDT system (Fig. 3), developed by project partner TWI Ltd (UK), is a shearography system that can inspect composite wind turbine blades. The critical defects found in composite wind turbine blades include cracks, voids, disbonds, delaminations, and impact damage [11]. As a rule of thumb, the minimal characteristic dimension of a defect is equivalent to its depth (the distance from the surface of the material). For a volumetric defect such as a void, the characteristic dimension is the diameter of the defect. For a planar or linear defect such as a delamination or a crack, the characteristic dimension is much complex to quantify since the effect of the defect on the surface deformation is highly dependent on the orientation of the defect.

The technique measures and qualifies both large and small complex geometries quickly and efficiently, yielding near real-time results. A thermal pulse is applied to the inspected part of the blade, distribution of the stress on the tested area is recorded using a shearing image interferometer, included in the digital shearography camera. By analyzing the derivatives of the stress distribution, any discontinuities are observed, thus every critical defect is revealed.

The single shearography camera is capable of detecting surface and sub-surface anomalies as small as 3 nm and scan the blade with an efficiency of up to 130 m^2 per hour. A wind turbine can be inspected within a day on average.

2.3 Design of End-Effector to Enable Shearography in the Presence of 3-D Blade Vibration

Shearography offers the advantage of non-contact inspection with the Shearography Unit (SU) deployed at one meter from the blade. However, vibration of the blade caused by wind creates a relative motion between the blade and the shearography system which renders the inspection data invalid. To overcome this motion of the blade, an end-effector was developed to place and firmly attach the shearography system to the blade with sufficient passive compliance in the system to enable the SU and blade to move together.

Blade oscillation data was gathered from a wind turbine representative of the targeted market, to design and dimension the whole robotic system.

The end-effector is composed of the following main components:

- Shearography Unit Holder holds the shearography unit and uses suction cups to attach to the surface of the blade.
- A mechanism to keep the Shearography Unit Holder connected to the rest of the robot climbing up the wind tower for the duration of the inspection session.
- A passive compliance mechanism between the Shearography Unit Holder and the robot arm that allows the Shearography Unit Holder and hence the Shearography Unit to remain stationary with respect to the vibrating blade.
- A control and communication system that ensures the correct sequence of steps are followed for inspection.

For the design and implementation of the mechanical and control system of the end-effector, the oscillation of a blade was characterized which would be representative of the targeted population of wind turbines and under inspection conditions. This characterization enables the design and dimensioning of the Shearography Holder Unit (SUH) and defines the specifications of the control system for the end-effector as well as the manipulator of the robot climbing up the wind tower.

Characterization of the Oscillation on a Gamesa G90 (2 MW) Wind Turbine

The amplitude and frequency of the oscillation of large onshore wind turbine blades was measured directly on the surface of the blade and under maintenance/inspection conditions. The deflection of the blade was measured at three different radii (distance from the nacelle along the surface of the blade). The chosen wind turbine was a G90 (2 MW) located in Barchín del Hoyo (Spain) wind farm.

The conditions of the test were representative of an inspection session, with the blade stationary at the four possible pitch angles used in a G90 blade, test lasting for about four hours, wind speed at the farm was between 1 and 6 m/s, being within the 10 m/s maximum wind speed allowed for safety reasons for inspection and maintenance work on site.

Real-time data on the oscillation of the blade was obtained using three IMU sensors (NGIMU, X-IO Technologies) placed on the upper shell near the root of the blade (radius R = 3000 mm), near the middle of the blade (R = 22500 mm), and near the tip of the blade (R = 44500 mm).

Deflection of the Blade

From the collected data of the variation of the pitch, roll and yaw angles of the IMU sensor, the deflection of the blade at the three selected radii was calculated. The pitch, roll and yaw angles of the IMU sensor correspond to the three axes of rotation of the sensor:

- X axis, which corresponds to the axis from the tip to the root along the length of the blade.

- Y axis, which corresponds to the out of plane axis, that is, the movement of the blade towards and away from the wind tower.
- Z axis, which corresponds to the in-plane axis, that is, the movement of the blade along the plane defined by the surface of the blade.

Figure 4 shows the data on deflection for the three radii, the four pitch angles of the blade and the entire duration of the test. The out of plane and in-plane deflection are of a comparable magnitude while the deflection along the X axis of the sensor is much smaller – this is the case for the measurement taken near the middle and the tip of the blade; near the root the deflection along the three axes are of a comparable magnitude (see Fig. 4). The maximum deflection calculated from the measurements on a G90 blade is 2.6 mm near the root, 9.6 mm near the middle, and 94.3 mm near the tip. Therefore, the maximum deflection caused by the wind to be expected on a G90 blade un-der inspection conditions is approximately 100 mm.

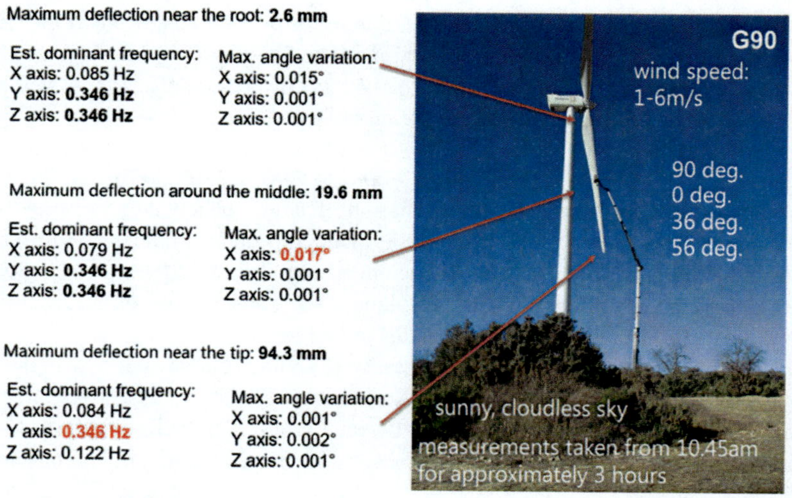

Fig. 4. Deflection of blade near the root, middle and tip of G90 blade calculated from experimental data for pitch angles of (90°, 0°, 36°, 56°), around the three directions of rotation.

The estimated dominant frequency, calculated from experimental data on a G90 blade for each axis and each radius of the blade is summarized in Table 1, the dominant frequency of blade under inspection conditions is approximately 0.35 Hz.

Angle Variation

The angle variation measured by the IMU sensors determines the misalignment to be expected between the blade and the surface of the cups while the robotic manipulator attempts to press the cups on to the blade. This angle variation is measured directly by the IMU sensors at each radius along the length of the blade. Table 2 summarizes the maximum angle variation detected at each radius, for each axis of rotation.

Table 1. Estimated dominant frequency calculated from experimental data on a G90 blade for each of the three radii considered and each axis of rotation.

Axis	Position along the blade		
	Root (R = 3000 mm)	Middle (R = 22,500 mm)	Tip (R = 44,500 mm)
X (tip to root)	0.085 Hz	0.029 Hz	0.084 Hz
Y (out of plane)	0.346 Hz	0.346 Hz	0.346 Hz
Z(in plane)	0.346 Hz	0.346 Hz	0.122 Hz

Table 2. Maximum angle variation directly measured for each radius considered on a G90 blade and each direction of rotation.

Axis	Position along the blade		
	Root (R = 3000 mm)	Middle (R = 22,500 mm)	Tip (R = 44,500 mm)
X (tip to root)	0.015°	0.017°	0.001°
Y (out of plane)	0.001°	0.001°	0.002°
Z (in plane)	0.001°	0.001°	0.001°

The maximum angle variation measured by the sensors at each location along the blade is one hundredth of a degree. This means that the end-effector looking at the area of inspection (a circle of 0.5 m diameter) will hardly notice any change of orientation of the blade due to the wind. The end-effector will see the area of inspection getting closer and away and moving from left to right by as much as 100 mm as previously shown.

Curvature of Gamesa's G90 Blade

Data provided by Gamesa for the G90 wind turbine blade was used to determine whether the footprint of the Shearography Holder Unit (SUH) (the area between the three pairs of suction cups) will be small enough to allow the suction cups to attach to the blade surface along the length of a blade (from its root to tip) and between the leading edge and the trailing edge of the blade. Figure 5 shows how the blade thickness is measured. Defining the longitudinal blade curvature as the radius of the blade surface from the blade root to its tip and the transversal curvature as the radius of the blade from its leading edge to its trailing edge. The transversal curvature is the radius of the surface across the chord. The radius changes depending on blade thickness, with blade thickness being higher at the root of a blade (small radius) and becoming thinner towards the tip (large radius). If the suction cups are too far apart, either longitudinally or transversally, they might not be able to attach to some areas of the blade as the compliance on the cups might not be able to adapt to the relative change of curvature between cups.

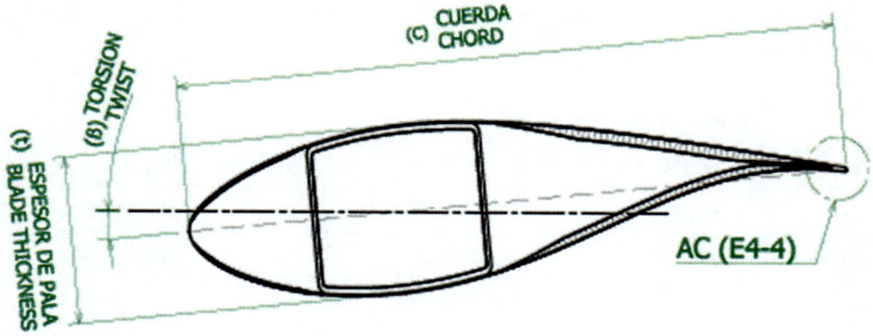

Fig. 5. Profile of Gamesa G90 blade. Note that the distance along the cross section of the blade is called its chord. The torsion line divides the profile in two sides, top and bottom.

3 Design of the End-Effector to Stabilize Shearography

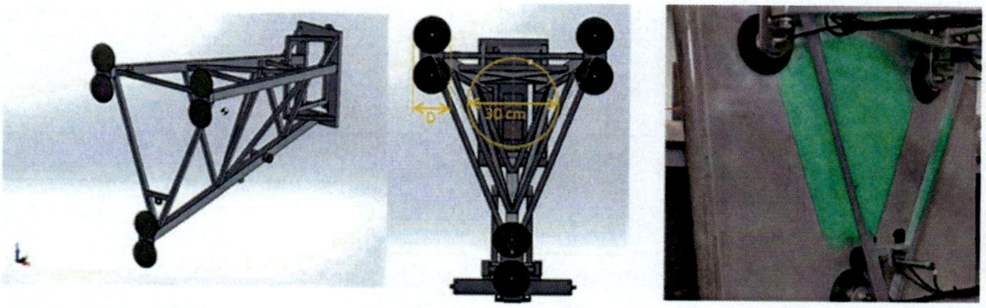

Fig. 6. (Left) Frame design of end-effector with arrangement of suction cups. (Middle) Front view shows the circle limiting the inspection area. (Right) Shearography laser imaging area.

Figure 6 shows the design of the end-effector. The pyramid-like end-effector structure is attached to a wind blade surface with suction cups located at the three vertices of a triangle. The shearography unit is placed at the top end of the pyramid so that its laser projection area is between the three triangle vertices and its optics have clear access to a measurement area within a circle of 30 cm diameter with its center at the optics focal point, as shown in Fig. 6, middle image. At each attachment vertex two suction cups aligned along the vertical, z axis, establish the connection between the end-effector and the wind blade. As the curvature of the blade is mainly along the chord (x axis), with vertical alignment of the suction cups we minimize its footprint along x direction and maximize the area of the blade we can inspect. The suction cups above the center of gravity of the end-effector will be in traction and the ones bellow will be in compression, meaning that more suction pressure is required at the upper part. To place the end-effector in the most stable configuration to withstand its own weight, the triangular base has been oriented pointing downwards.

The three-point contact configuration improves the stability and the rigidity of the end-effector when it is subjected to crosswinds or other side forces. The connection used in the two upper vertices has one-degree of freedom – z rotation, allowing for the suction cups to locally adjust to the blade curvature. This has the advantage that it is easier to attach the end-effector to the wind blade.

The end-effector structure was analyzed with the Finite Element Method (FEM) to perform: 1) Static Load Analysis; and 2) Modal Analysis. Suction cups and shearography unit were modelled with simplified versions that retain the most relevant geometrical dimensions and weight of the original components. Figure 7 shows the end-effector geometry, as well as the connections and boundary conditions for FEM. Fixed support condition applied to the suction cups at the bottom. Free z-rotation condition applied to the two connections between the frame and the suction cups.

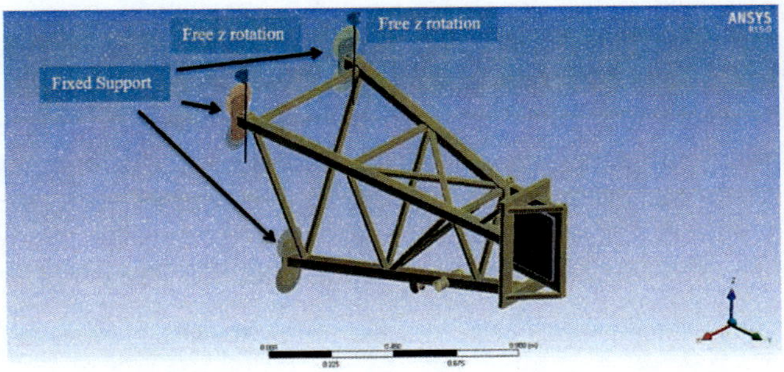

Fig. 7. Boundary conditions and connections for FEM analysis

Results from the FEM analysis showed the deformation in the x direction of the end-effector when submitted to 30N force in the x and y direction at the connection point with the compliant base. Additionally, also considered was the effect of the end-effector weight (190N). The maximum absolute value of the deformation is 0.079 mm.

The maximum absolute value of the deformation in the y direction is 0.021 mm. According to these results the maximum deformations are observed for the x (maximum of 0.079 mm), and z (maximum of 0.078 mm) directions, with the y direction showing the smallest deformation, with a maximum of 0.012 mm. The total deformation accounts for a maximum of 0.1 mm. In all these cases the maximum deformation falls below the maximum acceptable of 0.5 mm, meaning that the end-effector is rigid enough to cope with forces in all three direction.

Modal Analysis

Considering that the blade is an oscillatory body and that the end-effector will be attached to it, it is important to ensure that the last is not excited at its natural/resonant frequency. If such a scenario would occur, not only would the measurements be com-promised but also the integrity of the whole system, which could raise some safety issues. It is therefore important to estimate the resonant modes associated to the end-effector and ensure that these are far from the oscillations imposed by the wind blade. Additionally, evaluating a structure based on its resonant frequencies can be very helpful to optimize the relation between rigidity and weight. To evaluate the resonant frequencies of the end-effector a modal analysis was performed. The characterization of the oscillations of a wind blade type G90 determined that the dominant frequencies are between 0.085 Hz and 0.346 Hz - for wind conditions between 1 and 6 m/s with 10 m/s being the maximum allowed for inspection. Considering that lowest resonant frequency for the end-effector is 58.77 Hz and the highest dominant frequency of excitation is below 0.5 Hz, the oscillations are not expected to compromise the end-effector performance.

End-Effector Compliance Support

Figure 8 shows the end-effector and a detailed view from the top of the compliant connection which consists of two sliding platforms, one aligned with the blade in-plane direction and the other with the out-plane direction, which are centered by a set of springs shown in Fig. 8. These extension springs are pre-loaded by adjusting a set of screws attaching them to the sliding platform. When extended, within the expected oscillation limits (3 cm), the force of the spring applied to the end-effector should be around 30 N, as considered in the static analysis. The fact that the force of the spring must be limited implies that an auxiliary method is needed to guarantee a steady position of the end-effector during the deployment phase. To this end, a compressed air braking system is built into each of the sliding parts to hold the end-effector steady during its deployment. The air brakes are released after the end-effector is attached to the blade.

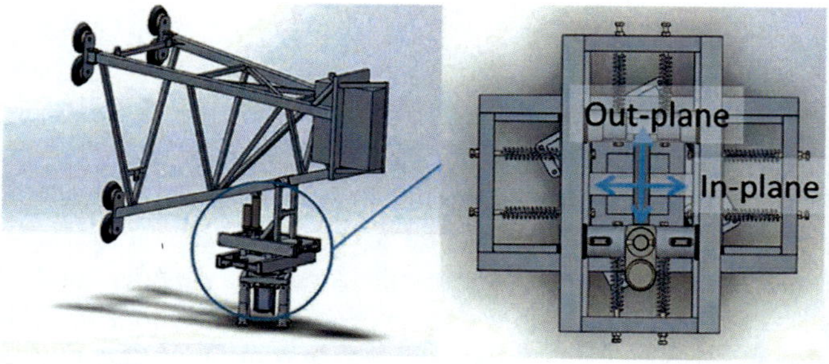

Fig. 8. Compliant connection of two sliding platforms showing the sliding platforms for the in-plane and out-plane oscillations. Springs in the sliding platforms center the end-effector.

Control, Power and Communication System for the End-Effector

A control system was implemented to orientate the Shearography Holder Unit (SUH) to match the inclination of the blade surface. Distance sensors on the end-effector calculate the pan and tilt angles required to align the end-effector to the blade at an inspection site. The five-axis robot arm performs the required orientation before presenting the end-effector suction cups to the blade for attachment. A passive compliance mechanism between the end-effector and the robot arm ensures that the end-effector remains attached to the blade despite the maximum measured amplitude of blade vibrations.

The state machine of overall control system of the end-effector is described in Fig. 9. An error-handling state (not shown) manages all contingencies that might happen during the control sequence. This state is accessible from and can communicate with all states of the machine as well as the ground station via Ethernet.

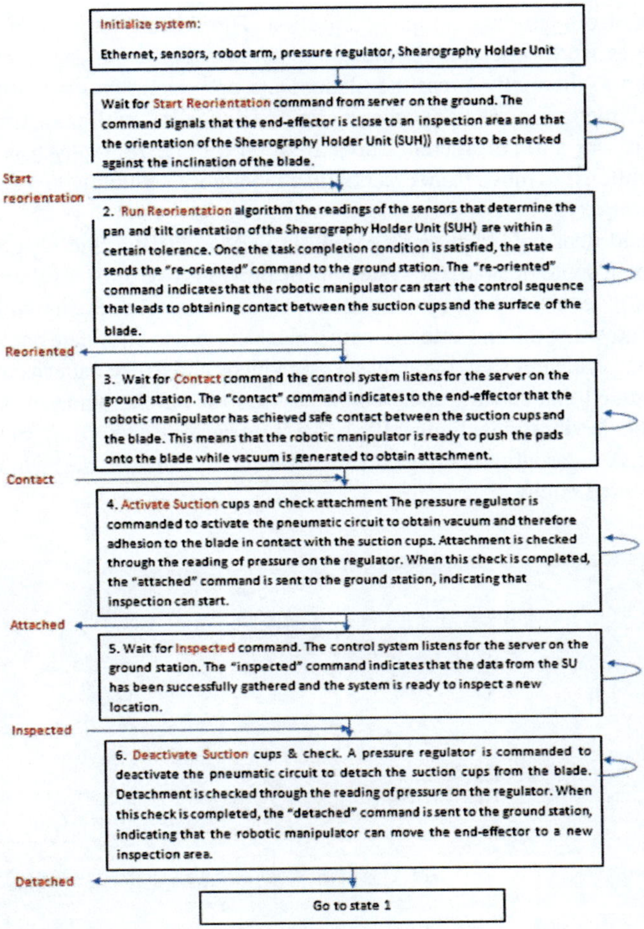

Fig. 9. State chart to control the Winspector system

4 Field Trials in Greece

Capabilities of WInspector were first tested in laboratory environments. Three separate field trials were conducted at the Centre for Renewable Energy Systems (CRES) wind farm near Athens, Greece. At the first trial, a problem identified was related to the attachment of the climber to the tower with the magnet system on the platform scratching the tower paintwork and not providing sufficient adhesion to remain attached. It was difficult to manhandle the system to bring it to the base of a tower and put it on to the tower.

The magnet system was modified by using stronger permanent magnets and increasing the air gap between the magnets and tower surface. A second field trial was conducted with no problems attaching the platform and winching it up the tower. Adverse weather conditions with wind speeds exceeding 8 m/s at times gave very limited time to collect data from the shearography inspection. Three 100-s-long video clips of laser speckle images were collected but there was some doubt about the capability of the heating set-up to thermally stress the blade. Data processing was performed by careful image subtraction to see some shearography fringes but with poor fringe contrast. However, this was still considered a breakthrough for shearography (operated in bad weather conditions) to move from essentially a laboratory technique to a field technique operating in the sky.

A third and final field trial was carried out in May 2019 at the CRES wind farm. Experience with deployment of the system in the previous two trials resulted in faster and more assured placement of the system on the tower. Thermal stressing with a powerful heat gun and covering the end-effector with a black curtain to eliminate background daylight interference, clearer shearography speckle videos and fringe patterns were obtained that demonstrated that shearography could be used for on-site blade inspection in the presence of blade vibrations. Figure 10 shows Winspector under-going trials on a wind turbine blade. A shearography image obtained on site is shown on the right with a defect marked with a red circle.

Fig. 10. Field trials. Manhandling of the platform into place at the bottom of the tower; winching it up; end-effector attached to the blade to perform shearography; RHS: a fringe pattern indicating defect (circled red) found in the blade.

5 Conclusions

Winspector is an innovative remotely controlled robotic NDT system for on-site WTB inspection. Three field trial demonstrations at the CRES wind farm in Greece have shown that shearography stabilized with the end-effector can overcome the effect of blade vibrations that would otherwise result in no useful shearography data. To our knowledge, this is the first time that shearography is proved to be working remotely on a wind tower blade.

Acknowledgements. The Winspector project was funded by the European H2020 FTI program, 1/09/2015, Reference: 700986, with lead partner WRS Marine (Netherlands), and partners IKnowHow (formerly Innora) (Greece), TWI Ltd (UK), London South Bank University (UK), Simens Gamesa (Spain).

References

1. Wind in Power: 2014 European statistics (report), EWEA (2015)
2. Global Eind Energy Council (GWEC) (2014)
3. 2014 JRC wind status report (2014)
4. Clark, D.: Cube Special Report: Breaking blades: Global trends in wind turbine downtime events (2014). http://www.gcube-insurance.com/reports/breaking-blades/
5. Summary of wind turbine accident data to 31 May 2015 – Caithness windfarm information forum (2015)
6. Sattar, T.P., et al.: Climbing robot to perform radiography of wind blades. In: Chugo, D., Tokhi, M.O., Silva, M.F., Nakamura, T., Goher, K. (eds.) Robotics for Sustainable Future, CLAWAR 2021. Lecture Notes in Networks and Systems, vol. 324, pp. 165–176. Springer, Cham (2022). https://doi.org/10.1007/978-3-030-86294-7_15
7. Sahbel, A., Abbas, A., Sattar, T.P.: System design and implementation of wall climbing robot for wind turbine blade inspection. In: ITCE 2019, Aswan, Egypt, pp. 242–247. IEEE Xplore Digital library (2019)
8. Li, Z., Tokhi, M.O., Zhao, Z., Gao, J., Zheng, H.: A compact laser shearography system integrated with robotic climber for on-site inspection of wind turbine blades. In: Proceedings of the CLAWAR2020, Moscow, p. 212–219 (2020)
9. Li, Z., Tokhi, M.O., Zhao, Z., Zheng, H.: A compact laser shearography system for on-site robotic inspection of wind turbine blades. J. Artif. Intell. Technol. **1**(3), 166–173 (2021)
10. Li, Z., Tokhi, M.O., Marks, R., Zheng, H., Zhao, Z.: Dynamic wind turbine blade inspection using micro-polarization spatial phase shift digital shearography. Appl. Sci. **11**, 10700 (2021)
11. Liu, Z.W., Gao, J.X., et al.: NDT capability of digital shearography for different materials. Opt. Lasers Eng. **49**(12), 1462–1469 (2011)

Robots in Education and Educational Robotics

Machine Learning Application with AI-Robotics Tool - What We Learned from Pilot Studies

Amy Eguchi(✉)

University of California San Diego, La Jolla 92093, CA, USA
a2eguchi@ucsd.edu

Abstract. The paper presents pilot studies exploring students' understanding of machine learning with an AI-Robotics tool. This is a follow-up to the "Promoting AI Literacy through AI-Robotics" [1] presented at the SITE 2021 conference. Artificial Intelligence (AI) is influencing practically every aspect of our lives as it has rapidly been developed to increase its complexities. Children are not an exception. What is specifically urgent is to prepare primary and secondary students for their future professions, which might not currently exist, and become citizens capable of understanding and utilizing AI-enhanced technologies in ways that do not benefit some populations over others. There have been several AI-tools focusing on teaching students about AI developed and used in education settings. However, how deep their understanding of AI becomes one of the remaining questions. This paper explores and reports how using Teachable Machine, one of the AI-tools, with a robotic tool deepens the understanding of AI among students. Two pilot studies using Teachable Machine and AI-Robotics tool are explained and how the process that a group of undergraduate students went through in developing AI-powered educational robotics lessons helped them deepen their own understanding of machine learning and its application in physical computing is reported.

Keywords: Machine learning · AI-powered robotics · Artificial Intelligence literacy

1 Introduction

A recent Washington Post's story about a Google employee named Blake Lemoine claiming that one of Google's Artificial Intelligence (AI) language models, LaMDA (Language Models for Dialogue Applications) [2], has become sentient started a public discourse about AI's future capabilities. The influences that AI has on our everyday life is growing rapidly. AI has become a topic of our everyday conversations. Machine learning models have grown their complexity and improved their capabilities to the level that even AI developers have difficulty understanding them in full. People are now heavily influenced by chatbots, recommender systems, and other AI-empowered technologies with their decisions on various aspects of their lives including purchases of products, services, and which contents to watch/listen to/read while there are significant ethical concerns raised by various experts [3].

© The Author(s), under exclusive license to Springer Nature Switzerland AG 2023
J. M. Cascalho et al. (Eds.): CLAWAR 2022, LNNS 530, pp. 383–394, 2023.
https://doi.org/10.1007/978-3-031-15226-9_36

The One Hundred Year Study on Artificial Intelligence (AI100) which was launched in the fall of 2014 provides a long-term investigation of the field of Artificial Intelligence (AI), including its influences on people, their communities, and society (https://ai100.stanford.edu/). To accomplish its aim, a Study Panel is formed every five years to evaluate the current state of AI. The 2021 Report, put together by the Study Panel composed of a diverse set of experts in AI, from academia to industry research laboratories, representing various fields such as computer science, engineering, law, political science, policy, sociology, and economics, highlights the importance of AI approaches that augment human capabilities, by seamless human-machine collaboration and cooperation which enables the decision making aligned with dynamic and complex human values and preferences [3]. It emphasizes that "an AI system might be better at synthesizing available data and making decisions in well-characterized parts of a problem, while a human may be better at understanding the implications of the data" (p. 9). The report highlights the importance of involving all stakeholders in the development of AI assistance that cooperates and collaborates with humans to produce a "human-AI team that outperforms either alone" (p. 9). It is expected that any routine and/or potentially dangerous tasks for humans to carry out will be taken over by AI systems. Such applications are possible and effective only when the collection and application of appropriate and inclusive data with the effective integration of algorithms into broader socio-technical systems are achieved.

In order to accomplish the goal where humans and AI technologies effectively collaborate to enhance each other's capabilities, we need to promote humans' understanding of AI. There is a broader understanding among politicians, industries, and educators across fields from computer science, AI, to education that it is crucial to promote AI literacy among primary and secondary students. AI literacy includes the understanding of the science behind AI, its limits, and potential societal impacts in our everyday lives as well as in the future. The urgent task for educators is to prepare students for their future professions which might not currently exist, and become citizens capable of understanding and utilizing AI-enhanced technologies effectively and in unharmful ways so that they will not benefit specific populations over others [4]. This is of particular importance to involve underrepresented minorities in STEM and CS, especially those of low socioeconomic status who have the potential to be left behind due to the lack of resources and opportunities to explore computing in their schools and lack of exposure to AI-technologies in their homes and communities.

Promoting the integration of AI literacy in school has become the focus of educators from around the world. Multiple initiatives have already been started. UNESCO started a global initiative focusing on equitable AI. Its Unit for Technology and Artificial Intelligence in Education reported three AI curriculum frameworks: AI Literacy: Competencies and Design Considerations, the AI4K12: K-12 AI Guidelines, and the Machine Learning Education Framework [5] representing such initiatives.

The Five Big Ideas of AI, developed in the US, is AI guidelines for students and teachers developed by the AI4K12 initiative [4]. The Five Big Ideas in AI include (Fig. 1):

1. Perception: Computers perceive the world using sensors,
2. Representation & Reasoning: Agents maintain representations of the world and use them for reasoning,
3. Learning: Computers can learn from data,

4. Natural Interaction: Intelligent agents require many kinds of knowledge to interact naturally with humans, and
5. Societal Impact: AI can impact society in both positive and negative ways.

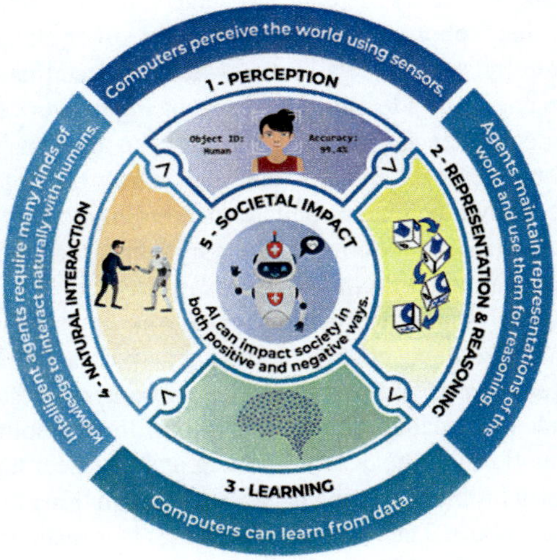

Fig. 1. AI4K12 Five Big Ideas in AI

AI for K-12 initiative's national guidelines provides a framework to assist standards and curricula, developers, on AI concepts, essential knowledge, and skills by grade band.

This paper reports the learning from two small pilot projects evaluating the learning experience of AI-powered educational robotics activities including which Big Ideas in AI could be introduced through the tool. The first pilot was with middle school students to test the AI-powered robotics tool regarding the age appropriateness of the tool. The second pilot was with a group of undergraduate students developing AI-powered educational robotics lessons introducing AI literacy, specifically machine learning and training data. It specifically focused on their experience creating training data for object recognition that they used with their robot and, as the result, it has deepened their understanding of what they intend to teach. First, it explains why we believe AI-powered educational robotics as a learning tool can promote a deeper understanding of AI. Then, it will provide an anecdotal and observational report of the undergraduate students' experience with Teachable Machine and training data.

2 Why AI-Powered Educational Robotics for AI, Computer Science, and Computational Thinking Learning?

Whenever we bring our robotics creations to classrooms, it generates excitement and creative effects among students [6]. In December 2018, the college students in the author's robotics class presented their robotics creations using recycling materials and boxes to pre-school students, where we observed the same reaction's from the students. Our visit triggered them to create their imaginary box robots as their end-of-the-year project. Similarly, in December 2019, 3rd graders (two 3rd grade classes) from a local elementary school serving underrepresented minorities visited our campus for the showcase of robotics creations by the author's undergraduate students. The 3rd graders were eager to learn about the robots created by the undergraduate students who were tasked to teach the 3rd graders how their robots were programmed by either explaining the concept or showing the code step-by-step. Since all robots were required to use at least one sensor, the 3rd graders were engaged in learning how sensor inputs work to control the robot's actions. The 3rd-grade teachers shared the feedback from the 3rd graders, which were all very positive including their excitement to continue learning about robotics and coding, and tips to make robots that the undergraduate students created better. Moreover, teachers who witnessed the effect of robotics on their students asked the author to help them start incorporating educational robotics in their lessons.

Educational robotics is an effective tool for promoting students' learning [6–8]. It becomes effective when students engage in making activities with the hands-on robotic tool, creating a fun, engaging hands-on learning environment for students. Making with robotics engages students in manipulating, assembling, disassembling, and reassembling materials while going through the design learning and engineering iteration processes. Students also engage in problem-solving processes through trial and error while improving the codes for their robots. The learning approaches that make learning with robotics successful are project-based, problem-based, learning by design, student-centered, and constructionist learning approaches where the focus is on the process of learning, rather than the final product [7].

Like robotics, when students are engaged in activities using AI technology, they become curious and wonder how it works. AI provides students with excitement and more learning opportunities to explore the new technology. It is necessary to provide AI literacy lessons to *all* students. To reach all students, it is crucial to incorporate AI learning early. To help the early introduction of AI, there are several tools available for young students. For example, *AI for Oceans* [9], designed for grades 3 and up, introduces AI, machine learning, training data, and AI bias. It aligns with some concepts of the United States' Computer Science Teachers Association Computer Science Standards (Data and Analysis and Impacts of Computing) as well as Five Big Ideas in AI (representation and reasoning and learning). There are also AI-enhanced online tools and resources that introduce AI which is fun to use. For example, there are several AI-tools accessible to K-5, including Scratch Lab's Face Sensing [10], Scratch extensions (i.e. text to speech, translate, video sensing, face sensing), and Machine Learning for Kids [11], and Cognimates [12].

Although there are several AI-enhanced online tools that support students' learning of AI, computer science, and computational thinking concepts, the concepts remain rather abstract for students since their work stays inside of the computer (virtual and abstract). Students, especially younger ones, tend to struggle to fully grasp abstract concepts and ideas without applying them in real-life situations using manipulatives. Seymour Papert's constructionism theory as its foundation, the learning environment that robotics tools create promotes students' learning of abstract concepts and ideas [7,

13–18]. Incorporating AI-powered robotics tools as manipulatives could make abstract AI, computer science, and computational thinking concepts and their ideas visible while students apply the concepts in real-life situations. Since the subject and concepts that AI introduces are foreign to many, making AI and its concepts visible through AI-robotics tools could help their construction of new knowledge, make their understanding of AI deeper, and help them retain the newly constructed knowledge.

3 CogBots: AI-Powered Educational Robotics Tool

Cogbots - CogBot and CogMini (Figs. 2 and 3) are open-source robotics kits developed by CogLabs in collaboration with Google and UNESCO. They are do-it-yourself-type robotics kits that spark students' creativity and imaginations. CogBots empower students through creating designs, assembling, and programming their own "thinking" robot with AI technology. It uses sustainable and readily available materials like recycled boxes and cardboards (Fig. 4), 3D printed parts and recycled smartphones. For the controller, CogBots uses an Arduino ESP32-base controller board that communicates with motors and a smartphone. ESP32 board is a low-cost, low-power microcontroller board with integrated WiFi and dual-mode Bluetooth communication capabilities, which is robust enough for the AI-robotics tool. Its motors, micro 360-degree continuous rotation servo motors, are also low-cost (US$5 each), which contributes to keeping the price of the robot affordable.

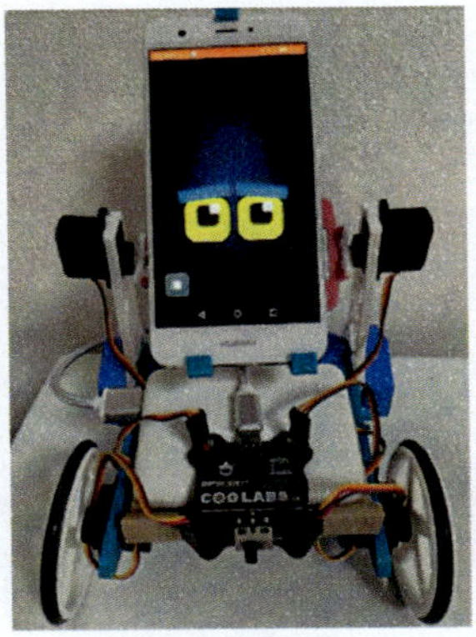

Fig. 2. CogBot

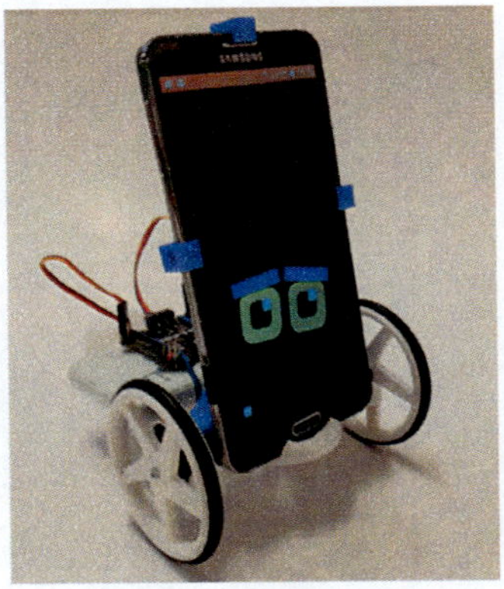

Fig. 3. CogMini

CogBot uses a smartphone as its sensor. A smartphone can be a recycled one without a cellphone data subscription (Android 5.0 onwards). It communicates with a computer via a local WIFI network and with the ESP32 board via Bluetooth. The current model works with Android phones only. However, there is a plan to make the iOS app available in the future.

Scratch is recognized as one of the most powerful, popular, and accessible block-coding tools available in many countries around the world, which makes it an ideal

Fig. 4. CogBot with cardboard and craft materials

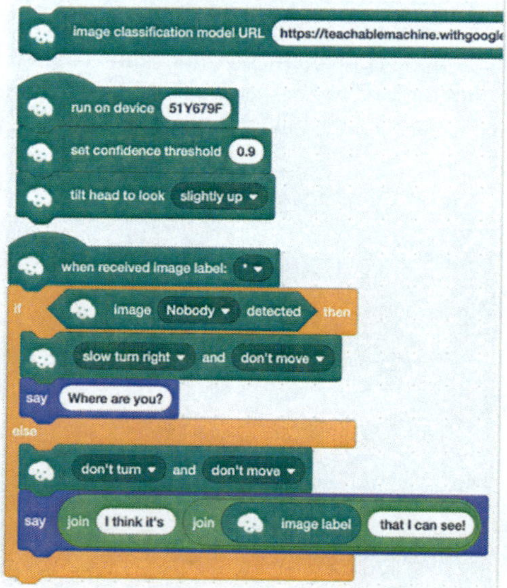

Fig. 5. ThinkBot Scratch extension

application for CogBots. Using the Scratch extension to make AI-CS experience accessible for students with little or no coding experience (Fig. 5). Scratch is beneficial because it is a web-based programming environment where no installation is required, making it easy for teachers to use in their classrooms since any installation on their classroom computers usually requires IT support. Since Scratch is widely used in classrooms around the world, it will make programming with CogBots accessible to many students who have experience coding with Scratch. Teachable Machine [19], by Google Creative Lab, is an additional tool that enables CogBot to provide a machine learning experience. ThinkBot

Scratch extension allows students to use an image classification model created with Teachable Machine in their codes to control the CogBots.

4 Pilot Study with Middle Schoolers

Since CogMini targets middle school students, the first pilot was done with middle school students to test its usability. In December 2020, a pilot study with two middle schoolers was conducted. The participants were 6th- and 8th-grade female students from a private school. The 6th grader had no text coding experience but has done some block-coding. She had some robotic experience; however, it was not enough to create her own program from the scratch. The 8th grader had some block-coding and text coding experience and could tinker with an existing program to create her own version.

The pilot study was conducted as a CogMini AI lesson, which lasted for 90 min. Prior to the CogMini AI lesson, students were asked if they knew what AI is. Both students heard about AI and had their own understanding of what it is. During the lesson, some AI and machine learning concepts were introduced to the students through a slide presentation (http://bit.ly/cogmini-1220).

4.1 Building CogMini Robot

After the lesson, the students were tasked to build a CogMini. The small screws were a little difficult to manipulate at the beginning for both. However, after the instructor showed them how to hold the motor and a wheel when screwing them together, they could manage to screw them up. Other parts were easy for them to assemble. This showed that the kit was an appropriate level of difficulty for middle school students without extensive robotics experience.

4.2 Training a Model with Teachable Machine

After the instructor demonstrated how to use the Teachable Machine's Image Project (https://teachablemachine.withgoogle.com/), the students could train a model independently. They quickly figured out that the model needs many images to make it accurately identify an object/face. The instructor showed them how the number of images influences the accuracy of the trained model (i.e. If the number of images is lower, the trained model has difficulty identifying an object accurately.) The students could integrate the trained model into the scratch program that they were presented with. Once they have succeeded, they were asked to tinker with the code to work on a project of their choice. The 9th grader continued to tinker with the original program by adding small modifications. The 6th grader was very interested in creating more complex training models and continued to play with the Teachable Machine. She figured out that it could detect different faces and differentiate between her face only, her father's face, both of their faces in one frame, and no face (four classes). Their robots could successfully identify the objects/faces that they used to train the model. It was because the robot was sitting on the same table where their laptops were used to train the model on Teachable Machine. This means that the testing data (the objects/faces) were in similar backgrounds as those of the training data. None of them explored if the background in the image frame could influence the results of the trained model regarding identifying their faces. Overall, the students spent more than 90 min going through the AI lesson and working on the CogMini project with Teachable Machine. The students' oral feedback after the lesson confirmed that they gained a basic understanding of AI and machine learning including the 1–3 Big Ideas of AI (Perception, Representation & Reasoning, and Learning).

5 Pilot Studies with Undergraduates

After the initial pilot study, the project moved to the lesson development phase. In May 2021, a group of six undergraduate students at the University of California San Diego (UCSD) joined the author to work on the development of AI-powered educational robotics lessons using CogMini. They came up with three project ideas (Gym Buddy, Sign Language Reader, Model Road) and worked on the project for about one month before going into a summer break. Due to COVID-19, they could not get together to work on the projects. Some of the students were back in their country when they join the group. The distance restricted them from accessing the CogMini robots to work on. In Fall 2021, only three students continued to work on their project (Model Road). The students were all senior students. Two students major in computer science and another major in data science. One of the students who majored in computer science has two additional minors (education studies and cognitive science). Each student received a CogMini to work with because COVID-19 continued to hinder them to gather in the lab to work on the project. Thus, they continued to work remotely on the project. Only two students out of the three are currently working on the project.

5.1 Model Road

Model Road is a robot maze project where a CogMini will be guided by different signs to go through a maze. A CogMini in a maze will be shown different signs. Through the vision/camera, the CogMini *decide*s its movement based on which sign is shown. For example, if a sign is an arrow, it will move forward. When it sees a stop sign, it will stop.

Step 1: First, the students used a "Stop" sign and no sign to control a robot. They used Teachable Machine to create training data to train a model. They could successfully train a model using their computer. Then they used the model to control a CogMini successfully. The initial learning goal of the project was to teach students how to train a model using Teachable Machine and to apply the trained model in Scratch to control the robot.
Step 2: Since the two-class control (stop or go) was successful. The next step was to add more classes using different signs. They added "Turn Right" and "Turn Left". They also decided to make the Model Road project into a game that students can play. The Model Road uses a set course that has two rooms. There will be a target (goal) in one room, which a robot can see from the other room. Students need to control their robot with the signs to navigate it through the rooms without crashing into anything. There will be obstacles in its way. They will lose some points when a robot hits a wall or

the obstacle(s). However, they struggled to train a successful model for the CogMini to identify the three different signs. They reported that the model worked fine when tested using the Teachable Machine platform on the computer. But when the model was used in the code that they developed for the CogMini to react to different signs, the model became less reliable.

Step 3: They switched to using a deck of poker playing cards, thinking that the prints on the cards were clearer and easier for a machine to identify than hand-drawn signs. They used four cards (7 diamond, 2 Club, 8 Club). However, that did not solve the problem. They reported that when the model was used with the robot, the model had difficulty identifying the cards, especially "8". They suspected that the difference between the background of the cards was the problem.

Step 4: Instead of using the numbers, they decided to identify the colors, hoping that the model would not be influenced so much by the background. They used Blue, Yellow, and Red Squares. In addition, they used a green screen background. Since the solution did not work, they added the "nothing" class when there was no card and moved the green screen to only cover 1/3 of the screen, which technically became the "nothing" class. Though it worked slightly better, the robot had difficulty identifying "blue".

Step 5: Since their thought was that the background was causing the model not to work, they added cardboard as the background when training a model. And the rooms/mazes were created using the same cardboard. They reported that it detected the background well but was confused between "blue" and "green". (The author suspected that it was the different lighting conditions causing the trouble.)

Step 6: They seemed to be stuck with the issue between the background with which the computer trained a model and the background/environment that a robot was in. The discussion focused on how to train a model with different backgrounds. They seemed to realize, then, that a machine is "sensing a whole range of different things that may contribute to a range of different outputs for no reason (from the lab report)". The suggested solution was to train a model with pictures with various backgrounds/environments, not just with one static background by moving a laptop/ smartphone around while taking pictures. They reported that this method seemed to work better. The next plan was to collect as many pictures of the cards (Uno cards instead of poker playing cards to simplify the images on the cards since they focused on using the colors) as possible. Then, they would combine all pictures into one training data to train a model to test on the robot.

Step 7: Since the previous method worked, the students plan to explore how many images it needs to train the model that works with CogMini, which will provide ideas for their lesson planning/learning outcomes for the project. They created an algorithm to extract image files from a video file. This makes it easy for them to collect as many images as possible while changing the background and images. They concluded that a model consisting of ~25 s videos with ~600 frames for training would work well. Their plan is to let students go through a similar experience that they went through while scaffolding their learning using the tips that they create including the algorithm for image extraction from a video.

5.2 Red Light, Green Light, One, Two, Three! (Inspired by the Squid Game) and More

After being satisfied with the Model Road project, they decided to create a project that could attract middle/high school students. Their idea is to create a robot that can detect a human body (either facing front or back) to play "Red light, Green light, One, Two, Three!" inspired by the Squid Game, a popular TV program. They used the algorithm for the image extraction for this project to train a posture model. At this point, they are more interested in training an almost-perfect model to use with the CogMini robot so that their focus is on exploring how to make an AI model better. At the end of the 2021–2022 academic year, they worked on a model that could identify how much water a clear plastic bottle has, which was not successful due to the transparency of water in the bottle. However, they could create a model that detects the amount of coke in a bottle (full, half, or empty).

6 Discussion and Next Step

The paper explored the idea of promoting computing education, including computer science and computational thinking skills and AI literacy through AI-powered educational robotics as a motivating learning tool. It, then, reported two pilot studies using the CogMini robot as a learning tool. The first pilot study indicated that the middle schoolers understood AI concepts (Perception, Representation & Reasoning, and Learning) and how machine learning works. However, through the pilot study with undergraduate students working on the development of AI-powered educational robotics lessons, UCSD students faced the issue that simply using a Teachable Machine on a computer to train a model using training data students collected does not necessarily work on the robot. This promoted a deeper understanding of training data and how a set of training data influence how well a model is trained. Through the project, the undergraduate students became aware of the issues that future students could face while training a model to be used with the CogMini robot. Being computer science and data science majors, the undergraduate students had an understanding of machine learning and how it works. What contributed to their struggle was their lack of experience applying machine learning using physical computing tools (CogMini robot). What contributed to their struggle was their lack of experience applying machine learning in a robotics project. They understand that the larger a training data set is, the better the model that it trains becomes. They understood that a machine looks for patterns to create a class. However, since humans can focus on the objects that they are interested in while ignoring everything else, they did not realize that what they thought they saw (either the numbers or colors) was not the only thing captured by the machine (everything else in the camera frame). The misunderstanding led to the problems that they faced while working on the project. They understand, in theory, that the model trained on the machine that Teachable Machine was running does not work on a CogMini robot because of the background. However, it took multiple tries for them to solve the problem. The months of struggles that the students ratify the author's belief that the application of AI in projects reinforces and deepens students' learning and understanding of AI. The undergraduate students have already started the discussion about how to turn their experiences and struggles into effective lessons for upper-elementary to high school students to promote AI literacy. What is interesting from the instructor's perspective is that their experience with the Teachable Machine and creating an AI model for CogMini robots triggered their interest in perfecting the AI model than coding the robot. Since the goal of the project is to promote their interest in learning AI, although their current focus is not on AI-robot programming, it can be considered to have a successful influence on the UCSD students and inspired them to continue learning about how to create an effective AI model. Thus, the educational robotics tool did inspire them to further their AI and machine learning knowledge. Based on the undergraduate students' experience through the project, we consider that the lessons will cover the Big Ideas in AI 1) Perception, 2) Representation & Reasoning, and 3) Learning, especially the issue with training data and a real-life application of AI model. We plan to integrate the Bid Idea 5) Social Impact in the final lessons. The lesson development will continue into the next academic year.

References

1. Eguchi, A.: Promoting AI literacy through AI-robotics. In: Presented at the SITE 2021 - Society for Information Technology & Teacher Education International Conference (2021)
2. Tiku, N.: The Google engineer who thinks the company's AI has come to life. The Washington Post (2022)
3. Littman, M., et al.: Gathering Strength, Gathering Storms: The One Hundred Year Study on Artificial Intelligence (AI100) 2021 Study Panel Report, Stanford University, Stanford, CA, 2021. http://ai100.stanford.edu/2021-report. Accessed 16 Sep 2021
4. Touretzky, D.S., Gardner-McCune, C., Martin, F., Seehorn, D.: Envisioning AI for K-12: what should every child know about AI? In: Presented at the AAAI 2019, Palo Alto, CA (2019)
5. UNESCO's Unit for Technology and Artificial Intelligence in Education. K-12 AI curricula - a mapping of government-endorsed AI curricula. United Nations Educational, Scientific and Cultural Organization, Paris, France (2022)
6. Eguchi, A.: Keeping students engaged in robotics creations while learning from the experience. In: Presented at the 11th International Conference on Robotics in Education (2020)
7. Eguchi, A.: Bringing robotics in classrooms. In: Khine, M.S. (ed.) Robotics in STEM Education, pp. 3–31. Springer, Cham (2017). https://doi.org/10.1007/978-3-319-57786-9_1
8. Eguchi, A.: Educational robotics for elementary school classroom. In: Proceedings of the Society for Information Technology and Education (SITE), pp. 2542–2549 (2007)
9. Code.org: AI for Oceans (2022). https://code.org/oceans. Accessed 01 May 2022
10. MI:T: Scratch Lab's Face Sensing (2022). https://lab.scratch.mit.edu/face/. Accessed 01 May 2022
11. Lane, D.: Machine Learning for Kids (2021). https://machinelearningforkids.co.uk/. Accessed 01 May 2022
12. Druga, S.: Cognimates (n.a.). https://cognimates.me/. Accessed 01 May 2022
13. Eguchi, A.: Student learning experience through CoSpace educational robotics. In: Proceedings of the Society for Information Technology & Teacher Education International Conference (2012)
14. Eguchi, A.: Educational robotics as a learning tool for promoting rich environments for active learning (REALs). In: Keengwe, J., (ed.) Handbook of Research on Educational Technology Integration and Active Learning, pp. 19–47. Information Science Reference (IGI Global), Hershey (2015)

15. Papert, S.: The Children's Machine: Rethinking School in the Age of the Computer. Basic Books, New York (1993)
16. Papert, S.: Mindstorms: Children, Computers, and Powerful Ideas, 2nd edn. Basic Books, New York (1993)
17. Bers, M.U.: Using robotic manipulatives to develop technological fluency in early childhood. In: Saracho, O.N., Spodek, B. (eds.) Contemporary Perspectives on Science and Technology in Early Childhood Education, pp. 105–125. Information Age Publishing Inc., Charlotte (2008)
18. Bers, M.U.: The TangibleK robotics program: applied computational thinking for young children. Early Child. Res. Pract. **12**(2), 20 (2010)
19. Google Creative Lab. Teachable Machine. https://teachablemachine.withgoogle.com/. Accessed 01 May 2022

Performing a Planetary Exploration Mission from the Classroom

Elsa Maria Alfonso Sanchez[1], Anestis Mavridis[2](✉), Monica Talevi[1], and Hugo Maree[1]

[1] European Space Agency, ESTEC, Keplerlaan 1, PO Box 299, 2200 AG Noordwijk, The Netherlands
Elsa.Maria.Sanchez@ext.esa.int, {Monica.Talevi, Hugo.Maree}@esa.int

[2] European Space Agency, ESEC-Galaxia, 2 rue devant les hêtres, 6890 Transinne, Belgium
Anestis.Mavridis@ext.esa.int

Abstract. The concept of a newly developed online teacher training course in robotics by the European Space Agency (ESA) Education Office is presented. The course aims at both primary and secondary school teachers, as the activities proposed are appropriate for students from 8 to 14 years old. Teaching scenario and design strategy of online activities are presented along with the space context used to wrap-up the whole activity. A separate paragraph is dedicated in presenting the activity of Remote Coding. This activity contributes to increase the interactivity of the course and gives participants the feeling of operating a real planetary exploration mission. Finally, a statistical analysis from participants' feedback is presented. Results show a continuous effort of teachers to implement new Science, Technology, Engineering and Mathematics (STEM) tools to their classes along with a clear preference on more interactive activities concerning the online content delivery method.

Keywords: Planetary · Exploration · Space · Robotics · STEM · Block-based · Programming

1 Introduction

Since the COVID-19 pandemic stroke, the ESA Education Office started implementing its new Online trainings portfolio. Whereas on online courses supporting the current school projects, such as CanSat, Astro Pi, Climate Detectives and Moon Camp, developing an online workshop on coding and robotics was prioritized, to go in line with ESA Education office's objective of enforcing youngster's technological competences, such as programming.

An exploration of the available coding online tools was done in the framework of the ESA/ESERO working groups, a collaboration established between ESA and its delegated offices in the member states, called ESERO (European Space Education Resource Office), [1].

© The Author(s), under exclusive license to Springer Nature Switzerland AG 2023
J. M. Cascalho et al. (Eds.): CLAWAR 2022, LNNS 530, pp. 395–403, 2023.
https://doi.org/10.1007/978-3-031-15226-9_37

The Open Roberta Lab environment, developed by the Fraunhofer IAIS (copyright Fraunhofer) [2], was selected as it is a free, open source, block-based programming platform that simulates the operation of many educational robots available in the market. As shown in Fig. 1, the environment of Open Roberta Lab is similar to other popular block-based programming tools like for example MIT Scratch [3].

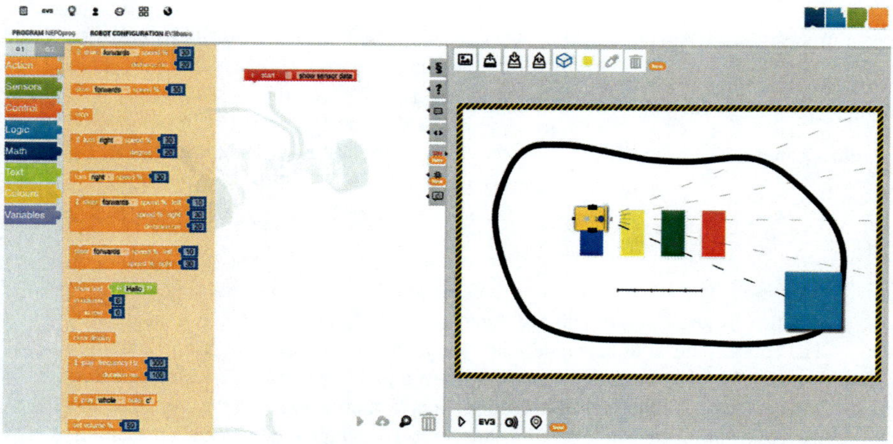

Fig. 1. The Open Roberta Lab environment.

The initial focus of the training was primary teachers, as block-based programming is considered an appropriate tool for introduction to programming for this age range [4–6]. However, the training was later extended to secondary teachers, as in some countries' curriculum coding may not start until a more advanced age.

Apart from providing a unique space context, the online course presented here includes an activity that connects the online environment of a simulator with a real educational robotics set located at ESA Education facilities. Its scope is twofold, to increase interactivity with the participants joining online and to connect the virtual environment of a programming tool with a real-world device.

2 The Online Training: From Its Origin to Its Refinement

The scope of designing this online training course was to introduce teachers to the basic concepts of robotics and coding, using the exploration of planet Mars as the context. Practically speaking, getting to know the environment of the Martian surface and a few simple commands of navigating a planetary rover was the initial target of the course.

As the course was still under development, the advantages offered by the use of the simulation window, located at the right-hand side of Open Roberta Lab environment shown in Fig. 1, led to extend the activities to include some interaction with the environment of the simulated rover. So, a number of activities using sensors were added.

The final addition to the course was a simple line follower algorithm. Although this can become a quite advanced and complex concept, our intention was to explain the basic idea of following a line using a light sensor and try to develop a working algorithm rather than a perfect one, in terms of rover's smooth movement.

2.1 Course Development from the Didactics Point of View

The training was created having in mind three aspects: Using innovative teaching methodologies, following progressive learning, and keeping an enquiry-based approach.

To achieve the first goal, challenge-based learning and game-based learning were chosen as the methodologies. In this way, the participants, who would be appointed as ESA Astronauts who just landed on Mars, would have to follow a series of challenges while navigating a rover on the Martian surface. While doing so, participants would be learning to use different features of their Mars rover, such as the ultrasonic sensor, the color sensor or the light sensor.

To keep progressive learning and enquiry in mind, the challenges are presented in order of difficulty. First challenges require the use of actuators only, rover's motors in this case, to navigate through the virtual Martian environment. Next, the use of sensors is required to take a decision on the route followed by the rover. Finally, the line follower challenge, requires the synthesis of actuators, sensors and taking decisions.

Additionally, as presented in Fig. 2, different levels of help are provided to participants in each of the challenges, so that they can use as few or as many tips as they need through their learning process.

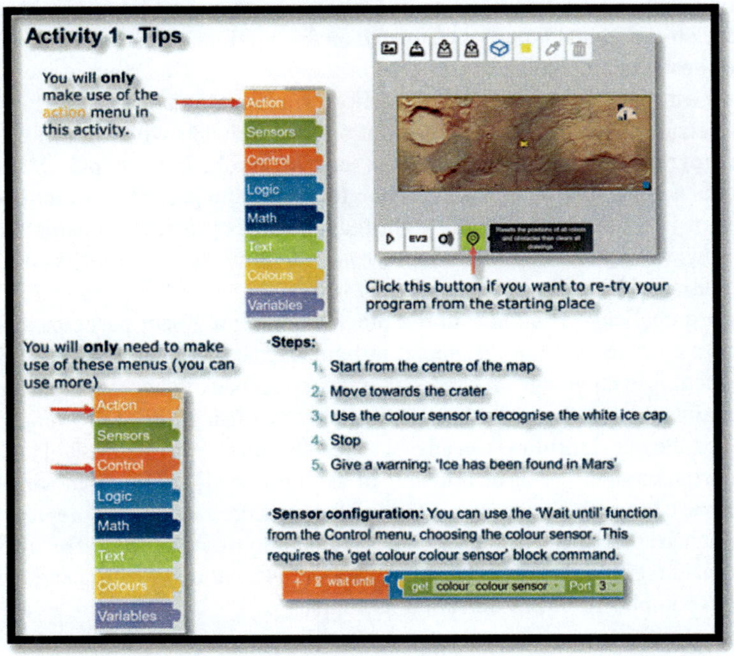

Fig. 2. Support provided for different difficulty levels.

2.2 The Course Activities

As a first step, the image of Fig. 3 from Mars Express [7] was selected as the environment for the training to take place in. The reason for its selection was the richness appreciated in one image: A valley, two craters and a mountain are visible, offering interesting exploration possibilities for participants.

Fig. 3. The "Dried out river valley network on Mars" image from Mars Express orbiter [7], was selected as the area to be explored, for the training.

The first challenge, called 'Back to home base', focuses on driving around the Mars terrain (back to a fictional home base situated on the top right corner, as shown in Fig. 3) using the most basic commands available in the platform. This way, participants get acquainted with the environment and the action programming blocks.

In the second challenge, 'detecting elevations', participants will need to make use of two new programming categories: 'Sensors' and 'Control'. In the coding blocks from those sections, they need to build their code following the procedure described in the flow chart of Fig. 4. In this way, the algorithm is always provided to participants in a readable way, so they can translate it into the corresponding coding blocks, even if they are not familiar with the flow chart tool.

The third challenge is similar to the previous one, only here participants need to find a crater by detecting the difference in light intensity, which is higher inside the crater. In the fourth challenge, they need to draw a path back to home base in the image using the paint tool, and then drive back to the home base using a line follower algorithm. Since a line follower algorithm is an advanced subject, teachers are not asked to build the algorithm from scratch. The basic principle of the simplest algorithm, of moving left and right over one edge of the line is presented and then teachers are given a preprogrammed algorithm where they must complete some numbers to make it work. The analog and Proportional-Integral-Derivative (PID) line follower algorithms are not mentioned since they are too complex to be presented in this course.

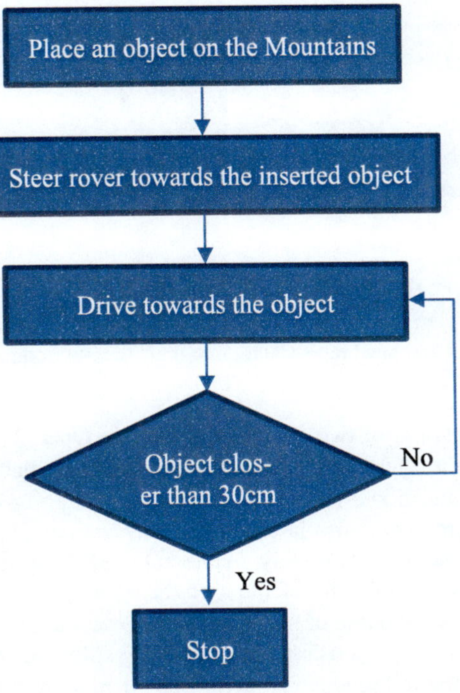

Fig. 4. Flow chart of program for detecting elevations.

3 From Simulation to Reality

Based on the number of applications received and the positive comments of the participants of the first pilot course, ESA Education Office decided to continue the delivery of the online course adding one more element to its structure. The basic idea was to find a way to make the course more interactive, giving the participants the opportunity to run their code on a real educational robot, even though participating in an online course.

To achieve this goal, the structure of the course was refined, introducing a new activity called "Remote Coding". During this activity the code developed by the participants online, is downloaded in a Lego Mindstorms EV3 [8] robot. This is performed by an ESA Education team member being present during the delivery of the course, at the e-technology lab, situated at the ESA ESEC site. The e-technology lab is ESA Education permanent teacher training facility, and it is equipped with a Martian Terrain Simulation table. So, the robot is driving on this table, executing the code developed remotely by the participants. A graphical representation of the activity's concept is presented in Fig. 5.

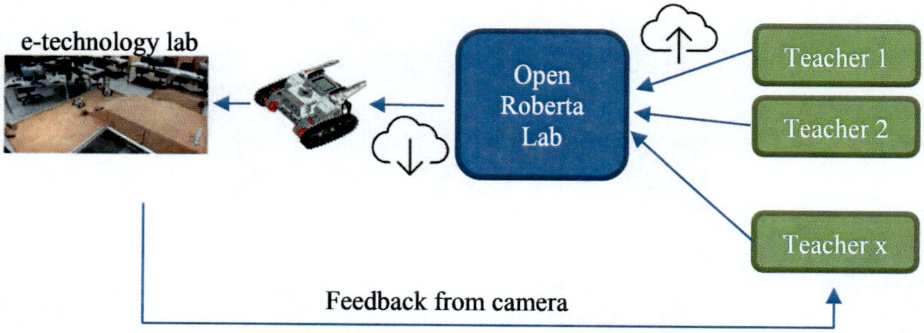

Fig. 5. A graphical representation of the "Remote Coding" activity.

The activity makes use of two features available in Open Roberta Lab. The first is the connectivity that it offers with several commercially available educational robots, including the Lego Education EV3. The second is the "user groups" option which is used to group participants in a common classroom-like environment. In this way any program created within the group is immediately visible to the ESA trainers at the e-technology lab who can then download it and execute it on the EV3 rover.

During the preparation phase of this activity, an exact model of the Martian Simulation Terrain table was designed using a 3D modelling software. A top view of the table was exported and imported to Open Roberta Lab simulation window. Both the actual table and the model inserted to Open Roberta Lab are shown in Fig. 6.

Special attention was given to the configuration of the simulated robot, in order to have exactly the same behavior as the real one. It is of critical importance that a distance covered in the simulation window corresponds to the same distance traveled by the real rover on the table. A high-definition web camera pointing at the table is used to provide feedback of the robot's actions to the participants.

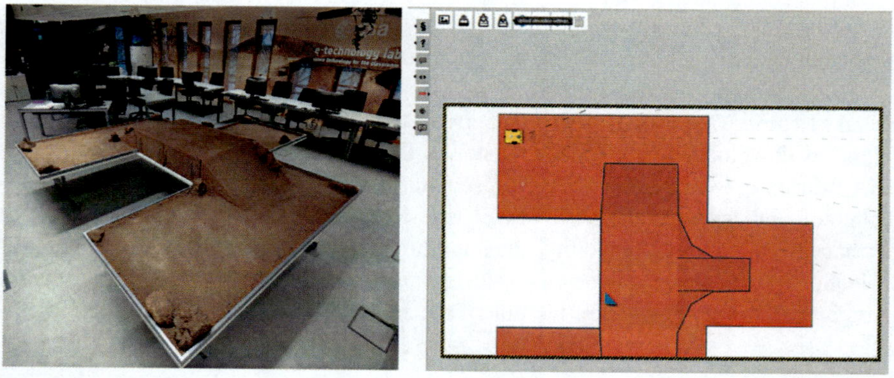

Fig. 6. A view of the e-technology lab Martian terrain table and the corresponding background image used in the Open Roberta Lab simulation window.

Two challenges were developed for this session, following the same progressive level of difficulty as the rest of the activities. During the first challenge, participants are asked to program their rover to drive from a certain point of the table to another. The second challenge is more advanced as it requires the use of a sensor to detect and approach an object placed on the surface of the table.

The new enriched version of the course has been delivered twice by the time that these lines are being written, getting very positive comments from the participants. What was mostly appreciated was the interactivity it offers. Moreover, it introduces the space context of operating a planetary exploration robotic vehicle.

4 Useful Conclusions from Participants' Feedback Analysis

Each teacher training course delivered by ESA Education is followed by a survey where all participants are kindly asked to provide their feedback. The course presented here, was delivered three times from May 2021 until April 2022. It was attended by 83 teachers in total, selected among the 345 who applied. The survey after the courses was submitted by 46 teachers.

The first worth mentioning conclusion is that the majority of participants, had already had some experience in block-based programming. This is illustrated in the graph of Fig. 7. Nevertheless, it seems that they are always exploring new tools that they could add in their lessons and implement in the classroom.

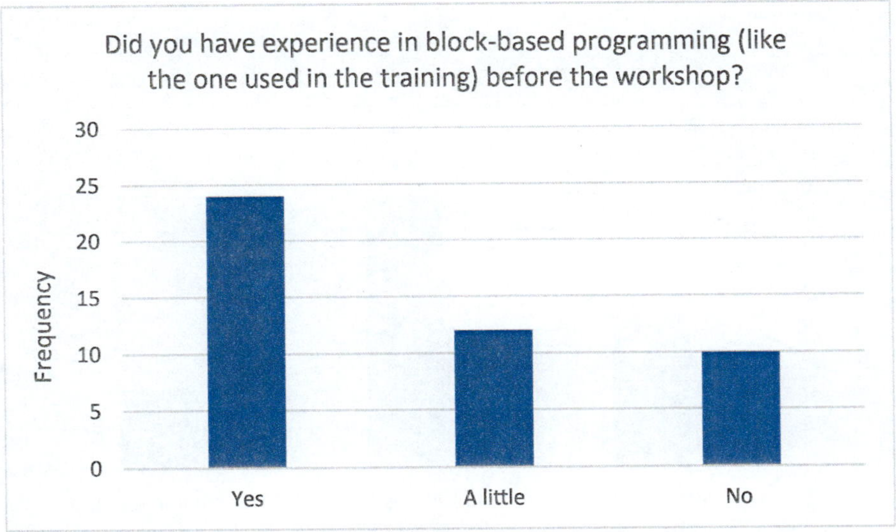

Fig. 7. Most participants had already had some experience with block-based programming.

What is most impressive is that, as presented in Fig. 8, all teachers have responded positively in the possibility of using Open Roberta Lab tool in their classroom.

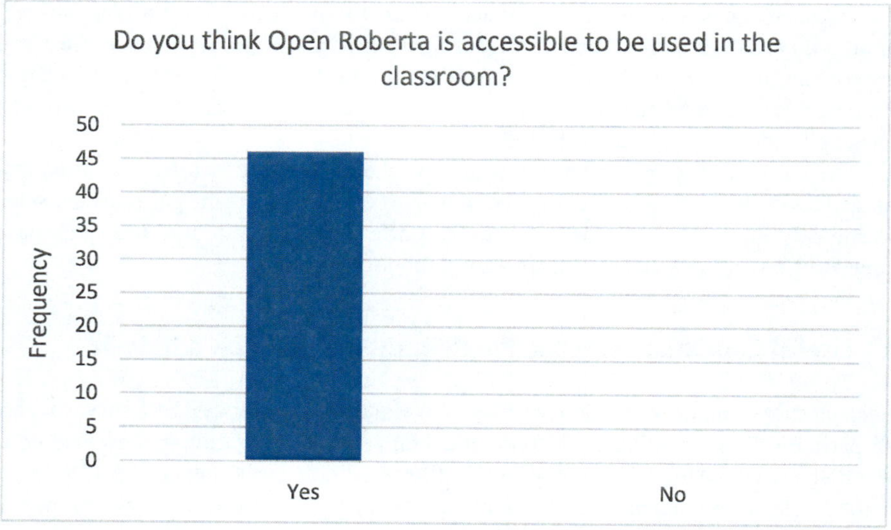

Fig. 8. 100% of participants think Open Roberta Lab can be used in the classroom.

Finally, the graph presented in Fig. 9 clearly shows that most of the teachers prefer the "Activity sessions" as a method of content delivery, in comparison to PowerPoint or discussion sessions.

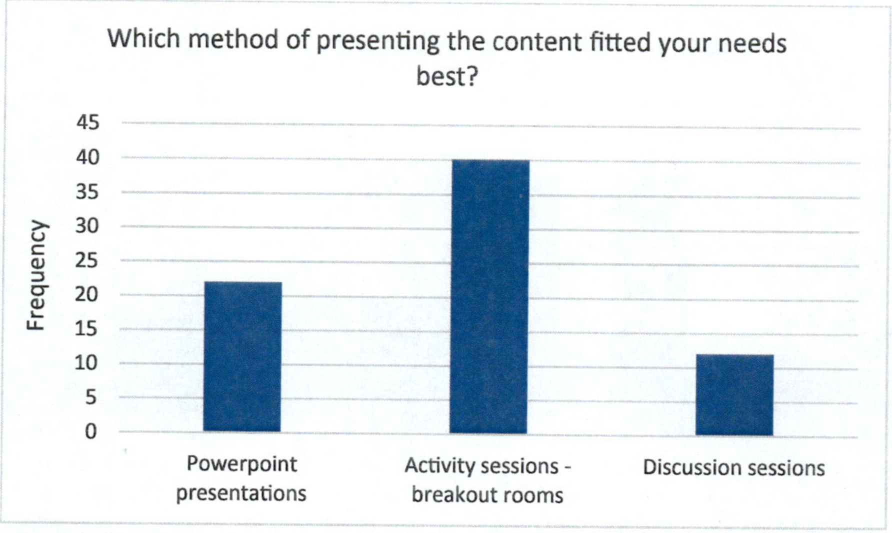

Fig. 9. Most of the participants prefer the "Activity sessions" as content delivery method.

5 Conclusions

The concept of a newly developed online robotics course for primary and secondary school teachers by the ESA Education Office was presented. The feedback received from participating teachers has been very positive even if most of them had already had some experience with block-based programming. This clearly shows that teachers are investigating new tools to implement in their STEM lessons in the classroom.

The remote coding activity that was added after the first delivery, increased the interactivity of the course. This activity is differentiating the course from a classical online course, since it gives participants the sense of performing a real mission. Feedback from participants showed a clear positive acceptance of this type of activity, compared to classical delivery methods like PowerPoint presentations or discussion sessions.

References

1. ESA, European Space Education Resource Office (ESERO) homepage. https://www.esa.int/Education/Teachers_Corner/European_Space_Education_Resource_Office. Accessed 01 Apr 2022
2. Open Roberta Lab homepage. https://lab.open-roberta.org
3. MIT scratch homepage. https://scratch.mit.edu/
4. Computing our Future, Computer Programming and Coding: Priorities, School Curricula and Initiatives Across Europe, Update 2015, European Schoolnet (2015)
5. Lindberg, R.S.N., Laine, T.H., Haaranen, L.: Gamifying programming education in K-12: a review of programming curricula in seven countries and programming games. Br. J. Educ. Technol. **50**(4), 1979–1995 (2019)
6. Bers, M.U.: Coding and computational thinking in early childhood: the impact of ScratchJr in Europe. Eur. J. STEM Educ. **3**(3) (2018). https://doi.org/10.20897/ejsteme/3868
7. ESA, Science and Exploration, Dried out river valley network on Mars. https://www.esa.int/ESA_Multimedia/Images/2019/02/Dried_out_river_valley_network_on_Mars#.YjR_TW8R4BY.link
8. LEGO Education Mindstorms EV3 set. https://education.lego.com/en-us/products/lego-mindstorms-education-ev3-core-set/5003400#lego-mindstorms-education-ev3

Learning Hands-On Electronics from Home: A Simulator for Fritzing

Andres Faiña(✉)

Robotics, Evolution and Art Lab (REAL), IT University of Copenhagen, 2300 Copenhagen, Denmark
anfv@itu.dk

Abstract. The recent pandemic has forced us to teach online, which is especially difficult for hand-on courses in robotics, like basic electronics. In this paper, I present a simulator which tries to replicate the same experience students will encounter during the exercises in the laboratory. The simulator has been developed in Fritzing, uses realistic multimeters for measuring and checks for common mistakes. Results show not only that the simulator was extremely useful during the pandemic, but also that it can supplement laboratory exercises when teaching in-classroom.

Keywords: Learning electronics · Electronics simulator · Fritzing · Educational tools

1 Introduction

Robotics is an interdisciplinary area which overlaps with the fields of mechanical engineering, electronics, and computer science. A basic understanding of these three fields is necessary to have a global understanding of robotics. However, learning to design and build electronic circuits is a difficult task for students without an engineering background. Students need to learn the basic principles of electric circuits which include electronic units, basic rules of circuits and their representation in schematics. Nevertheless, this knowledge is not enough to implement them.

To implement circuits, it is also necessary to learn practical skills that cannot be taught with just a schematic circuit. For example, a breadboard is often used to implement prototypes, and printed circuit boards (PCBs) are used for production. New electronic practitioners often struggle to build implementations of schematic circuits [1,2]. This happens as the schematic view is a symbolic representation of the circuit while circuits made on protoboards or PCBs are implementations with their own specific particularities. Moreover, debugging a circuit after implementing it on a breadboard or a PCB is usually necessary and requires knowing how to use multimeters and oscilloscopes. Therefore, implementation and debugging skills need to be learned as well.

© The Author(s), under exclusive license to Springer Nature Switzerland AG 2023
J. M. Cascalho et al. (Eds.): CLAWAR 2022, LNNS 530, pp. 404–413, 2023.
https://doi.org/10.1007/978-3-031-15226-9_38

The traditionally approach to teach electronics is to combine theoretical explanations about circuits using schematics with a practical session where students implement circuits in breadboards [3,4]. And after they become more experienced, they learn how to produce PCBs [5]. However, practical sessions are not always possible as they require (1) that the students have the electronic components and hardware tools and (2) that a teacher is available during the practical sessions.

During the current pandemic, a lot of electronic courses were taught online, and students did not have access to these practical sessions. In my case, I teach the first course of the robotic specialization, which is called "How to Make (Almost) Anything", where M.Sc. Computer Science students without engineering background learn mechanics, electronics and microcontroller programming by prototyping robots and machines. There are only four lectures about electronics, but the students should hand in a functional prototype. Thus, during the pandemic, the only solution was to use an electronic simulator to teach both the theory and the practicalities of implementing electronic circuits.

There are several electronic simulators available, but most of them focus on only one view of the circuit: the advanced simulators focus on schematics and simulators for learners use a breadboard (and wiring) view. In addition, few of them provide feedback on important mistakes: short-circuit detection, voltage, current or power exceeded, etc. Finally, debugging tools are often simplified and do not represent real tools accurately. For example, some simulators allow to measure a voltage with just one probe (the other one is assumed to be the ground terminal), which obfuscates that voltage is a relative measurement. Other simulators use specific parts for voltmeters, ammeters and ohmmeters, while in a real-life scenario it is standard to use a multimeter. Using this tool, students need to change the mode of the multimeter and connect its probes correctly to be able to measure. In order to overcome these limitations and provide a "practical" session with a simulator as close to reality as possible, I implemented a simulator in Fritzing, an open-source electronic design automation program.

2 Related Work

One of the most popular electronic simulators is SPICE [6] and its different forks such as LTspice [7] and Ngspice [8]. These simulators take the definition of an electronic circuit through a netlist and can perform different types of analyses: operating point, transient, noise, etc. They simulate electronic circuits accurately, but they are difficult to use for beginners. For example, SPICE and Ngspice do not offer a schematic view of the circuit. Ngspice can be used inside KiCad [9], a popular suite for electronic design automation, which provides a schematic editor and basic interaction with the simulator. However, it only has a very limited set of parts that can be simulated as SPICE models usually have proprietary licenses. Thus, users need to select a specific component, find its SPICE model and add it to the simulator. In contrast, LTspice comes with a large set of specific devices to simulate.

Table 1. Comparison of different electronic circuit simulators.

Simulator	View		Simulation			Real multimeters	Failure detection	Create components	Platform	License
	Breadboard	Schematic	DC	AC	Arduino					
Fritzing	x	x	x	–[a]	–	x	x	x	All[b]	GPL
TinkerCAD	x	–	x	x	x	–	Few[c]	–	Web	Closed
KiCad	–	x	–	x	–	–	–	x	All[b]	GPL
CircuitLab	–	x	x	x	–	–	–	x	Web	Closed
VirtualBreadboard	x	–	x	–	x	–	–	–	Win	Closed
SimulIDE	x[d]		x	x	x	–	–[e]	x	All[b]	GPL
Wokwi	x	–	–	–	x	–	–	x	Web	MIT
Proteus VSM	–	x	x	x	x	–	–	x	Win	Closed

[a]It will be implemented in future versions.
[b]Windows, MacOs and Linux.
[c]Polarity in electrolytic capacitors and, in some items, maximum voltage and maximum current.
[d]Mixes schematic symbols and images of Arduinos and sensors.
[e]Only for current in LEDs.

There are several simulators targeted to electronic learners such as TinkerCAD [10], SimulIDE [11] and WokWi [12]. These simulators are more oriented towards simulation of microcontrollers with limited capabilities of analog simulation. They provide a rich set of generic components, and the user can simulate them in a few clicks: drag and drop components, draw the wires and press run. However, they do not offer realistic multimeters and most of them cannot detect common failures: e.g., short-circuits or wrong polarity applied to electrolytic capacitors.

The most similar work to the simulator presented here is probably TinkerCAD, which offers some failure detections. However, the fact that it is closed source and does not allow creating new components makes it less flexible. In addition, it only provides a breadboard view, which does not encourage students to learn how to read and draw schematics.

There are plenty of electronic simulators and each of them has its own features. I have summarized the main features of the most popular ones in Table 1.

The simulator introduced in this paper is implemented inside Fritzing [13]. Fritzing is an electronic design automation software, which provides three concurrent views of the same circuit: breadboard, schematic and PCB. It is targeted to people without previous electronic knowledge, and it was developed to be easy-to use. Spice models are optional in Fritzing's part format and Fritzing can export a spice netlist. However, there were extremely few components with spice models (basically, only resistors, inductors and capacitors).

3 Simulator

The following section describes the main features of the simulator and its implementation. The simulator has extended Fritzing by adding simulation capabil-

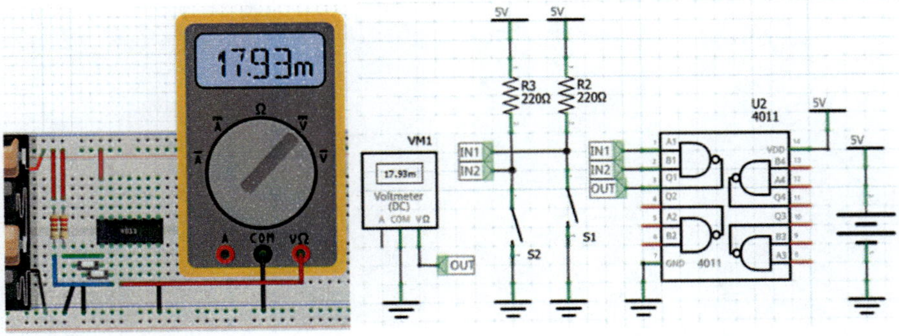

Fig. 1. The simulator allows students to use the breadboard view (left) or schematics view (right). In both views, a NAND gate (CD4011) is being tested. Two switches allow changing the input voltages and the output is measured with a multimeter. In the case shown, both inputs are 5 V and the voltage at the output is 17.93 mV.

ities.[1] Every time the simulator is triggered, a spice netlist is generated and simulated in a background thread using Ngspice [8]. Using the results of the spice simulation, the graphical user interface of Fritzing is updated.

The simulator tries to replicate what a student would observe when working with real hardware. Thus, it is not possible to measure anything without using a realistic multimeter. In this first implementation, the simulator only performs a direct current (DC) operation point analysis, which is enough for teaching basic electronic concepts. The simulator checks for failures and, if there are any, a smoke symbol is drawn on top of the damaged component.

The simulator comes disabled by default as it is still a beta feature in Fritzing, but it can be enabled in the properties menu. After that, a new button appears in the graphical user interface where the simulator can be started or stopped. Once the simulator has been started, the simulator will trigger a simulation each time that the user changes a property on one of the components (e.g., change the value of a resistor) or the wiring changes.

The main features of the simulator are described below.

Multi View. Fritzing has three different views (breadboard, schematics and PCB) that are synchronized: changes in one view are propagated to the other views. E.g., when adding a component, the component is added in all the views, and when adding a wire, ratsnests are generated in the other views. Thus, the students can use any of them to build a circuit and simulate it. This allows students to build the circuit in the view they feel more comfortable. Once a circuit has been built in a view, ratsnest lines help to wire other views. Currently,

[1] The simulator code is available at https://github.com/fritzing/fritzing-app. Fritzing code is open source, but currently official binaries cost €8 (the payments allow hiring a maintainer and developers). In addition, the simulator needs components with spice models, which are available at https://github.com/fritzing/fritzing-parts.

(a) Ohmmeter (b) Voltmeter (c) Ammeter (d) Bad wiring

Fig. 2. The realistic multimeters measuring different properties: (a) resistance, (b) voltage and (c) current. If a user connects the probes incorrectly, an error is reported. In (d), the multimeter is set to measure a resistance but the current probe ("A") is used instead of the common probe ("COM").

multimeters cannot be connected and LEDs do not light up in the PCB view. However, breadboard and schematic views are fully functional, see Fig. 1.

Multimeters. A realistic multimeter part has been designed with five different variations: a voltmeter (DC and AC), an ammeter (DC and AC), and an ohmmeter. The user can change the type of the multimeter from the user interface (inspector pane). As most real multimeters, the part has three different connectors for the probes: a common one (COM), one for measuring voltages and resistances (VΩ) and one for measuring currents (A). The user needs to connect the probes to the points to measure and change the wiring of the circuit to read currents.

Once the simulation is performed, the multimeter will show the value measured in its screen. If the probes connected to the multimeter do not match with the measuring mode of the multimeter, an error message is shown on the screen (ERR). Figure 2 shows three different multimeters measuring a resistance, a DC voltage and a DC current, and a fourth one where the probes are not connected correctly.

Failure Detection. The simulator performs several checks to analyse if any of the components of the circuit is working outside its specifications. If so, a smoke image is added on top of the component to indicate it, see Fig. 3. Currently, the simulator performs the following checks:

Fig. 3. Failure detection examples: the simulator checks for current exceeded (Led), reverse polarity (electrolytic capacitor), excess current in output pins (IR sensor), max voltage exceed (ceramic capacitor), and short-circuits (right battery).

- Short circuit in batteries and power sources
- Max current (diodes and LEDs, inductors, IO pins)
- Max power (resistors)
- Reverse voltage (tantalum and electrolytic capacitors, IR sensors)
- Max voltage (capacitors, motors, IR sensors).

Components. The parts need to have a spice model in order to simulate them. Fritzing comes with around one thousand parts, but only a few of them have spice models. Thus, parts that can and cannot be simulated can coexist in the same sketch. To avoid confusions, all the parts that cannot be simulated are greyed out when the simulator is active. In addition, the spice model is shown in the inspector pane when a part is selected. To find parts that can be simulated, a new extra bin called "Sim" has been created. In this bin, all the parts contain spice models and can be simulated.

The components represent generic versions of the most usual devices available at the laboratory, and they do not model a specific part. This simplifies the selection of components for the students, but it does not allow users to simulate circuits accurately. The spice model of a component cannot be changed through the user interface, but a spice model can use properties of the component. For example, the spice model of a potentiometer uses the maximum resistance property of the potentiometer, and this resistance can be changed by the user through the inspector pane. Advanced users can create new parts with specific spice models if they require accurate simulations.

Currently, the components that can be simulated are resistors, capacitors, inductors, batteries and power supplies, potentiometers, switches, diodes and LEDs, transistors (bipolar and MOSFET), IR sensors (analogue and digital), DC motors and NAND gates.

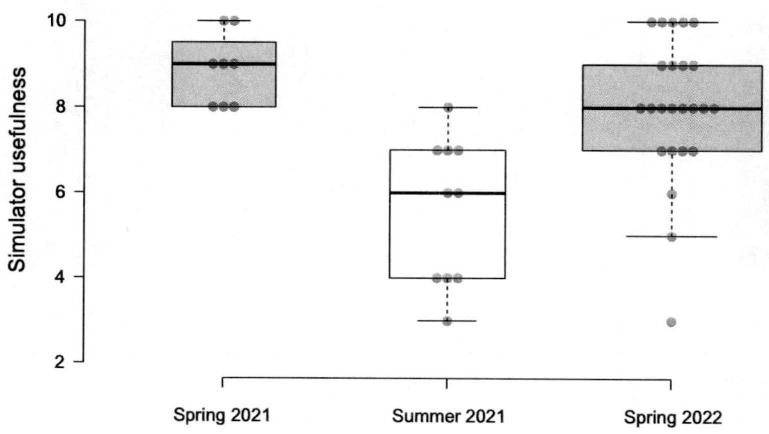

Fig. 4. Students graded the simulator usefulness from 0 to 10 during the course evaluation. Center lines show the medians; box limits indicate the 25th and 75th percentiles as determined by R software; whiskers extend 1.5 times the interquartile range from the 25th and 75th percentiles, outliers are represented by dots; data points are plotted as open circles. n = 8, 10, 24 sample points.

4 Teaching with the Simulator

The simulator has been used in three classes of the course "How to Make (Almost) Anything" at the IT University of Copenhagen in 2021 and 2022. The Spring 2021 course was taught completely online, and the Summer 2021 and Spring 2022 courses were taught physically at the university. In the 2021 courses, students used a preliminary (*alpha*) version of the simulator. The *alpha* version did not show the spice model in the user interface and did not have a simulation bin with all the parts that can be simulated. In addition, the simulation had to be triggered manually by pressing a button after any change in the circuit or in the properties of the components. In the 2022 course, students used the latest version of the simulator.

In the lectures, students learned basic electronics, saw some examples and had a brief introduction to Fritzing and its simulator. During the following exercise session, students had to work on some mandatory assignments. These assignments consisted on building circuits with switches, potentiometers, IR sensors, transistors and DC motors. The students had to hand in the circuits assembled in the breadboard view and their schematics. In courses taught at the university, students also had to build the circuits in a physical breadboard.

After each course, an anonymous and online survey about the simulator was carried out. The students were asked to answer "Was the Fritzing simulator useful?" and they could grade it from 0 to 10, where 0 meant "No, it was not useful" and 10 meant "Yes, it helped me to learn electronics". After the grading, they also could write in a text box what worked well in Fritzing and what could be improved. The survey was answered by 8 out of 30 students, 10 out of 18 and 24 out of 60 in the Spring 2021, Summer 2021 and Spring 2022 courses respectively.

Looking at the results of the survey, Fig. 4, we can say that the simulator helps them to learn electronics. However, we can observe a big difference between the Spring and Summer courses in 2021. In the Summer course survey, I found three comments from students who reported that it was not useful as they could not use it in their computers (Mac users had to use a virtual machine to run the *alpha* version), which probably explains the three lowest scores. Even discarding these three low scores, the simulator seems less useful when the students have access to practical exercises. In the Spring 2022 course, the simulator improvements have raised its usefulness, but it is still slightly below the Spring 2021 course.

I classified the answers of what could be improved in the Spring 2022 course (the one that used the latest version of the simulator). Five votes were for better support for MacOS (the Mac binary needed to install a library through the command line), four votes for more parts with simulation models, and two votes for making the brightness of the LEDs more noticeable, simulate Arduinos, and adding interactive components (quickly change the state of buttons and potentiometers). Finally, two persons also mentioned that working in both views, schematic and breadboard, at the same time could be confusing if unintentional connections are made.

In the surveys, I found some comments on the ability of the simulator to help them to learn electronics. Only one of them is negative ("*Fritzing was very helpful with creating schematics. Other than that, it was just easier to look at the real life electronics.*"). The rest are positive and can be classified in (1) better understanding of electronics (e.g., "*The simulation aspect of Fritzing was very nice. It was a good way of learning about switches and H-bridges and resistors, which can be quite abstract to learn.*" or "*I really liked how it helped me gain a good understanding about basic electronics.*"), (2) reducing stress of breaking components (e.g., "*I really enjoyed the simulation feature because it allowed me to be more confident in my circuits. I was very worried about damaging components so I liked to draw everything in Fritzing first just to make sure I wasn't gonna blow up any capacitors or the likes.*" or "*The parts that were there worked pretty well, and I liked being able to simulate circuits and getting some kind of security before actually building it in real life.*") and (3) helping to translate schematics into implementations (e.g., "*The wiring guidance when you had already done it in either schematics or breadboard.*" or "*I think the wiring and the connection between breadboard and schematic view was great.*").

5 Discussion

The simulator seems that it was extremely useful when the students did not have access to the hardware and less useful when they have access to practical sessions. This result is in line with other experiences using simulators [14].

From my personal experience, I have noticed that using Fritzing and the simulator has improved the learning of the students. Using Fritzing, they can work in three different views (breadboard, schematic and PCB). As the exercises forced the students to implement the circuits in a breadboard and draw the schematics for each exercise, they learn to draw and interpret schematics. In addition, the functionality of highlighting all the terminals that are connected together (clicking on a terminal), helps them to understand how breadboards, net labels and power symbols work.

While I think that working with hardware is necessary, the simulator has a couple of advantages. In some occasions, detecting a failure in hardware is not straight forward and it is easy to see this in the simulator. As an example, my students need to read the status of a switch. While the exercises tries to force them to use a pull-up or pull-down resistor, some of them solve this task by creating a short-circuit when the switch is pressed. While this solution is not acceptable, they cannot notice it as the short-circuit does not produce any noticeable effect except that there is no power (e.g., when taking 5 V from a voltage regulator with short-circuit protection). Instead, it is easy to see that this solution is not valid in the simulator, as they can see a clear smoke coming out of the battery. These kinds of issues also appear when a circuit is working outside of its specifications.

The feedback of the students in 2021 was used to improve the simulator. However, the simulator still has some limitations. It can only simulate one circuit at a time. Several circuits need to be connected to each other to be able to simulate them in Ngspice or two independent simulations need to be performed. There are no interactive parts. The users can change the properties of the parts in the inspector pane, but they cannot drag the knob of the potentiometers. This makes the experience less realistic. And finally, the simulator only allows for DC operating point analysis. While this is enough for basic electronic teaching, an AC simulation is still missing.

6 Conclusion and Future Work

Teaching electronics online poses some challenges on how to replace lab sessions with real hardware. In this paper, I have described how this shortcoming can be addressed with a simulator that tries to replicate the experiences of working with real circuits. The simulator has been implemented in Fritzing and allows students to implement the circuits on a breadboard and, at the same time, draw the schematics. Results shown that students find the simulator useful, and my personal experience suggests that a simulator can still be useful even if students have access to real hardware.

In the short term, I will focus on improving the simulator: increasing the parts with spice models, developing interactive parts and adding AC simulation and oscilloscopes while maintaining ease of use. Long term efforts include integrating the simulator in a holistic robot simulator (HoRoSim) [15] or introducing game-based learning in the simulator.

Acknowledgements. Thanks you to Kjell Morgenstern, who compiled the simulator, and to Peter Van Epp and Phil Duby, who fixed the original multimeter part. Finally, I would also express my gratitude to the students who tested the simulator and gave valuable feedback.

References

1. Kim, Y., et al.: SchemaBoard: supporting correct assembly of schematic circuits using dynamic in-situ visualization. In: Proceedings of the 33rd Annual ACM Symposium on User Interface Software and Technology, Association for Computing Machinery, New York, NY, USA, pp. 987–998 (2020)
2. Pedersen, B.K.M.K., Larsen, J.C., Nielsen, J.: From diagram to breadboard: limiting the gap and strengthening the understanding. In: Lepuschitz, W., Merdan, M., Koppensteiner, G., Balogh, R., Obdržálek, D. (eds.) RiE 2020. AISC, vol. 1316, pp. 339–353. Springer, Cham (2021). https://doi.org/10.1007/978-3-030-67411-3_31
3. Rosner, D., Sârbu, G., Tătăroiu, R., Deaconescu, R.: Applied electronics curriculum for computer science students. In: 2011 IEEE EUROCON - International Conference on Computer as a Tool, pp. 1–4 (2011). https://doi.org/10.1109/EUROCON.2011.5929184
4. Menendez, L.M., Salaverria, A., Mandado, E., Dacosta, J.G.: Virtual electronics laboratory: a new tool to improve industrial electronics learning. In: IECON 2006–32nd Annual Conference on IEEE Industrial Electronics, pp. 5445–5448 (2006). https://doi.org/10.1109/IECON.2006.348039
5. Braun, C.: Making things real in electronics laboratories. In: Proceedings Frontiers in Education 1995 25th Annual Conference. Engineering Education for the 21st Century, vol. 2, pp. 4c2.10–4c2.13 (1995). https://doi.org/10.1109/FIE.1995.483212
6. Nagel, L., Pederson, D.: Simulation program with integrated circuit emphasis. In: Midwest Symposium on Circuit Theory (1973)
7. AnalogDevices, LTspice
8. Vogt, H., Hendrix, M., Nenzi, P., Warning, D.: Ngspice users manual version 33 (2020)
9. KiCad EDA. https://www.kicad.org/
10. TinkerCAD. https://www.tinkercad.com
11. SimulIDE. https://www.simulide.com
12. Wokwi - Online Arduino and ESP32 Simulator. https://wokwi.com/
13. Knörig, A., Wettach, R., Cohen, J.: Fritzing: a tool for advancing electronic prototyping for designers. In: Proceedings of the 3rd International Conference on Tangible and Embedded Interaction, pp. 351–358 (2009)
14. Shalannanda, W.: Digital logic design laboratory using autodesk Tinkercad and Google classroom. In: 2020 14th International Conference on Telecommunication Systems, Services, and Applications (TSSA), pp. 1–5 (2020). https://doi.org/10.1109/TSSA51342.2020.9310842
15. Faiña, A.: HoRoSim, a holistic robot simulator: arduino code, electronic circuits and physics. In: Merdan, M., Lepuschitz, W., Koppensteiner, G., Balogh, R., Obdržálek, D. (eds.) RiE 2021. AISC, vol. 1359, pp. 256–267. Springer, Cham (2022). https://doi.org/10.1007/978-3-030-82544-7_24

Climbing a Volcano: A New Robotic Competition

Matthias G. Funk[1,3(✉)], José Cascalho[1,3], Francisco Pedro[2,3], André Behr[1], Paulo Medeiros[1,3], and Armando B. Mendes[1,3]

[1] Universidade dos Açores, FCT - UAc, Ponta Delgada, Portugal
{gunther.ma.funk,jose.mv.cascalho,andre.r.behpaulo.jf.medeiros}@uac.pt
[2] Escola Básica Integrada de Rabo de Peixe, Ribeira Grande, Portugal
francisco@acores.net
[3] Grupo de Robótica e Inteligência Artificial da Universidade dos Açores
GRIA-LIACC, Ponta Delgada, Portugal
https://gria.uac.pt

Abstract. In recent years, there has been a growing trend toward the use of educational robots to foster interest in STEAM and support collaborative learning. On the other hand, competitions are considered to be powerful incentives for humans to become strongly involved in a specific topic. In particular, in robotics, several competitions have been proposed in the context of Open Robotics where teams must create their own robots and test them in proposed environments. In this work, we present a new competition for a robot where a volcano scenario is used as a support for the activity, explaining the objectives and why using this scenario can be an interesting challenge for the educational robotics activity.

Keywords: Educational robotics · Competition · Volcanic environment

1 Introduction

In recent years, there has been a considerably growing trend to use educational robots in schools, even for young children [10,13]. This is seen as an incentive for collaborative learning, as suggested by Kamga et al. [12]. But the robots could also foster the interest for STEAM (*i.e.* for science, technology, engineering, art and mathematics) [4,18]. While Papert [14] considered robots as one of the best tools for implementing constructivist learning principles, Robinson [17] showed an increase in students' motivation toward mathematics and science courses.

It is common sense that competitions are a powerful incentive for humans to engage strongly with a specific topic, like in the RoboCup and RoboCup Junior. In [7], the authors, revisit the 20 years of this initiative as "a great vehicle to educate future academic and technical staff, allows for imparting broad knowledge in the field of robotics, multi-agent systems and artificial intelligence", and

© The Author(s), under exclusive license to Springer Nature Switzerland AG 2023
J. M. Cascalho et al. (Eds.): CLAWAR 2022, LNNS 530, pp. 414–422, 2023.
https://doi.org/10.1007/978-3-031-15226-9_39

referring to it as an initiative where "cutting-edge research and development do not stay abstract, but rather follow the '*get-your-fingers-dirty*' paradigm". On the other hand, Ribeiro and Lopes [16] presents an overview of robotics competitions for different levels of difficulty and with different learning targets. Other initiatives in technological areas have been proving the interest in the competition as a learning activity. One example is the International Computer Olympiad Informatics, which has for more than 30 years attracted many secondary school students to deal more intensively with topics in computer science and nowadays even robotics, as Kaloti-Hallak et al. [11] shows. Dagiene et al. [5] well document the benefits of learning competitions in different countries.

In the context of RoboCup and initiatives of a similar kind of RoboCup, some challenges, mentioned as *application leagues* in [7], have been applied to societal and economic relevant problems, promoting the use of robots and fostering robotics learning. Examples of these challenges are the Fire-Fighting Robot International Competitions [1] or the competitions related to rescue activities in RoboCup.

Following this idea, in this paper, we describe the Volcano Challenge, targeting the idea of using robotic devices to measure and explore the environment of volcanic activity. The use of robots in volcano monitoring is a well-known research area in robotics [6].

This proposal is in line with recent events related to volcanic activity where robotic technology has been used or could have been used. They are the volcanic activity verified in the autumn of 2021 on the island of Las Palmas and the intense seismic activity caused on the island of S. Jorge in the Azores archipelago during the year 2022.

The Volcano Challenge will be tested in two upcoming competitions. In the first one, a predefined set of teams will use a previously tested robot and scenario. The teams are free to modify the initial robot and program its behaviour. During this meeting, we expect to assess how the challenge fits with the learning goals (described in Sect. 2.2) and to adjust it to fit these objectives better, if necessary. In the latter, we intend to have teams building their own robots and use them in the previously tested scenario of the competition.

As both events have not yet taken place, this document presents the preparation process for the competition. In the next section, we present the competition and its rules. We then discuss some of the pedagogical issues related to the challenge. Finally, we make some concluding remarks.

2 The Competition Environment

The scenario intends to simulate an active volcano where the crater with outgoing lava emits high heat radiation represented by a radiator of infrared light. As the neighbourhood of the crater has a accumulation of toxic gas, no human could do any measuring. A robot should substitute the human inability, but cannot go beyond the point where the temperature would cause its destruction.

The challenge is to climb an irregular, sloping terrain towards an infrared light and stop at a certain stipulated light distance. The ground simulates a volcanic hill where the hot eruption point is localized as an infrared light. The robots must explore the upper area, identifying the colours on the ground as if they were taking samples, and then safely return to the starting area.

The basic component to simulate this environment is the ramp with three meters in length and approximately 90 cm in width. The inclination of the ramp is approximately 12 °C. This structure will be embedded with some obstacles (*i.e.* a couple of big solid stones) and covered by expansive foam simulating the uneven pavement. Figure 1 presents the prototype example of the ramp without the colours on the top and without the obstacles.

Fig. 1. A prototype of a ramp in the volcano competition.

The robot starting area, or base station, is marked by a line at the base of the ramp. The resistance limit of the robot is identified by an area where the robot should not cross or stay for long (*e.g.* in the Fig. 1 is represented as a red area on the upper edge of the structure). Before returning to the base station, the robot should detect colours on the ground, simulating the data being collected near the eruption site.

As a second environmental element, there is an infrared light radiator (700–1100 nm). One option is to install a radiator like the one in the picture 2. This infrared light should be located at the top end of the ramp and is used as a directional orientation marker for the robot using an infrared sensor. In future competitions, this element could be an infrared medical lamp which should better fit our proposal.

Fig. 2. Radiator

2.1 The Challenge Tasks

This challenge targets ages 13+ with teams of up to four participants and one tutor. The competitors will be separated into two levels depending on the age of their oldest students: one for students under 15 years old and another for students over this age.

The robot has to execute four tasks in a maximum time of 10 min, following the steps:

1. Climbing the ramp beginning from the designated blue starting.
2. Not surpass the given return line.
3. Detecting colours on the ground.
4. Going back to the starting area.

Successfully executing these four tasks gives the team 5 points each. The best score establishes the ranking. In the case of a draw, the execution time (till the last successful task) will be the decisive factor.

The robot can find obstacles on the ramp. So, it must use the ultrasonic sensor to detect and avoid them. Also, a temperature sensor is used to detect the minimum distance to the infrared light source. The teams should detect this distance heuristically.

The teams can complete only part of the tasks proposed. For example, a team scores for the safe return even if it does not detect the colours on the ground.

2.2 Learning Goals

Volcanic environments are tough for robots [3]. The use of robots on a rough ground where it is difficult to move, avoid obstacles and try to achieve a target is one of the project's goals. Another goal is explicitly using legs to achieve this task (*e.g.* using Whegs).

One of the difficulties concerning the task is to keep the robot on the target. The movement using the Whegs will balance up and down and to the left and right. The robot must use an ultrasonic sensor to avoid obstacles. At the same time, it must detect the direction where the light is stronger. The other difficulty is finding the way back to the initial area, descending the ramp. Finally, detecting colours on the ground could be tricky because of the interference from the light radiator.

The main immediate learning goals are the following:

- Learn to program the movement of the robot.
- Tuning the robot's activity using the sensors.
- Redesign some of the components (e.g. the Whegs of the robot).
- Fine tuning of the new physical components.

The first two goals are related to general foster spatial reasoning, as Ribeiro [15] shows. It should be noted that this is a cross-cutting aspect of such a kind of competition, the importance of which is underlined by Francis et al. [8]. The last two goals are aimed at improving the design of the Wheg wheels and could be related to a product or experimental design.

2.3 Robot Hardware

The hardware will be created in the spirit of digital fabrication described by Blikstein [2]. For the first event, organizers will provide a basic prototype that is depicted in the Fig. 3 with a different set of wheels/Whegs. The intention is to help teams concentrate on controlling the robot up and down the ramp.

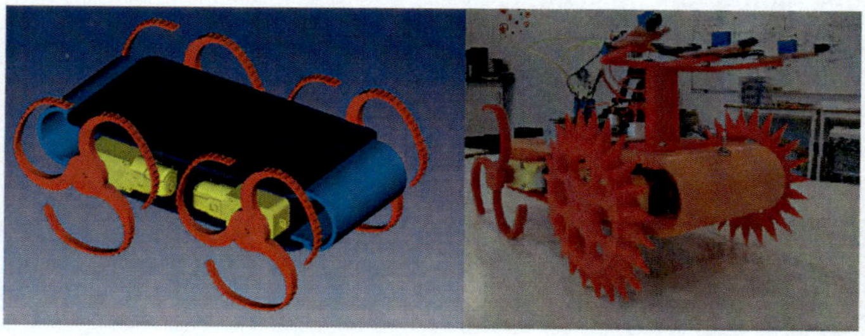

Fig. 3. Prototype **(a)** with Whegs wheels and **(b)** with Whegs and spike wheels as well three infrared sensors on top

There is a restriction on the size of the robot, with a 30 cm maximum length and 25 cm width. The components used in the prototype model are the following:

- Controller: ESP32 or Arduino (Uno, Nano, or Mega).
- Motor Driver: L298, DRV8835, or TB6612FNG.
- Motor: Micro Motor DC 6 V.
- Sensors: Flame or infra-red sensor, Ultrasonic sensor, Temperature sensor, and Colour sensor.

Using a robot as a prototype is also an invitation to the teams to adopt and modify the robot for better performance. Thus, teams are expected to modify the robot and create adapted or new components, even when the proposed components, shown in the Fig. 3 (a) and (b), are freely available in STL format.

3 Some Considerations Related to the Project Goals

This challenge is aimed at people over 13 years old, so the teams must be composed of school students with robotic activities in clubs or some educational pilot projects. We intend to apply, firstly, the spirit of *do-it-yourself* (DIY) and the Maker philosophy advocated by Blikstein [2], which promotes the use of 3D-Print and advocates the banning of all superfluous commercial equipment. It is important to remember that neither the teachers nor the students have experience in building robots. This means that we hope to have a context where teachers and students learn together (*e.g.* Robo-Party [15]).

In these circumstances, the delivered prototypes will give the opportunity for the teams to begin immediately with some of the programming issues.

Although the prototypes suggest the use of Whegs, depending on the obstacles in the environment, the robot could be still unable to climb the ramp. Thus, a modification of the Whegs should be prepared by the teams.

The construction of an adequate leg could be considered the main task to execute on the hardware side. The competition could easily be restrained on the programming issue, but this would lack an important educational component, called by Gil et al. [9] "structionist" which is the underlying concept of the Lego bricks. We think such a competition should include a strong "learning by design" or "learning by doing" approach.

4 General Pedagogical Aspects of the Robot Competition

The contest addresses two different age groups, and everyone probably will focus on specific aspects. We expect the younger group to focus more on the programming task. On the other hand, the older ones would be able to think about improving the hardware. We think if we keep the same rules for both groups, it is necessary to design the competition in a way that each group could reach some goals. So we intend to split the challenge into the four sub-tasks already described, where the first one is easily achievable, *i.e.* climbing the ramp guided by the source of radiation and avoiding the obstacles. In this case, a minimum score will be achieved by reaching the red line. To justify the utility of this task, we could invoke that a real robot passing the red line, even if it is burned, could have transmitted essential information to the base station. This conceptual embedding is important to motivate students for such a project.

The other tasks imply the following:

i. To avoid the immediate destruction in halting at the critical point.
ii. Collect the data and keep it.
ii. To go back to the initial area, prepare for a new mission.

Fig. 4. Prototypes: (a) possible whegs (b) given whegs in the prototype

One of the aspects that could be discussed deeply with the teams is the improvement of the given legs/wheels. Both types of legs/wheels given by the prototypes in Fig. 3 are able to overpass the inclination, but not necessarily the obstacles in the form of rocks on their way. The spiked wheel does not help to climb a real obstacle. The Whegs have more potential to solve the problem.

The two Whegs wheels in Fig. 4 show an evolution process in the development of this issue. The three legs on each Wheg have a space between them, as marked by the grey rectangle in the figure, in which a part of rock could enter and be used as a lever to elevate the robot. The problem is to find the right size for this gap. Another critical point is the thickness of the legs: a thick one has no flexibility, whereas a thin one has no resistance to push up the robot.

Finally, the white circle shows the critical point of the power transmission between the lever and the legs, which should be as thick as possible, but needs a gradual decrease to give the leg the necessary elasticity. As a third interesting point to investigate, there is the grip of the Whegs.

Having given students some direction of investigation does not mean that there is no freedom for them to find their own solutions out of the box. Therefore, is it important to include maximum flexibility in the design of the competition.

5 Conclusion

In this article, we present a robot competition in a scenario simulating an erupting volcano's slope. We present the objectives and rules of this activity, explaining why we understand that using this scenario can be an exciting challenge as an educational robotics activity. We present the pedagogical context where we address some technical issues related to the robotic movement, the design of the robot prototype, and the respective scenario where the competition will occur. This competition brings some difficulties related to the type of scenario. It simulates an irregular terrain with the danger of robot destruction and the need to collect data in a hostile environment. With this contest, we hope to stimulate the participants' creativity by connecting them to a real environment where robots have been deployed for over 20 years.

A first evaluation competition was held where we had two teams for each of the two age groups. The older participants were able to conclude all goals, but the younger pupils only concluded the first two steps. Therefore, we consider it appropriate to define multiple objectives for different age groups.

Acknowledgment. We thanks to our colleagues Manuel Silva, José Lima and Vitor Pinto for the fruitfully discussions and their contribution for the idea of the Volcano competition. This work is part of the project supported by Governo Regional dos Açores: PRO-SCIENTIA M3.4.B/PRO ROBÓTICA/001/2021

References

1. Ahlgren, D.J., Verner, I.M.: Fire-fighting robot international competitions: education through interdisciplinary design. In: International Conference on Engineering Education, 6–10 August 2001. Oslo, Norway (2001)
2. Blikstein, P.: Digital fabrication and 'making' in education: the democratization of invention. In: Walter-Herrmann, J., Büching, C. (eds.) FabLabs: of Machines. Makers and Inventors. Transcript Verlag, Bielefeld (2014)
3. Borgese, A., Cantelli, L., Guastella, D.C., Sutera, G., Muscato, G.: Highlights on some robots and drones for hazardous environments. In: 2021 International Symposium on Electrical, Electronics and Information Engineering, ISEEIE 2021, pp. 45–49. Association for Computing Machinery, New York (2021)
4. Conde, M.Á., Rodríguez-Sedano, F.J., Fernández-Llamas, C., Gonçalves, J., Lima, J., García-Peñalvo, F.J.: Fostering steam through challenge-based learning, robotics, and physical devices: a systematic mapping literature review. Comput. Appl. Eng. Educ. **29**(1), 46–65 (2021)
5. Dagiene, V., Mannila, L., Poranen, T., Rolandsson, L., Stupuriene, G.: Reasoning on children's cognitive skills in an informatics contest: findings and discoveries from Finland, Lithuania, and Sweden. In: Gülbahar, Y., Karataş, E. (eds.) Informatics in Schools. Teaching and Learning Perspectives, pp. 66–77. Springer International Publishing, Cham (2014). https://doi.org/10.1007/978-3-319-09958-3_7
6. Evita, M., Djamal, M.: A review of mobile robot navigation system for volcano monitoring application. Indonesian J. Phys. **32**(1), 1–11 (2021)
7. Ferrein, A., Steinbauer, G.: 20 years of Robocup. KI - Künstliche Intelligenz **30**, 225–232 (2016)
8. Francis, K., Rothschuh, S., Poscente, D., Davis, B.: Malleability of spatial reasoning with short-term and long-term robotics interventions. Technol. Knowl. Learn. **27**, 927–956 (2021). https://doi.org/10.1007/s10758-021-09520-7
9. Gil, P., et al.: An experience on mechatronics teaching on undergraduate students by means of the Skybot platform: from classroom to robot competition. Int. J. Robot. Educ. Art **2**(2), 1–15 (2012)

10. Hudson, M.-A., Baek, Y.: Increasing elementary students' computational thinking skills using a multifaceted robotics-based intervention. Comput. Sch. **39**(1), 16–40 (2022)
11. Kaloti-Hallak, F., Armoni, M., BEN-ARI, M.M.: The effectiveness of robotics competitions on students' learning of computer science. Olympiads Inf. **9** (2015)
12. Kamga, R., Romero, M., Komis, V., Mirsili, A.: Design requirements for educational robotics activities for sustaining collaborative problem solving. In: Alimisis, D., Moro, M., Menegatti, E. (eds.) Edurobotics 2016 2016. AISC, vol. 560, pp. 225–228. Springer, Cham (2017). https://doi.org/10.1007/978-3-319-55553-9_18
13. Kubilinskiene, S., Zilinskiene, I., Dagiene, V., Sinkevičius, V.: Applying robotics in school education: a systematic review. Baltic J. Mod. Comput. **5**, 50–69 (2017)
14. Papert, S.: Mindstorms: Children, Computers, and Powerful Ideas. Basic Books Inc, New York (1993)
15. Ribeiro, A.F.: Building a robot to use in school - teachers and students learning together. http://repositorium.sdum.uminho.pt/bitstream/1822/10248/1/A%20Fernando%20Ribeiro.pdf. Accessed 29 Apr 2022
16. Ribeiro, A.F., Lopes, G.: Learning robotics: a review. Curr. Robot. Rep. **1**(1), 1–11 (2020)
17. Robinson, M.: Robotics-driven activities: can they improve middle school science learning? Bull. Sci. Technol. Soc. **25**(1), 73–84 (2005)
18. Sullivan, F.R., Heffernan, J.: Robotic construction kits as computational manipulatives for learning in the stem disciplines. J. Res. Technol. Educ. **48**(2), 105–128 (2016)

Sensing, Positioning and Localization

Fast Prototyping of a Low-Cost Three-Dimensional Force Sensor for the Six-Legged Walking Robot LAURON

Carsten Plasberg(✉), Sven Sauerbaum, Arne Roennau, and Rüdiger Dillmann

Department of Interactive Diagnosis and Service Systems (IDS),
FZI Research Center for Information Technology, 76131 Karlsruhe, Germany
Plasberg@fzi.de

Abstract. Dynamic walking with six-legged robots is a fascinating but also demanding challenge, both for hard- and software. Modern approaches for dynamic walking require high resolution input on time and measurement accuracy in their feet to keep balanced. Multiplied with the number of their feet, force sensors with an adequate measurement range are heavy, expensive and often also bulky. This problem can be addressed with a novel approach using the tubular leg structure itself as a measurement carrier, combined with a lid in the foot. The proposed sensor design is easily manufactured at any workshop, cheap, lightweight and can be used for quite heavy walking robots.

Keywords: Three-dimensional force sensor · Strain gauges · Walking robot · Dynamic locomotion

1 Introduction

1.1 Motivation

For biological inspired walking robots, as like LAURON [1], state of the art is a force based control to enable dynamic locomotion. Dynamic locomotion implies, that there are phases during movement, when the center of mass of the robot is outside its support polygon [2,3]. Most advanced walking robots use model predictive control to achieve dynamic walking, e.g. MIT Cheetah [4] or ANYmal [5]. As well as Garofalo et al. [6] they need either torques of joints or contact forces at their feet as input into their control cycle. Approaches using joint torques as input usually estimate contact forces at their feet. As contact forces are the variable to control for each leg, the measurement of those forces, to be more specific, the difference between measurement and target value is an indicator for disturbances and vital for a robust control [3]. Because of interacting in a 3D space it is absolutely essential to handle this three dimensional force-vector for each leg in magnitude and direction (Fig. 1).

© The Author(s), under exclusive license to Springer Nature Switzerland AG 2023
J. M. Cascalho et al. (Eds.): CLAWAR 2022, LNNS 530, pp. 425–435, 2023.
https://doi.org/10.1007/978-3-031-15226-9_40

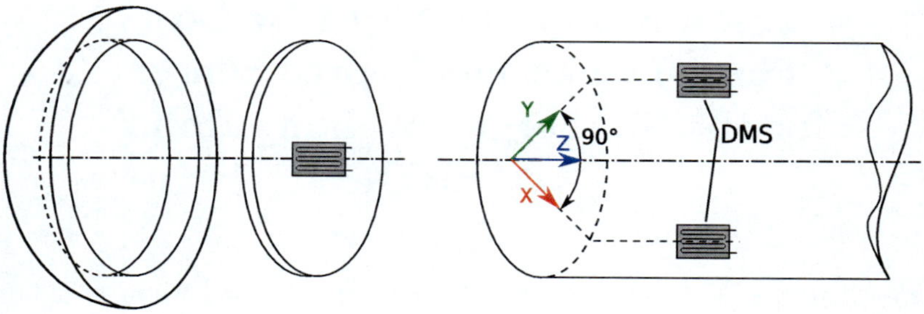

Fig. 1. Exemplary assembly of the strain gauges for x-, y- and z-direction, around the magnitude of a CFRP-tube with an offset of 90° and one at an aluminium sheet. From left to right, rubber food, aluminium lid with strain gauge and the CFRP-tube with its gauges.

To measure those forces the sensor should be located directly at the TCP (Tool Center Point), in case of a walking robot, its feet. At the same time this location applies the highest accelerations during walking on the sensor. To reduce energy consumption of the overall robot the new sensor is expected to be lightweight. Additionally this sensor is needed for each leg. In case of LAURON V, our primary target system, this means six sensors. Thus gives a further constraint in costs and weight for an individual sensor.

Prior to this work, industrial solutions for this problem have been investigated. Confronted with the same Problem as Wu et al. [7] in 2011 no appropriate solution with the constraints of weight, costs and measurement range was found.

1.2 Structure of the Paper

At first, requirements for the new sensor prototype are collected and different methods to measure forces are evaluated. This is done in Sect. 2. Section 3 focuses on the requirements and design of the newly developed sensor. Tests and results with the new sensor are presented in Sect. 4, while Sect. 5 concludes the work.

2 Force Measurement Methods

There are different possibilities to quantify the force. Since the force itself is not directly observable, the effect of an applied force into a predefined measurement body, the sensor [8,9], is usually measured. Here a short overview into some different principals is given.

Force Measurement Through Change of Resistance. Electric resistance of bodies can be described as shown in Eq. 1 [10].

$$R = \zeta \frac{4l}{\pi D^2} \tag{1}$$

The ohmic resistance R is influenced by with the specific resistance ζ, the length l and the diameter D of the body. Grote et al. [8] and Schleichert [11] show, that this equation can also be displayed in a differential form (Eq. 2). Here the lateral contraction $\mu = -\frac{\frac{dD}{D}}{\frac{dl}{l}}$ as in Park et al. [12] and the variation of length $\epsilon = \frac{\delta l}{l_0}$ from [8] are used.

$$\frac{\Delta R}{R} = (1 + 2\mu)\epsilon + \frac{\Delta\zeta}{\zeta} \tag{2}$$

Obviously this equation is composed as a sum of two terms. The first depends on the dimension of the body while the second part is directly related to the variation of the mechanical stress $\sigma = \frac{F}{A}$.

Force Measurement Through Optical Effects. Grote et al. [8] explain two optical methods. Both methods rely on optical fibers that are applied to the sensor-volumes surface. The Fabry-Pérot-Interferometrie [13] requires at least two fibers with an open end facing each other with a small distance. Due to deformation of the sensor the distance between the fibers varies. Illuminated this change of distance leads to a changing interference pattern which can be observed.

The Fiber Bragg Grating (FBG) method [14] works by shifting the spectral lines. By inserting a region with a different optical refractive index, applied forces can be observed as a shift of the spectral lines. The signal output reaction is relative to spectral shift of the signal input. This method is considered as an absolute measuring method.

Piezo Electrical Force Measurement. As described in [10], an interacting force to a piezo-crystal results in a shift of the electrical charge inside the crystal. This effect is called piezo-electrical effect. According to [15] an applied force to a piezoelectric bimorph can be detected as piezoelectric voltage V depending on geometry with width w, length l and thickness t as well as the piezoelectric stress constant g_{31}.

$$V = \frac{3}{2}\left(\frac{g_{31}l}{wt}\right)F \tag{3}$$

Inductive Force Measurement. The inductivity L of a coil is proportional to the product out of the relative permeability μ_0, the absolute permeability μ_r, the number of turns N, the cross-section area A as well as the inverse of the length of the coil l.

$$L \sim \frac{\mu_0 \mu_r N^2 A}{l} \tag{4}$$

As Giesbert et al. [16] and Grote et al. [8] declare, for inductive force measuring a non iron and electrical conductive metal is required, for example aluminium. This metal is placed inside or near the coil. By applying some force onto the sensor, the distance between the coil and the metal core decreases and results in a changing inductivity by changing the relative permeability.

Capacitive Force Measurement. The function of capacitive sensors is similar to the inductive method. Capacity can be evaluated with the permeability ϵ_0, relative permeability ϵ_r inside the dielectric, the plate distance D and the size of the overlapping area A.

$$C = \epsilon_0 \epsilon_r \frac{A}{D} \tag{5}$$

A force can either effect the distance of the contact or, as Al-Handarish et al. [17] describe, the overlapping area.

3 Design of the New Sensor

The newly designed three-dimensional force sensor will be used in Lauron V [18] (see Fig. 2) and its successor LAURON VI, which is currently under development. Both robots have or will have a weight of approx. 40 kg [18]. With LAURON VI it is one of the major aims to enable dynamic walking on a six-legged walking robot. In order to enable these dynamic movements, a feedback of actual forces and perturbations is needed.

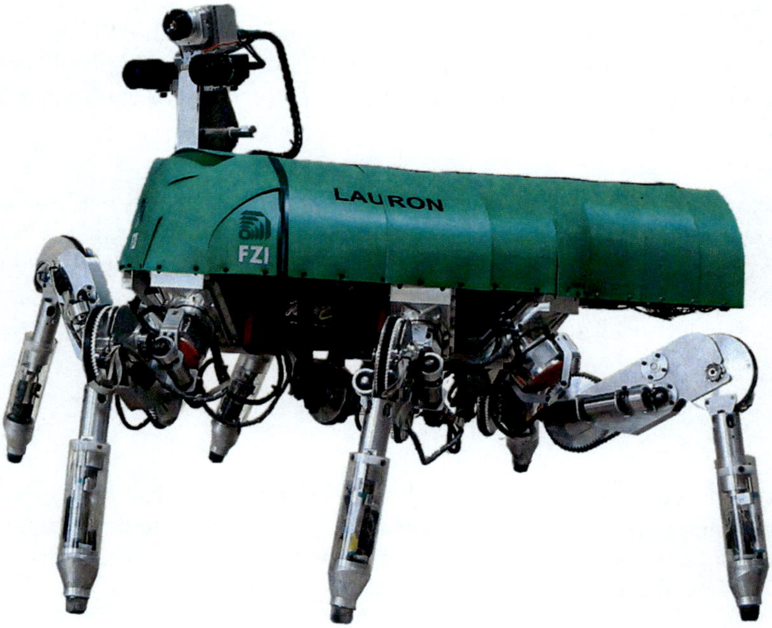

Fig. 2. The six-legged walking robot LAURON V, biologically inspired by the stick insect. This robot is the primary targeted system of the newly developed sensor prototype.

With LAURON VI it is planned to use serial elastic actuators, ANYdrives [19,20], that give feedback of torque in the rotational axis for each joint. There is no such option in LAURON V. Currently the forces at the TCP are calculated from measured motor currents. The resulting forces are heavily affected by noise as there are open gearboxes that can be influenced by the environment. Examples are grains of sand that jam the transmission but also grasses that increase friction in the joints.

Ground-Contact Forces are used in many different ways. They can improve mobility when traversing rough terrain [21] as quality of foot-placement can be assessed, but also allow to determine a leg's condition as suggested in [22]. This information can be used to compensate failures by sparing damaged legs [23].

With the newly proposed three-dimensional force sensor perception of perturbations during dynamic walking is possible also. The easy usage also allows to evaluate the drives internal measurements and to learn models of the legs' dynamics.

3.1 Requirements of the Sensor

To be used in a profitable way, the sensor has to fulfill some requirements. On a battery powered, non tethered mobile system there are several constraints for hardware and electronics. One of them is energy efficiency. The feet are exposed to high shocks during landing (of the feet) and repeatedly undergo the highest accelerations in the whole robot. Therefore they need to be robust and, for the state of energy efficiency, extremely lightweight. When used for feedback in the control loop which is running at 100 Hz to 500 Hz, this bandwidth shall be supported by the sensor as well as width a sufficient accuracy. In order to prevent the feet from getting caught on obstacles in unstructured terrain, the installation space is limited in width to the diameter of the feet which is 30 mm.

LAURONs legs can apply forces up to 500 N to hold the robot up (z-axis). Unlike a mammalian kinematics at the insects kinematics, loads exceeding the maximum power of the legs will rather result in lowering the body to the ground than in higher forces at the feet. In direction of x- and y-axis the maximum force limited by two factors. On the one hand there is the maximum force the drives can apply, while at the same time friction on the ground sets a maximum force, that, when exceeded, results in slipping away. For most of the activity ±30 N are not exceeded.

3.2 Sensor Principle

The basic idea behind the new sensor is to directly use the legs' structure at each foot as measurement carrier. An applied force deforms this structure. The selection of material for the leg, here a CFRP-tube, determines different moduli of elasticity depending on the direction of the applied force. For CFRP materials the longitudinal tensile modulus is an order of magnitude higher than the traverse tensile modulus [24,25]. Because of the very small crushing of CFRP-tubes, respectively the high modulus of elasticity on axial forces, these are measured with a aluminium lid on the tube.

Traverse forces on the feet result in a deformation of the CFRP-tube. These minimal changes in geometry can be measured with the applied strain gauges, their change is respectively in the resistance. More difficult is the measurement of longitudinal forces. This problem is solved with aluminium lid fixed to the bottom of the CFRP-tube. Applied forces in direction of the z-axis produce deformation of the rubber feet. The usage of an aluminium lid and a rubber-foot as shown in Fig. 3 allow the measurement of pressing forces.

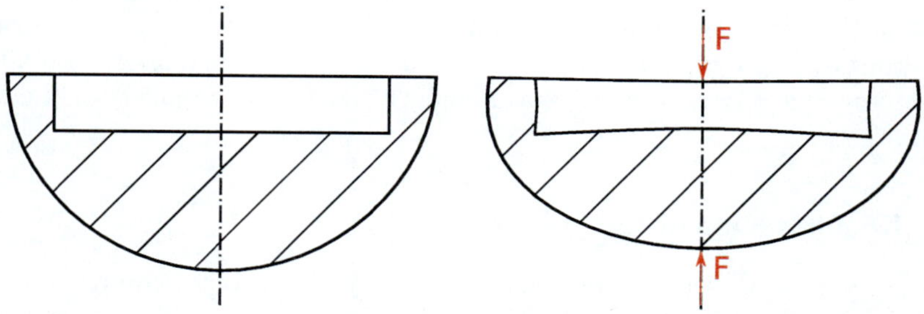

Fig. 3. Cross-section of the used rubber-feet. On the left side the no forces are applied. In contrast on the right a force in z-direction is applied. One can clearly see the widened top

3.3 Force Measurement

For prototyping, strain gauges of the type FAE2-A6251J-100-S6E by VISHAY [26], with a nominal resistance of 1 kΩ were used. They were applied to the CFRP-tube by using superglue. As a lid, an aluminium sheet with 1mm thickness was cut into shape. Here, the strain gauge is also glued on. Due to friction the rubber foot is put on the tube on holds. This has the advantage that occurring torques around the z-axis can be neglected as the foot can turn on the leg. Each of the strain gauges are connected to a HX711 board from SparkFun [27,28] for analog to digital conversion. The converted values get transmitted via SPI to an Arduino Nano [29]. Still in a raw format, these values are put in a ROS message and sent to a PC for processing and recording. To have only minimal noise, wiring between the strain gauges and the related ADCs should be as short as possible. Distance is not that critical anymore when using SPI or even USB to communicate with higher level devices.

This prototype setup offers room for several improvements. The selected strain gauges are not chosen optimally as their characteristic is optimized for steel instead of CFRP. Additionally, a second strain gauge should be used on the x- and y-axis, located directly vis-à-vis to get a differential measurement. This also helps to improve the separation of measurements in those axes as they are not completely independent. Also the load cell amplifiers should be replaced as they only offer a limited bandwidth not matching the previously identified requirements.

Still, the whole measurement setup can be placed (completely) inside the CFRP-tube. So it is well protected from any environmental impact on the leg. Because it is very lightweight (roughly 32 g) the influence on energy efficiency while walking is low.

With the proposed force sensor, a force measurement in three axes is possible. No special tools or environment is needed to built up the sensor. Superglue, sanding paper, and a soldering iron are just fine.

4 Experiments and Results

The Sensor-Prototype was tested, calibrated and compared to an industrial force-torque-sensor. For this a FTN-Delta-SI-660-60 from Schunk was used as reference system. It has a measurement range of ±660 N in x- and y-direction and a range of ±1980 N along the z-axis. Resolution in force is 0.125 N on x- and y-, 0.25 N on the z-axis [30]. Both sensors, the newly designed one as well as the reference sensor, are connected to the same PC. For testing the newly designed sensor is mounted on a robot leg which is attached to a fix test setup. By placing the foot onto the reference sensor and actuating the leg to apply forces the sensors data can be compared (Fig. 4).

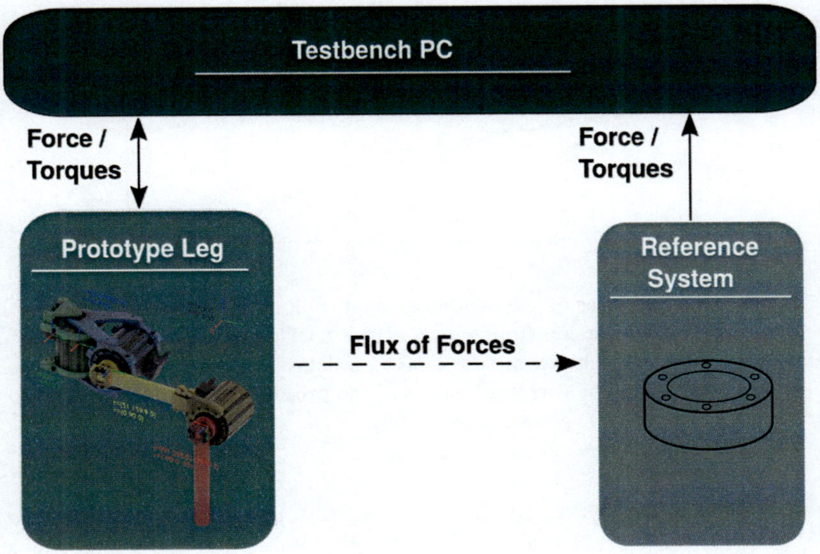

Fig. 4. Schematic representation of the setup to analyse and test the sensor prototype.

4.1 Linearity and Calibration

For a first test only a z-force was applied. Figure 5 shows in the upper graph the measured values from the reference system (blue) and unprocessed values $F_{z,raw}$ of the prototype (green). Most significant difference beside the deflection is a delay in time. A linear calibration model, as shown in Eq. 6, is assumed here.

$$F_{z,calib} = m \cdot (F_{z,raw} + c) \tag{6}$$

To calculate the calibration factor m the quotient of both curves is formed. The result can be seen in the middle graph of Fig. 5. The outlier at the beginning results from the delay in time which is caused by the low bandwidth of the used ADC. Nevertheless the quotient is constant for most of the time. In the bottom graph of Fig. 5 the corrected measurements of forces along the z-axis are shown. They match the measured values of the reference system pretty well.

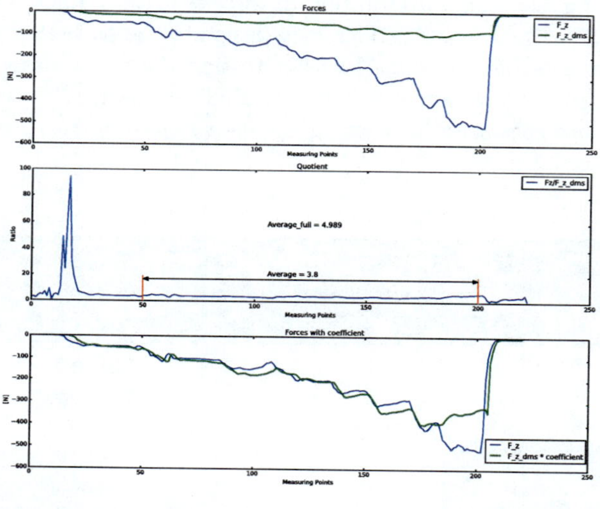

Fig. 5. Top: Measured force of the reference system (blue) and unprocessed raw values of the prototype. A conversion from values of the ADC to Force isn't done yet. Mid: Quotient of both graphs shown in top. Due to a delay in time there is an outlier in the beginning. Bottom: The corrected values of the prototype (green) compared to the measurement of the reference system (blue).

For the traverse axes a repetitive rectangular pattern was controlled (see Fig. 6). To provide enough friction for those measurements the z-axis was subjected to a force of 300 N. Due to some transformation of the coordinate systems between the prototype and the reference sensor the values of the x-axis have to be inverted. Here also calibration is done with the linear model from Eq. 6. Including the inversion the factor m can be calculated to -4.16.

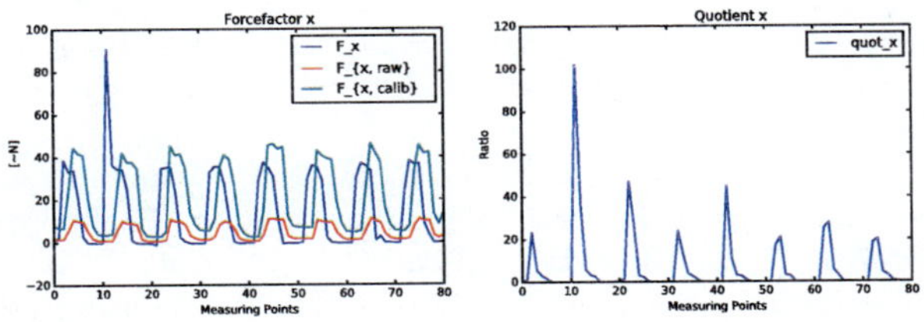

Fig. 6. The graphs show the measurement data of the x axis. On the left side the curves of the prototype sensor and the reference system are depicted. On the right side the quotient is shown again. As there are times without an applied force and due to the delay these plots show massive outliers.

4.2 Technical Specifications

The newly developed prototype sensor has been tested up to values of 500 N in longitudinal direction and 70 N for traverse axis. With the used hardware the maximum supported samplerate is 80SPS. Bottleneck here is the HX711 chip, which does not support higher samplerates. Torques can not me measured. With the round rubber feet, as used on LAURON, torques are a minor effect and can be neglected. All the Hardware can be placed inside the CFRP-tube and has a weight of only 32 g.

5 Conclusions and Future Works

In this paper a novel approach for a prototype of a low-cost three-dimensional force sensor has been introduced. This sensor is characterized by its very low weight, easy to purchase and cheap components as well as its mechanical durability. From the tests it can be seen, that the measurement range is well suited for walking robots like LAURON. Strain gauges are used to measure the forces on the leg, a CFPR-tube, itself. A novelty for force sensors in robots feet is the usage of a lid made of aluminum to evaluate longitudinal forces.

Possible future improvements include the usage of strain gauges optimized for CFRP in x- and y-axes. Beside that another ADC will be used to increase the samplerate. Also there are plans to use several strain gauges for each direction to improve temperature compensation.

Acknowledgement. The research leading to these results has received funding from the DLR Space Administration under grant agreement No. 50RA1730 by the German Bundestag.

References

1. Rönnau, A., Heppner, G., Pfotzer, L., Dillmann, R.: LAURON V: optimized leg configuration for the design of a bio-inspired walking robot. In: Nature-Inspired Mobile Robotics, pp. 563–570. World Scientific (2013)

2. Wieber, P.-B., Tedrake, R., Kuindersma, S.: Springer Handbook of Robotics. Ed. by B. Siciliano, O. Khatib. Springer, Cham (2016). https://doi.org/10.1007/978-3-540-30301-5
3. Vukobratovic, M., Surdilovic, D., Ekalo, Y., Katic, D.: Dynamics and Robust Control of Robot-Environment Interaction. World Scientific, Singapore (2009)
4. Bledt, G., Powell, M.J., Katz, B., Di Carlo, J., Wensing, P.M., Kim, S.: MIT Cheetah 3: Design and control of a robust, dynamic quadruped robot. In: 2018 IEEE/RSJ International Conference on Intelligent Robots and Systems (IROS), pp. 2245–2252. IEEE (2018)
5. Bellicoso, C.D., Jenelten, F., Fankhauser, P., Gehring, C., Hwangbo, J., Hutter, M.: Dynamic locomotion and whole-body control for quadrupedal robots. In: 2017 IEEE/RSJ International Conference on Intelligent Robots and Systems (IROS), pp. 3359–3365. IEEE (2017)
6. Garofalo, G., Ott, C., Albu-Schäffer, A.: Walking control of fully actuated robots based on the bipedal slip model. In: IEEE International Conference on Robotics and Automation, pp. 1456–1463 (2012)
7. Wu, B., Luo, J., Shen, F., Ren, Y., Wu, Z.: Optimum design method of multi-axis force sensor integrated in humanoid robot foot system. Measurement **44**(9), 1651–1660 (2011)
8. Bender, B., Göhlich, D. (eds.): Dubbel Taschenbuch für den Maschinenbau 1: Grundlagen und Tabellen. Springer, Heidelberg (2020). https://doi.org/10.1007/978-3-662-59711-8
9. Pott, P.P.: Entwurfskriterien feinwerktechnischer integrierter sensor-aktor-systeme. Darmstadt, April 2015. http://tuprints.ulb.tu-darmstadt.de/4616/. Accessed 25 Jan 2021
10. Bender, B., Göhlich, D. (eds.): Dubbel Taschenbuch für den Maschinenbau 1: Grundlagen und Tabellen. Springer, Heidelberg (2020). https://doi.org/10.1007/978-3-662-59711-8
11. Schleichert, J.: Entwicklung und untersuchung von mehrkomponentensensoren für kraft und drehmoment (2016)
12. Park, W.-T.: Piezoresistivity. In: Bhushan, B. (ed.) Encyclopedia of Nanotechnology, pp. 2111–2117. Springer, Dordrecht (2012). https://doi.org/10.1007/978-90-481-9751-4_222
13. Thurner, K.: Position sensing using a fiber-optic fabry-pérot interferometer. Dissertation, Technische Universität München, München (2018)
14. Roths, J., Kratzer, P.: Vergleich zwischen optischen faser-bragg-gitter-dehnungssensoren und elektrischen dehnungsmessstreifen intercomparison of optical fbg-based strain sensors and resistive strain gages. Tm-technisches Messen **75**, 647–654 (2008)
15. Tansock, J., Williams, C.: Force measurement with a piezoelectric cantilever in a scanning force microscope. Ultramicroscopy **42**, 1464–1469 (1992)
16. Giesberts, R., Sluiter, V., Verkerke, G.: Design and test of a new inductive force sensor. Sensors **18**, 2079 (2018)
17. Al-Handarish, Y.: A survey of tactile-sensing systems and their applications in biomedical engineering. In: Advances in Materials Science and Engineering, vol. 2020 (2020)
18. Rönnau, A., Heppner, G., Nowicki, M., Dillmann, R.: Lauron V: a versatile six-legged walking robot with advanced maneuverability. In: 2014 IEEE/ASME International Conference on Advanced Intelligent Mechatronics, pp. 82–87. IEEE (2014)
19. Fankhauser, P., Hutter, M.: ANYmal: a unique quadruped robot conquering harsh environments. Res. Features **126**, 54–57 (2018)

20. Prassler, E., Baroncelli, A.: 2017 IEEE/IFR award for IERA [Industrial activities]. IEEE Robot. Autom. Mag. **24**(3), 8–11 (2017)
21. Roennau, A., Kerscher, T., Ziegenmeyer, M., Zoellner, J., Dillmann, R.: Six-legged walking in rough terrain based on foot point planning. In: Mobile Robotics: Solutions and Challenges, pp. 591–598. World Scientific (2010)
22. Schnell, T., et al.: Robot health estimation through unsupervised anomaly detection using Gaussian mixture models. In: 2020 IEEE 16th International Conference on Automation Science and Engineering (CASE), pp. 1037–1042. IEEE (2020)
23. Plasberg, C., Kogel, D., Roennau, A., Dillmann, R.: Adaptive walking pattern generation with damaged legs for the six-legged robot Lauron V. In: 2021 20th International Conference on Advanced Robotics (ICAR), pp. 422–427. IEEE (2021)
24. Adams, R., Bacon, D.: Effect of fibre orientation and laminate geometry on the dynamic properties of CFRP. J. Compos. Mater. **7**(4), 402–428 (1973)
25. Kar, K.K. (ed.): Composite Materials. Springer, Heidelberg (2017). https://doi.org/10.1007/978-3-662-49514-8
26. Vishay Measurement Group GmbH: Transducer - Gauges and Accessories, Catalog-No. 306A, June 2010
27. SparkFun Electronics: SparkFun Load Cell Amplifier - HX711. https://www.sparkfun.com/products/13879. Accessed 30 Jan 2021
28. AVIA Semiconductor: HX711 - 24-Bit Analog-to-Digital Converter (ADC) for Weigh Scales. https://cdn.sparkfun.com/assets/learn_tutorials/5/4/6/hx711F_EN.pdf. Accessed 30 Jan 2021
29. Arduino SA: Arduino Nano. https://store.arduino.cc/arduino-nano. Accessed 30 Jan 2021
30. SCHUNK GmbH & Co. KG: Product information force/torque sensor FT delta. https://schunk.com/fileadmin/pim/docs/IM0019637.PDF. Accessed 30 Jan 2021

Position Estimator for a Follow Line Robot: Comparison of Least Squares and Machine Learning Approaches

Diogo Matos[1(✉)], João Mendes[2], José Lima[1,2], Ana I. Pereira[2], António Valente[1,3], Salviano Soares[3,4], Pedro Costa[1,5], and Paulo Costa[1,5]

[1] INESC TEC-Instituto de Engenharia de Sistemas e Computadores, Tecnologia e Ciência, Porto, Portugal
diogo.m.matos@inesctec.pt

[2] Research Centre in Digitalization and Intelligent Robotics CeDRI, Instituto Politécnico de Bragança, 5300-252 Bragança, Portugal
{joao.cmendes,jllima,apereira}@ipb.pt

[3] Universidade de Trás-os-Montes e Alto Douro, Vila Real, Portugal
{avalente,salblues}@utad.pt

[4] IEETA-Instituto de Engenharia Electrónica e Telemática de Aveiro, Aveiro, Portugal

[5] Faculty of Engineering of University of Porto, Porto, Portugal
{pedrogc,paco}@fe.up.pt

Abstract. Navigation is one of the most important tasks for a mobile robot and the localisation is one of its main requirements. There are several types of localisation solutions such as LiDAR, Radio-frequency and acoustic among others. The well-known line follower has been a solution used for a long time ago and still remains its application, especially in competitions for young researchers that should be captivated to the scientific and technological areas. This paper describes two methodologies to estimate the position of a robot placed on a gradient line and compares them. The Least Squares and the Machine Learning methods are used and the results applied to a real robot allow to validate the proposed approach.

Keywords: Line follower · Localisation · Estimation

1 Introduction

Line detection is a common problem for several types of robots. This kind of robot can be used for military purposes, delivery services, transportation systems, blind assisting applications [1]. Public transportation is also a focused research topic [2] as well as the high speed autonomous movable IoT devices such as line-following robotic vehicles [3].

Every year in Portugal a competition is held named Robot@Factory Lite in the Robotics Open [4], this competition consists in the development of a

© The Author(s), under exclusive license to Springer Nature Switzerland AG 2023
J. M. Cascalho et al. (Eds.): CLAWAR 2022, LNNS 530, pp. 436–447, 2023.
https://doi.org/10.1007/978-3-031-15226-9_41

small scale robot capable of transporting a small cargo from point A to point B. In this competition most robots use navigation based around the follow line concept, however in previous editions the degradation of the guidance lines in the map, due to the repeated passage of robots, presented a significant challenge to competitors.

To solve this problem, the implementation of a calibration methodology is necessary. Normally the calibration process is typically rather complex and time consuming. Therefore there is a need to simplify this process, in order to turn this competition more accessible to younger participants.

This paper addresses two methods to calibrate a Line Follower array sensor and compares them (Least Squares and Machine Learning methods). The used line is a large and gradient one. Large allows to be acquired by more than one sensor, whereas a gradient one allows to present different black levels to different sensors, as presented in Fig. 1.

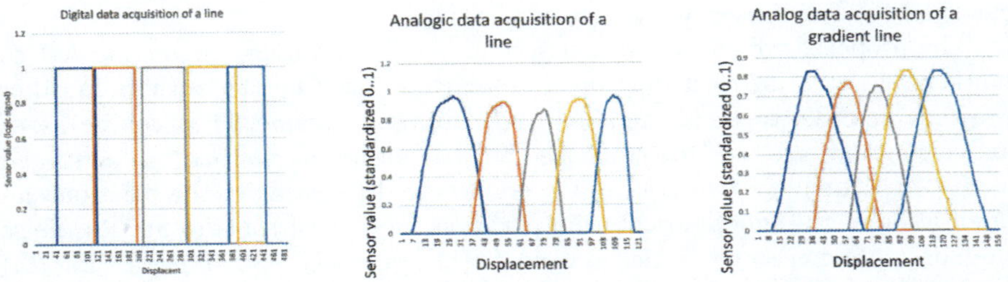

(a) Digital data acquisition of a line.

(b) Analogic data acquisition of a line.

(c) Analogic data acquisition of a gradient line.

Fig. 1. Different data acquisition and line types.

The proposed methodology addressed in this paper can also be implemented in other scenarios with aim of simplifying the calibration process of the optical sensors used in this type of robots.

There has been applied different methods and technologies for acquisition of this lines. There are approaches where a quadratic interpolation technique is used to calculate the centre of the line [5]. On the other hand, there is a research topic related to the minimisation of the line sensors number ensuring efficiency [6]. There is an increase of interest around Educational Robotics that requires a line following procedure avoiding the use of complex sensory systems [7]. Moreover, there are also research works that use learning methods to develop the control of these type of robots [8].

This paper is organised as follows: after a brief introduction where the problem is addressed and a short related work are presented, the Sect. 2 shows the Test-bed architecture used for the localisation of the robot and the data extraction developed application. Then, Sects. 3 and 4 present the Least square Calibration and the Machine Learning methods to estimate the position, respectively.

Section 5 brings the results and the comparison between the proposed methods whereas Sect. 6 finished the paper with conclusions and future work.

2 Testbed Architecture

The main architecture of the developed system is presented by Fig. 2a. There is a planar floor where the robot will move. At each corner, there is an arUco market to localise the arena without calibration need: just point a camera placed above to the arena. A rasberry pi is used to capture the image and identify the arUco markers. The position and orientation of the robot are extracted (as further explained in this section) and sent by ZeroMQ library using Ethernet. A main application runs the modelling and control tasks (and can also be added extra features, such as Digital-Twin among others) and connects the robot by WiFi protocol. The robot owns an ESP32 microcontroller that receives the motors left and right speed commands that using two H-bridges attaches the motors and can send the line sensor array data acquisition.

The proposed robot localisation system is based on image, detecting ArUco markers [9], as a low cost developed system also used by the authors in other projects. In order to avoid the apparatus calibration (camera positioning), four arUco markers are used to estimate the corners of the testbed, as presented in Fig. 2a. Moreover, two arUco markers are used to estimate the position and orientation of the robot (orientation performs more accurate and has less noise than one only marker). In order to validate the approach, a graph paper (Fig. 2b) is placed above the glass to place the robot markers on several positions and orientations.

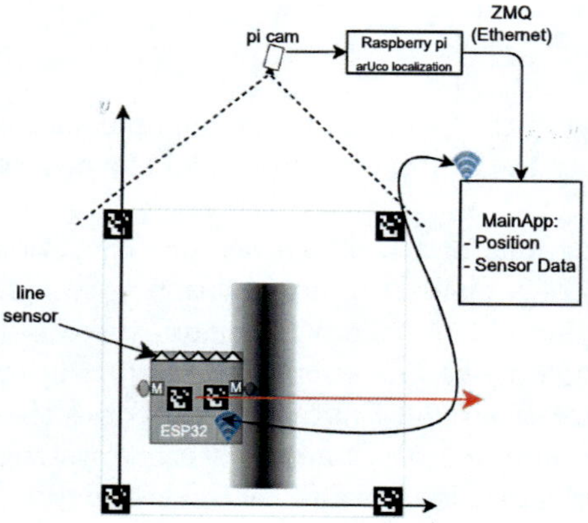

(a) Corner markers are used for the reference; coupled markers are used for robot position measurements.

(b) Testbed equipped with graph paper to measure the accuracy and the introduced error for the arUco markers.

Fig. 2. Used testbed for the experiment and its validation.

The library GSNCodes[1] is used to generate and localize the arUco markers allowing to obtain each marker coordinates, as presented on Fig. 2b. A set of Translation and Rotation should be done on the acquired image to measure the position and orientation of the robot. We will assume three referentials: the camera (O_c), the shifted one to have its origin aligned with the world origin (O_{c0}) and the world referential (O_w). By this way, the position of the camera remains free. According to Fig. 3, a translation is required to place the marker m_1^c on the O_w origin, as ${m_i}^{c0} = m_i^c - m_1^c, \forall i \in 1..6$.

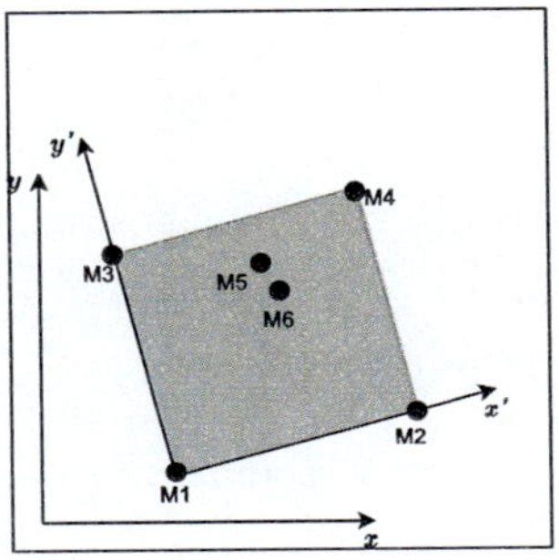

Fig. 3. Reference for the position of each detection.

Once it is not always true that the camera is parallel to the plane, a 3D rotation procedure will be needed. We will assume three referentials: the camera (O_c), the shifted one to have its origin aligned with the world origin (O_{c0}) and the world referential (O_w). Equations 1 and 4 calculate the vectors of x' and y' coordinates.

$$\boldsymbol{v_{12}} = m_2^{c0} - m_1^{c0} \tag{1}$$

$$\boldsymbol{v_{13}} = m_3^{c0} - m_1^{c0} \tag{2}$$

The unit vector for each vector can be calculated as:

$$\widehat{\boldsymbol{v_x}} = \frac{\boldsymbol{v_{12}}}{||\boldsymbol{v_{12}}||} \tag{3}$$

$$\widehat{\boldsymbol{v_y}} = \frac{\boldsymbol{v_{13}}}{||\boldsymbol{v_{13}}||} \tag{4}$$

The unit vector for z axis, can be calculated as the cross product of $\widehat{\boldsymbol{v_x}}$ and $\widehat{\boldsymbol{v_y}}$, according to Eq. 5.

$$\widehat{\boldsymbol{v_z}} = \widehat{\boldsymbol{v_x}} \times \widehat{\boldsymbol{v_y}} \tag{5}$$

[1] https://github.com/GSNCodes/ArUCo-Markers-Pose-Estimation-Generation-Python.

The rotation matrix, R, can be calculated as presented in Eq. 6.

$$R = [\widehat{\boldsymbol{v_x}} \mid \widehat{\boldsymbol{v_Y}} \mid \widehat{\boldsymbol{v_z}}] \tag{6}$$

All the markers positions in space can now be calculated as:

$$m_i^w = R^T.m_i^{c0}, \forall i \in 2..6 \tag{7}$$

Finally, the robot position (Rob^w) be determined as expressed in Eq. 8 whereas the angle of the robot, θ can be expressed in Eq. 9.

$$Rob^w = \frac{m_5^w + m_6^w}{2} \tag{8}$$

$$\theta = atan2({m_6}_y^w - {m_5}_y^w, {m_6}_x^w - {m_5}_x^w) + pi/2 \tag{9}$$

where $atan2(a, b)$ represents the angle that the vectors a and b do. The value $atan2(a, b)$ is located between $-\pi$ and π.

The developed acquisition and localization system is presented in Fig. 4a whereas the developed application to communicate with the robot is presented in Fig. 4b.

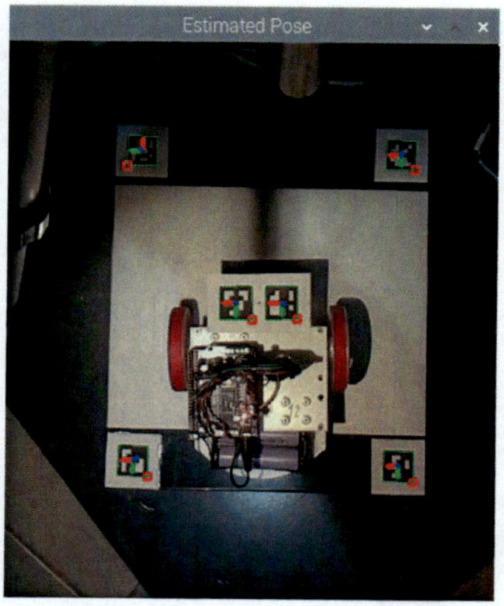

(a) Acquisition system: detection of ArUcos.

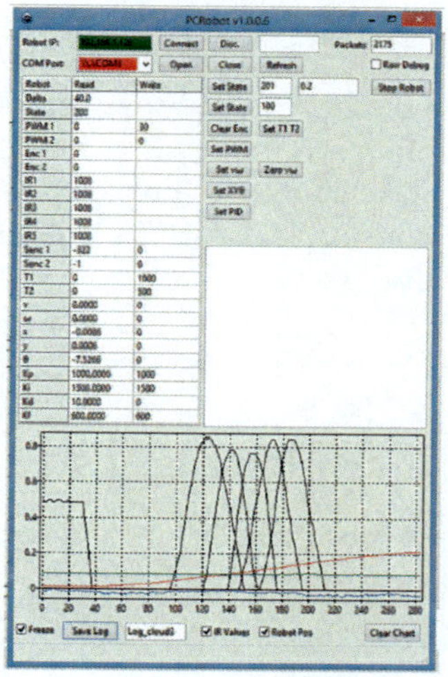

(b) PCRobot - Developed application to communicate with the robot and log generator.

Fig. 4. Used testbed and PCRobot application for the experiment and its validation.

In order to validate the robot localisation system, 256 samples were measured on well established positions (0.05, 0.05), (0.05, 0.15), (0.2, 0.05) with $\theta = 0$ and (0.15, 0.1) with orientations −45, −90, −135 and −180°. Table 1 shows the error on localisation (position and orientation).

Table 1. Error on measuring X, Y and Theta with arUco markers and acquisition software. Number of samples: 256 for each position.

REAL			Average Error			
X (m)	Y (m)	Theta (deg)	X (m)	Y (m)	Theta (deg)	Euclidean Dist (m)
0.05	0.05	0	−3.72E−04	1.28E−03	−1.67E+00	1.34E−03
0.05	0.15	0	7.95E−04	2.39E−03	−7.30E−01	2.65E−03
0.2	0.05	0	3.23E−03	1.62E−03	3.58E−01	3.62E−03
0.15	0.1	0	1.10E−03	1.57E−03	−6.12E−01	1.92E−03
0.15	0.1	−45	2.09E−04	1.57E−03	−1.17E−01	1.60E−03
0.15	0.1	−90	1.97E−03	1.32E−03	−1.00E+00	2.37E−03
0.15	0.1	−135	2.49E−04	1.12E−03	6.81E−01	1.26E−03
0.15	0.1	−180	9.13E−04	4.66E−04	1.63E−01	1.03E−03

As result, it is possible to verify that the average error on X axis is 0.001020 m, on y axis is 0.001436 m, theta average error is −0.394214 degrees and the Euclidean distance error is 0.001967 m.

This application allows to create a log from the sensors data that can be further processed, as detailed in next sections.

3 Least Square Calibration

The Least Square method was first design and developed to solve navigation problems in the astronomy field [10,11]. Nowadays, the Least Square method is a regression analysis method, it is typically used for approximate the solution of overdetermined systems [12]. The implementations based around this method are usually characterise by belonging to one of two classes: Linear Least Squares [13] and Non-linear Least Squares [14].

The selection of the type of Least Square regression method to used is intrinsically connected to the nature of the data available for the approximation of a estimation function. This approximation is commonly designated as the fitting process. This process consists in adjusting the parameters of a model function to best fit a certain data set [15]. Therefore if a certain dataset belongs to system that is controled/modelled using a linear function, an implementation from the linear least squares class should be used. The same applies to the non-linear least squares class of implementations. Independently of the type of implementation used, this methodology has as its core principal the minimisation of the sum of the squares of the residuals. The residuals are calculated as being the difference between the observed value and the the fitted value generated by a model [16]. By minimising the sum of these squared residuals the precision of the estimating function is increased, therefore its estimation will generate values closer to those observed in the real system.

Given that the main target of this study is to estimate the position of the robot in relation to the gradient line using the data provided by five sensors. Since all movement equations of this robot reside within the linear function domain, a linear implementation of the least squares methodology was chosen.

Within the Linear least squares (LLS) class their are several diferent types of possible implementations. The three most commonly used implementations are:

1. Ordinary least squares (OLS), which is commonly used to analyse both experimental and observational data [17].
2. Weighted least squares (WLS) are used when heteroscedasticity is present in the error terms of the model [17].
3. Generalized least squares (GLS) used mainly when both heteroscedasticity and correlations are present in the error terms of the model [17].

3.1 Methodology

For the propose of this study a Ordinary least squares (OLS) implementation was used. This implementation uses two matrices, matrix X and matrix Y to generate the estimator parameters under the form of matrix K.

Matrix Y is generated by compiling the data obtained from the ground truth system into a number of samples x 3 matrix. In this matrix each line is comprised by the X coordinate, the Y coordinate and the orientation (θ) value of the robot, computed by the vision localisation system. This is represented in Eq. 10, where m is the total number of samples.

$$Y = \begin{bmatrix} X_0 & Y_0 & \theta_0 \\ ... & ... & ... \\ X_m & Y_m & \theta_m \end{bmatrix} \tag{10}$$

Matrix X is generated by compiling the data obtained from the five sensors into a matrix with the dimensions: number of samples x 5. Similar to the Y matrix, each line is comprised by the data of each of the sensors, this matrix is represented in Eq. 11.

$$X = \begin{bmatrix} S1_0 & S2_0 & S3_0 & S4_0 & S5_0 \\ ... & ... & ... & ... & ... \\ S1_m & S2_m & S3_m & S4_m & S5_m \end{bmatrix} \tag{11}$$

The estimator parameters are then generated using Eq. 12.

$$K = (X^T X)^{-1} X^T Y \tag{12}$$

The values from matrix K can then be used to estimate the behaviour of the system when it is presented with new data. This new data comes in the form of new values for the sensor readings, which then allow us to conversely calculate X coordinate, the Y coordinate and the orientation (θ) value of the robot, using the estimator obtained from the regression. This is represented in Eq. 13, an example of a comparison between the ground truth and the estimated results can be seen in Fig. 5.

$$Ye = XK \tag{13}$$

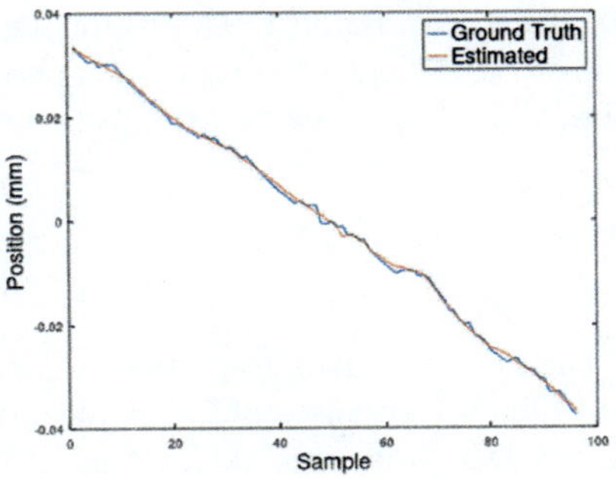

Fig. 5. Comparison between ground truth and the estimated values.

4 Machine Learning

In recent years there have been significant advances in the field of machine learning, allowing the use of this type of algorithms in practically all areas, robotics not being an exception [18]. Machine learning can be broadly defined as computational methods that use the experience to improve performance or make accurate predictions. This process has generalization as to its fundamental operating principle. This generalization is performed from the model training data (experience), data that provide essential characteristics of the problem. From these, a function of a set of hypotheses is formulated, which is a subset of the family of all functions. The selected function is subsequently used to label all instances, including unseen examples (test data) [19]. This process is applied to several standard tasks, the best-known being classification, regression, and clustering. Regression consists of predicting actual values for each item, the penalty for an incorrect prediction depends on the magnitude of the difference between the true and predicted values, in contrast to the classification problem, where there is usually no notion of proximity between various categories. The main objective of a regression problem is to generate accurate predictions for unseen items and design efficient and robust algorithms to produce these predictions even for large-scale problems.

In this way, and after analyzing several Machine Learning methods available for regression problems, the Decision Tree method was selected. This choice fell on this method due to its easy interpretation, quickly identifying which variables have the most significant influence on the predicted data. In addition to this advantage, it is a method that does not require a long training process and, therefore, can save a lot of modeling time [20].

Decision trees are one of several supervised machine learning methods. These are obtained by recursively partitioning the data space and fitting a simple prediction model within each partition. As a result, the partitioning can be graphically represented as a decision tree, which can be used for classification and regression. Classification trees are designed for dependent variables that assume a finite number of unordered values, with prediction error measured in terms of misclassification cost, in turn, regression trees are for dependent variables that assume continuous or discrete values ordered, and its prediction error is usually measured by the squared difference between the observed and predicted values [21].

4.1 Methodology

In order to implement the Decision Tree Regressor algorithm, the dataset presented above was used. This set consists of 137 values, referring to the robot position (Y) and the data of the line sensors (X) present in Eq. 10, 11, as shown in Fig. 2b.

Once the training data set to model the algorithm was established, some of its hyperparameters (max features, min samples leaf, ...) were also optimized using an optimization function (GridSearchCV[2]) implemented with ten times cross validation. By way of comparison, the prediction was also made using the same dataset combined with a simple Linear Regression method, without any optimization. To test the effectiveness of the two algorithms, a prediction of the robot's position (Y) was performed from a data set (X) totally new to the models. This new dataset, derived from another simulation, is composed of 114 entries and the predictions obtained will be presented below.

Observing the data regarding the prediction and the ground truth in the Fig. 6, it is possible to verify that the decision tree model presents better results, achieving a better prediction of the robot's position. Next section will stress and compare both methodologies.

5 Results

In addition to the results obtained with the test set presented above and taking into account the variation of readings given by the line sensors, four different situations were simulated to amplify the test set and thus test the effectiveness of the prediction models for new types of readings. To facilitate this evaluation, two usual metrics were used, the Root Mean Square Error (RMSE) and Coefficient of determination (R^2) [22].

[2] https://scikit-learn.org/stable/modules/generated/sklearn.model_selection.GridSearchCV.html.

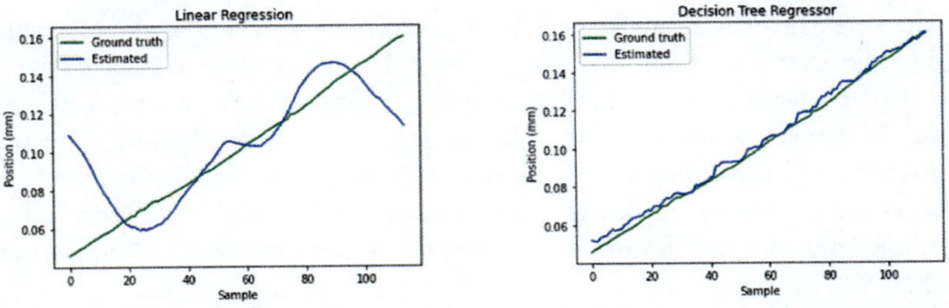

(a) Linear Regression model prediction.

(b) Decision tree regressor model prediction.

Fig. 6. Machine Learning Methods comparison

In this way, the matrix X Eq. 11 was modified, and after its modification, the robot positions were predicted. The changes tend to simulate different types of reading, thus testing their behaviour in various types of sensors and other reading conditions (luminosity, line intensity, ...). The first changes affect the entire length of X, for this, all data has been increased and decreased by 20%. Then, two new readings were simulated where only the central sensor of the matrix X was increased and decreased by 20%. This results in four different test sets that aim to cover the various possible readings of this type of sensors.

Table 2. Comparison of the results obtained from the implemented methods.

Method	Type of test set	RMSE	R^2
Least square regression	Decreased 20% on the center sensor	0.0032	0.9710
	Decreased 20% on all sensors	0.0071	0.9554
	Original Data	0.0025	0.9708
	Increased 20% on all sensors	0.0043	0.9809
	Increased 20% on the center sensor	0.0033	0.9621
Linear regression	Decreased 20% on the center sensor	0.0203	0.6299
	Decreased 20%	0.0210	0.6022
	Original Data	0.0203	0.6296
	Increased 20%	0.0211	0.5974
	Increased 20% on the center sensor	0.0203	0.6299
Decision tree regressor	Decreased 20% on the center sensor	0.0029	0.9923
	Decreased 20% on all sensors	0.0053	0.9743
	Original Data	0.0028	0.9929
	Increased 20% on all sensors	0.0034	0.9896
	Increased 20% on the center sensor	0.0029	0.9923

Observing the values in Table 2, it is possible to notice a clear advantage in using the Decision Tree Regressor method, noting that values closer to 0 in the RMSE metric are better, as there is a perfect forecast with RMSE=0 and, on the contrary, an R^2 =1 is also a 100% correct prediction. The results obtained through the simulations are also encouraging when we observe that, in general, results are very similar to those obtained with real data. Thus, allowing to mention that the predictions continued to be admissible for different sensor readings.

6 Conclusions and Future Work

This paper proposed two methods for position estimation of a robot using the data from an optic line sensor array. It was used an image based system to localize the robot in the testbed environment. Both methods (Least Squares and Machine Learning) were implemented and tested in a real scenario, where a gradient line is is used get data from the line sensor. Results are promising and showed that Decision Tree approach obtained better results.

As future work, the implementation of both methods in an embedded system will evaluate the required computational resources (memory and processing time). It also will address the availability of each method implementation with limited hardware resources.

Acknowledgements. This work has been supported by FCT - Fundação para a Ciência e Tecnologia within the Project Scope: UIDB/05757/2020 and also by National Funds through the Portuguese funding agency, FCT - Fundação para a Ciência e a Tecnologia, within project LA/P/0063/2020.

References

1. Pakdaman, M., Sanaatiyan, M.M., Ghahroudi, M.R.: A line follower robot from design to implementation: technical issues and problems. In: 2010 The 2nd International Conference on Computer and Automation Engineering (ICCAE), pp. 5–9 (2010). https://doi.org/10.1109/ICCAE.2010.5451881
2. Gumus, O., Topaloglu, M., Ozcelik, D.: The use of computer controlled line follower robots in public transport. Procedia Comput. Sci. **102**, 202–208 (2016). ISSN 1877-0509
3. Eleftheriou, G., Doitsidis, L., Zinonos, Z., Chatzichristofis, S.A.: A fuzzy rule-based control system for fast line-following robots. In: 2020 16th International Conference on Distributed Computing in Sensor Systems (DCOSS), pp. 388–395 (2020). https://doi.org/10.1109/DCOSS49796.2020.00068
4. Lima, J., et al.: An industry 4.0 approach for the robot@factory lite competition. In: 2020 IEEE International Conference on Autonomous Robot Systems and Competitions (ICARSC), pp. 239–244 (2020). https://doi.org/10.1109/ICARSC49921.2020.9096164
5. Engin, M., Engin, D.: Path planning of line follower robot. In: 2012 5th European DSP Education and Research Conference (EDERC), pp. 1–5 (2012). https://doi.org/10.1109/EDERC.2012.6532213

6. Chowdhury, N.H., Khushi, D., Rashid, M.M.: Algorithm for line follower robots to follow critical paths with minimum number of sensors. Int. J. Comput. (IJC) **24**(1), 13–22 (2017)
7. Evripidou, S., Georgiou, K., Doitsidis, L., Amanatiadis, A.A., Zinonos, Z., Chatzichristofis, S.A.: Educational robotics: platforms, competitions and expected learning outcomes. IEEE Access **8**, 219534–219562 (2020). https://doi.org/10.1109/ACCESS.2020.3042555
8. Saadatmand, S., Azizi, S., Kavousi, M., Wunsch, D.: Autonomous control of a line follower robot using a Q-learning controller. In: 2020 10th Annual Computing and Communication Workshop and Conference (CCWC), pp. 0556–0561 (2020) .https://doi.org/10.1109/CCWC47524.2020.9031160
9. Li, B., Wu, J., Tan, X., Wang, B.: ArUco marker detection under occlusion using convolutional neural network. In: 2020 5th International Conference on Automation, Control and Robotics Engineering (CACRE), pp. 706–711 (2020) https://doi.org/10.1109/CACRE50138.2020.9230250
10. Stigler, S.M.: The History of Statistics: The Measurement of Uncertainty Before 1900. Belknap Press of Harvard University Press, Cambridge, MA (1986).ISBN 978-0-674-40340-6
11. Legendre, A.M.: Nouvelles méthodes pour la détermination des orbites des comètes [New Methods for the Determination of the Orbits of Comets] (in French), Paris: F. Didot. hdl:2027/nyp.33433069112559 (1805)
12. Björck, Å.: Least Squares Methods, Handbook of Numerical Analysis, Elsevier, vol. 1, pp. 465–652 (1990). ISSN 1570–8659, ISBN 9780444703668. https://doi.org/10.1016/S1570-8659(05)80036-5
13. Tofallis, C.: Least squares percentage regression. J. Mod. Appl. Stat. Methods **7**, 526–534 (2009). https://doi.org/10.2139/ssrn.1406472. SSRN 1406472
14. Strutz, T.: Data Fitting and Uncertainty : A Practical Introduction to Weighted Least Squares and Beyond (2nd ed.). Springer Vieweg. (2016). ISBN 978-3-658-11455-8
15. Dekking, F.M., Kraaikamp, C., Lopuhaä, H.P., Meester, L.E.: A modern introduction to probability and statistics. STS, Springer, London (2005). https://doi.org/10.1007/1-84628-168-7. ISBN 978-1-85233-896-1. OCLC 262680588
16. Gregersen, E.: The Britannica guide to statistics and probability. Britannica Educational Pub. in association with Rosen Educational Services, New York, NY (2011)
17. Carroll, R.J.: Adapting for heteroscedasticity in linear models. Ann. Stat. **10**(4), 1224–1233 (1982). https://doi.org/10.1214/aos/1176345987. JSTOR 2240725
18. Xiao, X., Liu, B., Warnell, G., et al.: Motion planning and control for mobile robot navigation using machine learning: a survey. Auton. Robot. (2022). https://doi.org/10.1007/s10514-022-10039-8
19. Mohri, M., Rostamizadeh, A., Talwalkar, A.: Foundations of machine learning. MIT press (2018)
20. Zhao, Y., Zhang, Y.: Comparison of decision tree methods for finding active objects. Adv. Space Res. **41**(12), 1955–1959 (2008)
21. Loh, W.Y.: Classification and regression trees. Wiley Interdisc. Rev. Data Min. Knowl. Discovery **1**(1), 14–23 (2011)
22. Chicco, D., Warrens, M.J., Jurman, G.: In regression analysis evaluation, the coefficient of determination R-squared is more informative than SMAPE, MAE, MAPE MSE and RMSE. PeerJ Comput. Sci. **7**, e623 (2021). https://doi.org/10.7717/peerj-cs.623

Remote VDB-Mapping: A Level-Based Data Reduction Framework for Distributed Mapping

Marvin Grosse Besselmann(✉), Arne Rönnau, and Rüdiger Dillmann

Department of Interactive Diagnosis and Service Systems (IDS),
FZI Research Center for Information Technology, 76131 Karlsruhe, Germany
{GrosseBesselmann,roennau,dillmann}@fzi.de

Abstract. In recent times, the distribution of mobile walking robots has increased sharply due to their high availability. To utilize the increased mobility of these robots, autonomous navigation algorithms require fast and accurate volumetric maps, generated by the sensors of the robot. Usually these maps have to be transferred between different systems, like for example in a multi robot setup or if a operator needs access to the data on a remote control room. Since the memory footprint of the data scales indefinitely with the resolution and size of the environment, these maps can become quite large. This is especially the case in volumetric 3D maps due to the added third dimension. Sending the whole map over a network would quickly overburden a wireless network. In this paper we present a distributed mapping approach, which is able to efficiently share identical maps between different machines. Further this approach can be be used to combine data from various sources (e.g. collaborative robot teams) into a single map. This is archived by only exchanging memory efficient reduced update grids between the systems. We present a level based approach which is able to further reduce the memory footprint of the update grids depending on different use-cases. The algorithm can easily integrate various systems and was evaluated using an ANYmal robot operated over a remote base station. The presented experiments show, that this approach is able to significantly reduce the amount network traffic required in a remote mapping scenario.

Keywords: Robot mapping · Navigation · Robot teams · Distributed computing

1 Introduction

In shared or fully autonomous navigation is imperative for the robot to possess an accurate map representation of the present environment (Fig. 1). These maps are generated by evaluating the combined information of one or multiple range sensors (e.g. LIDAR). Based on these maps, it is possible to plan missions, search for possible paths between multiple locations or avoid obstacles while driving.

© The Author(s), under exclusive license to Springer Nature Switzerland AG 2023
J. M. Cascalho et al. (Eds.): CLAWAR 2022, LNNS 530, pp. 448–459, 2023.
https://doi.org/10.1007/978-3-031-15226-9_42

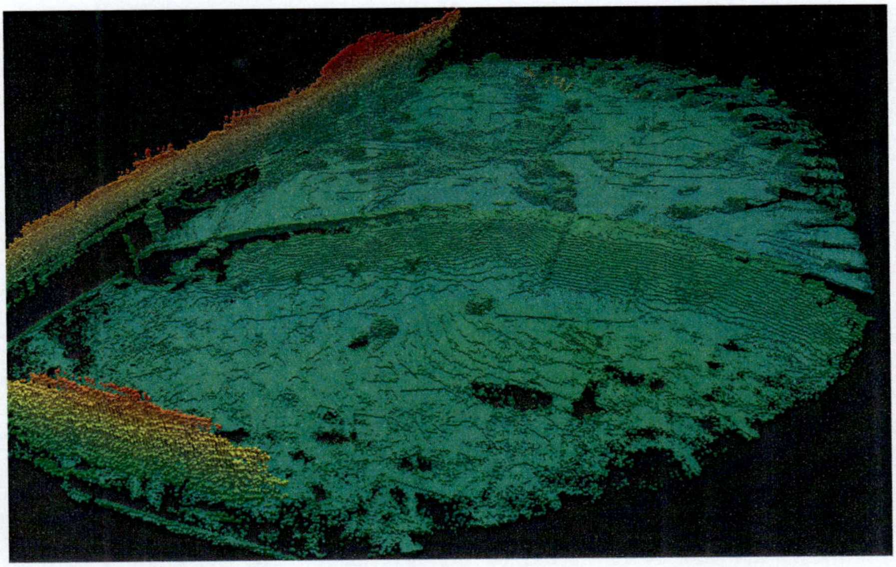

Fig. 1. Remote VDB-Map captured by an ANYmal with a mounted Velodyne VLP-16 during the ESA Space Resource Challenge.

Furthermore an operator on a remote base station can use the map to command high level functions of the robot based on the visible information. In order to efficiently utilize these maps, first they have to be shared over the network. However their required memory footprint can, depending on the enclosed space, grow quite large.

For simple environments (e.g. industrial environments like warehouses), these maps are often projections onto a 2D plane. Occupancy Grid Maps [1] discretise the world into a 2D grid representation in which each cell represents an area of the environment. To determine the state of the corresponding area these cell contain a probability value which specifies whether the area is free or occupied by obstacles. However in more complex environments a simple projection onto a 2D plane is no longer a valid option. Herbert et al. [2] propose the 2.5D maps, which also abstract the world as a 2D grid. Additionally each cell of the grid contains an value which encodes the height of the environment in this area. In Fankhauser et al. [3] this grid was extended with multiple layers which could contain additional information like for example surface normals. To decrease the memory footprint of 2.5D maps Heppner et al. [4] proposed to used a quadtree data structure instead of a full 2D grid. Triebel et al. [5] refined the concept in their Multi-Level Surface Maps, by stacking multiple layers of 2.5D maps onto each other to support environment where a single height value is no longer sufficient.

Recently, versatile mobile robotic platforms like walking robots [6,7] and unmanned aerial vehicles(UAVs) became much more accessible. These robots often operates in complex environments during search and rescue missions in which full 3D representations of the environment are necessary. Hornung et al. [8] proposed the OctoMap, which apply the idea of Occupancy Grid Maps onto a 3D octree. More recently Grosse Besselmann et al. [9] developed a similar approach using OpenVDB [10] as data layer, to increase the performance during the integrating new data into the map.

All these approaches are mainly focused on local map building and ignore the efficient transfer of data between different systems. The efficient transfer of large maps is here widely disregarded. In order to solve this shortcoming a popular approach is, to split the map into smaller parts, which have a much smaller memory footprint. In Hess et al. [11] the maps are split into sub-maps covering only a small portion of the environment. Here each submap contains a transformation which is continuously update by the graph SLAM back-end. This way the individual maps can be subsequently stitched together to create a single consistent map. Fox et al. [12] propose a similar approach in which complete parts of the map are send to remote instances. Here the map is subsequently aligned to a local map to estimate the relative transform between them. These approaches are mostly restricted to the 2D case, where the amount of data is comparably low due to the missing third dimension. In Petitti et al. [13] the distributed systems try to register if two robots are close to each other. If both operate in the same area, they periodically exchange their entire map as a Point Cloud to their neighbors. Usually the raw sensor data abstraction and the integration of the abstracted information into the map are tightly coupled into a single system. This leads to rather inflexible behavior since the only interface to the whole mapping system is the resulting map. As a result the processed information can only be used locally.

In this paper we present mapping approach extending VDB-Mapping [9] which can natively be run memory efficiently on a network of different machines to generate combined maps. To achieve this goal the mapping process is split into two distinct steps. First the raw sensor data is abstracted into a memory reduced map update, which can be transferred over the network. In order to support different use cases and scenarios the size and nature of the transferred update grids can be configure providing multiple levels of data reduction. Afterwards these updates can be applied to the local as well as remote mapping instances. As a result, each systems generates the same data, without sending the entire map over the network. The structure of this paper is as follows:

- In Sect. 2 a detailed description of the distributed mapping process and the different data reduction levels is given.
- Section 3 contains an in-depth evaluation of the mapping system. It presents multiple experiments evaluating the effect of the different level of data reduction.
- Section 4 highlights the prospects and limitations of the presented distributed mapping approach. Furthermore, a discussion about future extensions is presented.

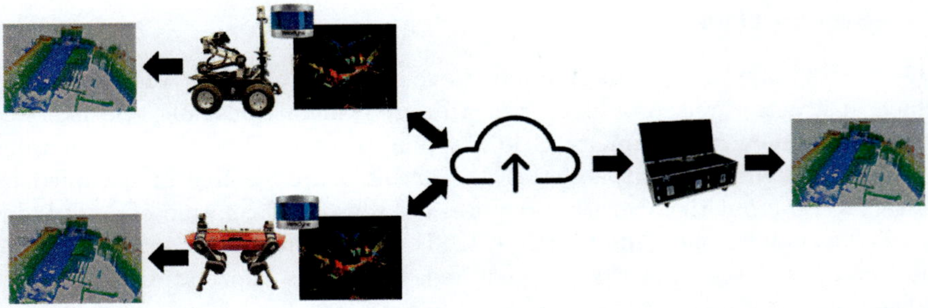

Fig. 2. System setup during remote operation. Both robots on the left gather data which is exchanged over the network.

2 Approach

Complex tasks often require robots to perform as a collaborative team. In our use-case, depicted in Fig. 2, the system setup consisted of multiple mobile robots, which during operation exchanged sensor information to generate a complete map of the surrounding environment. Additionally a remote base station was used which visualized the resulting map and where an operator was able control the robot using this information. The mapping approach described in this paper is based on VDB-Mapping and was extended with remote mapping capabilities. Up to now it was tested on a range of different walking robots (e.g. Lauron V, ANYmal, Spot), UGVs (e.g. Husky, Fetch Freight) and even flying drones (Matrice100). The environmental information is gathered using a LIDAR (in this case a Velodyne VLP-16 or HDL-32E, as well as multiple Intel Realsense RGB-D cameras per robot) and an additional inertial measurement unit (IMU). To fuse the data and estimate the current robot position we applied the Google Cartographer [11] SLAM algorithm. For the internal as well as external communication between the different systems the Robot Operating System (ROS) [14] was extensively used.

One of the main problems for remote mapping, is the huge amount of data required to be send over a wireless network. Sending the raw data, gathered by a LIDAR, results in huge amounts of 3D point data. Depending on the type, these kind of sensors generate about 400.000 to 1.39 million data points per second. Even considering efficient data structures like Point Clouds [15] sending this amount of data constantly over the network quickly becomes unfeasible. Alternatively sending the already generated map is equally unfeasible if not even worse. Since the size of the map scales directly with the enclosing space, and the desired resolution, the memory footprint is basically unbounded. In order to remedy this problem we propose a remote volumetric mapping framework based on small update VDB-Grids. Using a level based data reduction scheme, the memory footprint of these updates can be reduced to cope with the network bottleneck.

2.1 System Overview

The remote mapping system described in this work is based on VDB-Mapping, which abstracts raw sensor data into a discrete hierarchical 3D grid-like tree structure. Similar to Occupancy Grid Mapping [1] each cell of this grid contains a probability which specify whether the enclosed area is free or occupied by obstacles. As input the algorithm requires the sensor information of a 3D Lidar encoded as point cloud. Additionally the 3D pose of the robot corresponding to the location the scan was taken is required. Using this information, each point of the sensor data is raycasted through the data structure. This is done to find all free and occupied cells traversed by the sensors ray for a subsequent map update step.

In our approach we split the mapping process into two distinct steps. First the incoming raw sensor data is abstracted into an update which is subsequently applied to the map. Instead of raycasting each data point individually and directly updating each traversed cell, we first gather and combine the information of all raycasts. For this an additional update VDB-Grid is created, which is co-aligned with the actual map. Here for each cell traversed by the raycasting a flag is set specifying in which way the corresponding cell in the map has to be updated. The prior gathering of all necessary data has two significant advantages. First it prevents the integration of contradictory sensor data caused by the discretisation of the environment. And second, we only update each cell only once, instead of each time an additional beam crosses the same cell during raycasting. This effect is especially problematic in close proximity to the sensor where all beams are still closely bundled together. Both these effects would negatively impact the occupancy probability formulation and lead to incorrect map representation. After this process, each cell in the update grid is traversed and the corresponding cell in the map according to the set flag updated.

In this work we leverage this divided update process to create efficient map updates which can be shared between a host (i.e. the local robot which generates the data) and a remote system (i.e. the system which receives the data and also creates a copy of the map from it). In this case the creation of the map update is a byproduct of the local mapping process which does not require any further computation. To reduce the memory footprint of the send update grid it is compressed and serialized before sending it as a bit-stream over the network. On the remote side the bit-stream can then be unpacked again and applied as map update to the remote map just like it was on the host side. By using ROS as a communication layer, the update grid is published to the complete network and can be received by multiple instances (e.g. an additional robot and the remote base station) at once. The memory footprint for a map update is adjustable by a level-based data reduction scheme to further decrease the network load. In the following sections, the different data reduction levels and their use cases are closes described.

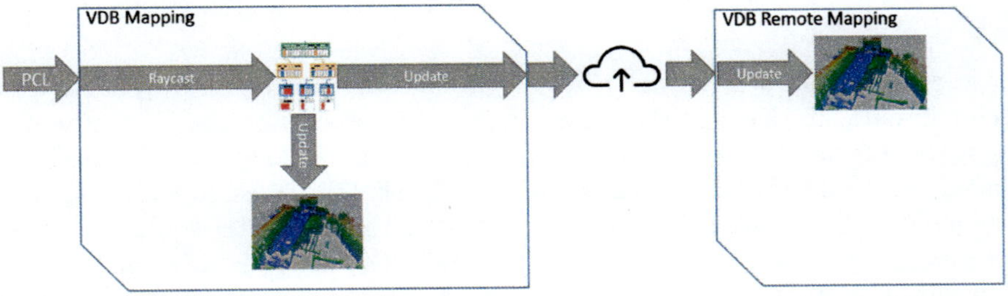

Fig. 3. Data reduction level 0: This is the default behavior of VDB-Mapping. The data is first raycasted into an update grid which is subsequently applied to the map. On the remote side the already raycasted data is directly applied to the map.

2.2 Level 0

Level 0 represents the standard operating behavior of VDB-Mapping which can be seen in Fig. 3. First all raycasting data is combined within the update grid. Afterwards the update grid is locally integrated into the current map. On the remote side, the identical integration process is executed to add the new data into the remote map. As a result we receive the same map on both systems without any loss of information. This level does not always reduce the amount of transferred data. In fact the transferred update grid is in some cases even larger than the input point cloud. This is due to the fact that the update grid not only contains information about the occupied cells but also about all free cells in between. Covering both cases results in a much larger number of relevant cells, than the amount of initial data points. However, the expensive operation of raycasting each point has already be done on the host system. Therefore the remote system has only to integrate the present data, which is comparatively cheap. This behavior is especially interesting for remote systems with low computational power, like for example flying drone.

2.3 Level 1

Level 1, depicted in Fig. 4 provides a medium data reduction compared to the point cloud size, without dropping any information. In order to archive this, we first integrate all raw data points into the update grid. This way we already filter out data points by discretising them due to the resolution of the map. As a result the total number of points that have to be send over the network is reduced without losing relevant information. Subsequently each data point is raycasted on both the host and remote system. In this case as well the data of all rays is first gathered within an update grid to prevent contradictory data. Afterwards the update grid is applied to create identical maps on both sides. Since the same operation is performed on both side no information contained in the sensor data is lost. This level is adapted for systems with similar computational specifications, where both are able to perform the raycasting.

2.4 Level 2

Data reduction level 2 shown in Fig. 5 provides the largest data reduction of the map data. In this case the complete mapping process on the host system is identical to level 0. However during the update process, for each updated cell the current and last state is compared. If the probability crossed the occupancy threshold, either above or below, a flag is set for this cell in the update grid. As a result the update grid that will be send over the network only contains the cells which occupancy state has changed in the last time step. This significantly reduces the number of relevant cells in the update grid. Similar to the other levels this grid is send over the network and applied to the remote map. However in this case we directly overwrite the binary occupancy state of the remote map. As a result we lose all information about about the probabilistic framework of the host map. Therefore this reduction scheme is more suited for a remote instance which only needs to visualize the data. Another use-case would be a system, which only relies on external data without integrating own data into a probabilistic grid.

3 Results and Experiments

This section provides an in-depth evaluation of the presented remote mapping approach. The performance of the algorithm can be directly measured by the reduction in data size send over the network. The second relevant property is the runtime of the algorithm. To evaluate the performance of the data reduction capacities of the algorithm, we first performed multiple tests using maps of different resolutions. This way we also gained an insight on how the size of the grid influences the used bandwidth. In the second part we took a closer look on the time it took to integrate new data into the map. Here as well VDB-Grids of various resolutions are used. The point cloud data used in this evaluation was created by a robot mounted Velodyne VLP-16. In order to align the incoming point clouds to each other Google Cartographer was used. All presented test cases were performed on a standard consumer laptop containing an Intel Core i7-6820HQ, 2.70 GHz.

3.1 Update Bandwidth Usage

The first experiment covers the emerging amount of used bandwidth of the generated map updates. These rates are evaluated for each level individually. As a general reference we can use the used bandwidth of the input point clouds used for the mapping. These pre-aligned clouds contain around 40000 data points per scan on a data rate 10 Hz resulting in roughly 400.000 data points per second. Data wise this gives us a network load of 2.79 MB/s as a comparatively baseline. Table 1 shows the bandwidth usage results of the different levels.

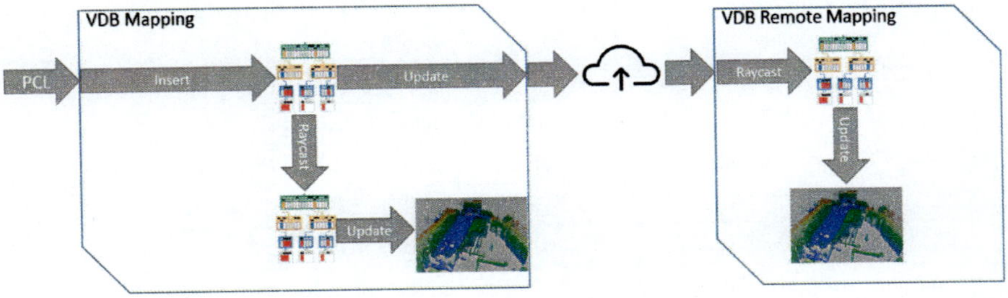

Fig. 4. Data reduction level 1: Here the raw data is directly inserted into the update grid structure. Afterwards both the local and remote side have to raycast the data and integrate the updates into the map.

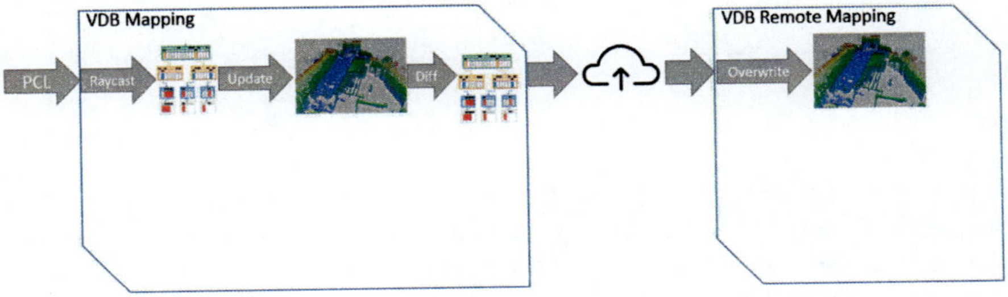

Fig. 5. Data reduction level 2: In this case, the complete mapping including raycasting and data integration is performed on the host system. During this process the algorithm tracks which grid cells changed their occupancy state and adds these cells to the update grid. On the remote system the update grid is used to overwrite the changed occupancy state. This will result in a loss of probability information.

From this data it is evident that the chosen grid resolution of the map has a huge impact on the amount of used bandwidth. Since the update grid has the same resolution as the actual map, it is obvious that a reduction of the resolution lead in general to smaller amount of data cells that have to be transmitted over the network. In case of reduction level 0 this even leads to higher use of bandwidth than of the input point cloud up to a resolution of 10 cm. This is due to the fact, that not only the end points of the LIDAR scan but also all raycasted points in between are encoded in the update. From a 10 cm resolution on however, we receive a data reduction even on this level. Level 1 on the other hand directly outperforms the input point cloud data wise. This is due to the fact that the amount of relevant data points is at most equal to the number of points in the input cloud. Additionally points are reduced due to the discretisation of the grid, leading to an even smaller amount of data as the resolution decreases. On the level 2 data reduction, only the cells which state has changed in the last time step are gathered in the update grid. Unexpectedly the data reduction of level 2 is larger than the input cloud when used on high resolutions. This effect can be explained by the low discretisation of the grid. The raycasting often traverses

Table 1. Resulting bandwidth usage for grids of different resolutions.

Map resolution	Used bandwidth (MB/s)		
	Level 0	Level 1	Level 2
2 cm	28.23	2.63	4.64
4 cm	12.65	2.34	1.29
6 cm	6.54	1.73	0.75
8 cm	3.17	1.08	0.45
10 cm	1.79	0.76	0.32
12 cm	1.16	0.52	0.26
14 cm	0.81	0.41	0.19
16 cm	0.62	0.35	0.17
18 cm	0.46	0.27	0.15
20 cm	0.38	0.24	0.13

through new, previously unobserved cells and immediately changes their occupancy cell. As a result the update grid contains all these free and occupied cells. However, as the resolution drops, the amount of necessary data for an update decreases significantly. Nevertheless this comes, as previously mentioned, at the cost of losing the probability data on the remote map.

3.2 Integration Time

In the second experiment we examine the required computation time for integrating the updates into the remote map. Here as well we evaluate the effect of differently sized VDB-Grids on the performance for each data reduction level. The desired goal of this mapping algorithm is to reach the data rate of the incoming sensor data. Otherwise it would drop data which in turn leads to incomplete maps of the environment. On the other hand the approach would also benefit from even higher rates as this would allow to integrate more data from different sensors simultaneously. The here used Velodyne VLP-16 offers a scan rate 10 Hz. Table 2 shows the result of the integration time experiments on various grid resolutions.

In this case as well the resolution directly influences the performance of the algorithm since it determines the number of cells which has to be updated. Apart from the 2 cm resolution the integration time of level 0 still stays below the sensor data rate. This is due to the effect that on the remote side only the probability update of each traversed cell is necessary. As previously mentioned this effect can be leveraged for systems with low computational power. The integration times of level 1 however are much higher, since in this case the raycasting has to be performed on the remote side as well. In contrast to the integration times of level 0 and level 1 the required time for updates on level 2 are nearly negligible. This

Table 2. Impact of grid resolution and data reduction levels to the integration time on remote maps.

	Integration time (ms)		
	Level 0	Level 1	Level 2
2 cm	297.10	552.71	2.920
4 cm	85.50	226.64	0.589
6 cm	39.61	135.82	0.281
8 cm	20.55	88.15	0.178
10 cm	11.88	65.23	0.122
12 cm	7.56	49.82	0.086
14 cm	5.21	41.62	0.067
16 cm	3.86	34.97	0.060
18 cm	2.83	30.15	0.040
20 cm	2.26	25.01	0.034

is due to the fact, that here only the state of the tree nodes has to be flipped. As a result the data stored within the leaf nodes of the VDB-tree do not have to be accessed.

In conclusion we have shown that our remote mapping approach can be used to significantly reduce the network traffic. Depending on the map resolution and the chosen data reduction level we are even able to reach the data rate of the sensor while still not overburden the network.

4 Conclusion and Future Works

In this paper we presented a level based remote mapping algorithm based on VDB-Mapping. The approach utilizes small update grids, which size can be tailored to different use cases and system properties. The reduction of size allows us to efficiently transfer these update grids over wireless networks. These update grids can subsequently be applied to local or remote mapping instances equally. As a result, the presented extension enables us the simultaneously create identical maps on different remote systems. The presented evaluation shows, that the network load compared to the input data and especially compared to sending the entire map could be reduced significantly. Furthermore this approach can be used to customize the time it takes to integrate new sensor data to support specific use-cases.

The key insights of this work are:

- **Reduced use of bandwidth:** Since we only send small updates instead of the entire map we were able to reduce the necessary network traffic for remote mapping. This effect could be further increased by applying different levels of data reduction.

- **Distributed mapping:** By applying the same update grids on different system, they all create identical maps of the surrounding area. These can subsequently be used to cooperatively work in a shared environment. Furthermore this can be used to cope for an occluded field of view of one system by integrated remote information.
- **Additional flexibility:** By not tightly coupling the data abstraction and the data integration we are able to create more customizable behavior.

This approach shows great potential for autonomous robot teams, which cooperatively explore and navigate through complex environments. All tests shown in this work are performed on a VDB-Grid containing 32-bit floating point data types. However current developments of VDB-Mapping enable us to use custom data types as leaf values of the VDB-tree. As these datatypes are more complex, the required memory footprint of the map as well as the update grids would increase. Future work should investigate, how custom data types in VDB-Maps can be better supported. As these can become quite large it should be examined, how the data necessary for an update can be better encoded. To further increase the data reduction performance, the impact of different applied compression algorithms on the grid bit-stream should be evaluated. This could result in even smaller update sizes to conserve even more network traffic.

Acknowledgement. The research leading to these results was supported by the ROBDEKON project funded by the German Federal Ministry of Education and Research under grant agreement No. 13N14679.

References

1. Moravec, H., Elfes, A.: High resolution maps from wide angle sonar. In: Proceedings. 1985 IEEE International Conference on Robotics and Automation, vol. 2, pp. 116–121. Institute of Electrical and Electronics Engineers (1985)
2. Kweon, I.S., Hebert, M., Krotkov, E., Kanade, T.: Terrain mapping for a roving planetary explorer. In: Proceedings, 1989 International Conference on Robotics and Automation, pp. 997–1002. IEEE Comput. Soc. Press (1989)
3. Fankhauser, P., Bloesch, M., Gehring, C., Hutter, M., Siegwart, R.: Robot-centric elevation mapping with uncertainty estimates. In: Mobile Service Robotics, pp. 433–440. World Scientific (2014)
4. Heppner, G., Roennau, A., Oberländer, J., Klemm, S.: Laurope-six legged walking robot for planetary exploration participating in the spacebot cup. WS Adv. Space Technol. Robot. Autom. **2**(13), 69–76 (2015)
5. Triebel, R., Pfaff, P., Burgard, W.: Multi-level surface maps for outdoor terrain mapping and loop closing. In: 2006 IEEE/RSJ International Conference on Intelligent Robots and Systems, pp. 2276–2282. IEEE (2006)
6. Roennau, A., Heppner, G., Nowicki, M., Dillmann, R.: LAURON V: a versatile six-legged walking robot with advanced maneuverability. In: 2014 IEEE/ASME International Conference on Advanced Intelligent Mechatronics, pp. 82–87. IEEE (2014)
7. Hutter, M., et al.: ANYmal-a highly mobile and dynamic quadrupedal robot. In: 2016 IEEE/RSJ International Conference on Intelligent Robots and Systems (IROS), pp. 38–44. IEEE (2016)

8. Hornung, A., Wurm, K.M., Bennewitz, M., Stachniss, C., Burgard, W.: OctoMap: an efficient probabilistic 3D mapping framework based on octrees. Auton. Robot. **34**(3), 189–206 (2013)
9. MBesselmann, M.G., Puck, L., Steffen, L., Roennau, A., Dillmann, R.: VDB-Mapping: a high resolution and real-time capable 3D mapping framework for versatile mobile robots. In: 2021 IEEE 17th International Conference on Automation Science and Engineering (CASE), pp. 448–454. Lyon, France. IEEE (2021). https://ieeexplore.ieee.org/document/9551430/
10. Museth, K.: VDB: high-resolution sparse volumes with dynamic topology. ACM Trans. Graph. **32**(3), 1–22 (2013)
11. Hess, W., Kohler, D., Rapp, H., Andor, D.: Real-time loop closure in 2D LIDAR SLAM. In: 2016 IEEE International Conference on Robotics and Automation (ICRA), pp. 1271–1278. IEEE (2016)
12. Fox, D., Ko, J., Konolige, K., Limketkai, B., Schulz, D., Stewart, B.: Distributed multirobot exploration and mapping. In: Proceedings of the IEEE, vol. 94, no. 7, pp. 1325–1339 (2006). http://ieeexplore.ieee.org/document/1677947/
13. Petitti, A., et al.: A distributed map building approach for mobile robotic networks. In: 2018 IEEE 14th International Conference on Automation Science and Engineering (CASE), pp. 116–121. Munich, Germany. IEEE (2018). https://ieeexplore.ieee.org/document/8560499/
14. Quigley, M., et al.: ROS: an open-source robot operating system. In: ICRA Workshop on Open Source Software, vol. 3, no. 3.2, p. 5. Kobe, Japan (2009)
15. Rusu, R.B., Cousins, S.: 3d is here: point cloud library (PCL) In: 2011 IEEE International Conference on Robotics and Automation, pp. 1–4. IEEE (2011)

Check for updates

Experiment of Social Robot Collaboration (SRC): Vision-Based Object Segmentation, Localization and Segregation in a Public Interaction Event

B. Kaushik[1] and Abhra Roy Chowdhury[2(✉)]

[1] Carnegie Mellon University, Pittsburgh, USA
[2] Indian Institute of Science, Bengaluru, India
abhra@iisc.ac.in

Abstract. To bridge the gap between humans and robots, there is a need to focus on the development of technology that can enhance the collaboration between the human and the robot (SRC). This research paper elaborates on a human interactive experiment of vision-based sorting of objects (candy) using a 3R serial spatial collaborative manipulator, used as an exhibit at the annual public 'Open Day' event at the Indian Institute of Science Bangalore. We collected feedback from the vast & variety of audience (mostly children and adult) and used this data to compute various human-robot collaboration parameters. 66.67% of the participants who were shown the demonstration demonstrated over 90% effectiveness in interaction during the sorting task as defined above. It was also statistically shown that there was no significant difference in the effectiveness between researchers and non-researchers who took part in the experiment.

Keywords: Human robot collaboration · Social robotics · Image localization · Image segmentation · Computer vision

1 Introduction

Robotics has made substantial impact in industrial settings and automated assembly lines, and only recently are they collaborating with humans in the factory [1]. However, they are yet to be integrated into our daily lives. A traditional computer vision approach is adopted in this paper to detect and sort various types of candy and is linked to real-world industrial applications [2] wherein defective parts are segregated [3] using similar techniques of distinguishing between objects' features by using contour, color and shape detection. As an extension to the work described above, quantitative methods for fault detection by measurement of dimensions of parts, such as gears and O rings, using traditional computer vision approaches have been presented in [4, 5]. These parameters will then be fused [7] to define metrics to determine the effectiveness and quality of the interaction. The metrics chosen in our work are based on the summary presented in [6] and from the early introduction of these metrics. In [8], the prestigious Royal

© The Author(s), under exclusive license to Springer Nature Switzerland AG 2023
J. M. Cascalho et al. (Eds.): CLAWAR 2022, LNNS 530, pp. 460–472, 2023.
https://doi.org/10.1007/978-3-031-15226-9_43

Society Summer Science Exhibition was used as a platform to allow public engagement with Robots to find out peoples' views about the future of robots in hazardous environments through questionnaires after the interaction. In [9] SoftBank Robotics' Pepper robot was introduced into a public shopping mall to determine the effectiveness of socially appropriate HRI and engaging behavior by combining speech-based interaction with non-verbal communication and human-aware navigation. Constant exposure and engagement between robots and the general public is a vital component for general social acceptance of robots in various fields [10]. Several attempts have been done to expose Robots to the general public. [12] illustrates experiments of Robots being used in nursing homes to alleviate the feelings of isolation and loneliness in addition to providing care to the elderly. [13] also have demonstrated an HRI system that can handle tasks even with no prior knowledge about the task or the language.

Furthermore, researchers in Japan [11] have conducted experiments to measure change in public perception of Robots before and after active engagement with Robots at Nagoya Science Hiroba Exhibition. This was based off a similar study, but the target audience was limited to teenagers visiting the Robo Life Museum. In coherence with these efforts, in our work we aim to inspire our audience to envision various applications for human society (social) robot collaboration or SRC in their daily lives. We use the annual 'Open Day' event at the Indian Institute of Science to demonstrate an entertaining and interactive task of identifying and sorting out various types of candy as commanded by the user. 'Open Day' is organized to exhibit research at the university as a gateway to and to communicate important concepts in science and engineering to the general public of varying age groups from primary school children, university students, working professions and academicians to interact with the various exhibits. The audience member can choose chocolate bars of their liking with voice commands to pick up only that item from the available assortment. The candy bars will be of different colors and sizes, and we will use these distinguishing features to carry out the sorting operation using computer vision. A selected group of audience members will then be asked to fill out a questionnaire, whose data forms the basis for further analysis to evaluate several Human Robot Interaction parameters. The rest of the paper is structured as follows: An overview of the method is described in Sect. 2. Sections 3 and 4 describe the steps of the method in detail. Section 5 describes the calibration process for obtaining the mapping function. The SRC experiments are described in Sect. 6, and the results are presented in Sect. 7. Finally, we conclude in Sect. 8.

2 Method Overview

2.1 Vision-Based Sorting of Objects (Candy)

In this section, we describe our assumptions and the steps involved in the implementation of our algorithm for vision-based sorting of candy. In the following sections, we elaborate each step of the algorithm in detail. The assumptions made are as listed below:

1. The camera coordinates are kept fixed.
2. The background behind each image is composed of a single colour which is different from the primary colour of the objects we are trying to sort.

The steps in our algorithm are as follows:

1. Retrieve voice commands from the user about the candy of their choice.
2. Capture the camera image and convert the input image into HSV colour space.
3. Colour Threshold the HSV image for the colour range of the background and the pre-determined set of candy to be sorted. The colour range of the background and the primary colour of each candy is calculated during the calibration process.
4. Apply Gaussian Blur on the image and create masks for each primary colour of the candies to be sorted.
5. Find the largest contour in the image and create a bounding box on it.
6. Find the coordinates of the centroid of the contours and convert those coordinates to the Robot's frame of reference using the linear regression-based mapping function.
7. Pick up the desired candy and place into the sorting box using the converted coordinates of the centroid of the objects.

The process flow is summarized in the flow chart as shown in Fig. 1. For the human and robot to collaborate with one another, we implement a voice-based communication interface to facilitate a natural and intuitive mode of communication with the robot. The user speaks their commands using a microphone and the robot responds back to the user using an external speaker. A speech recognition package is used for audio to text conversion for communication between the user and robot, pyttsx3 is used for text to audio conversion using which robot response is converted to speech and "pyaudio" is used for enabling the microphone. The knowledge engine that allows the generation of responses is Wolfram Alpha, whose API is imported into our script for creating a client for facilitating the human robot interaction. The voice of the robot is Microsoft Zira PRO, with a speech rate of 135 words per minute. The Google Web Search API (English: Indian version) is used to recognize speech from the audio input given by the user.

2.2 Voice Integration for Human-Robot Interaction

There are instances when the robot will be unable to identify the user's input commands. In such scenarios, in order to gracefully handle exceptions, we have included a provision for the user to type the command into the system manually. The control flow used for obtaining the user input is as demonstrated in Fig. 1. The robot first presents a generic greeting and then proceeds to ask their name to offer a personalized welcome message to the user. Once this is completed, the interaction enables the user to request the robot to perform the task of sorting the candy. Objects (Candies) of three different colored packages were included in the example. The algorithm has been programmed to either take the color of the candy or name of the candy bar as input so as to give the user the candy of their choice.

3 Image Segmentation

3.1 HSV Color Space

The first step in this process changing the colour space of the image from the default RGB colour space into the HSV colour space. A colour space is a mathematical model

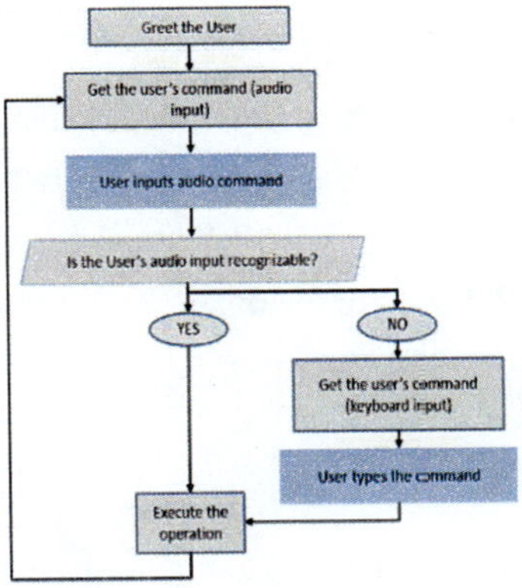

Fig. 1. Control flow for getting the user's input

of the description of various colours as a tuple of numbers, with an interpretation associated with each component of the tuple. We choose to transform the colour space into HSV because in the RGB colour space, the intensity and colour information cannot be discretely obtained (Fig. 2).

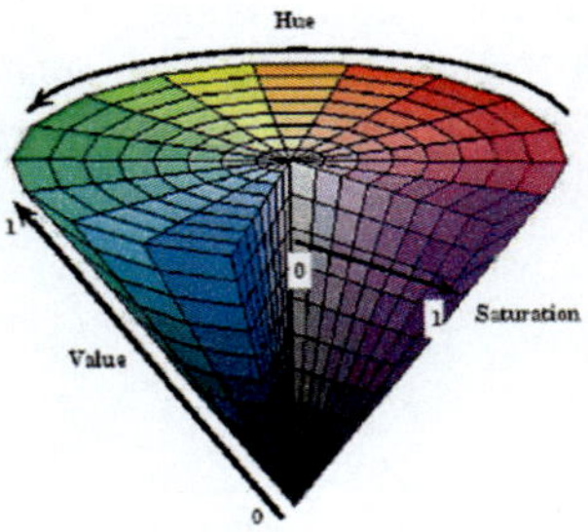

Fig. 2. HSV colour space: Source [14]

3.2 Defining Masks for Each Colour

In image processing applications, it is easier to analyse an image by breaking it down into meaningful segments, rather than processing an image in its entirety. This is because every image consists of a lot of unwanted information. Image segmentation works by grouping pixels having similar attributes into clusters, forming a distinct pixel-wise mask over certain objects in each image. Our algorithm uses colour thresholding to extract objects of a particular colour from the background. During the calibration process, we determine the upper and lower limits of HSV values for objects of different colours. A pixel is defined to be of an object of a particular colour if all three of its HSV components fall within the given the pre-calibrated range for that colour.

$$img(x, y) = \begin{cases} 0 \text{ if } Hmin \leq h \leq Hmax,\ Smin \leq s \leq Smax \\ Vmin \leq v \leq Vmax \\ 1 \text{ } otherwise \end{cases} \quad (1)$$

W: width of the image; *H*: height of the image; *h*: hue of the pixel*s*: saturation of the pixel; *v*: pixel val.

4 Object Localization

Once the image segmentation is completed and all objects of a given color have been extracted from the image, we proceed to find the centroid of each object. Prior to finding the centroid, we must find extract the contours of the objects having a specific color. The contours are used to infer the shape of the object by joining all continuous points present along the boundary that have the similar HSV values as described in Sect. 3.2 [6]. In order to avoid the detection of false positives, contours larger than a threshold only were considered, an approach as described in [7]. To ensure successful grasp acquisition, the end effector will grasp the object at its centroid. The centroid of a shape is the arithmetic mean of all points in a shape. With respect to image processing, each extracted shape is composed of pixels. Hence, the centroid is the weighted average of all the pixels constituting the shape. In this context, Image Moments are defined as the weighted average of pixel intensities using which we can determine the centroid among other parameters including the radius, area, etc. The centroid is given by the formula:

$$C_x = \frac{M_{10}}{M_{00}} \quad (2)$$

$$C_y = \frac{M_{01}}{M_{00}} \quad (3)$$

Here, C_x denotes the *X* coordinate of the centroid and C_y denotes the *Y* coordinate of the centroid. *M* denotes the Moment.

5 Mapping Function

In order for the end effector to grasp the object at its centroid as descried in Sect. 4, we calculate a mapping function which translates the coordinates of the centroid from the camera's frame of reference into equivalent coordinates in the Dobot Magician's frame of reference. We use a simple linear regression-based mapping function for this purpose. A set of seven points that cover entire camera frame and robot workspace were chosen.

$$X_{Dobot} = 0.5426\ X_{Camera} + 146.98 \quad (4)$$

$$Y_{Dobot} = -0.6249\ Y_{Camera} + 208.54 \quad (5)$$

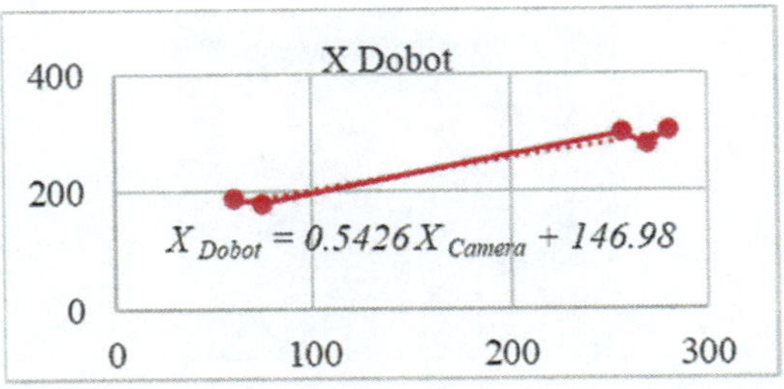

Fig. 3. Mapping function of Dobot Magician's X coordinates

The corresponding X and Y coordinates in the image and Robot frame of references were recorded and using linear regression, we determined the equation of the straight line that encapsulates this translation. The mapping equation for X and Y coordinates are given below. The Z-axis coordinates were pre-determined as the objects used in the experiments had similar thickness. The corresponding linear regression plots are as shown in Fig. 3 and Fig. 4.

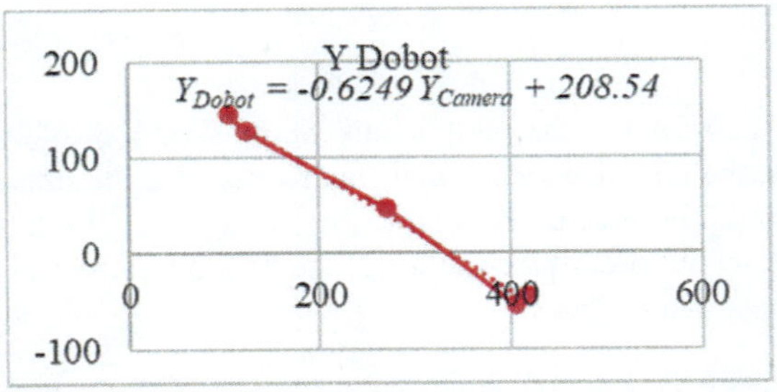

Fig. 4. Mapping function for Dobot Magician's Y coordinates

6 Experiments

Experiments were performed to sort 4 different types of candy based on their color. The experimental setup is as shown in Fig. 5 and 6. The experimental setup consisted of the 3R serial spatial manipulator of make a collaborative desktop robot (Dobot Magician), fitted with a suction gripper end effector to perform the sorting of the various objects. The camera used was mounted on a stand at a height of 35 cm and the objects were placed underneath it. A sorting box was placed in the workspace where the user's choice of candy will be sorted from the available assortment and placed. The algorithm was made to run on an Intel Core i7-8250U CPU with 16GB of RAM.

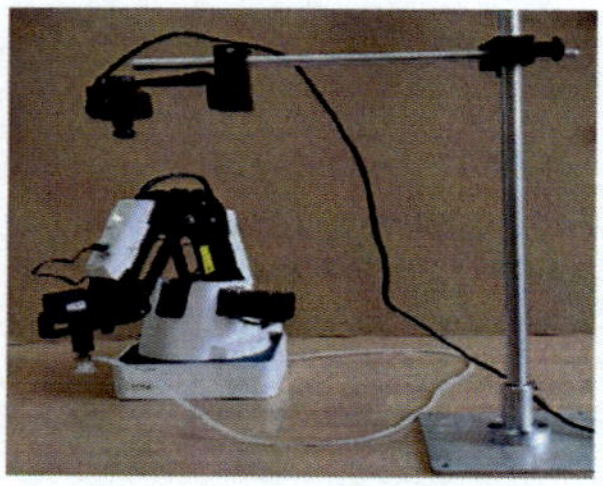

Fig. 5. Mounting of camera with respect to Desktop Cobot

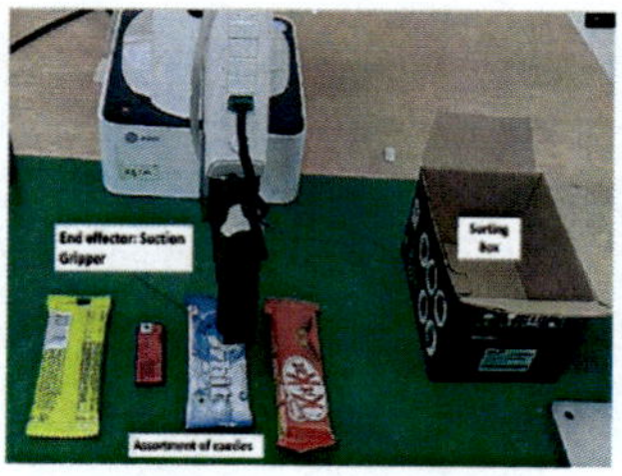

Fig. 6. Experimental setup

After each demonstration, the participants were asked the following questions in order to determine the effectiveness of their interaction and the impact the exhibit has had on them towards the role of Robots in their daily life. The first three questions evaluate the task effectiveness of the robot and the final question evaluates their overall perception of Robots in their lives.

1. Was the Robot able to understand and interpret your commands?
2. Did the Robot correctly identify the candy of your choice?
3. If correctly identified, was the Robot able to pick up and place the candy into the sorting box?
4. How was your overall experience in interacting with the Robot?
5. What other applications can you suggest for Robots in your life?

The responses to first three questions was given a quantitative score ranging from 1–5, with 5 being the highest positive response and 1 being the lowest negative response.

The input parameters chosen for this experiment and their average scores are as given in Table 1. The reasons for choosing these various input parameters explained as follows:

1. Comprehension of the user's commands **(Co):** This parameter is crucial to the effectiveness of the human robot interaction. It measures the ability of our algorithm to consistently understand the user's commands and take the right set of actions corresponding to the input command.
2. Wait Time **(WT):** In our interpretation, the higher the wait time score, the lesser the user had to wait for their commands to be comprehended and executed by the

Table 1. Input parameters for sorting of candy

Sl. No	Input parameter	Notation	Range	Avg. score
1	**Comprehension of the user's commands**	Co	{0, 0.2, 0.4, 0.6, 0.8, 1}	0.820
2	**Wait time***	WT	{0, 1}	0.796
3	**Success in correct identification of candy**	SI	{0, 1}	0.936
4	**Success in picking up the correct candy and segregating it**	SS	{0, 1}	0.879
5	**Inspiration Quotient: a measure of uniqueness of the suggested application domain for robotics**	IQ	{0, 0.5, 1}	0.451

*Here, Wait Time (WT) is a measure of the user's satisfaction with the time he/she waited for the Robot's response. So, higher values of WT are favorable.

robot. As with 'Co', this parameter is also crucial to the interaction as user interest and involvement in a conversation is dependent how quickly their commands are understood.

3. Success in correct **identification of candy (SI):** This parameter was used to infer whether after the correct comprehension of the user's voice commands, the vision algorithm consistently identified the right color as specified by the user with good repeatability.
4. Success in picking up the correct candy and segregating it ***(SS)*:** After the identification of the correct color, this parameter measures the ability of our system to correctly localize the correct candy at its centroid, effectively carry out the mapping from the camera's frame of reference to that of the Robot's coordinate system, and finally grasp and place the candy in the sorting box, whose coordinates have been predefined and calibrated at the beginning of the experiment.
5. Inspiration Quotient: A measure of uniqueness of the suggested application domain for Robotics ***(IQ)*:** The participants were also asked other tasks or applications they would recommend for this Robot. These participants were categorized as:

- *Category 1:* Participants who came up with new ideas that are different from the one demonstrated are given a score 1.
- *Category 2:* They are the people who gave suggestions, and improvisations of the demonstrated operations. They did not talk about any new and unique applications other than what was already demonstrated. They were thus given a score of 0.5.
- *Category 3:* This category includes the set of people who did not respond to the feedback question about alternate tasks, operations, and applications for Robotics in their daily lives. They were given a score of 0.

7 Observation

Out of the 51 participants, 18 chose not to answer the questions regarding the sorting operation. As a result, the operation was analyzed for 33 participants only. The successful identification (SI) of a candy of a color as specified by the user through a voice command is 93.55%. Furthermore, the mean score for the successful grasp acquisition and segregation of a candy (SS) into the sorting box provided of any color is 87.87%. There were instances observed when the identification of the candy was done successfully as per the user's commands and the grasp acquisition and placing into the sorting box was not successful.

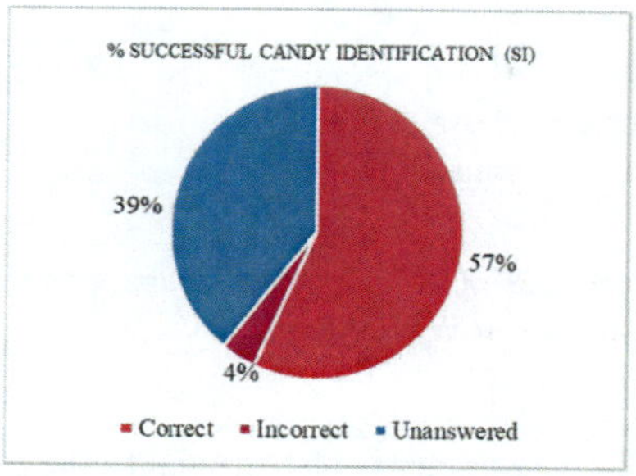

Fig. 7. Summary of SI candy sorting task

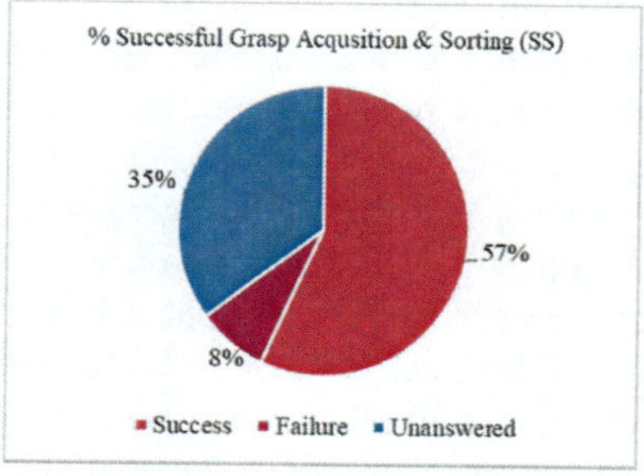

Fig. 8. Summary of SS candy sorting task

The summary of the characteristic metrics (SI & SS) for judging the success of the sorting operation are summarized in Figs. 7 and 8. As a metric to evaluate the overall success of the sorting operation, we introduce the Effectiveness in Sorting (ES) as the output parameter:

$$\boldsymbol{ES} = \boldsymbol{Co} + \boldsymbol{WT} + \boldsymbol{SG} + \boldsymbol{SS} \tag{6}$$

We have further segregated the participants into two groups constituting researchers and non-researchers. The researchers are from engineering background but with little or no practical experience in the domain of robotics. This ensures unbiased behavioral response from them. Based on the responses of the participants in these groups, we also present two hypotheses we wish to test based on these observations:

H_{0A}: There is no difference between the effectiveness of sorting parameter between the two groups.
H_{1A}: The group of researchers will have a higher effectiveness of sorting interaction with the robot.

The null hypothesis A suggests that there will not be any considerable difference between the group of researchers and non-researchers in the **effectiveness of the sorting (ES)** interaction between the human and the robot. ES frequency distribution plot is shown in Figs. 9 and 10.

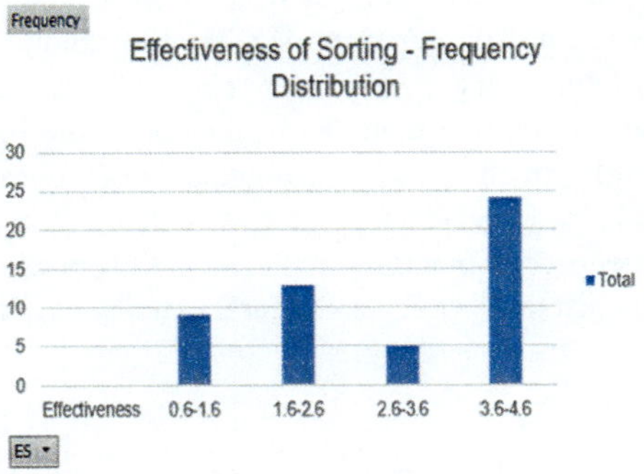

Fig. 9. Plot of frequency distribution of range of ES

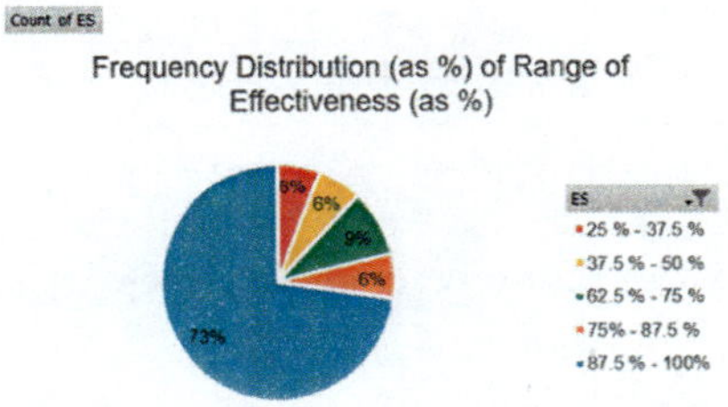

Fig. 10. Plot of frequency distribution (as %) of range of ES (as %)

The alternate hypothesis A is that the group of researchers will have better interaction with the robots and thus demonstrate a higher ES parameter than non-researchers. The thinking behind this alternative hypothesis is that due to research's familiarity in using similar technology beforehand it will naturally give them an edge over the non-researchers in asking questions or performing tasks in a way which is more predictable. For instance, we predict that researchers will be inherently attuned to speak relatively slowly and towards the direction of the microphone to allow the voice engine to process and respond to their commands. They will be more likely to follow along with the conversation and thus demonstrate better ability to interact with the Robot to perform the task. On the other hand, the non-researchers we assume will be less effective in their interaction due to the novelty in their interaction and the associated unpredictability.

Most people who are interacting with a Robot for the first time might not be familiar with its limitations and the nature of the commands it is capable of executing.

H_{0B}: There is no difference between the inspiration quotients of the ideas generated between the two groups.
H_{1B}: Researchers will have more innovative ideas to present for uses of Robotics in their daily lives than non-researchers.

The null hypothesis B suggests that there will not be any significant difference between the researchers and non-researchers in their ability to come up with radically new ideas for applications of Robotics in their daily lives. The alternate hypothesis B predicts that the researchers will be able to come with more unique ideas. Since reasoning behind the alternate hypothesis is that, since most researchers are likely to be exposed to the academic innovations in cutting-edge fields, they are more likely to understand the implications of the rapid growth of Social Robotics Collaboration (SRC) in everyday life, and thus will be able to suggest task domains which are far more creative than what was exhibited. We further predict that most audience members who are seeing this Robot for the first time will suggest tasks which are natural extensions of the task demonstrated.

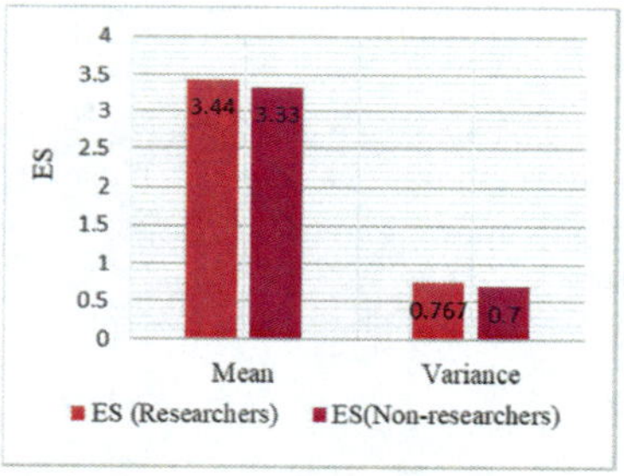

Fig. 11. Plot of mean & variance between for both groups (Hypotheses H_{0A} & H_{1A})

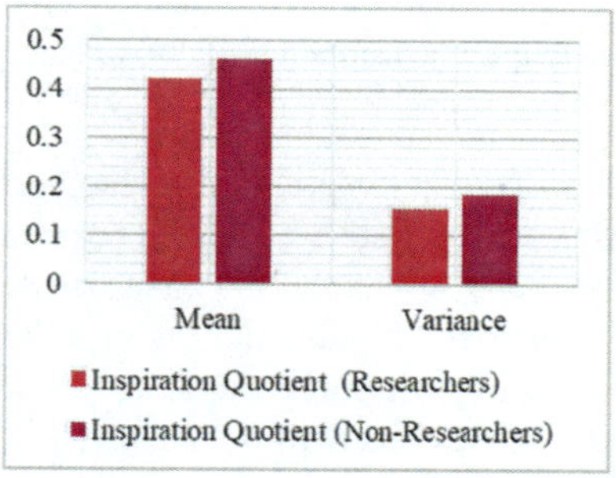

Fig. 12. Plot of mean & variance for both groups (Hypotheses H_{0B} & H_{1B})

The variance in the data was not significant for both data sets on 'ES' and 'IQ', shown in plots as Figs. 11 and 12.

8 Discussions and Conclusion

The social interaction experiment using audio and visual modalities by a collaborative robot with human subjects (children and adults) in a publicly organized event was reported (see Fig. 13) to be successful by a large majority of those that gave feedback. 66.67% of the participants who were shown the demonstration demonstrated over 90% effectiveness in interaction during the sorting task as defined above. The unanswered queries are largely attributed to the audience not being shown the demonstration. However, there were a small portion of anomalous instances where the query on success of sorting the desired candy was left unanswered or given a negative remark when the identification of the candy was done successfully but the grasp acquisition and sorting of candy was unsuccessful. This was due to calibration errors and inherent hardware limitations of the gripper. It was further shown that the effectiveness in sorting and inspiration quotient was not significantly different for researchers and non-researchers who were a part of the experiment. In future work, we will also incorporate eye-gaze tracking and real time audio processing to infer the success of the interactions from a person's attentiveness to the demonstration and their verbal feedback on the experiments. The collected audio can then be converted to text and a Natural Language Processing system will be able to gauge the overall response of the audience. In addition, to improve the task success rate and thereby achieve better HRI effectiveness, we will also include 3D cameras for depth perception.

Fig. 13. Social robot experiment with kids and adults on IISc Bangalore open day

Acknowledgment. This project is funded by Samarth Udyog, Common Engineering Facility Centre (CEFC) CPDM, IISc, Bangalore; Department of Heavy Industries (DHI), Govt. of India.

References

1. Robla-Gómez, S., Becerra, V.M., Llata, J.R., González-Sarabia, E., Torre-Ferrero, C., Pérez-Oria, J.: Working together: a review on safe human-robot collaboration in industrial environments. IEEE Access **5**, 26754–26773 (2017)
2. Anuar, S., Baharuddin, A.B., Mohammad, A.: Lego bricks color sorting machine. In: 2016 2nd International Symposium on Agent, Multi-agent Systems and Robotics (ISAMSR) (2016)
3. Wang, J., Hu, H., Chen, L., He, C.: Assembly defect detection of atomizers based on machine vision. In: CACRE 2019, Shenzhen, China (2019)
4. Wang, W., et al.: Measurement system of gear parameters based on machine vision. Meas. Control **48**(8), 242–248 (2015)
5. Peng, G., Zhang, Z., Li, W.: Computer vision algorithm for measurement and inspection of O-rings. Measurement **94**, 828–836 (2016)
6. Murphy, R.R., Schreckenghost, D.: Survey of metrics for human-robot interaction. In: 2013 8th ACM/IEEE International Conference on Human-Robot Interaction (HRI). IEEE (2013)
7. Bensch, S., Jevtic, A., Hellström, T.: On interaction quality in human-robot interaction. In: ICAART 2017 Proceedings of the 9th International Conference on Agents and Artificial Intelligence, vol. 1. SciTePress (2017)
8. Robb, D.A., et al.: Robots in the danger zone: exploring public perception through engagement. In: Proceedings of the 2020 ACM/IEEE International Conference on Human-Robot Interaction (2020)
9. Foster, M.E. et al.: The MuMMER project: engaging human-robot interaction in real-world public spaces. In: Agah, A., Cabibihan, J.J., Howard, A., Salichs, M., He, H. (eds.) Social Robotics. ICSR 2016. LNCS, vol. 9979, pp. 753–763. Springer, Cham (2016). https://doi.org/10.1007/978-3-319-47437-3_74
10. Savela, N., Turja, T., Oksanen, A.: Social acceptance of robots in different occupational fields: a systematic literature review. Int. J. Soc. Robot. **10**(4), 493–502 (2018)
11. Kim, M.-G., et al.: Effectiveness of robot exhibition through visitors experience: a case study of Nagoya Science Hiroba exhibition in Japan. In: 2016 International Symposium on Micro-NanoMechatronics and Human Science (MHS). IEEE (2016)
12. Kim, M.G., Lee, H., Lee, J., Kwak, S.S., Joo, Y.: Effectiveness and service quality of robot museum through visitors experience: a case study of RoboLife Museum in South Korea. In: 2015 International Symposium on Micro-NanoMechatronics and Human Science (MHS), pp.1–5. IEEE (2015)
13. Moro, C., et al.: Social robots and seniors: a comparative study on the influence of dynamic social features on human–robot interaction. Int. J. Soc. Robot. **11**(1), 5–24 (2019)
14. Erdogan, K., Yilmaz, N.: Shifting colors to overcome not realizing objects problem due to color vision deficiency.In: 2nd International Conference on Advances in Computing, Electronics and Electrical Technology - CEET. Kuala Lumpur, Malaysia (2014). https://doi.org/10.15224/ 978-1-63248-034-7-27

Simulation in Robotics Applications

Modelling of a Vibration Robot Using Localization Ground Truth Assisted by ArUCo Markers

Diogo Matos[1(✉)], José Lima[1,2], Ronnier Rohrich[3], André Oliveira[3], António Valente[1,4], Pedro Costa[1,5], and Paulo Costa[1,5]

[1] Centre for Robotics in Industry and Intelligent Systems, INESC TEC, Porto, Portugal
diogo.m.matos@inesctec.pt

[2] Research Centre in Digitalization and Intelligent Robotics CeDRI, Instituto Politécnico de Bragança, 5300-252 Bragança, Portugal
jllima@ipb.pt

[3] Universidade Tecnológica Federal do Paraná (UTFPR), Curitiba, Paraná, Brazil
{rohrich,andreoliveira}@utfpr.edu.br

[4] Universidade de Trás-os-Montes e Alto Douro, Vila Real, Portugal
avalente@utad.pt

[5] Faculty of Engineering, University of Porto, Porto, Portugal
{pedrogc,paco}@fe.up.pt

Abstract. Simulators have been increasingly used on development and tests on several areas. They allow to speed up the development without damage and no extra costs. On realistic simulators, where kinematics play an important role, the modelling process should be imported for each component to be accurately simulated. Some robots are not yet modelled, as for example the Monera. This paper presents a model of a small vibration robot (Monera) that is acquired in a developed test-bed. A localisation ground truth is used to acquire the position of the Monera with actuating it. Linear and angular speeds acquired from real experiments allow to validate the proposed methodology.

Keywords: Vibration robot · Modelling · ArUco markers · Localization

1 Introduction

Among the various characteristics that a robotic system can present, movement deserves to be highlighted, as it increases the ability to navigate and explore previously unknown environments. Thus, different solutions can be approached in different contexts and environments. The most common models for moving mobile robots involve complex mechanical systems of gears coupled to wheels or legs. An approach that also exists, but is not widespread, is vibration movement.

© The Author(s), under exclusive license to Springer Nature Switzerland AG 2023
J. M. Cascalho et al. (Eds.): CLAWAR 2022, LNNS 530, pp. 475–486, 2023.
https://doi.org/10.1007/978-3-031-15226-9_44

The basic concept of this method involves performing a robot design that generates a force resulting from displacement that is capable of being controlled and selected as needed. Therefore, vibration robots bring constructive simplicity, one of their greatest virtues, associated with evident exploratory capacity in this area of research. To develop control algorithms and evaluate them, simulation can be used. It is a tool that have been increasingly used. However, its performance has several dependencies such as the simulation step time and the model accuracy with special focus in Simulations with dynamic characteristics. For an accurate simulation, the model of the robot should be extracted. This paper addresses a model of a vibration robot that was developed with real components. A dynamic model of the robot was based on real experiments where the excitation signals and robot positions were acquired allowing to estimate the dynamic model.

This paper is organised as follows: after a brief introduction, Sect. 2 presents a related work on vibration robots. Then, Sect. 3 presents the test bed architecture that was developed to localise the robot during the experiments based on ArUco markers. Section 4 addresses the modelling and presents the results. Finally, Sect. 5 ends the paper with conclusions and future work.

2 Related Work

This section addresses the previous work on the developing the Monera Robot, as well as the modelling of vibration robots stressed on the community.

2.1 Monera Robot

Some different small-scale mobile robot typologies have already been developed for various purposes. In this work, some characteristics of the mobile robot Monera will be explored to provide contributions in the simulation of these robots.

This typology of robot has the same locomotion principle as the Kilobot robot, presented in [1]. The Kilobot was designed to make tests of collective algorithms on hundreds or thousands of robots accessible to robotics researchers. The Monera robot slightly increases the robot scale in order to use more common components in its design making it easily replicable.

That is, the movement occurs through the vibration of two motors fixed to the "chassis". The physical principles that involve the movement of the Monera robot are related to the direction generated after the activation carried out by each motor.

The first difference between these two topologies is that the Monera is built with components found in the retail market, allowing anyone to build their mobile robot. Another difference is that although Monera does not have odometry, all its information such as position, operating time, Pulse Width Modulation (PWM) levels, trajectory, among others, are accessible via ROS, providing flexibility in publishing and subscribing messages between any system connected to it. In Fig. 1, it is possible to have an overview of the Monera robot and its constituent parts.

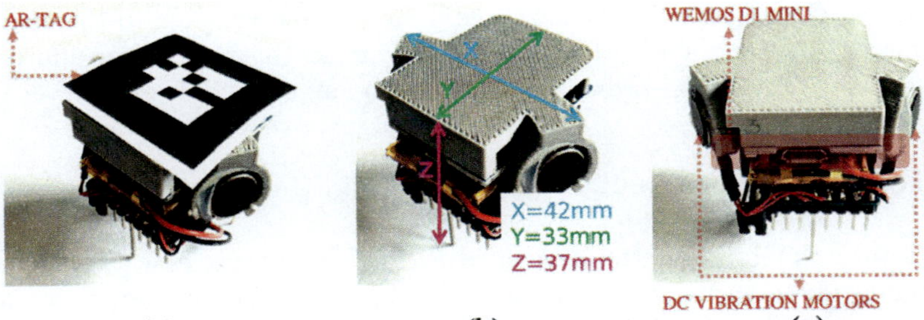

(a) (b) (c)

Fig. 1. Main components of the Monera robot. (a) ARTag - Fiducial marker system to support augmented reality. (b) Extreme dimensions of the robot Monera. (c) Location of vibration motors and control board.

The mechanical structure of the Monera robot was based on analyses carried out on the results of the work of [2], where the physical concepts for the movement and control of the position of the mobile robot through centripetal forces generated by vibrating micro-motors.

2.2 Modelling of Vibration Robots

The locomotion and movement of robots in small constricted spaces is a very complex task, most standard locomotion systems used nowadays, struggle to provide enough flexibility while at the same time providing an adequate level of miniaturisation [3,4]. This particular class of robots have the potential to enable several different implementations such as pipeline inspection and disaster rescue [5].

To deal with such complex environments the new studies into animal's locomotion systems, like snakes [6], caterpillars [7] and cockroaches [8] in order to establish new guidelines for modelling miniature robots.

The mechanical model of Shell [5] is based vibration-driven locomotion system based around the movement of earthworms [9,10]. A vibration-driven locomotion system is capable of without the need for a complex transmission system or the need of complex active components such as legs or wheels [5]. This makes this type of simple to design and suited for miniaturisation.

Several of these approaches are already in use on Active scope Cameras, such as proposed by HATAZAKI [11] to operate in the debris of collapsed structures for the purposes of search and rescue operations. These cameras use only one actuator to propel their body through the environment via the use of ciliary vibration. This idea was then further improved and adapted by [12] to create a locomotion system for autonomous robots, [12] uses Anisotropic friction and vibrations of the body to propel the robot forward using only an unbalanced rotor and a DC motor.

Similar to the Monera robot, which is the target of this paper, there is the Kilobot. The study by [13], already presents a simulation model for this types of robots in CoppeliaSim (also known as V-Rep). However, due to the changes implemented in Monera design, which aimed to make this robot more affordable and replicable, the model presented in [13] is no longer valid.

3 Test-Bed Architecture

The main architecture of the developed system is presented by Fig. 2. There is a planar glass where the robot should navigate. At each corner, there is an arUco market to localise the arena without calibration need: just point a camera placed above to the arena. A rasberry pi is used to capture the image, with a sample rate of 30 fps and identify the arUco markers. The position and orientation of the robot are extracted (as further explained in this section) and sent by ZeroMQ library using Ethernet. A main application runs the modelling and control tasks (and can also be added extra features, such as Digital-Twin among others) and connects the robot by WiFi protocol. The robot uses an ESP8266 microcontroller that receives the left and right motor speeds and command two H-bridges attached to the vibration motors.

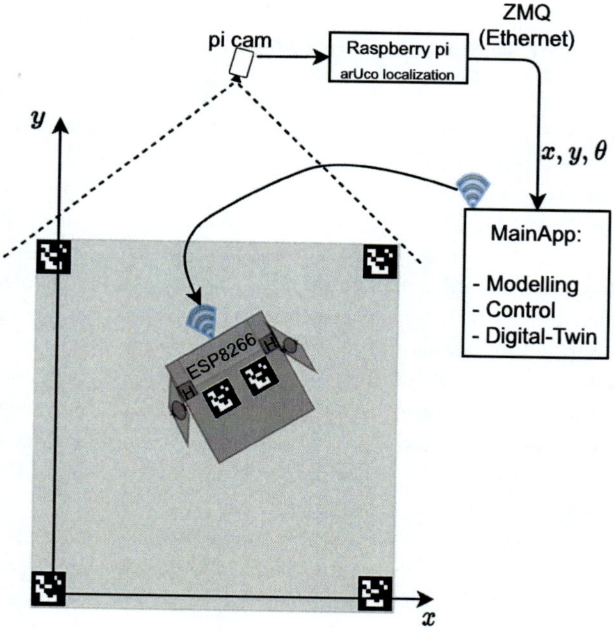

Fig. 2. System architecture.

The proposed localisation system is based on image, detecting ArUco markers [14]. In order to avoid the apparatus calibration (camera positioning), four arUco markers are used to estimate the corners of the testbed, as presented in Fig. 3a.

Moreover, two arUco markers are used to estimate the position and orientation of the robot (orientation performs more accurate and has less noise than one only marker). In order to validate the approach, a graph paper (Fig. 4a) is placed above the glass to place the robot markers on several positions and orientations.

(a) Corner markers are used for the reference; coupled markers are used for robot position measurements.

(b) Testbed equipped with graph paper to measure the accuracy and the introduced error for the arUco markers.

Fig. 3. Used testbed for the experiment and its validation.

The library GSNCodes[1] is used to generate and localize the arUco markers allowing to obtain each marker coordinates, as presented on Fig. 4a. A set of Translation and Rotation should be done on the acquired image to measure the position and orientation of the robot. We will assume three referentials: the camera (O_c), the shifted one to have its origin aligned with the world origin (O_c0) and the world referential (O_w). By this way, the position of the camera remains free. According to Fig. 4b, a translation is required to place the marker m_1^c on the O_w origin, as presented in Eq. 1.

$$m_i^c0 = m_i^c - m_1^c, \forall i \in 1..6 \tag{1}$$

Once it is not always true that the camera is parallel to the plane, the rotation procedure will be based on a tree dimensional operation. Equations 2 to 5 calculate the vectors of x' and y' coordinates.

$$\boldsymbol{v_{12}} = m_2^c0 - m_1^c0 \tag{2}$$

$$\boldsymbol{v_{13}} = m_3^c0 - m_1^c0 \tag{3}$$

[1] https://github.com/GSNCodes/ArUCo-Markers-Pose-Estimation-Generation-Python.

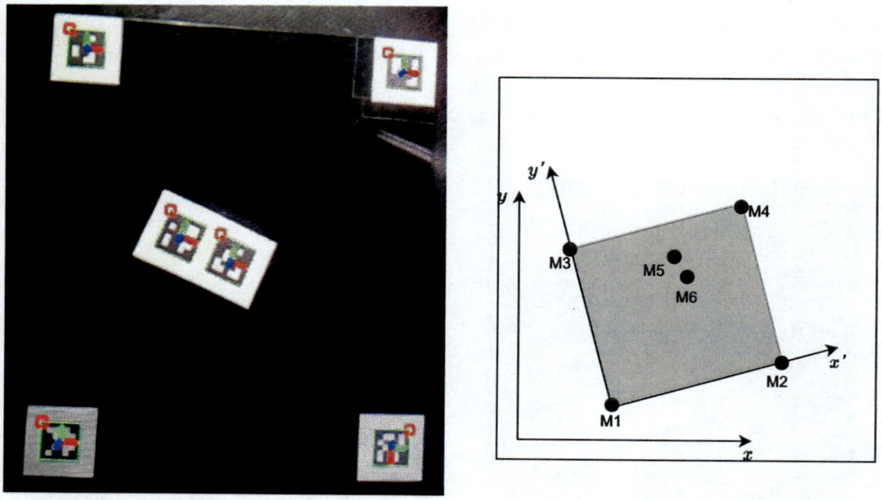

(a) arUco markers identification. (b) Acquisition image schematic.

Fig. 4. ArUco markers identification.

The unit vector for each vector can be calculated as:

$$\widehat{\boldsymbol{v_x}} = \frac{\boldsymbol{v_{12}}}{||\boldsymbol{v_{12}}||} \tag{4}$$

$$\widehat{\boldsymbol{v_y}} = \frac{\boldsymbol{v_{13}}}{||\boldsymbol{v_{13}}||} \tag{5}$$

The unit vector for z axis, can be calculated as the cross product of $\widehat{\boldsymbol{v_x}}$ and $\widehat{\boldsymbol{v_y}}$, according to Eq. 6.

$$\widehat{\boldsymbol{v_z}} = \widehat{\boldsymbol{v_x}} \times \widehat{\boldsymbol{v_y}} \tag{6}$$

The rotation matrix, R, can be calculated as presented in Eq. 7.

$$R = [\widehat{\boldsymbol{v_x}} \mid \widehat{\boldsymbol{v_Y}} \mid \widehat{\boldsymbol{v_z}}] \tag{7}$$

All the markers positions in space can now be calculated as:

$$m_i^w = R^T.m_i^c 0, \forall i \in 2..6 \tag{8}$$

Finally, the robot position (Rob^w) be determined as expressed in Eq. 9 whereas the angle of the robot, θ can be expressed in Eq. 10.

$$Rob^w = \frac{m_5^w + m_6^w}{2} \tag{9}$$

$$\theta = atan2(m_6 y^w - m_5 y^w, m_6 x^w - m_5 x^w) + pi/2 \tag{10}$$

In order to validate the robot localisation system, 256 samples were measured on well established positions (0.05,0.05), (0.05,0.15), (0.2,0.05) with $\theta = 0$ and (0.15,0.1) with orientations −45, −90, −135 and −180°. Table 1 shows the error on localisation (position and orientation).

Table 1. Error on measuring X, Y and Theta with arUco markers and acquisition software. Number of samples: 256 for each position.

REAL			Average error			
X (m)	Y (m)	Theta (deg)	X (m)	Y (m)	Theta (deg)	Euclidean Dist (m)
0.05	0.05	0	−3.72E−04	1.28E−03	−1.67E+00	1.34E−03
0.05	0.15	0	7.95E−04	2.39E−03	−7.30E−01	2.65E−03
0.2	0.05	0	3.23E−03	1.62E−03	3.58E−01	3.62E−03
0.15	0.1	0	1.10E−03	1.57E−03	−6.12E−01	1.92E−03
0.15	0.1	−45	2.09E−04	1.57E−03	−1.17E−01	1.60E−03
0.15	0.1	−90	1.97E−03	1.32E−03	−1.00E+00	2.37E−03
0.15	0.1	−135	2.49E−04	1.12E−03	6.81E−01	1.26E−03
0.15	0.1	−180	9.13E−04	4.66E−04	1.63E−01	1.03E−03

As result, it is possible to verify that the average error on X axis is 0.001020 m, on y axis is 0.001436 m, theta average error is −0.394214° and the Euclidean distance error is 0.001967 m. These errors are pretty acceptable when considering the robot velocity will be slow - few cm/s.

More detailed information in about the 2D Aruco Board Benchmark can be found in the [15] and [16].

4 Modelling Results

For this article the robot, represented in Fig. 5, was chosen in order to test the proposed methodology. This robot uses three static legs, as well as, two small electrical motors that generate the vibrations used for locomotion by this robot.

In order to determine the dynamic model in the overall model it is necessary to understand what is the effect of each of the motors in the movement of the robot. For this reason it is necessary to test and measure the robot movements, in these tests the time each motor is powered is varied, effectively in each of these tests the duty cycle of each of the motors is changed, this duty cycle was calculated using Eq. 11. After 5 s of operation the position of the robot is then measured and compared with the starting position. By analysing the displacement of the robot during the time interval of the test it is possible to calculate the angular and linear velocity of the robot in each of the tests.

In Eq. 11, T1 represents the time first motor is active and T2 the time second motor is active in milliseconds. The 0.5 offset is subtracted in order to center the duty values around zero. In this scale, the value 0 signifies that both motors are active the same amount of time and 0,5 and −0.5 signifies that only first motor and the second motor are active for the full duration of the tests. In Eq. 11 the total period, the sum of T1 and T2, is 100 ms.

$$Duty\_Cycle = T1/[T1 + T2] - 0.5 \tag{11}$$

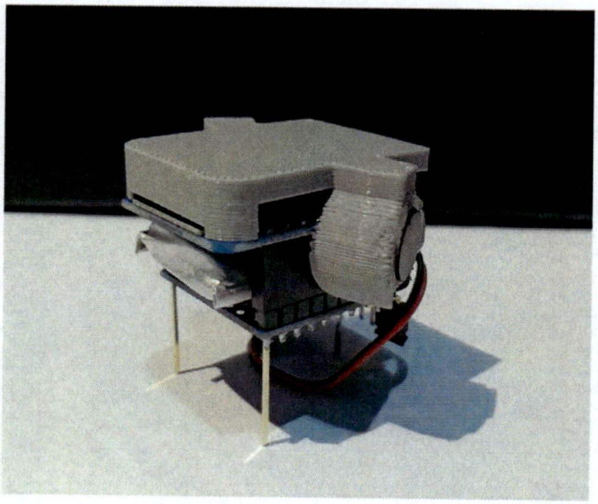

Fig. 5. Vibration developed robot under test.

In the following sub sections the results of these tests will be presented, as well as the proposed model for each of the velocity components. To accurately model these velocity components, several tests under the same scenario were executed. Since the results of these test were similar, for tests in the same scenario, the results of only one test were represented.

4.1 Angular Velocity

The first velocity component that will be analysed is the Angular Velocity, to obtain a broader specter of results the duty cycle of the motors was varied from −0,5 to 0,5. In order to identify the model for the angular velocity of the robot, 25 samples with different duty cycles were taken. The results of these tests were then represented via the chart in Fig. 6.

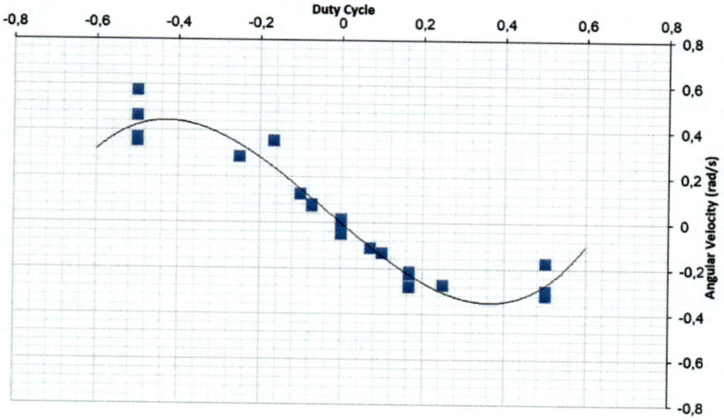

Fig. 6. Angular velocity in relation with the duty cycle of one motor.

Analysing Fig. 6 it is possible to correlate the test results using a third degree polynomial trendline. This polynomial trendline is modelled by Eq. 12 and has the fitness value showed in Eq. 13.

$$y = 3,1934x^3 + 0,3035x^2 - 1,4921x \tag{12}$$

$$R^2 = 0,9567 \tag{13}$$

From the results of these tests, it is possible to conclude that between the duty values of −0,3 and 0,3 the behaviour of the angular velocity can be easily modelled by a linear function as shown in Fig. 7. This linear function can be modelled by Eq. 14 and has the fitness value showed in Eq. 15.

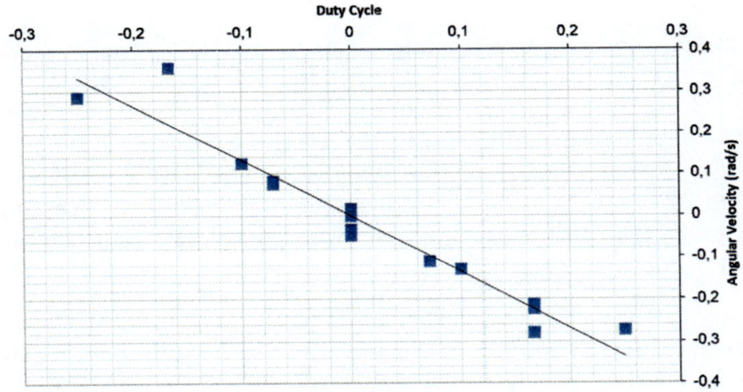

Fig. 7. Angular velocity in relation with the duty cycle of one motor between the duty values of −0.3 and 0.3, has seen in this interval the angular velocity can be modelled by a linear function.

$$y = -1,3392x - 0,0017(rad/s) \tag{14}$$

$$R^2 = 0,9375 \tag{15}$$

The tests executed with duty values below −0.3 and above 0.3 showed that there is a clear saturation situation in these intervals. Its also in these intervals where the peak value of angular velocity in both direction is reached. Has expected these values are reached when only one motor is active for the entirety of the test time interval.

4.2 Linear Velocity

To obtain the full model of the robot it is still necessary to model the linear component of its velocity. Similar to the tests executed for the angular component the same tests were executed for the linear component. The results of these tests were represented via the chart in Fig. 8.

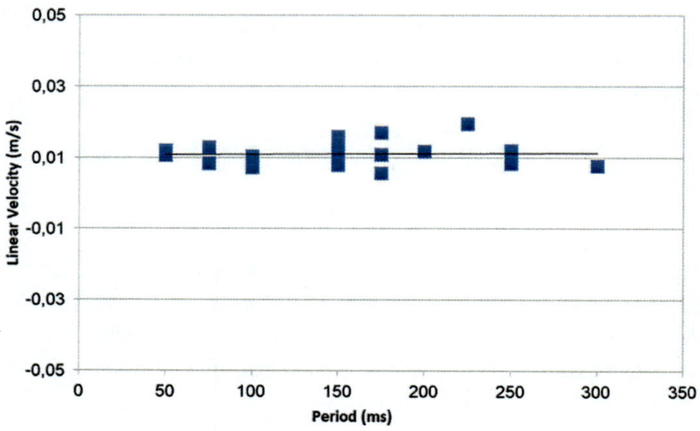

Fig. 8. Linear velocity in relation with the period that the motor is active.

Analysing the data, represented in the graph of Fig. 8, it is possible to conclude that the linear velocity of the robot is approximately constant and it is independent of the duty cycle for each motor. As long has at least one of the motors is active the robot will travel with a linear velocity that can be approximated by Eq. 16.

$$y = 0,011(m/s) \tag{16}$$

These tests also showed that, in this kind of robots, it is impossible to separate both velocity components, this means that this robot is not capable of rotating over it self since it will always have a linear constant velocity associated with any rotation.

As summary, after several tests it was possible to conclude that in terms of angular velocity the robot behaves similarly to a differential traction robot. However when it comes to the linear velocity, this robot behaves slightly differently from the differential ones having a near constant linear speed, as long as at least one of the motors is active.

5 Conclusions

Vibration robots present an affordable and more versatile alternative to the more classic differential traction approach.

This paper proposed a methodology to identify the dynamic model of a small vibration robot. A localisation apparatus was developed with the help of a small test bed, based on image analysis and the use ArUco markers in order to obtain the ground position of the robot in each moment. A model that is capable of predicting the robots movement based on the duty cycle applied to each of the robots electrical motors was developed, using the data of several different sample runs. This model can be further embedded in the simulation environment, as future work.

Despite the robot locomotion being optimized for glass surfaces, the authors of this paper plan to study the effect of other surface materials on the locomotion of the robot. Also in future work the authors plan to include the dynamics considering the torque and forces generated during locomotion.

Acknowledgements. This work has been supported by FCT - Fundação para a Ciência e Tecnologia within the Project Scope: UIDB/05757/2020 and also by National Funds through the Portuguese funding agency, FCT - Fundação para a Ciência e a Tecnologia, within project LA/P/0063/2020.

References

1. Rubenstein, M., Ahler, C., Nagpal, R.: Kilobot: a low cost scalable robot system for collective behaviors. In: 2012 IEEE International Conference on Robotics and Automation, pp. 3293–3298 (2012)
2. Vartholomeos, P., Papadopoulos, E.: Analysis, design and control of a planar microrobot driven by two centripetal-force actuators. In: Proceedings 2006 IEEE International Conference on Robotics and Automation, ICRA 2006, pp. 649–654 (2006)
3. Karydis, K., Liu, Y., Poulakakis, I., Tanner, H.G.: A template candidate for miniature legged robots in quasi-static motion. Auton. Robot. **38**(2), 193–209 (2014). https://doi.org/10.1007/s10514-014-9401-4
4. Transeth, A.A., Pettersen, K.Y., Liljebäck, P.: A survey on snake robot modeling and locomotion. Robotica **27**(7), 999–1015 (2009)
5. Zhan, X., Xu, J., Fang, H.: A vibration-driven planar locomotion robot–Shell. Robotica **36**, 1–19 (2018). https://doi.org/10.1017/S0263574718000383
6. Liljebäck, P., Pettersen, K.Y., Stavdahl, O., Gravdahl, J.T.: Snake Robots: Modelling, Mechatronics, and Control. Springer, London (2012). https://doi.org/10.1007/978-1-4471-2996-7
7. Lin, H.T., Leisk, G.G., Trimmer, B.: GoQBot: a caterpillar-inspired soft-bodied rolling robot. Bioinspir. Biomim. **6**(2), 026007 (2011)
8. Yao, J.J., Gao, S., Jiang, G.L., Hill, T.L., Yu, H., Shao, D.: Screw theory based motion analysis for an inchworm-like climbing robot. Robotica **33**(8), 1704–1717 (2015)
9. Bolotnik, N.N., Zeidis, I.M., Zimmermann, K., Yatsun, S.F.: Dynamics of controlled motion of vibration-driven systems. J. Comput. Syst. Sci. Int. **45**(5), 831–840 (2006)
10. Chernousko, F.L.: Analysis and optimization of the motion of a body controlled by means of a movable internal mass. PMM - J. Appl. Math. Mech. **72**(6), 819–842 (2006)
11. Hatazaki, K., Konyo, M., et al.: Active scope camera for urban search and rescue. In: Proceedings of the 2007 IEEE/RSJ International Conference on Intelligent Robots and Systems, San Diego, CA, USA, pp. 2596–2602 (2007)
12. Becker, F., Minchenya, V.T., Zeidis, I., Zimmermann, K.: Modeling and dynamical simulation of vibration-driven robots. In: 56th International Scientific Colloquium Ilmenau University of Technology (2011)
13. Pinciroli, C., Talamali, M.S., Reina, A., Marshall, J.A.R., Trianni, V.: Simulating Kilobots within ARGoS: models and experimental validation. In: Dorigo, M., Birattari, M., Blum, C., Christensen, A.L., Reina, A., Trianni, V. (eds.) ANTS 2018. LNCS, vol. 11172, pp. 176–187. Springer, Cham (2018). https://doi.org/10.1007/978-3-030-00533-7_14

14. Li, B., Wu, J., Tan, X., Wang, B.: ArUco marker detection under occlusion using convolutional neural network. In: 2020 5th International Conference on Automation, Control and Robotics Engineering (CACRE), pp. 706–711 (2020). https://doi.org/10.1109/CACRE50138.2020.9230250
15. Poroykov, A., Kalugin, P., Shitov, S., Lapitskaya, I.: Modeling ArUco markers images for accuracy analysis of their 3D pose estimation. In: Proceedings of the 30th International Conference on Computer Graphics and Machine Vision (GraphiCon 2020), Part 2, short14-1 (2020). https://doi.org/10.51130/graphicon-2020-2-4-14
16. Oščádal, P., et al.: Improved pose estimation of ArUco tags using a novel 3D placement strategy. Sensors (Basel, Switzerland) **20**(17), 4825 (2020). https://doi.org/10.3390/s20174825

Modelling and Simulation of Robotic Luggage Transport at OPO Airport

Miguel Pereira[1], Manuel F. Silva[1,2](✉), and André Siqueira[1]

[1] School of Engineering, Polytechnic Institute of Porto, Rua Dr. António Bernardino de Almeida, 431, Porto, Portugal
{mss,and}@isep.ipp.pt

[2] INESC TEC, Rua Dr. Roberto Frias, 4200-465 Porto, Portugal
http://ave.dee.isep.ipp.pt/~mss/

Abstract. Due to the lack of unskilled labour force that has been verified in the last years, several processes have been automated, both at industrial and services level. In terms of logistics tasks and transport of materials, it is increasingly common to use mobile robots, given the advantages that this equipment presents. This is also the case in airports, where the adoption of these vehicles to perform several tasks is becoming visible. Considering the possibility of using mobile robots to transport luggage at the Francisco Sá Carneiro Airport, this paper presents the development of a simulation model and the analysis of several scenarios, with different number of vehicles, in order to understand the time that passengers would have to wait for their luggage, in case this task is automated. The final objective is to determine the number of vehicles required and the changes that need to be made to the airport's operation in order to ensure a level of service identical to (or better than) that currently achieved, with these operations being carried out by human operators.

Keywords: Mobile robots · Simulation · Luggage transport · Airports

1 Introduction

Autonomous Guided Vehicles (AGV) are vehicles that can move autonomously after being programmed, travelling along a defined path, orientating themselves in space, thanks to the sensors and actuators with which they are equipped, and can operate indoors and outdoors. They have been mainly developed and programmed to deliver and collect products from factory premises or warehouses. Many companies have begun to realise that AGV can save labour, improve labour efficiency and save costs, which is the reason for their increased used in different logistic tasks. Typical AGV applications include the transportation of raw materials, in-process and/or finished products, support to production lines, support to the separation of materials in storage and distribution tasks, among others.

The main difference between AGV and autonomous mobile robots (AMR) lies in the use of pre-defined paths by AGV. While many AGV use fixed navigation

© The Author(s), under exclusive license to Springer Nature Switzerland AG 2023
J. M. Cascalho et al. (Eds.): CLAWAR 2022, LNNS 530, pp. 487–500, 2023.
https://doi.org/10.1007/978-3-031-15226-9_45

systems, such as wires or magnetic tape (usually on the ground), AMR are typically equipped with intelligent navigation features, *i.e.* sensors and camera systems, that allow them to detect and navigate through the workspace and even perform obstacle avoidance and contouring. For example, if an obstacle appears in the AGV workspace that interferes with its trajectory, it will require a physical change in the path definition to go around it and thus avoid the obstacle, whereas an AMR is able to perform the obstacle avoidance autonomously.

Currently, there are different categories of autonomous vehicles (either AGV or AMR) being used in industry, namely: Automated Guided Cart (AGC), Towing or Tugger, Unit Load and Stacker, among others [2]. AGC ride under the load to be transported, hence they are also called Under Ride, being flexible and easy to install, and generally have a lower cost than other types of autonomous vehicles. It is one of the solutions to perform repetitive tasks, such as transporting loads from a point A to a point B [5].

AMR are also starting to be used in airports for performing different tasks. One possible application is their use for transporting luggage between aircrafts and the passenger terminal. Considering this, the objective of the work described in this paper is to study, using simulation tools, the possibility of implementing a fleet of AMR, at the Oporto airport, to automate such transport of luggage. Considering the objectives for the transport of luggage in airports, the idea would be to use AGC vehicles for such purpose, able to autonomously transport the luggage cars presently used in airports.

Bearing these ideas in mind, the remainder of this paper is organised as follows. Sections 2 presents several applications of AMR on airports, followed by Sect. 3 that describes the airport considered for the possible introduction of AMR. Next, Sect. 4 describes the modelling of the airport flows of aircraft, AMR, and luggage, followed by Sect. 5 which presents the simulation scenarios considered and the obtained results. Finally, Sect. 6 presents the main conclusions of the work and some ideas for future developments.

2 Applications of AMR in Airports

Technology has been exploited to improve luggage flow performance at airports, also increasing labour productivity. Moreover, by automating airport traffic flows, operators can focus on management and decision-making actions that are more complex and difficult to automate, which saves time and improves efficiency in handling logistics equipment.

Dallas Fort Worth International Airport is among the pioneers to introduce a luggage handling robot system, known as FLEET (Fig. 1a). When the luggage robots are in action, passengers take their luggage to the automatic check-in stations, where they deposit their luggage in one of the robots, which after performing all the checks of the passenger's flight are delivered on the conveyor belt that transports them to the container corresponding to the flight. This solution from the Vanderlande company consumes 50 % less energy compared to traditional luggage handling systems and handles more bags per cubic metre [4].

(a) Vanderlande FLEET robot for luggage logistics [4]

(b) Autonomous robot in luggage container transport [10]

(c) Ottonomy autonomous delivery robot [6]

(d) Robot to bolster cleaning regime at London's Luton airport [12]

(e) AGV pallet trucks [8]

(f) Parked vehicles [8]

Fig. 1. Applications of AMR in Airports

By using intelligent autonomous technology, the need for fixed conveyors and sorting systems can be replaced. Each AMR in the fleet carries a single bag and determines the most optimal route through the airport. Thanks to its design, the FLEET is both scalable and flexible, and changes are easy to make. It is error-resistant and should one unit fail, only one AMR is affected as the other vehicles can bypass it, thus ensuring a degree of operational certainty.

Another example, is the autonomous luggage transportation for aircraft at Singapore's Changi International Airport [10]. Once passenger luggage arrives at the sorting area, the operator places it in containers and this autonomous luggage robot, depicted in Fig. 1b, intervenes. This robot is an intelligent vehicle that moves autonomously through the airport to reach the aircraft, its destination being pre-programmed through a touch screen. When approaching the aircraft, the robot stops and the worker takes control of the vehicle to subsequently load the luggage into the aircraft's hold. After that, the vehicle returns to the luggage sorting area following the same route in the opposite direction.

Ottonomy, illustrated on Fig. 1c, is part of a fleet of fully autonomous food and retail delivery robots launched at Cincinnati/Northern Kentucky International Airport. Passengers at this airport can use the orderatcvg.com website or the mobile app to order delivery of items sold at retail shops within the airport.

A different application is the one on which AMR equipped with cameras and sensors from Ice Cobotics (Fig. 1d) are being used in air cleaning and sanitation operations at Luton airport, being capable of cleaning vast areas of the airport floor unattended, freeing up operators who previously performed the same task

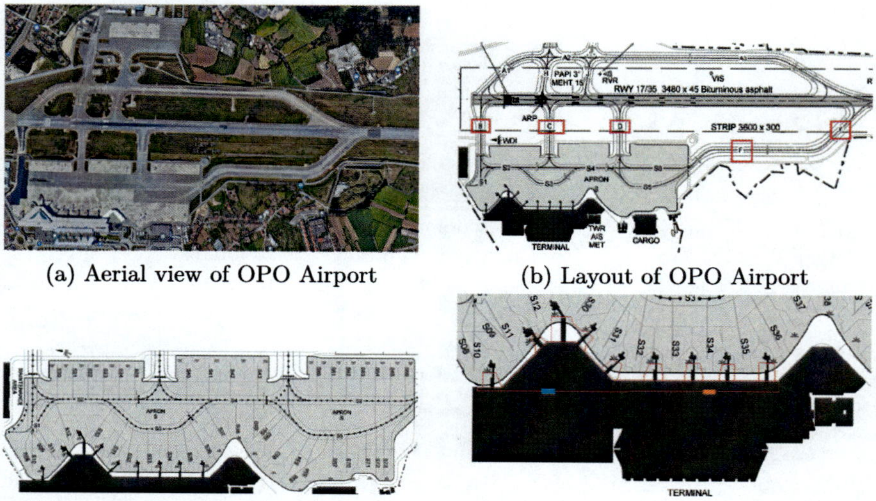

(a) Aerial view of OPO Airport

(b) Layout of OPO Airport

(c) Aircraft routes within the airport

(d) AGV routes within the airport

Fig. 2. Francisco Sá Carneiro Airport

manually [12]. This innovative technology is there to help cleaning operators who continue to refocus their attention on high-value tasks such as toilet hygiene and cleaning other touchpoints.

Finally, Stanley Robotics has installed a parking system at Gatwick Airport (London, UK), depicted on Fig. 1e. Using an AGV Pallet Carrier, whenever it detects a car it slides under the vehicle, lifts it and moves it to the storage (park) area. This way, it is expected that the current 6000 parking spaces will be enough to park 8500 vehicles, as parked with AGV does not need space for drivers to get in and out of the vehicle (see Fig. 1f) [8].

3 Francisco Sá Carneiro Airport

The Francisco Sá Carneiro Airport (OPO) (depicted in Fig. 2a) is located in the north of Portugal, in the district of Porto. It is the second largest airport in Portugal in terms of traffic, currently handling 20 movements/hour, with a capacity of up to 16 million passengers a year, and with 40 parking spaces (35 passenger aircraft plus 5 cargo aircraft). The Airport is developing an expansion plan in order to ensure a processing capacity of 32 movements/hour [1].

Figure 2b shows the runways (17/35) where the aircraft land and/or take off and the four aircraft circulation paths to the place where the loading and unloading of passengers and luggage is done; from left to right, these connections are named by: B, C, D and F. At the bottom of this image it is also possible to see the OPO airport terminal.

3.1 Aircraft and AGV Trajectories

The OPO airport has a runway of 3480 m, with various connections to the places where passengers (PAX) and luggage are loaded and unloaded. The places where the aircraft are positioned for assistance are called stand (or position) and are identified by a letter + number. They may be in contact with the terminal (through the telescopic bridges) or remote (accessible by transport, such as buses/minibuses, VIP transport or business aviation). There are 43 places where aircraft can be parked and 9 telescopic bridges where aircraft can pull over for loading and unloading.

Inside the airport, the aircraft must access one of the stands for assistance, if they are about to depart (exit from the airport), and to unload, if they are about to arrive (enter the airport). These accesses (zone S of Fig. 2b) can be through any of the channels (see Fig. 2c) B, C or D (the latter does not appear in the figure mentioned).

The AMR should make their journeys between the airport terminal and the respective stands in order to drop off/pick up luggage. They should leave the incoming luggage in the orange rectangle in Fig. 2d and load the luggage onto the aircraft in the blue rectangle. The line drawn in red is the AMR route; they only enter the stands if there is luggage to be picked up or delivered, otherwise they circulate on the main route. If they collect luggage at the terminal, they must take it to the desired stand for the aircraft to take off and leave it there. If the AGV is picking up luggage from an aircraft at one of the stands, it must carry it to the terminal and leave it there.

3.2 Aircraft, Passenger and Luggage Flow

The number of flights that pass through the airport is not fixed, varying every day and throughout the day. There are two types of flights: cargo flights and passenger flights. There was access to private information with statistics regarding the airport operations during the month of September 2016. It was possible to see that in that month there were 114 cargo flights and 943 passenger flights arriving and 115 cargo flights and 944 passenger flights departing. Therefore, although there are cargo flights, these are in smaller numbers. In total, there were 1057 flights into the airport and 1059 flights out of the airport.

During this month, the average number of passengers per flight was 152 and 89 486 bags were checked in. Of this total number of bags, an average of 79 bags was transported per flight and 0.5 bags per person. Obviously, these figures vary from month to month and according to the airline and the aircraft capacity.

3.3 Safety Rules

Airports have strict security rules that must be respected. First, nothing must pass under the aircraft. In fact, an imaginary perimeter is made where the AMR must not pass. Another rule that AMR must respect is not to obstruct the fuel supply trucks (orange cylinder in Fig. 3), which have priority over AMR.

The process for the arrival of an aircraft is as follows: first an aircraft arrives and, after its arrival, AMR move to the stand where the aircraft is. This rule is very important for the implementation as the waiting time of passengers for luggage is going to be evaluated, so the time it will take for the AMR to reach the aircraft stand is a must. When they arrive, the luggage is first unloaded from the aircraft and transported to the terminal and only after unloading is complete, the luggage loading process starts from the terminal to the aircraft.

The luggage entry and exit is always made on the right side of the aircraft and on the left side the passengers circulation is made. The circulation rules are depicted in Fig. 3. The fuel loading also takes place on the right side of the aircraft. It is also possible to see the Ground Power Unit (GPU) equipment that supplies power to the aircraft when they are parked at the airport.

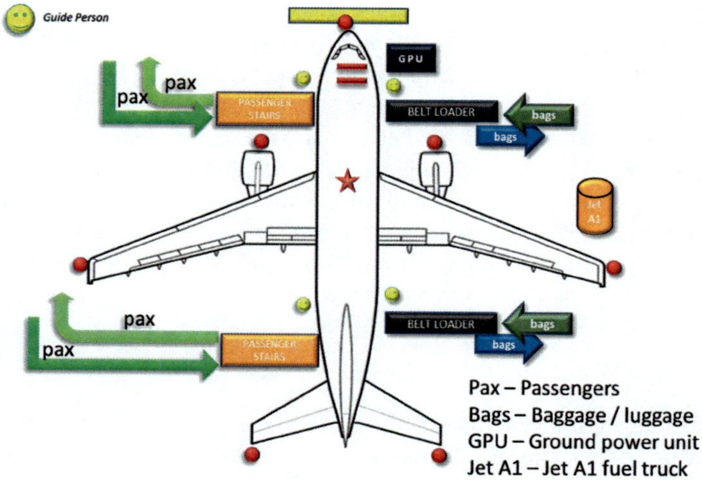

Fig. 3. Luggage and passenger circulation on the plane [11]

With this data it is possible to build a simulation model that is as close as possible to reality and to work with real values in order to have a simulation very close to reality.

4 OPO Airport Model

The software adopted for the airport modelling and simulation was FlexSim, version 21.0.10, from FlexSim Software Products, Inc. [7,9]. It is an object-oriented software, *i.e.* each object has a pre-built logic by performing various tasks, so that these are similar to the real world. Since it is a simulator, it is possible to create various "What If" scenarios with the possibility of importing custom Computer-Aided Design (CAD) drawings and without the need to make changes in the real world, thus being a software that allows to analyse, visualise and improve real world processes. In addition, it contains a flowchart tool (named *process flow*) to create the project model logic based on block diagrams.

For the implementation of this project functioning logic it was necessary to create a *process flow* (shown in Fig. 4), divided into three *containers* - divisions to improve the appearance and organisation of the developed *process flow*. The first *container* is aimed to the airport resources, the creation of luggage (both those of this airport and those of an outside airport) and the sending of these luggage to global lists of the system. The second one refers to the implementation of the Aircraft Flow and the third *container* to the implementation of the AMR Flow. The flowchart referring to the aircraft flow *container* is divided into two parts: on the left side, the process from arrival at the airport to unloading baggage at one of the *stands* and, on the right side, the process from loading departure baggage to leaving the airport. The flowchart referring to the AGV flow *container* is divided into two parts: on the left side, the process from arrival at the airport to luggage unloading at one of the *stands*, and on the right side, the process from loading departure luggage to departure from the airport.

4.1 OPO Airport Layout Modelling

The first step on developing the model was to create the layout of OPO Airport, following [3]. This document also allowed to understand the paths along which the aircraft could circulate. The resulting layout model is depicted on Fig. 5a. Along the routes of both aircraft and AMR, several control points (for both aircraft and AMR) are required. With these points are avoided collisions and are obtained reference points for travel. Figure 5b (magnification of the Flexim model layout) shows in detail the path that the AMR take to the stands (round trip). There is a main path and the appropriate "detours" that the AMR need to take to reach the aircraft stands, to load or unload luggage on them.

4.2 AMR Features

AMR will load and unload luggage between the terminal and the aircraft, and it was considered that the AMR would have no obstacle during its travel. The type of AMR chosen for this project was the AGC (under ride type). For an inbound flight, the AMR does not start moving until the aircraft has come to a stop. It then unloads the arriving luggage and, if there is already luggage at the terminal, loads the departing luggage. It was necessary to configure their properties, such as the speed at which they travel in different situations (straight line speed, cornering speed, loaded and unloaded speed (see Fig. 6a)), the load capacity and the time to load/ unload a piece of luggage. The load capacity of the AMR is 20 pieces of luggage, for volume reasons, as a hold luggage has a maximum total dimension $(a + b + c)$ of up to 158 cm (see Fig. 6b), or up to 300 cm if PAXs pay extra. The AMR take 10 s for each loading, and unloading, of luggage and it was defined that there would be no errors in these tasks.

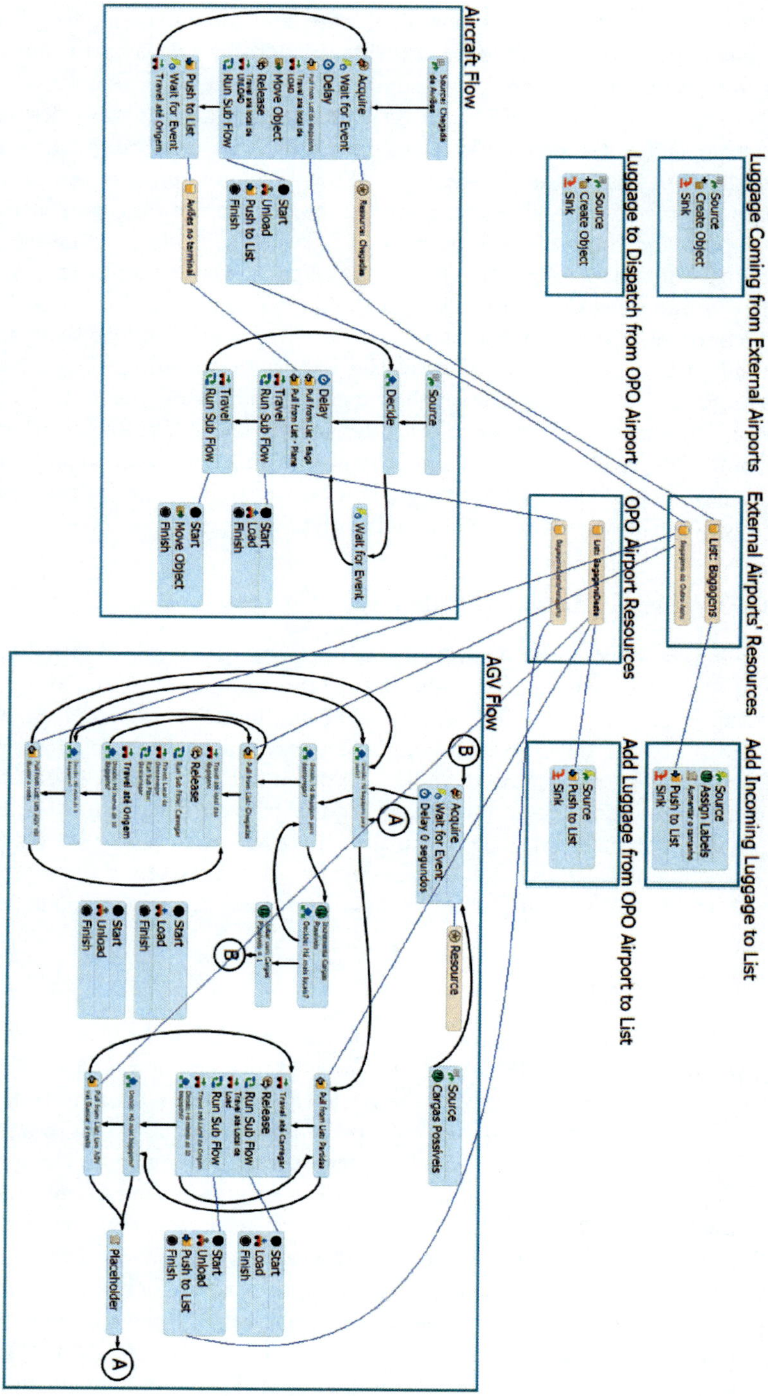

Fig. 4. Final aspect of the model process flow

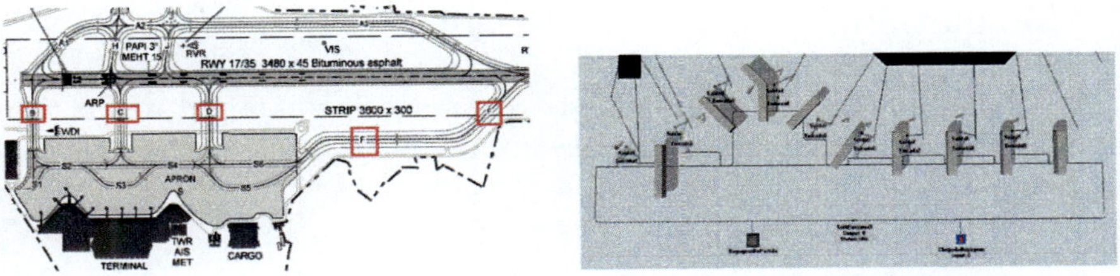

(a) OPO Airport Layout developed on Flexsim

(b) Layout of the AGV paths

Fig. 5. OPO Airport layout model

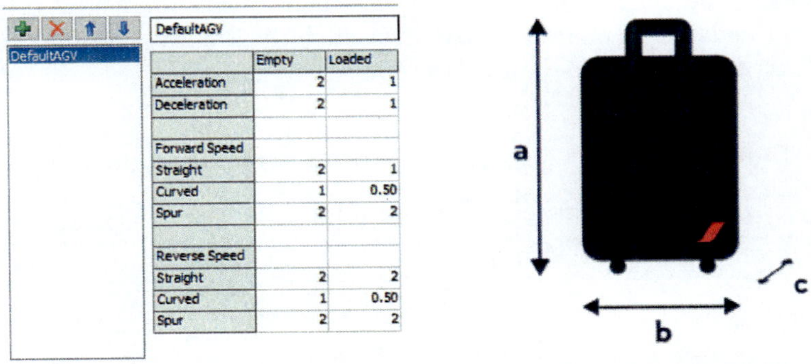

(a) AMR speed parameters

(b) Size of a hold luggage

Fig. 6. Simulation model parameters

4.3 Luggage Classification

Sources are used to create flow items (for example luggage). Each source creates a class. In this work two types of source were used: Schedule Source and Event-Triggered Source. In a schedule source it is possible to determine at what time and how many objects will be created (enter the simulation). For the luggage to appear at the intended instant (and location) it was necessary to create them through a schedule source. The event-triggered source is only active when an event occurs and one can associate a *token* with items from this source.

For luggage, different tokens were created: *token.BagsIn* refers to luggage arriving at OPO airport, while *token.BagsOut* refers to luggage departing from this airport to another one. Other references should be considered for priority luggage (5% of all luggage in the model, referring to 1st class luggage that will be carried at the start of the aircraft unloading) and transfer luggage (not considered in this study, referring to luggage that is transferred directly between aircraft). Sinks were used to simulate luggage departures (entry into the airport terminal or luggage departure from the airport).

5 Simulation Scenarios and Results

This section presents the considered scenarios, and the obtained results. The waiting time for luggage arriving at the airport (the time passengers will wait for their luggage to arrive at the terminal from the moment they leave the plane) was assessed and analysed, as a function of the number of AMR in operation.

5.1 Simulation Schedule Source Parameters

To obtain valid results based on actual data, a restricted circulation document listing the arrival and departure times of aircraft during a working day was used. Additionally, the number of luggage on each flight was estimated, as presented in Tables 1 and 2. This quantity is based on the historical average of luggage per flight. This data was entered into FlexSim.

Table 1. Number of luggage items arriving at each stand per flight

Time (s)	Quantity	Stand	Time (s)	Quantity	Stand	Time (s)	Quantity	Stand
50	78	1	37000	75	8	58800	79	4
350	76	2	40500	82	9	62400	78	5
650	75	3	42300	73	1	63600	77	6
950	68	4	42900	66	5	64800	76	7
1250	82	5	43500	73	3	65400	70	8
15000	71	6	44100	74	4	66000	83	9
19500	68	7	45600	76	2	67200	68	1
22800	74	8	46200	76	5	68400	72	2
23400	73	9	47400	74	6	69300	74	3
24000	73	1	48600	73	7	71400	76	4
24600	80	2	51600	72	8	74700	77	5
28800	79	3	54600	71	9	77700	78	6
30000	77	4	56100	76	1	78900	79	7
33600	81	6	57600	80	3	80700	76	8
35100	73	7						

Since FlexSim always starts the project simulation at 00:00, the planning started at that time, until the end of the flights. In total there were 86 flights performed on this day (43 arrival flights + 43 departure flights). So that no aircraft remained at the airport, the departure of the last flights in the planning was also simulated at the end of the simulation.

Table 2. Number of luggage items departing at each stand per flight

Time (s)	Quantity	Stand	Time (s)	Quantity	Stand	Time (s)	Quantity	Stand
18000	74	1	43200	68	9	64200	79	5
18900	76	6	44100	69	1	66000	78	6
19800	68	2	45000	72	2	67200	77	2
21600	83	3	45900	78	5	68400	78	7
22500	80	4	46800	79	3	69300	76	8
23400	71	7	48600	80	4	70200	78	1
26400	72	8	49500	74	5	71400	78	2
27600	73	9	51000	75	6	72600	68	3
28800	75	1	52500	72	7	74700	78	4
32400	74	4	54600	71	8	75600	67	9
34800	77	3	57000	67	9	82800	78	5
39600	78	6	58500	75	1	83700	72	6
40500	79	7	60000	74	3	84600	76	7
41400	80	8	61200	66	4	85500	77	8
42300	80	5						

5.2 Simulation Scenarios Considered

The tests started with a 13 AMR fleet in operation and with all of them going to all stands, *i.e.* no AMR is fixed to a certain stand. In total, 9 stands were considered. The data was extracted to verify the waiting times (average, minimum and maximum) at each stand (exit). The number of AMR in each simulation scenario was increased of 1 unit, until the AMR fleet had 17 vehicles. The 5 different scenarios considered were named Scenario A (13 AMR), Scenario B (14 AMR), Scenario C (15 AMR), Scenario D (16 AMR) and Scenario E (17 AMR).

In all the scenarios considered above, the AMRs are divided into two groups of equal size. Whenever a group of luggage arrives (at the airport or terminal), only half of the AMRs will collect that luggage. If another group of luggage arrives at almost the same moment, the second group of AMRs will collect that other luggage group. If all the AMRs had to collect the first set of luggage, they would only collect the second set at the end of the first one, increasing the associated waiting time. In the scenarios considered where the number of AMR is odd (Scenarios A, C and E) there will always be 1 AMR that is at rest and charging batteries during that time interval.

A test was still performed with 18 AMR, but the waiting times started to increase, as there were too many AMR in operation at the same time and there was no improvement in waiting times.

5.3 Simulation Results

To define which would be the best scenario for the day when this planning was done, Table 3 was made, presenting the maximum, minimum and average waiting times, for an AMR fleet with 13 to 18 vehicles. The minimum times are always the same since for priority luggage (PL), as they are the first to be picked up, the AMR that will load and unload them always takes the same time; while the minimum waiting time for non-priority luggage (NPL) is always the same due to the fact that on the last arrival flight, the AMR are free and thus, the luggage is unloaded immediately.

Table 3. Waiting time for luggage as a function of the AMR fleet size

Time (s)	13 AMR	14 AMR	15 AMR	16 AMR	17 AMR	18 AMR
Maximum	4680,32	4680,32	4319,43	3510,55	3248,97	4472,03
Minimum PL	766,98	766,98	766,98	766,98	766,98	766,98
Minimum NPL	1477,27	1477,27	1477,27	1477,27	1477,27	1477,27
Average	1932,98	1869,09	1825,43	1804,76	1788,75	1813,68

Analysing these results, the scenarios with 13 and 14 AMR can be discarded since both present a higher maximum waiting time than the others. Next, when comparing the remaining 3 scenarios, if the scenario with 15 AMR is separated from the others, it was found that the average of the minimum waiting times for non-priority luggage is lower in Scenario C than in Scenarios D and E, although the minimum values are the same. However, the maximum waiting time is much higher. The scenario with 16 AMR when compared to the scenario with 17 AMR differs by 16 s in average time, but in maximum time, scenario D (16 AMR) takes 262 s (4 min and 22 s) longer than scenario E (17 AMR). Whereas, if the average minimum waiting times for non-priority luggage are compared, the scenario with 17 AMR takes 13 s longer. Between these two scenarios, the scenario with 17 AMR, despite its minimum time for non-priority luggage being higher by a few seconds, gains 4 min in maximum waiting time, so it is better compared to the scenario with 16 AMR.

Therefore, on this day, the airport would have to choose the scenario according to which was the highest priority. Scenario C (15 AMR) would lose 1100 s seconds in maximum waiting time on one of the flights and 37 s in average waiting time throughout the day compared with Scenario E (17 AMR), but it would gain 15 s in minimum waiting time (for non-priority luggage). If the priority is to decrease the maximum and average waiting time throughout the day, the scenario to choose is Scenario E, with 17 AMR. However, if the priority is to decrease the minimum waiting time for non-priority luggage, the best scenario is Scenario C, with 15 AMR.

6 Conclusions and Future Developments

This work presented the modelling and simulation of the luggage transport between the airport terminals and the aircraft in the OPO Airport. Its main objective was to understand: (*i*) if it is possible to have this task performed by an AMR fleet in an automated way, without the intervention of humans, (*ii*) the inherent problems to this implementation. And (*iii*) the needed number of AMR to maintain the service levels to passengers. According to the simulation results, it was concluded that the best scenario would imply the use of 17 AMR.

Concerning future developments of this project, three main lines will be pursued. First, the AMR should reach more points in the airport: instead of reaching only the nine stands closest to the terminal, they should also reach the remaining points where the aircraft stop to load and unload passengers/luggage. This will also require the modelling and simulation of priorities between AMR and aircraft to avoid potential collisions. Second, there will be modelled and simulated the process of transfer luggage (bags that do not arrive at the airport terminal, on flights with stopovers, since they are immediately transferred to another aircraft). Third, and last, there is a detail that would improve the look of the simulation: changing the colour of priority luggage, to give simulation a better look and make it easier to interpret the arriving luggage.

Acknowledgements. This work was partially financed by National Funds through the Portuguese funding agency, Fundação para a Ciência e a Tecnologia (FCT), within project LA/P/0063/2020.

References

1. aerOPOrto: Plano de expansão (2017). https://aeroporto.news/plano-de-expansao/. Accessed 20 Mar 2022
2. AGV Network: Types of AGV vehicles with examples - what AGV should you buy? (2017). https://www.agvnetwork.com/types-of-automated-guided-vehicles. Accessed 20 Mar 2022
3. AIS - Portugal: Manual VFR (2021). https://www.nav.pt/docs/AIS/aerodromos/portoAD.pdf?sfvrsn=28. Accessed 09 Apr 2022
4. Bostick, N.: DFW International Airport Is Testing the World's First Robot Baggage Handlers (2019). https://www.dallasobserver.com/news/dfw-international-airport-is-testing-the-worlds-first-robot-baggage-handlers-11703209. Accessed 28 Feb 2022
5. DAIFUKU: Automated guided vehicles & carts (2022). https://www.daifuku.com/us/solution/vehiclecart/. Accessed 20 Mar 2022
6. Deletter, E.: CVG airport now has food delivery robots (2021). https://eu.cincinnati.com/story/news/2021/12/16/cvg-cincinnati-airport-ottobot-food-delivery-robots/8925637002/. Accessed 28 Feb 2022
7. FlexSim: Flexsim 3D simulation modeling software (2022). https://www.flexsim.com/flexsim/. Accessed 20 Mar 2022
8. London Daily Express (.K.): Fleet of robots to park cars at Gatwick airport (2019). https://cacm.acm.org/news/234461-fleet-of-robots-to-park-cars-at-gatwick-airport/fulltext. Accessed 20 Mar 2022

9. Palma, C.L.: FlexSim software. http://www.leansimulation.org/flexsim-software
10. Park, K.: The Airport of the Future Is Here. And It Doesn't Need Humans (2018). https://www.bloomberg.com/news/features/2018-09-16/singapore-s-changi-airport-is-partly-run-by-robots. Accessed 28 Feb 2022
11. Siqueira, A.: Gestão aeroportuária e turismo, January 2021
12. Thornton, J.: Robots enhance London Luton Airport cleaning regime (2020). https://www.roboticsandinnovation.co.uk/news/cleaning-sanitation/robots-enhance-london-luton-airport-cleaning-regime.html. Accessed 28 Feb 2022

Simulation and Digital Twin of a Robotic Sanitizing Process of Environments at Risk During the Pandemic

Francesco Cepolina and Elvezia Maria Cepolina(✉)

University of Genoa, Genoa, Italy
francesco.cepolina@edu.unige.it, elvezia.maria.cepolina@unige.it

Abstract. The goal of this work is the design and development of intelligent semi-autonomous processes conducted by a team of mobile robots for cleaning environments whose hygiene must comply with the law, such as restaurant and community environments, healthcare facilities, laboratories and hospital departments. The cleaning team is composed by mobile robots designed for wall cleaning tasks but they can also be used for floor and ceiling cleaning tasks.

The article proposes a simulator, based on the geometric model of the environment and functional model of the robot, that allows to define cleaning procedures adapted to needs and regulations with a good margin of flexibility.

This simulator is configured as a digital twin of the local cleaning system and can be used for cleaning procedure improvements, for the punctual definition of maintenance operations and for the review of the process in case of new regulations and or unexpected occurrences.

Keywords: Climbing robot · Healthcare environments · Robotic cleaning and disinfection · Logistics strategy · Digital twin · Industrial plants · Service plants

1 Introduction

The context of the research are plants for the production of food and healthcare services. The article proposes the definition of service plants for the creation of environmental hygiene conditions suitable for the production and preservation of food and to ensure the physiological well-being of workers and customers in the industrial facilities.

The problem of cleaning environments subject to legislation has always been present and automated solutions have been presented and applied for the execution of the processes and the documentation required by law [1]. Several companies are working in this sector with renewed sensitivity to the regulations that are published from time to time in different countries. Figure 1 shows the results published in the IFR World Robotics 2021 [2] about the statistics on professional cleaning service robots suppliers in different geographic regions.

Since the early days of the pandemic, national and international health organizations have defined and continue to update stringent cleaning methods for healthcare environments [3]. They claim that it is important to ensure that cleaning and disinfection

© The Author(s), under exclusive license to Springer Nature Switzerland AG 2023
J. M. Cascalho et al. (Eds.): CLAWAR 2022, LNNS 530, pp. 501–512, 2023.
https://doi.org/10.1007/978-3-031-15226-9_46

procedures are followed consistently and correctly with legal certification. All surfaces in healthcare facilities should be routinely cleaned and disinfected, especially high-touch surfaces, and whenever visibly soiled or if contaminated by body fluids.

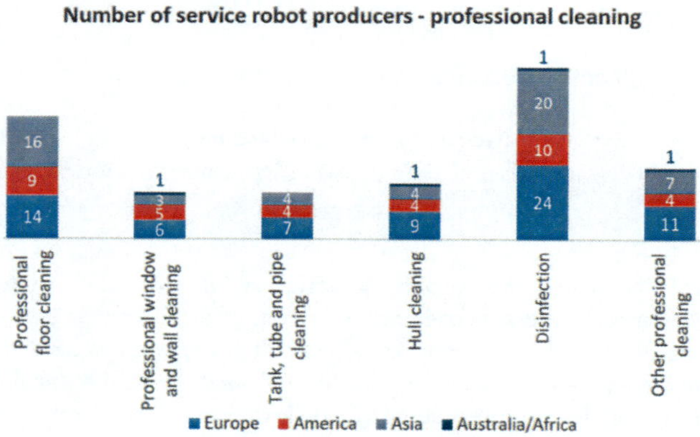

Fig. 1. Professional cleaning robot producers [2]

Detailed guidance on environmental cleaning and disinfection monitoring and auditing in the context of COVID-19 is available from The World Health Organization [4]. The selection of disinfectants should meet local authorities' requirements for market approval, including any regulations applicable to specific sectors, for example healthcare and food industries. Detailed standard operating procedures on the cleaning and disinfection of environmental surfaces in the context of COVID-19 are enacted and updated from the competent bodies. These procedures should be followed with a clear delineation of responsibilities regarding the type of surfaces and frequency of cleaning (Table 1).

The automation of cleaning processes in critical environments and with the use of materials that can be harmful to those who handle them offers undeniable advantages not only for automatically delivery of reporting procedures guaranteeing environment hygiene but also for safeguarding the health of cleaning personnel.

Certainly the management of the process must take into account the costs of materials, the energy consumption and the time that can be allocated to cleaning. Low energy consumption robots must be used [5], the team must be dimensioned with a number of agents and with effective logistics in order to guarantee the completion of the process within the time allowed: to this aim the same fleet optimization methodology proposed in [6] has been adopted. The proposed simulator is an effective tool to get a preview of the behavior and effectiveness of the cleaning system from these points of view.

The same simulator, if set as a digital twin (DT) with appropriate links between the virtual data of the simulation and the real data detected by the sensors on the process resources, is a great help in setting up and managing the cleaning procedures improving quality and timely delivery of the service [7] together with its sustainability performances [8].

Table 1. Healthcare setting: Recommended frequency of cleaning (WHO, 2020).

Patient area	Frequency [a]	Additional guidance
Screening/triage area	At least twice daily	• Focus on high-touch surfaces, then floors (last)
Inpatient rooms / cohort – occupied	At least twice daily, preferably three times daily, in particular for high-touch surfaces	• Focus on high-touch surfaces, starting with shared/common surfaces, then move to each patient bed; use new cloth for each bed if possible; then floors (last)
Inpatient rooms – unoccupied (terminal cleaning)	Upon discharge/transfer	• Low-touch surfaces, high-touch surfaces, floors (in that order); waste and linens removed, bed thoroughly cleaned and disinfected
Outpatient / ambulatory care rooms	After each patient visit (in particular for high-touch surfaces) and at least once daily terminal clean	• High-touch surfaces to be disinfected after each patient visit • Once daily low-touch surfaces, high-touch surfaces, floors (in that order); waste and linens removed, examination bed thoroughly cleaned and disinfected
Hallways / corridors	At least twice daily [b]	• High-touch surfaces including railings and equipment in hallways, then floors (last)
Patient bathrooms/ toilets	Private patient room toilet: at least twice daily Shared toilets: at least three times daily	• High-touch surfaces, including door handles, light switches, counters, faucets, then sink bowls, then toilets and finally floor (in that order) • Avoid sharing toilets between staff and patients

Next section is dedicated to the robotic cleaning system modelling including a brief presentation of the design characteristics of the reference cleaning robot, the description of typical healthcare environments, a look on currently suggested cleaning procedures. The third section presents the simulator architecture with objects, methods, procedures, and interfaces with particular view on cleaning team logistics. The last section concludes the paper with final discussions and a look to future developments.

2 Modelling and Simulation of the Cleaning Process

The purpose of the simulation is to study the cleaning process of smooth flat surfaces of the interiors of buildings, typically of hospitals, nursing environments, laboratories [8], community kitchen, restaurants [9]. In any case the geometric model of the considered environment is two-dimensional; it is seen developed in plan with the walls overturned on the floor level.

The simulator can be used to predefine the cleaning procedures: scuffing cycles with subsequent aspiration of residues, washing cycles and disinfection cycles for surface sanitization and sanitation. To organize the cleaning cycles, the simulator must implement the logistics of distributing the work on the different plane surfaces of the environment between the various robotic agents of the team [10]. The distribution of work among the various agents was approached by creating clusters of activities and assigning each cluster to an agent. A fuzzy k-means clustering algorithm has been adopted as in [11].

The environment can be considered a fairly structured environment and enough free from temporary encumbrances so that the procedures can be taught or assigned to the robots that will perform them independently, using algorithms for overcoming obstacles in the event that their sensors detect the presence of objects/people not foreseen on one's path [12]. The person avoidance algorithms assigned to the robots are defined with the aim of ensuring minimum disturbance to those working in the environment and maximum accessibility for the activities that people perform in the environment [13]. The simulator must also monitor the status of the robots and the resources for the direct performance of the cleaning functions: energy consumption, fluid and battery level. The tool thus set up allows to compare different cleaning procedures and logistics strategies from the point of view of efficiency (times), consumption and costs, providing information on the duration of completion of the procedures, on the rhythms of maintenance, on the stocks of consumables and of the battery recharging station (which affect refueling times).

2.1 Cleaning Agency

The cleaning agency is composed by a team of GeckoV robots that represent the last generation of robot Gecko [14], updated for the new pandemic needs and embedded with smart sensors for improving the automation functions. Because GeckoV has limited navigation capability being able only to move on plane surfaces by Cartesian coordinates, a human agent or a smart mobile robot assistant moving on the floor is in charge to put the single GeckoV on the planar surface it has to clean based on the logistics strategy adopted. The same agent is asked to change and/or refill the cleaning liquid in the GeckoV tank.

GeckoV is a simple low cost robot designed to accomplish wall and ceiling cleaning duties. But it can be adapted to cleaning also floors using a reduced movement depression, creating a sort of self-mop with mobility along two orthogonal axes.

Vacuum technique is chosen to sustain the robot [15]. The adherence to the wall is ensured by four, purposely depressurised, suction cups [16]. The vacuum is achieved in a central reservoir and distributed to each cup that can be activated independently.

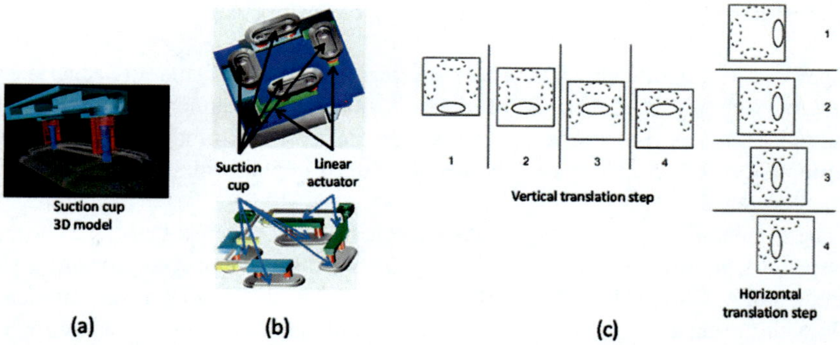

Fig. 2. Models of suction cup (a) and robot climbing motion system (b). On the right (c) the motion basic steps of the robot.

The actuation system allows GeckoV to move upward, downward, left and right, without the need to rotate the robot itself. Tidy up, typically, follows vertical tracks and horizontal displacements are used to reach a new track, avoiding gravity bias effect. Figure 2 shows the models of a suction cup and of the robot climbing motion system together with the mobility along horizontal and vertical steps.

The kinematic and dynamic behavior of GeckoV was analyzed using ProMechanica considering the simplified multibody system essentially consisting of rigid bodies, constraints (kinematic pairs), forces and moments. The positions of each member are identified with respect to the WCS (world coordinate system) by means of a vector of generalized coordinates $q = \lfloor \lceil q_1 \rceil^T \lceil q_2 \rceil^T \ldots . \lceil q_{nb} \rceil^T \rfloor^T$.

The number of components of this vector is nq = 6 × nb. Where nb is the number of rigid bodies making up the system and 6 the number of degrees of freedom of a body in the space.

The system of position constraint equations can be expressed in the system of non-linear equations in q, $\Phi(q, t) = 0$ which leads to the two linear constraint systems in velocity where $[\Phi_{,q}]$ is the Jacobian matrix of the system. The equations of dynamics, through the application of the virtual work theorem, take the following form:

$$\begin{bmatrix} [M] & [\Phi_{,q}]^T \\ [\Phi_{,q}] & [0] \end{bmatrix} \begin{bmatrix} \ddot{q} \\ \lambda \end{bmatrix} = \begin{bmatrix} Q^A \\ \gamma(q, \dot{q}, t) \end{bmatrix}$$

- $[M]$ is the matrix of the generalized inertias of the multibody system
- Q^A is the vector of the λ active forces acting on the respective components of
- λ is the vector of the Lagrange multipliers;

These models applied to GeckoV and solved in ProMechanica provided realistic evaluations of the kineto dynamic parameters of the robot that were used in the simulation. As an example, the results of the dynamic analysis of the GeckoV, Fig. 3 shows the forces exerted by the corners of the robot skirt on the wall and the skidding during a lateral step. Analogous results concerning the vertical step show that the skidding of the robot is practically nil; on a stroke of 154 mm, the robot moves laterally (with respect to the direction of travel) by 1.4 microns.

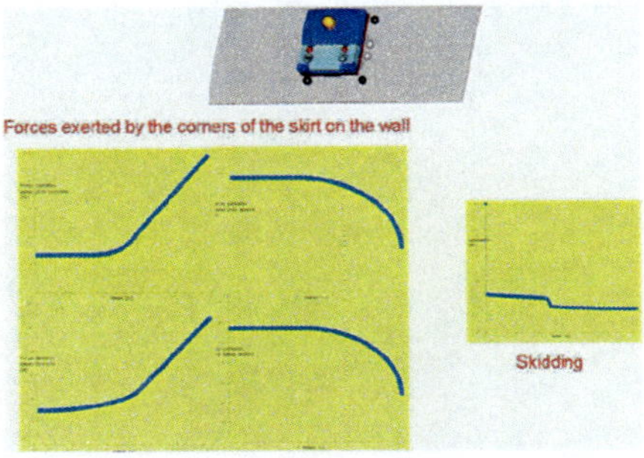

Fig. 3. GeckoV: dynamic analysis during the lateral step

The physical robot is embedded with low cost sensors that provide data useful for the knowledge of the real evolution of the cleaning procedure and to make comparison with the analogous data provided by the virtual digital twin [17].

2.2 Cleaning Process

The cleaning cycle consists of several phases depending on the needs of use and the legislation in place. It was decided to prepare a first phase of washing with steam and sucking up the dirty water before proceeding to subsequent phases of sanitation. The liquid tank on board is reduced in size to facilitate motion and handling between the different cleaning phases every time the tank must be emptied and filled with the reference liquids for the next phase. Suitable constructive measures are adopted in order to exploit all the liquid present in the tank, to vaporize it and for signalizing the low level (7–10%) of the volume of the liquid.

In relation to the process characteristics: dispensing of liquids, energy consumption, the hypothesis of uniformity was made, which is actually valid only in the case of normal operation; constant times were considered for maintenance, battery recharging and liquid restoring operations, a valid hypothesis in the case of the presence of an agent in charge of the service and of the related supplies of consumables.

2.3 Model of the Environment

The surface to be cleaned is thought to be quantized according to the size of the elementary cleaning area. This operation, referred to as mapping, constitutes an orientation constraint of the robot paths which, however, seems not limiting because it is set in an interactive phase and therefore under the direct control of the user.

To make the model more flexible, the mapping of the environment was organized in two different matrices. First of all, the map of the perimeter walls and fixed constraints must be provided; examples of fixed constraints are partition walls, windows and passage openings or work islands. Then, the user must introduce the overall dimensions of the local mobile equipment. Examples of mapped environments are given in Fig. 4.

The division of the cleaning area plan into two maps facilitates the updating of the map in case of relocation of the mobile structures, and allows their relocation from the point of view of ergonomics in the cleaning phases.

Fig. 4. Mapped plans of example environments to be cleaned.

3 Simulator Architecture

The program was developed using an object language to which modules, written in c, customized to the functional needs of simulation of the cleaning process and of the digital twin implementation mode, were added. The program is able to import graphic files (in bitmap format) and sound files (in wave format). In the field of work, these potentials have been fully exploited including also proprietary libraries [18].

Figure 5 shows a diagram of the organization of the simulation and digital twin program developed and fine-tuned. According to the authors, the concept and development of this simulator presents significant points of originality.

The different process actions, performed independently by the different robots of the cleaning agency, can take place simultaneously and, in any case, without a predefined time evolution.

To synchronize events, tools are used that allow to interrupt the execution of an activity until predefined conditions are met. By suitably combining these tools and various functional modules, it is possible to achieve the desired synchronization between processes occurring in parallel.

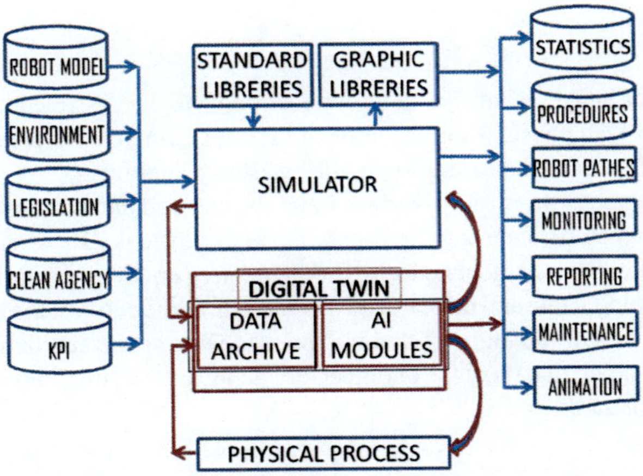

Fig. 5. Schematic of the developed simulation/digital twin

3.1 Objects

Objects are all entities that can change state and/or perform activities. Objects are characterized by state variables and by the methods that apply to them. They interact through actions, carried out by the methods, which can trigger events and change states which, in turn, activate other methods, determining the evolution of the process.

Objects can be static or dynamic depending on whether their representation is fixed or mobile in space. In our case, because the fields of an object can only be changed by the methods of the object itself, the development of appropriate modules was required for the creation of logics driving interaction between different objects. The main objects are the robots and the surface units of the environment.

Robot

The robot is a permanent dynamic object. Some of the characteristics are invariant attributes for the particular robot while others represent current values useful for the correct representation of the process and therefore for the interaction with other types of objects. There are also some statistical variables useful as information returned by the simulation. General features of the robot parent object include:

- constant characteristics that describe the robot: icon, travel, washing, sucking, sanitizing speeds, capacity of the detergent compartments, consumption for a unitary washing and sanitizing operation, energy consumption for a unitary transport operation, battery capacity, maintenance times related to changing and topping up detergents, recharging and replacing batteries
- characteristics that vary during the evolving simulation: actual position, current speed, current level of the detergent compartments, current battery level, current status: translation, washing, sucking, sanitation, maintenance.

Surface Unit

A record of color images of surface units corresponding to the possible levels of cleaning was matched to each physical surface units of the environment, in order to facilitate the visualization of the actual cleaning state: dirt, washing, sanitizing.

Constant characteristics of the surface units are the icon and its shape and size.

Variable characteristic during the simulation is the current state of cleaning.

Some of the variables associated with the elementary environment units are invariant while others are updatable and indicate the current state of the entities: the latter constitute the basis of choices of the logistics that will be made during the simulation.

It should be noted that this type of object, not being "alive", does not need to activate asynchronous methods.

3.2 Methods/Experience Modes

Each object is capable of carrying out multiple activities. The activity is launched from an object using a method that is able to pass the simulation time. There are also methods that are comparable to normal procedures for assigning parameters to agents. These methods make a synchronous call and do not allow the simulation time to pass.

The activities, even carried out in parallel, are governed by methods that, with asynchronous call, require the execution of the procedures to the objects and pass over without waiting for the invoked method to be finished or, for some reason, still to be started. In this way the simulation time is able to pass.

There is also another type of method that has characteristics of both previous methods: it makes a synchronous call but allows the simulation time to elapse.

The robot has a main method called MoveAround; it calls several methods which constitute synchronous calls for the execution of procedures and methods which allow the simultaneous carrying out of several parallel activities. This method, from the dynamic module of assignment of tasks to the single robots, reads the cleaning cycle that the individual robot must perform and uses this data to perform the following operations: - animate the robot, with the speed specific to each operation; - update the cleaning status mapping; - update in real time the current values of the battery energy indicators, the current speed, the current cleaned surface, the detergent level.

Path Planning

Some simple path planning algorithms have been implemented to solve the navigation problem in two-dimensional (2-D) near-static environments. A priori, the total area is divided into rectangular sub-areas that each constitute a block of the process. The passage from one block to the next is guided by heuristic criteria that are validated by the simulation. The proximity sensors of the GecoV activate appropriate variations to the trajectories set.

Digital Twin Rules

The developed methodology leads to the cleaning modes optimization through the evaluation of the sustainability, effectiveness and efficiency performance indexes.

There are many rules which, on the basis of the differences between simulated data and data collected on the real system, suggest actions for the improvement of both virtual and real twins. Among the rules for improving the model of the simulated system there are those for redefining simplifying hypotheses that lead to significant deviations between the evolution of the simulation and real cleaning process behavior. For example the hypothesis of constant speed or the hypothesized law of battery discharge or the consumption rate of cleaning liquids.

Among the rules that suggest actions to improve the management of the real process there are those that lead to an in-depth analysis and review of monitoring and diagnostic activities as well as reasoning and decision making on maintenance operations [19] or, at least, replacement of a robot considered unsatisfactory for its degraded state of health.

AI key technologies are applied to the large amount of data collected during the parallel evolution of the twins to make intelligent knowledge based decisions aimed to improve important aspects of the cleaning process such as lean agile resilient green aspects:

- Lean aspect concerns the improvement in cleaning performance by eliminating wastes and by enhancing the quality and timely delivery of process [20];
- Agile cleaning process is about the improvement of a rapid and cost-effective response to regulations demand fluctuation by looking at the minimization of costs;
- Resilient cleaning process looks for elimination of disturbances due to external risk as regulations turbulence, delay in resources procurement and internal risk such as robot breakdown;
- Green processing concerns the execution of environmental conscious cleaning practices to minimize ecological damage.

This is the most innovative part of the simulator. The DT algorithms are implemented in a modular way to give flexibility to the optimization procedures towards different key performance indexes (KPI). In our case only some of these aspects, mainly related to lean and green aspects, have been persecuted demonstrating the feasibility of the approach.

3.3 Interfaces

Libraries have been used that offer the possibility of representing both dialog-boxes and graphics in one, two or three dimensions during the simulation. These elements, provide a simple and immediate control of the evolution of the process, Standard graphic interfaces are also associated with useful sound indicators.

Input Interface

As input, the user must assign the characteristics of the environment and of the fixed encumbrances with the two-dimensional plans and their reference mapping for the simulation; the number of robots of the cleaning team; the dimensional and kinematic parameters of each robot, including the parameters of the tanks; the distinctive icons of the different robots for quick identification during the execution of the simulated process. Moreover, for the management of the cleaning process and for the planning of the path it is very important to discriminate between robots that can move in one direction only or bi-directionally in X and Y.

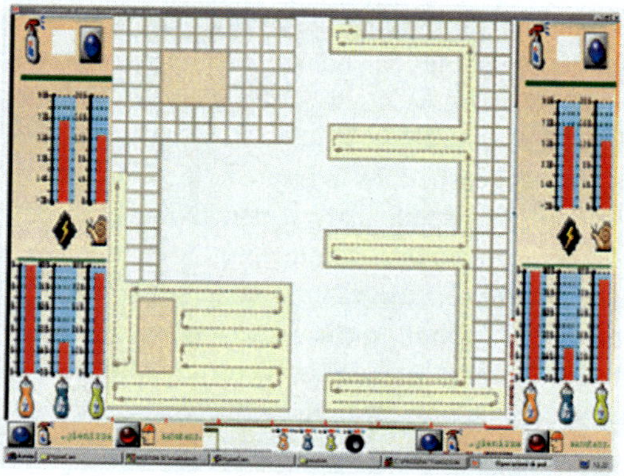

Fig. 6. The simulator main interface

Dynamic Digital Twin Interface

During the simulation the main process variables are monitored; to facilitate reading, they are associated with significant icons, see Fig. 6. Among these we remember:

- Indicator of simulation time that can be scaled with respect to real time.
- Energy consumption indicator for each robot.
- Speed indicator for each robot.
- Indicator of the level of the detergent currently for the cleaning operation for each robot.

- Low battery maintenance in progress indicator for each robot requiring replacement with backup battery
- Maintenance in progress indicator for emptying a detergent compartment that requires it to be filled for each robot.

These characteristic variables of the evolution of the simulated process, in addition to constituting useful information for the user, are the knowledge of the virtual process (a priori knowledge) and the data collected are compared with the analogous ones detected on the real physical process (a posteriori knowledge). This comparison [21] is the basis of the smart logic of improving updating the simulator and of careful management and optimization of the real cleaning process towards the performances selected from the aspects recalled in the previous section.

To allow the user to recognize more immediately the type of operation in progress, it has been chosen to assign a different sound to each operation that is performed during the simulation. Particularly useful is the acoustic warning which signals to the operator who supervises the cleaning process the need for a maintenance operation.

Output Interface
When the simulation is finished, a summary box appears containing the salient information of the simulation: for each robot the trace of the trajectory performed, the energy consumption and the number of battery changes, the consumption of liquids and the number of top-ups of detergent with the relative combined maintenance times, the execution time of each cleaning phase, the total cleaning time of the area assigned to it. Each simulation run generates a lot of data that constitutes a history sample of the cleaning process trial. All these trial samples through data smart processing corroborate the knowledge of the characteristics of the virtual twin of cleaning process.

4 Summary of Results and Conclusions

A flexible model of a cleaning process of environments at risk of contamination performed by a team of identical cleaning robots was presented. The cleaning agent robot is the updated version of a climbing robot developed years ago by the same research team. The upgraded robot able to perform sanitizing process has been successfully tested with a tank that supplies the cleaning liquid, be it simple water or sanitizer, to the vaporizer and nebulizer dispenser system.

A new simulator of the multi-robot cleaning process, embedded with digital twin modules, was developed and made it possible through virtual monitoring to evaluate alternative process logics. Digital twin supports the cleaning service life cycle management through monitoring, diagnostics, optimizing activities. The results are promising but the DT modules have yet to be validated.

In the current version, the developed system is able to suggest changes to logistics, inspection and equipment fault diagnosis and repair actions as well as control actions aimed at improving the sustainability and efficiency of the process and its compliance with current regulations.

The authors are designing a test campaign to evaluate the effectiveness of the cleaning process modalities suggested by the DT module.

References

1. Cepolina, F., Michelini, R.C., Molfino, R.M., Razzoli, R.P.: Gecko-Collie-home cleaning automation of floors walls and cupboards. In: Third International Conference on Climbing and Walking Robots, pp. 803–811 (2000)
2. IFR International Federation of Robotics: WorldRobotics 2021, Frankfurt, 04 November (2021)
3. Kraemer, M.U., et al.: The effect of human mobility and control measures on the COVID-19 epidemic in China. Science **368**(6490), 493–497 (2020)
4. WHO Cleaning and disinfection of environmental surfaces in the context of COVID-19. World Health, Geneva (2020)
5. Vaussard, F., Rétornaz, P., Hamel, D., Mondada, F.: Cutting down the energy consumed by domestic robots: insights from robotic vacuum cleaners. In: Herrmann, G., et al. (eds.) TAROS 2012. LNCS (LNAI), vol. 7429, pp. 128–139. Springer, Heidelberg (2012). https://doi.org/10.1007/978-3-642-32527-4_12
6. Cepolina, E.M., Farina, A., Holloway, C., Tyler, N.: Innovative strategies for urban car-sharing systems and a simulator to assess their performance. Transp. Plan. Technol. **38**(4), 375–391 (2015)
7. Coelho, F., Relvas, S., Barbosa-Póvoa, A.P.: Simulation-based decision support tool for in-house logistics: the basis for a digital twin. Comput. Ind. Eng. **153**, 107094 (2021)
8. Liu, W., Zhang, W., Dutta, B., Wu, Z., Goh, M.: Digital twinning for productivity improvement opportunities with robotic process automation: Case of greenfield hospital. Int. J. Mech. Eng. Robot. Res. **9**(2), 258–263 (2020)
9. Yang, Z., Zhang, X., Liu, X., Tang, J., Cao, H.: A conceptual framework of future home smart kitchen system with multiple experience modes. In: Rebelo, F. (ed.) AHFE 2021. LNNS, vol. 261, pp. 438–445. Springer, Cham (2021). https://doi.org/10.1007/978-3-030-79760-7_52
10. Cepolina, F., Zoppi, M., Zurlo, G., Molfino, R.: A robotic cleaning agency. In: Proceedings of IAS 2004–8th Conference on Intelligent Autonomous Systems, pp. 10–13 (2008)
11. Cepolina, E.M.: The packages clustering optimisation in the logistics of the last mile freight distribution. Int. J. Simul. Process Model. **11**(6), 468–478 (2016)
12. Joon, A., Kowalczyk, W.: Design of autonomous mobile robot for cleaning in the environment with obstacles. Appl. Sci. **11**(17), 8076 (2021)
13. Cepolina, E.M., Tyler, N.: Microscopic simulation of pedestrians in accessibility evaluation. Transp. Plan. Technol. **27**(3), 145–180 (2004)
14. Michelini, R.C., Cepolina, F.: Gecko, the walls cleaner. Ind. Robot. Int. J. **29**, 538–543 (2002)
15. Gradetsky, V.G., Knyazkov, M.M., Semenov, E.A., Sukhanov, A.N., Chashchukhin, V.G.: Vacuum generation for wall climbing robot adaptive contact to surface. Robots Hum. Life **80** (2020)
16. Asafa, T.B., Afonja, T.M., Olaniyan, E.A., Alade, H.O.: Development of a vacuum cleaner robot. Alex. Eng. J. **57**(4), 2911–2920 (2018)
17. Tao, F., et al.: Digital twin-driven product design framework. Int. J. Prod. Res. **57**(12), 3935–3953 (2019)
18. Cepolina, E.M., Cepolina, F.: Twin tools for intelligent manufacturing: a case study. In: Proceedings of the 23rd International Conference on Harbor, Maritime and Multimodal Logistic Modeling and Simulation (2021). https://doi.org/10.46354/i3m.2021.emss.059
19. Aivaliotis, P., et al.: Methodology for enabling digital twin using advanced physics-based modelling in predictive maintenance. Procedia Cirp **81**, 417–422 (2019)
20. Agostinelli, S., et al.: Cyber-physical systems improving building energy management: digital twin and artificial intelligence. Energies **14**(8), 2338 (2021)
21. Liu, J., et al.: Dynamic design method of digital twin process model driven by knowledge-evolution machining features. Int. J. Prod. Res. **60**, 2312–2330 (2021)

Common Information Model for Modules of Service Robots

Mi-Sook Kim[1] and Hong Seong Park[2(✉)]

[1] Kangwon University-Industry Cooperation Foundation, Chuncheon 24341, South Korea
[2] Kangwon National University, Chuncheon 24341, South Korea
hspark@kangwon.ac.kr

Abstract. In the research community, modularity in hardware and software has been practiced in the building of service robots. However, as yet, this concept hasn't made significant progress in the commercial market. The common information model (CIM) is a vehicle for interoperability and reusability for modular robots. Recently, the ISO published a standard about the general requirements for modularity of service robots in both the hardware and software sides including safety and security. This paper presents a common information model for modularity to make hardware modules, software modules, and hardware-software modules for modular robots. Along with that, the user's relationship with the CIM is discussed. Examples of detailed information models using CIM are addressed.

Keywords: Modularity · Robot module · Information model for modules

1 Introduction

In NASA's technical readiness levels [1], level 8 is the end of system development and lists a new reusable vehicle as an example. In the service robotics industry, a modular approach, providing reusability and composability in designing and building service robots, instead of a manufacturer dependent approach, is gaining traction with upcoming companies in the market [2]. In the research community, robot modular design has been an active topic for two decades. Some practitioners are OROCOS [3], OpenRTM [4], OPRoS [5], and ROS [6]. In the DARPA Robotics Challenge, the application of modularity has been reported in the robot, THOR-OP [7]. Hardware modularity led to the use of general purpose parts which lowers costs, and software modularity led to the use of software modules which also saves time and effort. Reusability has gained prominence, and modularity has been proven to be a good practice of this.

However, this new approach has not made a major impact in the service robot marketplace because modularity has mostly been applied for a company's particular use. To extend hardware and software modularity from an inter-company level to an intra-company levels, there are several endeavors.

Ministry of Trade, Industry and Energy and Korea Evaluation Institute of Industrial Technology supported this search (20006898) in 2022.

© The Author(s), under exclusive license to Springer Nature Switzerland AG 2023
J. M. Cascalho et al. (Eds.): CLAWAR 2022, LNNS 530, pp. 513–521, 2023.
https://doi.org/10.1007/978-3-031-15226-9_47

The Common Information Model (CIM) has a common definition of data being passed between applications [8] and its format can be either UML or XML style W3C RDF-based serialization.

HRIM [9] is an information model for robot component modules in ROS, but with hardware information being considered optional. Information about safety and security may also be included as an option.

There is an ISO group focusing on interoperability and reusability of service robot modularity. ISO TC299 WG6 has published ISO 22166-1 [10], and the standard introduced following nine main requirements: Composability, Integrability, Interoperability, Granularity, Platform Independence, Openness, Reusability, Safety, and Security. The standard also introduced a template of items to meet these elements. It applied modularity into not only software, but also hardware. It also introduced the concept that the development of the detailed interfaces of the template are in the family standards in the future.

This paper presents the information model of the template of the standard for interoperability and reusability between robot modules in the hardware and software sides. Section 2 defines three different types of robot modules and applies the items of the template into these types as CIM. Section 3 delivers the relationship of four types of information models: the CIM, information model for software modules, information model for hardware modules, and information model for hardware-software modules. Section 4 illustrates use of CIM.

2 Common Information Model for Robot Modules

To provide an overarching guide, ISO 22166-1, robot modules are specified for modular robots, and modularity is applied into not only software, but also hardware. The standard defines hardware modules including purely physical parts along with firmware as a software aspect. In this paper, hardware modules having a software aspect are named hardware-software modules.

The template for modules of ISO 22166-1 is transformed into the items shown in Table 1 as the CIM for modules. The symbols 'M' and 'O' in the table denote mandatory and optional, respectively. The CIM for modules is shown in Fig. 1.

Item No. 1, 'Module Name', is the name representing the module.

Item 2, 'Description', provides the overview of the module, what the module is, what it does, and how it can be used.

Item 3, 'Manufactures', provides contact information for designer, developer, and manufacturer of the module.

Item 4, 'Examples', provides typical use cases of the module.

Item 5, 'Information model version', is the version number of the information model.

Item 6, 'Module ID', is the unique identifier of the module within a system.

Item 7, 'Hardware Aspects' and Item 8, 'Software Aspects', are related only to composite modules. If the module is composed of two or more hardware-software modules, software modules, and/or hardware modules, their module IDs are listed in the Hardware Aspects (including hardware-software IDs) and the Software Aspects, respectively. System designers and/or module users are able to refer to the necessary information about

Table 1. CIM for robot modules and the corresponding group tags (used in Fig. 4)

No.	Items	Common information model[b]	Information models for modules[a]			Related group/Tag name (Abbreviation of each group)
			Hardware-software module	Hardware module	Software module	
1	Module name	M	M	M	M	GenInfo
2	Description	M	O	O	O	
3	Manufacture(s)	M	M	M	M	
4	Examples	M	O	O	O	
5	Information model version	M	M	M	M	IDnType
6	Module ID	M	M	M	M	
7	Hardware aspects	M	M	M	-	
8	Software aspects	M	M	-	M	
9	Module properties[c]	M	M	M	M	Properties
10	Inputs	M	M	M	M	IOVariables
11	Outputs	M	M	M	M	
12	Status	M	M	O	M	-
13	Services (capabilities)	M	M	M	M	Services
14	Infrastructure	M	M	M	M	Infra
15	Safety/security	M	M	M	M	SafeSecure
16	Modelling	M	O	O	O	Modelling
17	Executable form	M	O	O	M	ExecutableForm

[a]For information models such as hardware, software, and hardware-software, some types of items can be omitted depending on their functionalities. In particular, for information models of hardware modules and software modules, the items "software aspects" and "hardware aspects" are not included.
[b]All items for CIM are mandatory.
[c]It is mandatory only for those modules which can be influenced (set) from the outside or at least to those, which have an expected effect on other modules.

related modules from the module IDs listed in the Hardware Aspects and the Software Aspects.

Item 9, 'Module properties', are values that are generally used in initialization of modules. Module properties are classified into mandatory and optional. Environment constraints are also considered as properties, examples of which are operating temperature, operating humidity, and maximum allowed mechanical shock. Each coefficient used in the PID control algorithm is used once in initialization or is changed and used several times during execution of related software modules.

Items 10/11, 'Inputs/Outputs' describes names of variables for the interface. Inputs and outputs of a servo control software module are encoder values and motor control values, respectively. Properties are kinds of input values from the viewpoint of modules but have to be distinguished in that Inputs are related to the environment of the modules but properties are related to parameters contained in the modules. For example, Inputs of the servo control software module are encoder values, but its properties are Proportional (P), Integral (I), and Differential (D) coefficients.

Item 12, 'Status', describes the status of a module that is operating. It is not utilized at design time.

Item 13, 'Services (capabilities)', describes interfaces that a module provides and utilizes for robot services. Service means performing one or more functions of a module for other modules via a pre-described interface. Examples of function format for software aspects are shown in Table 2.

Table 2. Examples of function format for software aspects

Name	Arguments	Return value	Description
Initialize	int16 init_ival1, float64 init_rval2	int8	Initialization using 2 arguments Return value (0: success, negative value: error type)
	int16 init_ival1, float64 init_rval2, int32 init_ival3	int8	Initialization using 3 arguments Return value (0: success, negative value: error type)

An example of function format for electrical/electronic aspects is shown in Table 3, where the arguments 'connType', 'keying', and 'busProtocol' mean the type of connector, the gender, and the protocol type that the module uses. The function is used to check that the peer module utilizes the proper electrical aspects of the other module. Values for connector type can be USB-A, RJ45, DB-9, etc. The value for keying is male or female. Values of busProtocol can be USB, Ethernet, EtherCAT, RS232, etc.

Table 3. Example of function format for electrical/electronic adaptability

Name	Arguments	Return value	Description
checkElecConnectivity	String connType, String keying, String busProtocol	boolean	Check electrical/electronic adaptability using 3 arguments Return value (True: success, False: error)

Examples of function format for mechanical aspects is provided in Table 4, similar to the electrical/electronic aspects.

Table 4. Example of function format for mechanical adaptability

Name	Arguments	Return value	Description
linkConnectivity	array float64 origin, float64 mass, array float64 inertia, String shape, vector<float64>size, array float64 axis, String connection, array float64 collision	boolean	Check mechanical link adaptability using 8 arguments Return value (True: success, False: error)
jointConnectivity	array float64 origin, array float64 axis, float64 limits, float64 damping, float64 friction	boolean	Check mechanical joint adaptability using 5 arguments Return value (True: success, False: error)

Item 14, 'Infrastructure' lists hardware and/or software with which the modules commonly use or connect. These are Power type, middleware type, databus type, and database type that the modules are commonly utilizing.

Item 15, 'Safety/Security' describes the safety-related performance level and the security information provided by the module. For the Safety/Security of the general service robot, the safety-related performance level is used for the module, which is defined in ISO 13849-1 [11], and the security-related level is listed for the module, where the security level is 0–4. Security level 0 means that there are no security measures. Security levels 1–4 are defined in IEC 62443-4-2 [12]. For the specific robot types, such as a medical robot or a physical assistant robot, other safety-related standards and security standards should be utilized. A performance level is generally provided for each single safety function. If a module has several safety functions, it should provide a performance level of itself by using a combination of the performance levels of all of the safety functions of the module or via the verification and validation of the module's overall safety-related function.

Item 16, 'Modelling', provides different types of models for simulation and/or design purposes.

Item 17, 'ExecutableForm', provides program codes executed to achieve or support the modules' purpose.

3 Relationship Between CIM and Specific IMs for Robot Modules

The CIM for modules is used in information models for all types of modules, which are hardware-software, software, or hardware modules. Their relationship is shown in Fig. 1. Hardware-software modules are composed of hardware aspects and software aspects.

The relationships between information models of hardware modules, software modules, and hardware-software modules are illustrated in Fig. 2.

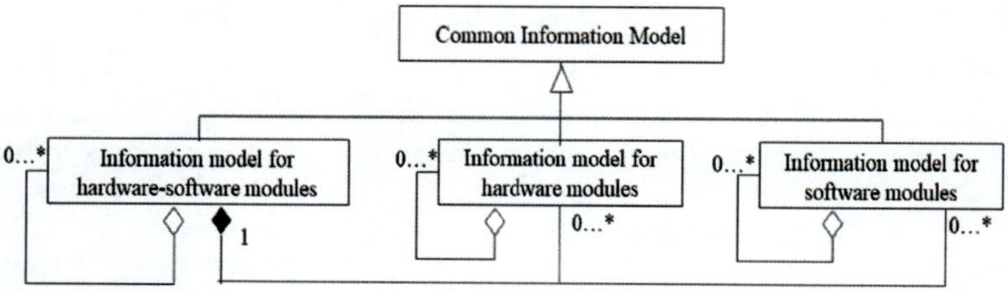

Fig. 1. Relationships among information models for module classes of CIM

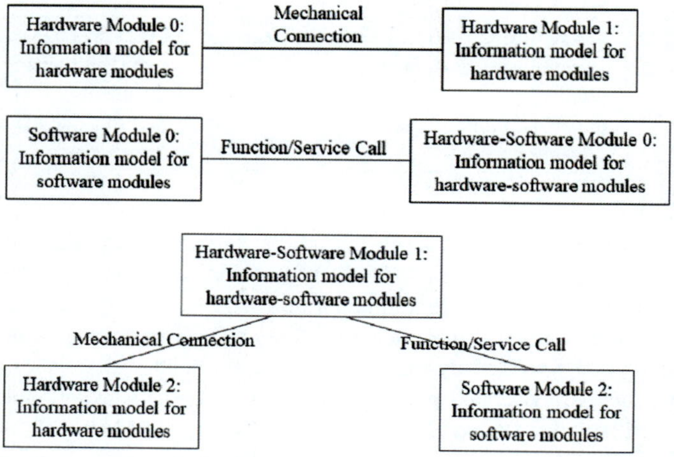

Fig. 2. Example of relationships between information models of modules

The CIM is be grouped into 9 inner classes as illustrated in Fig. 3. Eight classes are from the eight items of the Related Groups column of Table 1 and the additional class is the status-related information of a module. The four items from GenInfo, Module name, Description, Manufacturer, and Examples, become the member attributes of the class CIM.

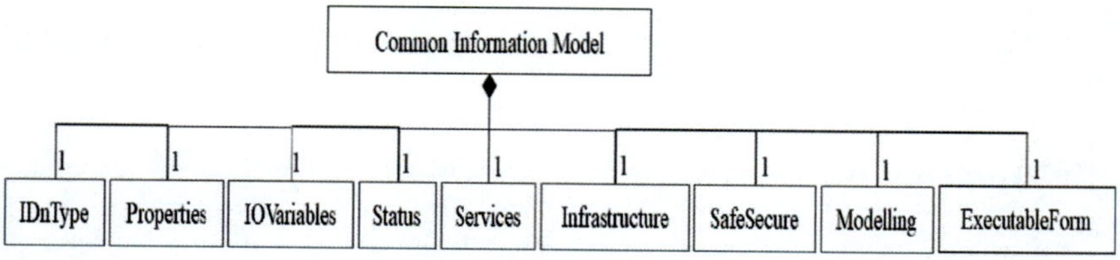

Fig. 3. CIM class for robot modules

The class of the CIM has attributes given in Table 5 and its attributes are based on subclauses 4–7 in ISO 22166-1[8]. IDnType, Properties, IOVariables, Status, Services, Infra, SafeSecure, and Modelling in Table 5 mean the classes used in CIM. Attributes can be assigned to one of following: mandatory (M), optional (O). The number of attributes are specified, with all in Table 5 being one (1). The four items of moduleName, description, manufacturer, and examples in the Table 1 are attributes of the class CIM.

Table 5. CIM of robot modules

Description: class of CIM for robot modules				
Derived From: none				
Attributes:				
moduleName	String	M	1	A name of the module in data type String
description	String	M	1	General usage of the module in data type String
manufacturer	String	M	1	A name of the module maker in data type String
examples	String	M	1	An example of the module in data type String
idnType	IDnType	M	1	- Unique id of the module in data type IDnType - In case of a composite module, the ids of sub modules' are included - A version number of CIM
properties	Properties	M	1	A list of properties of the module in data type Properties
ioVariables	IOVariables	M	1	A list of input and output data for module in data type IOVariables
status	Status	M	1	A status of the module in data type Status
services	Services	M	1	A list of services of the module in data type Services
infra	Infra	M	1	A list of infrastructures of the module in data type Infra
safeSecure	SafeSecure	M	1	A list of safety and security levels in data type SafeSecure

(*continued*)

Table 5. (*continued*)

modelling	Modelling	M	1	A modelling of the module in data type Modelling
execForm	ExecutableForm	M	1	A list of executable forms for the module in data type ExecutableForm

4 Description of Use of CIM for Robot Modules

Modularity of service robots is for interoperability and reusability. The very first users of the information model of modules are the makers of the initial module. For these modules to be broadly used, information needs to be provided about the module. Other users employing ready-made modules are secondary module makers who will create a new module by adding to the initial module and perhaps other secondary modules. System integrators will compose and integrate these initial and secondary modules in constructing a robot.

When all the modules provide their own CIM, system integrators can easily assemble modules based on this information for their robots. When a set of assembled modules is used as a module for a complex robot, then the information of this set is necessary so that it can be used again in several different kinds of robots. Additionally, safety and security can be assured to be are in accord with the users' robot design. Figure 4 displays three kinds of users' relations with the information model.

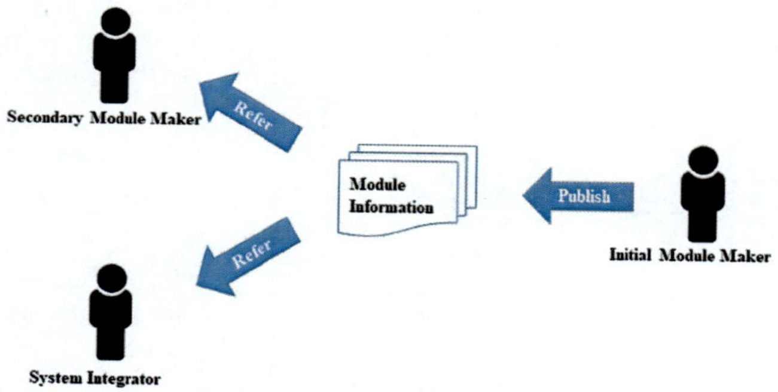

Fig. 4. Description of use of CIM of robot modules

5 Conclusion

Modularity for service robots is a globally shared topic for lowering costs and accelerating development and thus assist in the creation of numerous applications for robot markets. For these modular robots, CIM is a good driver for robot modules' interoperability and reusability. The CIM for robot modules in this paper supports not only the software side but also the hardware side providing a guide for the consistent, speedy construction of safe robots.

CIM is shared information of the three kind modules; hardware module, software module, and hardware-software module. Developing specific information for these modules will require research into enhanced information consistency. For example, information that hardware modules should provide is power types in Infra class of CIM, for hardware-software modules is firmware file name in ExecutableForm class, and for software modules, information about versions or type of operating systems that the software modules can be executed.

References

1. Zou, Y., Kim, D., Norman, P., Espinosa, J., Wang, J.-C., Virk, G.S.:Towards robot modularity—a review of international modularity standardization for service robots. Robot. Auton. Syst. **148** (2022)
2. Mankins, J.: Technology Readiness Level-A White Paper (1995). https://www.researchgate.net/publication/247705707_Technology_Readiness_Level_-_A_White_Paper. Accessed 13 June 2022
3. OROCOS. https://docs.orocos.org. Accessed 13 June 2022
4. OpenRTM. https://www.openrtm.org/openrtm/en. Accessed 13 June 2022
5. ROS. https://www.ros.org. Accessed 13 June 2022
6. OPRoS. https://github.com/opros-wiki/OPRoS_v1.1_Components/wiki/Open-Platform-for-Robotic-Services. Accessed 13 June 2022
7. Yi, S.-J., et al.: Modular low-cost humanoid platform for disaster response. In: 2014 IEEE/RSJ International Conference on Intelligent Robots and Systems (2014)
8. Uslar, M., Specht, M., Rohjans, S., Trefke, J., Gonzalez, J.M.: The Common Information Model CIM IEC 61968/61970 and 62325-A Practical Introduction to the CIM, 1st edn. Springer, Heidelberg (2012)
9. Zamalloa, I., Muguruza, I., Hernández, A., Kojcev, R., Mayoral, V.: An information model for modular robots: the Hardware Robot Information Model(HRIM). https://arxiv.org/abs/1802.01459. Accessed 13 June 2022
10. ISO 22166-1:2021 Robotics – Modularity for service robots – Part 1: General requirement
11. ISO, ISO 13849-1:2015 Safety of machinery – Safety-related parts of control systems – Part 1: General principles for design
12. IEC, IEC 62443-4-2 Security for industrial automation and control systems – Part 4-2: Technical security requirements for IACS components

Wearable Technology for Rehabilitation and Daily Assistance at Home and Work

Wearable Virtual Reality Tool for Balance Training: The Design and Validation on Healthy

Diana Rito[1,2(✉)], Cristiana Pinheiro[1,2], Joana Figueiredo[1,2], and Cristina P. Santos[1,2]

[1] Center for MicroElectroMechanical Systems (CMEMS), University of Minho, Braga, Portugal
a82528@alunos.uminho.pt
[2] LABBELS – Associate Laboratory, Braga/Guimarães, Portugal

Abstract. Balance disabilities affect the human quality of life. Current directions for balance rehabilitation require the inclusion of virtual reality (VR) tools as a complementary robotic tool to conventional physical therapies to accelerate balance recovery. This works aims to present the design and validation on healthy of a wearable and fully immersive VR-based tool following a user-centred design. This wearable VR tool comprises four Activities of Daily Living-based virtual challenges including nine motor tasks, chosen according to those most performed in the literature, as well as in the tasks of the Berg Balance Scale (BBS) and Timed Up and Go (TUG) clinical tests. This system comprises wearable VR technology, providing multimodal feedback (visual, sonorous, and vibrotactile cues), and integrates wearable inertial sensors for real-time motor assessment. The system's operability was validated with six healthy subjects executing BBS and TUG related motor tasks to assess balance performance after and before two training trials for each virtual challenge. The results showed statistically significant improvements regarding COM' displacement and velocity during "Cooking" and "Watch Tv" games and after VR training. The VR-based tool was rated with high IPQ and IMI scores. This wearable VR-based tool has the potential to effectively improve balance and walking, and to effectively increase the user's active participation and enthusiasm during balance training; thus, contributing to accelerate the user's balance and motor recovery.

Keywords: ADLs · Wearable inertial system · User-centred design

1 Introduction

Balance disability is one of the main risk factors for falling, being the second leading cause of unintentional injury deaths worldwide [1]. It is mainly caused by neurological

This work has been supported by the FEDER Funds through the Programa Operacional Re-gional do Norte and national funds from Fundação para a Ciência e Tecnologia with the Smar-tOs project under Grant NORTE-01-0145-FEDER-030386, under the national support to R&D units grant, through the reference project UIDB/04436/2020 and UIDP/04436/2020, under scholarship reference 2020.05709.BD, and under Stimulus of Scientific Employment with the grant 2020.03393.CEECIND.

© The Author(s), under exclusive license to Springer Nature Switzerland AG 2023
J. M. Cascalho et al. (Eds.): CLAWAR 2022, LNNS 530, pp. 525–536, 2023.
https://doi.org/10.1007/978-3-031-15226-9_48

conditions (i.e., stroke, Parkinson's disease, cerebral palsy), ear disorders (i.e., ear infections, vestibular problems as inner ear abnormalities or Meniere's disease), head injury, age (≥60 years old), or medication [2, 3]. The loss of balance itself and the associated fear of falling lead to loss of independence and increased difficulty to perform Activities of Daily Living (ADLs), compromising quality of life, and professional and social inclusion [4].

People left disabled can recover balance function and regain their motor independence through rehabilitation intervention, either physical or robotic [5, 6]. Physical therapy, as conventional rehabilitation intervention, has the disadvantages of being non-standard and dependent on the therapist's experience [5, 6]. Robotic therapy includes robotic devices to objectively assess user's motor condition and provide personalized intensive and repetitive training according to user's needs [7, 8]. Virtual reality (VR)-based tools are promising in robotic rehabilitation once they offer endless three-dimensional (3D) virtual reality environments (VREs), eliciting realistic perceptions and reactions in the user [9–11]. VR tools have the advantage of evolving the users in a multi-sensory, fully immersive, and enthusiast VREs designed and customised according to the user's imminent needs [12]. Thus, increasing user's active participation leading to accelerate balance recovery [12]. VR intervention has already been shown to be effective for improving balance [13–17]. A literature review was conducted on this topic [15], concluding that most of studies use non-wearable VR technology and biosensors, limiting the clinical practice of the VR tools to an indoor fixed facility [14–16]; there is a lack of fully immersive VR systems for balance training; and, none VR tool suggests a user-centred design not having into account the users' most performed and appreciated ADLs for designing the tool [15, 16]; and, no study mention to choose VR-encouraged motor tasks according to the clinically accepted BBS and TUG clinical tests for balance assessment [15, 16].

This work aims to present the user-centred design and validation on healthy of a wearable and fully immersive VR tool for balance training, having into account the most performed and appreciated ADLs and BBS and TUG-related motor tasks during the design. Thus, it contributes with a novel wearable and fully immersive VR-based tool for BBS and TUG-related motor tasks training while virtually performing the most performed and appreciated ADLs, and the evidence concerning its kinematic effects and user's experience evaluation on healthy. Firstly, the VR technology, sensor integration for objective motor assessment, four virtual challenges and related VRE and motor tasks are described. Secondly, six control strategies that allow the users to interact with the VRE and provide real-time feedback are shown. Thirdly, the experimental validation with healthy subjects and the related results are presented and discussed.

2 Methods

2.1 VR Technology

The VR commercial device HTC Vive Pro Full Kit (HTC VIVE Pro, HTC, Taiwan, Republic of China) was used to allow the fully immersion and interaction of the user with the VRE within a play area of 2 m × 1.5 m. Beyond being wearable, fully immersive, and portable, the system's ability to provide multimodal feedback (visual, auditory, and vibrotactile cues) also motivated its selection once improves the user's sense of presence [18]. A computer Republic of Gamers Strix Helios with a NVIDIA GeForce RTX 3080 Ti GPU was used to run the software needed by the VR-based tool. A monitor screen was also used so that the training can be followed by accompanying persons as physiotherapists, allowing to track in real-time the users' performance during the balance training and, consequently, provide additional support according to their imminent needs.

2.2 Sensor Integration

The wearable inertial measurement unit-based motion capture system MVN Awinda (Xsens, Enschede, Netherlands) was integrated into the VR-based tool given its reliability for body motion analysis in free-living conditions. This system was used in a full body suit configuration. After a successful calibration, the "Position + Orientation (Quaternion)" and "Centre of Mass" (COM) data were selected to be streamed in real-time to the third-party software used to develop the VRE at 100 Hz using UDP communication protocol. Its wearability and portability make it ideal for training in any location, potentiating daily practice and, thus, accelerating the recovery process.

2.3 Motor Tasks

The motor tasks encouraged by the VR tool were chosen based on the literature and BBS and TUG clinical tests which are the most performed tests for balance assessment in clinical practice [15]. So, while using this tool for balance training, the users are practicing the motor tasks clinically accepted and performed in conventional physical rehabilitation, allowing to increase the tool's acceptability. The chosen nine motor tasks were: 1) walking; 2) stepping; 3) sit-to-stand and stand-to-sit; 4) knee flexion and extension; 5) weight-shifting; 6) trunk and pelvic/hip movement; 7) look over the shoulder; 8) standing upright on one leg; 9) reaching.

2.4 Virtual Challenges and VRE

Four virtual challenges based on ADLs – "Fruit Catcher", "Cooking", "Take a Shower", and "Watch TV" - were proposed to enthusiasm the physical training. They match the most performed and appreciated ADLs that require balance according to the answers of a questionnaire completed by 64 persons described in [19]. Unity 3D software, Unity Technologies (Copenhagen, Denmark), was used to develop the VRE. It consists of a house with detailed indoor and outdoor spaces designed to be the most realistic and immersive as possible. Natural ambient sound and ambient sunlight were also created.

Fruit Catcher. Its VRE consists of a backyard with a tree, a ladder, and apples as main objects to carry out the virtual game (Fig. 1). The "Fruit Catcher" game aims that the user: 1) climbs six steps of the ladder perched on the tree through stepping with each foot alternately; 2) then, reach and catch nine apples, three forwards and three to each side left and right by performing weight shifting (with their feet together and without moving them) looking forwards and over the left and right shoulders, respectively. The nine apples were placed at predefined distances (5 cm, 12.5 cm, and 25 cm after the hand position with arms outstretched of a person with 1.70 m height) corresponding to the second, third, and fourth levels of the eighth BBS' task (entitled "Reaching forward with outstretched arm while standing"), respectively.

Cooking. For this virtual challenge, a kitchen division was created, with special emphasis on the four shelves full of ingredients (Fig. 1). The "Cooking" game aims that the user, with their feet together and without moving them, reaches six ingredients at specific points of the shelves: 1) three levels of height adjusted according to user's height (2 ingredients/level) through knee and hip flexion and extension; 2) on the left and right sides (3 ingredients/side) by performing weight shifting. The user is asked to put individually the ingredients in the basket beside her/him, keeping her/his balance.

Take a Shower. A bathroom division with special focus on the bathtub with a height of 40 cm (standard height of an undermount real bathtub [20]) was created (Fig. 1). The "Take a Shower" game aims that the user: 1) firstly, enters the bathtub with one foot at a time through knee flexion and extension without listening a sound of a kicking that is generated every time she/he touches the bathtub's rim; 2) once inside the bathtub, stands upright on one leg for listening to a sound of running water (auditory feedback) in case of success in lifting the leg enough, mimicking that she/he is washing her/himself. The user must lift the leg at least 25 cm and stand upright on one leg for at least 10 s (time to reach the maximum score on fourteenth BBS's task entitled "Standing on one leg") to achieve the maximum score.

Watch TV. A living room with special emphasis on the television and sofa, designed with a height of 47 cm according to the standards of real chairs [21], was created (Fig. 1). The "Watch Tv" game was inspired by the TUG-related motor tasks. In this game, the user, initially raised, must: 1) sit in the virtual sofa (supported by a real matched chair); 2) stand; 3) walk 3 m forward towards the television until it turns on; 4) rotate 180°; 5) walk again 3 m towards the sofa, ending the game. A video (visual and auditory feedback) plays on the television when the user reaches the 3 m forward the sofa. The

Fig. 1. Design of house backyard (upper left), kitchen (upper right), bathroom (lower left), and living room (lower right).

sit-to-stand and stand-to-sit tasks were performed in a different order when compared to the real TUG to ensure the user's safety.

2.5 Control Strategies

Six control strategies were created to allow the user to interact with the VRE and receive feedback from it, namely: 1) first and second person views (available during all serious games); 2) grab; 3) vibrotactile feedback; 4) visual and auditory feedback; 5) stepping up the ladder; 6) shelves height.

A **first-person perspective** was implemented by matching the position tracking from the headset and controllers with an avatar mimicking in real-time the user's body movement through the "MVN Live Animation" Unity plug-in (Fig. 2). A **second-person perspective** showing in real-time the user's feet with her/his COM projected was also created (Fig. 2). The COM is represented by one red ball (diameter of 1 cm) placed at feet level to easily visualize if it is within the base of support (area that includes every point of contact that the person makes with the supporting surface and, in the case of having more than one point contact, the area between them too) [22]. The position of the virtual red ball was controlled according to the streamed COM data post-transformed from the MVN to the Unity reference axes. Thus, the user can adjust her/his posture during VR training towards maintaining balance and preventing her/him from falling.

Fig. 2. First person view during "Watch Tv" virtual challenge with the second person view highlighted by the orange square.

Grab control strategy allows the apples on the tree and the ingredients placed on the shelves to be caught with the HTC controllers, if they are not static objects and have physical properties. This strategy was associated to both controllers' Trigger button by configuring "Open binding UI" to add "Grab" action on "SteamVR Input". **Vibrotactile feedback** occurs when user grabs or collides with virtual objects (as the apples on the tree and the food placed on the shelves) with "Box Colliders" and "OnTriggerEnter" components using HTC controllers. The vibrotactile cues are generated by the controllers with haptic amplitude, duration, and frequency of 88 g, 0.4 s, and 22 Hz, respectively, by configuring "Open binding UI" to add "Haptic" action.

In "Watch Tv" game, a video (**visual and auditory feedback**) plays in the Tv screen when the avatar's collider enters in the television's collider. For "Take a Shower" game, a sound (**auditory feedback**) imitating a kick against the wall is enabled every time the user touches the bathtub' wall, by creating colliders on the stone of the bathtub and on the user's right and left feet. Also, a sound of running water (**auditory feedback**) plays each time the height of user's foot (assessed through "Position + Orientation" streamed data) is above 25 cm from floor and stops 10 s after it or if the user puts her/his foot below the threshold value.

In "Fruit Catcher" game, the avatar **steps up the ladder** perched in the tree by moving up and forward the vertical and horizontal position of the avatar every time the height of right or left user's foot is above 25 cm from the floor and backs to the floor's height. The **height of the shelves** in "Cooking" game was automatically customised according to the user height by agreement with standard measures of kitchen shelves and ergonomic rules [23].

2.6 Experimental Validation

The validation protocol was conducted following the principles of the Declaration of Helsinki and Oviedo Convention, and approved by the local ethical committee CEICVS 006/2020 at University of Minho. Six healthy participants (gender: 3 females, age: 23.67 $\pm$ 0.94 years, height: 1.66 $\pm$ 0.09 m, body mass: 62.93 $\pm$ 10.65 kg) were recruited and signed an informed consent to participate in the validation protocol. They were instructed to execute BBS and TUG-related motor tasks to assess balance performance after and before VR training. The VR training included 2 trials in a row per virtual challenge (randomly ordered) with a maximum duration of 3 min each and 3 min of rest between trials. The participants self-selected the order to pick up the apples and the ingredients in the virtual challenges "Fruit Catcher" and "Cooking", respectively. The participants were accompanied during the entire protocol to ensure their safety.

Sensor-based assessment were performed before, during, and after the VR training using Xsens inertial system. The outcomes during VR intervention and the BBS and TUG-related motor tasks are the COM's position and velocity in the anteroposterior (AP) and mediolateral (ML) directions. After the intervention, all participants answered the Intrinsic Motivation Inventory (IMI) and Igroup Presence Questionnaire (IPQ) to evaluate the user's experience. The IPQ questionnaire evaluates the VR tool in terms of the VRE's realism and how engaged the participant feels by it and IMI questionnaire assesses the experience lived by the user while using the VR-based tool to complete the virtual challenges. After processing, COM displacement and minimum and maximum COM velocity on the AP and ML directions were obtained.

The statistical analysis included the two-tailed *t*-tests and Wilcoxon signed-rank tests for parametric and non-parametric metrics, respectively, using the IBM SPSS software version 26.0 (IMP Corp, Armonk, NY, USA). The statistical tests were conducted to evaluate the following null hypotheses: 1) there are no statistically significant differences between trial 1 and trial 2 of VR intervention; 2) there are no statistically significant differences between before and after VR intervention.

3 Results

3.1 During VR Intervention

During VR intervention, maximum COM's velocity in AP direction statistically significant decreased from T1 to T2 of "Cooking" game (Table 1). In "Watch Tv" game, the COM's displacement in ML direction and the minimum COM's velocity in AP direction during walking statistically significantly decreased and increased, respectively (Table 1). The virtual challenges "Fruit Catcher" and "Take a Shower" did not present statistically significant differences between T1 and T2 (p-value > 0.05).

Table 1. Mean and standard deviation (SD) of trial 1 (T1) and trial 2 (T2), and the p-value of the statistical test for COM displacement (Disp), and maximum (Max) and minimum (Min) COM velocity (Vel), on AP and ML directions revealing statistically significant differences, considering a significance level of 5%

Virtual challenge	Mean ± SD		p-value
Outcome	T1	T2	
Cooking			
VelMaxAP	11.008 ± 1.769	7.985 ± 0.857	**.027**
Watch Tv			
DispMLWalk	9.113 ± 0.982	7.483 ± 0.553	**.049**
VelMinAPWalk	42.647 ± 3.259	46.773 ± 3.836	**.020**

3.2 Before vs After VR Intervention

By comparing the balance performance before versus after the VR intervention, statistically significant differences were found for the following BBS's tasks: sit-to stand, stand-to-sit, transfers, standing with eyes closed, standing with feet together, reaching forward, and standing on one leg, and for TUG's task during walking (Table 2). The BBS's tasks standing unsupported, sitting unsupported, pick up an object, looking over shoulders, turn 360°, place alternate foot on step, and standing with one foot in front did not present statistically significant differences between pre and post VR training (p-value > 0.05).

3.3 User's Experience Evaluation

The VR-based tool proved to be well accepted by the users, who experienced a sense of presence, realism, and involvement by the created VRE (Table 3). It was also proved that the participants were interested during the VR intervention and felt highly competent while performing the virtual challenges (Table 4). For the three open-ended questions of the IMI questionnaire, participants reported that the tool seems useful for improving

Table 2. Mean and standard deviation (SD) of pre and post VR training, and the *p*-value of the statistical test for COM displacement (Disp), and maximum (Max) and minimum (Min) COM velocity (Vel), on AP and ML directions revealing statistically significant differences, considering a significance level of 5%

Motor task	Mean ± SD		*p*-value
Outcome	Pre	Post	
Sit-to-stand (BBS)			
DispAP	42.235 ± 1.445	37.548 ± 2.344	**.047**
Stand-to-sit (BBS)			
DispAP	40.058 ± 1.844	35.810 ± 2.045	**.011**
VelMaxAP	2.472 ± 0.453	3.663 ± 0.631	**.041**
Transfers (BBS)			
VelMinAP	−38.720 ± 2.442	−41.933 ± 2.678	**.030**
Standing with eyes closed (BBS)			
VelMinAP	−2.017 ± 0.610	−1.635 ± 0.576	**.018**
VelMaxAP	1.090 ± 0.179	1.270 ± 1.184	**.022**
Standing with feet together (BBS)			
DispML	1.313 ± 0.093	1.565 ± 0.145	**.046**
VelMaxML	1.143 ± 0.174	1.792 ± 0.517	**.046**
Reaching forward (BBS)			
DispAP	9.773 ± 0.949	8.365 ± 0.934	**.001**
Standing on one leg (BBS)			
VelMinAP	−3.853 ± 0.670	−2.853 ± 0.471	**.032**
Walking (TUG)			
VelMaxAP	120.197 ± 3.172	113.738 ± 3.866	**.047**

balance once it helps them to understand their limits in posture and balance (participant 6), to increase balance and trunk mobility (participant 5), and to recover from physical injuries of the lower body (participant 4).

Table 3. Mean and standard deviation (SD) of all participants and best score for each factor of IPQ

	Spatial presence	Involvement	Experienced realism	Global presence
Mean ± SD	4.25 ± 0.53	3.58 ± 0.45	3.88 ± 0.24	3.90 ± 0.24
Best score	6.33/7.00	6.00/7.00	7.00/7.00	6.44/7.00

Table 4. Mean and standard deviation (SD) of all participants and best score for each subscale of IMI

	Interest/ Enjoyment	Perceived Competence	Effort/ Importance	Pressure/ Tension	Value/ Usefulness
Mean ± SD	5.81 ± 0.61	5.53 ± 0.58	5.27 ± 0.98	3.03 ± 1.53	5.33 ± 0.84
Best score	7.00/7.00	7.00/7.00	2.20/7.00	1.00/7.00	7.00/7.00

4 Discussion

This work proposes a novel wearable VR-based tool for balance training allowing their use in a non-fixed facility in opposite to most current literature [14–16]. The VR technology included provides fully immersion of the user in the VRE enabling the most realistic training in opposite to [17]. Sensors were integrated into the VR-based tool for objective assessment of user's motor performance preventing dependence on physiotherapist's experience in clinical practice as in [17]. Nine motor tasks included in the clinically accepted most performed clinical tests for balance assessment (BBS and TUG) are encouraged during the proposed VR training so that the users can efficiently improve their balance ability [15, 16]. Following a user-centred design, the VR training implies virtual challenges that correspond to the most performed and appreciated ADLs, increasing the enthusiasm during training and the impact in users' daily life [15, 16, 20]. Six control strategies were developed to provide multimodal feedback and allow users' interaction with the VRE, fostering their sense of presence and active participation, respectively, as in [24].

The VR-based tool was preliminary validated with six healthy subjects evaluating the kinematic effects regarding COM's displacement and velocity before, during, and after VR training and the qualitative assessment of user's experience. During "Cooking" training, the maximum COM velocity in AP direction decreased from T1 to T2. However, COM velocity in ML direction did not statistically significantly decrease, being the ML direction more encouraged than AP direction due to lateral weight-shifting. Regarding the "Watch Tv" training, the results suggest a balance improvement in both ML and AP directions once the COM's displacement decreased and the minimum COM's velocity increased from T1 to T2, respectively. Therefore, the users improved their side-to-side movement control (more stable balance) and were able to move faster during walking.

Regarding the kinematic effects before versus after VR training, in sit-to-stand and stand-to-sit tasks (related to "Cooking" and "Watch Tv" games), the decrease of COM's displacement in the AP direction means that after VR training participants were able to perform these tasks with greater stability of the movement. The increase of maximum COM's velocity in stand-to-sit task represents a faster forward movement when participants lean before sitting. When performing transfers between chairs, participants shown an improvement of minimum COM's velocity, indicating that they were able to perform this motor task faster without losing balance. While standing with eyes closed, participants decreased in module the minimum but increased the maximum COM's velocity in the AP direction. Cho et al. [25] also showed improvements in postural sway velocity in the AP direction, although not significant. For standing with feet together, participants decreased the maximum COM's velocity but increased the COM's displacement in the ML direction. The negative results of standing with eyes closed and standing with feet together motor tasks can be explained by the lack of static motor task in the developed VR-based tool. For the task of reaching forward (related to "Fruit Catcher" game), the participants were able to reach the same object with greater stability and balance after VR training once the COM's displacement decreased in AP direction. Further, in standing on one leg (related to "Take a Shower" game), it was reported a module decrease of minimum COM velocity in ML direction, indicating greater stability. Regarding walking (related to "Watch Tv" game), the decrease of the maximum COM's velocity in AP direction may be related to the reduced space to walk (3 m each way) being difficult for people to increase speed and, thus, giving priority to a slower gait. Although the healthy ability of the participants and the reduced number of training trials, balance improvements were registered during and before vs after VR training.

The users experienced high (>3.50) spatial presence, involvement, realism, and global presence. Moreover, they reported high (>3.50) Interest/Enjoyment and Perceived Competence. Despite the high Effort/Importance, they reported feeling low Pressure/Tension. In addition, they also allocated it a high Value and Usefulness for improving balance. Thus, findings point out that the tool seems promising for balance rehabilitation, complementing conventional physical therapy.

5 Conclusion

This work presents a wearable and fully immersive VR-based tool for balance training following a user-centred design. The tool integrates inertial sensors for real-time objective assessment of users' motor performance. Furthermore, it includes multimodal feedback, providing visual, auditory, and vibrotactile cues, through HTC HMD, their built-in headphones, and controllers, respectively. The VR training comprises ADL-based four virtual challenges to encourage the execution of BBS and TUG-related motor tasks. The results from the preliminary validation with six healthy subjects demonstrate statistically significant improvements regarding COM' displacement and velocity during "Cooking" and "Watch Tv" games and after VR training. The VR-based tool was rated with high IPQ and IMI scores. Future work comprises a feasibility study with balance injured patients.

References

1. Falls - WHO | World Health Organization, 26 April 2021. https://www.who.int/news-room/fact-sheets/detail/falls. Accessed 31 Mar 2022
2. Balance problems - Symptoms and causes - Mayo Clinic, 18 June 2020. https://www.mayoclinic.org/diseases-conditions/balance-problems/symptoms-causes/syc-20350474. Accessed 31 Mar 2022
3. Balance Disorders—Causes, Types & Treatment | NIDCD, 06 March 2018. https://www.nidcd.nih.gov/health/balance-disorders. Accessed 31 Mar 2022

4. Kaveh, M.H., Bahadori, F., Doosti, A., Asadollahi, A.: The effect of balance exercise training on balance status, and quality of life in elderly women: a randomized controlled trial. Malaysian J. Med. Heal. Sci. **17**(2), 129–136 (2021)
5. Rahayu, U.B., Wibowo, S., Setyopranoto, I., Hibatullah Romli, M.: Effectiveness of physiotherapy interventions in brain plasticity, balance and functional ability in stroke survivors: a randomized controlled trial. NeuroRehabilitation **47**(4), 463–470 (2020). https://doi.org/10.3233/NRE-203210
6. Guo, S., Cai, H., Guo, J.: A novel fuzzy neural network-based rehabilitation stage classifying method for the upper limb rehabilitation robotic system. In: 2020 IEEE International Conference on Mechatronics and Automation, ICMA 2020, pp. 261–266 (2020). https://doi.org/10.1109/ICMA49215.2020.9233519
7. Gassert, R., Dietz, V.: Rehabilitation robots for the treatment of sensorimotor deficits: a neurophysiological perspective. J. Neuroeng. Rehabil. **15**(1), 1–15 (2018). https://doi.org/10.1186/s12984-018-0383-x
8. Klamroth-Marganska, V.: Stroke rehabilitation: therapy robots and assistive devices. In: Kerkhof, P.L.M., Miller, V.M. (eds.) Sex-Specific Analysis of Cardiovascular Function. AEMB, vol. 1065, pp. 579–587. Springer, Cham (2018). https://doi.org/10.1007/978-3-319-77932-4_35
9. Tieri, G., Morone, G., Paolucci, S., Iosa, M.: Virtual reality in cognitive and motor rehabilitation: facts, fiction and fallacies. Expert Rev. Med. Devices **15**(2), 107–117 (2018). https://doi.org/10.1080/17434440.2018.1425613
10. Arcuri, F., Porcaro, C., Ciancarelli, I., Tonin, P., Cerasa, A.: Electrophysiological correlates of virtual-reality applications in the rehabilitation setting: new perspectives for stroke patients. Electronics **10**(7) (2021). https://doi.org/10.3390/electronics10070836
11. Howard, M.C.: A meta-analysis and systematic literature review of virtual reality rehabilitation programs. Comput. Human Behav. **70**, 317–327 (2017). https://doi.org/10.1016/j.chb.2017.01.013
12. Marchionne, F.: Virtual Reality Rehabilitation - The Future of Physical Therapy, 17 December 2019. https://imotions.com/blog/virtual-reality-rehabilitation/. Accessed 31 Mar 2022
13. Asadzadeh, A., Samad-Soltani, T., Salahzadeh, Z., Rezaei-Hachesu, P.: Effectiveness of virtual reality-based exercise therapy in rehabilitation: a scoping review. Inform. Med. Unlocked **24**, 100562 (2021). https://doi.org/10.1016/j.imu.2021.100562
14. Corbetta, D., Imeri, F., Gatti, R.: Rehabilitation that incorporates virtual reality is more effective than standard rehabilitation for improving walking speed, balance and mobility after stroke: a systematic review. J. Physiother. **61**(3), 117–124 (2015). https://doi.org/10.1016/j.jphys.2015.05.017
15. Rito, D., Pinheiro, C., Figueiredo, J., Santos, C.P.: Virtual reality tools for post-stroke balance rehabilitation: a review and a solution proposal. In: 2021 IEEE International Conference on Autonomous Robot Systems and Competitions, ICARSC 2021, pp. 240–245 (2021). https://doi.org/10.1109/ICARSC52212.2021.9429777
16. Mohammadi, R., Semnani, A.V., Mirmohammadkhani, M., Grampurohit, N.: Effects of Virtual reality compared to conventional therapy on balance poststroke: a systematic review and meta-analysis. J. Stroke Cerebrovasc. Dis. **28**(7), 1787–1798 (2019). https://doi.org/10.1016/j.jstrokecerebrovasdis.2019.03.054
17. Xu, Y., et al.: A depth camera–based, task-specific virtual reality rehabilitation game for patients with stroke: pilot usability study. JMIR Serious Games **9**(1), 1–12 (2021). https://doi.org/10.2196/20916
18. VIVE Pro Full Kit | VIVE European Union. https://www.vive.com/eu/product/vive-pro-full-kit/. Accessed 29 Mar 2022
19. Rito, D., Figueiredo, J., Santos, C.P.: Virtual reality tool for balance training. University of Minho (2022)

20. Badeloft: Standard Bathtub Dimensions For Every Type of Tub | Badeloft, 22 November 2017. https://www.badeloftusa.com/ideas/bathtub-dimensions/. Accessed 31 Mar 2022
21. List of Standard Table & Chair Heights | How to Calculate the Ideal Height. https://parotas.com/en/standard-table-chair-heights-guide/. Accessed 13 Nov 2021
22. Lugade, V., Lin, V., Chou, L.S.: Center of mass and base of support interaction during gait. Gait Posture **33**(3), 406–411 (2011). https://doi.org/10.1016/j.gaitpost.2010.12.013
23. Ergonomia da Cozinha - Arquidicas. https://www.arquidicas.com.br/ergonomia-da-cozinha/. Accessed 22 Nov 2021
24. Jaffe, D.L., Brown, D.A., Pierson-Carey, C.D., Buckley, E.L., Lew, H.L.: Stepping over obstacles to improve walking in individuals with poststroke hemiplegia. J. Rehabil. Res. Dev. **41**(3A), 283 (2004). https://doi.org/10.1682/JRRD.2004.03.0283
25. Cho, K.H., Lee, W.H.: Effect of treadmill training based real-world video recording on balance and gait in chronic stroke patients: a randomized controlled trial. Gait Posture **39**(1), 523–528 (2014). https://doi.org/10.1016/j.gaitpost.2013.09.003

Effect on Jumping Height by Changing Jumping Powered Exoskeleton Attachment Position for Augmentation of Human Instantaneous Movements

Yusuke Ishii(✉), Fumio Ito, Shunichi Kurumaya, and Taro Nakamura

Precision Mechanics, Chuo University, Kasuga 1-13-27, Tokyo 112-8551, Japan
{y_ishii,f_ito,s_kurumaya}@bio.mech.chuo-u.ac.jp,
nakamura@mech.chuo-u.ac.jp

Abstract. This paper reports the construction of a robot model based on the human jumping mechanism and a powered exoskeleton, and simulation results on the effect of the powered exoskeleton on the jumping height. The objective of this study was to develop a powered exoskeleton that can follow human movements and significantly augment instantaneous movements. This paper describes the effect of the attachment position of the exoskeleton on human jumping height, which has not been clarified before, through the analysis of experiments using a jumping robot. Clarifying this effect is significant in that it will facilitate the development of jumping powered exoskeletons. Related studies include jumping exoskeletons using motors and artificial muscles. However, these exoskeletons have not been able to significantly augment the jumping height, and exoskeletons capable of significantly augmenting instantaneous movements have not yet been developed. This paper describes a comparison of the enhancement of the jumping height by different attachment methods when the exoskeleton is attached to a jumping robot in parallel or in series. It was estimated that the jump height was 1.06 and 1.18 times higher with the exoskeleton attached than without it, in parallel and in series, respectively. This confirms that the powered exoskeleton is expected to be significant, especially in the serial attachment, and that an enhanced jumping height can be expected. These results have the potential to significantly accelerate the development of powered exoskeletons that augment jumping motion by clarifying the effect of the exoskeleton's attachment method on the jumping height.

Keywords: Human augmentation · Powered exoskeleton · Jumping robot

1 Introduction

Many wearable robots have been developed for use in the medical and industrial industries [1–3], with particular interest in the development of the powered exoskeleton that augments the physical capabilities of the human body. The development of the exoskeleton that enables instantaneous movements will enable, for example, rapid movement over uneven terrain and overcoming obstacles that are difficult to overcome with human-derived movement force alone.

© The Author(s), under exclusive license to Springer Nature Switzerland AG 2023
J. M. Cascalho et al. (Eds.): CLAWAR 2022, LNNS 530, pp. 537–549, 2023.
https://doi.org/10.1007/978-3-031-15226-9_49

Related research includes a powered exoskeleton that augments human exertion through human-exoskeleton interaction in static movements [4], an exoskeleton that reduces metabolic costs [5], and powered exoskeletons designed to augment instantaneous movements [6–8]. The powered exoskeleton, which augments human exertion, achieves a large exertion force that would be difficult to achieve without the exoskeleton, such as when a wearer is required to lift a load of 22 kg or more with one arm. Additionally, the exoskeleton that reduces human metabolic costs reduces the wearer's oxygen consumption by approximately 15%, enabling long-distance travel. Thus, these exoskeletons have been developed to reduce slow movements and metabolic costs, such as carrying heavy loads.

However, there are jumping exoskeletons with motors [6] and pneumatic actuators [7, 8] that are designed to improve instantaneous movements. Because of the low compliance of motors, there was concern that the force of the exoskeleton would inhibit human movement, resulting in a lower rate of enhancement of physical performance. Additionally, when large forces and high speeds are required, the motors become heavier and larger, making it difficult to achieve instantaneous movements. Pneumatic actuators are considered to augment human motion efficiently because of their high compliance and output-to-weight ratio. However, these exoskeletons have not led to significant improvements in the jumping height. Therefore, powered exoskeletons that significantly augment human instantaneous movements have not been developed. This makes it impossible to rescue people at disaster sites by moving quickly or to develop highly realistic amusement facilities that allow people to fly through the air with their own power. The development of powered exoskeletons that can extend instantaneous movements will enable the realization of these things.

The purpose of this study is to develop a powered exoskeleton that can follow human movements and augment instantaneous movements. Instantaneous movements include running, jumping, and throwing. Among the instantaneous movements, jumping is not a periodic movement, and instantaneous force can be evaluated simply from the jumping height in a single movement. Therefore, the first step in achieving our goal is to augment the vertical jump height. We elucidate the interactive phenomena between the powered exoskeleton and human generated forces and develop a powered exoskeleton that can augment the jumping height of humans. Related studies aimed at increasing the jumping height can be broadly classified into (i) cases where the powered exoskeleton applies forces to joints in parallel [6–8] and (ii) those where the powered exoskeleton directly exerts upward forces on the body in series [9, 10]. However, no clear comparison has been made between these two methods, making it impossible to utilize them in the development of powered exoskeletons. Moreover, the hazards associated with jumping experiments make it difficult to construct a safe experimental environment, making the development of powered exoskeletons difficult. In the study and development process, the first step was to determine the basic design guidelines necessary for the exoskeleton by verifying the exoskeleton's effects on the body under various conditions in a real-world environment. This paper reports on the simulation results.

This paper describes the construction of a model of a small robot based on the human jumping mechanism and a powered exoskeleton, and a comparison of the enhancement of jumping height by the attachment method of the powered exoskeleton through dynamic analysis. Specifically, we compare the improvement in the jumping height by each attachment method: (i) when the powered exoskeleton applies forces to the joints in parallel, and (ii) when it directly exerts an upward force on the robot in series. Pneumatic actuators with high compliance and power-to-weight ratios are effective for powered exoskeletons that efficiently augment human motion. In this study, artificial muscles to which pneumatic pressure is applied are used as simple passive springs, and elastic energy is accumulated while the muscles are fixed with latches; force is instantaneously exerted by releasing the latches.

The contributions of this paper are as follows.

- We constructed a model of a jumping robot and powered exoskeleton based on the human jumping mechanism.
- From simulations using the constructed model, we compared the improvement in the jumping height depending on the attachment position of the exoskeleton and estimated the attachment method of the powered exoskeleton that improves the jumping height.
- We obtain insight into the effect of combining a high stiffness actuator with a low stiffness actuator on instantaneous movement.

2 Target Jumping Motion

The squat jump is widely used as a jumping motion when evaluating human vertical jumps; the jumping motion in this study is the squat jump. The squat jump is performed from a stationary standing posture to a semi-squat posture, which is held for approximately 3 s, and then an upward motion is initiated from the static state before releasing the ground [11]. In the static state of the semi-squat posture, gravity and the ground reaction force are in balance. With the upward motion, the ground reaction force becomes larger than gravity, and the ground reaction force curves convexly upward to 0 N at take-off. The larger the force product of the ground reaction force, the greater the initial velocity of the jump and the greater the jumping height. It should be noted that the squat jump operation time is 0.30–0.43 s (from the start of ascent to the take-off) [12]. The squat jump performance is also affected by the initial knee angle in the semi-squat position. High jumps are possible at initial knee angles of 80–90 [13], and the negative performance effects of static stretching before jumping can be ignored at 90–110 [14]. This suggests that when the initial knee angle is 90 , the negative effects of static stretching can be ignored, and high jumps are possible. Therefore, the initial knee angle of the jumping robot is set to 90 in this work.

3 Outline of the Organization

3.1 Humanoid Jumping Robot

Figure 1 shows a model of the jumping robot based on the human jumping mechanism. The constructed jumping model is based on the jumping model of Ingen Schenau et al. [15], in which the foot and thigh are connected by a non-stretchable constraint. From this, the rotation of the knee joint causes the ankle joint to rotate subordinately and play the role of the gastrocnemius muscle. The length of each link is determined based on the length ratio of the lower limb structure [16], and the weights of the non-lower limb are placed with reference to the weight ratio of each human body part [16]. The scale was set to approximately one-third of the actual size for future verification on an actual machine. A DC motor is mounted on the weight, and it drives the knee joint via a belt and pulley, connecting a series elastic element (SEE) based on human muscles to enhance the compliance characteristics of the knee joint. A torsion spring is added to the hip joint to apply torque in the extension direction. The floor is covered with a non-slip surface to prevent the slippage of the toes. The assumed DC motor is a DCX16L motor manufactured by Maxon, and the parameters were set as a GPX16 reduction gear manufactured by Maxon (Table 1).

Table 1. Parameters referenced to the prototype of the jumping robot and powered exoskeleton.

l_1	Length of the thigh	137	mm
l_2	Length of the leg	150	mm
l_3	Length of the foot	84	mm
l_A	Length of the ankle	18	mm
m_1	Mass of the thigh	18.8	g
m_2	Mass of the leg	17.3	g
m_3	Mass of the foot	12.7	g
m_W	Mass of the weight	139.9	g
m_J	Mass of the jumping robot	264.0	g
d_M	Distance between motor and hip	24.75	mm
d_K	Distance between knee and wire	44	mm
d_A	Distance between ankle and wire	40	mm
k_H	Spring constants of hip spring	1.74	N mm/deg
k_K	Spring constants of SEE	2.491	N/mm
R_M	Radius of motor pulley	10	mm
R_K	Radius of knee pulley	43	mm
V	Voltage applied to DC motor	24	V
R	Terminal resistance of DC motor	6.22	Ω
L	Inductance of DC motor	0.411	mH
k_t	Torque constant of DC motor	17.8	mN m/A
J	Rotor inertia of DC motor	2.28	g cm^2

(*continued*)

Table 1. (*continued*)

n	Reduction of gearhead	16:1	–
l_a	Length of the arm	100	mm
m_a	Mass of the arm	9.0	g
d_P	Maximum displacement of tension spring	10	mm

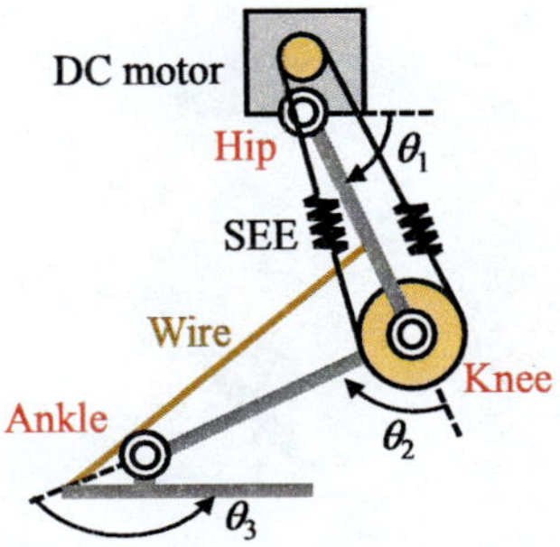

Fig. 1. Model based on the human jumping mechanism.

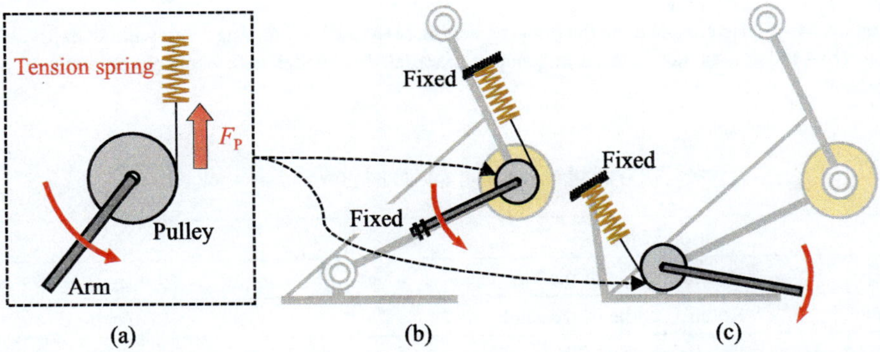

Fig. 2. Powered exoskeleton model and attachment. (a) Model of the powered exoskeleton. (b) Parallel attachment of the powered exoskeleton. (c) Serial attachment of the powered exoskeleton.

3.2 Powered Exoskeleton

A model of the powered exoskeleton is shown in Fig. 2(a). The exoskeleton's actuators use tension springs, which are driven independently of the jumping robot. Because the knee joint extension motion is the factor that most influences the jumping height among the joint extension motions [17], two methods of connecting the powered exoskeleton are considered: (i) in parallel with the knee joint (Fig. 2(b)) and (ii) in series with the knee joint (Fig. 2(c)). (i) is an active link that can rotate dependent on the knee joint, whereas (ii) is an active link that can rotate independently of the joint. In (i), the central axis of the pulley is attached to the knee joint, the two links are fixed to the body, and torque is applied to the joint in parallel. In (ii), the central axis of the pulley is attached to the ankle joint and one link is fixed to the body to transmit the upward force obtained when the arm strikes the ground to the body in series. A roller is attached to the end of the arm to eliminate the effect of friction and transmit only the upward force to the jumping model.

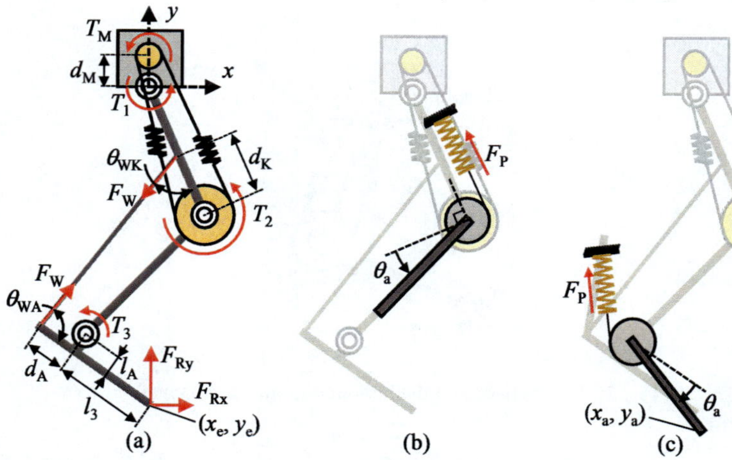

Fig. 3. Model used to calculate the jumping height. (a) Model of jumping robot without exoskeleton. (b) Model with exoskeleton attached in parallel. (c) Model with exoskeleton attached in series.

Table 2. Symbol of jumping robot and powered exoskeleton.

y	Position of the hip	[m]
θ_1	Rotation angle of the hip	[rad]
θ_2	Rotation angle of the knee	[rad]
θ_3	Rotation angle of the ankle	[rad]
T_1	Torque acting on the hip	[N m]
T_2	Torque acting on the knee	[N m]
T_3	Torque acting on the ankle	[N m]
T_H	Torque of the torsion spring added to the hip joint	[N m]
F_W	Wire pull force	[N]

(*continued*)

Table 2. (*continued*)

θ_{WK}	Angle between the thigh and the wire	[rad]
θ_{WA}	Angle between the foot and the wire	[rad]
(x_e, y_e)	Position of the toe	[m]
(x_a, y_a)	Position of the end of the arm	[m]
θ_a	Rotation angle of the arm	[rad]
I_a	Arm moment of inertia	[kg m^2]
F_P	Force of the tension spring	[N]
R_P	Radius of powered exoskeleton's pulley	[m]
F_{Rx}	Horizontal component of ground reaction force	[N]
F_{Ry}	Vertical component of ground reaction force	[N]
k_R	Coefficient of reaction force against sinking displacement	[N/m^2]
c_R	Viscosity coefficient for ground reaction force	[N s/m]
δ	Sinking displacement into the ground	[m]
t_0	Take off time	[s]
v_0	Velocity of the hip at takeoff	[m/s]
y_0	Position of the hip at takeoff	[m]
a	Gravity center acceleration	[m/s^2]
g	Acceleration of gravity	[m/s^2]
m_{total}	Overall mass including jumping robot and exoskeleton	[kg]
h_{max}	Maximum jump height	[m]

4 Dynamic Models of the Jumping Robot and Powered Exoskeleton

4.1 Conditions in Constructing the Model

In this chapter, we model a jumping robot based on the human jumping mechanism and powered exoskeleton. As shown in Fig. 3 (a), an *x-y* coordinate system with the origin at the initial position of the hip joint is established, the effects of friction and air resistance are ignored, and the maximum jumping height of the jumping model is set as the evaluation parameter h_{max}. The parameters used to construct the model are listed in Table 2.

4.2 Dynamic Model Construction

The rotational speed of a DC motor varies with the applied voltage, and the output torque T_M is inversely proportional to the rotational speed ω_M. The torque characteristic of the motor can be expressed using the stall torque T_S and the no load speed ω_N as shown in Eq. (1).

$$T_M = T_S(1 - \omega_M/\omega_N) \quad (1)$$

The toe positions of the jumping model (x_e, y_e) are given by Eqs. (2) and (3) from kinematics, and the Jacobian matrix $\boldsymbol{J}$ is expressed by Eq. (4).

$$x_e = l_1 cos\theta_1 + l_2 cos(\theta_1 + \theta_2) + l_3 cos(\theta_1 + \theta_2 + \theta_3) + l_A sin(\theta_1 + \theta_2 + \theta_3) \quad (2)$$

$$y_e = y + l_1 sin\theta_1 + l_2 sin(\theta_1 + \theta_2) + l_3 sin(\theta_1 + \theta_2 + \theta_3) - l_A cos(\theta_1 + \theta_2 + \theta_3) \quad (3)$$

$$\boldsymbol{J} = \begin{bmatrix} \frac{\partial x_e}{\partial \theta_1} & \frac{\partial x_e}{\partial \theta_2} & \frac{\partial x_e}{\partial \theta_3} \\ \frac{\partial y_e}{\partial \theta_1} & \frac{\partial y_e}{\partial \theta_2} & \frac{\partial y_e}{\partial \theta_3} \end{bmatrix} \quad (4)$$

Each joint torque $\boldsymbol{T}_{\mathbf{R}}$ owing to the ground reaction force acting on the toe is given by Eq. (5), and the total torque acting on each joint is given by Eqs. (6), (7), and (8).

$$\boldsymbol{T}_{\mathbf{R}} = \begin{bmatrix} T_{R1} \\ T_{R2} \\ T_{R3} \end{bmatrix} = \boldsymbol{J}^{\mathrm{T}} \begin{bmatrix} F_{Rx} \\ F_{Ry} \end{bmatrix} \quad (5)$$

$$T_1 = T_{R1} - T_H + T_M(R_M + d_M cos\theta_1)/R_M \quad (6)$$

$$T_2 = T_{R2} + T_M R_K/R_M - F_W d_K sin\theta_{WK} \quad (7)$$

$$T_3 = T_{R3} - F_W \sqrt{l_A^2 + d_A^2} sin(\theta_{WA} - tan^{-1}(l_A/d_A)) \quad (8)$$

Assuming that the ankle joint angle θ_3 is a function of the knee joint angle θ_2 and that the mass of the constraint connecting the foot and thigh is 0, the equation of motion for the jumping model is Eq. (9).

$$\boldsymbol{H}(\theta_1, \theta_2) \begin{bmatrix} \ddot{\theta}_1 \\ \ddot{\theta}_2 \\ \ddot{\theta}_3 \end{bmatrix} + \boldsymbol{h}(\theta_1, \theta_2, \dot{\theta}_1, \dot{\theta}_2) + \boldsymbol{g}(\theta_1, \theta_2) = \begin{bmatrix} T_1 \\ T_2 \\ T_3 \end{bmatrix} \quad (9)$$

In Eq. (9), the terms are, from left to right, the inertia term, the centrifugal and Coriolis force terms, and the gravity term, respectively. When the powered exoskeleton is connected in series, the arms are initially positioned parallel to the leg links, and the positions of the arm tip (x_a, y_a) are given by Eqs. (10) and (11) from kinematics. Additionally, the equations of motion are given by the ground reaction force and gravity in Eq. (12).

$$x_a = l_1 cos\theta_1 + l_2 cos(\theta_1 + \theta_2) + l_a cos(\theta_1 + \theta_2 + \theta_3 - \theta_a) \quad (10)$$

$$y_a = y + l_1 sin\theta_1 + l_2 sin(\theta_1 + \theta_2) + l_a sin(\theta_1 + \theta_2 + \theta_3 - \theta_a) \quad (11)$$

$$I_a \ddot{\theta}_a = F_P R_p - F_R l_a cos(\theta_1 + \theta_2 + \theta_3 - \theta_a) + m_a g l_a cos(\theta_1 + \theta_2 + \theta_3 - \theta_a)/2 \quad (12)$$

The center-of-gravity acceleration a of the jumping model is given by Eq. (13), and $F_{Ry} = 0$ when the jumping model or exoskeleton leaves the ground. The force product of the ground reaction force gives the initial velocity v_0 in Eq. (14) and the maximum jump height h_{max} in Eq. (15).

$$a = F_{Ry}/m_{total} - g \tag{13}$$

$$v_0 = \int_0^{t_0} F_{Ry} dt / m_{total} \tag{14}$$

$$h_{max} = y_0 + v_0^2/(2g) \tag{15}$$

In this study, the vertical ground reaction force is represented F_{Ry} by the hypothetical nonlinear viscoelastic model shown in Eq. (16).

$$F_{Ry} = k_R \delta^2 + c_R \dot{\delta} \tag{16}$$

5 Analysis

5.1 Analysis Purpose

In this analysis, the state of the jumping motion is estimated by a dynamic analysis considering the models of the jumping robot and the powered exoskeleton as shown in Fig. 4. The exoskeleton is then attached in parallel or in series to the jumping model to compare the effects of the attachment method on the jumping height.

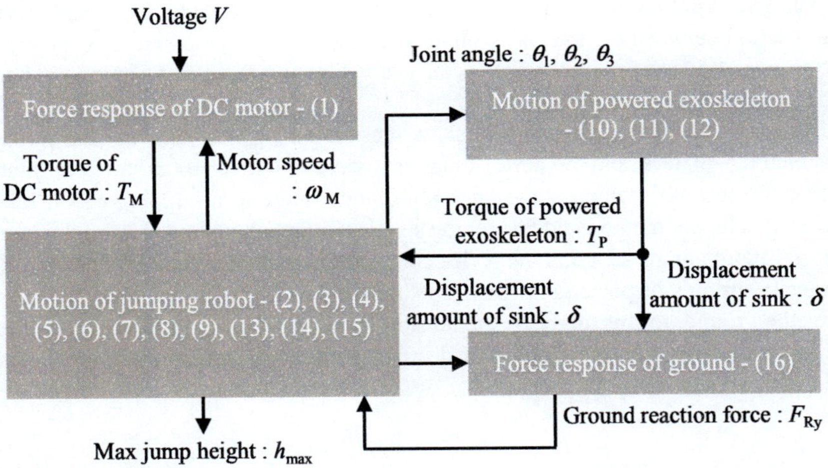

Fig. 4. Schematic diagram of jumping model and powered exoskeleton. Number in parentheses indicate the related equation

Table 3. Initial angle and range of motion of each joint.

Initial angle of the hip	$\theta_1 = -66.83°$
Initial angle of the knee	$\theta_2 = -90.0°$
Initial angle of the ankle	$\theta_3 = 156.83°$
Range of motion of the hip	$-96.1° < \theta_1 < 0°$
Range of motion of the knee	$-90.0° < \theta_2 < 6.1°$
Range of motion of the ankle	$<\theta_3 < 180°$

5.2 Analysis Environment

We used Simulation X (Ver. 4.1, NewtonWorks corporation) for this analysis. The analysis step was 0.1×10^{-9} s, and the sampling time was 0.05×10^{-3} s. The ground reaction force F_{Ry} was represented by the hypothetical nonlinear viscoelastic model shown in Eq. (16), with $k_R = 55$ N/mm^2 and $c_R = 0.10$ Ns/mm.

5.3 Analysis Conditions

The hip joint of the jumping robot is attached to a guide rail fixed vertically to the ground, and the robot's degrees of freedom are limited to the vertical translation and pitch rotation. The initial angles and ranges of motion of each joint are set as listed in Table 3, and the initial posture is a squat position with the toe grounded just below the axis of rotation of the hip joint. The DC motor is driven at a constant voltage (24 V) and the voltage is applied until the jumping robot takes off the ground. The coefficient of static friction between the toe and ground μ is 2.0, and the coefficient of the dynamic friction μ' is 1.5. The hip joint height is the evaluation parameter h. The hip joint height in the initial posture is $h = 0$ m.

For each attachment method, the same tension spring was used for the powered exoskeleton actuators, and the pulley diameters were designed so that the arm rotation angle range was 90°, which is the knee joint angle range of a jumping robot, for the parallel attachment method and 60° for the serial attachment method. The parameters of the arm rotation angle range in the series connection were empirically adjusted so that the tensile spring's displacement reaches its maximum displacement when the arm tip leaves the ground. In this analysis, the tension spring is initially in the tensioned state and varies linearly with the amount of displacement converted from the arm rotation angle θ_a as shown in Eq. (17).

$$F_P = 120\{1 - R_P\theta_a/(2d_P)\} \tag{17}$$

However, $R_P = 6.35$ mm for the parallel attachment and $R_P = 9.55$ mm for the serial attachment. The displacement of the tension spring was set to 0 m in the initial posture for the parallel attachment and to 0 m with the arm parallel to the foot link for the serial attachment. The force exerted by the tension spring at a displacement of 0 m is 120 N. The tension spring exerts force from the start of the drive.

In the serial attachment, if the exoskeleton is driven too quickly, the force of the exoskeleton causes the joints of the jumping model to flex. Therefore, the timing of the powered exoskeleton drive in the serial attachment was set to 0.095 s after the DC motor of the jumping model was driven.

6 Results and Discussion

Figure 5 (a) shows the maximum jump height results without the exoskeleton, parallel attachment, and serial attachment. In both attachment methods, the jumping height was approximately 1.06 times higher than that without the exoskeleton, and approximately 1.18 times higher than that with the parallel and serial attachment, respectively. From this, we estimated that the powered exoskeleton was effective. Additionally, the serial attachment contributed the most to the enhanced jumping height under the conditions set up in this study. As shown in Fig. 5 (b), the energy consumption of the DC motors was lower in the parallel attachment than in the serial attachment, with the parallel attachment consuming 7.1% less energy than without the exoskeleton. The above results are discussed. The parallel attachment reduced the workload of the motor because the exoskeleton force acted on the same joints simultaneously with the motor force. Thus, in the parallel attachment, the force of the exoskeleton acted as an assist exoskeleton such that the workload of the motor was reduced, whereas the improvement in the jumping height was small. In serial attachment, however, the exoskeleton and motor forces act on the joints at different times, so that the exoskeleton does the work after the motor. We hypothesized that this would not increase the workload of the motor but would add to the workload of the powered exoskeleton, resulting in a significant increase in the jumping height. The workload of the motor also decreased, suggesting that it also had the effect of assisting motion. Based on these results, the serial attachment of the powered exoskeleton is expected to extend instantaneous movements such as jumping.

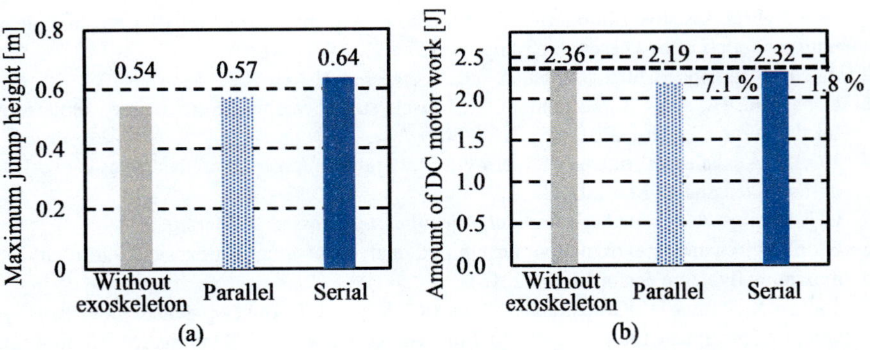

Fig. 5. Analysis results. (a) Comparison of the effect of powered exoskeleton attachment method on maximum jump height. (b) Effect of powered exoskeleton attachment methods on the amount of work done by the DC motor. The amount of work done by the DC motor without the powered exoskeleton is 2.36 J.

However, in serial attachment, changes in the drive timing of the powered exoskeleton may cause the joints of the jumping model to flex, or the model may not be able to receive the upward force of the exoskeleton. It is also necessary to consider the efficiency improvement of the power output due to the drive timing of the exoskeleton and the interactional factors between the jumping robot and exoskeleton. Furthermore, the effect of changing the arm length on the jumping height should also be considered, because the period during which the ground reaction force can be obtained changes as the arm length is changed. In addition, in this paper, the actuator of the powered exoskeleton is simply a passive spring but, in the future, the characteristics of a pneumatic actuator with a large output-to-weight ratio will also be considered.

7 Conclusion

This paper describes the construction of a model of a small robot based on the human jumping mechanism and a powered exoskeleton designed to enhance the jumping height, and a comparison of the effect of the attachment method of the powered exoskeleton on the jumping height by dynamic analysis. The maximum jumping height of the serial attachment was approximately 1.18 times higher than that without the exoskeleton. This indicates that the serial attachment is the most effective way to achieve the maximum jump height and is expected to be effective in extending instantaneous movements such as jumping.

In the future, we will clarify the effect of the drive timing of the powered exoskeleton on the jumping height considering the characteristics of the pneumatic actuators and confirm the validity of the model through experiments on actual equipment.

References

1. Kubota, S., et al.: Robotic rehabilitation training with a newly developed upper limb single-joint Hybrid Assistive Limb (HAL-SJ) for elbow flexor reconstruction after brachial plexus injury: a report of two cases. J. Orthop. Surg. **26**(2), 1–7 (2018)
2. ATOUN Homepage. https://atoun.co.jp/en/. Accessed 21 Mar 2022
3. Kobayashi, H., Aida, T., Hashimoto, T.: Muscle suit development and factory application. Int. J. Autom. Technol. **3**(6), 709–715 (2009)
4. XOS2 exoskeleton. https://www.army-technology.com/projects/raytheon-xos-2-exoskeleton-us/. Accessed 21 Mar 2022
5. HULC. https://bleex.me.berkeley.edu/project/hulc/. Accessed 21 Mar 2022
6. Farid, M.: Dynamic exoskeleton: design and analysis of a human exoskeleton to enhance maximum dynamic performance (2016)
7. Tadano, K., Araya, H., Kawashima, K., Youn, C., Kagawa, T.: Development of jump assist system using pneumatic rubber muscle. In: Proceedings of the 8th JFPS International Symposium on Fluid Power, pp. 677–682. JFPS, OKINAWA (2011)
8. Majoe, D., Widmer, L.: Power and endurance for comfortable wearable robotics. In: 2014 IEEE International Conference on Robotics and Automation (ICRA), pp. 4811–4818. IEEE, Hong Kong (2014)
9. Dick, G.J., Edwards, E.A.: Human bipedal locomotion device. US5,016,869A (1991)
10. Böck, A.: Device for helping a person to walk. US6,719,671B1 (2004)

11. Van Hooren, B., Zolotarjova, J.: The difference between countermovement and squat jump performances: a review of underlying mechanisms with practical applications. J. Strength Cond. Res. **31**(7), 2011–2020 (2017)
12. Van Hooren, B., Bosch, F.: Influence of muscle slack on high-intensity sport performance: a review. Strength Conditioning J. **38**(5), 75–87 (2016)
13. Mitchell, L.J., Argus, C.K., Taylor, K.L., Sheppard, J.M., Chapman, D.W.: The effect of initial knee angle on concentric-only squat jump performance. Res. Q. Exerc. Sport **88**(2), 184–192 (2017)
14. La Torre, A., et al.: Acute effects of static stretching on squat jump performance at different knee starting angles. J. Strength Cond. Res. **24**(3), 687–694 (2010)
15. van Ingen Schenau, G.J., Bobbert, M.F., Rozendal, R.H.: The unique action of bi-articular muscles in complex movements. J. Anat. **155**, 1–5 (1987)
16. de Leva, P.: Adjustments to Zatsiorsky-Seluyanov's segment inertia parameters. J. Biomech. **29**(9), 1223–1230 (1996)
17. Hubley, C.L., Wells, R.P.: Awork-energy approach to determine individual joint contributions to vertical jump performance. Eur. J. Appl. Physiol. **50**, 247–254 (1983)

Evaluation of Support Device for Manual Handling of Gas Cylinders

Kiyotaka Oba[1](✉), Manabu Okui[1], Rie Nishihama[2], and Taro Nakamura[1]

[1] Chuo University, Kasuga 1-13-27, Tokyo 112-8551, Japan
k_oobas@bio.mech.chuo-u.ac.jp, {m_okui, nakamura}@mech.chuo-u.ac.jp

[2] Research and Development Initiative, Chuo University, Kasuga 1-13-27, Tokyo 112-8551, Japan
r_nishihama@bio.mech.chuo-u.ac.jp

Abstract. In this study, a support device for liquefied petroleum (LP) gas cylinder delivery was developed and evaluated using questionnaires. This study contributes to developing a method for reducing the burden of gas cylinder delivery. LP gas cylinders are manually transported, and the physical burden on delivery personnel is high. Existing methods to minimize this burden include cart transportation. In the delivery environment, narrow spaces and steps exist, increasing the difficulty of cart transportation. Therefore, we propose a support device as a new method for reducing this burden. The support device was designed based on a survey of the LP gas cylinder delivery environment and a method for holding cylinders. The support device enhanced walking and was less physically demanding. The subjective evaluation showed that it exhibits the characteristics of the main method used for holding cylinders while improving the shortcomings. This device significantly reduces the load on the upper body during delivery.

Keywords: LP gas delivery · Work support · Assistive device

1 Introduction

Liquefied petroleum (LP) gas is used for heating and is a vital energy source that supports people's daily lives. Unlike the grid supply of electricity and city gas, the LP gas supply system is decentralized. LP gas is supplied to residences and facilities through installed LP gas cylinders. In Japan, 24 million households, representing approximately half of all households, use this service [1]. There are two types of LP gas supply methods: bulk and cylinder. In the bulk method, LP gas is filled directly from a tanker into a large bulk cylinder installed in advance at the facility. In the cylinder-type method, gas is supplied by transporting a gas cylinder to a facility where the gas is available and the cylinder is replaced. In Japan, the cylinder-type method is widely used in ordinary detached houses. Gas cylinders are transported by road from the filling station to the installation site and manually transported by the delivery personnel when they arrive near the installation

© The Author(s), under exclusive license to Springer Nature Switzerland AG 2023
J. M. Cascalho et al. (Eds.): CLAWAR 2022, LNNS 530, pp. 550–559, 2023.
https://doi.org/10.1007/978-3-031-15226-9_50

site. In Japan, the declining number of young workers and the aging of society have resulted in an aging and understaffed delivery workforce, placing a heavy labor burden on delivery workers.

Fig. 1. 20 kg cylinder

(a) Narrow space (b) Stairs

Fig. 2. Delivery environment

Efforts to improve work efficiency include optimization of delivery routes [2] and Internet-of-Things-based monitoring of gas cylinders [3]. However, the physical burden of manually handling cylinders remains high. Cylinders are heavy, weighing approximately 20 kg, even for lightweight models. The delivery person manually carries a heavy cylinder. The delivery environment contains narrow spaces and stairs, which require stooping. A survey [4] conducted in Taiwan, where gas cylinders are delivered manually, as in Japan, showed that more than half of the delivery workers experienced shoulder and back pains within a year. The prevalence of musculoskeletal disorders within a year exceeded 90%. In addition, work-related accidents during cylinder transportation occur more frequently when heavy cylinders are carried by hand [5]. Heavy loads and physically demanding delivery environments pose a significant risk of musculoskeletal disorders and are major accident hazards for gas cylinder delivery personnel. Therefore, there is a need to develop a method to minimize the burden of manually transporting gas cylinders.

The burden of gas cylinder delivery can be reduced by using a cart. Therefore, a cart for gas cylinders has been developed. Okunribido et al. [6] conducted an experiment and designed a handle for carts. Al Amin et al. [7] conducted a questionnaire survey to design a trolley for transporting LP gas. However, delivery environments often include narrow spaces and stairs, making it challenging to use trolleys. Therefore, the weight reduction of the cylinders and the development of orthotics to reduce the burden of handling cylinders have been conducted. The weights of cylinders can be reduced by replacing conventional steel cylinders with fiber-reinforced polymer (FRP) cylinders [8]. However, the application rate of FRP cylinders in Japan is low, owing to their high manufacturing costs. Based on workers' opinions, Devereux et al. [9] developed a device that imitates a steering wheel to reduce the burden during transportation. However, owing to the specifications of the device, it is difficult to apply it to heavy gas cylinders used in Japan because of the burden placed on the arms and shoulders.

In an initial study, we developed a support device to reduce the physical burden of LP gas delivery. The concept of the developed device was determined based on a survey of the actual conditions related to gas cylinder delivery using questionnaires. The proposed device is used to hold a cylinder at waist height.

Table 1. Techniques of holding cylinders

	Carry the cylinder on the shoulder	Hold the sideways cylinder with both hands	Hold the cylinder protector with both hands
		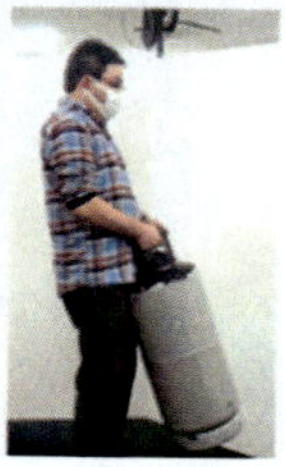	
(1) Walkable	○	×	×
(2) Well balanced	×	×	×
(3) Visibility	×	○	×
(4) Use in narrow space	×	×	○
(5) Difficult to lift	×	×	×
Physical burden	shoulder waist	arm waist	arm waist

By holding the device at the waist height, it can be smoothly transported, and the physical burden can be reduced. A questionnaire evaluation of the prototype device was conducted, and it was found that the physical burden when using the device was slight.

2 Proposed Assistive Device

2.1 On-the-Spot Investigation

Before developing the device, a survey of the gas cylinder delivery sites in Japan was conducted. The survey comprised the size and weight of frequently handled cylinders, method for transporting cylinders, and work environment during delivery. The most widely used transported cylinder type is shown in Fig. 1. This type is called a "20 kg cylinder" in Japan. The size of the cylinder is 300 mm × 300 mm × 850 mm. The weight of the cylinder is 39.5 kg when filled and 18.5 kg when empty.

The cylinder is lowered from the truck bed close to the delivery person's waist height and transported to the installation site. Cylinders are installed at various locations in the main delivery destination, that is, detached houses. They are often placed on the ground behind buildings or at inconspicuous locations. The route to the installation site includes stairs, steps, and narrow stoop locations, as shown in Fig. 2. When a cart is available, the delivery person uses it, but in most cases, the final product is transported by hand to the installation site.

Table 1 shows the main ways of holding the "20 kg cylinder" for manual handling. Delivery personnel use the three transportation techniques depicted in Table 1: "carry a cylinder on the shoulder," "pulling by top handles," and "lifting both ends," depending on the delivery environment and situation.

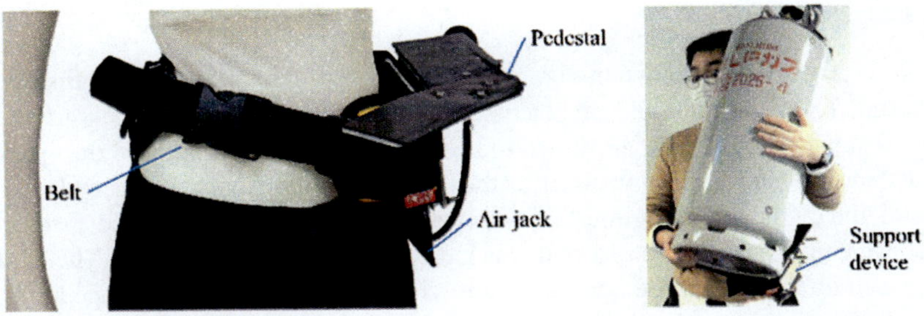

(a) Support device details (b) Attaching support device

Fig. 3. Overview of support device

Table 1 lists the advantages of the three holding styles. For each item, a "0" was indicated if its holding method was superior. "Carry a cylinder on the shoulder" is a method for carrying a cylinder by placing it on one's shoulder. Most delivery personnel use this holding method because of their superior walking performance. The weight is borne on the shoulders and hips. "Lifting both ends" is a transportation technique in which the cylinder is carried by grasping both ends. This method has excellent visibility but strains the arms and the lower back. "Pulling by top handles" is often used in environments that require discretion, such as narrow spaces. In this case, the arms and lower back are stressed. All the carrying methods were found to have disadvantages regarding the ease of balancing, indicating that the cylinders are sometimes carried under highly hazardous conditions.

2.2 Concept of Device

Each of the three holding styles depicted in Table 1 exhibits different characteristics and is applied in different ways, depending on the situation. In this study, we developed a new method for holding the support device instead of focusing on any of these ways of holding it. We aimed to develop a device to overcome the disadvantages of each holding method while utilizing the advantages of each method.

In this study, Items (1)–(4), related to transportation, were targeted for the initial study on development among the required specifications. We propose a new technique for holding a cylinder to satisfy these requirements: "hold the cylinder at the waist height."

If the cylinder is held at the waist height, walking is enhanced because the cylinder does not interfere with the feet. It is easier to move in tight spaces where the worker must stoop. In addition, the cylinder is lower when it is held at the waist height than when it is carried on the shoulder, reducing the lifting burden. However, using only the hands to hold the cylinder at the waist height burdens the arms and the upper body. In addition, the overlap between the face and cylinder may obstruct the carrier's view. Therefore, a device that supports the cylinder on the side of the lumbar region and can distribute the burden to body parts was considered. The lumbar region can provide improved visibility.

2.3 Proposed Device

The prototype device is shown in Fig. 3. The developed device was a belt equipped with a pedestal. The pedestal could be placed at the waist by wrapping the belt around the waist, as shown in Fig. 3(a). As shown in Fig. 3(b), a cylinder was placed on a pedestal to distribute the load to the lower back as the cylinder was transported. When placed on a pedestal, the cylinder did not interfere with walking movements. It enabled a stable gait and convenient balance. The pedestal was located on the side of the body. The cylinder was placed on the side of the body to ensure visibility. The surface of the pedestal was covered with a nonslip sheet to prevent the cylinder from sliding down. An air pad was placed inside the belt near the pedestal. The air pressure of the air pad prevented the pedestal from tilting, owing to the weight of the cylinder. This pressure prevented the pedestal from digging into the legs. The air pad was inflated using a hand pump. The belt length could be adjusted to fit the physique of the delivery worker.

3 Subjective Evaluation of Device

3.1 Evaluation Method

A questionnaire-based evaluation was conducted to determine whether the developed equipment satisfied the required specifications. This experiment was conducted to compare the three methods of holding presented in Table 1 when the device was used and not used. The cylinder used in this study was an empty "20 kg cylinder" weighing 15 kg. Each subject performed the target movement while holding a cylinder. The target movements were walking movements in three environments (a flat road, stairs, and a narrow space), as shown in Fig. 4. The frame dimensions of the narrow space were 75 mm × 180 mm. The subjects answered the questionnaire presented in Table 2 regarding the required specifications (Q1–Q3) and body burden (Q4–Q7) at the end of the walking activity.

Table 2. Questionnaire

Require specification	
Q1 Is it easy to walk?	1.Excellent
Q2 Is it easy to balance?	2.Very Good
Q3 Is visibility good?	3.Fair
	4.Poor
	5.Unacceptable
Physical burden	
Q4 shoulder	1.Slight burden
Q5 arm	↕
Q6 waist	
Q7 knee	5.Significant burden

(a) Flat road (b) Stairs (c) Narrow place

Fig. 4. Aspects of experiment

The questions were scored from 1 to 5 based on a five-point scale for items related to requirement specifications. Body burden was scored on a scale of 1 to 5 for each body part. The scores for the items related to the requirement specifications were higher when the responses were positive, and the scores for the physical burden were higher when the burden was greater. The scores for each item were used to test for differences using Wilcoxon's signed-rank test, with the proposed device as the target group and the three conventional holding methods as the comparison group. The results were further corrected using the Bonferroni correction method. The subjects were ten males in their 20 s (height = 172 ± 7.02 cm; weight = 60.4 ± 12.8 kg). The Ethical Review Committee of Chuo University approved this study (No. 2021–58).

3.2 Results and Discussion

The results for the required specifications and physical burden are presented in Figs. 5 and 6, respectively. Figures 5(a) and 6(a) show radar charts of the results, and Figs. 5(b) and 6(b) show the box plots for each item. The proposed device scored higher than the three conventional methods for all items in all the environments. Figure 6(a) shows that the smaller the rectangle, the lower the body burden. The proposed device caused less arm and shoulder strains in both environments. The details of each item are presented below.

Q1: Is Walking Easy? The median score for the proposed device in both environments was five points, significantly higher than those of "lifting both ends" and "pulling by top handles." The median value of the "carry a cylinder on the shoulder" device was high because of its "easy to walk" feature, and it did not differ from that of the proposed device. In this survey, "carry a cylinder on the shoulder" had the best walking performance. However, scores for "carry a cylinder on the shoulder" varied significantly for walking on stairs and in narrow spaces, and some subjects felt that walking was challenging. However, all subjects scored higher on the proposed device than on the "carry a cylinder on the shoulder" device. Therefore, although no significant difference was observed, the proposed device may yield a better walking performance than the "carry a cylinder on the shoulder" device. Based on the above, a person can walk with the proposed device without the person's feet interfering with the cylinder because the cylinder is held at waist height.

Q2: Is Maintaining Balance Easy? The median score of the proposed device was high in all environments and significantly exceeded those of "carry a cylinder on the shoulder" and "pulling by top handles." The score for "lifting both ends" was also high, and no significant difference with the proposed device was confirmed. However, a more significant intersubject variability was observed for "lifting both ends" than for the device. For the "lifting both ends" option, the subject holds the cylinder horizontally, supported by the front of the thighs, which causes the center of gravity of the cylinder to sway more during walking. Depending on the subject's muscle strength, it may be impossible to suppress the sway. In the case of the proposed device, the median value was high, and the variation was minimal in all environments. The survey revealed few positive comments regarding the "ease of balancing" for any holding styles. The proposed device demonstrated the possibility of carrying cylinders more than existing holding methods.

Q3: Is the Visibility Good? The median score for the proposed device was high in both environments, but the conventional holding methods had good scores with no significant differences. We confirmed that all the conventional holding methods are simple to apply on flat roads, and the proposed device enhances visibility on a flat road. Conventional holding methods are highly variable in stairways and narrow spaces. These environments require the user to keep an eye on their feet, and the conventional technique of holding the cylinder places a heavy burden on the arms, shoulders, and waist, which may make it difficult to see the cylinder while maintaining a stable grasp.

In the case of the proposed device, a minimal variation was observed because the feet could be checked in any environment. These results indicate that the proposed device can provide visibility similar to or better than conventional devices.

Q4–Q7: Physical Burden. For the shoulders and arms of the upper body, the proposed device had significantly less burden than "carry a cylinder on the shoulder" and "pulling

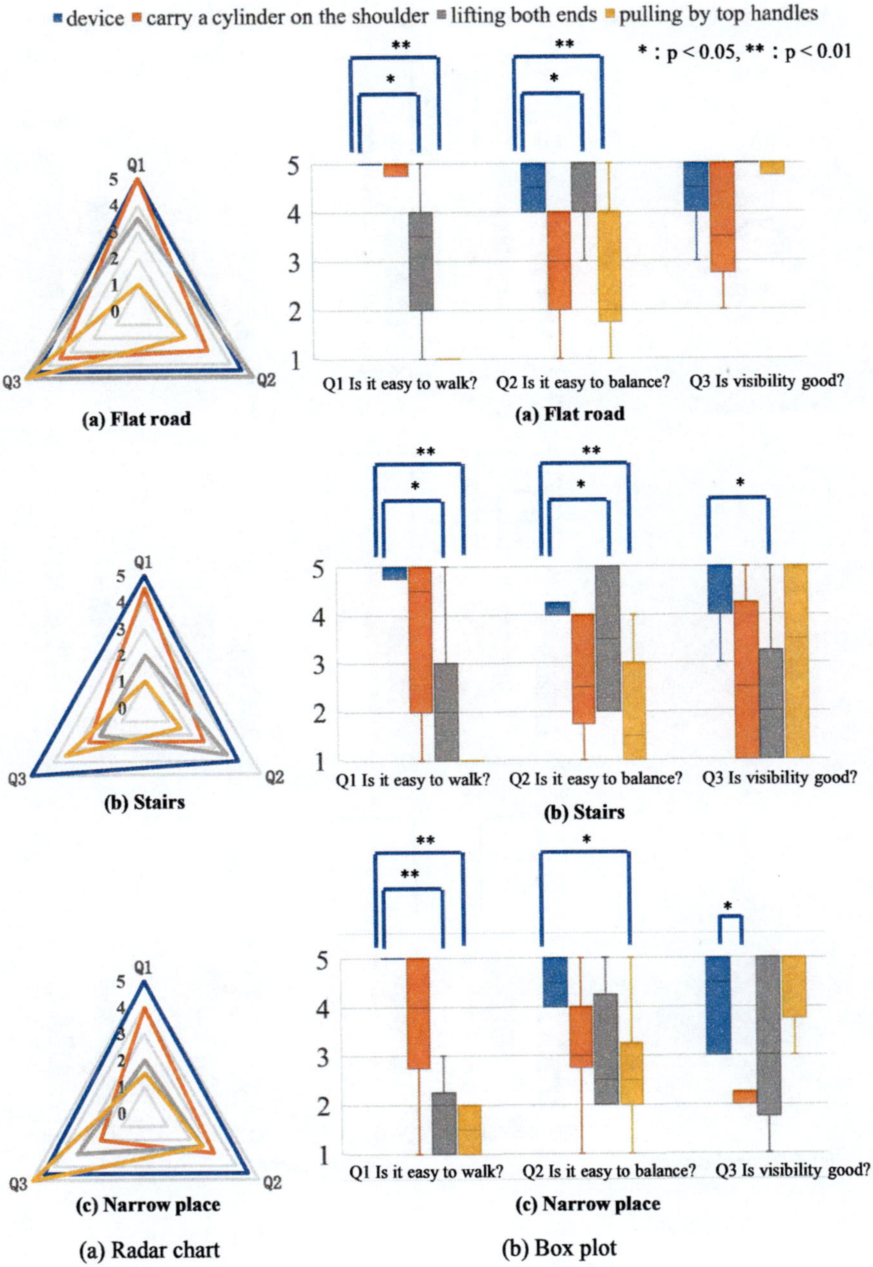

Fig. 5. Results for required specification

by top handles," which are techniques of holding the device that places a significant burden on the body of the upper body. At the waist, the median value for the device

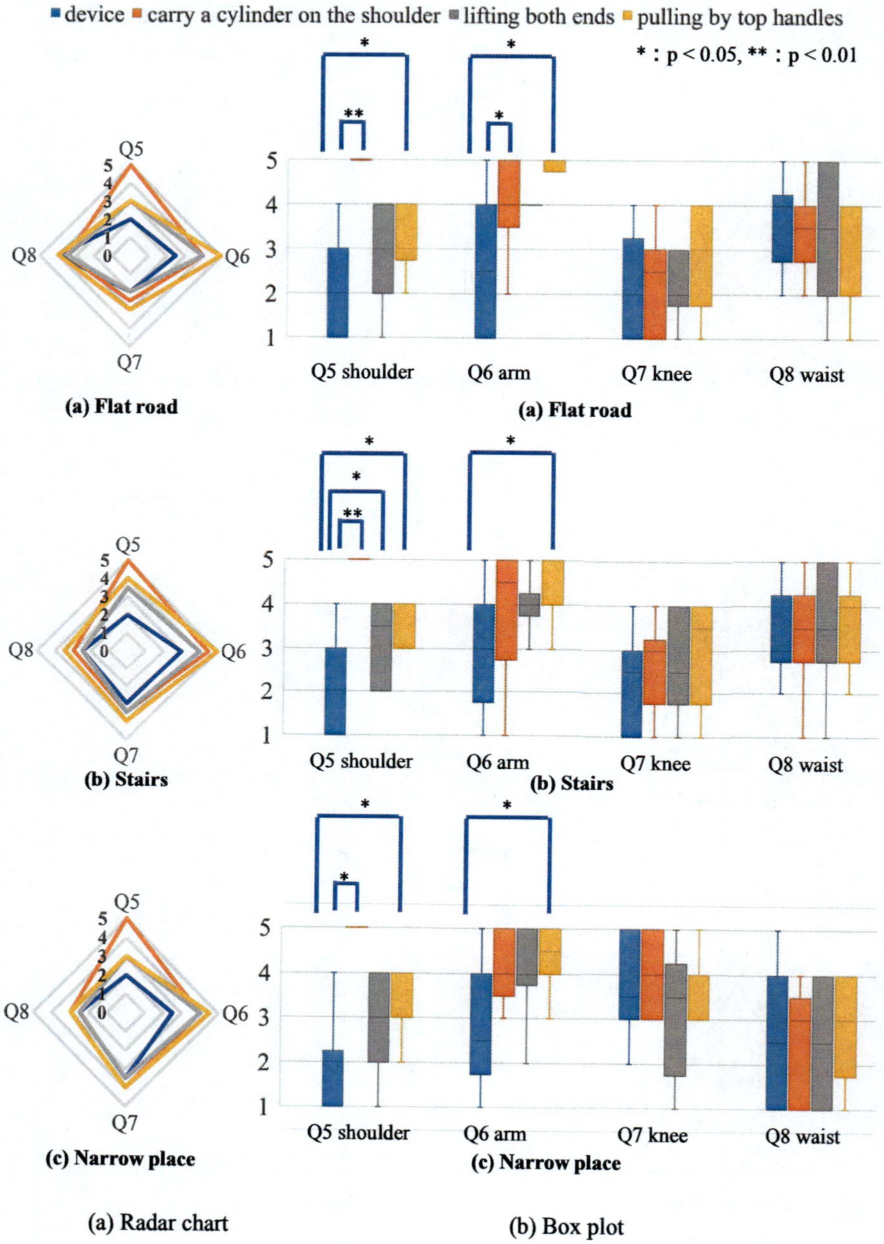

Fig. 6. Results for physical burden

was four points, which was not significantly different from other holding methods. This was because the cylinder was heavy, and any holding technique with or without the device placed a heavy burden on the lower back. Moreover, no significant difference was observed in knee strength between any holding styles. The proposed device was designed to mainly reduce the load on the arms and shoulders, and the results were expected. However, the lower half of the body also receives a significant load, and we plan to reduce the load on the lower half in future studies.

4 Conclusion

A support device was developed to reduce the physical burden of transporting LP gas cylinders. The developed device places an LP gas cylinder on a platform attached to a waist belt, making it easy to walk with the cylinder in one's hand and reducing the load on the body. Based on the gait evaluation, it was confirmed that the support device has the advantages of the existing holding methods (ease of walking and good visibility) while overcoming their disadvantages (significant load on the upper body and difficulty in maintaining balance). In the future, we will quantify physical load reduction using muscle potentials. In addition, we plan to develop a device to support the lifting and lowering of cylinders, which is outside the scope of this study.

Acknowledgments. This study was conducted in collaboration with Astomos Energy Corporation. The authors express their gratitude to the following:

References

1. METI-Japan: Energy White Paper 2020 (2020). (in Japanese)
2. Hsieh, Y.-C., You, P.-S., Chen, C.-S.: Scheduling the periodic delivery of liquefied petroleum gas tank with time window by using artificial intelligence approaches: an example in Taiwan. Sci. Prog. **104**(S3), 1–26 (2021)
3. Srivastava, A.K., Thakur, S., Kumar, A., Raj, A.: IoT based LPG cylinder monitoring system. In: 2019 IEEE International Symposium on Smart Electronic Systems (iSES, formerly iNiS), Rourkela, India, pp. 268–271 (2019)
4. Chen, Y.L., Yu, C.Y., Lin, D.Y.: Musculoskeletal disorders caused by gas cylinder handling tasks: a case study report. Work **56**(3), 403–407 (2017)
5. Kim, J.N., Jeong, B.Y., Park, M.H.: Accident analysis of gas cylinder handling work based on occupational injuries data. Hum. Factors Ergon. Manuf. Serv. Ind. **27**(6), 280–288 (2017)
6. Okunribido, O.O., Haslegrave, C.M.: Effect of handle design for cylinder trolleys. Appl. Ergon. **30**(5), 407–419 (1999)
7. Al Amin, M.S., Isa, H., Febrian, I., Nuradilah, Z.T., Nor, A.M.: Application of quality function deployment to design a liquefied petroleum gas trolley. Appl. Mech. Mater. **761**, 698–702 (2015)
8. Moketla, M.B., Shukla, M.: Design and finite element analysis of FRP LPG cylinder. Int. J. Instrum. Control Autom. **2**(1), 121 – 124 (2012). Article 9
9. Devereux, J., Buckle, P., Haisman, M.: The evaluation of a hand-handle interface tool (HHIT) for reducing musculoskeletal discomfort. Int. J. Ind. Ergon. **21**(1), 23–24 (1998)

Rehabilitation Device for Lower Limbs Through Virtual Training and Electrical Acupuncture Stimulation

Pavel Venev(✉), Ivanka Veneva(✉), Georgi Katsarov, and Dimitar Chakarov

Institute of Mechanics, Bulgarian Academy of Sciences, "Acad. G. Bonchev" Street Bl.4., 1113 Sofia, Bulgaria
{pavelvenev,veneva,mit}@imbm.bas.bg, goshokat@gmail.com

Abstract. The aim of this paper is to present the design and development of a device for rehabilitation for lower limbs through electrical acupuncture stimulation. The rehabilitation system uses an active lower extremity orthosis (exoskeleton) that can act as an assistive (or resistive) device that interacts with virtual reality and provides force feedback. The system includes a multi-electrode device for stimulating various acupuncture points of the legs by applying electrical impulses. A user-friendly interface has been created to adjust the strength and duration of the electrical pulses. The designed system provides visualization, adjustment and feedback in real time, guiding the operator in the steps of the therapeutic procedure. An interactive communication has been created between the electrical stimulating device and the therapist, which makes the rehabilitation therapy more adaptable to the personal characteristics of the patient without creating inconveniences. The hardware of the system and the construction of the matrix with electrodes for electrical stimulation are described. Finally, the experimental results show the applicability and precision of the designed system.

Keywords: Rehabilitation · Electrical acupuncture stimulation · Virtual training · Exoskeleton for lower limbs

1 Introduction

Paralysis is one of the main consequences of a cerebrovascular accident or spinal cord injury. The recovery process is usually time consuming and can cause extensive medical complications. That is why the rapid development of a special rehabilitation program, timely solutions for moving paralyzed limbs and maintaining muscle trophism play an important role.

Electrical stimulation (ES) involves the use of electric current to stimulate nerves or nerve endings (NMES) that innervate the muscles under the skin. Electrical stimulation can be applied superficially to the skin (transcutaneous electrical stimulation, TES) or intramuscularly to improve muscle function. The basic theoretical premise is that if the peripheral nerve can be stimulated, the resulting excitation impulse will be transmitted along the nerve, causing muscle contraction [1, 2].

© The Author(s), under exclusive license to Springer Nature Switzerland AG 2023
J. M. Cascalho et al. (Eds.): CLAWAR 2022, LNNS 530, pp. 560–571, 2023.
https://doi.org/10.1007/978-3-031-15226-9_51

Electrical stimulators provide direct, alternating, pulsating or pulsed forms of energy. The devices are used to exercise muscles, demonstrate muscle response to nerve stimulation, relieve pain, relieve incontinence, and provide test measurements. Electrical stimulators typically have controls for adjusting pulse length, pulse repetition rate, pulse amplitude and trigger modes. The electrodes are permanent, implanted transcutaneously or superficially. Therapeutic FES (Functional Electrical Stimulation) allows typically resistive exercises to prevent muscle atrophy and promote cardiovascular conditioning. Functional FES allows or improves standing, movement, gripping, pinching, reaching, breathing and more. [3]. Neuromuscular electrical stimulation (NMES) uses transcutaneous application of electrical currents to induce muscle contractions to promote reinnervation, prevent atrophy, relieve muscle spasms, and promote voluntary muscle control in individuals who have lost muscle function due to surgery, neurologically injury or disabling condition.

The American Occupational Therapy Association practice guidelines for adults with stroke state that for improved occupational performance of individuals with motor impairments, there is high certainty based on evidence that the use of ES has a moderate net benefit. The guidelines also state that the evidence is weak regarding whether or not this therapy improves patient outcomes [4].

There is some evidence that the use of NMES may be an effective rehabilitation regimen to prevent muscle atrophy associated with prolonged immobilization of the knee after ligament repair surgery or injury; however, controlled clinical trials are needed to determine whether adding NMES to the rehabilitation program will improve health outcomes.

There are various solutions and developments, such as a hybrid orthosis [5, 6] for performing repetitive movements with a closed FES control cycle such as a foot press and cycling. FES stimulates muscles in accordance with the sequence of muscle contraction for a given movement, kinematics and dynamics of movement. A lower limb training system activated by electrical stimulation has been developed (called "MotionMake"), composed of two orthoses with 3 DOF comprising a multi-channel electrostimulator, sensors, a control unit managing the electrical stimulation and motors with real-time regulation [7].

Knutson et al. [8] evaluated the effect on upper limb rehabilitation after stroke and concluded that 12 weeks of FES therapy resulted in improved manual dexterity. Lin and Hsu [9, 10] concluded that both higher and lower doses of NMES lead to improvements in motor function, but Shen [11] demonstrates that compared to conventional NMES, FES provides better restoration of function in stroke patients. Negm concluded that low-frequency ($\leq$100 Hz) pulsed electrical stimulation is effective for improving physical function [12]. Meng conducted a study on the effects of electroacupuncture (EA) and concluded that transcutaneous electrical acupuncture stimulation (TEAS) can improve the quality of recovery in the early period [13].

Several different models of EMC devices have recently appeared on the Japanese market, which are used for sports, dietary purposes and for weight loss [14]. These websites do not provide enough information about the devices to compare their features. There are also no customer reviews to provide us with feedback. Therefore, they cannot be used as a basis for comparison with the device we developed. What can be seen from the site is that the device with insole for electrical stimulation is used for EMS foot massage, leg weight loss, muscle training, fat burning, swelling of the legs, sensitivity to cold and more.

The results show that electrical stimulation can improve muscle properties, but with certain limits and precautions that can be useful in a rehabilitation program.

We want to understand the effect and possible benefits of electrical stimulation on the acupuncture points of the foot, as well as whether it can improve rehabilitation.

The aim of our project is to develop an innovative non-invasive neurorehabilitation multi-electrode device to stimulate various acupuncture points on the leg by applying electrical impulses to restore the natural movement of the patient's limbs in cases of spinal cord injury. The developed neurorehabilitation system affects the spinal cord pathways (afferent and efferent part) sending impulses to the brain and back to the peripheral nerve and muscles presenting an innovative solution for rehabilitation of paralysis due to spinal cord injury (SCI).

The developed prototype of a device for rehabilitation through electrical stimulation includes a control unit, a matrix for electrical stimulation of acupuncture points, a sensor circuit and an interface for communication between the microcontroller and the computer application. Advanced and intuitive software technology provides real-time visualization, setup and feedback capabilities, guiding the operator on the steps of the therapeutic procedure.

The addition of an active orthosis and a virtual scene makes the system multifunctional. The advantage of using the active orthosis is that it is possible to perform active and passive mobilization exercises with the desired position, speed and torque, which will be of great benefit to therapists and patients.

2 Material and Methods

According to the Sensory Nervous System, the receptors that respond to the stimulus and initiate the sensation process receive various physical stimuli and transduce the signal into a potential for electrical action. This action potential then travels along afferent neurons to specific brain regions, where it is processed and interpreted. [15, 16].

The device developed under the project (Contract 11IF-02–14/03.12.2020, NIF) is intended to affect the nerve pathways through the sensors of the foot. After applying a low-voltage electric current, the impulse is transmitted to the afferent part of the peripheral neuron and through the posterior horns of the spinal cord the information reaches the centres of the cerebral cortex, where the information is analysed (depending on spinal cord injury). After the analysis in the cortex, impulses are sent back to the efferent part of the spinal cord, to the peripheral nerve and are transmitted to the muscles of the limbs. The developed software application allows to modulate the intensity of stimulation during rehabilitation and to monitor in real time the change in the behaviour of immobilized limbs. All stimulation parameters can be set and changed via a graphical interface: frequency, minimum and maximum intensity, pulse duration, synchronization and order of stimulations, type of interactions and number of repetitions. Maximum thresholds should be set according to the characteristics of the individual patient's sensory system.

In the process of work and during the experiments, the available active orthoses (exoskeletons) are used to support the movements of the human limbs. Orthoses support the movement of individual joints and provide all anatomical degrees of freedom. Each joint of the orthosis has a specific range of motion, corresponding to the natural movements of the limb during exercises (flexion/extension, abduction/adduction). The movements of the participant are monitored in real time through a system of sensors for position, force and pressure, as well as electomyographic sensors, which give a measure of the muscles force. The actuation of the exoskeleton is realized with pneumatic actuators, position and pressure sensors, controllers and appropriate control algorithms.

The most important result from a practical point of view is the creation of a convenient interface for communication with the rehabilitation device, as well as visual and audio feedback from virtual reality scenes. A software application for visualization of signals in work areas, settings and control of parameters and human communication with the virtual environment has been developed and tested [17–19]. In the created virtual scene, researchers and users are given the opportunity to analyse scenarios of human-avatar interactions, to simulate them, to measure contact parameters and reactions during rehabilitation.

Unlike the most widely used rehabilitation methods that affect the muscles to improve their range of motion and strength, in this paper, we present a new type of rehabilitation device with adaptive regulation of electrical stimulation, with changes in the contact surface and highly effective rehabilitation technics that help people to achieve safe physical interaction. This goal will be achieved in real demonstrations of interaction. The subject of future activities will be conducting long-term experiments to test the functionality of the device with patients (volunteers) in a hospital environment.

3 Description of the System

The system consists of an exoskeleton for lower limbs, a control unit that performs electrical stimulation and an electrode matrix for electrical stimulation of acupuncture points. Electromyographic system is used for measuring muscle strength. The software modules are the following:

- Graphical user interface for calibration of the orthosis parameters;
- Graphical user interface for setting the mode of electrical stimulation;
- Virtual scene, representing the movement of the avatar in real time;

The orthosis is an exoskeleton with four degrees of freedom, placed on the outside of the human limb. It is designed to drive three joints: hip, knee and ankle with the help of artificial pneumatic muscles. The joints of the orthosis are coaxial with the joints of the human limb and provide two rotations of the hip joint and rotations of the knee and ankle in the sagittal plane. The lengths of the orthosis segments can be adjusted manually. The range of motion of the joints and the applied torque are calculated individually for each person using a specially designed graphical interface for calibration of the exoskeleton [17–19]. The orthosis can work in two modes, depending on whether we want to support the movement of the limb or create resistance forces to perform active exercises. The visualization of the movements is presented in scenes created in virtual reality, which serve as feedback for evaluation and as a psychological stimulus for progress in the process of rehabilitation. The person sees himself as a virtual avatar who exercises in a gym or walks in a forest area (Fig. 1).

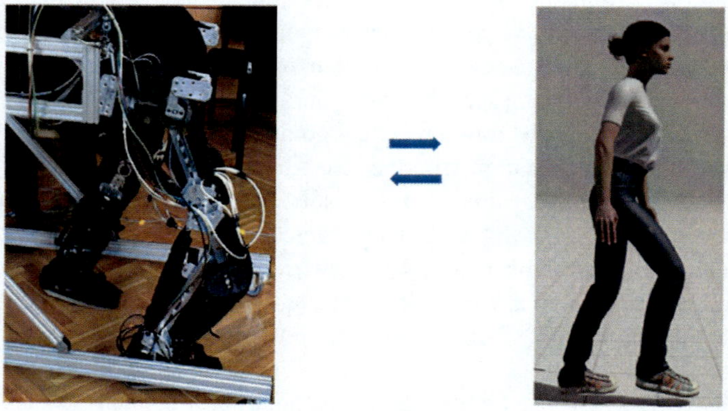

Fig. 1. Visualization of the movements presented in virtual reality

The aim of the paper is to present an approach to the management of an electrical stimulator, which allows the stimulation of selected acupuncture points on the human foot, using pulse width modulation. The graphical user interface for setting the electrical stimulation mode will be described below.

In this work, the effect of ordinary electrical stimulation on the acupuncture points of the foot is tested. The experiment and data collection were performed on 5 subjects and the obtained results were analyzed. This study aims to understand the full potential of electrical stimulation and to identify its possible benefits or disadvantages for muscle properties.

4 Results

Rehabilitation device for lower limb with electrical stimulation of acupuncture points has been developed, which includes:

- A Control Unit with microcontroller, sensor circuit and control algorithm,
- An algorithm for controlling the device for electrical stimulation,
- Matrix with electrodes for electrical stimulation of acupuncture points,
- Graphical user interface for communication between rehabilitation device and computer application.

Modeling and specification of the main parameters of the device for electrical stimulation of acupuncture points has been performed:

- The electrical stimulation will be performed on the surface of the skin by means of flat flexible electrodes, fixed on the matrix and with a terminal for connection to the electrical stimulation device.
- The developed device will affect the nerve pathways through the sensors of the foot at the appropriate acupuncture points. Excited by electric current, the pulse is transmitted to the afferent part of the peripheral neuron.
- A model of a scheme for acupuncture impact through electrical stimulation has been created, which enables the matrix to act on selected acupuncture points by distributing the electrodes.
- The software application allows setting all stimulation parameters: frequency, minimum and maximum intensity, pulse duration, synchronization and order of stimulations, type of user interactions and number of repetitions.
- In order to ensure maximum protection and comfort of the patient, it is possible to stop the procedure and turn off the equipment at any time.
- With the help of a graphical interface, a switching frequency and pulse width to the microcontroller are set via USB communication.

4.1 Control Unit Design

Figure 2 shows a diagram of a control unit, which consists of a microcontroller STM32, driver unit, power supply, controller and interface for communication with the computer. The device is powered by an external 5 V power supply, which is plugged into a special power connector. When powered, the device regulates the 5 V to 3.3 V using the LM3940IS LDO regulator. This regulated voltage supplies both the microcontroller and the high-voltage haptic driver.

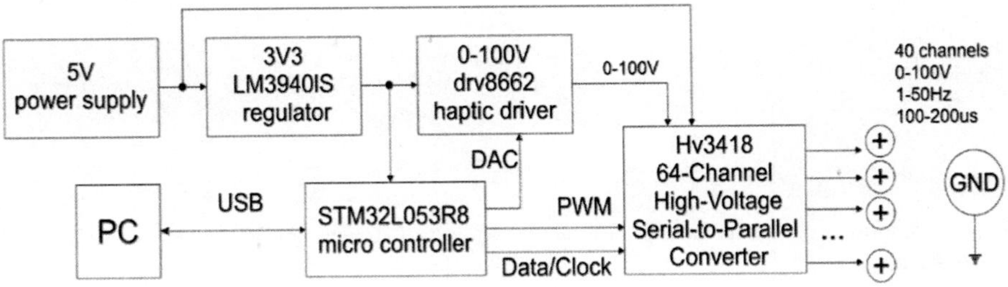

Fig. 2. Control unit and communication with the computer

The microcontroller performs a transceiver function by communicating with the computer software and controlling the output voltage of the haptic driver. The control is performed by a digital-to-analog converter which signal enters the input of the integrated amplifier. There, with the help of the internal boost converter and several external elements, the input voltage from 0 to 3 V is amplified with a gain of 28.8 dB, 34.8 dB, 38.4 dB, or 40.7 dB. This voltage is applied to the power input of the high-voltage 64-channel serial-to-parallel converter, where with the help of pulse-width modulation and a certain signal frequency is distributed to the corresponding output electrode. The controller can control up to 64 electrodes simultaneously.

4.2 Graphical User Interface

A Graphical user interface for adjusting the electrical stimulation of acupuncture points has been created (Fig. 3). The user interface contains a step diagram with marked acupuncture zones and provides the ability to select the position of the electrodes on the diagram, as well as to activate certain electrodes (1 to 40) for electrical stimulation. The interface has the following functionalities:

- Fields for entering modes and parameters of the electrical stimulation program;
- Buttons for connecting to the device, selecting a limb diagram, starting and stopping electrical stimulation.
- Diagram of the foot with distributed acupuncture zones;
- Interface elements for activating electrical stimulation at certain acupuncture points (from 1 to 40);
- Positioning of the interface elements on the limb diagram.

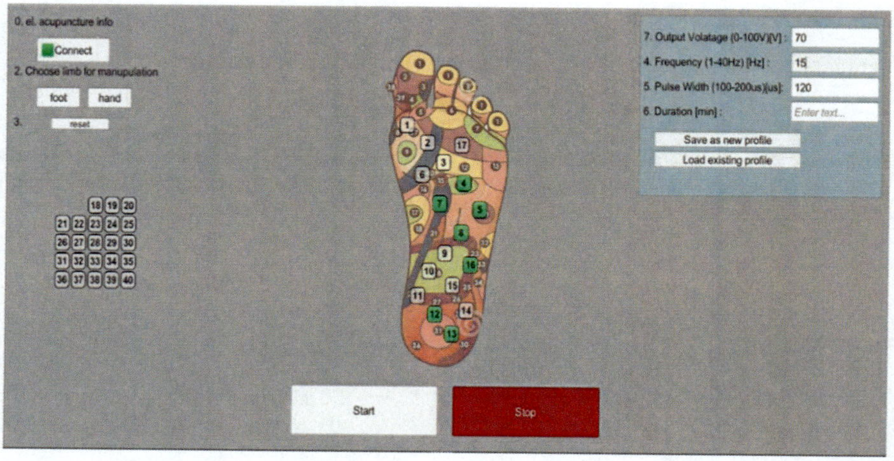

Fig. 3. Graphical user interface for interactive setting of modes and parameters for electrical stimulation.

4.3 Control Algorithm

An algorithm for controlling the device for electrical stimulation of acupuncture points has been created (Fig. 4). The firmware of the device and the accompanying computer interface include the following functionalities: *Graphical interface with interactive setting of the device parameters; Communication via serial interface;*

Control and communication with the device:

– Making a direct connection with the device,
– Setting parameters of the electrical stimulation program,
– Send and receive data to and from the device,
– Setting the control pins of the device,
– Setting the operating mode according to the received information,
– Setting the program of the operation to the peripheries of the microcontroller,
– Output constant signal with set voltage level,
– Output voltage control (controlled with DAC),
– Setting a series of output pins according to the choice of active electrodes,
– Implementation of electrical stimulation and control of electrical impulses,
– Status monitoring (via device status pins and PWM for signal control),
– Setting control levels for LED display.
– Starting electrical stimulation;
– Stopping the electrical stimulation.

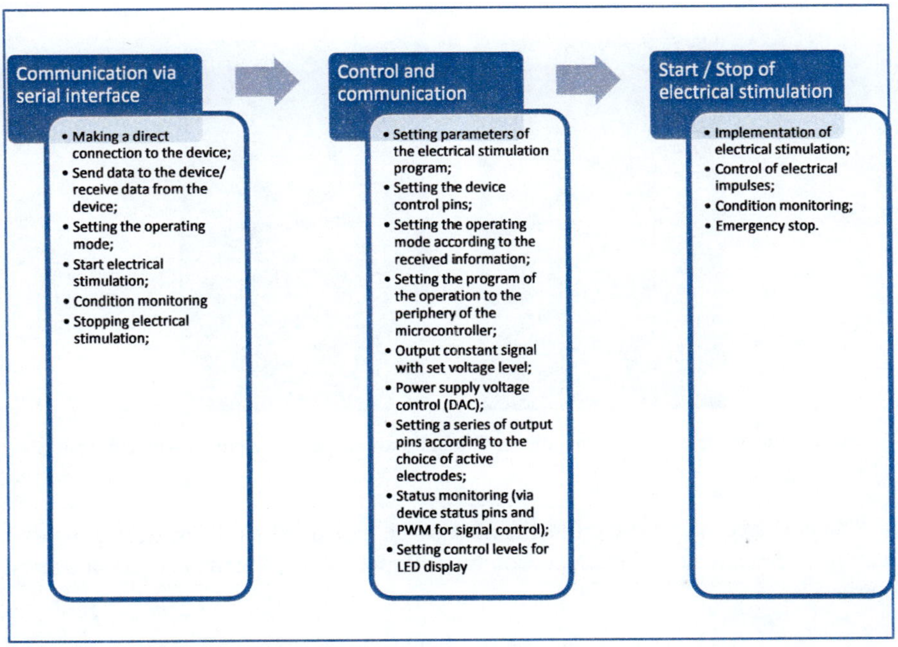

Fig. 4. Control algorithm

The electrostimulator consists of 40 independent modular channels. The parameters of each channel are controlled in real time by the central device via a serial connection. Each channel is capable of generating current pulses up to 0.5 mA. The voltage can reach up to 100 V. The pulse frequency and width are regulated in the range from 10 to

40 Hz and from 100 to 200 μs, respectively. Each output of the channel is electrically floating to avoid current between the electrodes.

The main concern is patient safety, which is ensured through hardware and software security systems, such as programmable power limitation and emergency shutdown. For proper use of the electrical stimulation device, an output voltage of around 100 V is recommended. This unit must be able to personalise and regulate the output voltage so that the patient can set the appropriate degree of stimulation without experiencing any discomfort. The voltage can be set via the graphical interface. As a result of the integrated low current control, the current depends on the different impedance of the patient's skin.

4.4 Electrode Matrix for Electrical Stimulation of Acupuncture Points

A prototype of a matrix for acupuncture by electrical stimulation has been developed (Fig. 5). An insole-shaped matrix with 40 electrodes is placed on the sole of the limb. The location of the electrodes can be personalized and change depending on the size of the foot, which allows us to position them depending on which acupuncture points we want to influence.

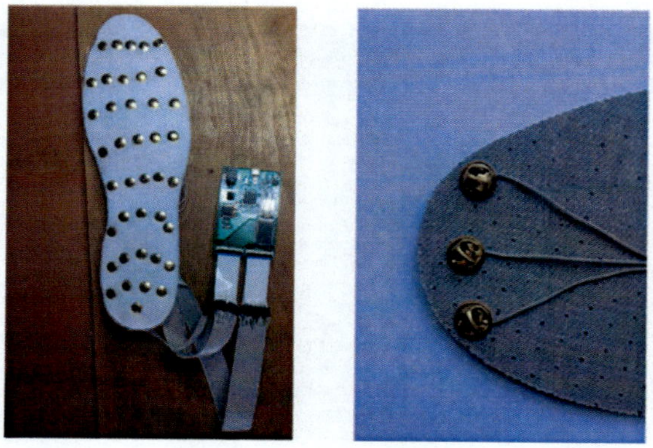

Fig. 5. Electrode matrix for electrical stimulation of acupuncture points with controller

The next step is to test the functionality of the individual modules: testing the user interface, firmware code for connection to a computer and communication via a serial interface.

4.5 Experimentation, Testing and Settings of the Electrical Stimulation Device

Experimental testing of electrical stimulation device settings was performed. Several experiments with volunteers have been conducted. The pulse width is adjusted at 100–200 μs to ensure optimal stimulation, the frequency is set depending on the individual tolerance of the patient and the muscular response of the subject. The current amplitude is adjusted to 100 mA so that it does not cause unpleasant sensations. The user interface has arranged steps to indicate to the user the methodology, procedures and modes of operation of the device. It also provides protection and emergency shutdown. Figure 6 shows the output signals controlled by the graphical user interface:

- Voltage regulation from 0 to 100 V,
- Frequency adjustment from 1 to 40 Hz,
- Duty cycle adjustment from 100 to 200 us (20 us resolution),
- Setting the time for stimulation (pulse duration).

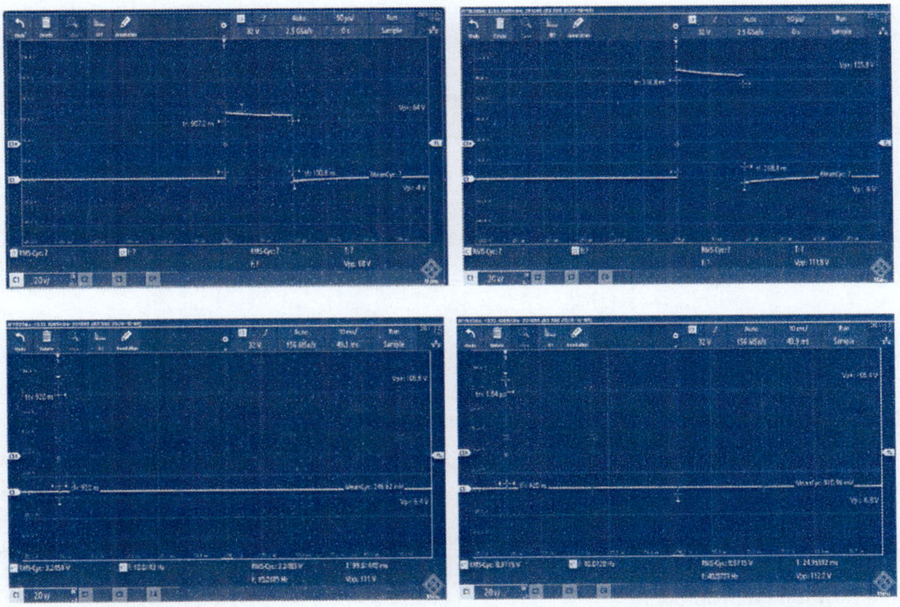

Fig. 6. Experimental testing. Output signals controlled by the computer's graphical interface

In the experiment, the set parameters were achieved: frequency in 1–40 Hz and pulse duration of 100–200 μs, with minimal delay and instruction execution time of 5–20 μs.

The instructions are sent to the device when START is pressed in the graphical interface. The device starts the assigned operations immediately. The STOP button sets an instruction that stops the signal of all electrodes for up to 5 ms.

The experiment and data collection were performed on 5 subjects and the obtained results were analyzed. During the experiments, the exoskeleton are used to support the movements of the lower limbs. The movements of the participants are monitored in real time through a system of sensors for position, force and pressure and electomyographic sensors. The first experiment was performed without external assistive force applied by the exoskeleton. Figure 7 on the left shows a rotation in the hip joint while the person raises his left leg and holds it at about 90°. We can see that the patient lifts his leg with great effort, and the first rotation is relatively slow and uncontrolled. In the next two rotations, he increased his speed to reduce the required muscular endurance. The second experiment was performed with external assistive force of 15 Nm applied by the exoskeleton or 4.5 bar of the pneumatic actuator (Fig. 7 on the right). The patient showed much more control of his movements, the rotations were clean, and he reported that it was much easier. Accordingly, Fig. 7b shows that the measured muscle strength from the electromyographic sensors was greater during the first experiment.

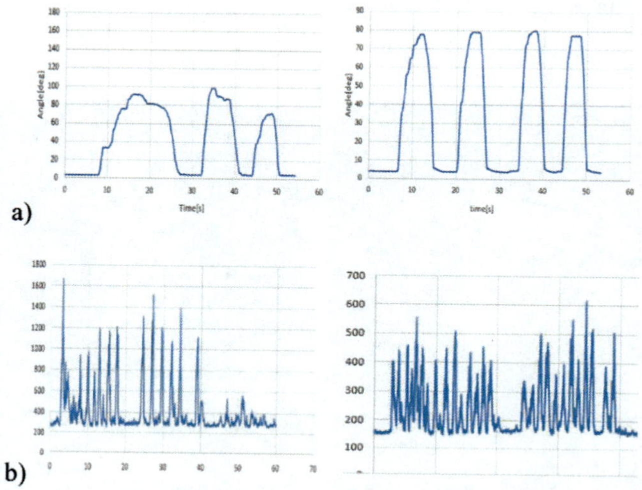

Fig. 7. Hip rotation with (right) and without (left) active assistance from the exoskeleton; b) measured muscle strength by electromyographic sensors

In conclusion, we can say that the subject of future activities will be conducting long-term experiments to test the functionality of the device with patients (volunteers) in a hospital environment.

5 Conclusion

The aim of the paper is to present an approach to the management of a functional electrical stimulator, which allows the stimulation of selected acupuncture points on the human foot, using pulse width modulation during the stimulation cycle. A rehabilitation device for electroacupuncture with a controller for elctostimulation and electrode matrix which signals are received from two sensor grids built into the insoles of both legs of the user has been developed. A graphical user interface for adjusting the electrical stimulation of acupuncture points has been created.

The version of the device described here is the first prototype, which will be used to study healthy subjects in our laboratory and people with SCI in a specialized clinic that cooperates in the project. The results of clinical trials of patients will be presented in the next article. The results will be analyzed in order to optimize the device and thus ensure further development.

Acknowledgements. This research was supported by the National Innovation Fund (project 11IF-02–14/03.12.2020, Development of an innovative solution for rehabilitation after a cerebrovascular accident or spinal cord injury), the Operational Program "Science and Education for Smart Growth" through the project "MIRACle"- Grant No BG05M2OP001–1.002–0011-C02 and co-financed by the European Union through the European structural and Investment funds.

References

1. ASIA neurological standards committee (eds.): Int. standards for neurological classification of SCI (2000)
2. Electrical stimulation for the treatment of pain and muscle rehabilitation, united healthcare commercial medical policy No: 2021T0126DD, Effective, 1 January 2021
3. Hayes: Inc. medical technology directory. Functional Electrical Stimulation (FES) Rehabilitation following Spinal Cord Injury. Lansdale, PA: Hayes, Inc.; November 2017, November 2018
4. Wolf, T.J., Nilsen, D.M.: Occupational therapy practice guidelines for adults with stroke. Bethesda (MD): American Occupational Therapy Association (AOTA) (2015)
5. Schmitt, C., Métrailler, P., Al-Khodairy, A., Brodard, R., Fournier, J.: The motion maker: a rehabilitation system combining an orthosis with closed-loop electrical muscle stimulation. In: 8th International Workshop on Functional Electrical Stimulation, Vienna, Austria (2004)
6. Zejdlik, C.P.: Management of Spinal Cord Injury, 2nd (edn.) Jones & Bartlett, Boston (1992)
7. Popovic, D., Sinkjear, T.: Control of Movement for the Physically Disabled, pp. 277–281. Springer, London (2000)
8. Knutson, J.S., Gunzler, D.D., Wilson, R.D., et al.: Contralaterally controlled functional electrical stimulation improves hand dexterity in chronic hemiparesis: a randomized trial. Stroke **47**(10), 2596–2602 (2016)
9. Lin, Z., Yan, T.: Long-term effectiveness of neuromuscular electrical stimulation for promoting motor recovery of the upper extremity after stroke. J Rehabil Med. **43**(6), 506–510 (2011)
10. Hsu, S.S., Hu, M.H., Wang, Y.H., et al.: Dose-response relation between neuromuscular electrical stimulation and upper-extremity function in patients with stroke. Stroke **41**(4), 821–824 (2010)
11. Shen, Y., Yin, Z., Fan, Y., et al.: Comparison of the effects of contralaterally controlled functional electrical stimulation and neuromuscular electrical stimulation on upper extremity functions in patients with stroke. CNS Neurol Disord Drug Targets **14**(10), 1260–1266 (2015)
12. Meng, J.B., Jiao, Y.N., Zhang, G., et al.: Electroacupuncture improves intestinal dysfunction in septic patients: a randomised controlled trial. Biomed. Res. Int. **26**(2018), 8293594 (2018)
13. Mi, Z., Gao, J., Chen, X., et al.: Effects of transcutaneous electrical acupoint stimulation on quality of recovery during early period after laparoscopic cholecystectomy. Zhongguo Zhen Jiu **38**(3), 256–260 (2018)
14. EMS Foot Massage Foldea. https://shopping.dmkt-sp.jp/products/06793173?sku=06793173
15. Kolb, B., Whishaw, I.Q.: Fundamentals of Human Neuropsychology (2003)
16. Sensory nervous system, Wikipedia, the free encyclopedia (2022). https://en.wikipedia.org/wiki/Sensory_nervous_system
17. Veneva, I., Chakarov, D. Tsveov, M., Zlatanov, E., Venev, P.: Active orthotic system for assistance and rehabilitation. In: 20th International Conference CLAWAR 2017, Porto, Spain, pp. 597–604 (2017)
18. Veneva, I., Chakarov, D. Tsveov, M., Trifonov, D., Zlatanov, E., Venev, P.: Active assistive orthotic system: (exoskeleton) enhancing movement. In: Handbook of Research on Biomimetics and Biomedical Robotics. IGI Global eEditorial Discovery, pp. 48–75 (2017)
19. Chakarov, D., Veneva, I., Tsveov, M., Venev, P.: Powered upper limb orthosis actuation system based on pneumatic artificial muscles. J. Theor. App. Mech. **48**(1), 23–36 (2018)

Check for updates

Wearable Lower Limb Neuroprosthesis: System Architecture and Control Tuning

Simão P. Carvalho[1,2], Joana Figueiredo[1,2](✉), and Cristina P. Santos[1,2]

[1] Center for MicroElectroMechanical Systems (CMEMS), University of Minho, Braga, Portugal
joana.figueiredo@dei.uminho.pt
[2] LABBELS – Associate Laboratory, Braga/Guimarães, Portugal

Abstract. The use of functional electrical stimulation (FES) through neuroprosthesis is becoming a promising solution in lower limb neurorehabilitation. However, the wearability constraints and time-consuming tuning of stimulation parameters still limit the daily use of neuroprostheses. This work proposes two major contributions, namely: (i) a conceptual design and technical architecture of a fully wearable lower limb neuroprosthesis; and (ii) a Matlab-OpenSim framework that enables fast subject-and muscle-specific tuning of FES controllers based on OpenSim musculoskeletal models. The validation procedures for this study were divided into three phases: (i) Verification of the system architecture real-time requirements; (ii) evaluation of the reliability of the MATLAB-OpenSim framework for tuning PID controller; and (iii) its subsequent use in the neuroprosthesis control with a healthy subject. The obtained results demonstrated that the neuroprosthesis system was able to meet the real-time requirements, with control and data acquisition call periods below 10 ms. Further findings indicated reliable and stable behavior of the simulation-tuned PID controller with an overshoot of 9.82% and a rise time of 0.063 s. The trajectory tracking control results with the neuroprosthesis corroborated the robustness of the tuned PID controller in tracking the desired ankle trajectory (RMSE = 17.23 ± 2.97° and time delay = 0.21 ± 0.070 s).

Keywords: Closed-loop control · Functional electrical stimulation · Wearable rehabilitation robot

1 Introduction

Around 250 k to 500 k people suffer a spinal cord injury (SCI) every year [1]. Sequalae of SCI may include loss of sensory and/or motor control of lower limbs, spasticity, muscle weakness, and pain [2]. Robot-based therapies are a promising complementary

This work has been supported by the Fundação para a Ciência e Tecnologia (FCT) through the Reference Scholarship under Grant SFRH/BD/147878/2019, the Stimulus of Scientific Employment under Grant 2020.03393.CEECIND, and in part by the FEDER Funds through the Programa Operacional Regional do Norte and national funds from FCT with the SmartOs project under Grant NORTE-01-0145-FEDER-030386. It is also supported under the national support to R & D units grant, through the reference project UIDB/04436/2020 and UIDP/04436/2020.

© The Author(s), under exclusive license to Springer Nature Switzerland AG 2023
J. M. Cascalho et al. (Eds.): CLAWAR 2022, LNNS 530, pp. 572–583, 2023.
https://doi.org/10.1007/978-3-031-15226-9_52

possibility to enhance rehabilitation outcomes [3]. Robotic assistive devices such as functional electrical stimulation (FES) stand out as the only intervention that can re-establish muscle function after paralysis and its ability to induce physiological benefits (e.g., muscle strength, counteract muscle atrophy, improved cardiovascular health) [4, 5]. FES acts towards the re-establishment of sensorimotor control by creating alternative neural pathways through the use of devices named neuroprosthesis [6].

During the last decades, several neuroprostheses have been developed targeting different body locations and joint motion (e.g., ankle [7], knee [8], shoulder and elbow [9], knee and ankle [10]). Despite advancing control algorithms, the proposed solutions [7–10] are still not compliant for long-term use in daily assistance, given their lack of wearability. The level of motor function recovery of SCI patients obtained from robotics-based rehabilitation is directly dependent on the execution of activities of daily living (ADLs) [4]. In this sense, the development of wearable rehabilitation robots is an untended challenge that is fundamental to reaching task-oriented locomotion assistance adapted to the patient's ADLs and motor progress [11].

Recent studies have proposed promising trajectory tracking control strategies for neuroprosthesis, commonly implemented through PID controllers, to elicit different types of movement [8, 9, 12]. Despite the major advancements suggested in these studies [8, 9, 12], one important challenge remains in the way the controller is tuned to generate the desired stimulation. It requires a subject- and muscle-specific tuning to achieve muscle-specific stimulation profiles, which should be regularly updated given time-variable muscle response. Thus, the process of tuning an FES controller may take a considerable number of experiments, being time-consuming. Further, the tuning is limited by the subject's ability to stand high levels of stimulation for long-lasting experiments [13], being a factor that may cause muscle fatigue.

This work aims to contribute to these two challenges. First, it presents the conceptual design and technological description of a modular and real-time architecture for a fully wearable lower limb neuroprosthesis. The system is formed by a stimulation unit that can electrically stimulate up to 8 muscles, an inertial based-sensor network able to track full lower limb kinematics [14], and a wearable central processing unit (CPU). The technological innovation, when compared to [8, 9, 12, 15], lies in using a wearable CPU and a modular architecture that enables both stand-alone use and direct integration into third-party systems. Overall, the technological description may guide future developments in wearable neuroprosthesis. Additionally, this study provides a Matlab-OpenSim framework that enables fast subject- and muscle-specific tuning of FES controllers, considering the open-source OpenSim musculoskeletal models. These models can be customized to the subject's anthropometric data and include all relevant muscles for lower limb motion. To demonstrate the reliability of the Matlab-OpenSim framework, this study describes the implementation and the tuning of the neuroprosthesis-oriented PID controller for the ankle joint. Experimental procedures, considering the controller parameters obtained from the framework, were conducted with a healthy subject during ankle joint motion evoked by a two-channel neuroprosthesis. The findings demonstrated the potential of the Matlab-OpenSim framework for FES controller tuning.

2 Methods

2.1 Neuroprosthesis Architecture

The neuroprosthesis architecture is composed of a Raspberry Pi 4B (CPU), motionstim8 (stimulation unit), InertialLab [14] (inertial sensor network), and the power supply, as demonstrated in Fig. 1a). Figure 1b) depicts the fully wearable neuroprosthesis placed on a subject. The power supply, CPU, stimulation unit, and the InertialLab master board are fixed inside the backpack, while the Inertial Measurement Units (IMUs) and electrodes are placed on the lower limbs. The connectivity between the backpack and the electrodes and IMUs is made through cables guided alongside the user's legs. Additionally, an Android App, running on a smartphone, was developed to configure, start, and stop the system. A description of each module is presented below.

2.1.1 Central Processing Unit

The CPU is a Raspberry Pi 4 Model B (Raspberry Pi Foundation, UK), a single-board computer with a Quad-core Cortex-A72 (ARM v8) 64-bit SoC @ 1.5 GHz and 4 GB of RAM. It has 56.5 × 85.6 × 11 mm and weighs 46 g. The InertialLAB and the stimulation unit communicate with the CPU at 100 Hz through Universal Asynchronous Receiver/Transmitter (UART) using a USB converter (FT232RL FTDI). The CPU uses the Ubuntu Mate OS to run the control strategy at 100 Hz, using the C++ programming language.

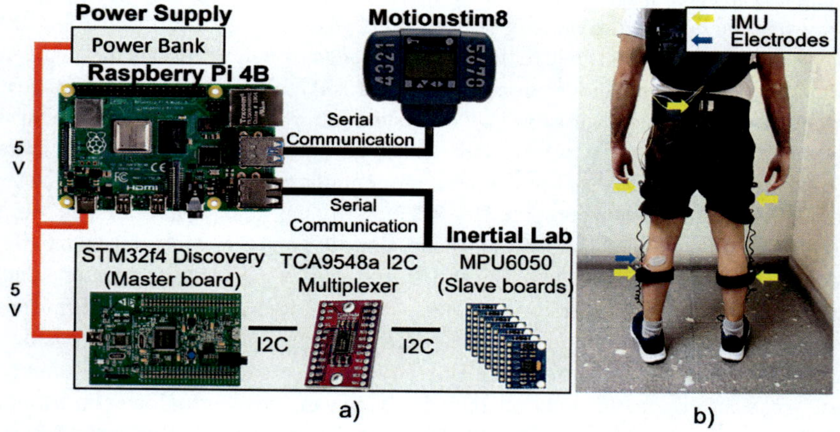

Fig. 1. a) Schematic overview of the neuroprosthesis system. b) Subject wearing the system with 7 IMUs and 2 stimulation channels.

The software architecture of the CPU (depicted in Fig. 2) was organized into four main software modules (classes), namely: (i) *CentralController* (in charge of setup, start, and stop all configurable modules), (ii) *ExternalDevice* (manages the communication between the CPU and external devices), (iii) *Monitor* (handles the data log), and (iv) *NPController* (manages the assistive control strategy). Each class has a thread based on POSIX Pthread Library. The *CentralController* is the main class and, therefore, inherits all the other classes. Through an Android APP, which communicates with the *CentralController* via Bluetooth, the user can easily configure the system (i.e., number and location of sensors, number and location of channels, gait speed, type of control, control settings, user's anthropometric data, among others), and start and stop the therapy. Given the system modularity, the user can select any sensor and channel as well as limit the therapy to data monitoring (i.e., without stimulation). Further, the APP can also provide feedback to the user in case of unexpected system errors. The *ExternalDevice* class receives the user commands and configures the external devices accordantly (i.e., the inertial sensor network and stimulation unit). During the therapy, the *ExternalDevice* handles the communication with these two devices through tasks of Pthread set at 100 Hz. The *Monitor* class guarantees that all sensor and control-specific variables are saved, every 10 ms, in the CPU memory in text files for posterior data analysis.

The *NPController* class has three hierarchically organized subclasses, namely: HLController, MLController, and LLController. This control architecture is inspired by the principles and organization of the human motion-control system [16]. The high-level, perception layer was designed to decode the user's locomotion intentions, locomotion status, and disability level. The mid-level, the translation layer, generates and adapts position reference trajectories to the user's needs and intentions. Lastly, the low-level layer runs PID-based position tracking controllers to generate assistive commands to the stimulation unit. This architecture presents a modular design to be expandable for including further assistive control strategies. A timer sets up the *NPController* tasks execution at 100 Hz.

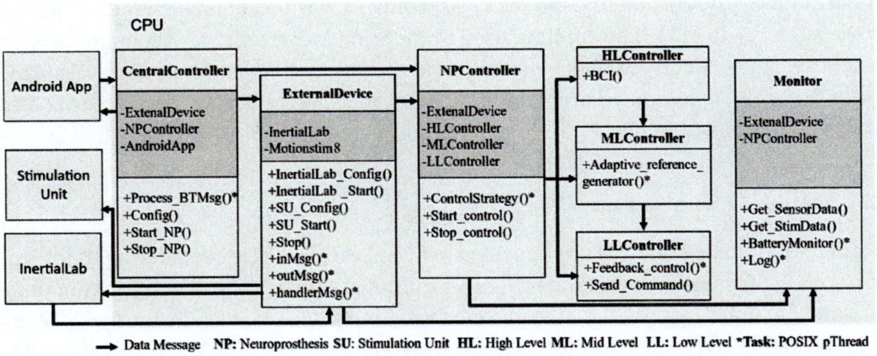

Fig. 2. Neuroprosthesis software architecture diagram - classes and their interconnectivity.

2.1.2 Power Supply

The power supply system includes two standard power banks (10000 mAh, 5 V) to power the InertialLab and CPU and allows at least 4 h of autonomy. Together, the body mass (360 g) and dimensions (92 × 60 × 44 mm) of the power banks match the requirements for system wearability. The stimulation unit has its own-integrated battery.

2.1.3 Inertial Sensor Network

The inertial sensor network is the InertialLab [14], which allows to set up from 1 up to 7 IMUs for real-time acquisition at 100 Hz. It includes the STM32F4-Discovery development board (STMicroelectronics, Switzerland) that communicates with the CPU through a UART interface. Each IMU consists of the MPU-6050 (InvenSense, Boston, MA, USA) that combines a 3-axis MEMS accelerometer (±8 g) and a 3-axis MEMS gyroscope (±2000/s) for the kinematic data acquisition. The MPU-6050 sensor communicates with the STM32F4-Discovery via I^2C through the TCA9548A multiplexer to manage the multi-channel data collection [14].

The InertialLab provides the raw data for each IMU (i.e., XYZ acceleration and XYZ angular velocity), up to seven segment angles (i.e., trunk, right/left thigh, right/left shank and right/left foot), and up to six joint angles (i.e., right/left hip, right/left knee, and right/left ankle). The segment and joint angle estimation are described in [14].

2.1.4 Stimulation Unit

The Motionstim8 (MEDEL GmbH, Medicine Electronics, Hamburg, Germany) is an electrical stimulation device that allows the individual control of up to 8 stimulation channels simultaneously. This device may be manually controlled or may receive commands from a third-party device (i.e., science mode). For the present work, the Motionstim8 was used in the science mode. It communicates with the CPU using a serial communication protocol (baud rate = 115200). The electrical stimulation pulses are rectangular and biphasic and can be controlled regarding their intensity (1−125 mA), pulse width (10−500 μsec), and frequency (1−99 Hz). The stimulation intensity and frequency were found empirically, considering the pain threshold of the subject and the range of motion of the ankle, as in [17]. The configuration of the stimulation unit was set up to 30 Hz of pulse frequency, 30 mA of pulse intensity, and the pulse width as the control command. The pulse width is computed in the low-level control using the trajectory tracking control strategy, as depicted in Fig. 4.

2.2 Matlab-OpenSim Framework

A simulation framework was implemented in MATLAB R2021a, considering its integration with the OpenSim environment. The OpenSim platform allows the execution of the forward dynamics by providing the muscle excitations that generate the musculoskeletal model's motion [18]. The developed framework is represented in Fig. 3. It is formed by the *Main* script, a recursive *Simulation* function, and the musculoskeletal model.

The *Main* script starts by including the OpenSim libraries (*org.opensim.modeling*). Then, it loads the *Gait10dof18musc* OpenSim model and allows the user to configure the stimulation frequency, activate/deactivate the PID tuning, change the reference trajectory signal, and set up PID gains. To conduct the simulations, the *Gait10dof18musc* OpenSim model was pinned to prevent it from being pulled by gravity, as if it was hanging by a harness. In addition, all joints were locked except for the ankles to prevent residual movements. The model was selected considering its computation efficiency, given its simplified muscle anatomy, which includes the main nine muscle groups in each leg, namely: *Hamstrings, Biceps Femoris Short Head, Gluteus, Iliopsoas, Rectus*

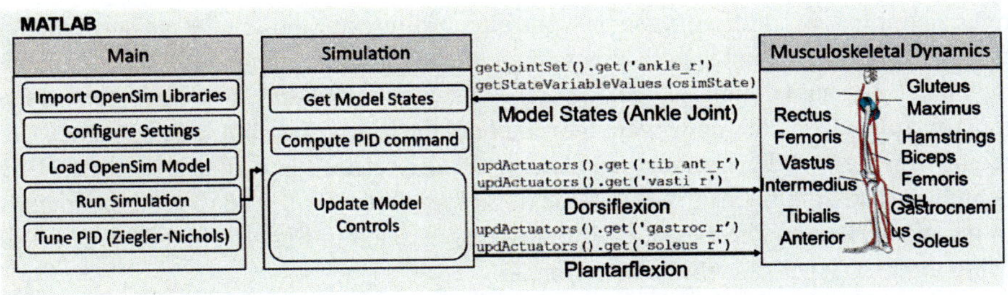

Fig. 3. Schematic overview of the Matlab-OpenSim framework.

Femoris, *Vastus Lateralis*, *Gastrocnemius*, *Soleus* and *Tibialis Anterior*. At this stage, the forward dynamics simulation is configured. The simulation function is defined with the *controlFunctionHandle* and the integration function is the *ode15s*. At the end of the simulation, the PID tunning function executes (when activated), automatically providing the PID gains.

The *Simulation* function executes recursively at 100 Hz, and each iteration considers the model states (i.e., joint angles) and the reference trajectory to compute the position error. The position error is used to compute the PID command, which is saturated between -1 and 1. The function *Update Model Control* sends the excitation signals (i.e., PID commands) to the dorsiflexion and plantarflexion muscles of the *Gait10dof18musc* OpenSim model.

The tunning process was conducted for the ankle joint, targeting the *Vastus Lateralis*, *Tibialis Anterior* for ankle dorsiflexion, and *Gastrocnemius*, *Soleus* for ankle plantarflexion. For this purpose, a step trajectory of 50° was used. The integral and derivative gains were set to zero and the ultimate (proportional) gain was obtained after achieving a stable oscillation of the ankle around the reference position (Kp = 5, Ki = 0, Kd = 0). Subsequently, the PID gains were computed using the Ziegler Nichols method.

2.3 Neuroprosthesis Control

The neuroprosthesis control architecture is hierarchically structured into three layers, namely: (i) High-level; (ii) Mid-level; and (iii) Low-level, as mentioned in Sect. 2.1.1. The current neuroprosthesis development stage only includes the mid and low-level layers; however, the architecture is prepared for the addition of the high-level in further developments. The high-level layer will be composed of a brain-computer interface capable of decoding the user's lower limbs motion intention, which will subsequently and automatically trigger the electrical stimulation. The mid-level includes the regression model proposed in [19] that generates user-oriented position trajectories for the hip, knee, and ankle joints according to the user's height and gait speed (both indicated through the Android App). This regression model scales the amplitude of the trajectory attending to the user's height, and the amplitude and timing of the trajectory based on the gait speed [19]. The low-lever controller covers a trajectory tracking control strategy relying on a feedback controller (PID). The PID includes an anti-windup strategy to reset the integral component when a zero-crossing occurs between the measured angle and the reference angle. In addition, when the controller hits its saturation limits, an anti-windup back-calculation method is used to discharge the integrator component. In this context, the plant is the human muscle, and to control the dorsiflexion (upward movement of the foot) and plantarflexion (downward movement of the foot), two stimulation channels are required in the *Tibialis Anterior,* and *Gastrocnemius,* respectively. In this sense, the PID is built to send commands to the dorsiflexion channel if $e_k > 0$, and to send commands to the plantarflexion channel if $e_k < 0$, as demonstrated in Fig. 4. The PID was tuned using the MATLAB-OpenSim framework.

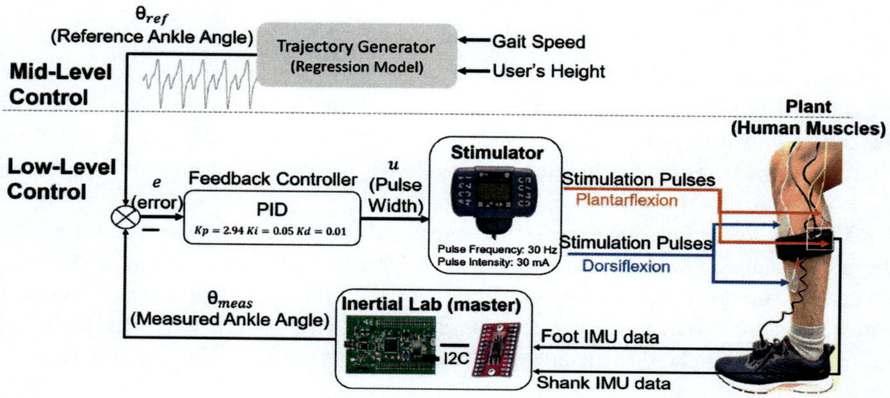

Fig. 4. Neuroprosthesis' control architecture, depicting the low- and mid-level layers.

2.4 Validation

This study has issued three validation phases. First, we conducted bench tests to verify whether the implemented architecture is able to fulfill the real-time requirements for inertial data acquisition and control every 10 ms. The validation was performed for 10 min when acquiring data from the 7 IMUs and simulating the 8 channels. During the validation, the timing of the call periods of each class was collected with a digital oscilloscope for a posterior analysis (mean and standard deviation).

The second and third phases aimed at evaluating the reliability of MATLAB-OpenSim framework for tuning the PID controller, and its subsequent use in the neuroprosthesis control in a real environment, respectively. In the second phase, the response of the PID controller obtained in the simulation was tested with a step reference of −50° and a sinusoidal reference trajectory (frequency of 1Hz, a peak amplitude of 20°, and an offset of −20°). The step response was analyzed with overshoot percentage, rise time, and settling time, considering the maximum requirements (i.e., overshoot <10%, rise time <0.2 s, settling time <2 s). On the other hand, the response of the controller to a variable trajectory (sinusoidal) was analyzed with the RMSE, normalized RMSE, and time delay, to determine its reliability to be used in the neuroprosthesis.

In the third phase, we tested if the PID gains found in the MATLAB-OpenSim framework enable a reliable stimulation profile for controlling the human ankle motion.

This validation analysis was based on the RMSE, time delay, and their normalized values. This phase involved a healthy male subject 27 years old, 70 kg, and 182 cm in height. The subject gave his informed consent according to the ethical conduct defined by the University of Minho Ethics Committee (CEICVS 006/2020) following the standards set by the declaration of Helsinki and the Oviedo Convention.

The participant was instrumented as follows. For the dorsiflexion channel, one oval electrode (4 × 6 cm) was positioned just below the fibular head, above the *tibialis anterior* muscle belly. The second oval electrode (4 × 6 cm) was placed about 2/3 of the way down the shin with the leads facing toward the midline of the body. Regarding the plantarflexion channel, one oval electrode (4 × 6 cm) was placed just below the knee, and the second electrode (4 × 6 cm) two fingers below the first electrode, in the belly of the *soleus* muscle (see Fig. 4). Further, he wore an IMU on the center of the left foot and another IMU on the lateral side of the left shank (see Fig. 4) by using straps.

Before initializing the experiment, a procedure was conducted to identify the maximum painless stimulation pulse width values. The limits for the dorsiflexion and plantarflexion channels were set to 250 μs and 100 μs, respectively. Afterwards, the PID testing experiment with the neuroprosthesis was conducted with the subject in the upright position while having his weight partially supported by a harness system. The experiment included 8 trials, in which the neuroprosthesis control was actively stimulating the subject for 5 min. Between each trial, the subject rested for 2 min.

3 Results and Discussion

3.1 Phase I – System Architecture

Table 1 presents the results of the computational performance of the neuroprosthesis architecture, for the inertial data acquisition and control. This evaluation considered the timers and tasks evocated repeatedly in real-time to investigate the effects of a non-centralized architecture during real-time therapy.

Table 1. Computational performance evaluation

Task	Requirement (ms)	Measures	
		Mean ± STD (ms)	RMSE (ms)
Control	10	9.999 ± 0.002	0.003
Data acquisition	10	9.998 ± 0.005	0.007

Table 1 reports RMSE values lower than 0.003 ms and 0.007 ms regarding the control and data acquisition tasks, respectively. These results indicate that the time requirements (10 ms) for the system architecture execution were met. Further, there is variability in the call period times along the therapy execution, as shown by STD values presented in Table 1. This variability was more evident during the system initialization and is higher for the data acquisition task, given the higher processing complexity for the InertialLab.

3.2 Phase II - Matlab-OpenSim Framework

The tunning process using the MATLAB-Opensim framework revealed that the ankle joint of the musculoskeletal model started to present a constant oscillatory response when the proportional gain (K_p) reached the value 5 (ultimate gain). Figure 5(a) demonstrates the oscillatory angle response to a step reference of −50°.

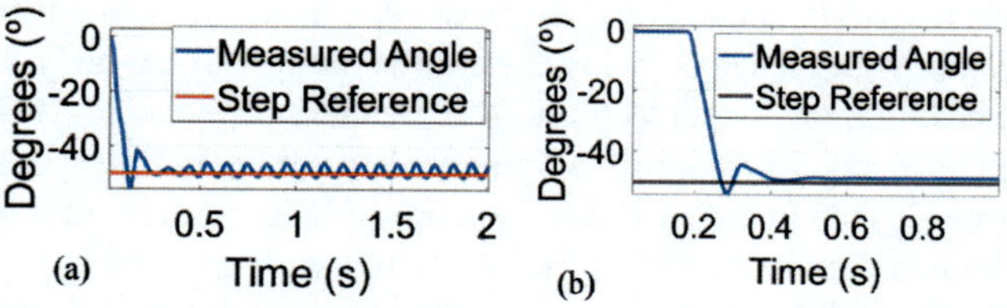

Fig. 5. a) Simulated oscillatory response of the ankle under a proportional controller ($K_p = 5$). b) Simulated step response of the ankle using the tunned PID ($Kp = 2.94$, $Ki = 0.05$, $Kd = 0.01$)

From the Ziegler Nichols classic tunning rules, the final PID gains were determined and corresponded to $Kp = 2.94$, $Ki = 0.05$, $Kd = 0.01$. The PID controller was subsequently tested with these gains, considering a step response of −50°. The resulting ankle joint response of the musculoskeletal model is demonstrated in Fig. 5(b).

The step response analysis reported an overshoot percentage of 9.82%, a rise time of 0.063 s, and a settling time of 1.7042 s, indicating that the system was able to meet the specified requirements (overshoot <10%, rise time <0.2 s, and settling time <2 s). Further, the reduced oscillatory response indicates the high stability of the system. The obtained step response suggests that the controller is suited to be used in the real setting, considering the improved results in overshoot (<8%) and rise time (<1 s), regarding a similar PID tuning approach for a different human joint [13].

Further validation procedures were conducted in simulation for the PID controller, considering a sinusoidal reference trajectory.

Figure 6 presents the results obtained including the trajectory tracking response, the tracking error, and the PID command sent to the OpenSim musculoskeletal model.Results reported a RMSE of 0.029°, a Normalized RMSE of 3.905%, and a time delay of 0.01 s. These results suggest a robust tracking performance using a variable reference trajectory, demonstrating reduced muscle excitation values associated with reduced RMSE and time delay, which stands as an ideal scenario.

3.3 Phase III - Neuroprosthesis Control

The neuroprosthesis trajectory tracking control was subsequently validated using the PID gains determined with the MATLAB-Opensim framework. Figure 7a) represents the reference trajectory (blue line) and the measured ankle joint angle (red line) considering the execution of three gait cycles. For this experiment, a healthy ankle angle trajectory was used as the reference. Figure 7b) shows the generated pulse width commands for the dorsiflexion channel (blue line) and the plantarflexion channel (red line).

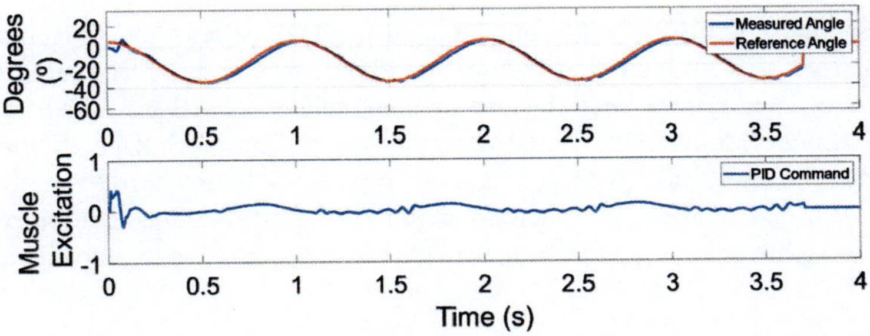

Fig. 6. Simulated ankle behaviour for the sinusoidal reference using the tuned PID.

From the neuroprosthesis control validation, performance results were obtained, namely, a RMSE of 17.23 ± 2.97°, a normalized RMSE of 3.97 ± 0.69%, a time delay of 0.21 ± 0.070 s, and a normalized time delay of 0.0044 ± 0.0014 s. The pulse width values tended to increase with time as a compensation for the increased muscle fatigue. Similar behaviors were reported in previous studies using FES [12]. These preliminary results suggest a promising reference tracking, presenting normalized RMSE values bellow literature (3.97% < 16.7%) reported values under a similar control approach for the knee joint [13]. Additionally, the neuroprosthesis controller was able to maintain the tracking delay below literature values (250 ms) [12], which stands as a challenge in FES controllers due to the existing natural electromechanical muscle delay (10–100 ms) [20].

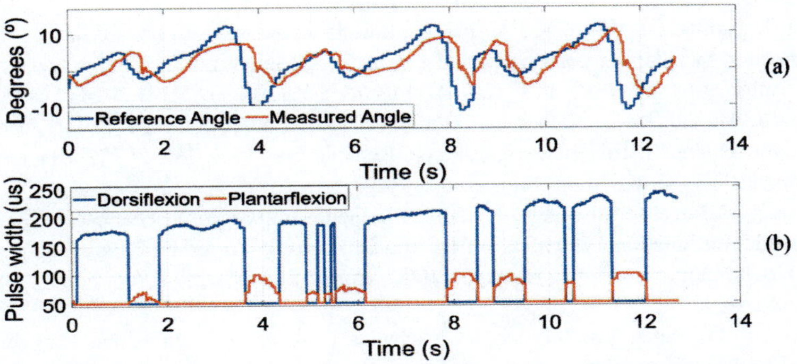

Fig. 7. Example of three gait cycles of Neuroprosthesis tracking performance with the healthy subject using the PID tuned by the MATLAB-OpenSim framework.

4 Conclusions

In this paper, we presented the modular architecture of a neuroprosthesis that tackles the lack of wearable neuroprosthetic solutions. The tests indicated that the neuroprosthesis architecture successfully meets the real-time requirements set for gait analysis and rehabilitation. Further, we proposed a MATLAB-OpenSim-based framework to enable a fast subject-and muscle-specific tuning of FES controllers. To demonstrate the reliability of the framework, this study evaluated the tuned PID controller in a real neuroprosthesis when controlling the ankle joint. Results reported reduced RMSE and time delay of the tuned PID controller in the MATLAB-OpenSim-based framework. Overall, this shows the potential of the MATLAB-OpenSim-based framework to tune further controllers of wearable neuroprostheses. Future developments include scaling the neuroprosthesis to the remaining lower limb joint as well as extending the validation to more subjects.

References

1. W. H. Organization, "WHO | International perspectives on spinal cord injury," WHO (2013). https://www.who.int/disabilities/policies/spinal_cord_injury/en/. Accessed 22 Apr 2020
2. Marquez-Chin, C., Bolivar-Tellería, I., Popovic, M.R.: Brain–computer interfaces for neurorehabilitation: enhancing functional electrical stimulation, 2nd edn. Elsevier BV. Amsterdam
3. Moreno, J.C., Figueiredo, J., Pons, J. L.: Exoskeletons for lower-limb rehabilitation. In: Rehabilitation Robotics: Technology and Application, London, United Kingdom. Elsevier, pp. 89–100 (2018)
4. Moreno, J.C., Mohammed, S., Sharma, N., Del-Ama, A.J.: Hybrid wearable robotic exoskeletons for human walking. INC (2019)
5. Molazadeh, V., Zhang, Q., Bao, X., Dicianno, B.E., Sharma, N.: Shared control of a powered exoskeleton and functional electrical stimulation using iterative learning. Front. Robot. AI **8**(11), 1–13 (2021). https://doi.org/10.3389/frobt.2021.711388
6. Shah, M., Peterson, C., Yilmaz, E., Halalmeh, D.R., Moisi, M.: Current advancements in the management of spinal cord injury: a comprehensive review of literature. Surg. Neurol. Int. **10**(174), 1–4 (2019). https://doi.org/10.25259/SNI
7. Seel, T., Laidig, D., Valtin, M., Werner, C., Raisch, J., Schauer, T.: Feedback control of foot eversion in the adaptive peroneal stimulator. In: 22nd Mediterranean Conference on Control and Automation, pp. 1482–1487 (2014). https://doi.org/10.1109/MED.2014.6961585
8. Ferrarin, M., Palazzo, F., Riener, R., Quintern, J.: Model-based control of FES-induced single joint movements. IEEE Trans. Neural Syst. Rehabil. Eng. **9**(3), 245–257 (2001). https://doi.org/10.1109/7333.948452
9. Resquín, F., Pons, J.L., Brunetti, F., Ibáñez, J., Gonzalez-Vargas, J.: Feedback error learning controller for functional electrical stimulation assistance in a hybrid robotic system for reaching rehabilitation. Eur. J. Transl. Myol. **26**(3), 255–261 (2016). https://doi.org/10.4081/ejtm.2016.6164
10. Müller, P., Del Ama, A.J., Moreno, J.C., Schauer, T.: Adaptive multichannel FES neuroprosthesis with learning control and automatic gait assessment. J. Neuroeng. Rehabil. **17**(1), 1–20 (2020). https://doi.org/10.1186/s12984-020-0640-7
11. Kim, H., Park, G., Shin, J.H., You, J.H.: Neuroplastic effects of end-effector robotic gait training for hemiparetic stroke: a randomised controlled trial. Sci. Rep. **10**(1), 1–9 (2020). https://doi.org/10.1038/s41598-020-69367-3
12. Pedrocchi, A., Ferrante, S., De Momi, E., Ferrigno, G.: Error mapping controller: a closed loop neuroprosthesis controlled by artificial neural networks. J. Neuroeng. Rehabil. (2006). https://doi.org/10.1186/1743-0003-3-25
13. Cardoso, A.C., et al.: A comparative study on control strategies for FES cycling using a detailed musculoskeletal model. IFAC-PapersOnLine **49**(32), 204–209 (2016). https://doi.org/10.1016/j.ifacol.2016.12.215

14. Figueiredo, J., Carvalho, S.P., Vilas-Boas, J.P., Gonçalves, L.M., Moreno, J.C., Santos, C.P.: Wearable inertial sensor system towards daily human kinematic gait analysis: benchmarking analysis to MVN BIOMECH. Sensors (Switzerland) **20**(8), 2185 (2020). https://doi.org/10.3390/s20082185
15. Müller, P., et al.: Adaptive multichannel FES neuroprosthesis with learning control and automatic gait assessment. J. Neuroeng. Rehabil. **17**(1), 1–20 (2020). https://doi.org/10.1186/s12984-020-0640-7
16. Tucker, M.R., et al.: Control strategies for active lower extremity prosthetics and orthotics: a review. J. Neuro Eng. Rehabil. **12**(1), 1 (2015). https://doi.org/10.1186/1743-0003-12-1
17. Bouri, M., Selfslagh, A., Campos, D., Yonamine, S., Donati, A.R.C., Shokur, S.: Closed-loop functional electrical stimulation for gait training for patients with paraplegia. In: 2018 IEEE International Conference on Robotics and Biomimetics, ROBIO, pp. 1489–1495 (2018). https://doi.org/10.1109/ROBIO.2018.8665270
18. Seth, A., et al.: OpenSim: simulating musculoskeletal dynamics and neuromuscular control to study human and animal movement. PLoS Comput. Biol. **14**(7), e1006223 (2018). https://doi.org/10.1371/JOURNAL.PCBI.1006223
19. Koopman, B., van Asseldonk, E.H.F., Van der Kooij, H.: Speed-dependent reference joint trajectory generation for robotic gait support. J. Biomech. **47**(6), 1447–1458 (2014). https://doi.org/10.1016/j.jbiomech.2014.01.037
20. Cavanagh, P.R., Komi, P.V.: Electromechanical delay in human skeletal muscle under concentric and eccentric contractions. Eur. J. Appl. Physiol. Occup. Physiol. **42**(3), 159–163 (1979). https://doi.org/10.1007/BF00431022

Comparative Analysis of Walking in the Lower Limbs' Exoskeleton with Various Strategies for Constructing the Ankle Joint' Trajectories

Sergey Jatsun, Andrei Malchikov(✉), Andrey Yatsun, and Andres Santiago Martinez Leon

Southwest State University, Kursk 305040, Russia
teormeh@inbox.ru

Abstract. The paper explores various approaches to obtaining the trajectories of the ankle joint' movements of a exoskeleton walking machine. One of the options for setting the movement of the exoskeleton foot is the use of piecewise polynomial laws that provides a shockless cyclic mode of movement of the links during walking. The second option considered in the work is the use of natural movement characteristics obtained during the analysis of walking videos with subsequent approximation of the results by an artificial neural network. As part of the work, a mathematical model is proposed that allows a comparative analysis of the characteristics of the controlled movement of the executive links with different strategies for setting the trajectory of the movement of the ankle joint of the exoskeleton. The article shows the assessment of the positioning error during the development of trajectories, as well as an assessment of the energy efficiency of the methods and an assessment of their applicability in real rehabilitation devices.

Keywords: Walking robot · Mathematical modeling · Trajectory planning

1 Introduction

According to official statistics, more than 400 thousand cases of stroke occur annually in the Russian Federation, while disability due to stroke ranks first among all causes of primary disability. Among the most likely consequences of a stroke: aphasia, movement disorder and apraxia of walking. According to the Stroke Registry of the Research Institute of Neurology of the Russian Academy of Medical Sciences, by the end of the acute period of stroke, hemiparesis was observed in 81.2% of surviving patients, including hemiplegia in 11.2%, severe and severe hemiparesis in 11.1%, mild and moderate hemiparesis in 58.9% of patients [1]. At the same time, complications are characterized by a decrease in strength, limitation of the movements volume, changes in muscle tone, spasticity [2, 3]. Among the most effective therapeutic measures: medical treatment, physiotherapy, massage and therapeutic gymnastics. In the latter case, rehabilitation mechanotherapeutic complexes built on the basis of the lower limbs' exoskeleton are often used [4–7].

© The Author(s), under exclusive license to Springer Nature Switzerland AG 2023
J. M. Cascalho et al. (Eds.): CLAWAR 2022, LNNS 530, pp. 584–593, 2023.
https://doi.org/10.1007/978-3-031-15226-9_53

When developing rehabilitation devices, including devices that provide imitation of walking with the support of the operator's torso, the most important role is played by the planning of the trajectories of movement of the executive links of the rehabilitation apparatus – the exoskeleton. As a rule, such devices implement walking on a treadmill, which differs from natural walking on a fixed base.

In [8, 9], the method of constructing a closed trajectory is considered, the feature of which is to ensure shock-free interaction with the movable surface of the treadmill. Ensuring synchronization of the exoskeleton foot movement and the treadmill canvas through the use of a closed synthetic trajectory reduces the impact of dynamic processes of contact with the surface, both on the patient and on the design of the exoskeleton electromechanical system. The method will be discussed in detail in the second section of this article. It should be noted that the form of the synthetic trajectory considered in the work differs from the natural one characteristic of a healthy person walking on a fixed base, therefore, the task of comparative analysis of the synthetic and natural trajectories is relevant.

The purpose of the article is to compare the kinematic and dynamic parameters of the movement of the executive links of the exoskeleton with various strategies for constructing the trajectories of the ankle joint, through mathematical modeling. The proposed analysis will make it possible to establish the requirements for the electromechanical system of the exoskeleton and the applicability of the proposed techniques for solving a variety of tasks.

2 Mathematical Model and Formulation of the Simulation Problem

The paper considers an experimental stand designed to study the features of walking in a robotic lower limbs' exoskeleton. The robot is a fixed body equipped with two-link legs that perform movements imitating a human gait. Schematically, the device under consideration is shown in Fig. 1.

The paper deals with the lower limbs' exoskeleton during the walk implementation along the moving belt of a treadmill (Fig. 1). The upper part of the exoskeleton 3 is attached to a fixed base – the frame. The femoral link 1 by means of active joint that contain electric drives creating moment, that can have an assisting effect on the patient, are attached to the body. To control the relative position of the links, the active hinges are equipped with angle sensors and torque sensing systems. In a similar way, the femoral links are connected to the lower leg links 2.

The task of the system that controls drives of the hip and knee joints of the device is to realize this relative motion of links, at which point O_3 move along the desired trajectory 4, implementing a walking motion of the exoskeleton.

For mathematical description of the kinematics of motion, there is a system of equations:

$$\begin{array}{ll} x_{O2} = x_O + l_1 \cos(\varphi_1) \\ y_{O2} = y_O + l_1 \sin(\varphi_1) \end{array}, \quad \begin{array}{ll} x_{O3} = x_O + l_1 \cos(\varphi_1) + l_2 \cos(\varphi_2) \\ y_{O3} = y_O + l_1 \sin(\varphi_1) + l_2 \sin(\varphi_2) \end{array}, \tag{1}$$

where l_1 and l_2 – long links, respectively: thighs and shins, x_O, y_O – coordinates of the femoral joint in the inertial reference system, φ_1, φ_2 – absolute rotation angles of the leg links.

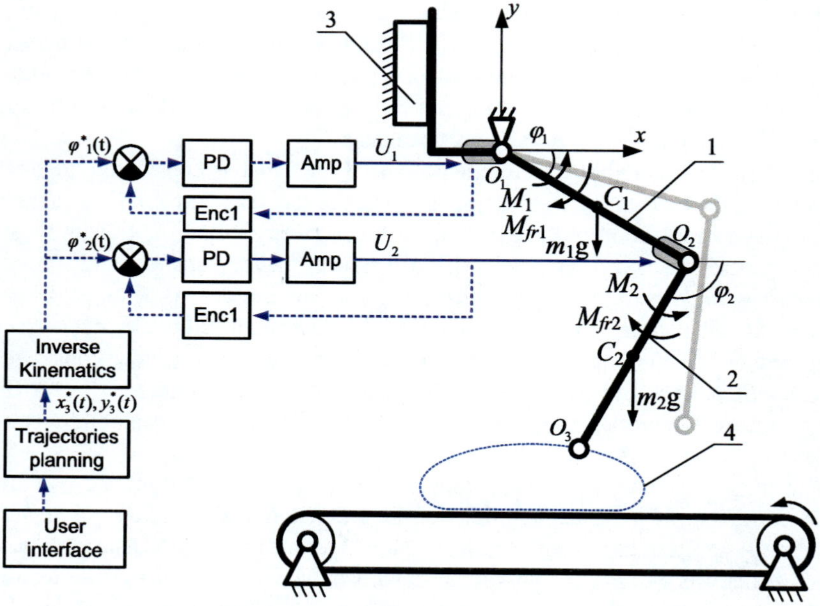

Fig. 1. Scheme of the laboratory stand and the structure of the control system

The links' rotation angles can be determined in various ways, in the framework of this work, they are defined as follows:

$$\varphi_1 = \begin{cases} arctng\left(\frac{y_{O3}}{x_{O3}}\right) + \arccos\left(\frac{l_1^2 - l_2^2 + (x_{O3}^2 + y_{O3}^2)}{2l_1\sqrt{x_{O3}^2 + y_{O3}^2}}\right) & \text{if } (x_{O3} > 0) \\ arctng\left(\frac{y_{O3}}{x_{O3}}\right) + \arccos\left(\frac{l_1^2 - l_2^2 + (x_{O3}^2 + y_{O3}^2)}{2l_1\sqrt{x_{O3}^2 + y_{O3}^2}}\right) - \pi & \text{if } (x_{O3} \le 0) \end{cases}, \quad (2)$$

$$\varphi_2 = \varphi_1 + \left(\pi + \arccos\left(\frac{l_1^2 + l_2^2 - (x_{O3}^2 + y_{O3}^2)}{2l_1 l_2}\right)\right) \quad (3)$$

Using Eqs. (2) and (3), it is possible to determine the absolute rotation angles of links when a point O_3 moves in the lower quadrants of the coordinate plane ($y_{O3} < 0$).

Thus, the leg of the exoskeleton can be represented by two-coordinate pendulums driven by the torques of electric drives equipped with a position control system.

In this case the system of differential equations describing the leg of the exoskeleton will take the form:

$$\begin{cases} \ddot{\varphi}_1 l_1^2(m_1+m_2)+\ddot{\varphi}_2 m_2 l_1 l_2\cos(\varphi_1-\varphi_2)+\dot{\varphi}_2^2 m_2 l_1 l_2\sin(\varphi_1-\varphi_2) \\ \quad = M_1 - M_{fr1} - \mu\dot{\varphi}_1 - (m_1+m_2)gl_1\sin\varphi_1, \\ \ddot{\varphi}_2 l_2^2 m_2+\ddot{\varphi}_1 m_2 l_1 l_2\cos(\varphi_1-\varphi_2)-\dot{\varphi}_1^2 m_2 l_1 l_2\sin(\varphi_1-\varphi_2) \\ \quad = M_2 - M_{fr2} - \mu\dot{\varphi}_2 - m_2 g l_2\sin\varphi_2, \\ L\dfrac{dI_1}{dt}+RI_1+k_E\varphi_1 i_G = U_1,\quad M_1 = k_m I_1, \\ L\dfrac{dI_2}{dt}+RI_2+k_E\varphi_2 i_G = U_2,\quad M_2 = k_m I_2, \end{cases} \tag{4}$$

where, m_1, m_2 – mass of the links of the exoskeleton, M_1, M_2 – torques generated by the drive, M_{fr1}, M_{fr2}– torques of friction forces, $\mu\dot{\varphi}_1$, $\mu\dot{\varphi}_2$ – torques of viscous friction forces, $k_m k_E$ – torque and speed coefficients of the motor, L and R– inductance and active resistance of the motor armature winding, i_G – total gear ratio of the drive gearbox.

U_1, U_2 – voltages generated by the drive system regulators in the following form:

$$U_i = k_p\left(\varphi_i^* - \varphi_i\right) + k_d\left(\dot{\varphi}_i^* - \dot{\varphi}_i\right) \tag{5}$$

where, k_p, k_d – coefficients of the PD controller. The coefficients of the regulator were selected as result of computational experiments.

The proposed mathematical apparatus makes it possible to obtain kinematic and dynamic parameters of the executive links' controlled movement with various strategies for constructing the movement trajectories of the exoskeleton ankle joint. The patterns obtained in the course of modeling will allow performing a comparative analysis of the movement task of synthetic and natural motion trajectories.

3 Trajectory Planning

The method of constructing the trajectory of the exoskeleton executive links on a movable base determines both the proximity of the gait to the natural for a person, and the requirements for the drive system and control system regulators. Within the framework of this work, it is proposed to compare two different methods for obtaining the driving forces for exoskeleton drives. Both approaches will be considered in detail.

The first approach considered in this paper is the creation of a trajectory that corresponds to a set of step parameters and boundary conditions determined by the treadmill movement. Such trajectory is a piecewise polynomial smooth function of the ankle joint movement, as an argument, which is the vector of gait parameters: $\bar{s} = (s, h_s, H_s, k_v, T_s)^T$ in which s – is the step length, h_s – the height of the leg when walking, H_s – the distance from the hip joint to the treadmill running, T_s – the duration of one step [8]. Thus, the type of trajectory is determined only by the boundary conditions (synchronization of the movement of the foot and the treadmill canvas) and the step parameters, while the movement between the characteristic points is determined by the nature of the initially selected polynomial function. Within the framework of the article, we will call such a trajectory – synthetic. This technique is described in detail in [8].

The advantage of the synthetic gait is the guaranteed provision of the boundary conditions (zero relative speed of the track and the foot of the exoskeleton) and the smoothness of changing the setting angles of the control system. In [10], it was proposed to use an artificial neural network that allows obtaining the required trajectory based on the motion parameters.

The second approach to the formation of driving influences for the drive system is characterized by the desire to imitate the natural trajectory of the human ankle joint in the process of walking as much as possible. Further, for convenience, such a trajectory will be called natural.

To obtain natural trajectories of human movement, various methods are used, including the method of analyzing the video-file of walking with markers attached to the body [8], which is used in this work. To obtain a set of points of the natural trajectory, a video recording of walking with various step parameters was performed, followed by analysis and modeling of video fragments using the "tracker" program [11]. The process of obtaining a natural trajectory is shown in Fig. 2.

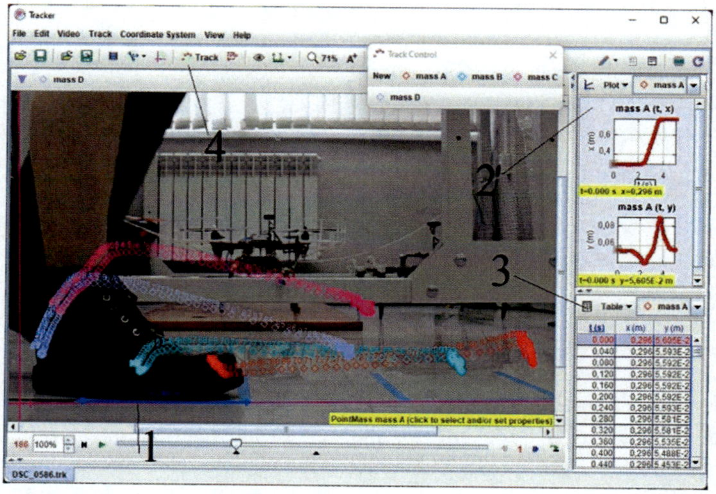

Fig. 2. Obtaining the natural trajectory of foot movement using the "tracker" program

A fully connected artificial neural network is used to approximate the results of experimental measurements. When training a neural network, the step parameters similar to those used when setting a synthetic trajectory will be used in the following ranges as initial (input) data:

- step length s,
- foot height related to the movable base h_s,
- trajectory coefficient k_υ.

It should be noted that the trajectory coefficient for the natural trajectory was determined experimentally as the ratio of the human movement speed when walking to the maximum movement speed of the ankle joint.

Thus, we can write the input vector x for the neural network as follows:

$$x = \left[t^{\mathrm{T}}\ s^{\mathrm{T}}\ h_s^{\mathrm{T}}\ k_v^{\mathrm{T}} \right]. \quad (6)$$

In the process of training the neural network, the results of about 80 experiments were used, while the results of leg movements are represented by the coordinates of the ankle joint:

$$y = \left[x_1^{*T},\ y_2^{*T} \right] \quad (7)$$

All data was randomly divided into 3 groups: Training (70%) – this data is used by the network during training, Validation (15%) – this data is be used to evaluate the results of the network and stop training, Testing (15%) - this data is used to determine the accuracy and performance of the network.

The used basic network model was a two-layer feed-forward network trained with the Levenberg-Marquardt method with a number of neurons in the hidden 25 layer.

The results of the trajectory generator operation with step parameters $\bar{\mathbf{s}} = (0.6,\ 0.155,\ 0,\ 0.43,\ 1.4)^T$ are shown in Fig. 3.

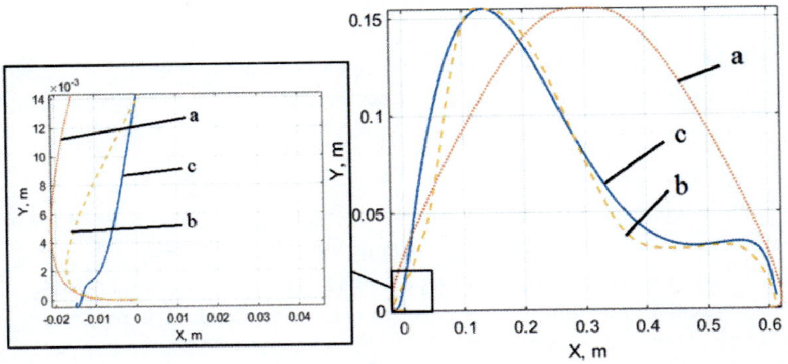

Fig. 3. Planning the trajectory of moving links of a walking machine: a – synthetic trajectory, b – natural trajectory, c – experimentally obtained trajectory

This figure clearly shows that the symmetrical synthetic trajectory (Fig. 3, a) differs significantly from the natural one, while the neural network approximation (Fig. 3, b) has minor deviations from the original experimental curve (Fig. 3, c).

It should be noted that the trajectory generator for natural and synthetic trajectories provides impact-free contact with the surface, due to the observance of the boundary conditions – synchronization of the movement of the treadmill and the exoskeleton foot.

The hodograph and the velocities values of the ankle joint center for various strategies for constructing motion trajectories will be shown in Fig. 4.

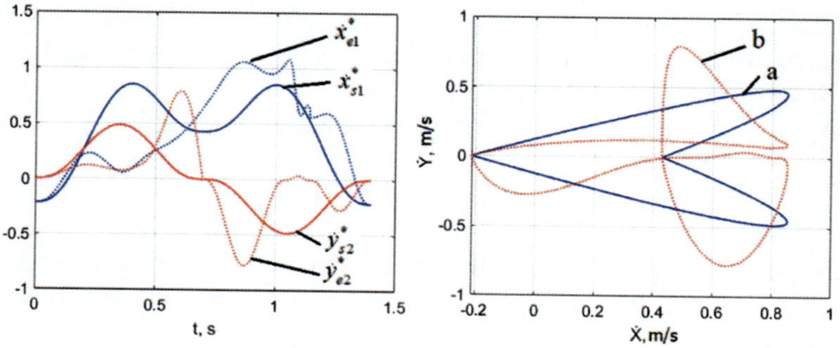

Fig. 4. Time laws of speed change and the hodograph for the ankle joint: a – for synthetic trajectory $(\dot{x}^*_{s1}, \dot{y}^*_{s2})$, b – for natural trajectory $(\dot{x}^*_{e1}, \dot{y}^*_{e2})$

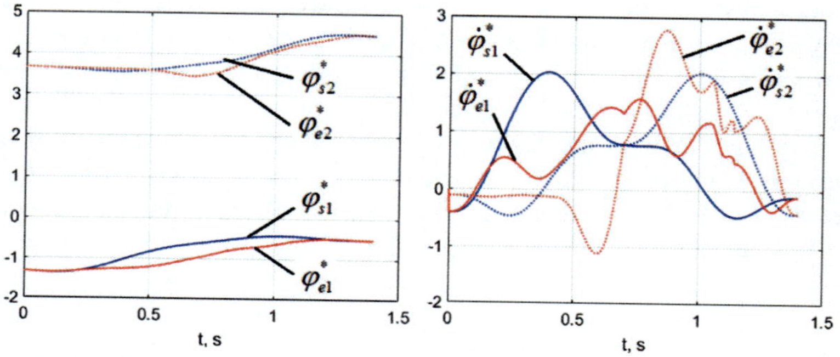

Fig. 5. Laws of change of rotation and velocities angles of the ankle joint: a – absolute rotational angles for synthetic $(\varphi^*_{s1}, \varphi^*_{s2})$ and natural trajectories $(\varphi^*_{e1}, \varphi^*_{e2})$, b – angular velocities for synthetic $(\dot{\varphi}^*_{s1}, \dot{\varphi}^*_{s2})$ and natural trajectories $(\dot{\varphi}^*_{e1}, \dot{\varphi}^*_{e2})$

The laws of change in the angles of rotation and angular velocities of the walking machine links can be represented in the form of time diagrams (Fig. 5).

The analysis of the graphs allows speaking about a more complex nature of the system movement when simulating the natural gait of a person. The maximum angular velocity when turning the second link is 2.8 rad/s, which is 30% higher than when setting a synthetic trajectory. It should be noted that the natural trajectory is also characterized by significant required accelerations up to 40 rad/s^2, which is more than 4 times higher than the analogous values for the synthetic task.

Reproduction of the natural trajectory by the drive system will require more power costs due to higher required speeds and accelerations of the movement of the executive links, and the complex nature of changing the setting angles requires a high quality of the drive control system.

4 Simulation Results of Controlled Motion Dynamics

To assess the influence of the trajectory setting method on the dynamics of the exoskeleton links' controlled movement, a mathematical model of the electromechanical exoskeleton system equipped with a closed control system was developed, the scheme of which is shown in Fig. 1.

To obtain a solution to a system of nonlinear differential equations that allow taking into account the moments of dry and viscous friction, the measurement error of the rotation angle sensors, the transport delay in the control loop, a program was developed in the Matlab environment.

The system parameters of the mathematical model are presented in the form of a Table 1.

Table 1. Parameters of the mathematical model

Parameter description	Symbol	Used in the model value
Mass of the femoral link of the exoskeleton	m_1	3.5 kg
Mass of the lower leg link of the exoskeleton	m_2	4.5 kg
Length of the femoral link of the exoskeleton	l_1	0,52 m
Length of the lower leg link of the exoskeleton	l_2	0,52 m
Armature winding inductance	L	0,072 mH
Armature winding resistance	R	0,103 Ω
Torque coefficient	k_m	0,35 Nm/A
Speed coefficient	k_E	25,9 rad/Vs

The results of numerical simulation will be shown in the form of time diagrams. Figure 6 shows the time dependences of the links rotation angles, when working out synthetic and natural trajectories. It should be noted that the settings of PD regulators are the same for both experiments ($k_p = 12.3$, $k_d = 0.25$). The method used in the work for selecting the coefficients of the regulator is described in detail in [9].

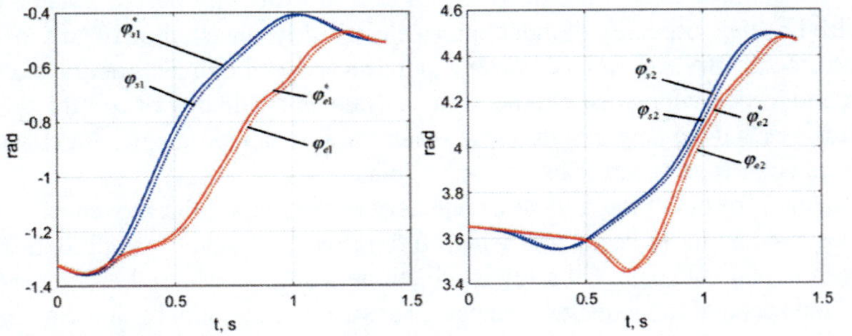

Fig. 6. Time diagrams of the links' rotation angles during the development of natural $\left(\varphi_{e1}, \varphi_{e2}\right)$ and synthetic trajectories $\left(\varphi_{s1}, \varphi_{s2}\right)$

To illustrate the differences between the motion task strategies, time diagrams of the instantaneous values of the errors $E_{s1} = \left(\varphi_{s1}^* - \varphi_{s1}\right)$, $E_{s2} = \left(\varphi_{s2}^* - \varphi_{s2}\right)$, $E_{e1} =$

$(\varphi_{e1}^{*} - \varphi_{e1}), E_{e2} = (\varphi_{e2}^{*} - \varphi_{e2})$ and the torque $M_{s1}.M_{s2}$, M_{e1}, M_{e2} created by the drives during trajectory processing will be shown (Fig. 7).

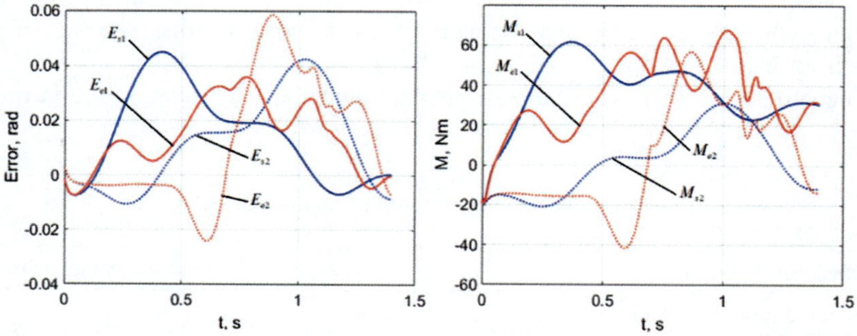

Fig. 7. Time diagrams of the errors when working out the specified rotation angles and torque of the drives

According to the simulation results, when the synthetic trajectory was worked out by a drive system equipped with a PD-controller, the average error was 0.0172 (rad) for the first drive and 0.0162 (rad) for the second one. When working out a natural trajectory, these values were 0.0188 (rad) and 0.0192 (rad), respectively. Thus, it can be concluded that when working out a synthetic trajectory, the positioning accuracy is on average 12% higher than with a natural trajectory. The performed estimation of the required power when working out one step with the exoskeleton system allows us to speak about an increase in the average energy consumption by 9.8% when working out a natural trajectory compared to a synthetic one.

5 Conclusions

1. A mathematical model of the lower limbs' exoskeleton is proposed, which is characterized by the possibility of modeling an electric drive installation of an exoskeleton with electric drives of limited power and the presence of moments of dry and viscous friction forces, which makes it possible to evaluate the quality of control, the values of torques and the power of the drive system with various strategies for constructing the trajectories of the exoskeleton ankle joint.
2. A method for obtaining a natural trajectory of the ankle joint movement based on video processing is described, which differs by the method of approximation of experimental data using an artificial neural network, for which the step parameters and the speed of movement of the device are used as initial data, allowing to obtain motion trajectories with a non-monotonic character imitating the natural gait of a healthy person.
3. The results of a comparative analysis of the kinematic characteristics of motion for synthetic and natural trajectories of motion are presented, showing an increase in the speeds and accelerations of the exoskeleton executive links when using a natural trajectory.

4. The results of a comparative analysis of the dynamic characteristics of controlled motion for synthetic and natural motion trajectories are presented, which showed a large control error when working out the natural trajectory, explained by the power limitation and inertia of the drive system, as well as increased energy consumption.
5. Based on a comparative analysis of trajectory setting methods, it has been established that the use of a natural trajectory is advisable for stationary devices in which the maximum approximation of the movement of the exoskeleton legs to the natural movements of a healthy person is fundamental. The synthetic trajectory is less demanding on the control system and the power of the exoskeleton power plant can be used for transport mobile exoskeleton systems.

Acknowledgments. The work was supported by RSF, research project No. 22-21-00464.

References

1. Shakhparonova, N.V., Kadykov, A.S.: Reabilitaciya bol'nyh s postinsul'tnymi spasticheskimi parezami. RMZH **20**(31), 1533–1535 (2012)
2. Carlsson, G.E., Möller, A., Blomstrand, C.: Consequences of mild stroke in persons. Cerebrovasc. Dis. **16**(4), 383–388 (2003)
3. Forster, A., Young, J.: Incidence and consequences of falls due to stroke: a systematic inquiry. BMJ **311**(6997), 83–86 (1995)
4. Shi, D., Zhang, W., Zhang, W., Ding, X.: A review on lower limb rehabilitation exoskeleton robots. Chin. J. Mech. Eng. **32**(1), 1–11 (2019)
5. Veneman, J.F., Kruidhof, R., Hekman, E.E., Ekkelenkamp, R., Van Asseldonk, E.H., Van Der Kooij, H.: Design and evaluation of the LOPES exoskeleton robot for interactive gait rehabilitation. IEEE Trans. Neural Syst. Rehab. Eng. **15**(3), 379–386 (2007)
6. Beyl, P., Van Damme, M., Van Ham, R., Versluys, R., Vanderborght, B., Lefeber, D.: An exoskeleton for gait rehabilitation: prototype design and control principle. In: 2008 IEEE International Conference on Robotics and Automation, pp. 2037–2042. IEEE (2008)
7. Pinto-Fernandez, D., et al.: Performance evaluation of lower limb exoskeletons: a systematic review. IEEE Trans. Neural Syst. Rehab. Eng. **28**(7), 1573–1583 (2020)
8. Jatsun, S., Malchikov, A., Yatsun, A., Leon, A.S.M.: Simulation of a walking robot-exoskeleton movement on a movable base. J. Artif. Intell. Technol. **1**(4), 207–213 (2021)
9. Jatsun, S., Malchikov, A., Postolniy, A., Yatsun, A.: Trajectory planning strategy for the links of a walking human-machine system using a neural network. In: Chugo, D., Tokhi, M.O., Silva, M.F., Nakamura, T., Goher, K. (eds.) CLAWAR 2021. LNNS, vol. 324, pp. 255–261. Springer, Cham (2022). https://doi.org/10.1007/978-3-030-86294-7_22
10. Malchikov, A., Yatsun, A., Bezmen, P.: Control features of the electromechanical system with end-effector considering the regulated torque. In: MATEC Web of Conferences, vol. 113, p. 02001. EDP Sciences (2017)
11. Tracker Homepage. https://physlets.org/tracker/

Adaptive Navigation Control of an Active Smart Walker

Andrea Borgese[1], Dario C. Guastella[1(✉)], Giuseppe Sutera[1], Alessia Biondo[1], and Giovanni Muscato[1,2]

[1] Dipartimento di Ingegneria Elettrica Elettronica e Informatica, University of Catania, Catania, Italy
andrea.borgese@phd.unict.it, {dario.guastella,gmuscato}@dieei.unict.it, giuseppe.sutera@unict.it

[2] Istituto di Calcolo e Reti ad Alte Prestazione, Consiglio Nazionale delle Ricerche, Palermo, Italy

Abstract. The rapid increase of elderly population all over the world has determined an enhanced attention to their needs: among the others, mobility is one of the most important human skills to support. In this work we propose a prototype of an active smart walker featuring obstacle avoidance and autonomous navigation capabilities, which is meant to assist the elderly people during the walk. The proposed system is able to plan and move in a known environment, adapting its behavior according to the presence of the user. The proposed solution has been widely tested in its different operating modes in a real setting.

Keywords: Smart walkers · Elderly care · Assistive robots · Human-robot interaction · Autonomous navigation · Obstacle avoidance

1 Introduction

Assistive devices are employed to improve balance and correct postural stability. Such devices are widely used also for the prevention of falls in patients with soft instability problems. Patients using these appliances, besides regaining individual mobility, reported also benefits in psychological attitude, such as increased self-confidence and self-esteem.

A large body of research on assistive technologies for the elderly deals with the design of solutions to help those suffering from walking abnormalities [4,10]. Several types of assistive devices have been proposed; the most popular are canes and walkers. In particular, the latter reduce the load on the lower limbs in patients with weight problems and facilitate the mobility of those suffering from poor balance.

The aim of this work is to develop a smart walker, including sensing and control systems, in order to enhance its usability by making it able to move in the environment, thus relieving the user from the effort of pushing the walker, while

© The Author(s), under exclusive license to Springer Nature Switzerland AG 2023
J. M. Cascalho et al. (Eds.): CLAWAR 2022, LNNS 530, pp. 594–603, 2023.
https://doi.org/10.1007/978-3-031-15226-9_54

avoiding collisions with obstacles. We propose a solution that adapts the walker control to either a fully autonomous behavior or alternatively to an assistive semi-autonomous behavior, depending on the presence of the user.

The main contributions of our solution compared to those already available in the related literature are:

- the adaptive behavior of the navigation control
- the autonomous navigation capability, that allows the sharing of the platform among different people, as it may occur in nursing homes or hospitals.

2 Related Work

Among the solutions proposed in the literature, smart walkers can be divided into passive and active devices, depending on their propulsion. Passive walkers exploit the automatic actuation of brakes [1,2] or steered wheels [3] for obstacle avoidance, while the user is in charge of pushing the walker.

In [1] the user is guided away from the obstacles through a proper braking strategy that is activated according to the onboard 2D LIDAR (Light Detection and Ranging). The solution proposed in [3] is a passive system that can steer by adjusting the facing direction of its front wheel. The walker attempts to infer the user's driving intention in order to properly avoid any potential obstacle along the user's path.

Active walkers usually adopt actuated wheels to drive the platform within the environment. In [7] a walker is equipped with infrared sensors to detect the user's legs in order to control the walking speed. The authors present an omnidirectional platform which ensures good maneuverability. Sierra et al. [11] developed a solution on top of a differential drive platform with a human-robot interface for smart walkers. The proposed walker includes a people detection system and can be both remotely operated by medical staff and controlled by the user's intentions of movement.

3 Proposed Solution

Our solution consists of an active smart walker which overcomes some of the main limits of the proposals available in the literature. Walkers in the literature usually propose only simple obstacle avoidance in order to prevent collisions with the surrounding objects. Our system employs a 360° LIDAR capable of monitoring both the user walking parameters and the surrounding environment. In particular, the former are the user's stride and the distance from the walker.

In addition to obstacle avoidance, the map of the operating environment is used for autonomous navigation and can be employed to locate some *points of interest* (e.g. charging station, kitchen, living room, bedroom etc.) that the platform could reach autonomously, i.e. without any user intervention.

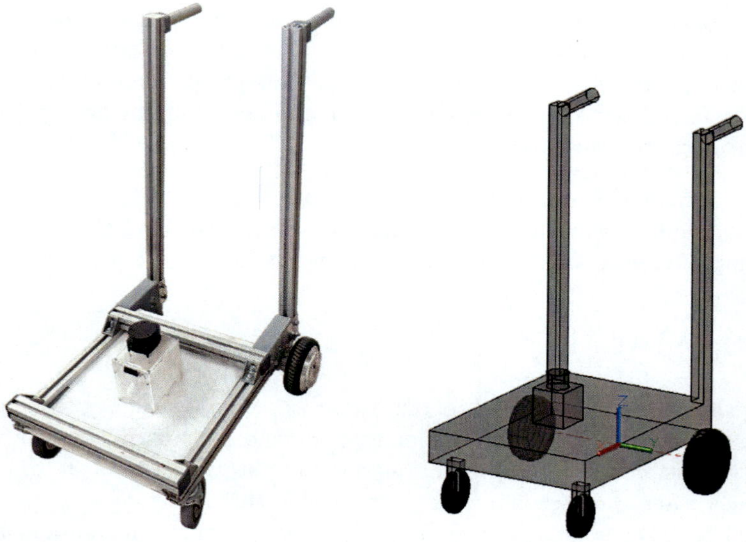

(a) Developed prototype. (b) 3D model.

Fig. 1. The proposed smart walker.

3.1 Mobile Platform

The proposed smart walker is a prototype entirely built in our laboratories with aluminum profiles and measures $0.6 \times 0.65 \times 1$ m (L × W × H). It is a differential drive platform with two castor wheels on the front and two rear drive wheels rigidly fixed to the frame (Fig. 1a). We define the platform reference system attached as shown in Fig. 1b, with its origin at the midpoint between the two wheels, the x axis pointing forward and the z axis pointing upward. Wheel odometry is derived from magnetic encoders.

3.2 Sensing and Actuation

A 2D 360° laser rangefinder, an RPLIDAR A1 by Slamtec, is placed on the platform. It can perform scans within a 6 m range with a scanning frequency of 5.5 Hz when sampling 360 points per rotation. The onboard computing unit of the smart walker is a Raspberry Pi 3 Model B+, which is connected to the motor drivers and provides them the control signals to actuate the drive wheels. ROS (Robot Operating System) has been used as the communication middleware for our software system [9].

3.3 Operating Modes

The proposed walker envisions two operating modes: an *autonomous navigation* mode, that allows the navigation towards a goal, and an *assistive semi-autonomous* mode, during which the walker proceeds forward while guiding the user to avoid obstacles along the path.

The platform autonomously adapts its behavior, i.e. switches from one operating mode to the other, by processing LIDAR readings to detect the presence or not of the assisted person. In particular, since the laser scanner detects the user's legs, the average distance value (d_{mean}) of the related scan points is considered.

Autonomous Navigation Mode. When the walker detects the absence of the user, it is allowed to autonomously navigate, given the environment occupancy grid and a target position. Autonomous navigation of the smart walker has been achieved by integrating the ROS Navigation Stack [8] in our solution, after a proper tuning and configuration for our specific platform.

In particular, the navigation stack features LIDAR-based adaptive Monte Carlo localization (AMCL) algorithm [6], while the Dijkstra algorithm[1] and the dynamic window approach [5] are used for global path planning and obstacle avoidance respectively. The latter take as an input the information from the LIDAR and wheel odometry and compute linear and angular speed commands. When in autonomous mode the platform is allowed to rotate in place, by assigning opposite angular speeds to the drive wheels, since it is assumed that the user is not present.

If needed, laser-based map building can be preliminary performed when the smart walker is employed in a new operating environment. In this case the walker is manually operated in order to explore the area of interest, while the *Gmapping* SLAM process[2] [6] incrementally builds the environment occupancy grid from the LIDAR data, filtering out the portion of scans affected by the users presence resulting in partial occlusions of the environment.

Assistive Semi-autonomous Mode. This operating mode is activated once a user is detected. In this case the platform will start to slowly move forward, taking into account the user's stride. In this manner, the user can still rely on the support of the walker without the effort of pushing it forward.

In particular, recalling the mean value, d_{mean}, introduced in Sect. 3.3, three possible cases are considered:

- if the mean value is greater than an *activation distance*, $d_a = 0.9$ m, the user is not present and linear velocity v is set to zero
- if the mean value is smaller than the *activation distance* and greater than a *minimum distance*, d_{min}, the walker moves at a linear velocity that is inversely proportional to the user's distance from the platform
- if the mean value is smaller than the minimum distance, the walker proceeds at a maximum linear speed allowed for the walker, v_{max}, that is experimentally tuned depending on the user

[1] http://wiki.ros.org/navfn.

[2] http://wiki.ros.org/gmapping.

The three cases described above can be summarized as follows:

$$\begin{cases} v = 0 & if \quad d_{mean} > d_a \\ v = \dfrac{k_v}{d_{mean}} & if \; d_{min} < d_{mean} < d_a \\ v = v_{max} & if \quad d_{mean} < d_{min} \end{cases} \tag{1}$$

Moreover, a control action is defined in order to avoid any obstacle along the user's path, by taking into account both the distance to closest obstacle and the angle between the walker's forward direction and the obstacle itself. Any obstacle detected at a distance larger than a *detection threshold*, $d_{th} = 2$ m, is neglected. This threshold has been experimentally determined after several tests performed in the environments which the platform was designed for (i.e. rooms, hallways, and, more generally, indoor areas). On the other hand if the obstacle is within a critical distance, $d_{crit} = 0.6$ m, the walker stops. Two different behaviors are defined otherwise:

- the closest obstacle is at a distance smaller than the threshold d_{th} but greater than the *intermediate distance*, $d_{int} = 0.9$ m: since in this case the obstacle is not yet too close, the walker can steer with a large radius of curvature, i.e. the linear velocity is set to the minimum, while the angular velocity is inversely proportional to the distance from the obstacle. The sign of the angular velocity depends on the position of the obstacle
- the closest obstacle is at a distance smaller than the intermediate distance but greater than the critical distance d_{crit}: the walker will have to steer more urgently and carefully, i.e. the linear velocity is set to zero, while the angular velocity is the minimum to make the platform move. Also in this case the sign of the angular velocity depends on the position of the obstacle

4 Experimental Results

In the following sections, the experimental results of the tests performed to validate the different functions of the walker are shown. The tested functions are: 1) linear velocity modulation, 2) obstacle avoidance, and 3) autonomous navigation.

A video showing the overall behavior and performance of the smart walker can be found in the link below[3].

4.1 Assistive Semi-autonomous Mode

Linear Velocity Control. An example of linear speed control depending on the user's distance is depicted in Fig. 2, where the chart on the top shows the user's distance, while the one on the bottom reports the forward velocity. When the user is at a distance less than $d_a = 0.9$ m, the linear velocity increases inversely proportional to the distance.

[3] https://youtu.be/cRwUxL_2V9U.

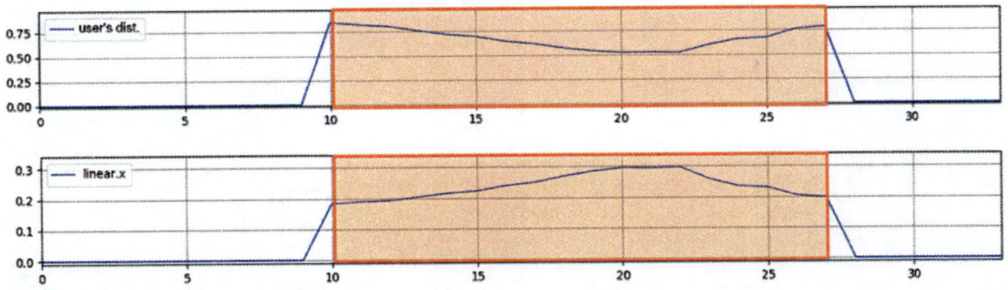

Fig. 2. Linear velocity control: on the top the detected user distance and on the bottom the forward velocity over time

As shown in Fig. 2, initially the walker is stationary because there is no user closer than 90 cm: the forward velocity is zero in the time interval before the highlighted portion. At the same time, in the top chart, the distance is forced to zero because the value of the distance is used only when it is less than 90 cm (and no target pose is given).

At the beginning of the highlighted portion of Fig. 2, the distance takes a value different from zero because a user has been detected and the speed starts to increase inversely proportional to the distance. The two charts are perfectly mirrored each other as expected.

Obstacle Avoidance. Several examples of obstacle avoidance are shown in Figs. 3, 4, 5 and 6. Each figure is composed of three charts:

- the chart on the top represents the distance to the closest obstacle over time;
- the chart in the middle represents the angle over time related to the closest obstacle;
- the chart on the bottom represents the angular velocity over time.

Taking into account how the onboard 2D LIDAR stores the detected ranges, a negative angle in the chart in the middle corresponds to an obstacle on the left of the walker, whereas a positive angle corresponds to an obstacle on the right. Therefore, to avoid an obstacle on the left, a right turn is needed, i.e. a negative angular velocity is assigned (given the reference system of the platform in Fig. 1). In a similar way, an obstacle on the right of the walker is avoided with a left turn, i.e. by assigning a positive angular velocity.

In Figs. 3, 4, 5 and 6 the actions related to an obstacle on the right (positive angle) with consequent left steering (positive angular speed) are highlighted in yellow, whereas the actions related to an obstacle to the left (negative angle) with consequent right steering (negative angular velocity) in red. In all the examples it can be observed that the walker begins to steer only when the closest obstacle distance is smaller than $d_{th} = 2$ m.

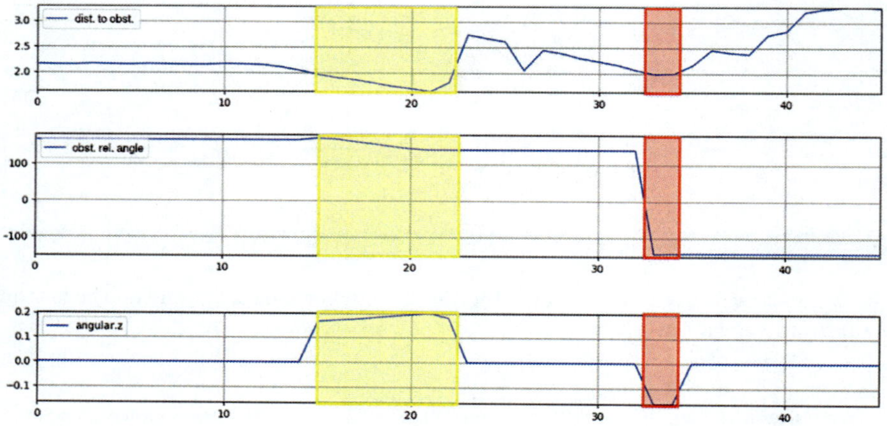

Fig. 3. Obstacle avoidance example: the first obstacle detected is on the right, the second is on the left.

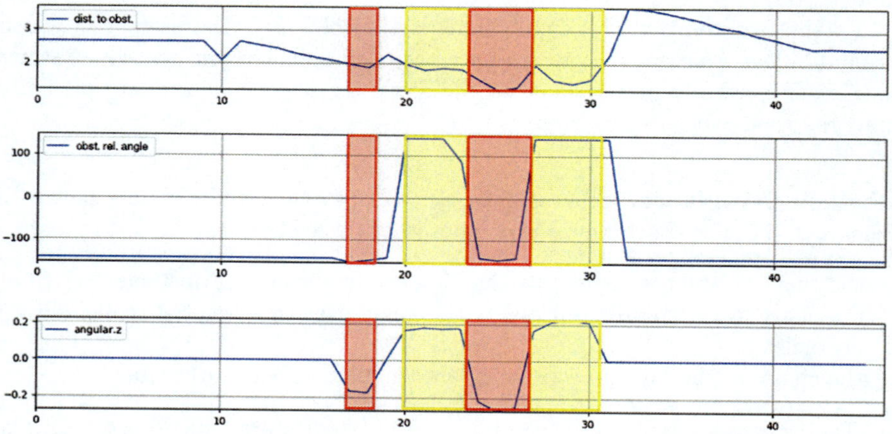

Fig. 4. Obstacle avoidance example: the first obstacle detected is on the left, the second is on the right, the third on the left and the fourth on the right.

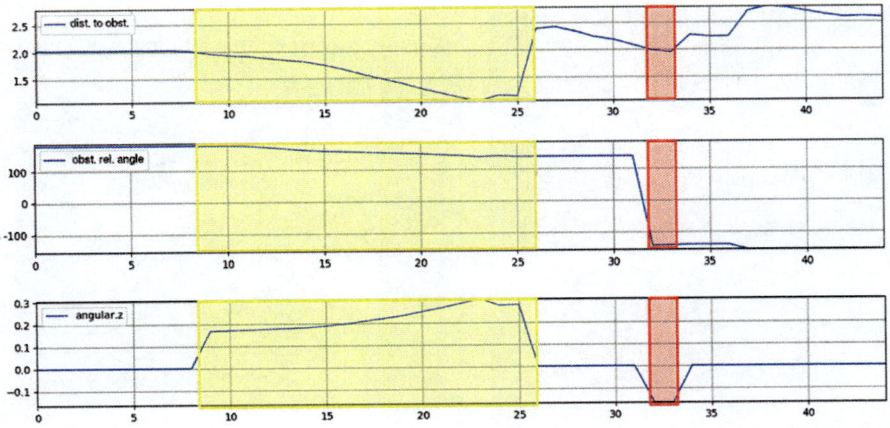

Fig. 5. Obstacle avoidance example: the first obstacle detected is on the right, the second is on the left.

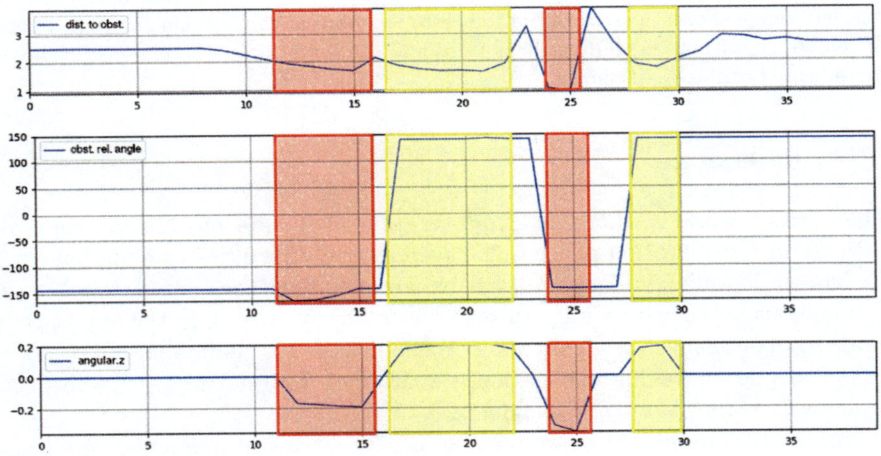

Fig. 6. Obstacle avoidance example: the first obstacle detected is on the left, the second is on the right, the third on the left and the fourth on the right.

4.2 Autonomous Navigation

Thanks to the careful tuning and the proper integration of the Navigation Stack, the smart walker proved to offer a remarkable autonomous navigation capability. The platform was capable of smoothly navigating also through narrow passages and to avoid sudden obstacles (hence not included in the map). Both these features are crucial for a smart walker which is assumed to move in indoor environments and in presence of people (Fig. 7).

The performance achieved in the autonomous navigation can be better appreciated in the video link reported above.

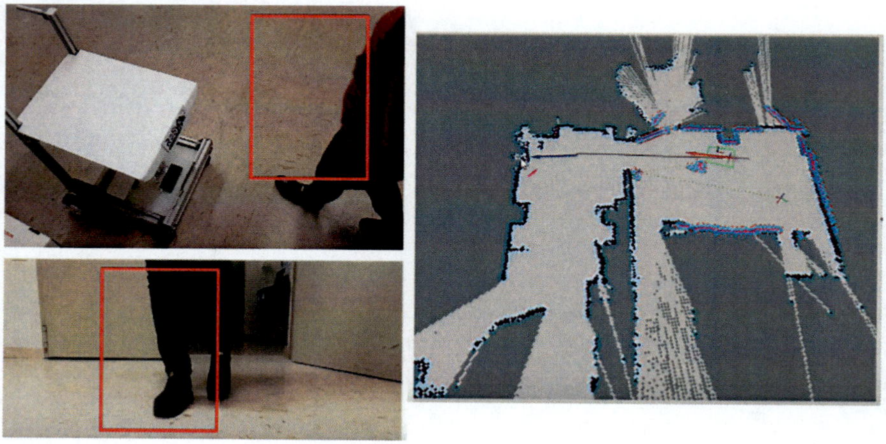

Fig. 7. An image from the autonomous navigation experiment. On the top left a third-person perspective image highlighting a walking person with a red rectangle. The bottom left image shows the same situation from an onboard camera. On the right figure, the environment map, including the current and the target pose, the planned path and the detected obstacles, is shown.

5 Conclusions

The present work is aimed at improving the daily life of elderly people or patients in rehabilitation phase that have physical difficulties to maintain equilibrium while walking. Our smart walker can operate either in an assistive semi-autonomous mode or in a fully autonomous mode, switching to the appropriate mode by detecting the user's presence. When in assistive mode, the walker moves at a proper speed, taking into account the distance from the user, while ensuring a safe walk through obstacle avoidance.

As a future development we plan to include force sensors on the handles of the walker in order to sense when the user is holding them. In this manner it would be possible to perform posture monitoring by checking the distance of the legs with respect to the ground projection of hands position, thus being able to correct the distance between the user and the walker itself and even to prevent potential falls.

References

1. Andò, B., et al.: An obstacle avoidance strategy using an indoor localization system. In: Monteriù, A., Freddi, A., Longhi, S. (eds.) ForItAAL 2019. LNEE, vol. 725, pp. 191–196. Springer, Cham (2021). https://doi.org/10.1007/978-3-030-63107-9_15
2. Andò, B., et al.: An introduction to patterns for the internet of robotic things in the ambient assisted living scenario. Robotics **10**(2) (2021). https://doi.org/10.3390/robotics10020056

3. Annicchiarico, R., Barrué, C., Benedico, T., Campana, F., Cortés, U., Martínez-Velasco, A.: The i-walker: an intelligent pedestrian mobility aid. In: Bichindaritz, I., Vaidya, S., Jain, A., Jain, L.C. (eds.) Computational Intelligence in Healthcare. Studies in Computational Intelligence, vol. 178, pp. 708–712. Springer, Heidelberg (2008). https://doi.org/10.1007/978-3-642-14464-6_6
4. Chaparro-Cárdenas, S.L., Lozano-Guzmán, A.A., Ramirez-Bautista, J.A., Hernández-Zavala, A.: A review in gait rehabilitation devices and applied control techniques. Disabil. Rehabil. Assist. Technol. **13**(8), 819–834 (2018). https://doi.org/10.1080/17483107.2018.1447611
5. Fox, D., Burgard, W., Thrun, S.: The dynamic window approach to collision avoidance. IEEE Robot. Autom. Mag. **4**(1), 23–33 (1997). https://doi.org/10.1109/100.580977
6. Grisetti, G., Stachniss, C., Burgard, W.: Improved techniques for grid mapping with Rao-Blackwellized particle filters. IEEE Trans. Rob. **23**(1), 34–46 (2007). https://doi.org/10.1109/TRO.2006.889486
7. Lee, G., Ohnuma, T., Chong, N.Y.: Design and control of JAIST active robotic walker. Intel. Serv. Robot. **3**(3), 125–135 (2010)
8. Marder-Eppstein, E., Berger, E., Foote, T., Gerkey, B., Konolige, K.: The office marathon: robust navigation in an indoor office environment. In: 2010 IEEE International Conference on Robotics and Automation, pp. 300–307 (2010)
9. Quigley, M., et al.: ROS: an open-source robot operating system. In: ICRA Workshop on Open Source Software, 3.2, p. 5. Kobe, Japan (2009)
10. Rajavenkatanarayanan, A., et al.: A survey of assistive technologies for assessment and rehabilitation of motor impairments in multiple sclerosis. Multimodal Technol. Interact. **3**(1), 6 (2019)
11. Sierra, S.D., Molina, J.F., Gómez, D.A., Múnera, M.C., Cifuentes, C.A.: Development of an interface for human-robot interaction on a robotic platform for gait assistance: AGoRA smart walker. In: 2018 IEEE ANDESCON (2018). https://doi.org/10.1109/ANDESCON.2018.8564594

Human Models Simulating the Physical Conditions of the Elderly Individual and Standing Assistance Method Based on These Models

Daisuke Chugo[1] (✉), Yuya Miyazaki[1], Satoshi Muramatsu[2], Sho Yokota[3], Jin-Hua She[4], and Hiroshi Hashimoto[5]

[1] Kwansei Gakuin University, Sanda, Hyogo, Japan
chugo@kwansei.ac.jp
[2] Tokai University. Hiratsuka, Kanagawa, Japan
[3] Toyo University, Kawagoe, Saitama, Japan
[4] Tokyo University of Technology, Hachioji, Tokyo, Japan
[5] Advanced Institute of Industrial Technology, Shinagawa, Tokyo, Japan

Abstract. This paper proposes a new estimation method for the lower limb muscle strength of the elderly that can be easily performed in nursing care facilities. Furthermore, based on the estimated lower limb muscle strength information, this paper proposes a human model that simulates the physical conditions of the elderly and, using this model, proposes a new design method for standing assistance strategy adapted to the physical conditions of the elderly. The main advantage of the proposed method is that it can estimate and model leg muscle strength, which depends on the individual elderly person, in a simple way that can done by staff in nursing care facilities who are not engineering specialists, without the use of specialized equipment such as electromyography.

Keywords: Standing assistance · Muscle strength estimation · Safety tolerance

1 Introduction

1.1 Background and Motivation

Standing is one of the most serious and important activities of daily living because of the possible lack of strength and stability in the elderly [1, 2]. Typically, in bad cases, an incapacitated elderly person may not be able to stand up and subsequently become limited to wheelchair living or bedridden [3]. Furthermore, when an elderly assumes such lifestyle, the lack of exercise and consequent decline in physical fitness becomes more pronounced [4]. Therefore, there is a need for an assistive robot to help the use of residual muscle strength during voluntary movements in order to maintain the individual's muscle strength.

To achieve this goal, the assistive robot must be designed to accept some degree of variability in the individual's body movements, as human movement does not always

© The Author(s), under exclusive license to Springer Nature Switzerland AG 2023
J. M. Cascalho et al. (Eds.): CLAWAR 2022, LNNS 530, pp. 604–616, 2023.
https://doi.org/10.1007/978-3-031-15226-9_55

conform to the established reference pathway [5]. During the standing motion, these robots interfere with the individual's body movements and are adjusted to conform to the reference pathway. Such external interventions would prevent the elderly from using his or her own strength in the process of standing up, which would result in a decrease in the individual's muscle strength.

1.2 Difficulties in the Corona Disaster and the Objectives of This Study

In order to address this problem, we defined the "safety tolerance" as the range in which a person can continuously perform standing motions from the standpoint of physical balance and muscle strength [6]. Furthermore, we developed a standing assistance robot which allowed its user's voluntary movements as long as its user's posture was within the safety tolerance [7]. Using this idea, our robot could use its user's physical strength as much as possible.

As muscle weakness and the level of weakness are different in individuals, the safety tolerance needs to be set based on the physical condition of the targeted elderly individually. Usually, the physical condition of the target elderly is obtained by the robot developer visiting the elderly directly and measuring the elderly's physical condition using myoelectric measurement sensors or similar devices.

However, the global corona pandemic prevents robot developers and elderly people from directly seeing each other. As the elderly have a high death risk from corona infection, in Japan only residents and facility staff are allowed in nursing care facilities. This prohibition has continued from 2020, when the corona pandemic occurred, to the present day, and there is no prospect of the prohibition being lifted. There is an urgent need to develop technology that enables robotics researchers to estimate the physical condition of elderly people and develop assistive robots suited to individual elderly people, without the researchers having to meet the elderly patient directly.

Therefore, the objectives of this study are to develop a human model that simulates an elderly person, using only data which is measurable by nursing facility staff who are not engineering experts, and to use this human model to design a reference path for our standing assistance robot that is suitable for that elderly person.

2 Estimating the Physical Condition of Elderly People

2.1 Overview of the Proposed Estimate Scheme

In order to design methods to assist standing up, the load on its user needs to be considered. Generally, joint traction is widely used as an index of the elderly's load in the linkage model of a human body [8, 9]. However, this index does not consider the posture of its user, and in some cases, this index diverges from the experience of nursing specialists, especially when its user is in a half-sitting posture. When an elderly stands, the muscles shown in Fig. 1 generate a lifting motion. Many muscles (shown in Fig. 1) are used to accomplish the standing movement, and the traction, which muscles can maximally generate, changes according to the relative position between frames and muscles.

In our previous research [10], we already developed the load estimation method with musculoskeletal model considering the role of the antagonistic and biarticular muscles. This estimation technique was achieved by using a musculoskeletal model to represent the relationship between the force that each muscle can generate and the force the elderly can finally perform at the toes in the posture he or she is in at the time. Using this model, we could estimate whether each elderly could perform the force required for the standing movement with their toes based on their muscle strength, and if not, how much force they should be assisted with individually.

This study reverses this model. In this study, we propose to estimate individual's muscle strength using our musculoskeletal model. By inputting force vectors that the elderly can produce maximum force on the toes, our musculoskeletal model can estimate capacities of each muscle strength. To estimate them, simply ask the elderly to produce a force on their toes in six or more directions and measure the amount and direction of the force. These force measurements can be made easily using simple force sensors and therefore any nursing facility staff can do this.

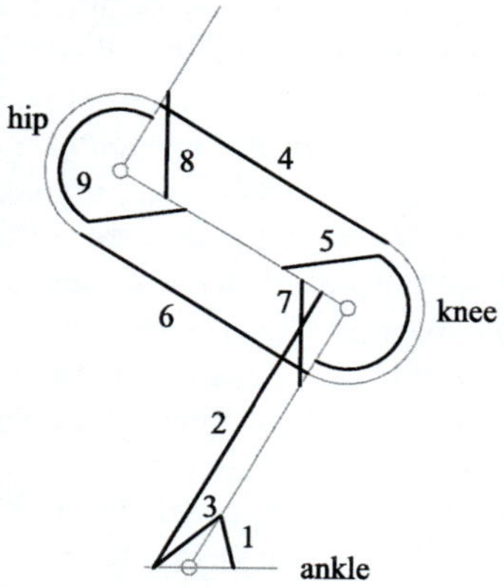

No	Muscle actuator	Corresponding muscle model
1	Tibialis anterior (TA)	
2	Gastrocnemius (GAS)	
3	Soleus (SOL)	
4	Rectus femoris (RF)	e3
5	Vastus lateralis (VAS)	e2
6	Semimenbranosus (SM)	f3
7	Biceps femoris and short head (BFSH)	f2
8	Iliopsoas (IL)	e1
9	Gluteus masimum (GMAX)	f1

Fig. 1. Muscle arrangements at the human leg. The corresponding muscle model shows the mapping to the effective leg muscles shown within the models in Fig. 2.

2.2 Muscle Strength Estimation Using a Bi-articular Muscle Model

Previous researches [11, 12] have shown that the human lower leg can be represented as a musculoskeletal model consisting of six bi-articular muscles, as shown in Fig. 2. The musculoskeletal model consists of a base representing the trunk, a first link representing the thighs and a second link representing the lower legs, and is driven by six muscles (f1, e1, f2, e2, f3, e3). Muscles f1 and e1 are monoarticular muscles acting antagonistically on the waist joint and muscles f2 and e2 on the knee joint. Muscles f3 and e3 are also bi-articular muscles, acting simultaneously and antagonistically on the waist and knee joints. r_1 and r_2 are the moment arms of the waist and knee joints, l_1 and l_2 are the lengths of the first and second links, and θ_1, θ_2 and θ_3 are the angles by link postures.

The six directions (a, b, c, d, e, f) shown in Fig. 2 indicate the force output direction at the ankle joint. Directions a and d are directions connecting the first and second joints, directions b and e are parallel to the second link and directions c and f are parallel to the first link.

The muscles that output forces in the directions a, b, c, d, e and f are muscles e2, e1, f3, f2, f1 and e3, respectively, depending on their kinematical relationship. The relationship between the muscular force generated by these muscles and the force resulting at the ankle joint by the muscular force generated can be expressed by (1). Based on the previous study [11], we assumed $r = r_1 = r_2$. Relationship bewteen each muscle, the muscular force and the force resulting from it are also shown in Table 1.

$$\begin{aligned} F'_{mf1} &= \frac{F_{mf1} r}{l_1 \sin\theta_2} \quad F'_{mf2} = \frac{F_{mf2} r}{l_2 \sin\theta_3} \quad F'_{mf3} = \frac{F_{mf3} r}{l_2 \sin\theta_2} \\ F'_{me1} &= \frac{F_{me1} r}{l_1 \sin\theta_2} \quad F'_{me2} = \frac{F_{me2} r}{l_2 \sin\theta_3} \quad F'_{me3} = \frac{F_{me3} r}{l_2 \sin\theta_2} \end{aligned} \tag{1}$$

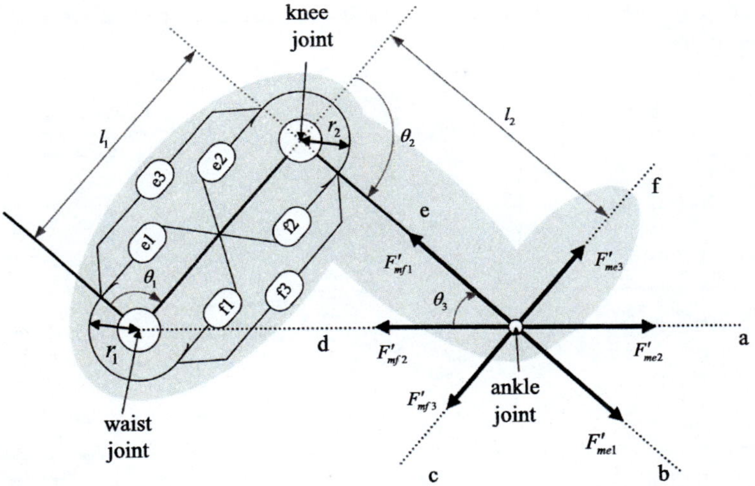

Fig. 2. Musculoskeletal model considering the role of the antagonistic and biarticular muscles.

As shown in Fig. 3, the direction of the force output of the ankle joints is as follows: the force output by muscles f1 and e1 is in direction b and e, parallel to the second link; the force output by muscles f2 and e2 is in direction a and d, connecting the waist joint and ankle joint; the force output by muscles f3 and e3 is in direction c and f, parallel to the first link. Therefore, force output in the six directions from a to f can be achieved by activating only one of the six muscles.

When the two muscles are exercised simultaneously, force output will be possible for the grey area in Fig. 3. Furthermore, when three muscles are active simultaneously, force output is possible within the hexagon ABCDEF. For example, as shown in Fig. 3, the combined force vector of $\mathbf{F'}_{\mathbf{me2}}$ and $\mathbf{F'}_{\mathbf{me3}}$ plus the force vector of $\mathbf{F'}_{\mathbf{me1}}$ results in the

Table 1. Functional effective muscles and their strengths.

Functional effective muscle	Maximum muscle strength	Generated force by maximum muscle strength	Direction
f1	F_{mf1}	F'_{mf1}	b
e1	F_{me1}	F'_{me1}	e
f2	F_{mf2}	F'_{mf2}	d
e2	F_{me2}	F'_{me2}	a
f3	F_{mf3}	F'_{mf3}	c
e3	F_{me3}	F'_{me3}	f

A vector, which represents the position of vertex A. The vertices of the hexagon are not necessarily on the a, b, c, d, e, f axes, because each muscle force is not same. Note that muscles e1 and f1, e2 and f2, e3 and f3 are antagonist muscles, so they cannot be worked simultaneously. Therefore, the maximum number of muscles used in the proposed model is limited to three.

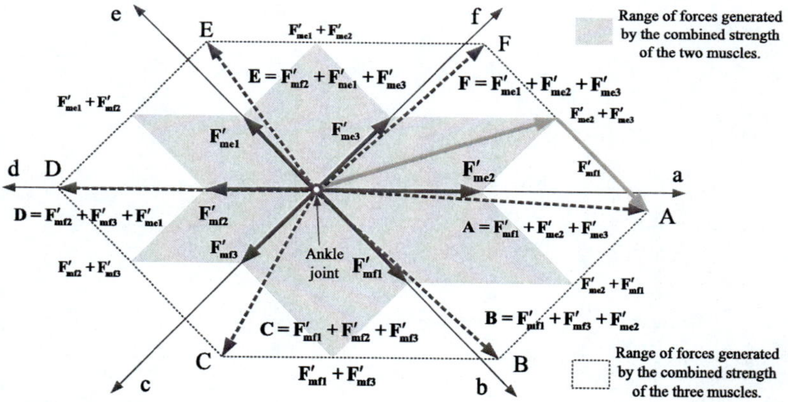

Fig. 3. Relationship between the range of force output and each muscle force at the ankle joint.

If the force output vector corresponding to each vertex of the hexagon in the proposed musculoskeletal model can be measured, the force output in the direction from a to f can be estimated and also the maximum muscular performance of each muscle can be estimated using (1). The combined force vector representing each vertex of the hexagon can be expressed by (2).

$$\begin{aligned} \mathbf{A} &= \mathbf{F'}_{\mathbf{mf1}} + \mathbf{F'}_{\mathbf{me2}} + \mathbf{F'}_{\mathbf{me3}} \quad \mathbf{B} = \mathbf{F'}_{\mathbf{mf1}} + \mathbf{F'}_{\mathbf{mf3}} + \mathbf{F'}_{\mathbf{me2}} \\ \mathbf{C} &= \mathbf{F'}_{\mathbf{mf1}} + \mathbf{F'}_{\mathbf{mf2}} + \mathbf{F'}_{\mathbf{mf3}} \quad \mathbf{D} = \mathbf{F'}_{\mathbf{mf2}} + \mathbf{F'}_{\mathbf{mf3}} + \mathbf{F'}_{\mathbf{me1}} \\ \mathbf{E} &= \mathbf{F'}_{\mathbf{mf2}} + \mathbf{F'}_{\mathbf{me1}} + \mathbf{F'}_{\mathbf{me3}} \quad \mathbf{F} = \mathbf{F'}_{\mathbf{me1}} + \mathbf{F'}_{\mathbf{me2}} + \mathbf{F'}_{\mathbf{me3}} \end{aligned} \tag{2}$$

Equation (2) cannot be solved simply because they are linearly dependent. Therefore, we use the relationships that muscles f1 and e1, f2 and e2, and f3 and e3 are antagonist muscles. From (2), we obtain (3), which represents the relationship between the antagonist muscles only.

$$\mathbf{A}-\mathbf{B}=\mathbf{F'}_{\mathbf{me3}}-\mathbf{F'}_{\mathbf{mf3}} \quad \mathbf{C}-\mathbf{D}=\mathbf{F'}_{\mathbf{mf1}}-\mathbf{F'}_{\mathbf{me1}} \quad \mathbf{E}-\mathbf{F}=\mathbf{F'}_{\mathbf{mf2}}-\mathbf{F'}_{\mathbf{me2}} \tag{3}$$

The force vectors generated by the antagonist muscles f1 and e1, f2 and e2, f3 and e3 are opposite vectors, so (4) is fulfilled. Note that k_1, k_2 and k_3 are coefficients.

$$\mathbf{F'}_{\mathbf{me1}}=-k_1\cdot\mathbf{F'}_{\mathbf{mf1}} \quad \mathbf{F'}_{\mathbf{me2}}=-k_2\cdot\mathbf{F'}_{\mathbf{mf2}} \quad \mathbf{F'}_{\mathbf{me3}}=-k_3\cdot\mathbf{F'}_{\mathbf{mf3}} \tag{4}$$

From the above, the force that the patient can perform in the direction from a to f at the ankle joint is given by (5).

$$\begin{aligned}
F'_{mf1}&=\left|\mathbf{F'}_{\mathbf{mf1}}\right|=\frac{|\mathbf{C}-\mathbf{D}|}{1+k_1} & F'_{me1}&=\left|\mathbf{F'}_{\mathbf{me1}}\right|=\frac{k_1|\mathbf{C}-\mathbf{D}|}{1+k_1}\\
F'_{mf2}&=\left|\mathbf{F'}_{\mathbf{mf2}}\right|=\frac{|\mathbf{E}-\mathbf{F}|}{1+k_2} & F'_{me2}&=\left|\mathbf{F'}_{\mathbf{me2}}\right|=\frac{k_2|\mathbf{E}-\mathbf{F}|}{1+k_2}\\
F'_{mf3}&=\left|\mathbf{F'}_{\mathbf{mf3}}\right|=\frac{|\mathbf{B}-\mathbf{A}|}{1+k_3} & F'_{me3}&=\left|\mathbf{F'}_{\mathbf{me3}}\right|=\frac{k_3|\mathbf{B}-\mathbf{A}|}{1+k_3}
\end{aligned} \tag{5}$$

Finally, from (1) and (5), the maximum muscle strength of each muscle (F_{mf1}, F_{me1}, F_{mf2}, F_{me2}, F_{mf3} and F_{me3}) can be estimated.

2.3 Measuring Method of Muscle Strength and Its Analysis

The measurement of the toe force vector generated by the elderly should be done by the nursing facility staff. Each of the muscle strength estimation methods presented in the previous subsections requires accurate measurement of the six force vectors performed at the ankle joint. However, it is considered difficult to measure in a complex way with limited staff in nursing care facilities. This chapter therefore proposes a method for estimating the maximum physical strength of each muscle using a rough measurement method.

As shown in Fig. 3, when the maximum muscle strength is generated, the force vectors originating at the ankle joint are on the sides of the hexagon. So, if at least six force vectors pointing to each side of the hexagon can be measured, the hexagon which shows the range of possible forces can be determined kinematically. Because the AB and DE sides are parallel to the first link, the BC and EF sides are in the direction connecting the hip and ankle joints, and the CD and FA sides are parallel to the second link. Therefore, if a point on each side can be identified, the straight lines that make up each side can be found. Furthermore, as there are six intersections of these lines, which can be easily obtained, thus, we can derive this hexagon.

We therefore asked staff in care facilities to measure the following.

- Elderlies are asked to sit in a chair and position their posture with $\theta_1 = 45°$ and $\theta_2 = 90°$. (In practice, this was explained using illustrations showing the elderly in a sitting position.) The chair should have a vertical backrest and should not recline.

- Ask the elderly to apply maximum force to the toes in the (1) upward direction, (2) forward upward direction, (3) forward downward direction, (4) downward direction, (5) backward downward direction and (6) backward upward direction. Measure and record the direction and amount of the force with a force sensor with a peak hold function. If possible, this should be done for each foot.
- Measure the elderly's height, weight, lower leg and thigh lengths.

With the cooperation of nursing facilities, facility staff recorded force data that two subjects could perform on their toes, as shown in Table 2. Subject A is able to stand on her own, but is at high risk of falling when standing, as shown in Fig. 4, which is a typical standing procedure for elderly people with reduced muscle strength [13]. Subject B is unable to stand up on her own and always requires assistance from nursing staff during standing. Subject B is able to maintain a standing position on her own after finishing standing, but needs to use a walker to walk.

The kinematical analysis of the force output range at the ankle joint, based on the data measured by the facility side, is shown in Fig. 5. We kinematically derived vectors **A**, **B**, **C**, **D**, **E** and **F**, which represent the maximum drivable capacity in each direction, based on the force data provided by the nursing facility side. In addition, we used these analytical results to estimate the maximum potential capacity of each muscle which is shown in Table 3. Values from previous studies were used for r, representing the moment arm [14], and k_1, k_2 and k_3, representing the output ratio of the antagonist muscle [15], respectively.

Table 2. Elderly subjects.

Subject	Height [cm]	Weight [kg]	Age	Gender	Care level
A	157	54.9	83	Female	2
B	148	38.5	80	Female	3

Fig. 4. Standing procedure of subject A.

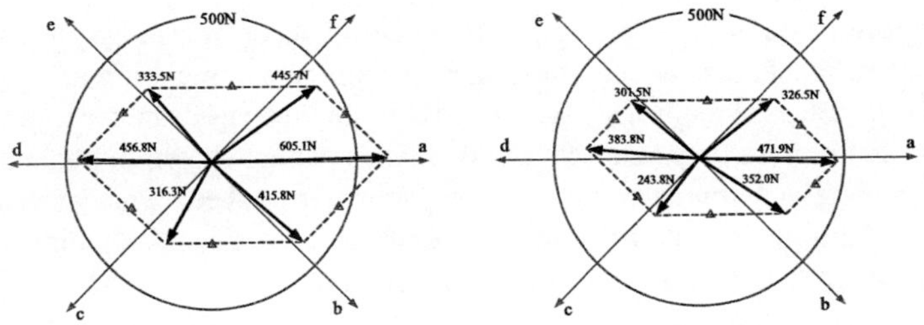

(a) Subject A (b) Subject B

Fig. 5. The kinematical analysis results of the force output range at the ankle joint of a right foot. The points in the red triangle represent the force data measured by the nursing facility.

Table 3. Potential capacity of each muscle. [N]

	Subject A		Subject B	
	Right foot	Left foot	Right foot	Left foot
F_{mf1}	694	689	432	429
F_{me1}	486	482	303	300
F_{mf2}	71	65	30	26
F_{me2}	532	491	222	191
F_{mf3}	617	615	407	403
F_{me3}	247	246	163	161

3 Designing Individualized Standing Assistance Measures

3.1 Conditions for a Successful Standing Movement

In terms of body mechanics, the safety standing motion must meet two conditions. The first condition is stability condition. The elderly should be able to maintain body balance during standing motion. We define this condition as: the center of gravity position (COG) is within range of the elderly's feet, while maintaining body balance when standing.

The second condition is muscle condition. The elderly should be able to control his or her body movements when standing. In general, the positional relationship between muscle and bones changes depending on a subject's posture [5], so the output force generated by a muscle changes with posture. In other words, in an unsuitable posture, it is not possible to generate enough upward output force to advance the standing motion even if the elderly has enough physical strength. In this study, this muscle condition is defined as the required output force of the muscle shown in Fig. 1 should not exceed the maximum output of the muscle during standing up.

In our previous studies [6, 7], we have examined this acceptable range (we call "safety tolerance") that satisfies these conditions through computer simulation using OpenSim [16], a human motion dynamics simulator package. In this paper, the maximum muscle capacity, which represents the muscle status of each patient shown in Table 3, is input into a musculoskeletal simulator as a muscle parameter to check what movements are possible according to the elderly's muscle strength, and to discuss standing assistance strategies adapted to the individual's physical situation.

3.2 Simulation Setup

In this simulation, as Fig. 6, 3DGait-Model 2392 was used as the human body model and the body parameters were changed to match a target subject. The standing movements were based on references recommended by physical therapist [17], as shown in Fig. 7(a). The angular changes in each joint of the patient during this movement are shown in Fig. 7(b). In Fig. 7(b), the Y-axis shows the angular values of the pelvis and trunk and the knees and ankles, and the X-axis shows the movement pattern [2], which is the ratio of the standing movement, as shown in (6). t_s is the time required for completion of the standing operation and t is the present time.

$$\hat{s} = \frac{t}{t_s} \tag{6}$$

In this computer simulation, the variation of the movements was increased by adding fluctuations to the basic movements by the physical therapist, and it was verified whether the human model could satisfy the stability and muscular conditions in each movement. Each phase was therefore subjected to a computer simulation with a variation of 30 [deg] width in the reference posture (postures (B)–(D)) at the end of each phase. Note that since posture (A) is the start posture and he sits on his chair, thus, no variations were set in posture (A).

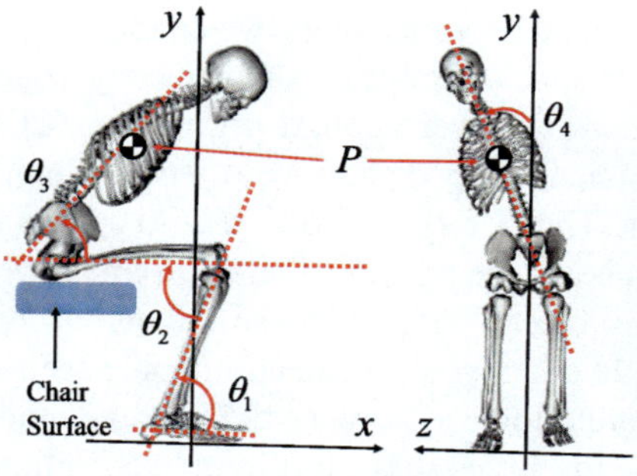

(a) Side view (b) Front view

Fig. 6. Human model and its coordination.

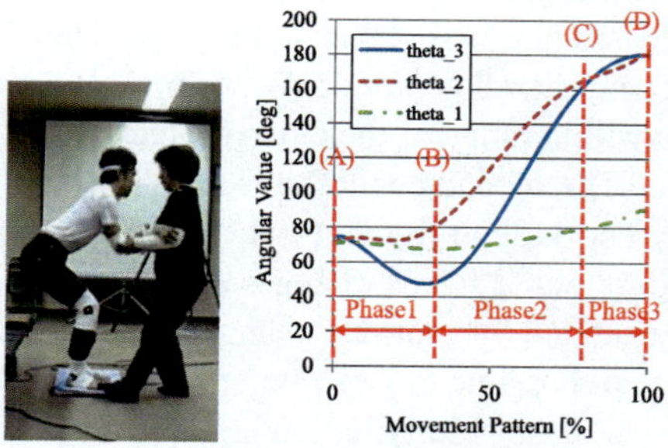

(a) Its motion (b) Angular values

Fig. 7. Recommended motion by PT.

3.3 Simulation Results and Discussion

Subject A: The simulation results are shown in Fig. 8. Figure 8 shows the position where the subject A's center of gravity, P, can satisfy the stability and muscular conditions. In other words, the gray-filled area in Fig. 8 indicates the safety tolerance, which extends before and after the standing trajectory recommended by the physical therapist. If the patient adopts a posture in which the P point is within the safety tolerance, the patient's postural stability is maintained and guaranteed to be able to continue standing with his or her own muscle strength.

From Fig. 8, we can conclude that subject A can stand up using only her own muscular strength if she shifts her body's center of gravity appropriately during the standing motion. Although the standing movement is a sagittal movement, the posture is tilted from the center to the Z direction, indicating that the safety tolerance is reduced in Fig. 8. Figure 9 is therefore shown to verify in the sagittal plane how the safety tolerance changes as the posture inclines in the z-direction. From Fig. 9(b), we can see that a subject A has a high risk of failing in the standing motion with her muscle strength, especially if her trunk is tilted to the left or right when lifting her buttocks from the chair. In general, elderly people with weak muscles become stiff due to fear of falling [13], and as a result, they cannot take the necessary posture for standing motion.

From the above, our robot is required to lead subject A to an appropriate posture during the standing motion. On the other hand, as Subject A is considered to have the necessary muscle strength, the robot does not need to provide force assistance.

Subject B: The simulation results are shown in Fig. 10. Figure 10(b) shows that subject B is unable to lift her buttocks from the chair and lift her trunk upwards using only her own muscle strength. Therefore, we can see that subject B needs a standing assistance robot if she wants to stand up.

Therefore, we assumed that the standing assistance manipulator of our robot [6, 7, 10] assists the upward force and input a 200 N vertical upward force to the human model during 10%–80% of the standing motion, the safety tolerance is shown in Fig. 10(b). This indicates that subject B can stand up with force assistance from our robot. However,

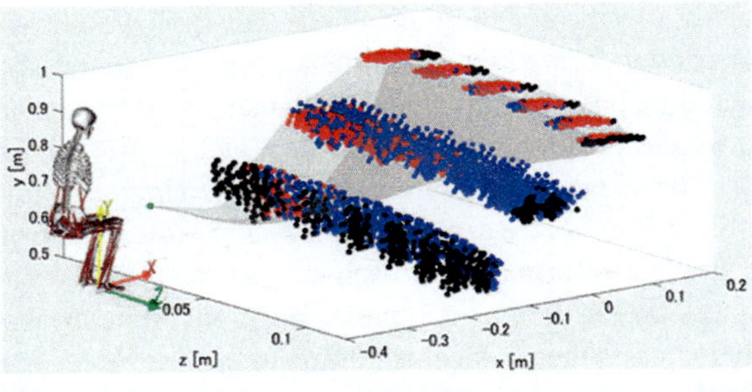

Fig. 8. Investigated safety tolerance of Subject A. Each point represents the position of the P-point when the subject A is in a posture. The red dots represent postures that satisfy both stability and muscle conditions. Blue dots represent postures in which the stability condition was satisfied but the muscle condition was not. Black dots indicate postures in which both the stability and muscle conditions were not satisfied.

the stability margin is narrower than for Subject A. This means that Subject B needs to maintain an appropriate posture at all times during the standing motion.

From the above, our robot is required to lead subject B to an appropriate posture during the standing motion. Furthermore, our robot needs to provide force assistance.

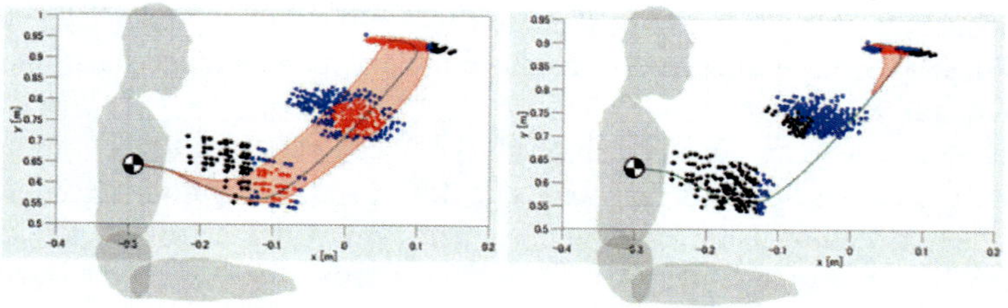

(a) Z=0[m] (b) Z=0.1[m]

Fig. 9. Investigated Safety tolerance of subject A on sagittal plane. The green line is the trajectory of the P point when the subject stands with a recommended way by the nursing specialist, and the stability margin range is within the red-filled area.

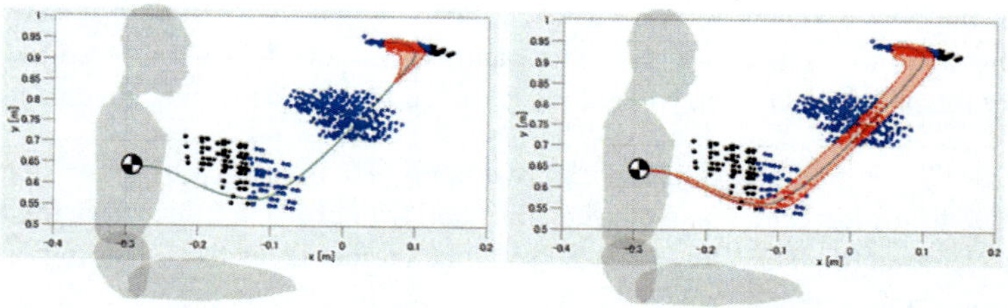

(a) Without assistance (b) With assistance

Fig. 10. Investigated Safety tolerance of subject B on sagittal plane, $Z = 0$[m].

Validation of Our Proposed Scheme: To confirm the validity of the above results, subjects need to actually use the standing assistance robot. However, as this is currently difficult by corona pandemic, we asked nursing facility staffs to measure the fluctuation of their COG at standing position by own strength using a stabilometer.

Figure 11 shows the position of the subject's COG in the standing position. The safety tolerance for the standing position is equal to the simulated final posture of the standing motion. From Fig. 11, we can see that the safety tolerance from the simulation using the muscle strength information of each subject matches the range in which the COG of each subject changes during the standing position. From the above, the proposed method accurately indicates the physical condition of each subject and the effectiveness of the proposed method can be verified.

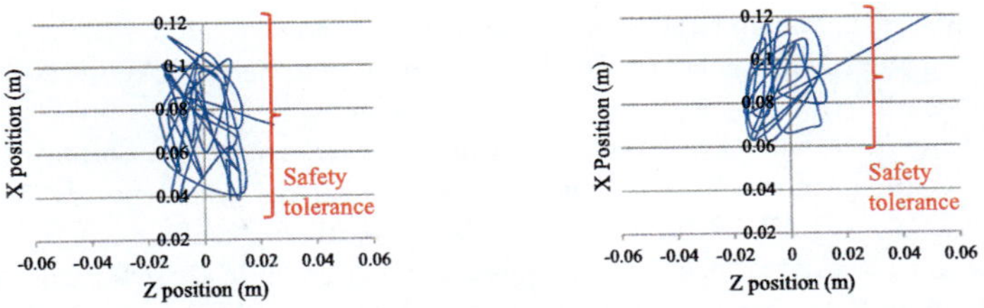

(a) Subject A (Comparison with Fig. 9(a)) (b) Subject B (Comparison with Fig. 10(a))

Fig. 11. The position of COG measured by stabilometer. Subject B lost her balance and leaned on the handrail during the measurement, which caused the values to extend outside the graph.

4 Conclusion

In this paper, we propose muscle strength estimation method for the lower limb muscles of the elderly which is based on the bi-articular muscle model. The muscle strength of each part can be estimated by asking the elderly to simply generate force on their feet in six different directions. Furthermore, by inputting the estimated muscle strength of each part as a parameter of the musculoskeletal simulator, our proposed method obtains a safety motion tolerance, which is the individual range of postures required for this elderly person to maintain body stability and to stand up using his or her own muscle strength. Finally, by designing the reference trajectory of the standing assistance robot to guide the posture of the elderly person using the derived safety motion tolerance, a standing assistance robot suitable for the target elderly person can be realized without the robot developer directly measuring the physical condition of the elderly person. We compared the derived safety motion tolerance with the actual movement range of the elderly. The results showed that the two were in close agreement, thus we can verify the effectiveness of our proposed method.

Acknowledgments. This study received a generous support on the nursing standing assistance way from Prof. Tsuyoshi Uematsu (occupational therapist), Mr. Yoichi Kondo (occupational therapist) and staffs on Kunitachi Visiting Nursing Station, Tokyo, Japan.

This work is supported in part by a Grant-in-Aid for Scientific Research B (20H04566) from the Japan Society for the Promotion of Science, the Matching Planner Program (VP29117940231) from Japan Science and Technology Agency, JST and Individual Special Research Subsidy from Kwansei Gakuin University.

References

1. Alexander, N.B., et al.: J. Geom. Med. Sci. **46**, M91 (1991)
2. Hughes, M.A., et al.: J. Rehab. Res. Dev. **133**, 409 (1996)
3. C. Office, KOUREISHA HAKUSHO (The whitepaper on the aged society, written in Japanese). Government of Japan (2021)
4. Hirvensalo, M., et al.: J. Am. Geriatric Soc. **48**, 493 (2000)
5. Ito, T., Azuma, T., Yamashita, N.: Eur. J. Appl. Physiol. **92**, 186 (2004)

6. Yokota, M., et al.: Pattern based standing assistance for a low level of care. In: Proceedings of the International Conference on Climbing and Walking Robots (CLAWAR 2018), Panama City, Panama (2018)
7. Chugo, D., et al.: Standing assistance which realizes voluntary movements of the patient within a safety motion tolerance. In: Proceedings of the International Conference on Climbing and Walking Robots (CLAWAR 2020), Virtually, Moscow, Russia (2020)
8. Hatsukari, T., et al.: Self-help standing-up method based on quasi-static motion. In: Proceedings of the IEEE International Conference on Robotics and Biomimetics (Robio 2009), Guilin, China (2009)
9. Nishida, I., et al.: J. Jpn. Ergon. Soc. **47**, 244 (2011)
10. Yamada, T., Sakaguchi, M., Chugo, D., et al.: A load estimation of a patient considering with a posture during standing motion. In: Proceedings of the International Conference on Climbing and Walking Robots (CLAWAR 2014), Poznań, Poland (2014)
11. Oshima, T., Fujikawa, T., Kumamoto, M.: J. Precision Eng. **65**, 1772 (1999)
12. Spector, S.A., et al.: J. Neurophysiol. **44**, 951 (1980)
13. Maki, E., Holliday, P.J., Topper, A.K.: J. Gerontol. **46**, 123 (1991)
14. Hoy, M.G., Zajac, F.E., Gordon, M.E.: J. Biomech. **23**, 157 (1990)
15. Oshima, T., Fujikawa, T., Kumamoto, M.: J. Precision Eng. **67**, 1824 (2001)
16. Opensim documentation gait 2392 and 2354 models (2018). https://simtk-confluence.stanford.edu/display/OpenSim/Gait+2392+and+2354+Models
17. Kamiya, K.: Development and evaluation of life support technology in nursing. In: Proceedings of the 7th RACE Symposium, Chiba, Japan (2005)

Author Index

© The Editor(s) (if applicable) and The Author(s), under exclusive license
to Springer Nature Switzerland AG 2023
J. M. Cascalho et al. (Eds.): CLAWAR 2022, LNNS 530, pp. 617–619, 2023.
https://doi.org/10.1007/978-3-031-15226-9